DEBURRING and EDGE FINISHING HANDBOOK

DEBURRING and EDGE FINISHING HANDBOOK

LaRoux K. Gillespie, PE, CMfgE

Society of Manufacturing Engineers
Dearborn, Michigan

American Society of Mechanical Engineers
New York, New York

98765432

Library of Congress Catalog Card Number: 98-061356
International Standard Book Number: 0-87263-501-5

Additional copies may be obtained by contacting:

Society of Manufacturing Engineers
Customer Service
One SME Drive, P.O. Box 930
Dearborn, Michigan 48121
1-800-733-4763
www.sme.org

SME staff who participated in producing this book:

Millicent Treloar, Senior Editor
Rosemary Csizmadia, Production Supervisor
Frances Kania, Production Assistant
Kathye Quirk, Production Assistant/Cover Design

Cover photo courtesy of Rex-Cut® Products, Inc., U.S.A.

Printed in the United States of America

To Koya Takazawa, whose unswerving dedication to burr technology and leadership is changing the world. His teacher became the student and he became the teacher. He focuses the attention of the world on deburring, builds a staff, and moves to repeat the process in yet another country. Burr, edge and surface technology groups in Japan, China, Korea, Germany, Russia, and Taiwan owe their current start to his pioneering efforts. The Worldwide Burr Technology Committee also owes its existence to him. He is the strength of and the force behind international burr technology efforts.

In the immortal words of a *Star Wars* leader, "May the force be with you!"

About the Author

LaRoux Gillespie is the world-renowned leader of the deburring "industry." He knows more about the formation and removal of burrs than anyone else. He has identified at least 100 deburring processes, 57 of which are commonly used. A recent search of the the Society of Manufacturing Engineers (SME) database for credits to LaRoux turned up 97 citations, most having to do with burrs, burr formation, burr tools, and deburring machinery. He has written, or edited 14 books, including many well-recognized publications of SME. He is chairman, or has chaired, virtually every SME committee that deals with machining or deburring. LaRoux is Manager, Engineering Products with a major manufacturer where he has been employed for more than 30 years. He is chairperson of the Worldwide Burr Technology Committee, an international standards body. LaRoux and his wife Judy have five children.

LaRoux earned his Bachelor of Science in Mechanical Engineering from Kansas University and his Master of Science in Manufacturing Engineering from Utah State University. He is a Registered Professional Engineer in Missouri and California, and a Certified Manufacturing Engineer.

In spite of LaRoux's great accomplishments, he is a modest man. He learned the business from the ground up—"hands on engineering" I call it. To quote him, "I learned from Mabel on the floor who had to remove difficult burrs."

Being great in one's field means more than knowing a tremendous amount about the field. It means sharing that knowledge with others; it means encouraging young people to enter the field and seek their own greatness. LaRoux willingly shares all he knows with anyone interested in furthering his or her own knowledge. He encourages young engineers to aspire to similar greatness. His motivational lecture entitled, "Seeds of Greatness—The Million Dollar Lecture" is frequently requested at annual events for young engineers.

While LaRoux stands alone in our industry, he works hard to change that. He is active in promoting the certification program of SME. He encourages others, especially young engineers, to pursue knowledge, both academically and with hands-on experience. He set the standard for dedication, and most of us would do well to emulate him in some way, in some facet, of our personal and professional lives.

All of us owe a debt of thanks to LaRoux. I have attempted to thank him for all of us, though I stopped short because I imagine him over my shoulder saying, "'nuff said, already."

R. Steven Marcus
Markee International Corporation
Columbus, Ohio

Table of Contents

Preface

As manufacturer's increasingly compete across national borders, the cost and quality of deburring and edge finishing methods, like all other areas of manufacturing, have come under scrutiny. The costs of deburring, deflashing, and surface finishing can range from being a very small percentage of the part cost to being the single largest cost. If an inappropriate method is used, it can, in extreme cases, totally eliminate the economic justification for the product. It is well known that a manufacturer in a country with very low manual labor costs can provide a fine hand finish at costs difficult to match by machines in industrialized countries. This challenges all manufacturers to find the most cost-effective methods. As is the case with materials and machines, the manufacturer who understands as much as possible about the options available for burr removal and surface conditioning will be better able to compete, and survive, in the future.

The removal of burrs and treatment of edges affect the economics of such diverse industries as automobiles, bearings, needles, fasteners, eyeglass frames, toys, computers, and clothing. Consequently, the methods of deburring and edge finishing encompass so many different tools and applications that no single work has previously attempted to cover the subject in its entirety. This book does. It is the most complete work ever presented on the subject.

This book begins with an overview of deburring. It addresses the aesthetic, functional, and economic justifications. It identifies five basic approaches to consider when one is charged with the cost reduction or process improvement to a product that requires any kind of deburring or edge treatment. Two approaches focus on burr removal. Three approaches focus on product design, tool design, process design, and the minimization or avoidance of burrs. Subsequently, Chapters 2–5 provide considerable insight into reducing or eliminating deburring and edge finishing costs through product design, tooling, and manufacturing procedures. Many tactics are covered, such as intentionally generating the burrs in areas where they do not need to be removed. An evolving system of standards is described in Chapter 2. Research areas promising breakthroughs that greatly reduce, or even eliminate, the burr problem for many product lines are identified. Finally, the book addresses today's environmental and safety concerns by providing basic information necessary to launch responsible planning. This information is introduced in Chapter 1 and is built upon in the last section of each of the core chapters on deburring methods.

Many readers will approach this book looking for definitions of deburring methods. They will not be disappointed. In Chapters 6–19, they will get that information in greater detail than ever before. It even provides descriptions of a few patented methods that have never come to market, perhaps motivating some reader, somewhere, to consider ideas that did not bear fruit in a previous manufacturing environment but that may solve problems in the future. Further, this book provides global coverage of deburring methods. For instance, what has been tried in Russia, China, or Japan? Some quite exotic methods are presented.

Many tables in the book detail the efficiency of different deburring methods. The reader will find ample technical information on radii generation, surface finish, burr dimensions, and other measurements pertinent to selecting a deburring method.

This book belongs in the library of every manufacturer of hard goods. It belongs in every college, university, trade, or technical school dealing with manufacturing technology at any level. Within the confines of manufacturing companies, the contents of this book should be known to such diverse people as presidents, controllers, safety directors, product designers, tool designers, manufacturing engineers, process engineers, plant managers, production supervisors, and even production workers. There is knowledge in this book for everyone.

The accumulation of deburring and edge finishing knowledge contained in this book could scarcely have been accomplished in a single lifetime. It brings together the knowledge and expertise of many diverse writers, researchers, engineers, manufacturers, machine operators, students, and professors. Dozens of people have contributed directly or indirectly to the information found in this book. It stands alone in its field.

R. Steven Markus
Markee International Corporation
Columbus, Ohio

Acknowledgements

Dozens of individuals have directly contributed to this book and behind them hundreds of researchers, practitioners, and equipment manufacturers generated material that helped provide the original developments and ideas. Much of the material has previously been published in SME papers. The author has also brought much of his own experience to this handbook and is indebted to the many persons who helped him as he developed his background in this international field. Dr. Y.T. Lin of General Motors and Steve Marcus of the Markee Corporation provided excellent technical suggestions for improving this book. Dr. David Dornfeld of the University of California at Berkeley also reviewed the manuscript and has been a fellow champion for better understanding of burr formation mechanics. The work of Dr. Dornfeld and his students would fill a book of this size by itself, and while we have captured the basics of his work in this field, there just is not space to include the full extent of the work developed at Berkeley.

A special acknowledgment goes to John Kittredge, a leader in the field of mass finishing, for permission to use many fact-filled papers and articles. John's works are one of the few places readers could go and find new information critical to making informed selections of deburring processes. In addition, John's persistent support for others, such as the author, made our job much easier over the years in learning more of the capabilities and nuances of burrs and deburring.

Many of the ideas in this book are reprinted with permission from a variety of company and association publications. Grateful acknowledgment is extended for permission to use material from the following sources: Ultramatic Equipment; U.S. Filter/Wheelabrator; Fraunhofer-Institute fur Producktionstechnik und Autmoatisierung (Germany); Anocut, Inc.; S.S. White Technologies; 3M; Polygrat GmbH (Germany); Allied Signal Aerospace, Inc.; Chiba Institute for Technology (Japan); Senn Co.; Weiler Corp.; Klaus Przyklenk (Germany); Surf/Tran Division Robert Bosch Corp.; Royal Products; Klingspor Abrasives; Osborn Manufacturing Co.; Extrude Hone; Brush Research Manufacturing Co.; Koya Takazawa (Japan); Superior Abrasives; Empire Abrasive Equipment Corp.; Comco, Inc.; E.I. du Pont de Nemours & Co.; Dynetics; Dynabrade; Timesavers, Inc.; Maxi-Blast, Inc.; Shoreview Metalcraft, Inc.; Kramer Industries; Precision Machined Products Association; *Advances in Machine Tool Design and Research,* Pergamon Press; M. Matsunaga (Japan); Canadair Ltd. (Canada); Masao Murakawa (Japan); Akihiko Shimizu (Japan); J.J. Claus; Pines Manufacturing, Inc.; Albert Madwed; Robert W. Johannesen; Dr. Friedrich Shäfer (Germany); Acme Manufacturing Co.; Harper Surface Finishing Systems.

Lastly, SME has continued its efforts to provide a top quality publication by assigning a very excellent book editor to improve style, format and understanding. Millicent Treloar has added more than a decade of publishing skills to make this a more usable tool for each of its users. Rosemary Csizmadia has done an exemplary job of managing the myriad production details. Joni Harlan has done an excellent job as copy editor. Thanks are due to Donna Hicks for working miracles with the electronic art illustrations and to Kathye Quirk for the clean, easy-to-follow layout of the text.

Thanks to all of those mentioned here and those many persons who also helped put this book in print.

LaRoux K. Gillespie
Kansas City, Missouri

Symbols and Abbreviations

The following is a list of symbols and abbreviations in use throughout this handbook.

A

A	ampere
AA	arithmetic average
ABS	acrylonitrile-butadiene-styrene
ADCI	American Die Casting Institute
AFM	abrasive flow machining
AISI	American Iron and Steel Institute
AJM	abrasive-jet machining
AMS	aerospace material specification
AMSA	American Metal Stamping Association
ANSI	American National Standards Institute
aq.	aqueous
ASTM	American Society for Testing and Materials

B

Bhn	Brinell hardness number

C

C	Celsius
CAMI	Coated Abrasive Manufacturers' Institute
CFR	Code of Federal Regulations
CHM	chemical machining
CLR	circuit-loop resistance
cm	centimeter
CNC	computer numerical control

D

dB	decibel
DC	direct current
DOE	design of experiments
dm	dyne per meter

E

EBM	electron beam machining
ECD	electrochemical discharge, electrochemical deburring
ECG	electrochemical grinding
ECH	electrochemical honing
ECM	electrochemical machining
EDM	electrical discharge machining
ELP	electropolishing
EPA	Environmental Protection Agency
EPDM	ethylene-propylene rubber
Eq.	equation
ESM	electrostream machining

F

F	Fahrenheit
FEPA	Federation of European Producers of Abrasives
fig.	figure
FMS	flexible manufacturing system
ft	foot/feet

G

g	gram
gal	gallon

H

h	helical
HCG	hot chlorine gas
HDM	hydrodynamic machining
hp	horsepower
hr	hour
Hz	hertz

I

IBM	ion beam machining
IC	integrated circuit
ID	inside diameter
in.	inch
μin.	microinch
ipm	inches per minute
ipr	inches per revolution
ISO	International Standards Organization

K

kg	kilogram
kHz	kilo Hertz
kPa	kilo Pascal
ksi	1,000 pounds per square inch
kW	kilowatt

L

L	liter
lb	pound
lbf	pound force
LBM	laser beam machining
LED	light-emitting diode

M

m	meter
μm	micrometer
mg	milligram
mi	mile
MIL	military specification
min	minute
mm	millimeter
MPa	mega Pascal
ms	millisecond

N

N	Newton
N/C or NC	numerical control
NEMA	National Electrical Manufacturers Association
NIST	National Institute for Science and Technology
nm	nanometer
N/m	Newton/meter
NSMPA	National Screw Machine Products Association

O

OD	outside diameter
OHD	overhead drive
ohm	unit of electrical resistance
OSHA	Occupational Safety and Health Administration
oz	ounce

P

PAM	plasma arc machining
PCM	photochemical machining
PH	precipitation hardening
ppm	parts per million
psi	pound per square inch
PVC	polyvinyl chloride
PWB	printed wiring boards

Q

qt	quart

R

r	rotation
R	radius
RCRA	Resource Conservation and Recovery Act
rev	revolution
RMS	root mean square
rpm	revolution per minute

S

s or sec	second
SAE	Society of Automotive Engineers
sfm	surface feet per minute
SiC	silicon carbide
SME	Society of Manufacturing Engineers
SPC	statistical process control

T

3-D	three-dimensional
TCLP	toxicity characteristic leaching procedure
TEM	thermal energy method

U

U.K.	United Kingdom
UL	Underwriters' Laboratories
U.S.	United States
USM	ultrasonic machining

V

V	volt

W

W	watt
WAM	waterjet machining
WBTC	Worldwide Burr Technology Committee
WJM	waterjet machining

Y

yd	yard

MATH SYMBOLS

$\sim$	about equal to
$^\circ$	degree
/	divided by or per
$>$	greater than
$\geq$	greater than or equal to
α	Greek alpha
β	Greek beta
Δ	Greek delta
ϵ	Greek epsilon
γ	Greek gamma
λ	Greek lambda
μ	Greek mu
Ω	Greek omega
ϕ	Greek phi
π	Greek pi
σ	Greek sigma
θ	Greek theta
ϑ	Greek theta (lower case)
$<$	less than
$\leq$	less than or equal to
$-$	minus
%	percent
$+$	plus
$\pm$	plus or minus
$\times$	times

1 Overview of Deburring and Edge Finishing Technology

One dictionary defines a burr as "a thin ridge or area of roughness produced in cutting or shaping metal." For most people, this definition conveys the basic idea of a burr. However, for engineers charged with removing burrs from manufactured parts, a process known as "deburring," this definition is inadequate.

What constitutes a "burr-free" part varies among companies and quality control departments. For some, it means having no loose materials at an edge. For others, it means having nothing visible to the naked eye or an edge condition that will not cause any functional problem in the next assembly process. Missing material or a hump of rounded metal at an edge may or may not be called a burr.

Burrs and sharp edges create many problems. Sharp edges can be the result of inadvertently leaving a sharp edge or can come from producing a burr that typically has many sharp facets. Burrs on sheet metal parts, for example, cause premature tearing during forming. Plating over burrs and sharp edges allows early corrosion of the material or a poor fit during assembly. Fine burrs left by grinding automotive cylinders can cause engine failure. Undetected burrs on life safety devices can undermine performance. Every automotive mechanic has received cuts and bruises from burrs and sharp edges left on automotive components. Edge quality is of concern for the performance, safety, cost, and appearance of a part. The following is a reasonably complete list of the problems caused by improperly finished edges:

- Cut hands in assembly or disassembly.
- Interference fits in assemblies.
- Jammed mechanisms.
- Scratched mating surfaces that allow seals to leak.
- Increased or changed friction (not allowable in some assemblies).
- Increased wear on moving or stressed parts.
- Electrical short circuits.
- Cut wires from sharp edges and sharp burrs.
- Unacceptable high-voltage breakdown of dielectric.
- Irregular electrical and magnetic fields.
- "Detuning" of microwave systems.
- Metal contamination in aerospace assemblies.
- Clogged filters and ports from loose burr accumulation.
- Cut rubber seals and O-rings.
- Excessive stress concentrations.
- Plating buildup at edges.
- Paint buildup from electrostatic spray over burrs.
- Paint thinout over sharp edges from liquid paints.
- Edge craters, fractures, and crumbling from initially nonsmooth edges.
- Turbulence and nonlaminar flow.
- Reduced formability.
- Inaccurate dimensional measurements.

After examining the preceding list, it is not surprising that burr technology and edge finishing is a vast and complex world. Burrs have physical properties and thus have various acceptable deburring processes. Furthermore, deburring is a manufacturing cost, and there are trade-offs between cost and product quality. The process of developing common definitions and standards across plants, industries, and countries is still in its early stages.

Although this handbook largely concerns burr removal, many of the following chapters also discuss edge finishing. It is important to recognize that a product may need more than simply having the burr removed. Typically, some specific edge configuration is also required. This handbook covers the major elements that designers must consider in the deburring and edge finishing process, including the requirements of typical industrial parts and processes applicable from microscopic features to earth-moving equipment. The single missing aspect of edge finishing that is not covered is edge sharpening, such as for knife and razor-blade edges.

This chapter provides an overview of burr technology, and specific topics are covered in more detail in later chapters. The chapter begins with a discussion of burr properties such as material properties, machining properties, and part configurations. It proceeds to an introduction of burr standards. Industrially significant and research deburring processes are introduced, and typical costs and operation are discussed.

1.1 FUNDAMENTALS

The fundamental principles of burr technology rely on a few simple concepts. The first principle encompasses the following:

- Burr properties are a function of material properties, machining and blanking process, and part configuration.
- Acceptable deburring is a function of burr properties, part configuration, acceptance standards, and deburring process parameters.
- Cost-effective deburring is a function of acceptable deburring quality; scheduled quantities; cycle time; and environmental, safety, and health issues.
- All deburring processes have side effects.

The second principle involves five basic approaches to reducing deburring costs:

- Improving product design.
- Preventing burrs.
- Minimizing burr properties.

- Removing burrs during the machining and blanking cycle.
- Developing or obtaining better deburring processes.

The third principle recognizes that edge finishing and edge quality are two different aspects of deburring.

The fourth principle recognizes the vast number of processes and process variations used for deburring. Users have over 100 deburring and edge finishing processes from which to choose.

The fifth principle recognizes that subtle tricks of the trade can produce major savings by eliminating the need for new machines, training, and maintenance of high-tech equipment. In deburring and edge finishing, innovation is the key to success.

1.2 BURR PROPERTIES

The first key to reducing edge issues is minimizing burr size. When burrs are small, deburring requires little thought and effort. If a burr is only .0001 in. thick × .0001 in. tall (2.5 × 2.5 μm), it can be removed in a few seconds on any part using any process. In contrast, when a burr is .005 in. thick × .005 in. tall (127 × 127 μm) and part tolerances are critical, carefully conceived approaches to removal are required. Figure 1-1 illustrates how manual deburring time increases as burr thickness increases for a simple shape on precision parts. A typical burr size in many operations is .003 in. thick at the root × .010 in. high (76.2 × 254 μm). Chapter 4 discusses the different types of burrs, describes how they form, and provides essential information on minimizing them.

1.2.1 Material Properties

Two factors related to workpiece material are directly linked to burr size: (1) the ductility of the workpiece material and (2) the strain-hardening exponent of the material.

Large burrs cannot form in brittle materials. Cast irons, for example, often have edges with no visible burrs. These materials have values of elongation of 0.5–3.0% in a 2 in. (50 mm) gage length. Since the material has little capacity for plastic deformation, large burrs cannot form. If, however, the cutting tool heats the cast iron enough to change its structure and the material is no longer brittle at the edges of machined surfaces, a noticeable burr can form.

Burr size is also a function of the strain-hardening exponent (or strain-hardening coefficient) (Datsko 1966). Nonstrain-hardening materials will form burrs, but they will be considerably smaller than those formed on materials having large strain-hardening tendencies. As the strain-hardening exponent increases, burr thickness generally increases, but the relationship is not usually directly proportional. (The terms *strain hardening* and *work hardening* are synonymous.) Table 1-1 presents typical data on material properties related to burr size that are useful in estimating burr-forming tendencies.

As a general rule, so-called aerospace materials (high nickel-content materials) form large burrs. The key factor, again, is high ductility in the material. Thus, stainless steel that work hardens easily will have high strain-hardening exponents and will easily form burrs—very large burrs in some instances.

Dull tools can significantly heat a part while being machined, causing normally small burrs to become monstrous.

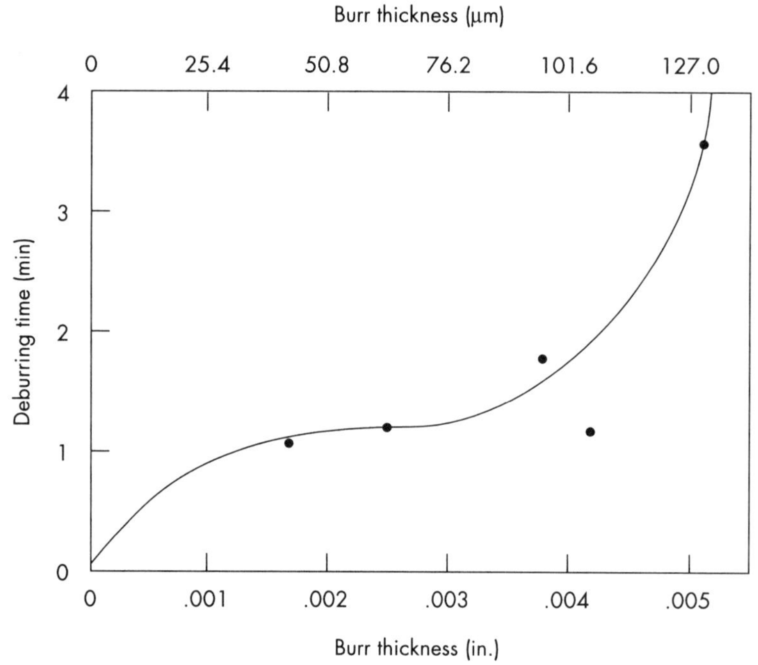

Figure 1-1. Hand-deburring time as a function of burr thickness on precision miniature parts (Wick and Veilleux 1985).

Table 1-1. Material properties related to burr size

Material	Yield strength (ksi)	Tensile strength (ksi)	Elongation (%)	Strain-hardening exponent	Burr tendency
Cast iron	55	80	6	0	Low
1020 steel	30	55	25	0.22	Medium
303 Se Stainless	60	180	50	0.56	High
Kovar	50	105	72	0.42	High
Hiperco 50	57	57	0.78	0	Low
BeCu	95	102	26	0.10	Medium
2024 Aluminum	11	27	20	0.15	Medium
Copper (soft)	10	32	45	0.50	High
4340 Steel	69	108	22	0.09	Medium
Vanadium (annealed)	66	78	20	0.35	High

(Datsko 1966; Gillespie 1977a; Wick and Veilleux 1982)

The underlying cause of these monster burrs is poor control of machining, but dull tools increase part temperatures, which further increases ductility and subsequently burr thickness.

1.2.2 Effects of Machining and Blanking Processes

Typical burrs are not the result of poor planning or poor engineering. They are a natural result of machining and blanking processes. Large burrs, however, may be the result of poor planning. For example, the cost of burr removal may be increased when certain machining or blanking processes are selected. They may also result from choices made in other manufacturing operations. The sequence in which dimensions are machined or blanked affects the location of burrs and the effort required to remove them. Inattention to tool sharpness can make deburring by traditional means impossible. Feeds, speeds, depths of cut, cutter geometry, sequence of cutter paths, and even machine tool design and repair affect burr sizes and consequently the cost of burr removal. Chapter 4 discusses the impact of process variables on burr sizes. Chapter 5 provides a complete look at how several products can be machined to reduce deburring costs. Clearly, high feed rates will typically, but not always, increase burr dimensions. Processes that normally employ rough-and-finish passes to hold tolerances also tend to have smaller burrs, because the finish passes are taken at smaller tooth or chip loads.

The geometry of cutters also plays an important role in burr production. With over 50 designs of countersink cutters alone, not surprisingly some cutters produce smaller burrs than others do. The challenge is to find the most cost-effective cutters. From a business viewpoint, cost effectiveness includes the costs of removing burrs and finishing the surface as well as producing the specific feature. Chapter 3 provides insight into preventing burrs.

1.2.3 Effects of Part Configuration

A variety of strategies and some software exists for designing parts and processes so that burrs are a less-costly problem. Part configuration affects the bottom line in three ways:

1. It defines geometry conditions that do not produce burrs.
2. It defines geometry conditions that produce smaller burrs.
3. It defines simple approaches to putting burrs where they can be easily removed at the least cost.

Chapter 5 provides basic guidance on the effects of part geometry on burr formation. The chapter provides an example of how the angles at which a milling cutter traverses a part (and the angles on the part) can affect burr size and shape. Other examples show how simple design changes can prevent the need for removing burrs. Although the topic of part configuration is not simple, the many examples provided in later chapters may prevent hundreds of hours of unnecessary effort.

Part geometry affects not only burr size but also the ease of burr removal. Figure 1-2 provides a simple analysis of the impact of angles. The ratchet wheel shown in Figure 1-2 has milled teeth; a blanked contour would involve similar deburring issues. Note that the angle at different teeth ranges from R_1 to R_4, and each angle is significantly different from the others. Mechanical deburring, such as tumbling processes, work on each edge for the same amount of time (at least in this example). Because of the angle differences, each edge after deburring will have a different radius, though it may not have had a burr initially. The center hole will also have a somewhat different radii than the other edges. If each edge begins with a different-sized burr (a common occurrence), the final edges will have even more differences. These differences are the result of part-geometry

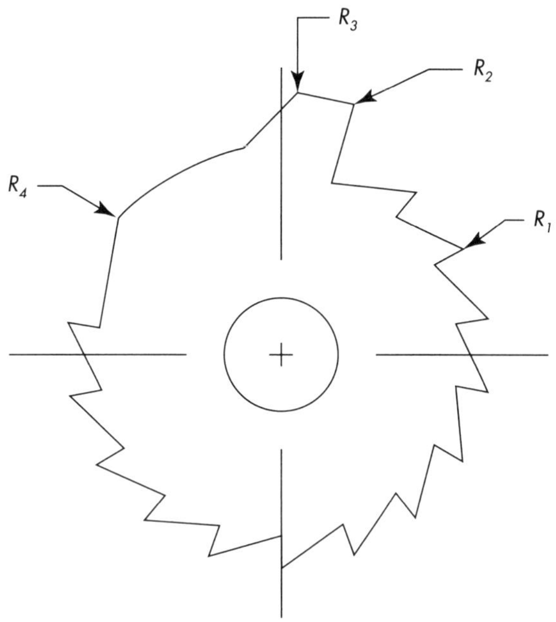

Radii produced on three edges of part while maintaining tolerance of radius R_2 on phosphor bronze:

Radius feature	Edge angle (degrees)	Radius produced (in.)	Radius produced (μm)
R_1	60	.0018 ±.001	45.7 ±25.4
R_2	90	.0050 ±.001	127.0 ±25.4
R_3	125	.0113 ±.001	287.0 ±25.4
R_4	140	.0121 ±.001	307.3 ±25.4

Figure 1-2. Phosphor bronze ratchet and edge radii produced by centrifugal barrel finishing (Gillespie 1978).

effects. For some products, the differences in edge results are not critical. For critical applications, the choice of deburring approaches becomes much more difficult.

The number of edges on a part and the total length of edges to be deburred on a single part may not be apparent. Both of these factors provide some measure of the difficulty of obtaining perfect edges when high precision is required. Figure 1-3 shows that a single gear tooth has 10 different line segments that must be burr-free. For a miniature instrument gear small enough to fit under a fingernail and having 12 teeth, 120 edge segments must be burr-free.

Figure 1-4 illustrates a simple fine-pitch 5.8 × 0.35-5h mm screw that is .320 in. (8 mm) long. It has 17 threads with 24.4 in. (620 mm) of burrs on the crests. This small part has over 24.8 in. (629.9 mm) of thread to deburr! A burr left anywhere on its crest could jam its mating part. The burrs on a single part of this design could, if broken off into loose particles of .00076 in. (0.019 mm) diameter or length, produce 32,632 particles.

1.3 EDGE STANDARDS

Figure 1-5 illustrates two edge conditions: a burr-laden edge and a rounded edge after removal of a burr. Deburring has many levels and variations, as shown in Table 1-2. This variation suggests the need for some form of edge standard, including a definition of "burr-free." The edge quality requirements listed in Table 1-2 are examples of product needs and illustrate why a simple "burr-free" note on a drawing does not adequately reflect product needs.

Clearly articulating desired edge conditions is a low-cost strategy for reducing deburring and edge finishing costs. As obvious as it sounds, however, companies continue to omit this simple step in their battle against the burr. Nevertheless, as customer expectations increase, clear standards are becoming more important.

To illustrate the range of standards, consider the case of manufacturers and researchers who identify the microscopic slivers left on ground surfaces (surfaces, not edges) as burrs. One company reportedly checks for burrs at 400× magnification. The idea of burr-free parts is meaningless without some reference to inspection approaches. Omitting reference to magnification levels of inspection does not necessarily or legally imply "when viewed with the aided eye." One company uses no magnification, another 10×, and another 40–100×.

What exactly does the term *burr-free* mean? Without clearly written, detailed, and uniform burr and edge finishing standards, we will continue to face unexpected scrap and functional problems. The lack of uniform standards creates three problems today:

1. *The lack of detailed definitions of what is and is not a burr cause many parts to be rejected when they should not be rejected and some to pass when they should not pass.* The cost of both actions is enormous. Every company has faced the problem of an inspector, floor supervisor, or engineer arguing about whether the small speck on the part is a burr or only "raised material." Needless to say, the lost time, unnecessary rework, and confusion is expensive.
2. *The lack of a concrete and detailed understanding of edge needs (as opposed to merely knowing word definitions) causes overzealous deburring,* which can waste tens of millions of dollars each year. Failing to meet real needs causes scratches, cuts, tears, product failures, assembly-line stoppages, excessive scrap, cycle time delays, and widespread frustration on the part of everyone in manufacturing.
3. *Constantly changing, undocumented standards create unnecessary expense and delays.* People may argue about whether a burr is normal or atypical or whether it is the same size as it was yesterday. "You accepted it last week, why won't you accept it today?" is a common refrain. Even if the words in the specifications do not change, if they exist, the implementation of standards may change daily. How does this happen? Every

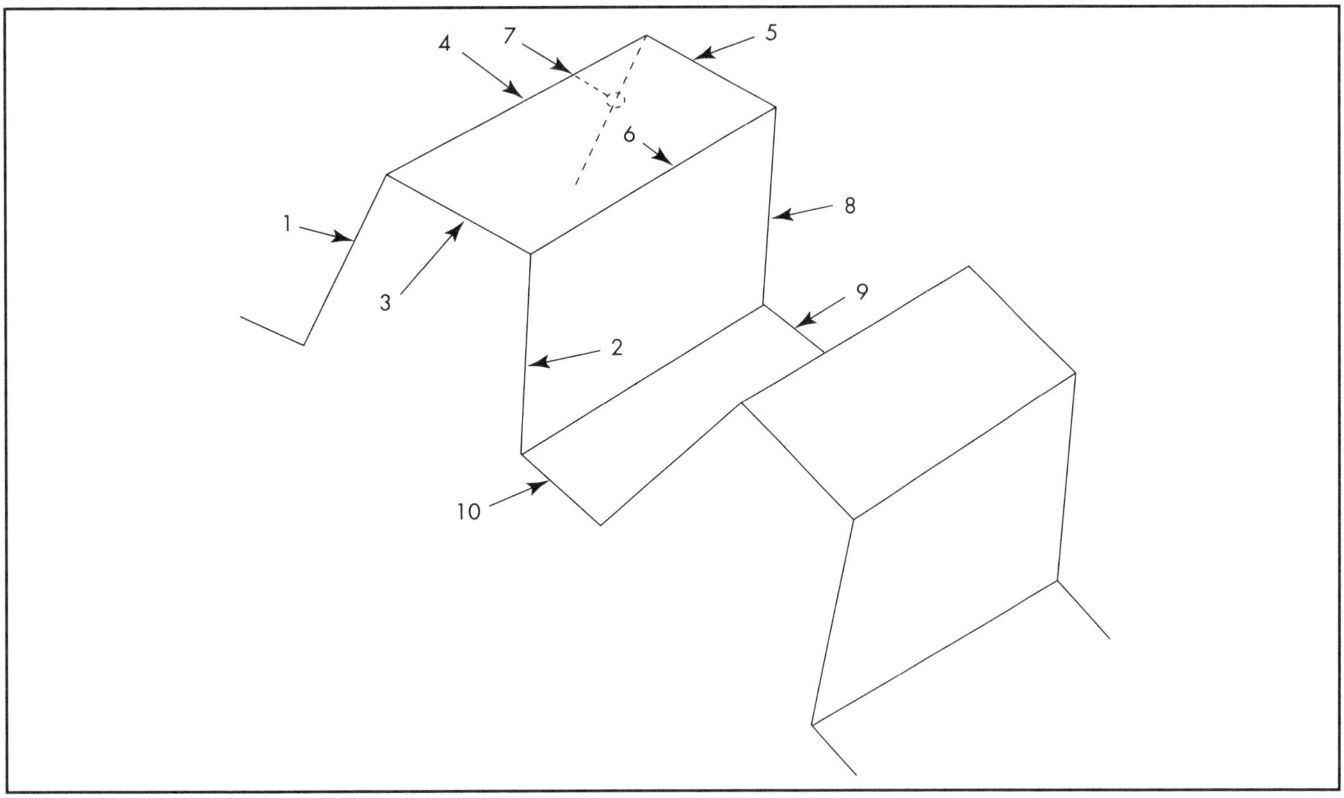

Figure 1-3. Line segments on a gear tooth (Gillespie 1982).

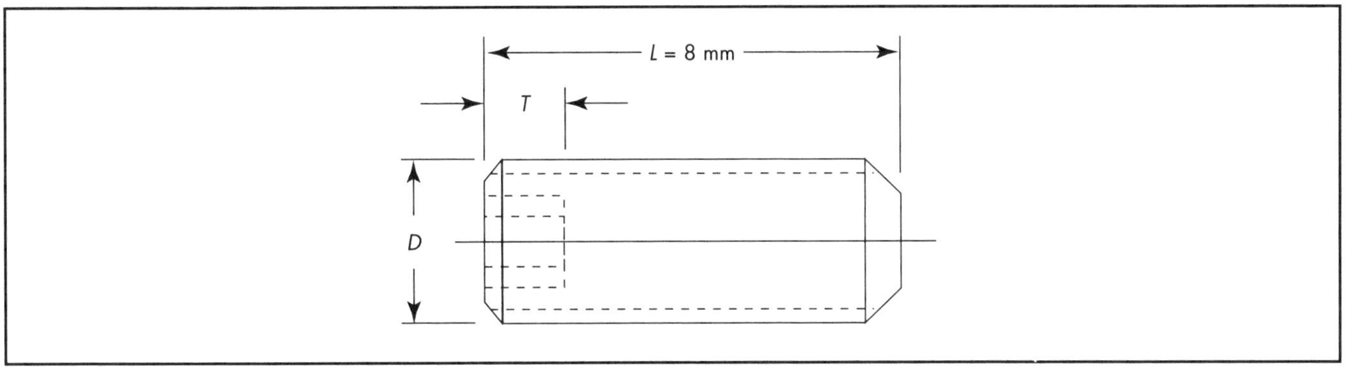

Figure 1-4. Fine-pitch screw (5.8 × 0.35-5h) has over 2 ft of burrs on its crests.

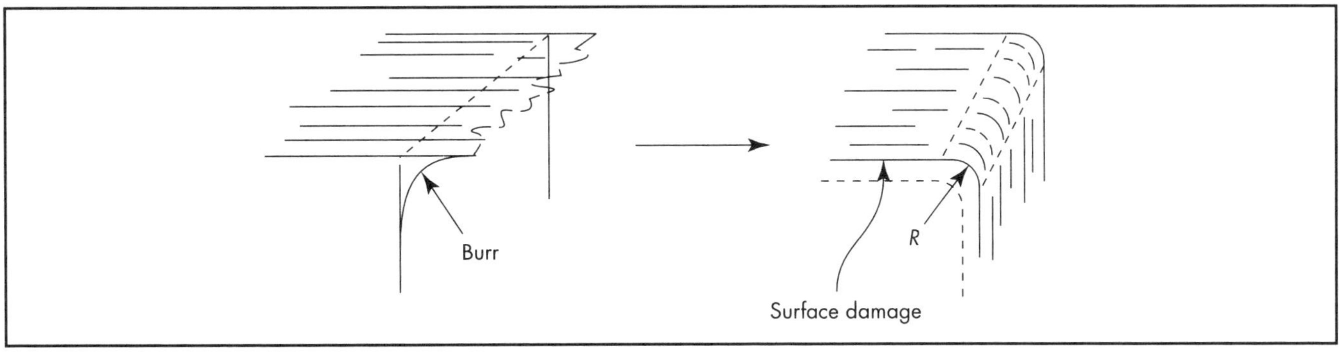

Figure 1-5. Definition of edge conditions (Takazawa and Kato 1997).

Table 1-2. Overview of edge quality requirements

Class	Grade	Drawing definition	Radius tolerance	Qualitative evaluation	Quantitative evaluation	Typical application
E0	Exceptional high-quality edge	0.0002*R*	0.01–0.02 mm		Interference microscope SEM	Diamond microtome knife edge
E1	High-quality edge	0.002*R*	0.3–5 mm	Cuts paper	Universal tool microscope, profile measuring machine, light section	Edge of cutting tools, edge of dies
E2	Sharp edge	0.02*R*	8–30 mm	Cuts fingernail	Same as above	Hydraulic orifice edge
E3	Rounded edge	0.2*R* or chamfer	0.08–0.3 mm	Will not cut finger	Stereo microscope Replica measurements	Mechanical parts, gyro pivots, piston rings, hydraulic spools
E4	Chamfered edge	0.5*R* or chamfer	0.4–0.6 mm	Naked eye, magnifying glass	Optical comparator	Mechanical parts
E5	Dull edge			No cut fingers	UL sharpness gage	Some automotive parts

(Takazawa and Kato 1997)

time a personnel change occurs in inspection, engineering, or manufacturing, a wealth of knowledge regarding intent and practice is lost. Furthermore, memories fade over time on some long-approved issues. *Written standards, when consistently used, can prevent this problem.*

Chapter 2 provides an overview of standards as well as references to additional works on this topic.

1.4 DEBURRING PROCESSES

Figure 1-6 illustrates the most commonly used deburring processes. Unfortunately, no single machine or process produces all the required edge quality on every edge for every burr without side effects. Table 1-3 outlines the known deburring processes in use worldwide. Each process has a segment of the edge finishing business to which it is particularly well suited. Some of the more common approaches, as well as some less common but novel approaches, are outlined in the following section.

1.4.1 Mass-finishing Processes

Vibratory deburring has been a mainstay of finishing for several decades, but today centrifugal disk machines (also called *roll flow* by some manufacturers) are encroaching on the popularity of vibratory machines because of their faster action. *Barrel tumbling*, another mass-finishing process, is still a major force in finishing some plastic eyeglass parts and other metal operations. *Centrifugal barrel finishing* is the fastest of the mass-finishing processes, but it is a batch process, whereas others, such as vibratory deburring, can be continuous-flow processes.

Which type of mass-finishing machine is the most cost effective continues to be debated and may depend on how knowledgeable the vendors and users are. Almost any existing process can be made more effective, and valid comparisons of output or costs depend on a clear understanding of how the comparison is being made. Comparing a poorly operated process with a new, optimized process will produce a different result than comparing an existing process that has been optimized with the new process.

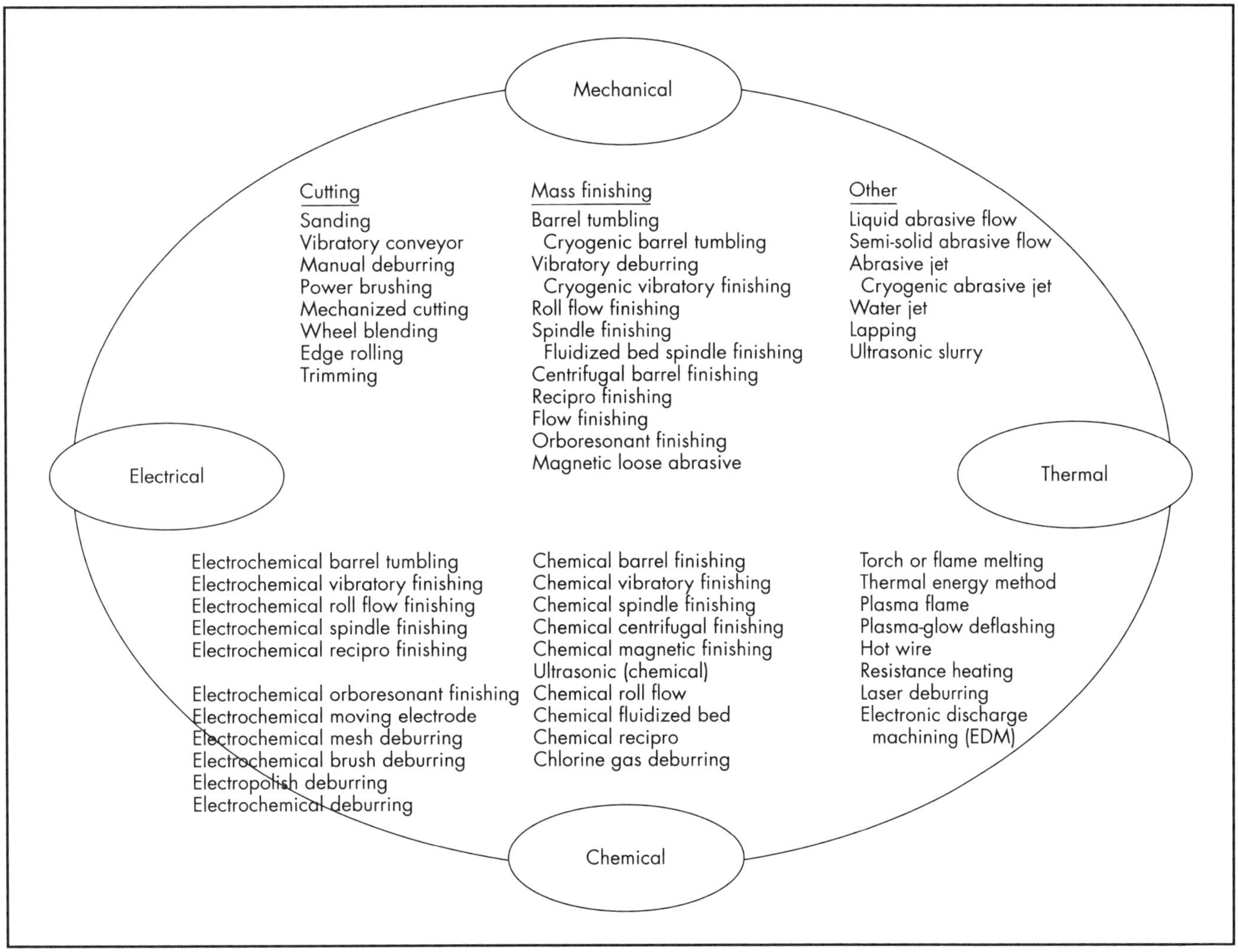

Figure 1-6. Principal deburring processes and their removal mechanisms.

Table 1-3. Known deburring processes (1997)[a]

Process		Quantity in U.S.	Quantity worldwide
Abrasive finishing (A)[b]	Barrel tumbling (Al)	8,000	16,000
	Loose belt tumbling (A1A)	10	50
	Vibratory finishing (A2)	12,000	30,000
	Vibratory shaker mixer finishing (A2s)	15	30
	Roll-flow (centrifugal disk) finishing (A3)	300	500
	Centrifugal barrel finishing (A4)	1,000	2,000
	Spindle finishing (A5)	500	1,200
	Vibratory spindle finishing (A5a)	0	20
	Fluidized bed spindle finishing (A5a)	1	40
	Recipro finishing (A6)	0	5
	Orboresonant finishing (A7)	5	10
	Flow finishing (A8)	5	10
	Cascading media (A9)	35	50
Chemical loose abrasive finishing (AC)	Chemical barrel tumbling (AC I)	100	200
	Chemical vibratory finishing (AC2)	200	400
	Chemical roll-flow (centrifugal disk) finishing (AC3)	5	10
	Chemical centrifugal barrel finishing (AC4)	10	20
	Chemical spindle finishing (AC5)	0	0
	Chemical fluidized bed spindle finishing (AC5a)	0	0
	Chemical recipro finishing (AC6)	0	0
	Chemical orboresonant finishing (AC7)	0	0
	Chemical flow finishing (AC8)	0	0
Cryogenic loose abrasive finishing (ACRY)	Cryogenic barrel tumbling (ACRYL)	50	90
	Cryogenic vibratory finishing (ACRY2)	50	90
	Cryogenic vibratory shaker mixer finishing (ACRY2s)	0	0
	Cryogenic roll-flow (centrifugal disk) finishing (ACRY3)	0	0
	Cryogenic centrifugal barrel finishing (ACRY4)	0	0
	Cryogenic spindle finishing (ACRY5)	0	0
	Cryogenic fluidized bed spindle finishing (ACRY5a)	0	0
	Cryogenic recipro finishing (ACRY6)	0	0
	Cryogenic orboresonant finishing (ACRY7)	0	0
	Cryogenic flow finishing (ACRY8)	0	0
Magnetic loose abrasive finishing (AM)	Magnetic abrasive barrel finishing (Aml)	5	400
	Magnetic abrasive vibratory finishing (AM2)	0	0
	Magnetic abrasive spindle finishing (AM5)	0	5
	Magnetic abrasive cylindrical finishing (AM5a)	0	5
	Magnetic abrasive tube-ID finishing (AM5b)	0	5
	Magnetic abrasive ball finishing (AM5c)	0	5
	Magnetic abrasive special shape finishing (AM5d)	0	5
	Magnetic abrasive prismatic finishing (AM7)	0	5
	Mixed metal fiber magnetic finishing (AM8)	0	1
Chemical magnetic loose abrasive finishing (AMC)	Chemical magnetic abrasive barrel finishing (AMCL)	0	0
	Chemical magnetic abrasive vibratory finishing (AMC2)	0	0
	Chemical magnetic abrasive spindle finishing (AMC5)	0	0
	Chemical magnetic abrasive cylindrical finishing (AMC5a)	0	0

Table 1-3. Known Deburring Processes (1997)[a] *(continued)*

Process		Quantity in U.S.	Quantity worldwide
Chemical magnetic loose abrasive finishing (AMC)	Chemical magnetic abrasive tube-ID finishing (AMC5b)	0	0
	Chemical magnetic abrasive ball finishing (AMC5c)	0	0
	Chemical magnetic abrasive special shape finishing (AMC5d)	0	0
	Chemical magnetic abrasive prismatic finishing (AMC7)	0	0
Electrochemical loose abrasive finishing (EAC)	Electrochemical barrel tumbling (EACL)	0	0
	Electrochemical vibratory finishing (EAC2)	1	1
	Electrochemical roll-flow (centrifugal disk) finishing (EAC7)	0	0
	Electrochemical centrifugal barrel finishing (EAC3)	0	0
	Electrochemical spindle finishing (EAC4)	0	0
	Electrochemical fluidized bed spindle finishing (EAC4a)	0	0
	Electrochemical recipro finishing (EAC6)	0	0
	Electrochemical orboresonant finishing (EAC7)	0	0
	Electrochemical flow finishing (EAC8)	0	0
Lapping (A9)		10	20
Abrasive jet (AB1)		20,000	40,000
Cryogenic abrasive jet (ABCl)		5	0
Liquid hone abrasive flow deburring (AB2)		1	5
Abrasive flow finishing (AB3)		500	700
Abrasive flow orbital finishing (AB3O)		2	4
Abrasive flow stream finishing (AB3s)		1	2
Ultrasonic liquid deburring (AUl)		20	30
Ultrasonic abrasive flow deburring (AB3U)		1	2
Ultrasonic slurry finishing (AU2)		10	15
Chemical deburring (Cl)		20	40
EDM deburring (El)		1	2
Electrochemical deburring processes (EC)	Electrochemical deburring (salt) (ECl)	600	1,200
	Electrochemical rotary electrode (EC1R)	1	2
	Electrochemical deburring (glycol) (EC2)	50	90
	Electrochemical mesh deburring (EC5)	0	0
	Electrochemical moving electrode deburring (EC4)	0	0
	Electrochemical brush deburring (EC3)	0	0
	Electrochemical nonwoven abrasive magnetic finishing (ECM6)	0	1
Electropolish (EL3)		50	100
Manual deburring (Ml)		300,000	800,000
Mechanized cutting (M2)		10,000	20,000
Skiving (M2s)		50	100
Robotic deburring (M2R)		300	1500
Brushing (M3)		50,000	100,000
Buffing (M3b)		1,000	2,000
Wheel blending (M4)		300	600
Sanding (M5)		25,000	50,000
Vibratory conveyor deburring (M5a)		1	1
Abrasive chemical finishing (MC1)		1	5
Edge rolling (M7)		400	600

Table 1-3. Known Deburring Processes (1997)[a] *(continued)*

Process	Quantity in U.S.	Quantity worldwide
Edge burnishing (M8)	300	300
Edge peening (M13)	1,000	2,000
Trimming (press work) (M9)	800	1,600
Edge coining (M10)	100	150
Water-jet deburring (M6)	500	800
Ballizing (M11)	1	1
Tearing (M12)	10	50
Torch-cut deburring (T1)	500	3,000
Plasma deburring (T3)	50	60
Plasma-glow deburring (T5)	5	5
Thermal energy method (T2)	500	700
Hot-wire deburring (T4)	1	1
Hot-blade deburring (T4b)	100	200
Hot-tool deburring (T4c)	100	200
Resistance heat deburring (T6)	1	1
Laser deburring (T7)	1	5
Chlorine deburring (TC1)	0	0

Note: Items in parentheses are codes for the specific process.

[a] These estimates are based on the intentional use of the process for deburring. Several of these processes also perform other finishing functions. Some of the processes have been studied in the laboratory but not used industrially. The quantities shown for manual deburring refer to the number of people estimated to perform deburring using some form of hand-held tool.

[b] All processes in this category, except fluidized bed spindle finishing, can use steel or metal media, which peens rather than abrades edges.

1.4.2 Spindle Finishing

Spindle finishing is widely used for some cylindrical parts. It employs many of the same elements of mass finishing and provides complete deburring in 2–5 minutes, but it requires individual part handling.

1.4.3 Less Common Loose Abrasive Processes

Some of the less well known loose abrasive processes, which have their own markets and applications, include *chemically accelerated mass finishing*, *cryogenic tumbling processes* for plastics and die castings, and a host of other minor variations. Adding acidic chemicals to mass-finishing processes will increase removal rates. *Magnetic abrasive finishing* still has not caught on in the United States despite its ability to produce surface finishes of 0.1 μm while removing .0012 in.-thick (30 μm) grinding burrs. Edge radii up to .0016 in. (40 μm) are produced in 4 minutes or less while finishing to these lustrous surfaces. The equipment is commercially available in Japan and is used for finishing camshafts, roller bearings, ball valves, sewing-machine components, hydraulic sleeves and valves, and molds; and for polishing taps, cookware, tubes, and holes to brilliant finishes.

1.4.4 Blasting

Blasting with abrasive particles or metal preforms is widely used for breaking off loose material at edges, cleaning, dulling edges, and removing sharpness. The steady impingement of particles on the surfaces is an advantage for some highly stressed parts, whereas that same effect is not tolerated for other high-precision parts because of potential damage or incomplete burr removal. Blasting cuts burrs or beats them over depending on the media used and the process parameters.

1.4.5 Bonded Abrasive Finishing

Bonded abrasive finishing, previously called *sanding,* is a major process for both mechanized and manual applications. For finishing minute crevices on single parts and robot-loaded machines, bonded abrasives continue to be effective in a variety of situations.

1.4.6 Brushing

Brushing plays a major role in effective edge finishing. The number of brush materials continues to increase, as do the brush designs themselves. Recent research on brushing dynamics has provided users with even better predictions of the effects of this process.

1.4.7 Abrasive Flow Processes

Abrasive flow processes extrude a semisolid over part edges, allowing inexpensive tooling to be used to produce smooth, burr-free edges. The process works best when edges allow media to freely flow over them. This process is ideal for intersecting holes and complex contours.

1.4.8 Electrochemical Deburring

Electrochemical deburring is ideal for hard-to-reach locations, intersecting holes, and precision parts. This process relies on salt or glycol solutions and electricity applied with a specially designed tool to concentrate material dissolution at the part edges. The burrs literally dissolve in 5–10 seconds with little effect on the rest of the part.

1.4.9 Electropolish

Electropolish combines an electrical current with solutions such as phosphoric acid to remove burrs. Burrs attract the electrical action much like a lightning rod, resulting in more material removed at edges than on surfaces. This process is often used for very thin burrs and highly polished surfaces.

1.4.10 Thermal Energy Method

The *thermal energy method (TEM)* is the fastest deburring process, relying on a high-temperature shock wave to burn or evaporate burrs before the heat dissipates into the part.

1.4.11 Mechanized Cutting

Mechanized cutting—whether on machining centers, special design machines, or robots—cuts burrs off with small rotary burrs or countersinks. Several companies produce mechanized machines that combine brushing, sanding, and cutting. Robot-deburring capabilities continue to improve, particularly with force feedback. Automotive and similar high-production manufacturers are principle users of this technology. One of its advantages is the ability of the robot to move either the part to the tool or the tool to the part.

1.4.12 Manual Deburring

Manual deburring is still the most widely used process for many reasons, including extreme flexibility, an inherent added inspection effect, use of low-cost tools, lack of technology needed, and ability to accommodate one part or thousands.

Sheet metal parts allow the use of many of the processes already listed, but processes such as skiving, edge rolling, and coining are particularly suited to these parts.

1.4.13 Less Commonly Used Processes

A variety of processes described in later chapters are also applicable to specific situations.

Water-jet deburring cuts off burrs with a fine stream of high-pressure water. Pressures exceeding 15,000 psi (103 MPa) provide the best cutting, but pressures as low as 5,000 psi (34.5 MPa) will knock off or beat down thin burrs.

Chemical deburring uses buffered acids to remove burrs, polish surfaces, and change part size.

Laser deburring promises fast, selective edge finishing, but the literature is still absent of any reference to actual applications other than trimming of plastic.

For very thin burrs and "hair-like" projections on plastics, *plasma-glow deburring* (a low-temperature plasma using argon or oxygen) can be effective. This process is used primarily for cleaning aerospace and electronic components, and deburring is a secondary application.

Ultrasonic abrasive-flow finishing uses an ultrasonic polisher to remove very fine burrs from critical parts. The 20,000-cycles per second vibration polishes surfaces and edges. Although only small burrs are removed with this polishing process, unlike ultrasonic machining, it requires no special tools. A blank stick of graphite and a slurry of 200–800-grit abrasive acts somewhat like a gentle jitterbug sander. The total stock removed by the process is typically .0005 in. (12 μm), so the process is most applicable to grinding, honing, lapping, and similar burr-removal processes.

Table 1-4 summarizes the key properties of each of the preceding processes. Although these are typical properties, the limitations and advantages shown in the table are generalizations, since nearly any process can be mechanized to meet a specific need. It is important to note that a part may be effectively deburred and radiused by a variety of processes at a similar cost. Visitors to major vendor facilities will see the same or similar parts in the showcases of many of these leading deburring equipment manufacturers. Each process does have its own product, geometry, cycle time, and cost portion of the finishing spectra, but that spectra overlaps many other process spectra.

1.4.14 Deburring Process Parameters

Like machining processes, deburring and edge finishing processes can be improved by varying many of the parameters. Some processes have several adaptations or tooling changes that can significantly alter the results. Later chapters describing each process present in detail the variables most often employed, but other variables can also be successfully used in specific circumstances.

Consider vibratory deburring. As shown in Table 1-5, which shows key processes used on precision parts, users can adjust 12–15 variables to obtain the desired results, thus providing a great breadth of capabilities. The amount of time in a vibratory deburring machine is a one-dimensional variable—more time provides more deburring. In contrast, there are at least 29 different shapes of vibratory finishing media, and each shape affects part features differently. For example, changing from one shape to another can affect three different portions of a single part such that one portion receives greater cutting, another less, and another the same. It is

Table 1-4. Key properties of commonly used deburring processes

Process	Tooling	Batch or continuous process	Typical applications	Typical limitations
Abrasive flow	Simple	Batch or single part/load	Intersecting holes, irregular contours, drilled plates, fine finishes	Blind features not readily deburred
Abrasive blasting	None, or simple	Batch, single part/load, or continuous	External burrs, some fine work	Abrasive or metal particles beaten into surface, residual stresses
Barrel tumbling	None	Batch	External features	Part geometry, internal features not deburred
Bonded abrasive (sanding)	None	Single part/load or continuous	Flat surfaces, accessible edges	Internal features not deburred
Brushing	None	Batch, continuous or single part/load	External edges, internal holes, blind features	Large burrs in hard to reach areas
Centrifugal barrel	None	Batch	External features	Internal features not deburred
Chemical	None	Batch or continuous	Parts that can tolerate size changes	Can cause some pitting, limited to metals
Cryogenic vibratory	None	Batch or continuous	Plastic deflashing or thin flash on conductive metals	Not applicable for normal burrs or most metals
Edge coining	Yes	Single part/load	High-production parts whose edges are readily accessible	Limited to sheet metal parts
Edge rolling/ burnishing/skiving	Yes, simple	Continuous	Sheet metal strips/coils	Not applicable to individual components
Electrochemical	Yes, simple to complex	Single part/load	Holes, intersecting holes, slots, gears, intricate contours	Total part with complex shapes
Electropolish	Yes, contacts	Batch or continuous	High-volume, high-finish parts with thin burrs	Will not remove thick burrs without distorting dimensions
Hand (manual)	Yes, hand tools	Single part/load	Low-volume, any configuration and burr size	Time consuming, error prone
Magnetic abrasive	Yes	Single part/load	High-finish parts with small thin burrs	Normal burrs, each part requires its own setup
Mechanized cutting	Yes, fixtures, cutters	Batch, continuous, or single part/load	External burrs or high-volume internal work	Varying burr size or location
Robotic	Yes	Single part/load	External burrs, normal edge breaks	Intricate features, low-volume
Roll flow	No	Continuous or batch	External edges	Not applicable to internal features
Spindle finishing	No	Single part/load	External cylindrical parts	Not suited to internal features or prismatic parts
Thermal energy	Yes, simple	Batch	External thin burrs and some internal thin burrs	Not suited to heavy burrs, some metals, precision finishes
Vibratory	No	Batch or continuous	External edges of most parts	Not suited to internal features
Water jet	Yes, fixturing	Single part/load	Thin burrs, metal or plastic deflashing	Not suited for thick burrs, hard-to-reach areas

simply inaccurate to say that changing shapes increases cutting, since the shape has varying effects depending on the part geometry. The variable "media shape" is multidimensional, and combined with other multilevel variables can result in hundreds and even thousands of combinations for improving deburring processes. This phenomenon is one reason that loose abrasive tumbling processes, such as vibratory finishing, are so widely used.

1.4.5 Deburring Side Effects Influencing Process Selection

A crucial factor in selecting a deburring process is knowing how the deburring process itself affects dimensions, finishes, cleanliness, flatness, plating, soldering, welding, residual stresses, surface imperfections, corrosion rates, luster, and color. The challenge to manufacturing engineers is

Table 1-5. Overview of key processes used on precision parts

Process	Number of key variables	Typical burr thickness removed (µm)	Typical effect on part size (µm)	Typical edge break (µm)	Normal best repeatability of edge radii (µm)	Surface finish range (µm)	Typical part size (mm)	Typical part materials
Abrasive blasting	7–10	25–75	None	75–250	±75	1–7	up to 1–1,000	Zinc, steel,plastics; no soft metals
Abrasive flow	5–7	75	5	75–250	±38	0.4	20–500	All metals
Barrel tumbling	13–15	75	5	25–750	±25	0.2–0.8	20–300	All metals, plastics, rubber
Bonded abrasive (sanding)	8–12	25–75	2–50	75–500	—	0.8–1.6	50–500	All metals
Brushing	5–10	5–75	0–2	50–250	±50	0.4	20–500	All metals, ceramics, some plastics
Centrifugal barrel	14–15	5–75	0–12	75–500	—	0.05–0.8	1.5–200	All metals, ceramics, some plastics
Chemical	4–5	2–12	1–50	25	—	0.15–0.2	20–500	Some limit on metals
Cryogenic vibratory	15–16	—	—	—	—	—	20-200	—
Edge coining	4–5	2-12	0	—	—	—	10-100	—
Edge rolling/ burnishing/skiving	4–5	2-12	0	25–75	—	0.05–1.0	—	—
Electrochemical	5–6	7–127	0	50–250	±50	0.4	20–500	Metals other than titanium
Electropolish	5–7	2–25	5–12	25–125	±1	0.05	20–200	Many metals
Hand (manual)	20–50	5–75	0	50–250	±50	—	5–500	All metals, ceramics, plastics, rubber
Magnetic abrasive	5–10	5–25	1–12	5–40	—	0.1	50–200	Metals
Mechanized cutting	5–10	75–250	0	125–750	—	—	50–500	All metals, plastics
Robotic	5–10	75–250	0	125–750	—	—	50–500	All metals, plastics
Roll flow	10–15	25–125	1–12	75–750	±30	0.2–1.0	20–200	All metals
Spindle finishing	10–15	25–125	—	75–750	±30	0.2–1.0	20–200	All metals
Thermal energy	3–6	5–50	0–10	25–75	—	0.8–2.0	20–200	Many metals, a few plastics
Vibratory	12–15	25–125	1–12	75–750	±30	0.2–1.0	20–200	All metals, some plastics
Water jet	4–6	12–50	0	50–125	—	—	50–500	All metals, plastics

Note: Blanks indicate a lack of published data.

(Gillespie 1976)

to remove burrs without adversely affecting part definition and function, to do it quickly at a reasonable cost, and to produce the desired edge definition at every edge. Table 1-5 outlines some of the predominant side effects of deburring processes such as part-size changes, edge breaks produced and their repeatability, and surface-finish results.

1.4.16 Deburring versus Part Finishing

It is important to recognize that in many plants the deburring operation typically encompasses removing burrs; taking out broken taps and drills; improving the finish; hiding scratches, nicks, and dings; completing the paperwork; removing stains and discoloration; as well as other functions. In other words, *"deburring" often means more than just removing the burr.* These additional, often unspecified functions are clearly beyond the capabilities of many deburring processes, although some processes can accommodate some of these added functions. When the intent is to reduce the time required to do any form of deburring, recognizing the complete breadth of work to be performed by those doing deburring is essential. Many so-called cost savings never occur after equipment is bought because those purchasing the equipment never take into account that those doing the deburring only spend a small part of their time actually doing deburring. For the purposes of this book, deburring means only to remove burrs and finish edges.

1.4.17 Burr Prevention

Some burrs can be prevented, and this should always be considered before investing any resources for burr removal. Chapter 3 provides a complete overview of burr prevention. Most users will find that although prevention is possible for some edges, it is not for others, so some deburring will still be required. A basic tenet of deburring is *if you do not make the burr, you do not have to take it off.* Burr minimization is always possible and is one path used to reduce burr problems and expense.

1.4.18 Data-driven Deburring

A strategy gaining increasing exposure as a result of total quality efforts is to quantitatively characterize the processes that produce burrs as well as the deburring processes. Manufacturing engineers are taking on the tasks, once held by researchers, of documenting how production variables affect burr size and deburring economics (Gillespie 1994). Simply put, improving any process requires a detailed knowledge of process mechanics and reactions. Statistical process control (SPC) data from a process and design of experiments are the two most prevalent tools used to provide production-line data. Characterization efforts include gathering data on the size of burrs produced as well as data on deburring effectiveness. Increasing drill feed rate, for example, significantly increases burr height (length) as well as thickness. For most processes, burr thickness is the most important variable in removal. Burr height has little to do with the economics of some processes; however, if a person does not know how thickness increases and how removal efforts are affected by thickness, making consistent improvements is difficult. Measuring edge radii and its repeatability after finishing, documenting the size of the "little" or inconsequential burrs left after vibratory finishing, and noting where they occur, even when not a problem, is a part of process characterization.

Deburring is complex and involves part geometry, process details, edge tolerances, and size effects. Broad generalizations are impossible. This is why it is essential to understand the effect of each process and how consistently it will be produced. How a process is controlled is also important. Detailed process knowledge allows engineers to calculate deburring times before the part is made. Processes can be changed and optimized to minimize deburring costs before committing time and resources to them.

1.4.19 Deburring versus Edge Finishing

The removal of burrs is one function of edge finishing. Edge finishing, however, is much broader in scope and has a more positive connotation than deburring. Consider that the edge of a broken glass has no burrs; it fractures in a brittle mode and leaves very sharp edges. It is erroneous to discuss deburring glass—there simply are no burrs. However, edge finishing encompasses every aspect of finishing and protecting an edge, as shown in the list of comparisons in Table 1-6.

Table 1-6. Comparison of attitudes toward edge finishing and deburring

Edge finishing	Deburring conceptually	Deburring in practice
Adding quality and function	Wasted effort	Removing problems
Removing burrs	Removing burrs	Removing burrs
Rounding edges	Rounding edges	Rounding edges
Removing sharpness	Fixing mistakes	Removing sharpness
Blending torn edges		Polishing
Generating knife edges		Cleaning
Generating longitudinal edge profiles		Completing paperwork
		Informal inspection

Deburring in some companies has a negative connotation or is referred to as cleanup, poor engineering, and lax machining, rather than as building in quality. To promote better deburring and to obtain the necessary resources, advocates may need to consider how the terms *deburring* and *edge finishing* are perceived by management. To some professionals *edge finishing* has a more positive sound. In some facilities deburring is perceived as removing someone else's mistakes. Furthermore, often it has been considered too unimportant to receive necessary resources or planning. In actuality, the problem of burrs is too complex to solve easily, and there are few experts in the field who can attack a company's complete burr problem and solve the system-wide problems within a company.

1.5 COST-EFFECTIVE DEBURRING

The steps involved in reducing the costs of deburring are:

1. Design to minimize deburring costs.
2. Practice burr prevention.
3. Minimize burr size.
4. Improve deburring processes.

As shown in Figure 1-7, these elements are part of an interlocking puzzle to reduce overall costs.

Designing to minimize deburring costs requires an understanding of how part geometry influences burr size and deburring costs and an understanding of what each edge contributes to and costs in part function. Geometry influences are discussed further in Chapter 5.

An understanding of how each edge affects product function and manufacturing costs is important in the design phase. Tables 1-7 and 1-8 are checklists for qualitatively assessing the impact of each edge. If these checklists are carefully filled out by the designer in conjunction with the manufacturing engineer, the key issues will have been addressed. To use these checklists, each edge of a part is marked with a number, and an assessment is made for each edge. The data support decisions about the requirements of each edge. Designers can then use the approaches discussed in Chapter 2 to provide edge-related drawing notes and details. Chapters 3 and 4 describe the basics of burr prevention and minimizing burr size.

Selecting a deburring process and documenting the details of its operation are steps in either an iterative process or a "this is good enough" decision-making process. The iterative process may be essential for those not familiar with the field and those searching for noticeably more economical solutions. Generally, consideration of schedule, work force, and equipment constraints combined with some previous experience will result in acceptable costs from the first iteration. The result may not be optimum, but the user will be able to determine if the cost is in the range of allowable costs. Purchasing new types of equipment and new processes requires experimentation and data-driven solutions—solutions that document the size of burrs removed, numerical values of edge conditions produced, time to obtain, and associated indirect costs.

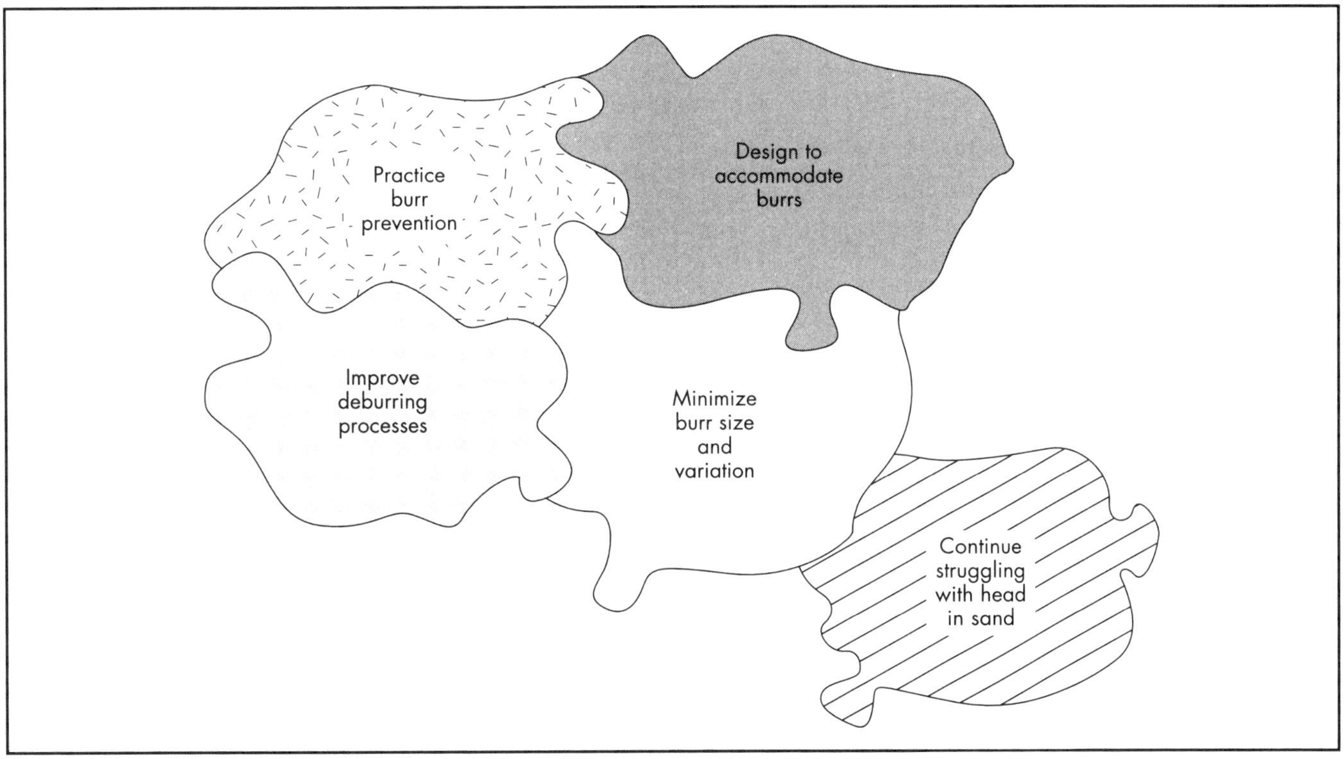

Figure 1-7. Key Elements in deburring technology (Gillespie 1972).

Table 1-7. Designer's checklist for assessing edge function

Edge function	Edge 1	Edge 2	Edge 3	Edge 4
None	X			
Clearance	X			
Meshing				
Cutting		X		
Electrical contact		X		
Minimize contact stresses			X	
Aesthetics				
Prevent cutting				
Reduce residual stresses				X
Reduce high-voltage breakdown				
Prevent microwave detuning				
Reduce plating buildup				
Prevent edge cratering, fractures, crumbling				
Reduce turbulence				
Reduce forming deformation				

Table 1-8. Designer's checklist for assessing edge concerns

Would burrs:	Edge 1	Edge 2	Edge 3
Cut hands?	no	yes	
Cause an assembly interference?	yes		
Jam mechanisms?	yes		
Scratch mating surfaces?	yes		
Increase friction?	yes		
Increase part wear?	yes		
Short-circuit device?	no		
Cut wires?	yes		
Cause high-voltage breakdown?	no		
Disrupt magnetic/electric fields?	no		
Detune microwaves?	no		
Contaminate assembly?	no		
Clog filters and ports?	no		
Cut rubber seals?	yes		
Create stress concentrations?	yes		
Cause plating buildup?	no		
Cause paint buildup?	no		
Cause edge fractures/ crumbing?	no		
Create turbulence/nonlaminar flow?	no		
Reduce sheet metal formability?	no		
Come loose during use causing other problems?	no		

Tables 1-9 and 1-10 provide specific details on the capabilities of the various deburring processes. The information in these tables, and in Tables 1-3 and 1-5, is provided because deburring requires a knowledge of all of the following factors:

- Burr properties—thickness, height, and hardness in relation to part hardness.
- Allowable dimensional changes in the part resulting from the deburring operation.
- Required edge radius.
- Required final surface finish as well as the surface finish of the part before deburring.

Although conditions such as burr location, part geometry, and part material are important, a knowledge of the preceding factors is essential for the preliminary selection of deburring processes for precision parts. Figure 1-8 illustrates these basic requirements. For an accurate selection of a deburring process for the simple cube shown in in the figure, information that must be known includes the approximate hole size, the amount of stock loss from the hole resulting from deburring, the repeatability of the stock loss, the initial surface finish, the required final surface finish, and the required edge radius at the entrance and exit of the hole.

The size of the hole is important because some deburring processes will not work well on small holes and others will not work well on large holes. When precision holes are involved, both the average stock loss and the repeatability of the stock loss must be ascertained. Although the holes can be drilled smaller to accommodate the stock loss from deburring, this action is of little help if the stock removal is not repeatable.

Similar data pertaining to the external edges of the part are also needed. The selection of deburring processes for a complex part involving many edges, surfaces, and burr sizes may require considerable thought. Although the availability of actual data is highly desirable, previous experience may allow users to select the appropriate process. However, when the allowable stock loss must be kept below .0001 in. (2.5 μm), there is little margin for error.

Table 1-9 indicates the processes that can be used to deburr external edges while maintaining a maximum allowable edge break, an allowable stock loss, and a given stock-loss repeatability. The data in this table are valid only for burrs .003 in. thick × .003 in. high (76.2 × 76.2 μm) on a stainless steel workpiece requiring a surface finish of 32 μin. (0.81 μm) or better. An initial surface finish of 32 μin. (0.81 μm) is assumed, although some deburring processes would improve a rougher finish.

Table 1-9. Typical processes for precision deburring of external metal edges

Maximum allowable edge break (μm)	Allowable thickness change on external surfaces (μm)				Allowable thickness change repeatability (μm)		
	1.25	2.50	25.0	125	0.5	5.0	50
2.5	—	—	—		—	—	—
25	—	—	AB3	AB3	—	AB3	AB3
50	—	M2, M5, EC	AB3, M2, M5, EC	AB3, M2, M5, EC	EC	AB3, EC, M2, M5	AB3, EC, M2, M5
125	EC, M1–3, M5–8	EC, M1–3, M5	A, AM, AB1, AB3, E1, M1–5, M7, T1, T2	Most	EC, M1–3, M5	A, AB3, EC, M1–5	Most
250	E1, EC, M1–8	E1, EC, M1–M10	A, M, AB1, AB3, E1, M1–10, T1, T2	All	E1, EC, M1–8	A, AB3, EC, M1–5, T2	All

Note: Letters are codes for processes—see Table 1-3 for definitions. To use this table select the maximum edge break that a drawing allows and then move horizontally across the row that has that edge break or the next largest value. Move until you reach a column that has an allowable stock loss (part-thickness change). The processes shown in that cell will most likely yield acceptable effects on stainless steel parts having a surface finish requirement of 32 μin. (0.8 μm) or better (1 μm = 39.37 μin.). Data are based on complete removal of burrs having an initial thickness and height of 0.003 in. (76 μm). The last three columns provide an estimate of size-change repeatability for the various processes and conditions stated. Dashes indicate that no process will produce the desired result for the specified burr size, edge break, surface finish, part material, or stock loss. In this case the burr must be hand worked or machined to a smaller size before edge-finishing processes can provide complete burr removal.

Table 1-10. Typical processes for precision deburring of intersecting holes

Maximum allowable edge break (μm)	Allowable hole diameter change repeatability (μm)				Allowable hole diameter change (μm)		
	1.25	2.50	25.0	125	0.5	5.0	50
2.5	—	—	—	—	—	—	—
25	—	—	—	—	—	—	—
50	EC1–2, T2	EC1-2, T2	EC1–2, T2	EC1–2, T2	—	EC1–2, T2	EC1–2, T2
125	EC1–2, T2	EC1-2, T2, M1	EC1–2, M1–2, T2, AB3, E1, M5	C1, EC1–2, EL3, M1–2, T2, AB3, E1, M5M2	EC1–2	EC1–2, T2, M1, M2	Most
250	EC1–2, M2, T2	E1, EC, M1-2	EC1–2, M1–2, T2,AB3,E1, M5	C1, EC1–2, 2, EL3, M1–2, 5, T2, AB3, E1, M5	EC1–2, M2	AB3, C1, EC1–2, EL3, M1–2, 5, T2	Most

Note: To use this table select the maximum edge break that a drawing allows and then move horizontally across the row that has that edge break or the next largest value. Move until you reach a column that has an allowable stock loss (hole-diameter change). The letters in that cell define the processes most likely to yield acceptable effects on stainless steel parts having a surface finish requirement of 32 μin. (0.8 μm) or better (1 μm = 39.37 μin.). The data are based on complete removal of burrs having an initial thickness and height of .003 in. (76 μm). Letters are codes for processes—see Table 1-3 for definitions. The last three columns provide an estimate of size-change repeatability for the various processes and conditions stated. Dashes indicate that no process will produce the desired result for the specified burr size, edge break, surface finish, part material, or stock loss. In this case the burr must be hand worked or machined to a smaller size before edge finishing processes can provide complete burr removal.

The information shown in the tables is based on the best available data at this time. Additional research may necessitate changes and rearrangements in some of the tabular material.

When two holes intersect, the angles between the hole surfaces vary considerably around the intersection. Because many, if not all, deburring processes are affected by the edge angle, deburring tolerances are also affected. As shown in Table 1-10, no process will maintain a .002 in. (50.8 μm) radius for the burr size and surface finish just mentioned when the allowable hole-size repeatability must be held to 19.6 μin. (0.5 μm) or better. In this instance the cause of the process inability is an effect of the varying angles of intersection as well as natural process capabilities.

When the burr size is reduced to a thickness and height of .001 in. (25.4 μm), several processes can be used to produce the desired results (Gillespie 1977b). Precision parts from the automatic screw machines typically have burrs of this size. Consequently, they generally can be deburred by most processes without excessive loss in stock or radii. As parts become smaller, however, some processes such as centrifugal barrel finishing begin to lose their effectiveness. For example, a .020 in. (0.5 mm) pin, .125 in. (3.18 mm) long, may require twice the time to remove the same size burr as that on a .125 in. (3.18 mm) pin of the same length. Threaded parts represent a problem to many deburring processes when the thread features are small and have close tolerances.

The significance of Tables 1-9 and 1-10 can be appreciated by briefly scanning the data on typical burr sizes described in each of the later chapters and elsewhere (Gillespie 1972). As seen in Tables 1-9 and 1-10, a typical burr by most processes is .002–.003 in. (50.8–76.2 μm) thick. No single process appears capable of removing these size burrs while maintaining stock losses less than .0005 in. (12.7 μm) and guaranteeing final edge radii of .002 in. (50.8 μm) or less. If edge radii of .005 in. (12.7 μm) are allowable, several processes can be used.

When parts require small precision radii and low loss conditions, only two possibilities exist: (1) develop nonstandard variations of the deburring processes, or (2) make smaller burrs. The following is a rational approach to selecting a deburring process (Gillespie 1976). Variations of this approach also work.

1. Determine typical burr sizes produced today or in the past.
2. Determine drawing edge requirements.
3. Eliminate those processes, from Table 1-4, whose limitations make them inappropriate.
4. Using Tables 1-5, 1-9, and 1-10, determine which processes meet the requirements of the part drawing.
5. Using the list resulting from step 4, review Chapters 6–19 in this book for more details.
6. Estimate production rates and deburring costs of the processes remaining from step 5.
7. Select the best process from step 6 or continue obtaining more details from manufacturers, distributors, job shops, or other sources, such as those listed in the references and bibliography.
8. Verify the selection with experienced plant staff.
9. Implement the best process.
10. Fine-tune the experimental results.

Note that the data shown in Tables 1-5, 1-9, and 1-10 are typical for hand-sized and smaller parts, but the processes also can be used for larger parts, nonprecision parts, and bigger burrs. As shown in Tables 1-9 and 1-10, normal burr size, sharp edges, and near-zero part-size changes severely constrain the potential deburring processes.

If the preceding approach fails to yield an acceptable deburring process, users must carefully consider using nonstandard processes (which generally are slower or more costly), changing part tolerances or specifications, making smaller burrs, or contacting deburring specialists to determine alternatives.

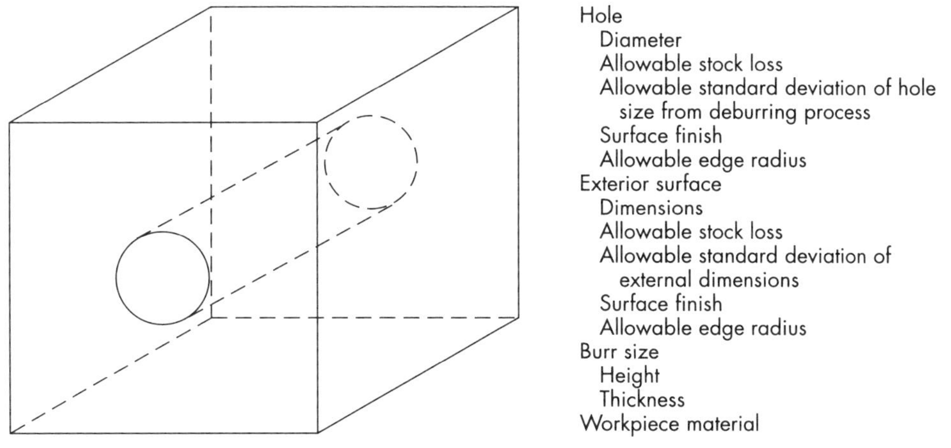

Figure 1-8. Considerations for selecting a deburring process for a workpiece with hole (Gillespie 1978).

1.5.1 Deburring Economics

Few examples of comparative deburring economics have been published. The following equations provide a reasonable estimate of costs using various processes. The equations use the following terminology:

C = Deburring cost per part.
C_D = Depreciation cost per hour = machine cost per operating hours.
C_M = Maintenance cost per hour of operation.
C_L = Labor cost per hour to run machine.
C_p = Cost of power used ($ per kilowatt-hour).
C_A = Cost of cleaning per hour after deburring (labor + material).
C_E = Cost of media per hour = media cost × percentage of hourly attrition.
C_C = Cost of compound per hour.
C_W = Cost of water per hour.
D_o = Overhead as percentage of labor rate.
N = Number of parts run per hour = n/t.
n = Number of parts run per cycle.
t = Time (hours) per cycle.
C_g = Cost of gas per cycle.
C_b = Cost of brush.
C_B = Cost of cleaning materials per hour.
N_p = Total number of parts run.
N_{p1} = Number of parts run for a given quantity of solution or tool life.
C_t = Total tool cost.
C_s = Total cost of solution.
W = Power used, in kilowatts (1 hp = 0.75 kW).
K_1 = Percentage of cycle time that operator actually spends controlling deburring operation.
K_2 = Percentage of cycle time that operator spends cleaning parts.

For vibratory (loose abrasive) processes, use

$$C = [C_D + C_M + WC_P + C_B + C_E + C_C + C_W]/N + [C_L(1 + D_O)(K_1 + K_2)]/N \quad (1\text{-}1)$$

For thermal energy methods, use

$$C = [C_D + C_M + C_L(1 + D_O)\,WC_P + C_A]/N + C_g/n + C_t/N_p \quad (1\text{-}2)$$

For brush deburring, use

$$C = [C_D + C_M + C_L(1 + D_O)\,WC_P + C_A]/N + C_b/N_{p1} \quad (1\text{-}3)$$

For flame deburring, use

$$C = [C_D + C_M + C_L(1 + D_O)\,WC_P + C_A]/N + C_g/n \quad (1\text{-}4)$$

For manual deburring, use

$$C = [C_L(1 + D_O) + C_A]/N + C_t/N_p \quad (1\text{-}5)$$

For mechanical deburring, use

$$C = [C_D + C_M + WC_P + C_A + C_L(1 + D_O)]/N + C_t/N_p \quad (1\text{-}6)$$

For chemical deburring, use

$$C = [C_D + C_M + WC_P + C_A + C_L(1 + D_O)]/N + C_s/N_p \quad (1\text{-}7)$$

For electrochemical deburring, use

$$C = [C_D + C_M + WC_P + C_A + C_L(1 + D_O)]/N + C_t/N_p + C_s/N_{p1} \quad (1\text{-}8)$$

For electropolish deburring, use

$$C = [C_D + C_M + WC_P + C_A + C_L(1 + D_O)]/N + C_t/N_p + C_s/N_{p1} \quad (1\text{-}9)$$

These equations have some important limitations. First, they assume that the conventional form of the deburring process is used. As mentioned earlier, it is frequently possible to alter the process slightly to obtain faster or better results. Such alterations may insert another cost term into the equation, however. Unless the conventional approach is used, these equations provide only initial estimates of cost.

A second limitation of these equations is that they assume that one knows the value of each component and the time required to remove the burr. Although it is possible to use "rule of thumb" costs for media, compounds, and the like, only a few publications provide any information on the time required to remove a burr of a specific size. As additional research is reported, this will become less of a limitation. In the interim, information can be extrapolated from the results produced by other parts subjected to the same process.

A third limitation of these equations is that they ignore the costs of floor space, area heating, lighting, maintenance, insurance, and supervision. These costs can add $1–2 more per hour than indicated by Equations (1-1)–(1-9).

For example, assume that 400,000 parts are to be deburred and that the machine used must have a life that will accommodate that many parts. Also assume that the values shown in Table 1-11, which shows examples of costs for various processes, are representative. For the values shown in the table, the thermal energy method would be the least expensive process (calculated as C = $.02 per part), whereas manual deburring would obviously be an undesirable approach ($1.21 per part). These calculations are based

Table 1-11. Typical costs of various processes (in U.S.$ per unit)

Cost item	Process Vibratory	Thermal Energy Method (TEM)	Manual	Chemical
C_D	.40	5.00	—	.20
C_M	.04	1.00	—	.02
C_L	5.00	5.00	5.00	5.00
C_p	.04	.04	.04	.04
C_A	—	5.80	5.50	5.50
C_E	.60	—	—	—
C_C	.30	—	—	—
C_W	.15	—		—
D_o	.8	.8	.8	.8
N	50	1,500	12	400
n	100	6	1	100
t	2	.004	.08	.25
C_g	—	.024	—	—
W	4	4	0	0
C_t	—	1,000	1,000	—
N_p	400,000	400,000	400,000	400,000
C_s	—	—	—	6,000
Calculated C	.106	.020	1.210	.052
K_l	.17	—	—	—
K_2	.17	—	—	—
C_B	.60	.80	.60	.50

Note: This table assumes that no automatic load/unload is used.

on the assumption that the process will remove the burrs without adversely affecting parts.

The data shown in Table 1-11 are reasonable estimates of costs for working with precision aerospace components, but before making decisions the values should be discussed with knowledgeable vendors or users. The number of parts per cycle and the cycle duration are functions of part size, burr size, and other variables.

Note that vibratory deburring is typically one of the least expensive processes that can be used if the burrs are accessible. However, three factors can increase the cost of this process:

1. After deburring, special cleaning is required to remove all traces of compound and loose particles (common on precision aerospace parts but often not on commercial parts).
2. A relatively long run time and small quantity of parts per cycle.
3. The thermal energy method and chemical processes are automated, whereas the vibratory method is not.

In most cases these three factors would not be as pronounced. In all equations except (1-1), it is assumed that the operator is devoting all of his or her time to the deburring operation.

A computer program to estimate finishing costs using loose abrasive processes is available and has been described in the literature (Kittredge 1985).

As mentioned earlier, almost any process can be made more effective than it is, so valid comparisons of output or costs of different machines depends on a clear understanding of how the comparison is being made. For example, are you comparing how you use the machine or how another expert would use it? If workers that are more knowledgeable were hired, would the same process result in lower costs? In addition, accounting procedures vary and can provide conflicting results; for example, one accounting procedure used by some companies assigns costs simply by the number of days a part or assembly is in house and ignores all other variables. Needless to say, it is important for users to understand their company's practices for establishing process and part costs.

1.6 ENVIRONMENTAL, HEALTH, AND SAFETY CONCERNS

Some deburring processes involve environmental, health, or safety concerns that must be addressed. It is important to investigate these concerns before purchasing new processes and supplies. Table 1-12 provides an overview of these considerations. The list of issues summarized in the table is not meant to be comprehensive, but it does provide some initial insight. Every process listed is currently in industrial use, and means have been found to accommodate all environmental, health, and safety issues while meeting production demands. As local, regional, national, and international regulations and laws change, this list will need to be updated.

1.7 TRICKS OF THE TRADE

Every machinist, inspector, engineer, and deburring operator has developed methods for reducing burr-related problems. These "tricks of the trade" are presented throughout this book, and references to other sources of ideas are given in the reference lists and bibliographies at the end of each chapter. These innovative ideas have reduced costs for many companies—they often cost little yet have benefits for years.

References

Datsko, Joseph. 1966. *Material Properties and Manufacturing Processes.* New York: Wiley & Sons.

Table 1-12. Environmental, health, and safety concerns in deburring

Process	Environmental issues	Safety issues	Health issues
Abrasive blasting			Breathing dust particles
Abrasive flow		Hydraulic clamping	
Barrel tumbling	Disposal of compounds in water		
Bonded abrasive (sanding)		Explosions from excessive dust in air	Breathing dust
		Accumulations of metal dust can cause fires	Nickel dust from stainless steel is suspected carcinogen
Brushing		Flying metal bristles can injure	
Centrifugal barrel	Disposal of compounds in water		
Chemical	Metal compound wastes	Acid splashes	Breathing acid fumes
Cryogenic vibratory			Low temperatures can damage body tissue
Edge coining		Hands in pinch points	
Edge rolling/burnishing/skiving		Hands in pinch points	
Electrochemical	Disposal of metallic compounds in solutions	Skin irritation from salt solutions	Long-term effects of glycol breakdown may be an issue
Electropolish	Disposal of metallic compounds	Acid on skin	Skin irritations
			Fumes from acids
Hand (manual)		Parts or tools thrown by hand motors	Ergometric damage such as repetitive motion damage
		Brush bristles hitting skin or face	
Mechanized cutting		Flying chips	
Robotic		Unexpected robot motion hitting worker	
Roll flow	Same as vibratory		
Spindle finishing	Same as vibratory		
Thermal energy		Controlled explosions must meet fire codes	
		Explosive gases	
Vibratory	Disposal of compounds in water		
Water jet		High pressures can cut skin	

Gillespie, LaRoux K. 1972. *Deburring Capabilities and Limitations*. Dearborn, MI: Society of Manufacturing Engineers (SME).

———. 1976. "Deburring Capabilities for Miniature Precision Parts." Report BDX-613-1604. Kansas City, MO: Bendix (available from NTIS).

———. 1977a. "Machinability of Metals as Related to Precision Miniature Parts." Division report BDX-613-1723. Kansas City, MO: Bendix (available from NTIS).

———. 1977b. "Side Effects of Deburring Processes." Technical paper MRR77-17. Dearborn, MI: Society of Manufacturing Engineers (SME).

———. 1978. *Advances in Deburring*. Dearborn, MI: Society of Manufacturing Engineers (SME).

———. 1982. "The Management of Deburring." Technical paper MRR82-04. Dearborn, MI: Society of Manufacturing Engineers (SME).

———. 1994. "Process Control for Burrs and Deburring." Report TR94-1. Kansas City, MO: Deburring Technology International.

Kittredge, John B. 1985. "Computer-aided Mass Finishing Process Development." Technical paper MR85-836 and computer disk. Dearborn, MI: Society of Manufacturing Engineers (SME).

Takazawa, K., and Kato, Y. 1997. "Edge Quality Classification and Grade." *Proceedings of BEST-J*. Tokyo, Japan: Best J: 3(17): p. 45.

Wick, C., and Veilleux, R., eds. 1985. *Tool and Manufacturing Engineer's Handbook (TMEH)*, Vol. 3, 4th ed. Dearborn, MI: Society of Manufacturing Engineers (SME).

Bibliography

Alting, L., and Nygaard, F. 1980. "Systematic Generation of Deburring Processes." Technical paper MR80-339. Dearborn, MI: Society of Manufacturing Engineers (SME).

Gillespie, LaRoux K. 1982. "The Management of Deburring." Technical paper MRR82-04. Dearborn, MI: Society of Manufacturing Engineers (SME).

Ioi, Toshihiro, Kobayashi, Hisamine and Matsunaga, Masahisa. 1981. "Computer-aided Selection of Deburring Methods." Technical paper MR81-389. Dearborn, MI: Society of Manufacturing Engineers (SME).

Narayanaswami, Rangarajan, and Dornfeld, David A. 1994. "Design and Process Planning Strategies for Burr Minimization and Deburring." *Transactions of NAMRI*. Dearborn, MI: Society of Manufacturing Engineers (SME).

2 Plant Standards and Edge Needs

The processes of deburring and edge finishing require some understanding of what the finished edges should be. Generally, users desire edges that are free of all burrs and have some small radius or chamfer left to allow easy assembly. The term *burr-free*, however, includes many different levels of edge quality. In microelectronics a single loose metal particle, such as a burr only .0002–.0004 in. (5–10 μm) long, can short circuit a device, causing erroneous function and perhaps tens of thousands of dollars of equipment loss. In contrast, automotive manufacturers tolerate burrs and sharp edges that can require mechanics to wear gloves. Design drawings of most devices typically stipulate or imply that the product shall be "burr-free," yet within a manufacturing plant workers often know that this specification does not really mean that every edge of every part must be totally devoid of burrs or other material. Obviously, each manufacturer has a standard, but these standards are not uniform across industries or, generally, even within companies.

This chapter presents an overview of existing standards and a sample standard suitable for companies to adapt for their own applications.

2.1 TYPES OF EDGE STANDARDS

Every plant has standards for burrs and edge finishing. In some plants standards may not be in written form, and they may vary from day to day and between individuals. In such plants there is some general understanding of allowable edge conditions. Some plants have comprehensive written standards. Typically, large companies will have detailed written standards because consistency is important, and written standards are the only way to ensure some measure of consistency throughout the plant. A few companies may have such stringent process control that the control provides the consistency.

Standards are meant to make life easier for those who work in a specific field. They help to assure a common level of performance and understanding. *They do not, however, necessarily restrict users to a single solution*. Many standards define *multiple allowable levels* or states. In addition, the multiplicity of standards in use testifies to the breadth of alternatives that can be tolerated when a single standard is not adequate.

Most manufacturing and design personnel have encountered instances in which their assumptions about edge quality differed markedly from those of the others. Such instances highlight the need for a common language within plants, between plants, and among individuals. If a design drawing only contains the statement "deburr" or "burr-free," does this mean "visible with the unaided eye" or "under magnification"? Does the fact that magnification is used on some parts in the plant imply any legal requirement that all parts must be finished to that same level of quality? The engineering answer is often different from the legal answer. The existence of and reference to a national standard or to comprehensive company standards prevents or greatly reduces such problems.

The need to establish a standard that can be communicated in multiple languages, with multiple cultural and practice differences, further reinforces the need for common terminology and definitions. Words take on different meanings in different countries, and in some countries the concepts implied by some words are so powerful that explicit definitions may be unnecessary. In most countries, however, this is not the case. With the current trend toward international manufacturing liaisons, electronic communications, and rapid decision making, clear standards become even more important for meeting delivery deadlines, quality standards, and cost expectations.

The ideal standard is one that can be widely used and understood. Currently, industry is still diverse enough that many companies need only simple standards to conduct their dialog on edge conditions. Existing specialty standards in the United States include the American Metal Stamping Association (AMSA) standard for allowable burrs on stamped parts (AMSA 1976), the Washer Division of AMSA standard for stamped washers (AMSA 1977), and the American Iron and Steel Institute (AISI) standard for allowable edge conditions of sheared sheet metal, as defined by the American Society for Testing and Materials (ASTM 1974). The American Die Casting Institute (ADCI) has two standards for allowable flash on die-cast parts (ADCI 1955). Allowable edge sharpness is also defined by Underwriters' Laboratories (UL 1976, 1993) and the Consumer Product Safety Commission (Consumer Product Safety Commission 1975, 1976). Powder coated edge quality is defined by the American Society for Testing and Materials ASTM (ASTM 1977).

The National Institute for Science and Technology (formerly the National Bureau of Standards) proposed a standard for detecting hazardous edges (Sorrells and Berger 1974) that Underwriters' Laboratories formalized (UL 1993). The standards listed here cover only a small part of the need, and each is summarized in SME's *Deburring Technology for Improved Manufacturing* (Gillespie 1981).

The Worldwide Burr Technology Committee (WBTC), a group of leading deburring authorities from the major industrialized countries, provides seven different standards for use by small and large shops around the world. The most simple standard is two pages long and is most suitable for very small shops (Gillespie 1978, 1995, 1996a–e). The German DIN 6784 standard (used by some companies around the world) also provides a means to define edge conditions.[1]

2.1.1 Criteria for Good Industry Standards

For a standard to be useful, it must contain the following elements:

1. Clear guidelines about when to apply the standard.
2. Clear rules about how to specify the standard.
3. Clear expression of the standard itself, including unequivocal language and terminology.
4. A business reason for the specifications in the standard.
5. Awareness that the standard exists.
6. A mechanism for periodically reviewing the standard.
7. A mechanism for incorporating user comments.

In the administration of national standards, some countries require the governing agency generating a standard to assure that those affected by it have a legitimate voice in its establishment.

2.1.2 International Standards for Burrs and Edge Finishing

At least three levels or forms for specifying standards are required by most design and manufacturing firms worldwide:

1. Design specifications.
2. Manufacturing specifications.
3. Inspection specifications and practices.

Design Specifications

Design requirements can be specified by notes, graphical icons, and/or textual (numerical) references to other documents. Drawings are expected to describe the complete requirements of the part or assembly, functionally and aesthetically. Edges are part of that description. Therefore, clear expectations of edge configuration are a part of drawing definitions. Currently, U.S. national standards for drawings do not adequately define edge quality needs.

Drafting personnel and designers generally concentrate on design needs and intent. The drawing standard (document) visually or textually describes the edge requirements and assures that the words used are defined such that the intent of the requirement is clear and unambiguous. Special symbols for edges are defined in this document or referenced in the drawing to another document.

Manufacturing Specifications

The manufacturing staff turns drawings into products through one or more processes, each of which can produce different edge characteristics. To clearly convey instructions to machine operators, various forms of further instructions and specifications are necessary. One company's manufacturing specification for adequate edge quality may state "Blast surfaces and edges with 600 mesh aluminum oxide to remove burrs and lose particles"; another company may provide more detailed edge finishing instructions. Instructions typically contain many illustrations, enabling operators to clearly understand the needs of each edge. Thus, manufacturing specifications differ from drawing specifications. Operators need more detailed, and in some cases different, information than drawings contain. For example, operators sometimes produce edges that are later machined away; drawings may not address such edges, but they are important to those manufacturing the parts. Thus, staff responsible for manufacturing sequences and operations must recognize final as well as intermediate edge needs.

Edge standards may require several quality levels, and standards in production may need to be higher than those specified by drawings or the plant's general needs. Edge standards in manufacturing include process routing instructions, manufacturing specifications (departmental, product related, or plant-wide procedures), edge finishing machine operation instructions, and machine calibration instructions.

Inspection Specifications and Practices

Inspection standards answer such questions as "Explicitly, how will the inspector check for the presence or absence of burrs?" One company may use the unaided eye alone, and another may use a number 2 pencil to probe edges. Some will paint part edges to illustrate particular areas to check more closely. Each approach may produce the same result. A company may choose to produce and inspect edges at a higher quality level than defined in drawings in order to set itself apart as a precision facility capable of extra quality. In some cases edges can only be monitored by a unique device such as a borescope. Inspection standards or specifications help assure that both the intent of the drawing and company policy are met.

Inspection standards include instructions for inspectors describing what and how to inspect, quality levels for acceptance, tools and equipment to use, calibration instructions, and record-keeping details. There are over 40 different ways to measure burrs, and, consequently, different edges, different products, and different materials may dictate using many different approaches to determine the presence of burrs.

With three levels or forms of edge standards, a company can assure complete conformance to edge expectations; communicate clearly to customers, vendors, and in-house staff; and consistently, year after year, guarantee conformance despite changes in personnel.

2.2 PLANT STANDARDS FOR EDGES AND BURRS

In the following sections, a sample standard is presented that may be adapted to suit a particular company's needs.

[1] DIN 6784, available from Document Center, 1504 Industrial Way, Unit 9, Belmont, CA 94002.

This standard provides a framework for integrating all the requirements within a company. The sample standard is based on the Worldwide Burr Technology Committee's Standard WBTC—STD14.1997 draft (Gillespie 1997). It is compatible with ISO 9000 requirements for clear definition of expectations. Although the sample standard may be used as is, sections may also be adapted or intermingled with a company's existing standards.

2.2.1 Sample Standard

Table 2-1 summarizes the elements of the sample standard and defines the hierarchy of requirements. The table uses a numbering system to identify the various elements of the standard and how the elements are organized. For example, section 01.01 is subordinate to section 01. Each successive level refers back to its parent level to provide a coordinated structure.

Section 01: Corporate Quality Expectations (for Edges and Burrs)

Products designed or produced by this company shall consider the effects of burrs and edges and provide appropriate actions and controls. Product designs shall identify edge requirements. Technical staff shall be aware of edge requirements.

Section 01.01: Design Requirements (for Edges and Burrs)

This requirement is driven by corporate quality expectations and provides a complete approach to defining edge conditions on product drawings. All new and current product designs shall conform to this system. Manufacturing and inspection instructions and practice will assure that design requirements are met.

Burr and edge terminology used in this standard are defined in the Appendix at the end of the chapter. Edge finishing quality levels "A–G" are defined in the following list of edge quality definitions. These levels should be specified when describing desired edge qualities. When a specific level is *not* specified on a drawing or purchasing document, "deburring is not required" is implied. The level designated (including "deburring is not required") applies to the entire part or component unless specifically stated otherwise.

Table 2-1. Outline of burr standards using an ISO 9000 approach

Standard no.	Directed at the corporate level	Directed at design engineering	Directed at manufacturing engineering	Directed at the shop floor
01	Quality expectations			
01.01		Design requirements		
01.01.01			Manufacturing requirements	
01.01.01.01				Work instruction details
01.01.01.02				Machine operation details
01.01.01.03				Training details
01.01.01.04				Calibration details
01.01.01.05				Documentation details
01.01.02			Inspection requirements	
01.01.02.01				Inspection instruction details
01.01.02.02				Inspection techniques and equipment details
01.01.02.03				Training details
01.01.02.04				Calibration details
01.01.02.05				Documentation details
01.01.03			Applied edge and burr research requirements	
01.02		Microscopic and scientific considerations		

Edge Quality Definitions

Seven levels of edge finishing are defined here. The design drawing specifies which level to use for a specific part or assembly.

A. *Deburring is not required.* All edges may be left as produced by the sequence of production processes used. Burrs or similar projections produced at edges may cause some dimensions to fall outside their normal limits. Whenever a drawing, purchase order, or specification fails to indicate specific edge finishing requirements, this level of edge finishing is implied. This level of edge quality allows producers to remove burrs and finish edges if they desire to do so.

B. *Remove sharp edges.* Edges defined by this level of finishing will be smoothed to the extent that hands will not be cut, nor would electrical wires or mating parts. Burrs may still remain on the products; they may be beaten over, flattened, completely removed, or rounded. The size of remaining material at an edge shall not cause the material dimensions to exceed drawing limits.

C. *Remove all visible burrs.* No projections visible to the unaided eye are permitted beyond the normal plane of adjacent surfaces. Small projections that are too small to be detected by normal unaided vision may remain. Any remaining material shall not cause dimensions to fall out of drawing specifications. This specification level also requires that edges shall not be sharp to the extent that they could cut hands, wiring cables, or mating parts.

D. *Remove all burrs visible at* _× *magnification (user must fill in magnification level on drawing).* No projections visible at ___× magnification are permitted beyond the normal plane of adjacent surfaces. Small projections that are too small to be detected by the indicated power of magnification may remain. Any remaining material shall not cause dimensions to fall out of drawing specifications. This specification level also requires that edges shall not be sharp to the extent that they could cut hands, wiring cables, or mating parts. This level of edge quality allows inspection by any quality of optics and any form of lighting. Lighting type and optics quality can greatly affect the ability to detect minute particles and burrs. Tactile or other nonoptical inspection approaches are not allowed in this level of deburring.

E. *Break edges* ___ × ___ *mm minimum (user must fill in values on drawing).* Edges shall be chamfered, blunted, or smoothed such that no material falls above a chamfer of the indicated minimum dimensions. Small burrs may remain on edges of the chamfers, and some raised material may remain near edges. Any material left at edges shall not cause product dimensions to fall out of their drawing tolerances.

F. *Round edges* ___ *to* ___ *mm radius (user must fill in values).* Edges shall have a curvature falling within the indicated limits. Chamfers are not acceptable. Projections at the edge, regardless of their radius, are not allowable when viewed with the unaided eye.

G. *Do not deburr.* Edges shall be left as produced by the sequence that produced them. This statement, in contrast to level A, explicitly prohibits deburring.

Specifying the Desired Edge Finish

Normal practice requires that the desired edge finish be specified on design drawings or other forms of specifications. For example, the note "Break all edges .004–.010 in. (0.1–0.25 mm) as specified in WBTC STD-14.1997 draft (*company inserts its own standard number here, if one exists, in place of the WBTC reference*)" is adequate for defining expectations. Alternately, designers may indicate the standard on the "bill of materials"—for example, "Note 1: Break all edges .004–.010 in. (0.1–0.25 mm)." In this case, the use of the level identifier (that is, "A, B, C," and so on) is not required on the drawings.

Other methods of specifying edge finishing levels are acceptable as long as they leave no doubt about the intent of the requirement. Drawings for components requiring more than one edge finishing level must clearly indicate which requirement applies to which edges. Providing explicit notes on the drawing with arrows pointing to the edges involved is the most commonly used approach, but other approaches are acceptable.

Figures 2-1–2-6 illustrate common burr notes when exceptions to the standard are required. These particular notes also illustrate approaches to identifying requirements unique to specific edges.

Inadequate Notes

Although notes such as "small burr satisfactory" or "burr raised in slotting operation is acceptable" (as in Figure 2-1) may be adequate for parts made within a specific plant, they should be avoided if parts are to be made by outside vendors. Sooner or later the product designer will be asked to define "small burr." One company faced a $9 million lawsuit because of a lack of detail in design specifications.

Preferred Notes

An illustration of a preferred note when deburring is not required as shown in Figure 2-2. The preferred practice for specifying edge quality, with allowable burr sizes described, is illustrated in Figure 2-3. Chamfering produces a generally smaller burr than those produced by other processes and may be the only deburring required. Either drawing notes or an in-plant standard should be used to indicate whether chamfering represents adequate deburring. When a smooth blend is required, the edge should be specified with a radius.

Edge Breaks

Edge breaks should be specified so that either a chamfered or a radiused condition is allowable. This specification permits the manufacturing engineer to determine whether a machining or a deburring process will provide the most economical edge condition. Typical corner breaks are .010 × .010 in. (0.25 × 0.25 mm) minimum. Radii should not normally be specified larger than .010 in. (254 μm) nor smaller than .003 in. (76.2 μm), unless required by function or stress concentrations, because of the extra expense in producing both large and small radii. Normal deburring processes readily provide these median-value edges. Chamfering or corner-rounding operations on machine tools may be required for larger values. Figure 2-4 illustrates the use of a maximum rather than a minimum edge condition.

Chamfers

When machined chamfers are explicitly required, they can be defined as chamfers by drawing notes (Figures 2-5 and 2-6). Typically, chamfers are specified as a dimension and an angle. A chamfer of .010–.015 in. (254–381 μm) × 45° is typical of small parts, whereas a chamfer of .020–.050 in. (0.5–1.27 mm) × 45° may be common for earth-moving equipment. A 45° angle is implied by this standard when not otherwise stated, but good practice requires including the angle. The angle of chamfer is not normally inspected unless it is obviously visually wrong. Very small and very large chamfers can significantly increase machining times.

Burr Direction

The direction a burr faces is sometimes more critical than its actual size. If this is the case, the correct orientation of the burr should be noted on the part drawing (Figures 2-5 and 2-6). With symmetrical threaded parts, it is helpful to the manufacturer if the designer indicates which end of the part the thread is started from, eliminating the need to deburr both ends.

Edge Breakout

When edges fracture, instead of forming burrs, the edge typically has a nonsmooth fractured surface with a geometry that varies along the edge. Allowable conditions may be expressed by the standard edge break with special notes, or with a separate specification.

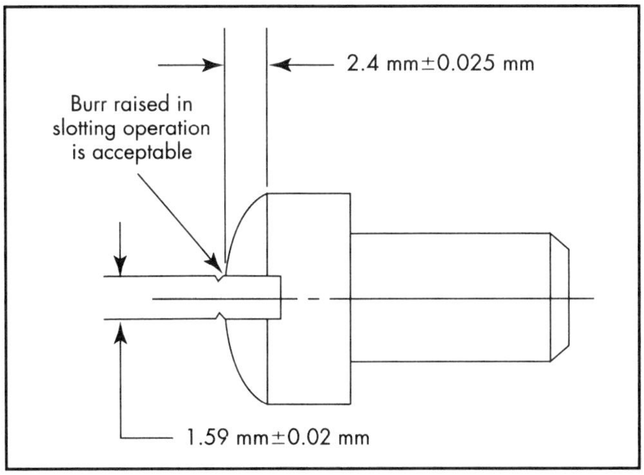

Figure 2-1. Example of note for slotting burr (NSMPA 1955).

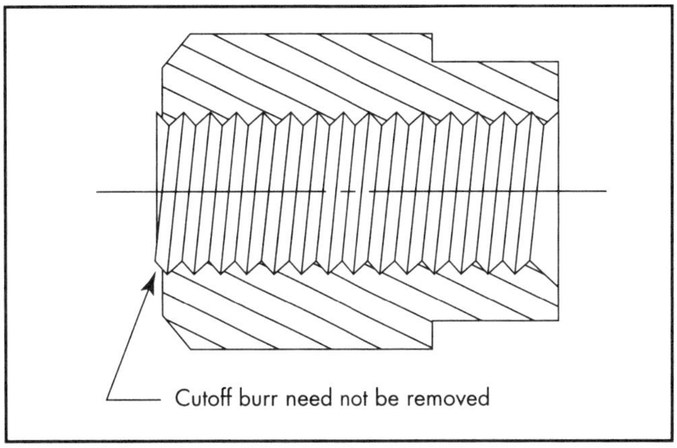

Figure 2-2. Example of note for cutoff burr (NSMPA 1955).

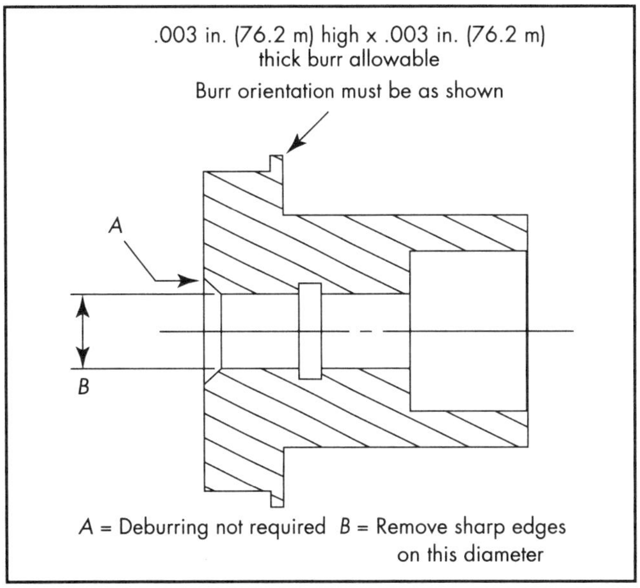

Figure 2-3. Allowable burr size on one part.

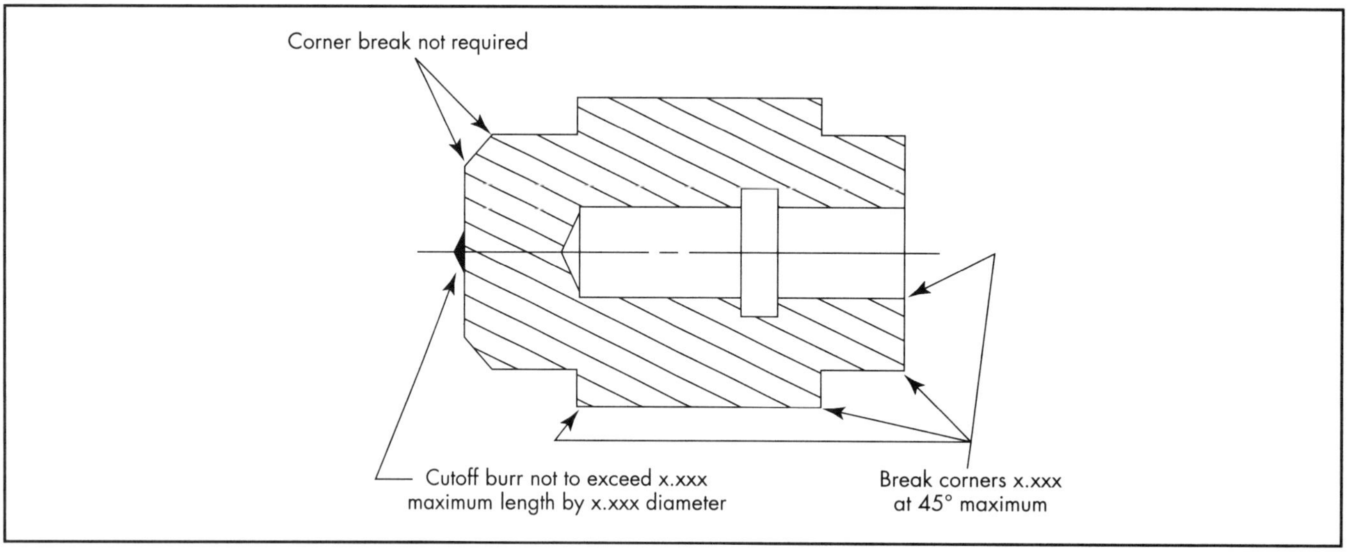

Figure 2-4. Typical burr notes for external edges (NSMPA 1965).

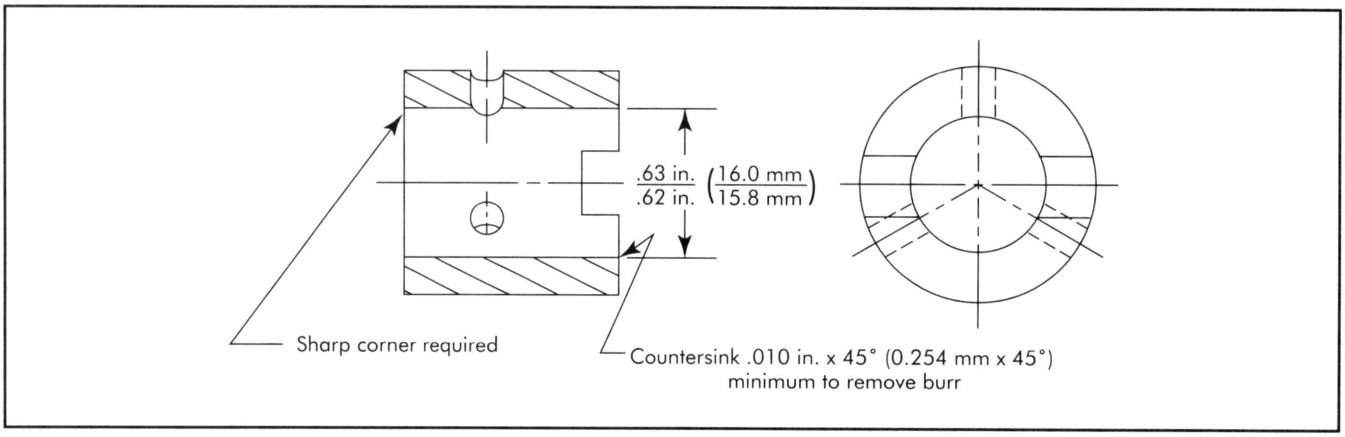

Figure 2-5. Example of note for sharp corner and countersink edges (NSMPA 1955).

Specific Notes

A burr always forms at the intersection of two holes. When a burr cannot be tolerated in one hole but can be in another, this should be noted on the drawing (Figure 2-6). Defining where burrs can exist on formed parts may eliminate the need to deburr the sheet stock. With proper thought and communication between product designer, tool designer, and manufacturing engineer, forming dies can be designed so that burrs on the blank will be in an out-of-the-way location on the finished part.

Removing Sharp Edges

On many parts, the only significant edge requirement is that all sharp edges be removed. In such cases, beating over burrs and dulling edges is adequate. Designers can handle this requirement in at least two ways:

1. By specifying the process that provides an acceptable edge.
2. By defining the actual edge quality needed.

When it is desirable to specify a specific edge finishing process rather than an edge requirement, the documentation must clearly indicate the process to be used and, preferably, the process details. Relatively minor changes to processes can have major effects on some components. Minor changes are routinely made in the course of doing business. Failure to specify process details leaves the selection to the producer.

When sharpness is desired, it may be defined by a note on the drawing, as shown in Figures 2-5 and 2-6. In such instances it is important to note precisely what "sharp" means. (See the Appendix at the end of this chapter for three definitions of "sharp" edges.) Any other terminology must be defined on the drawing as well.

In addition to the product drawing note, plant manufacturing and quality documents can, for example, contain a statement that "the plant adheres to Worldwide Burr Technology Committee Standard WBTC STD-14.1997, The Integrated International Standard for Burr and Edge Finishing."

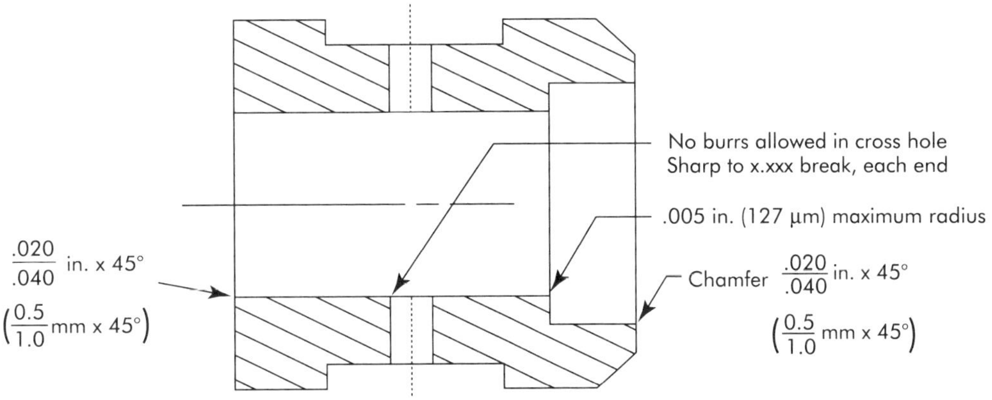

Figure 2-6. Typical burr notes for internal edges (NSMPA 1965).

Section 01.01.01: Manufacturing Requirements (for Edges and Burrs)

Manufacturing requirements are driven by design requirements and provide a complete approach to defining edge conditions for manufacturing operations. All new and current manufacturing instructions and practices shall conform to this requirement.

Section 01.01.01.01: Work Instruction Details

Work instructions for burrs, deburring, and edge finishing shall be written and provide adequate detail so that those who perform any of the operations can understand the requirements and how to perform the work. Work instructions shall assure that drawing callouts and specifications are met or exceeded.

When work instructions refer to machine operations, written details describing machine operation shall be provided. Operators shall follow work instructions. Operators who deburr or finish edges shall be trained, and a record shall be kept of their training. The training provided shall include all the elements specified on work instructions, machine operation procedures, plant general specifications, and expectations and requirements. All equipment dials, gages, and measuring devices and controls shall be calibrated to a set of written calibration standards, and a record shall be kept of such calibration. Documentation of any issues required by work instructions shall be maintained.

If not explicitly specified by drawings or manufacturing instructions, all edges entering acceptance inspection shall be broken to a maximum of .015 in. (381 µm), and no burrs or sharp edges shall be visible with the unaided eye. Instructions such as "deburr complete" are adequate for many operations and shall imply a maximum break of .015 in. (381 µm). (Inspection can only accept or reject to drawing requirements.)

Figure 2-7 illustrates one acceptable approach for defining edge requirements in work instructions. This approach provides explicit instructions of edges to be deburred, size of final edge configuration, and the equipment to be used. Figure 2-8 illustrates another approach, using color coding, that may be used. In this example, a sample part is painted white all over, then each feature to be deburred is marked with a different color, indicating the deburring tool required. Desired radii or chamfers may be hand painted where requirements differ from the plant standard.

Section 01.01.01.02: Machine Operation Details

In the absence of any specific deburring process described in manufacturing instructions, hand deburring shall be used. Whenever a technique is critical or its costs are clearly advantageous, the work instructions shall specify them. Table 2-2 defines the processes, calibration, and training requirements for deburring at this plant. When specified, a process may be identified by name or by the codes ("A, B, C," and so on) shown in Table 2-2. Specific requirements of each process are described in the following list. Operators shall verify their own work.

A. *Hand deburring:* An operator may use any tools, materials, and techniques on hand unless otherwise specified or prevented by work instructions.
B. *Vibratory finishing:* (company inserts description of machine operation here).
C. *Blasting:* (company inserts description of machine operation here).

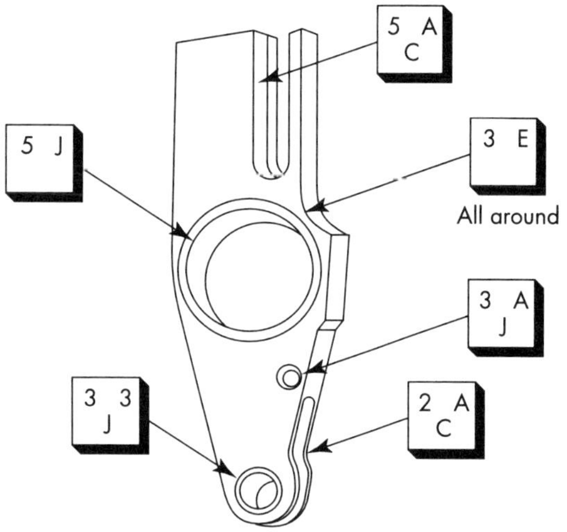

Code	Instruction
1	Deburr and break edge .015–.025 in. radius
2	Deburr and break edge .005–.010 in. radius
3	Deburr and break edge .005 in. maximum radius
4	Deburr and break edge .002 in./.003 in. radius
5	Deburr and break edge .002 in. maximum radius
11	Remove heavy burr only
12	Remove feather edge
13	Chamfer first and last thread
A	Burr knife
B	Wire brush
C	Nylon brush with AL_2O_3
D	240 grit paper
E	String brush with AL_2O_3
G	File
J	Cratex "bullet"
S	Burr ball

Figure 2-7. Example of work instructions that define edge requirements.

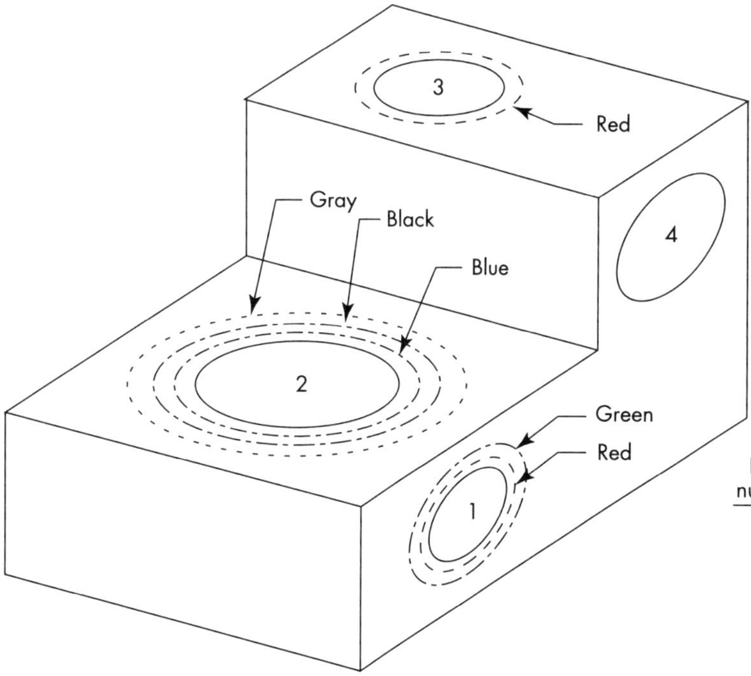

Hole number	Color band	Hand deburring approach
1	Red	Flat sand
1	Green	MX wheel in air motor
2	Blue	Swivel blade knife
2	Black	Triangular knife
2	Gray	Abrasive-filled rubber in motor
3	Red	Flat sand
4	None	None

Figure 2-8. Example of work instructions using color-coded approach to define edge requirements.

D. *Sanding:* (company inserts description of machine operation here).
E. *Brushing:* (company inserts description of machine operation here).
F. *ECD:* (company inserts description of machine operation here).
G. *Centrifugal barrel:* (company inserts description of machine operation here).

Section 01.01.02 Inspection Requirements (for Edges and Burrs)

Inspection requirements are driven by design requirements and provide a complete approach to defining edge conditions for inspection. All new and current inspection instructions and practices shall conform to this requirement.

Table 2-2. Summary of manufacturing process approaches for deburring and edge finishing*

Code	Process		Calibration		Training	
	Name	Spec. no.	Type	Spec. no.	Name	Spec. no.
A	Hand deburring	01.01.01.01A	None	N/A	Hand deburring	01.01.01.01 t-1
B	Vibratory finishing	01.01.01.01B	Vibratory machine cal-1 & cal-17	01.01.01.01 basics	Vibratory finishing	01.01.01.01 t-2
C	Blasting	01.01.01.01C	Abrasive blasting	01.01.01.01 cal-2	Blasting basics	01.01.01.01 t-3
D	Sanding	01.01.01.01D	Sanding machines	01.01.01.01 cal-3 sander	Principles of flat belt	01.01.01.01 t-4
E	Brushing	01.01.01.01E	Brushing machine no. 1	01.01.01.01 cal-4 machine operation	Three-headed brush	01.01.01.01 t-5
F	ECD	01.01.01.01F	ECD machine	01.01.01.01 cal-5 ECD machine	Operating the dynetics	01.01.01.01 t-6
G	Centrifugal barrel	01.01.01.01G	Barrel finisher	01.01.01.01 cal-6	Harperizer operation	01.01.01.01 t-7

*Company inserts its own specification numbers in this table.

Section 01.01.02.01 Inspection Instruction Details

Instructions for inspecting for burrs, deburring, and edge finishing shall be written and provide adequate detail so that those performing any inspections can understand the requirements, the measuring and recording devices, and how to perform the work. Instructions for inspection shall assure that drawing callouts and specifications are met or exceeded.

When instructions for inspection refer to operation of equipment, written details describing equipment operation shall be provided by some means, as indicated by the specification number shown. Inspectors shall follow inspection instructions. Inspectors who inspect for deburring and edge finishes shall be trained and a record kept of their training. The training provided shall include all the elements specified on inspection instructions, equipment operation procedures, plant general specifications, and expectations and requirements. All equipment dials, gages, and measuring devices and controls shall be calibrated to a set of written calibration standards, and a record shall be kept of such calibration. Documentation that parts conform or do not conform to edge and deburring requirements shall be maintained.

Section 01.01.02.02: Inspection Techniques and Equipment Details

The presence of burrs on parts, the measurements of burr properties, edge configuration, and related items can be documented by many approaches. Any of the inspection approaches shown in Table 2-3 may be used, but the specific approaches required for specific edges or parts shall be defined by inspection instructions. By default, in the absence of any specific instructions, procedure "A" shall be used. Operational details, calibration, and training requirements for each piece of equipment and machinery is described in the indicated procedures or subsections. Inspection shall be performed on all parts and assemblies unless otherwise stated. All features shall be inspected on all parts chosen out of a lot for inspection. Some of the inspection approaches shown in Table 2-3 are defined in the following list:

A. *Unaided visual inspection:* Unaided visual inspection recognizes that some personnel use eyeglasses, and if these are normally required by the personnel, they shall be classed as unaided. Visual acuity varies significantly among individuals, so absolute determination of burr-free conditions is not assured.

B. *Toothpick (tactile) inspection:* Wooden toothpicks can generally detect burrs and steps as small as .0005 in. (12.7 μm). Draw the end of the toothpick across the edge(s) in question, slowly searching for a slight upward movement or a momentary stopping. Sharpness is detected by drawing the body of the wooden toothpick lightly over the edge—if any shavings appear, the edge is sharp. Shavings would generally indicate that edge radii was approximately .001 in. (25.4 μm) or smaller.

C. *Number 2 pencil point inspection:* Using the point of a sharpened number 2 pencil, jab at the burr. If the point breaks off before the

Table 2-3. Summary of inspection approaches for edges and burrs[a]

Code	Process		Calibration		Training	
	Name	Spec. no.	Machine/gage no.	Spec. no.	Type	Spec. no.
A	Visual (unaided)	01.01.02.01A	None	None	Burr inspection	01.01.02.01.01
B	Toothpick	01.01.02.01B	None	N/A	Burr inspection	01.01.02.01.01
C	No. 2 pencil point	01.01.02.01C	None	N/A	Burr inspection	01.01.02.01.01
D	Fingernail	01.01.02.01D	None	N/A	Burr Inspection	01.01.02.01.01
E	4× glass	01.01.02.01E	None	N/A	Burr inspection	01.01.02.01.01
F	Microscope	01.01.02.01F	M1, M2, M3s	cal821	Microscope	01.01.02.01.02
G	Borescope	01.01.02.01G	Bore 1, 2, 3	cal986	Borescope	01.01.02.01.03
H	Metallurgical mount	01.01.02.01H	LTR 347	cal555	Mount preparation	01.01.02.01.04
J	Height gage	01.01.02.01J	CE4235	cal345	Height gage use	01.01.02.01.05
K	Profilometer	01.01.02.01K	CE3981	cal941	Profilometer use	01.01.02.01.06
L	3-D comparator[b]	01.01.02.01L	CE125	cal767	Laser comparator use	01.01.02.01.07
M	Light comparator	01.01.02.01M	CE45	cal4532	Comparator use	01.01.02.01.08
N	Toolmaker's microscope	01.01.02.01N	CE3247	N/A	Toolmaker's microscope use	01.01.02.01.09
P	Light section	01.01.02.01P	CE127	cal 887	Light section gage use	01.01.02.01.10
Q	Diode scan	01.01.02.01Q	CE128	cal 969	Diode scan gage use	01.01.02.01.11
R	Taper section mount	01.01.02.01R	None	N/A	Mount preparation	01.01.02.01.04
S	UL sharpness gage	01.01.02.01S	ga76	OI-ga76	UL sharpness gage use	01.01.02.01.12
T	Capacitance gage	01.01.02.01T	ga120	OI-ga73i	Capacitance gage	01.01.02.01.14
U	Water path	01.01.02.01U	FM125342	OI-ga43	Water flow use	01.01.02.01.13
V	Laser line vision system[c]	01.01.02.01V	CE321	OI-ga435	Disruption in line	01.01.02.01.15

[a] Company inserts its own equipment, specification, and document numbers in this table.
[b] Flores 1992.
[c] Stein et al. 1993.

burr or any other material is broken free, the burr is assumed to be rigidly adhered.

D. *Fingernail inspection:* Draw the end of the fingernail over the edges in question. If the fingernail hangs up, slows, or experiences a slight upward motion, a burr or raised metal exists at the edge. This technique can often detect burrs as small as .0005 in. (12.7 μm) high. Sharpness is detected by drawing the *top surface* of the fingernail lightly over the edge. If any shavings appear, the edge is sharp. Shavings would generally indicate that edge radii is approximately .001 in. (25.4 μm) or smaller.

E. *4× magnification inspection:* A 4× magnifying glass, with or without accompanying lighting, will be used. Inspectors may use eyeglasses in addition if these are normally required by the inspector. Visual acuity varies significantly among individuals, so absolute determination of burr-free conditions is not assured.

F. *Microscope inspection:* A stereozoom microscope shall be used with associated lighting of any kind. Magnification of 10× shall be used if not otherwise specified. Microscopes shall be in good working order and be maintained under a defined maintenance plan to assure adequate optics quality.

G. *Borescope inspection:* Borescopes shall be used to inspect the interior of holes or other cavities not otherwise readily seen. Borescope inspection must be done with operators qualified for detecting burrs and associated edge conditions. Acceptance may be made by borescope optic eyepiece or by borescope videoscreen unless otherwise defined.

H. *Metallurgical mounts inspection:* Metallurgical cross-sectional mounts will be made of edge conditions only when specified or when other means are inconclusive. This is a destructive inspection.

J. *Height gage inspection:* Raised material and burrs may be detected by passing a height gage and indicator over the edge in question. This approach requires several measurements along each edge. Careful use can detect burrs only .0002 in. (5.08 μm) tall, but it cannot determine burr thickness.

(Company inserts here definitions of any additional techniques it uses.)

Section 01.01.03: Applied Edge and Burr Research Requirements

Research standards are driven by corporate requirements and provide a means to better define edge conditions for design, manufacturing, and inspection. All new and current research efforts shall conform to this requirement.

When applied research is performed on burrs, deburring, and allied edge conditions, terminology shall be used that is consistent with both plant terminology and international practice. When new terms are coined, an attempt will be made to clearly define the term and relate it to existing terms so that its meaning is clearly understood by other users of the research. It is plant practice to use existing terminology where possible. For research purposes, burr dimensions or parameters will be measured and identified in a manner that allows various company personnel to analyze the data. A record of research results shall be maintained.

Section 01.02: Microscopic and Scientific Considerations (for Edges and Burrs)

Some scientific studies driven by corporate requirements provide a means for better understanding and defining minute details of edges, burrs, and related effects. This section concerns edge effects seen or believed to exist when viewed at magnifications of 400–100,000×. All new and current scientific and theoretical research efforts shall conform to this requirement.

When advanced scientific research is performed on burrs, deburring, and allied edge conditions, terminology shall be used that is consistent with both plant terminology and international practice. When new terms are coined, an attempt will be made to clearly define the term and relate it to existing terms so that its meaning is clearly understood by other users of the research. It is plant practice to use existing terminology where possible.

2.3 DICTIONARIES FOR EDGE AND BURR TERMINOLOGY

Because of the vast differences in burrs, material other than burrs at edges, and edge configurations, researchers must have a way to clearly and precisely communicate research results. They use specialized dictionaries for this purpose. The beginnings of such a dictionary exist in order to explain the various taxonomies of burrs (Gillespie 1996e). Although machinists, inspectors, and engineers need common terminology, dictionaries used by personnel at manufacturing sites need not be at the same level of sophistication and comprehensiveness as those used by researchers.

Researchers studying edges and surfaces at the theoretical, and microscopic (nanometer) levels often work with equations and advanced analytical lab equipment. They see a different universe than those working at the applied level, and consequently their needs for language, terms, and applications also differ. Their work includes the physics of materials and is not immediately applicable by those on the production floor. Nevertheless, although their need for standards may be minimal, and different from that of other workers, research standards still exist. For example, consider the surface of an atomic force microscope probe. It is literally scraping across molecules of material. The end of the probe may need to be described in normal vectors by B-splines or other definitions, but since it is the minute end of the tool—the edge—that is important in its use, the quality of the end edge is critical.

APPENDIX: BURR AND EDGE TERMINOLOGY

The edge-related terms defined here are presented in the following sequence:

- Edge characteristics.
- Burr-related terms.
- Flash-related terms.
- Miscellaneous edge-related terms.

Edge Characteristics

burr—Plastic deformation is produced at workpiece edges as a result of machining or shearing. A burr includes all metal extending past the theoretical intersection of the two surfaces that surround the burr (Figure 2-9a). In some instances, a burr can lie inside the theoretical intersection (Figure 2-9b). A burr can be a sharp, ragged projection, firmly adhered, or a loosely hanging projection. It also can be a swell of material at an edge. In Germany and other European countries, the word *burr* also means flash from castings and sintered metal operations. To provide more specificity for the standard contained in this chapter, this definition is limited to cutting and shearing processes.

chamfer—A machined inclined surface cut at an edge, typically in the order of .005–.050 in. (0.12–1.27 mm). Sometimes a chamfer is produced to remove burrs. A chamfer may have small burrs at its edges.

corner—The intersection of three or more edges.

edge—The intersection of two surfaces.

edge break— The amount of material removed from the theoretical intersection of two surfaces. Typically identified as if a uniform chamfer or radius were produced, such as ".05 in. (1.27 mm) edge break minimum required."

edge breakout—The condition left when a brittle material fractures at the edge and leaves an irregular fractured surface. Cast iron and ceramics particularly display this condition. It is an absence of material at an edge, whereas burrs and raised material are extra material left at an edge.

edge sharpness—See sharpness.

sharp edges—Edges that can readily cut or tear. To clearly illustrate what is meant, the following three classes of sharp edges are defined:

- Class 1: For thin materials .010-in. (0.25-mm) thick or less, a burr-free edge (when viewed at 5× magnification) with a radius, chamfer, or edge break less than $t/10$, where t is workpiece thickness at the edge to be measured.
- Class 2: For a workpiece thickness greater than .010 in. (0.25 mm) at the edge of interest, a burr-free edge (when viewed without magnification) with a radius, chamfer, or edge break of .001 in. (0.025 mm) or less.
- Class 3: For normal commercial use, any edge that would cut hands, containers, or nearby components in normal handling, installation, or repair. Sharp edges in this class include burr-laden edges as well as burr-free edges that cut. The determination of whether or not an edge will cut can be made by actual experiment

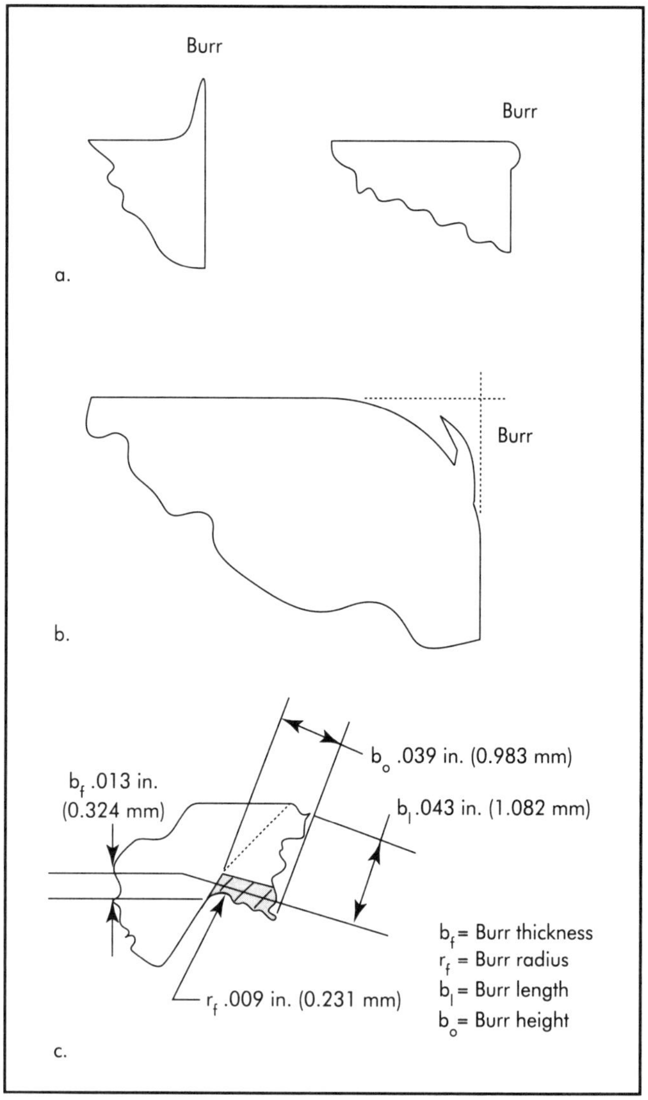

Figure 2-9. Examples of burrs and burr properties. (a) Metal extending outside the theoretical intersection of two surfaces is a burr; (b) example of a burr that lies inside the theoretical intersection of two surfaces; (c) definition of burr length, thickness, and radius, and actual measurements.

or by using the Underwriters' Laboratories sharpness testing unit.

Burr-related Terms

burr hardness—The hardness at the root of the burr at the theoretical intersection of the two surfaces adjoining the burr.

burr height—The burr dimension measured perpendicular to the edge from the theoretical intersection of surfaces to the most distant point of the burr in its cross section.

burr length—The longitudinal length of a burr along an edge (see Figure 2-9c).

burr radius—The radius of the burr profile at its root. For a rollover burr, the radius occurs on the side of the burr opposite that touched last by the cutting or shearing tool.

burr thickness—The width or thickness of the burr at its root as measured along one surface.

cutoff projection—The portion of the material left on the workpiece when the workpiece is allowed to fall before a cutting tool completely cuts through. On a turned workpiece, the cutoff projection is normally a short, small diameter projection from the cutoff surface. On milled stock, it may be a thin, rectangular piece attached to one edge. The material is not affected by plastic deformation. It is not material raised above a surface as a result of cutting. It is material left uncut.

entrance burr—Burr formed on the surface where the cutting tool enters the workpiece.

exit burr—Burr formed on the surface where the cutting tool leaves the workpiece.

feather burr—A very fine or thin burr. It can look like a feather with many fine spines.

feather edge—Sometimes used to indicate a feather burr except that a feather edge may also refer to the ends of a lead-in or lead-out thread (a very thin, machined ridge and not a burr).

hanging burr—Loose portions of a burr that are not firmly attached to a workpiece (that is, they are hanging from the workpiece).

Poisson burr—A burr formed predominantly by the phenomena responsible for Poisson's ratio (see Chapter 4).

rollover burr—A burr formed by the cutter when it exits over a surface and allows the chip to be rolled away, rather than sheared, from the cutter.

tear burr—A burr formed by the sides of a cutter as the cutter tears a chip from the workpiece.

tensile burr—A burr produced in blanking or piercing operations in which the slug separates from the stock as a result of tensile stresses.

tensile plus compressive burr—A burr produced in blanking or piercing operations in which the slug separates from the stock in a stress field that is initially tensile but changes to compressive at actual separation. This burr is usually associated with small die clearances and short die life.

wire edge—A very fine or thin burr that typically is sharp, looks like a wire, and is parallel to an edge.

Flash-related Terms

fin—Portion of flash that flows into the small gap between the movable workpieces of a mold.

flash—Excess material squeezed out of the mold cavity as a compression mold closes or as pressure is applied to a transfer or injection mold. Flash includes both the fin and the fragments left in the mold.

flash extension—The portion of flash remaining on the workpiece after trimming. Flash extension is measured from the intersection of the draft and the flash at the body of the forging or molded workpiece to the trimmed edge of the stock.

Miscellaneous Edge-related Terms

dross—Molten metal that resolidifies on the workpiece. This material occurs at the edges of material cut by thermal processes, such as plasma arc, oxyacetylene, and heliarc cutting.

raised metal—Material raised above the surrounding surface. This may be the root of a burr after most, but not all, of the burr has been removed, or it may be material raised when something bumps against the part or edge, such as another part. Raised material or raised metal is generally a small hump or bump of material as opposed to a sharp burr. In some instances raised material may be as objectionable as burrs.

References

American Die Casting Institute (ADCI). 1955. "Flash Removal." Product Standard ADCI-EIO-65. "Flash Formed by Ejector Pins." Product Standard ADCI-E9. New York: ADCI.

American Metal Stamping Association (AMSA). 1976. "Design Data Sheets." *Stamping Buyer's Guide*. Richmond Heights, OH: AMSA.

———. 1977. *Facts About Washers*. Richmond Heights, OH: AMSA.

American Society for Testing and Materials (ASTM). 1974. "General Requirements for Steel Sheet and Strip, Alloy, Hot Rolled, and Cold Rolled." ASTM Standard A505-73. *1974 Annual Book of ASTM Standards*. West Conshohocken, PA: ASTM: 349-64.

———. 1977. "Standard Method of Test for Edge Coverage of Coating Powders." ASTM Standard D-296. *Annual Book of ASTM Standards*. Vol. 39. Conshohocken, PA: ASTM: 685–87.

Consumer Product Safety Commission. 1975. "Bicycles, Toys, and Other Children's Articles." Proposed Rules. Federal Register 49, no. 4 (7 January): 372 (to be codified at 16 C.F.R., section 1500).

———. 1976. "Reproposed Rules for Sharp Edges: Working Draft of Revised Proposed Rules for Allowable Sharp Edges on Children's Toys or Related Articles." Draft No. 1. Code of Federal Regulations (16 C.F.R., Section 1500.48). 17 May. Washington, DC: Consumer Product Safety Commission.

Flores, Gerhard. 1992. "Mechanical Deburring: Process, Tools, Machines, and Applications." Technical paper MR95-272. Dearborn, MI: Society of Manufacturing Engineers.

Gillespie, LaRoux K. 1978. "Product Design Influences." *Advances in Deburring*. Dearborn, MI: Society of Manufacturing Engineers: 2: 21-29.

———. 1981. "Standards." *Deburring Technology for Improved Manufacturing*. Dearborn, MI: Society of Manufacturing Engineers: 3: 65–109, 605–46.

———. ed. 1995. "The Small Firm Standard for Burrs and Edge Finishing." WBTC-STD 11. 1995 draft. Kansas City, MO: Deburring Technology International.

———. ed. 1996a. "Burr and Edge Terminology: An International Dictionary." WBTC-STD 3.1996, draft. Kansas City, MO: Deburring Technology International.

———, ed. 1996b. "Burr and Edge Terminology: Definitions." WBTC-STD 1.1996, draft. Kansas City, MO: Deburring Technology International.

———, ed. 1996c. "Comments on Use of the DIN 6784 Standards for Burrs and Edge Conditions." WBTC-MS96-1. Kansas City, MO: Deburring Technology International.

———, ed. 1996d. "The International Standard for Burrs and Edge Finishing." WBTC-STD 13.1996, draft. Kansas City, MO: Deburring Technology International.

———, ed. 1996e. "Standard Nomenclature for Researchers in Burrs, Deburring, and Edge Finishing." WBTC-STD 2.1996, draft. Kansas City, MO: Deburring Technology International.

———, ed. 1997. "An Integrated International Standard for Burrs and Edge Conditions." WBTC-STD 14.1997, draft. Kansas City, MO: Deburring Technology International.

National Screw Machine Products Association (NSMPA). 1955. *Buyer's Guide for Design of Screw Machine Products*. Cleveland, OH: NSMPA (now Precision Machined Products Association).

———. 1965. *Draftsman's Guide for Screw Machine Parts*. Cleveland, Ohio: NSMPA (now Precision Machined Products Association).

Sorrells, John R., and Berger, Robert E. 1974. "An Inspection Procedure for Detecting Hazardous Edges." National Bureau of Standards Report NBSIR 74-428, April. Washington, DC: National Bureau of Standards.

Stein, Julie M., Narayanaswami, Ranga, Ho, Samuel, Lam, Anselm Y., Park, Ilwhan, Babu, Madhu, Afzal, Asim, and Dornfeld, David. 1993. "Intelligent Deburring of Precision Components." Technical paper MR93-319. Dearborn, MI: Society of Manufacturing Engineers.

Underwriters' Laboratories (UL). 1976. "Subject 1410: Proposed Revisions of the Standard for Television Receivers and Video Products—Proposed Effective Dates." UL Bulletin, 26 Oct. Melville, NY: Underwriters' Laboratories.

——— 1993. "Subject 1439: Standard for Sharpness of Edges on Equipment." UL Bulletin, 1439 93, 19 Feb. Northbrook, IL: Underwriters' Laboratories.

Bibliography

Gillespie, LaRoux K. 1977. "An Extension of Proposed Definitions for Burrs and Related Edge Conditions." SME paper MRR77-09. Dearborn, MI: Society of Manufacturing Engineers.

———. 1981. "Progress in the Battle with the Burr." SME paper MRR81-07. Dearborn, MI: Society of Manufacturing Engineers.

Przyklenk, Klaus, and Schlatter, M. 1987. *Entgraten von Werkstucken aus Aluminum*. Düsseldorf, Germany: Verlag-Aluminum: 5-9.

Seimens. 1982. "Prufeinrightung Fur Die Mechanische Messung Dunner Und Labiler Metallteile." Brochure (in German). München, Germany: Seimens.

3 Burr Prevention

Deburring problems disappear if parts are produced without burrs. It is possible to prevent burrs from forming, and every consideration should be given to doing so before spending time and money on removal efforts. The first axiom of deburring is always *if you do not make the burr, you do not have to remove it.*

Some techniques for preventing burrs are:

- Form instead of machining parts.
- Cast instead of machining edges.
- Use form tools instead of tools that face and turn.
- Use nontraditional machining processes (such as chemical machining).
- Employ radical machining processes (such as ultra high-speed machining).
- Prevent cutters from exiting from the part.
- Use selective hardening to reduce ductility at machined or blanked edges.
- Use sacrificial backup material.
- Employ large exit angles on part edges.
- Use push-back blanking sheet metal processes.

The practical restraints of burr prevention are considerable, but so are the benefits. The following are some of the limitations to burr prevention:

- With conventional machining, burrs may be prevented at one edge but not all edges because of the stresses and angles involved.
- Nontraditional processes may leave or produce other objectionable edge material.
- Sacrificial backup material may cost more than burr removal, and it may lead to other production problems.
- Exit-angle effects depend on material properties. Typically, most edges have a functional need for right angles.

3.1 PREVENTING BURRS VERSUS MINIMIZING BURRS

Recall from Chapter 1 that three approaches can be used to reduce deburring costs:

1. Prevent burrs.
2. Minimize the size of burrs.
3. Put burrs where they can be removed more easily.

It is easy to lump all three approaches into one category and miss the distinctions among them. This chapter concentrates on preventing burrs. True burr prevention means that no burr is formed. Chapter 4 describes how to minimize the size of burrs, and Chapter 5 describes how burrs can be placed for more effective deburring. Small burrs or "acceptable" burrs are still burrs, and to some "no burrs" may mean that no burrs exist when viewed with the naked eye. Throughout this handbook, the statement "no burrs" means that absolutely no burrs exist. In this handbook absolutes and numerical values will be presented so that readers will know exactly what occurs rather than having to determine what constitutes "small" or "acceptable."

A burr-free part may still require subsequent finishing, such as dulling its sharpness, removing molten metal and surface flaps, improving surface finishes, and other cosmetic treatments. Furthermore, "burr-free" does not necessarily imply cost reduction, cycle time improvement, or easier processing. Although many printed articles, reports, and other statements claim that a burr-free condition will result from a particular process, burrs may, in fact, still be produced by these processes. Most of these claims are the result of not having clear definitions of edge quality, a distinct disregard for the mechanics of cutting at an edge, and the belief that if a burr is smaller than normal, it is not a burr.

Unfortunately, altering speeds, feeds, and tool geometries will not prevent burrs. Both analytical and empirical studies have shown that although tool sharpness and cutting conditions can minimize burr size and control burr repeatability, they cannot prevent burrs. *With only a few exceptions, conventional machining techniques always produce burrs.*

Although burrs can be prevented by some conventional processes, they typically are prevented on only one edge, not all edges. This can be seen by looking closely at a workpiece. Most operations produce burrs at more than one location. In drilling, for example, a burr is produced at both the entrance and the exit holes. Techniques to prevent one burr may still allow the other burr to form. In milling operations, for example, burrs can be produced on up to 10 edges in two single cuts. Thus, at best, one burr may be prevented and others minimized; though minimizing burr size is a distinct advantage, it is not as desirable as burr prevention.

3.2 USING FORMED OR EXTRUDED PARTS INSTEAD OF MACHINED PARTS

Extruded sections prevent burrs from forming at part edges for two reasons: (1) Extrusion is a high-pressure forming operation rather than a chip-producing operation, so by definition anything at an edge is not a burr, and (2) the die is a continuous surface tool, which prevents any metal from squirting out past the edge. In one sense it wipes the part edges as the edges are produced. Note that when the extrusion is sliced into sections, a burr will form at the cutoff position, but the extruded edges will be free of burrs and will have the desired finished condition.

Other processes used to form parts include flashless forging, hot forming on closed-die machines, and wedge rolling. Care must be exercised in these processes if edges are to be truly free of undesired material. As mentioned earlier, some of these processes have been described as producing burr-free edges when, in fact, some metal may extend past the edges unless great care is taken to prevent such metal movement. These are not machining or shearing processes, so flash rather than burrs would form, but flash is also undesirable for many of the same reasons that burrs are undesirable.

Sheet metal forming also allows users to roll or form edges so that no burrs occur at the working edge. Although the burr still exists at the actual sheet metal edge, that edge is not an exposed edge, and thus the user sees no burr, only a nicely rounded edge. Additional approaches for sheet metal parts are presented at the end of this chapter and in Chapter 18.

3.3 CASTING PART EDGES

Casting saves finishing costs, such as deburring, and machining costs. When edges are cast, no burr forms. Flash, however, will form at die openings and generally must be removed for the same reasons that burrs must be removed. Since most parts have many edges and the flash will only form on a few of them, the net effect of cast parts is a big savings in edge finishing efforts and the lack of burrs or any other projection at most edges.

Powder metal forming and several other forming or molten metal operations that use dies all have these same "no burr" characteristics.

3.4 USING FORM TOOLS WHEN CUTTING

As an example of using form tools when cutting, consider automatic screw machines. Burr size can be controlled on these machines using three approaches:

- *Use a form tool to form diameters.* No burr will form when the tool produces the diameter and the adjoining face at the same time (Figure 3-1). Any intermediate burr formed is wiped off.
- *Break all edges with a chamfering or corner-rounding end mill.* Since tool position can be accurately controlled to .0002 in. (5 μm) on many machines, the rounding tool can remove the burrs and still assure small final edge breaks. Tools also have been designed to burnish these edges after chamfering, resulting in a burr-free, rounded edge. In this case the burr actually forms, but it is cut off in the edge-producing step. The user sees a burr-free part, although it was not initially produced as burr-free. Spindle growth from machine heating may limit this approach on close tolerance surfaces. It is possible to use corner-rounding cutters, which also produce the cavity or cutout at the same time. In this approach, the tool is actually producing a burr-free edge as one of the machining steps rather than just cutting it off. Some standard corner-rounding tools have a 90° corner design (Figure 3-2). For more tolerant edge rounding that allows some spindle axial variation, users would want to use a cutter that trails away from the edges and has an included angle of more than 90°.
- *Generate corner radii by N/C or CNC lathes or grinding.* Although the cutting tool can be programmed to cut radii at edges, on some machines this is difficult to do,

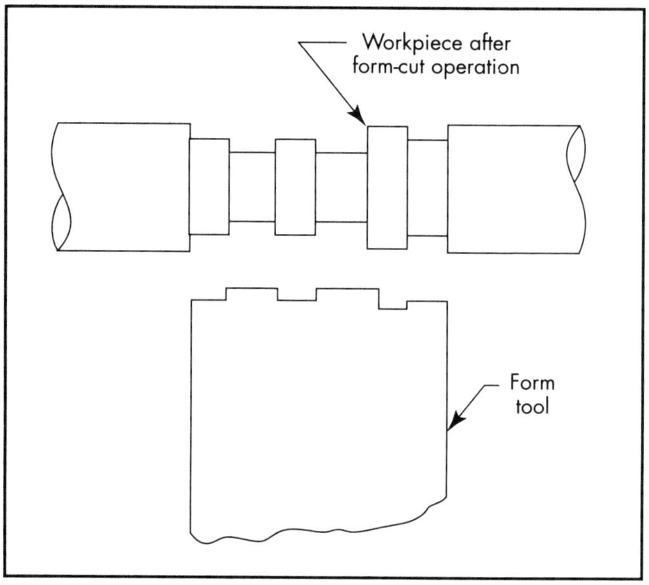

Figure 3-1. Form tool for turned parts.

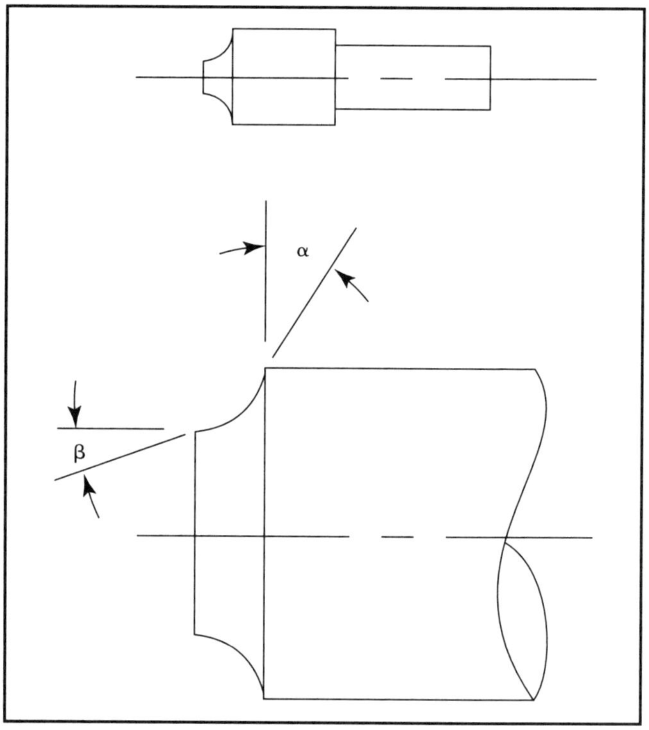

Figure 3-2. Standard corner-rounding cutter and cutter with flaired-out surfaces to reduce mismatch.

and it severely limits the adjustment capability of tools producing adjacent features. This approach is the lowest cost method for providing burr-free edges on many precision parts that cannot tolerate scratches or handling damage.

3.5 NONTRADITIONAL PROCESSES

Burrs can be prevented by employing some of the nontraditional processes. As shown in Table 3-1, most of the nontraditional processes do not produce burrs (when viewed at 30× magnification). Despite many statements to the contrary, electrical discharge machining (EDM), electron beam machining (EBM), and laser beam machining (LBM) do produce burr-like projections of recast material. Although little test data is available on the edge quality produced, two studies provide some insight into this controversy. One study (McBride 1969) documents the effects of EDM parameters on EDM burr size. Another discusses the edge conditions produced by EBM and LBM (Machinability Data Center 1980). When a high-velocity air blast is synchronized with the laser, the majority of the resolidified edge material is blown out before it can solidify on the workpiece.

Whenever possible, processes such as chemical machining (CHM), photochemical machining (PCM), electrochemical grinding (ECG), electrochemical machining (ECM), electrochemical honing (ECH), electropolishing (ELP), and electrostream machining (ESM) should be used. They not only eliminate deburring costs, but also provide excellent surface finishes and minimize welding, brazing, and plating problems caused by media impregnation or improper cleaning. In addition, eliminating unnecessary operations reduces paperwork costs and shortens production flow time.

The disadvantages of using the nontraditional processes include high equipment costs, limitations to certain geometries and workpiece materials, and workpiece tolerance and surface integrity problems. These factors are discussed in detail in a variety of other texts.

The nontraditional machining processes are generally the only machining approaches that can be used to prevent burrs; their use, however, depends on such factors as workpiece tolerance, geometry, surface integrity, and the availability of such equipment. For example, photochemical machining (PCM) can be used on cylinders, spheres, and complex shapes, as well as on more conventional flat parts. Ultrasonic machining also will not leave burrs since it does not generate chips at part edges; the ultrasonic tools in a sense form and slightly round the edges.

3.6 RADICAL MACHINING APPROACHES

Ultra high-speed machining (with cutting velocities in excess of 10,000 surface ft/min [3,048 m/min]) has been observed to produce no burrs with some metals. At present the general validity of this claim remains to be established (as noted earlier, to most users a claim of "no burrs" actually means smaller burrs or no burrs of consequence).

In the case of high-speed machining, there is good reason to believe that burrs might not be visible or detectable with the fingers. Figure 3-3 illustrates some typical curves of ductility and energy required to fracture a specimen as a function of testing velocity. In this case, the results are for

Table 3-1. Nontraditional machining processes

Process	Typically makes burr?*	Typical edge radii produced in./mm	Typical machining tolerance in./mm
Abrasive jet machining (AJM)	No	.003/0.08	—
Chemical machining (CHM)	No	Unknown	±.002/0.05
Photochemical machining (PCM)	No	.001/0.03	±.002/0.05
Electron beam machining (EBM)	Yes	—	±.001/0.03
Electrochemical discharge (ECD)	Unknown	Unknown	Unknown
Electrochemical grinding (ECG)	No	.003/0.08	±.002/0.05
Electrochemical machining (ECM)	No	.001/0.03	±.002/0.05
Electrochemical honing (ECH)	No	.0005/0.013	±.0002/0.005
Electrical discharge machining EDM	Yes	—	±.00006/0.015
Electropolishing (ELP)	No	.001/0.03	±.0005/0.013
Electrostream machining (ESM)	No	.002/0.05	±.001/0.03
Hot chlorine gas (HCG)	No	.002/0.05	±.003/0.08
Ion beam machining (IBM)	No	.00005/0.0013	±.0001/0.003
Laser beam machining (LBM)	Yes	—	±.001/0.03
Plasma arc machining (PAM)	Yes	—	±.003/0.08
Ultrasonic machining (USM)	No	.001/0.03	±.001/0.03
Water jet machining (WJM)	No	Unknown	±.003/0.08

*Where burr is visible under 30× magnification.

tensile testing of flat coupons. Similar results have been obtained for cylindrical specimens (Clark and Wood 1950; Mann 1936; Fukui et al. 1967). The same basic patterns have been observed to occur in blanking operations (HERF 1967; Johnson and Slater 1967). In each case, a metal has a critical velocity above which less energy is required to induce failure or separation. Lower energy requirements are the result of smaller amounts of plastic deformation, which is the cause of burrs. In the case of 17-7 PH stainless steel, at 600 ft/s (183 m/s), the ductility of the specimen tested was one-seventh of its normal amount. It was one-tenth of that required at 100 ft/s (30 m/s). Thus, it is highly plausible that burrs could be effectively prevented by high-speed machining.

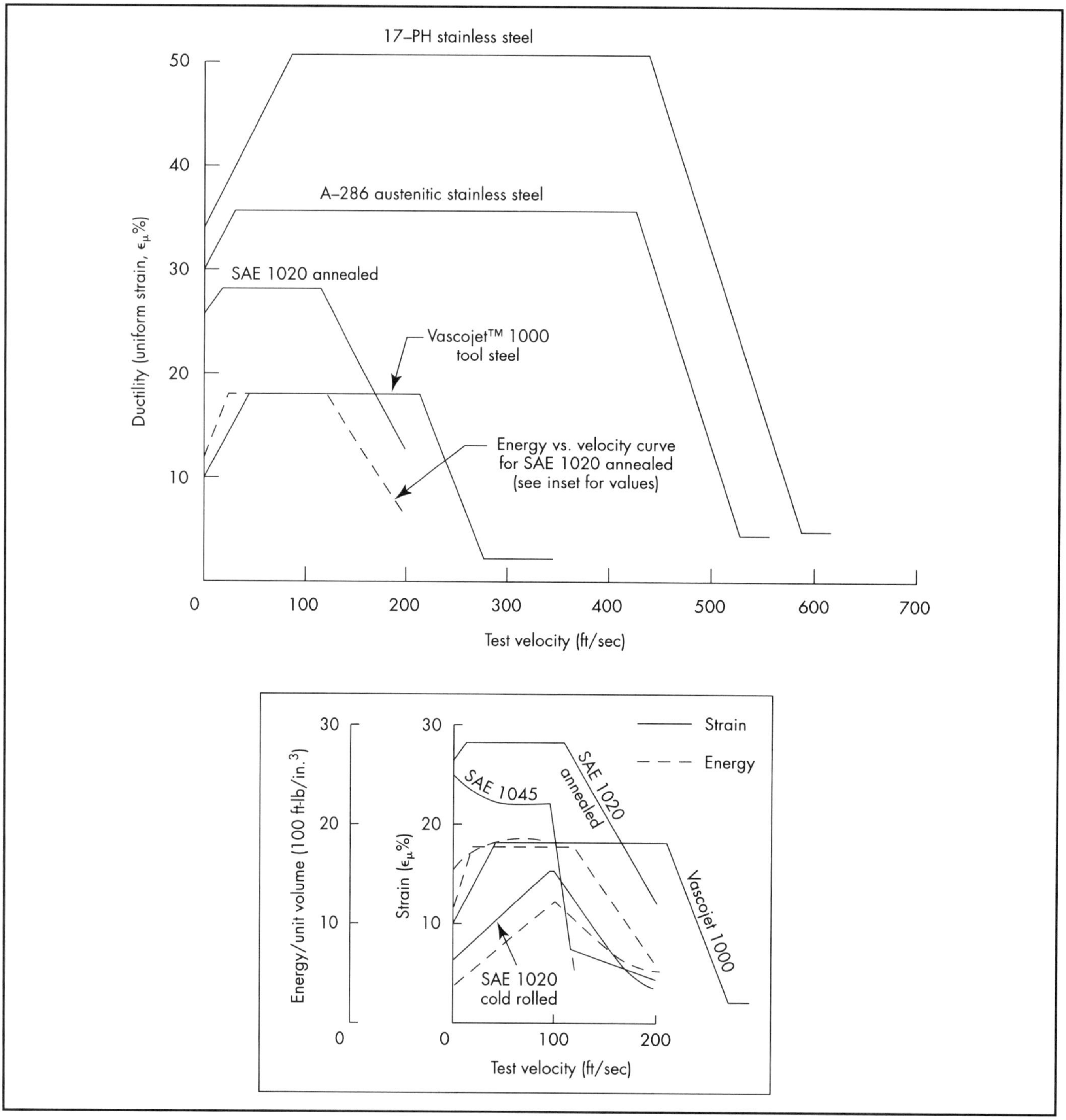

Figure 3-3. Energy requirements for fracturing a specimen as a function of testing velocity (Wood 1963).

3.7 PREVENTING THE CUTTER FROM EXITING OVER PART EDGES

Burrs may or may not occur where the cutter enters a part, but they always occur on ductile metals where the cutter exits from a part. Normally, the cutter must exit over edges to produce the desired features; however, this is not always the case. If a flat surface has to be machined, it is possible for the machinist to feed the part from one side to the center, rotate the part, and repeat this step until all edges have been entered. This procedure may not be economical, and it may require more process planning and N/C tape effort, but it can be done, and it saves a finishing operation. If one has special machines that automatically compensate and rotate, then the cutter will never exit the part except at the center of the surface. This procedure is in production use despite its apparent oddity. One company grinds parts such that the grinding wheel never exits the edge of the part—it exits at the middle of the part. Although this procedure requires complex planning and may require special machines, it saves tens of thousands of dollars each year.

3.8 SELECTIVE HARDENING

Selective hardening prevents plastic deformation from occurring macroscopically at part edges. Any environment that prevents this will minimize or prevent burrs. Suitable techniques for selective hardening include:

- Laser hardening.
- Mechanical localized hardening.
- Chemical hardening (liquid and gaseous).
- Thermal hardening.

In a unique application of localized hardening, researchers have shown that liquid mercury can improve machining and could reduce burrs because it tends to remove surface ductility and helps to break up chips (News and Trends 1975; Scanning for Ideas 1975). Any solution that accomplishes this same result will help reduce burrs.

Figure 3-4 illustrates burr prevention using a ball indenter. When a hole is drilled into a brass cartridge, as shown in Figure 3-4, an objectionable burr is produced. To alleviate this situation before drilling, one company pressed a ball into the area where the drill would exit. This left a spherical indentation that resulted in a 150° edge angle when the hole was drilled. Since the material in this particular example was brass, the indentation was accomplished quickly and easily. The indentation provided two other advantages: (1) If a burr were ever produced, it would be small and by the geometry involved would be recessed below the inner surface; and (2) the indentation process hardens the metal around the indentation, which in turn reduces most of the ductility in the area. Since large burrs cannot occur in nonductile materials, burr size is further reduced. Note that the ball size must be large enough to result in the desired angle.

Some plastic parts have traditionally been frozen in liquid nitrogen to temporarily embrittle flash to make it easier to remove. One enterprising company freezes its Teflon® sheet stock prior to blanking for the same purpose. However, in this instance, the low temperatures take out the ductility before the burr is made, and a burr-free part is produced. When the part warms up, it regains the normal Teflon properties.

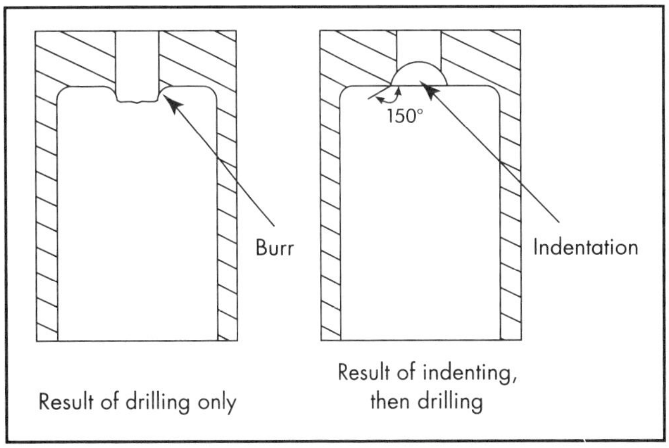

Figure 3-4. Preventing burrs by using a ball indenter prior to drilling.

3.9 SACRIFICIAL BACKUP

The production of printed wiring boards (PWB) has for many years required placing sheets of hard, thin plastic materials between layers of stacked bare boards. The backup material minimized or prevented burrs from forming in the copper-clad PWB. This is an easy solution when parts are flat or in sheet material.

The key to the success of this approach is to use a backup material that is harder than the material having the burr. Figure 3-5 provides test data on burr size when a block of backup material was inserted below the exit point of a drilled hole. Clearly, harder backup material greatly reduced the size of burrs produced, but it did not prevent them entirely in this case. For some applications, this process may be adequate. Another key to this approach is keeping the backup material in close contact with the part during machining. Any separation between the backup material and the part surface during machining will allow a burr to form. PWB drilling machines employ small pressure pads to push down on the stack around the hole as it is being drilled to keep the stack tightly together.

Keeping the materials tightly together becomes particularly challenging when machining stacks of material that have small variations in the dimensions or flatness of parts or backup material. Figure 3-6 illustrates a stack of miniature gears hobbed in a stack with a sacrificial backup between each gear (located around the gear hub). Each backup spacer and gear surface has to be flat within a few microns. The spacers must be thicker than the hub, and the spacers must be hard enough to prevent the burr. In addition, the hobbing fixture or machine must apply enough pressure to keep the stack together. Figure 3-6 also demonstrates a best

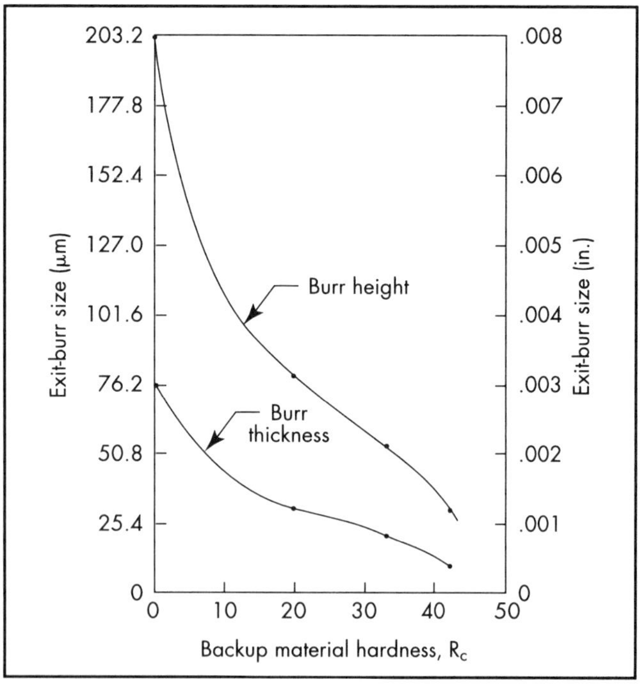

Figure 3-5. Effect of backup material hardness on burr size when drilling Type 303 Se stainless steel.

practice of always placing any burr on a flat, instead of a stepped, surface, which makes deburring easier and less expensive.

Theoretically, it would be possible to completely cover a part with "backup" material and prevent all burrs. From a practical standpoint, however, this procedure is not realistic because the backup material must be harder than the workpiece to prevent burr formation. Hard plastic coatings may help with isolated parts. Making a sandwich of parts with sacrificial backup material can work, but it adds to processing, planning, and costs.

3.10 EXIT EDGE ANGLES

The effect of exit edge angles on burr size and burr prevention is shown in Figure 3-7. For a specific material and machining condition, increasing the exit edge angle decreases the burr. In Figure 3-7, with this particular combination of material (unhardened low-carbon steel) and machining process, no burr was produced when the edge approached 150°. For other materials and machining processes, the burr is smaller but may still exist. When exit edge angles are small, a large burr typically forms. Small exit edge angles may also result in more displaced material at cutter entrances to the part.

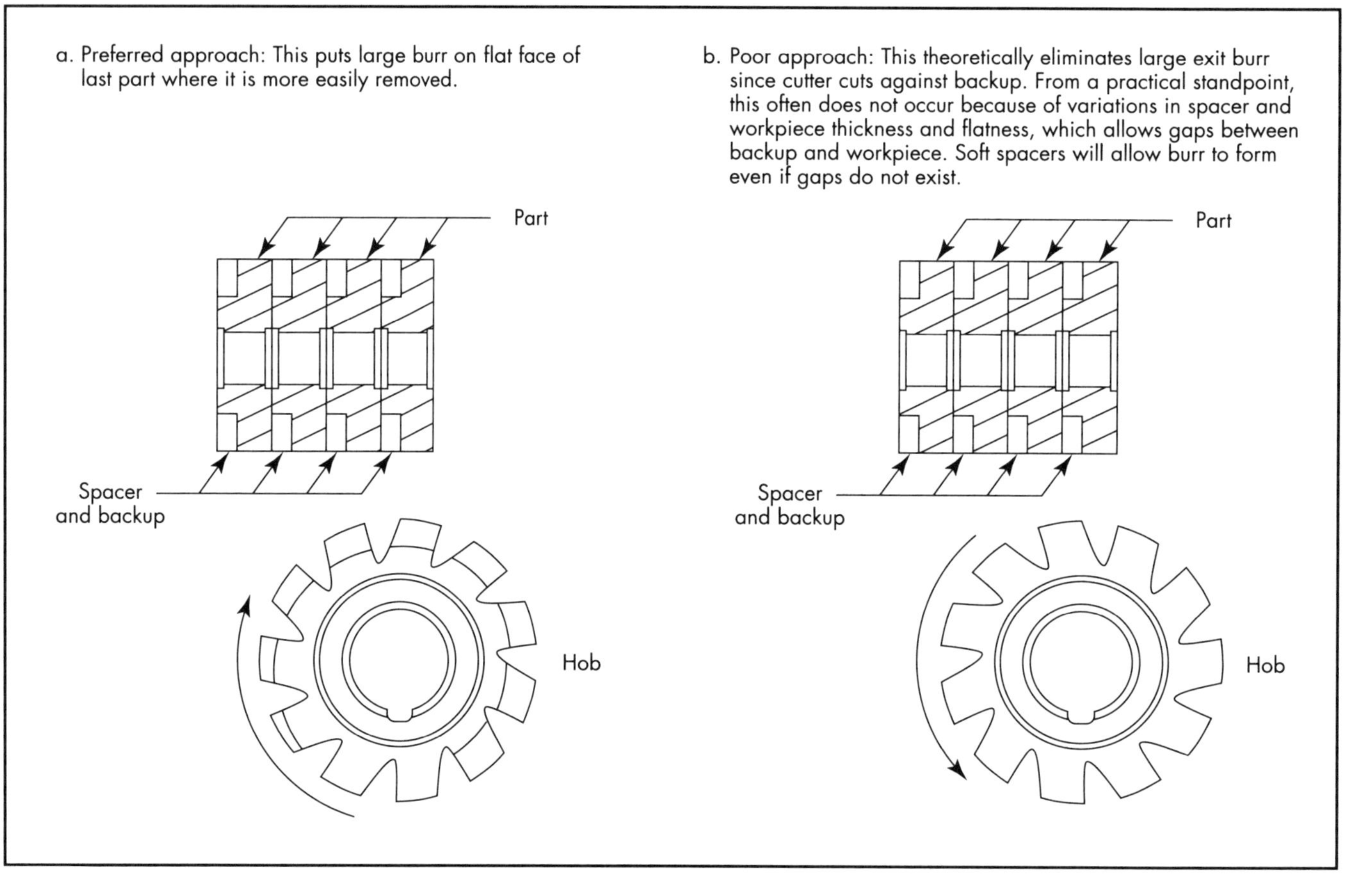

Figure 3-6. Sacrificial backup used when hobbing a stack of gears.

Most designs are based on 90° included edge angles; however, if the parts are machined castings, then the castings can employ any edge angles at no significant cost. As an alternative to using large edge angles, designers can take advantage of casting specific radii that result in the same effective edge angle (see Figure 3-8).

3.11 OPPOSED PUNCH PROCESSES

When burr-free edges are essential on sheet metal parts, the opposed-punch processes known as reciprocating blanking (Maeda 1958), push-back blanking, opposed-die blanking (Kondo et al. 1974), counter cutting, and roll-slit (Maeda and Murakawa 1977) should be considered. The basic principle behind these processes requires that a die roll be generated on both the top and bottom of the workpiece surfaces. The separation of the slug from the blank occurs in the interior of the cross section.

3.11.1 Reciprocating Blanking

Figure 3-9 illustrates the two-step reciprocating blanking process. In the first step (Figure 3-9b), the material is partially penetrated by the upper punch to a predetermined depth. The slug has, however, not yet separated. In the second step (Figure 3-9c), a lower punch forces the slug back into the sheet and eventually out the top surface (Figure 3-9d).

Figure 3-10 shows the sheared surface and cross section obtained by the reciprocating blanking process. Notice the die roll on both ends of the slug. Because of the die roll, no burrs can form on the sheared edges.

The key point of this method is to produce the first die roll and maintain the position of the punch until the lower punch can act. For this purpose, the first step must be stopped beyond the maximum shearing force in the punch load stroke curve shown in Figure 3-11. If the punch penetration advances too much in the first step, a crack will break out from its cutting edge, resulting in a sheared surface with burrs. In the most desirable blanking condition, the second step should be performed after a crack is formed from the die-cutting edge in the first step. For this purpose, it is desirable that the first-step punch-cutting edge have a small radius to initiate a small crack.

Because of the complicated tool arrangement, which inevitably leads to a special blanking press, this burr-free blanking method has very few practical applications that have been put to use. However, no other complete burr-free shearing method has yet been developed.

3.11.2 Push-back Blanking Method

Push-back blanking, shown in Figure 3-12, is designed to simplify the complicated tool arrangement found in reciprocating blanking in such a way that it can be adapted to a conventional press and tooling. In this method, no blanking punch and die are used in the second step; instead, a pair of flat plates is used to push back the half-sheared blank.

The operative feature of this method is similar to the so-called push-back feeding system used for transferring sheared blanks. The only difference is that whereas the blanks are completely sheared off in the first step in the ordinary

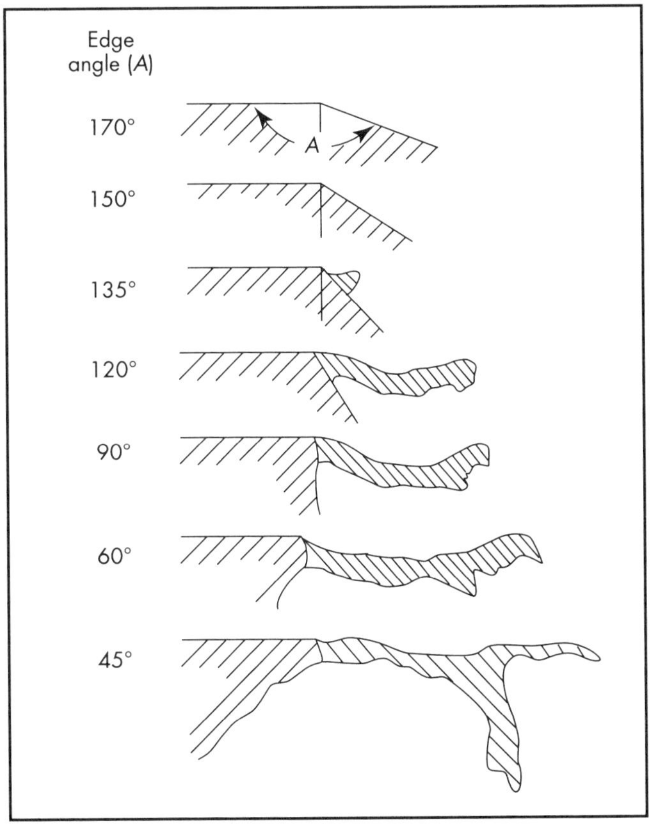

Figure 3-7. Effect of exit edge angle on burr size at end of cut (Shafer 1975).

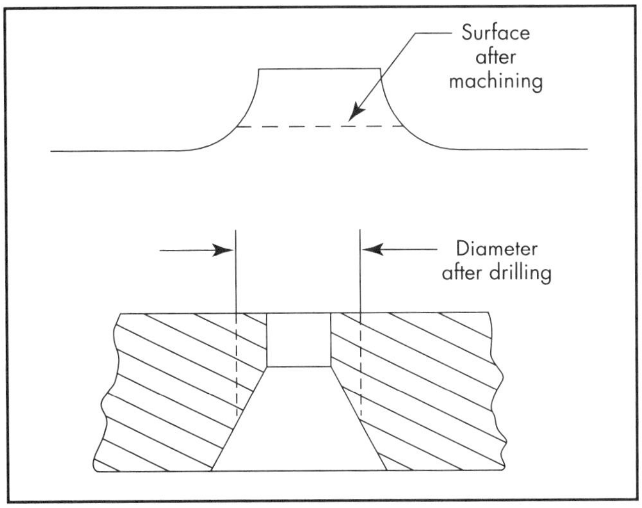

Figure 3-8. Machining through large cast radii and cast chamfers can result in large exit angles.

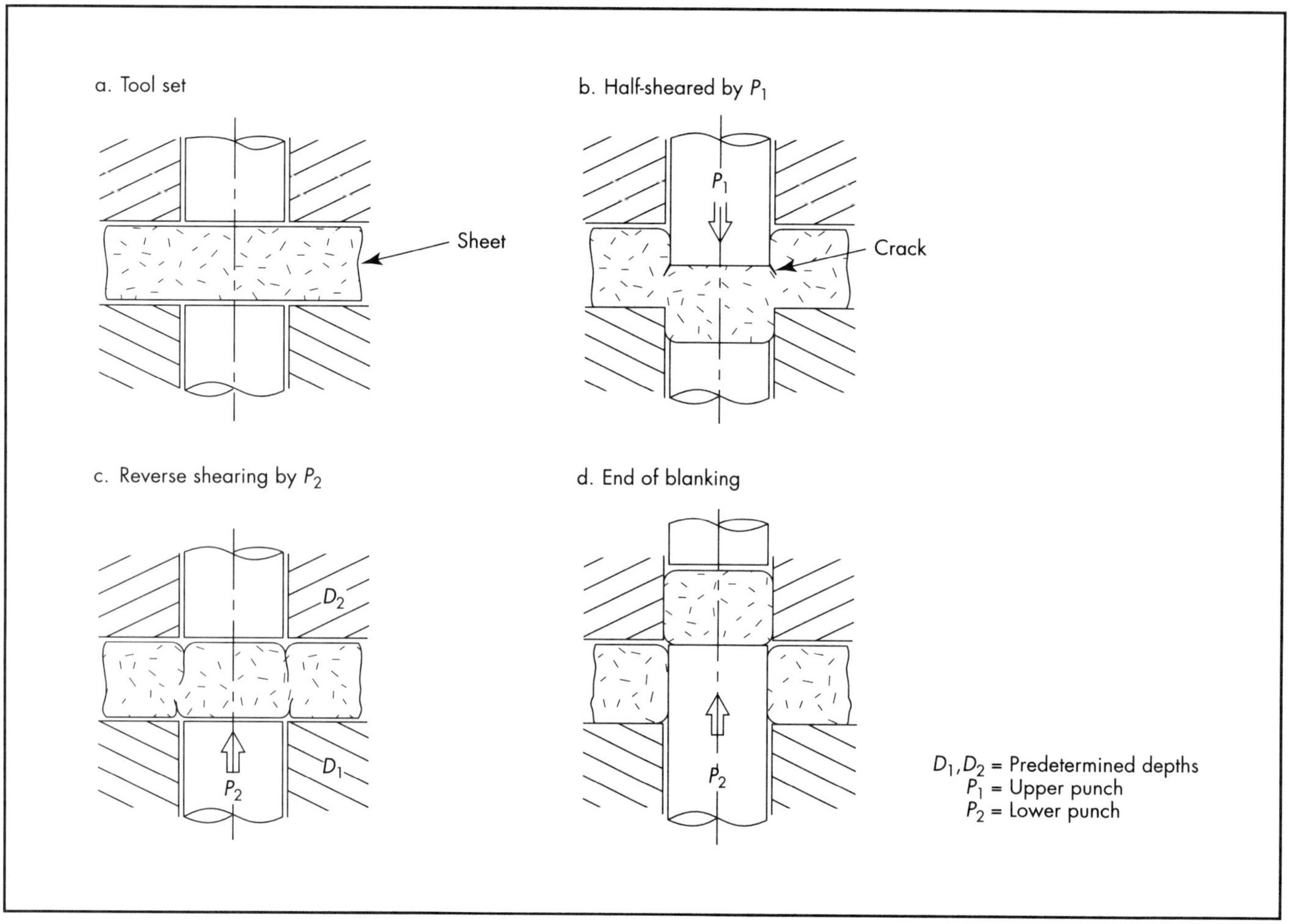

Figure 3-9. Process of two-step reciprocating blanking (Murakawa, Maedo, and Nakagawa 1977).

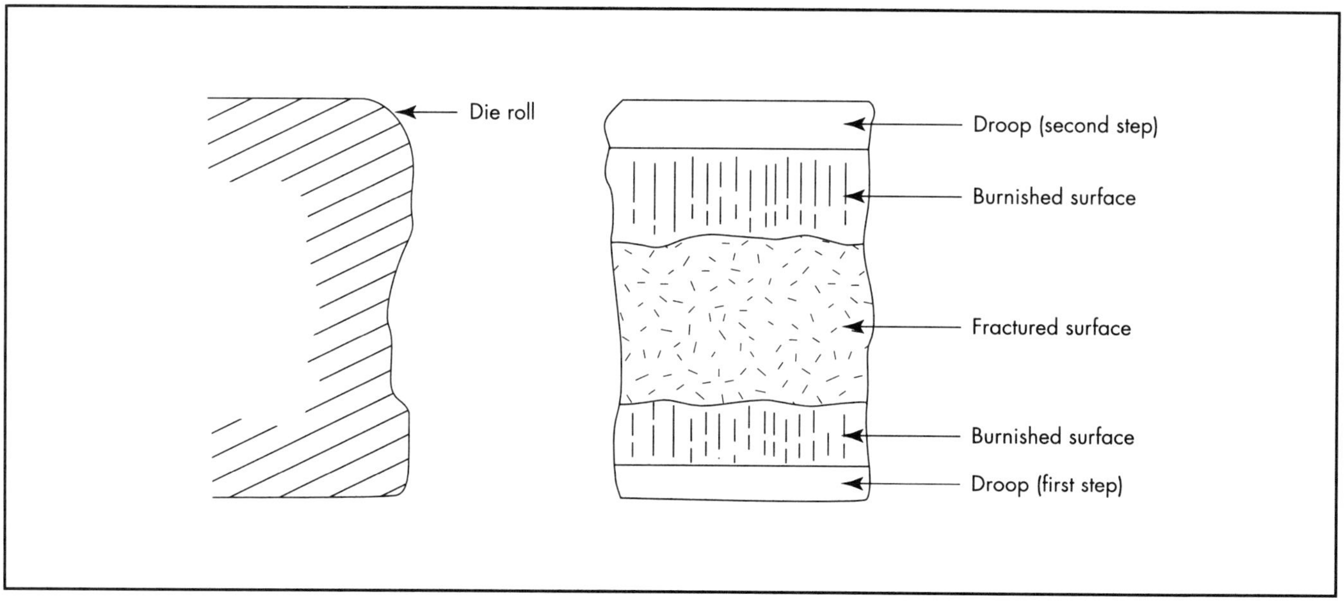

Figure 3-10. Surface blanked by reciprocating blanking (Murakawa, Maedo, and Nakagawa 1977).

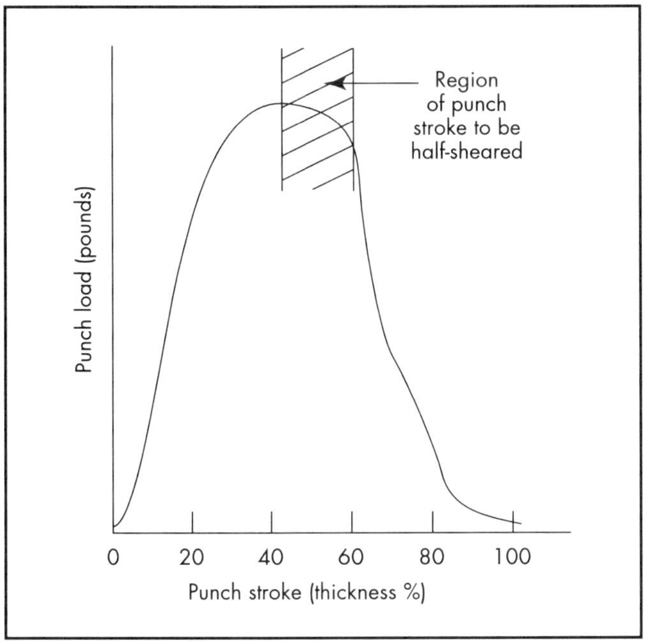

Figure 3-11. Region of punch stroke to be half-sheared in first step of reciprocating blanking process (Murakawa, Maedo, and Nakagawa 1977).

push-back system, the blanks in this case are half-sheared in the first step so that a relatively large push-back load is applied in the second step. A knock-out operation is necessary as a final step. With this method, it is not only possible to simplify the tooling and press design, but also to prevent burrs from forming, which could happen in the second step of the reciprocating blanking process if a poorly aligned blanking tool is used. This simplified reciprocating blanking method has been used successfully for blanking such material as aluminum, mild steel, and .060–.120-in. (1.5–3.0-mm) thick stainless steel.

3.11.3 Opposed-die Blanking Method

As illustrated in Figure 3-13, the main characteristic of opposed-die blanking is that a smooth sheared surface is obtained. This method can be applied to the blanking of thick and brittle material, and the blank die roll it produces is quite small. Opposed-die blanking also has a burr-free feature because of its shearing mechanism. Die roll is formed on both edges of the blanked surface before the blank is knocked out and separated from an adjacent scrap chip.

Figure 3-14 shows the factors affecting burr formation in opposed-die blanking. To produce burr-free blanked parts by this process, deeper penetration (*a*) of the protruding die and smaller clearance (*c*) between knock-out punch and

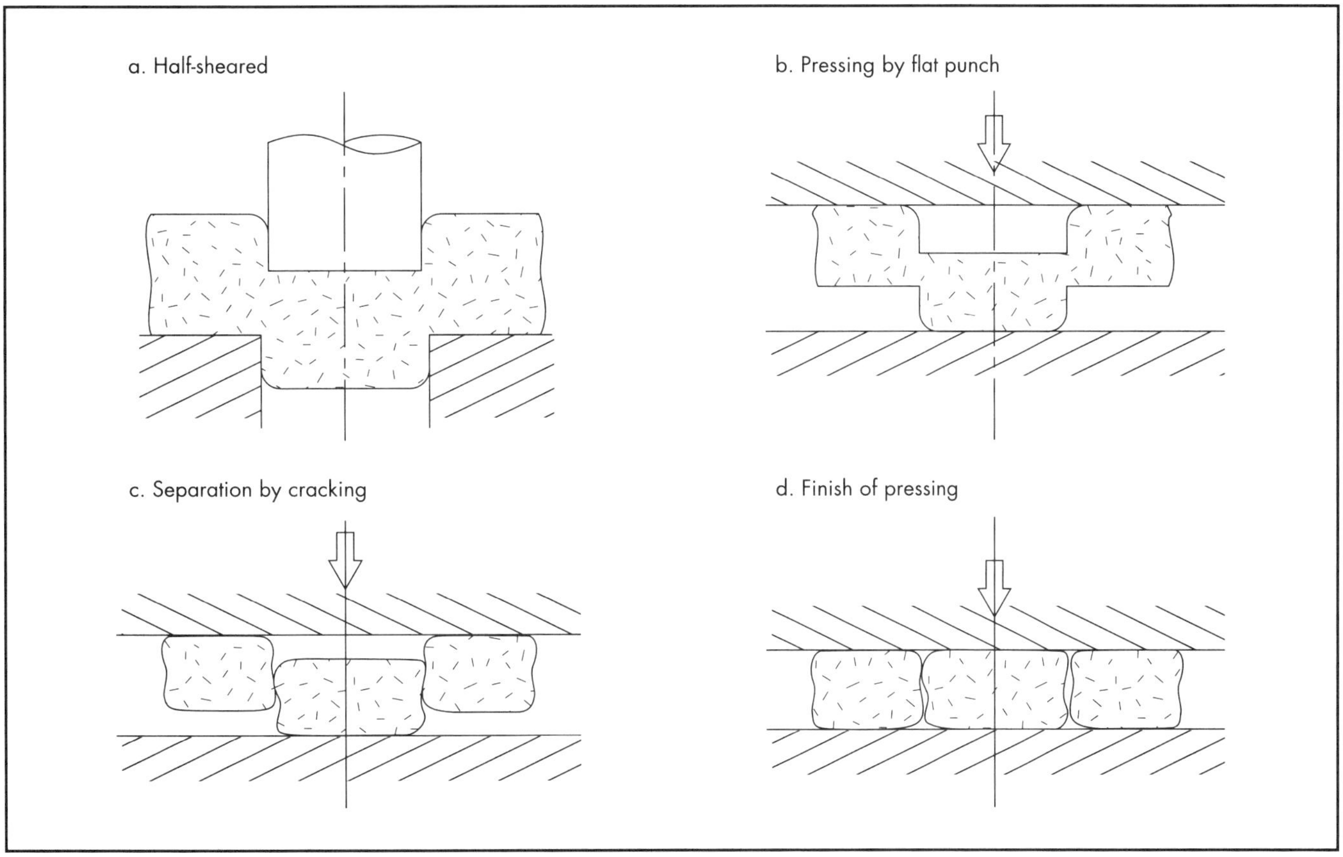

Figure 3-12. Push-back blanking (Murakawa, Maedo, and Nakagawa 1977).

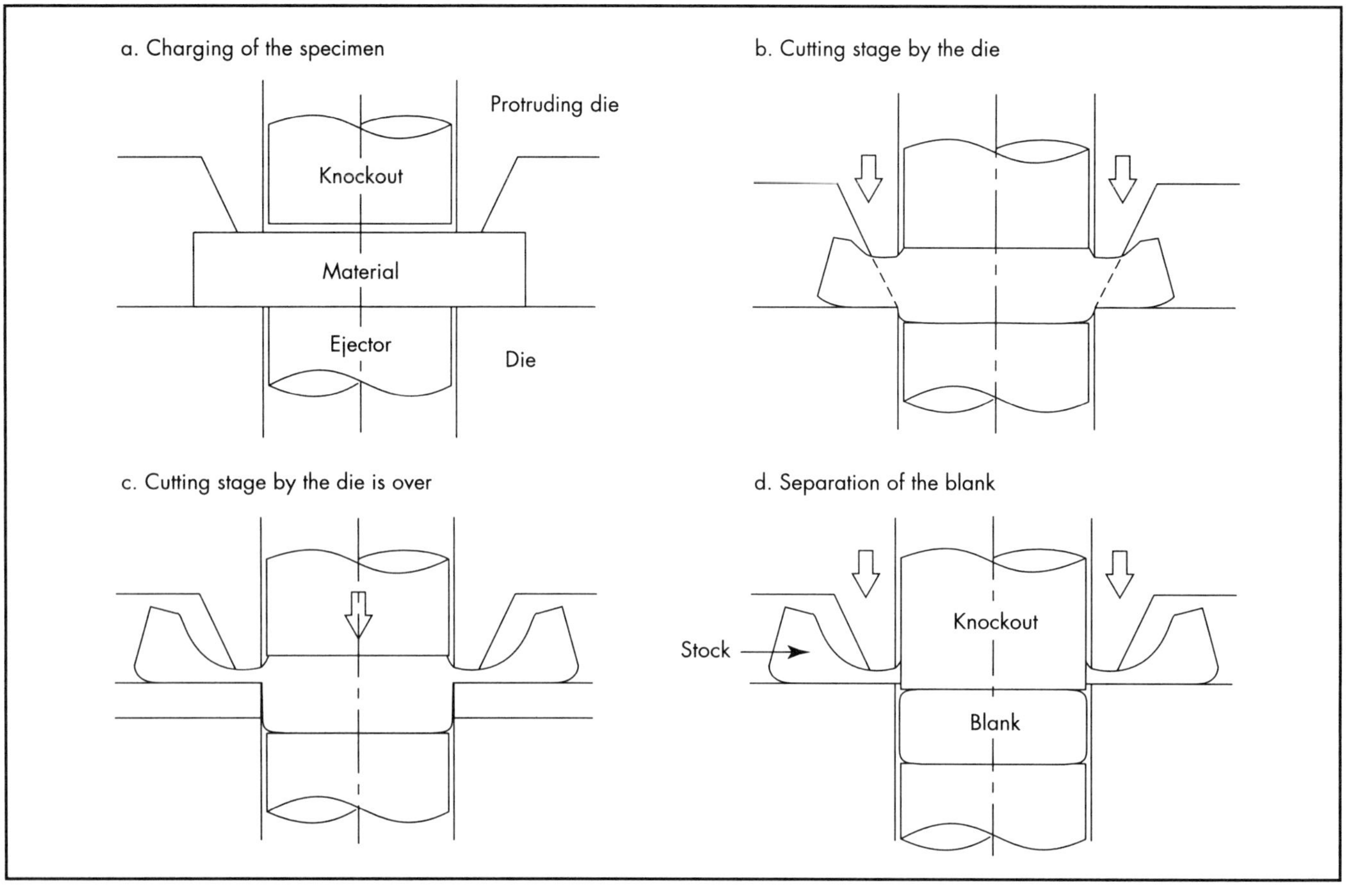

Figure 3-13. Opposed-die blanking (Murakawa, Maedo, and Nakagawa 1977).

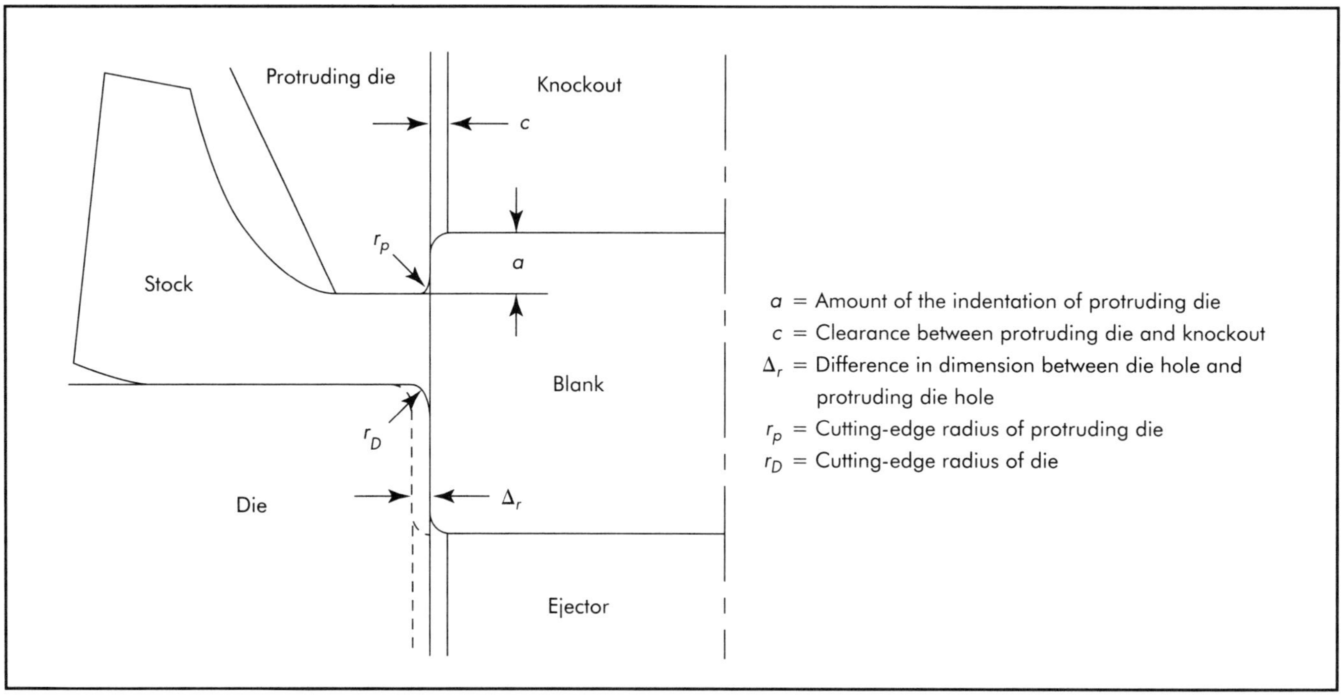

Figure 3-14. Factors affecting the burr in opposed-die blanking (Murakawa, Maedo, and Nakagawa 1977).

protruding die are desirable. This process has been successfully used for burr-free blanking of .032-in. (0.81-mm) thick silver-plated stator plates for electric motors.

3.11.4 Counter-cutting Method

Counter-cutting burr-free slitting was developed in 1968 in West Germany. Its mode of action is illustrated in Figure 3-15. In the initial cutting operation (Figure 3-15a), the cutters penetrate into the material to a specific depth. In the second step (Figure 3-15b), the circular cutters are arranged in the staggered position opposite the first pair of cutters, so that separation can be performed from the opposite side. Here the circular cutters are engaged as far as required for the complete separation of the material, thus producing burr-free slit strips.

Compared with conventional slitting, where the material is cut through only by the first-step cutter, the number of cutter disks in counter cutting is naturally doubled, and consequently substantially more maintenance (such as exchanging, regrinding, and realigning the cutter disks) is required. For counter cutting to be successful, both the alignment of the axes of the two pairs of cutter disks and the guide devices for the rolled material must be precisely maintained. Because of this requirement, the design and manufacture of the slitting stand and all the related components, as well as the daily operating and maintenance services, also must be precise. The production cost of this type of slitting machine is two or three times that of a conventional arrangement having a similar production capacity.

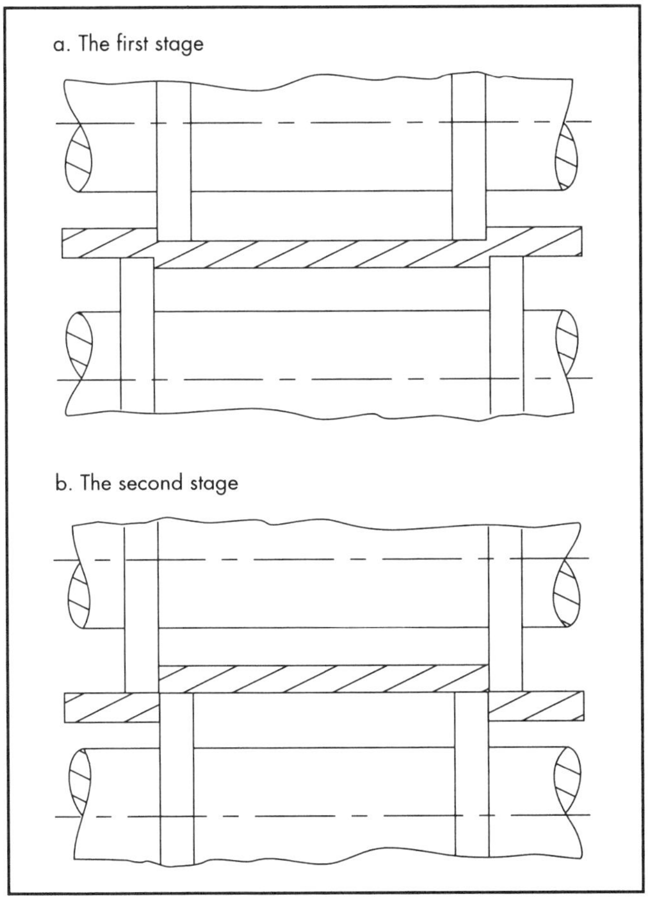

Figure 3-15. Counter-cutting method (Murakawa, Maedo, and Nakagawa 1977).

3.11.5 Roll-slit Method

The roll-slit method was designed to circumvent the need for highly accurate and costly slitting equipment. This method consists of two steps, as shown in Figure 3-16. The first step is essentially the same as in counter cutting: the flat material is partially sheared or penetrated but not separated from each adjacent part of the material (Figure 3-16a). The second step involves inserting this partially sheared material into the spacing between an opposing pair of rollers.

In the separating procedure of the roll-slit method shown in Figure 3-17, the sheared edge face, *a,* obtained from the first step, still remains at the initial stage of the second step. Meanwhile, a burnished surface, *b,* is formed adjacent to face *a*. In addition, die roll *a* formed on the strip edge is gradually reduced as the strip goes between the flattening rolls. Simultaneously, a new die roll, β, is gradually formed. As shown in Figure 3-17c, the plastic deformation of the half-sheared material will come to its limit during the squeezing operation of the rolls, and cracking will occur from the ends *(A, A')*. The partially sheared strip is then parted into the final slit products I, II, and III (Figure 3-16c). Figure 3-17d exemplifies the cut surface of a completely severed mild steel strip, where t = .128 in. (3.25 mm) from Figure 3-16c. A small die roll, *a*, exists, and a fracture surface is generated between the sheared face, *a,* and the burnished face, *b*. No burrs are formed.

In order to perform burr-free slitting, the selections of the cutter overlap L (Figure 3-16) and the cutter clearance C are key requirements. If the value of L is preset too small, the strip material will be sheared and severed completely in the first step, thus forming burrs. If L is too large, it will cause the half-sheared strip to be only pushed back and not completely separated in the second step. If the clearance C is too large, the complete separation of the strip without burrs is not obtained in the second step, regardless of suitable overlap value L in the first step. When C is too small, the sheared wall is not perpendicular to the surface.

The second key point for successful roll slitting is to set the clearance at or near zero. Tests on soft steel, aluminum, aluminum alloy, stainless steel, silicon steel, pickled hot-rolled strips, copper, and other materials, reveals that these two key points apply to all materials.

References

Clark, D. S., and Wood, D. S. 1950. "The Tensile Impact Properties of Some Metals and Alloys." *Transactions of the American Society of Metals.* Materials Park, OH: American Society for Metals: 45-74.

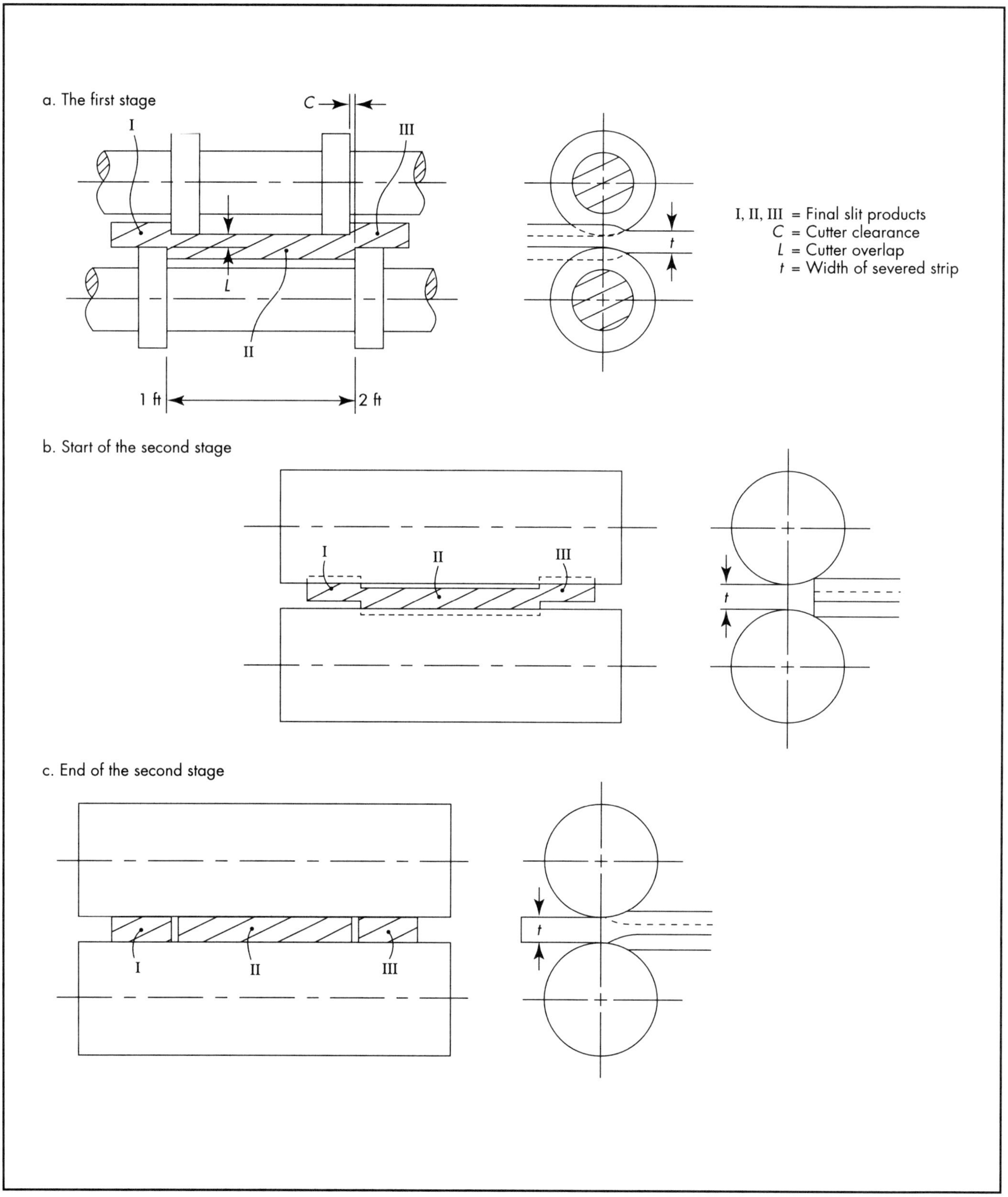

Figure 3-16. Roll-slit method (Murakawa, Maedo, and Nakagawa 1977).

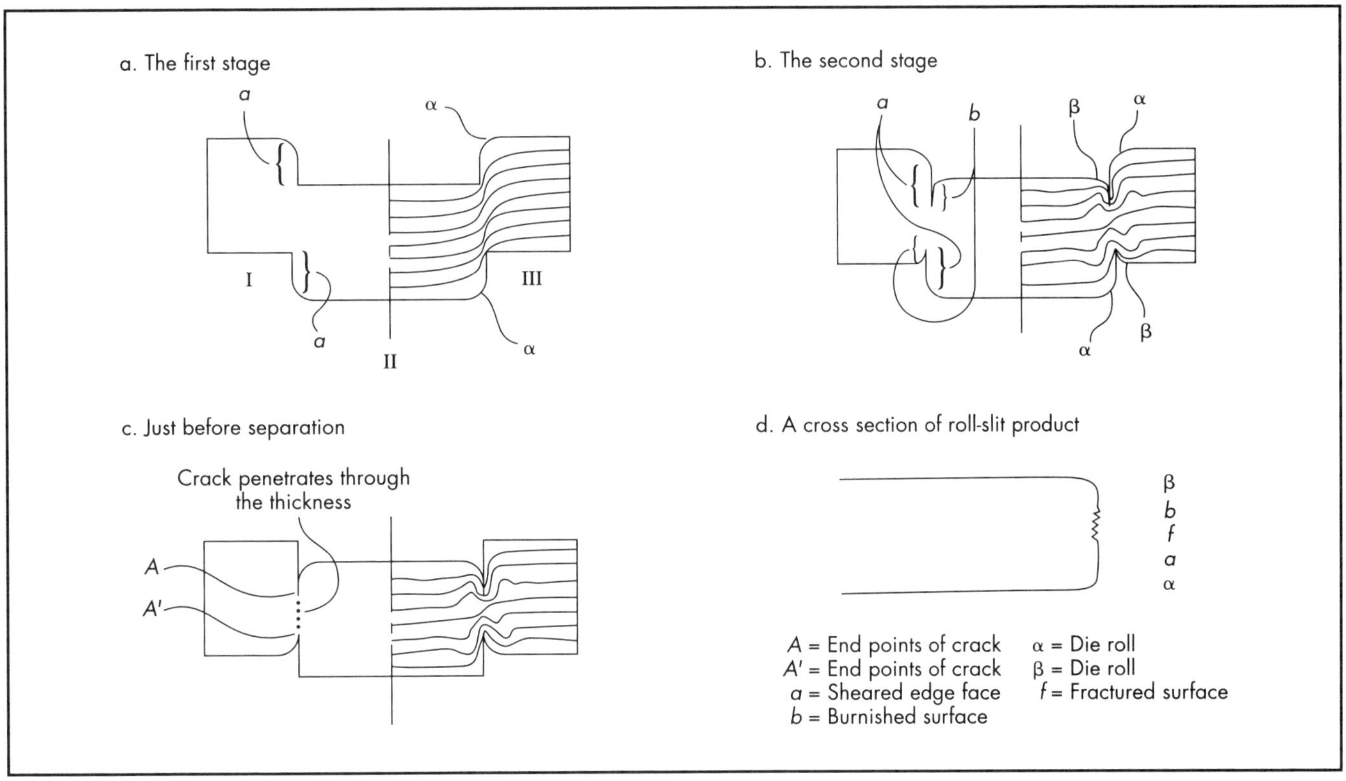

Figure 3-17. Separating procedure in roll slitting (Murakawa, Maedo, and Nakagawa 1977).

Fukui, S., Kawata, K., Takada, N., and Seino, J. 1967. "Effects of Tensile Speed on the Deformability of Sheet and Metals." *Proceedings of the 1967 CIRP Conference*. Detroit, MI: ASTME: 1163.

"HERF Shears Edges Clean." 1967. *Steel*, 28 November: 56-57.

Johnson, W., and Slater, R. A. C. 1967. "A Survey of the Slow and Fast Blanking of Metals at Ambient and High Temperatures." *Proceedings of the 1967 CIRP Conference*. Detroit, MI: ASTME: 825–31.

Kondo, K., Shikimyama, Tetsuol, Yoshida, Motoak, and Takada, Kunio. 1974. "Application of Opposed Dies Shearing Process." *Proceedings of the International Conference on Production Engineering,* Pt. 1, 251. Tokyo: Japan: Society of Precision Engineering.

Machinability Data Center. 1980. *Machining Data Handbook,* 3rd ed., vol. 3. Cincinnati, OH: Machinability Data Center.

Maeda, T. 1958. "Reciprocating Blanking Method." *Science of Machine,* 10(1): 140.

Maeda, T., and Murakawa, M. 1977. "The Development of Burr-free Slitting." *Journal of Japan Society for Technology of Plasticity,* 18(193): 114.

Mann, H. C. 1936. "High-velocity Tension Impact Tests." *Proceedings of the American Society for Testing Materials.* Philadelphia, PA: American Society for Testing Materials: vol. 36, Pt. II, 85–97.

McBride, James N. 1969. "The Magnitude of Burrs Caused by Electrical Discharge Machining." Masters thesis. Logan, UT: Utah State University.

Murakawa, Madao, Maedo, Teizo, and Nakagawa, Takeo. 1977. "Investigation on Burr-free Shearing in Japan." SME paper MR77-472. Dearborn, MI: Society of Manufacturing Engineers.

"News and Trends, Embrittlement Process Makes Hand Materials Easier to Machine." 1975. *Automation,* August: 14.

"Scanning for Ideas, Mercury Coating Makes Hand Materials Easier to Cut." 1975. *Machine Design,* September 18: 48.

Shafer, Fredrich. 1975. "Deburring Processes in Perspective." SME paper MR75-482. Dearborn, MI: Society of Manufacturing Engineers.

Wood, William W. 1963. "Combined Effects of High Velocity and Temperature on Metal Formability." *Proceedings of the 1963 Conference on International Research in Production Engineering.* New York: American Society of Mechanical Engineers: 311.

Bibliography

Allen, D. M., Brightman, J. R. C., Gillbanks, P. J., and Smith, P. M. 1987. "Three-dimensional Photochemical Machining." *CIRP Annals* 36(1): 91–93.

Bryan, Welsford J. 1986. "Chemical Machined Parts." *Handbook of Product Design for Manufacturing*. James D. Bralla, ed. New York: McGraw-Hill.

Coombs, Clyde F. 1988. *Printed Circuits Handbook,* 3rd ed. New York: McGraw-Hill.

Guo, Y. B., and Dornfeld, D. A. 1998. "Finite Element Analysis of Drilling Burr Minimization with a Backup Material." *Transactions of the North American Manufacturing Research Institute (NAMRI) of SME*. Vol. No. XXVI. Dearborn, MI: Society of Manufacturing Engineers: 207-212.

Shafer, Fredrich. 1975. *Entgraten*. Mainz, Germany: Krausskopf-Verlog.

Springborn, R. K., ed. 1967. *Nontraditional Machining Processes.* Dearborn, MI: Society of Manufacturing Engineers.

Takayama, H., and Kato S. 1991. "Burrless Drilling by Means of Ultrasonic Vibration." *CIRP Annals* 40(1): 83–86.

Weller, E. J. 1984. *Nontraditional Machining Processes.* Dearborn, MI: Society of Manufacturing Engineers.

4 Burr Formation

Reducing the costs of burrs requires knowledge of how burrs form. Preventing and minimizing burrs is technically challenging, and considerable research remains to be done. One must understand exactly which burrs form by which mechanism and recognize how the many machining variables affect the various burr properties. Changing one variable to reduce the thickness of a burr may increase the thickness or length of another burr on the same part. The issue is so complex that each type of burr may have several subcategories. Drilling burrs, for example, have at least six different appearances, and each appearance has its own name.

This chapter provides a framework for understanding burr formation and describes specific ways to minimize burrs in turning, drilling, milling, grinding, reaming, and press-working operations. An extensive reference section and bibliography at the end of the chapter provides sources for additional information on burr formation.

4.1 BURR FORMATION MECHANISMS

Burrs, flash, and related protrusions are formed by the six physical processes listed in Table 4-1. Burrs formed by the first three processes involve plastic deformation of the workpiece material. Solidification of material on the working edges, the fourth process of formation, forms a burr-like projection. The fifth process, incomplete cutoff, occurs when the workpiece is allowed to fall from the part before the cut is completed. Flash forms whenever the pressure on molten material is sufficient to force the material between the two halves of a die or mold.

4.1.1 Poisson Burr

A Poisson burr occurs whenever the cutting edge extends past an edge of the workpiece (see Figure 4-1). It is the result of lateral deformation that occurs whenever a solid is compressed. The extent of this deformation is a function of the workpiece material, the size and shape of the contacting components, and the applied load. The extent of the lateral deformation (burr size) is usually relatively small. The size of the burr is proportional to the cutting-edge radius and the applied pressure.

Table 4-1. Physical processes involved in the formation of burrs, flash, and related protrusions

Process	Name of protrusion
1. Lateral flow of material	Poisson burr
2. Bending of material (such as chip rollover)	Rollover burr
3. Tearing of chip from workpiece*	Tear burr
4. Redeposition of material	Recast bead
5. Incomplete cutoff	Cutoff projection
6. Flow of material into cracks	Flash

* A tear burr also forms in stamping operations when the punch tears the part from the stock.

4.1.2 Entrance Burr

When a cutting edge first indents a workpiece, as shown in Figure 4-2, another type of burr may form. This entrance burr is material that has flowed opposite to the direction of the tool. It is similar to the ridge that forms around the indentation made by a Brinell hardness tester. Whether a burr forms at this point depends on the workpiece properties and, probably, the shape of the cutting edge.

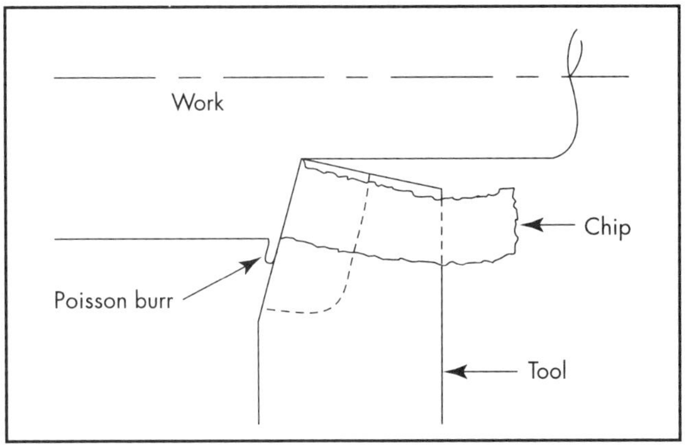

Figure 4-1. Poisson burr formed when cutting edge of tool extends past edge of workpiece.

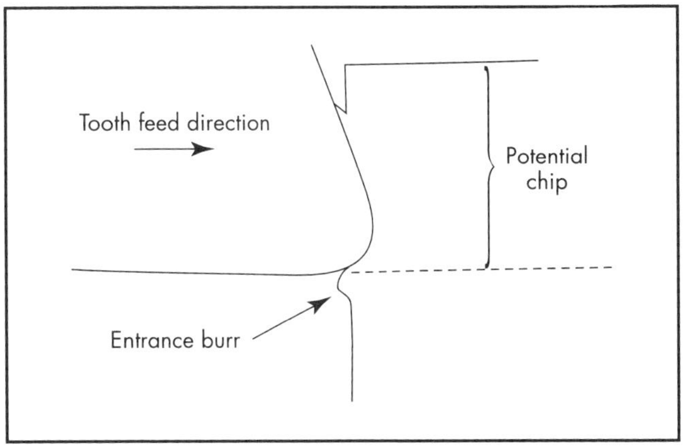

Figure 4-2. Cutting edge produces indentation burr as it enters workpiece.

Strain hardening also plays an important role in the formation of Poisson and entrance burrs. An entrance burr generally forms when the material has a low strain-hardening exponent (below 0.3). Materials having high strain-hardening exponents tend to cause a bulge, but not a sharp burr. Burrs will not form in materials unless they have some ductility. Material-ductility properties are generally expressed as a percentage of elongation. Ductile materials have high values of elongation; thus users can reasonably predict burr thickness problems by comparing elongation values. Materials that are exceptionally ductile will typically produce thick and long burrs.

4.1.3 Rollover Burrs

When a cutting edge exits from a workpiece, a rollover burr normally occurs. This happens when bending the chip is easier than cutting it or fracturing the edge (Figure 4-3). The thickness of a rollover burr (Figure 4-4) in a continuous depth-of-cut process can be expressed by Equation (4-1):

$$h = t(\operatorname{cosec} A - 1) \qquad (4\text{-}1)$$

where

h = Depth of plastic deformation, which also is thickness of a rollover burr.

t = Depth of cut.

A = Shear angle, expressed approximately by the following equation:

$$A = \tan^{1} [(t/t_2) \cos B]/[1 - (t/t_2) \sin B] \qquad (4\text{-}2)$$

or

$$A = \tan^{-1} [\cos B]/[t_2/t - \sin B] \qquad (4\text{-}3)$$

where

t_2 = Chip thickness.

B = Rake angle of cutting tool.

Shallow depths of cut and large shear angles minimize burr thickness. Large rake angles and small cutting forces minimize the chip thickness, thus reducing burr thickness.

The length of a rollover burr is a function of the cutting conditions and the plasticity of the workpiece material. With the exception of some milling operations, the length of the burr cannot be more than the total depth of the cut. If in bending the strain exceeds the strain required to fracture, the majority of the burr will break off, leaving only a short burr.

Note that the Poisson burr thickness at its root also can be estimated by Equations (4-1), (4-2), and (4-3), but they will be less accurate than for the rollover burr, since the equations ignore the effect of lateral flow. For the Poisson burr, however, the approximation is at least order of magnitude.

4.1.4 Tear Burrs

Tear burrs form when chips are torn rather than sheared from the workpiece (Figure 4-5). Although such burrs can form in most cutting processes, they are easiest to produce in side-milling operations. The milling cutter tooth forces the chip up and forward. As it does so, the sides of the chip are torn from the workpiece. The tear burr is the torn portion remaining on the workpiece. In punch-press operations, the burr formed is basically a tear burr. In early research on this subject, the burr from blanking was called a *tensile burr*.

4.2 COMMON CHARACTERISTICS OF BURRS

The Poisson burr, the rollover burr, and the tear burr all have one common characteristic: a radius occurs on the back side of the burrs (see Figure 4-6). All of these burrs are continuations of the parent material and have a grain structure similar to the parent material. However, the burr hardness will typically be somewhat different from the parent material.

4.3 OTHER PROJECTIONS AT EDGES

Burrs are only one form of edge projections. Projections, like burrs, are generally objectionable and typically must be removed. The removal process may be the same as for burrs, but recast metal may require more time because of the metallurgical properties of the recast, and cutoff projections are typically much thicker than burrs. These projections do not have the same characteristics commonly found in the various forms of burrs (see Section 4.2).

Some of the literature on burrs may not distinguish between the types of projections being removed, and industry personnel may consider any projection at an edge a burr. As mentioned in a preceding chapter, in this handbook burrs will be differentiated from other edge effects.

4.3.1 Recast Material

Recast material (see Table 4-1) is formed when molten metal resolidifies on an edge of a workpiece (Figure 4-7). This type of projection most commonly occurs on electrical discharge machined (EDM) features. The amount of recast material is a function of polarity, workpiece material, and frequency and amperage of the current defining the spark energy.

Most EDM-recast material forms on the tool-entrance side of the part. The size of the recast material can be reduced significantly by starting the cut with less than 20 amps, then slowly increasing the power. Finishing passes will also remove much of the recast material.

Laser machining leaves recast material, but little data is available to characterize the material. Modern machines are capable of leaving less than .004 in. (0.1 mm) of such recast thickness at edges during percussion laser drilling, and .002 in. (0.05 mm) during laser trepanning. The amount of material left on parts depends on material, time, energy, type of laser, and other factors.

4.3.2 Cutoff Projections

Cutoff projections (see Figure 4-8) occur when workpieces are severed from barstock (Gillespie 1977). In such cases, the workpiece is allowed to fall from the stock before the

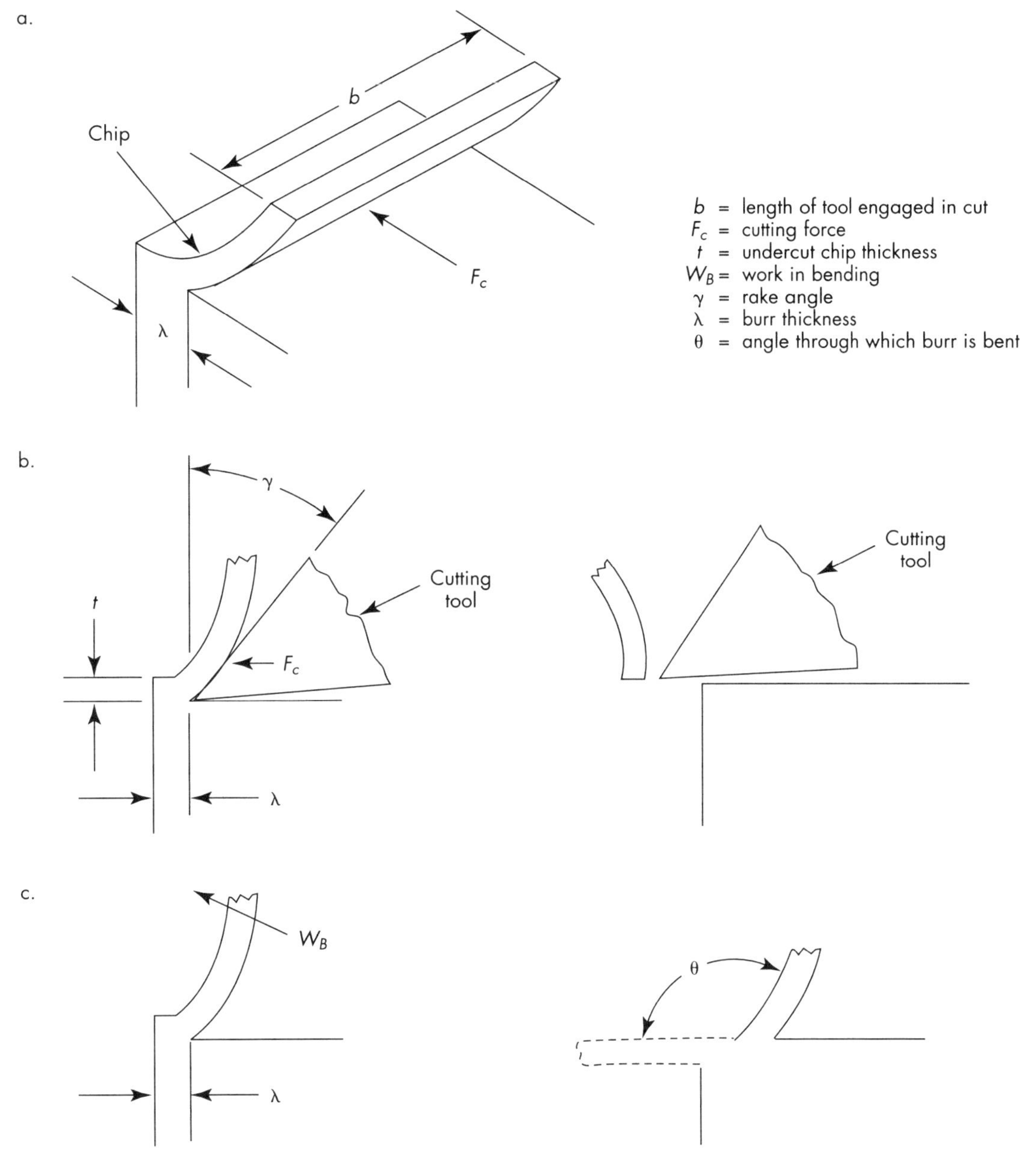

Figure 4-3. Rollover burr forms when the cutter nears the end of the cut.

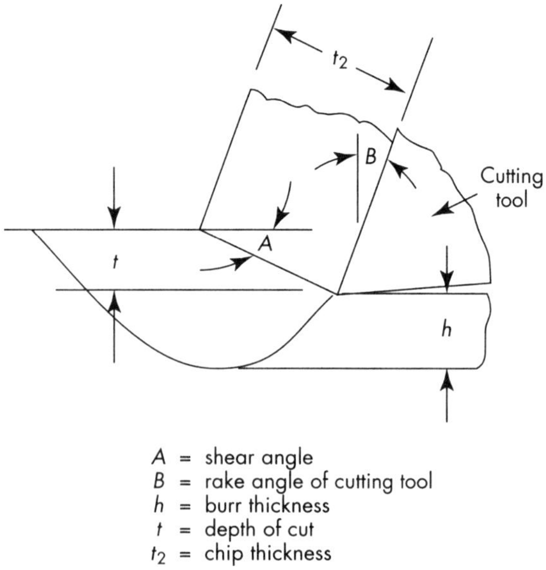

Figure 4-4. Rollover burr thickness, *h*, as a function of idealized conditions (Adapted from Barash and Schoech).

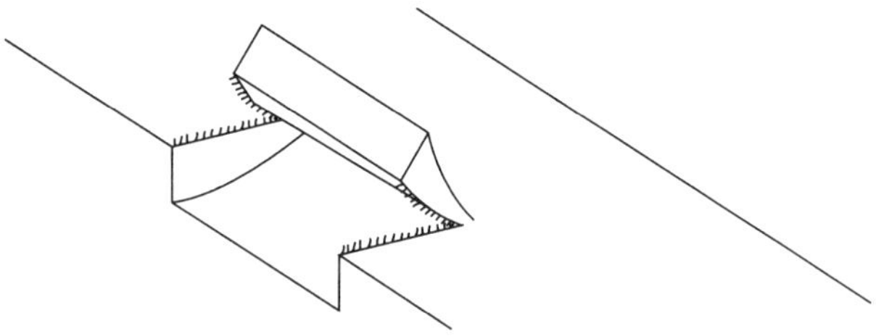

Figure 4-5. Tear burr forms when chip is torn from workpiece.

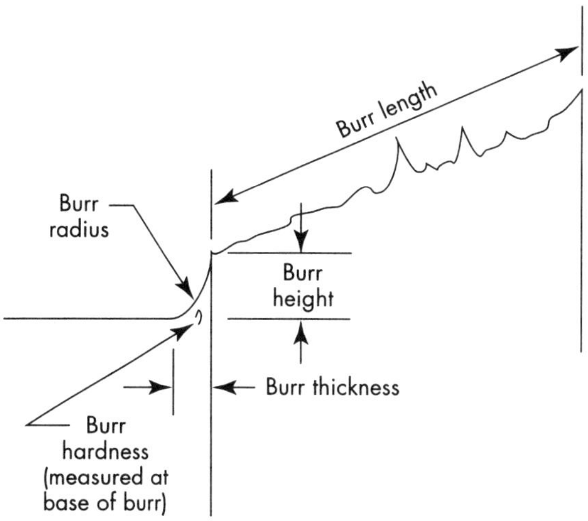

Figure 4-6. Definition of burr characteristics.

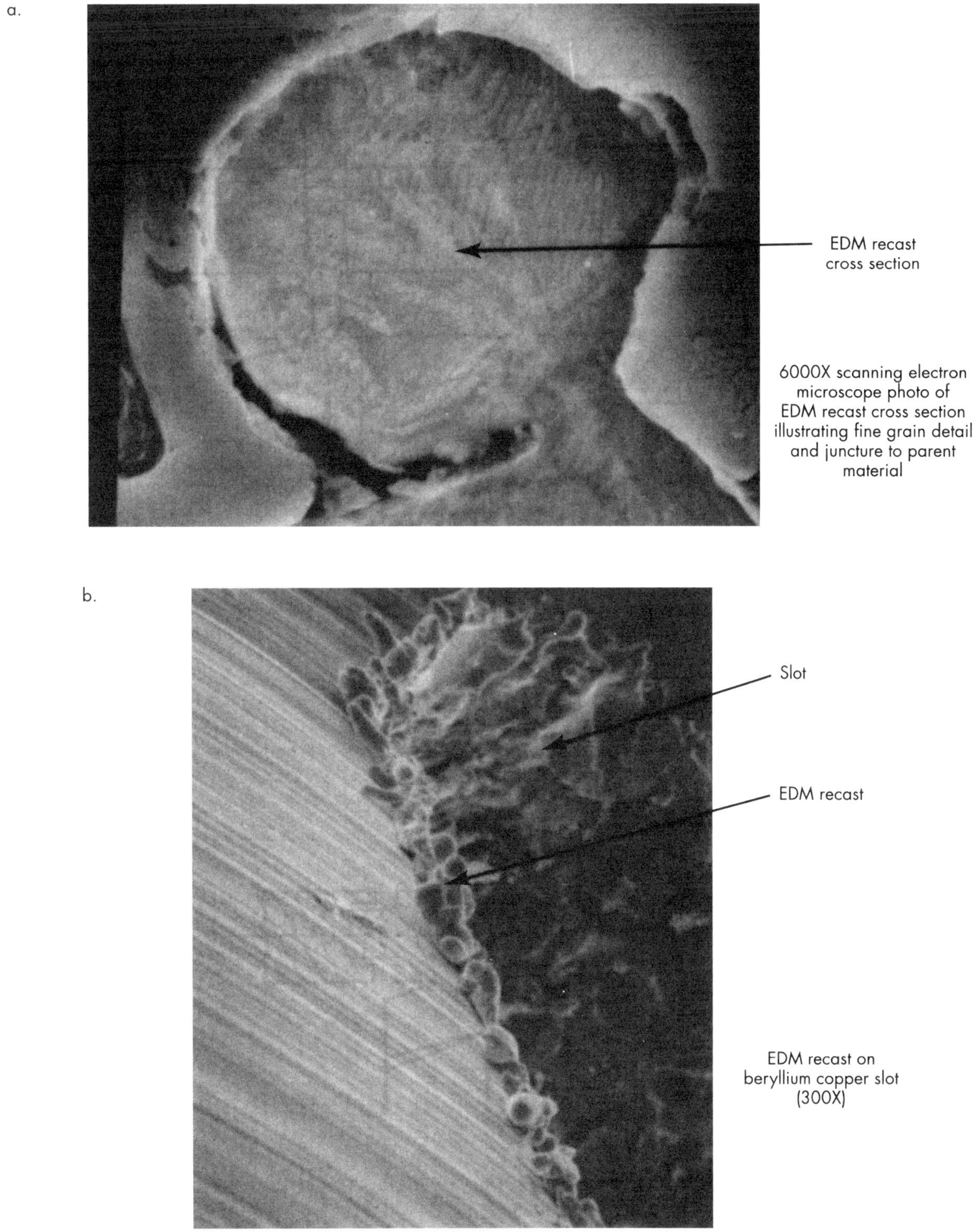

Figure 4-7. Recast material from EDM.

parting tool completely severs the material. Projections can occur in any parting operation, although they most commonly occur in turning operations.

4.3.3 Flash

If adequate clamping pressure is provided between casting dies or molds, flash will only occur when die or mold halves are not perfectly matched. In such cases, flash thickness is equal to the thickness of the voids between die or mold halves. Flash height is equal to or shorter than the length of these voids. The hardness of the flash is approximately the same as that of the workpiece. Flash is generally more difficult to remove than burrs because it is often thick and located on a surface rather than on an edge where removal is easier.

Another form of flash occurs when a hot tool is used to cut plastics. The tool locally softens the material and causes it to flow. Although no dies are involved, it forms by the same mechanism as conventional flash.

4.3.4 Cratered Edges

Brittle materials may crater, crack, tear, or crumble as the cutter passes over the edge. This leaves an edge with missing material. Although the absence of material does not require deburring, edges will often need to be smoothed to reduce stresses and prevent further fracture. Figure 4-9 provides photographs of the formation of a rollover burr (Figure 4-9a) and a fractured edge (Figure 4-9b).

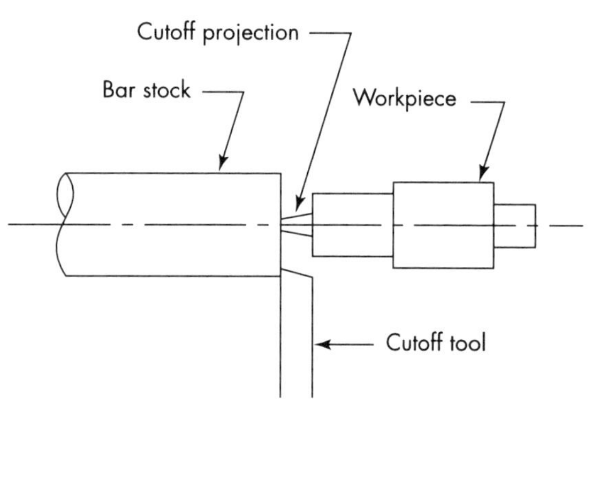

Figure 4-8. Projection left on workpiece after cutoff operation.

4.3.5 Projections Produced While Plastically Deforming Part Shapes

Consider for a moment the job of blacksmithing. The blacksmith picks up a piece of glowing orange steel, places it on an anvil, and hits it with a hammer. The part is smashed, forged, or otherwise permanently deformed. What was once a straight edge now has metal squirting past the old edge. Is that metal a burr? No, it is not. If the edge were smashed while cold, it would not be a burr. It would be forging flash—metal allowed to move past the edge. Some operations found in machining or other production facilities are similar to blacksmithing—for example, the process of producing rolled threads has some of the same characteristics as blacksmithing.

4.3.6 Nicks, Dings, Scratches, and Other Raised Material

When a tool, another part, a lid, or any other hard metal item hits with one edge against the edge of a part, some form of damage will occur. Nicks and dings are the two most common forms of damage. The indentation is only one form of the damage, however. Raised metal of some form normally also occurs at the edge of the ding or nick. Is this raised material a burr? In the normal sense that there was no shearing action or attempt to produce a chip, one might think not. However, raised material forms in the same way as entrance burrs when a metal cutting tool enters the cut. The only difference between this inadvertent damage and the "intended damage" resulting from cutting is that the raised material from accidents only occurs at sporadic positions on an edge.

When a sharp tool is drawn across a surface, a scratch is produced (Figure 4-10). If the material is ductile, small mounds of material can be seen at the sides of the scratch if it is viewed under high magnification. The scratch is a cutting action, and the burr occurring at the edge of the scratch forms in a manner similar to a tear burr or an entrance burr. In this case, the burr is at the edge of the scratch, not the edge of the part. There is a field of research that concentrates on scratches and the piling-up phenomena at the front and sides of scratches.

When one considers that a lathe tool is tearing out a long thin strand from the surface of a part, it might not be surprising to learn that even in conventional machining one can find hundreds of these fine fiber-like or wire-like strands of material on the surface of freshly machined parts. They are generally one or two orders of magnitude smaller than machining burrs, must be viewed with a microscope at 20–100× magnification, and have not been a concern on most products in the past. As product performance requirements increase for some products, such as those requiring clean room conditions and an absence of fibers and particles, so does the need to understand how these projections affect products and processes. Researchers are now beginning to study these "microslivers" that occur in grinding and other cutting processes. These particles are often naturally removed by common deburring processes.

a.

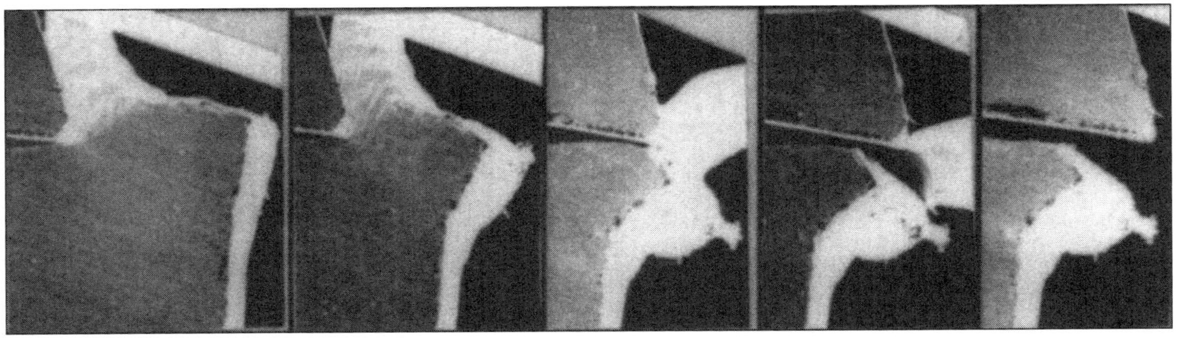

Feed 0.1 mm Cutting speed 0.025 mm/s Copper 0.1 mm

b.

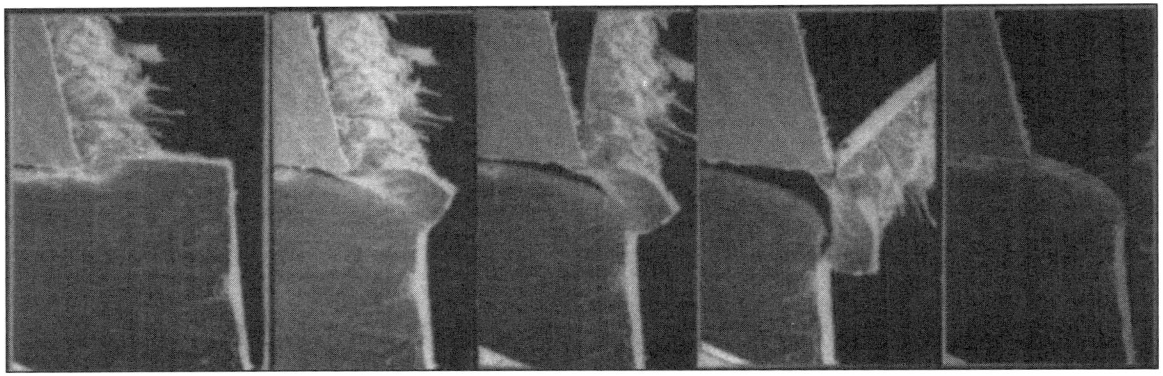

Feed 0.1 mm Cutting speed 0.025 mm/s Aluminum 20240 0.1 mm

Figure 4-9. Rollover burr formation. (a) In copper orthogonally machined at .004 in./s (0.1 mm/s) with cutting speed of .001 in./s (0.025 mm/s); (b) aluminum 20240 machined under the same conditions results in a cratered edge or absence of edge material (Hashimura and Dornfeld 1995).

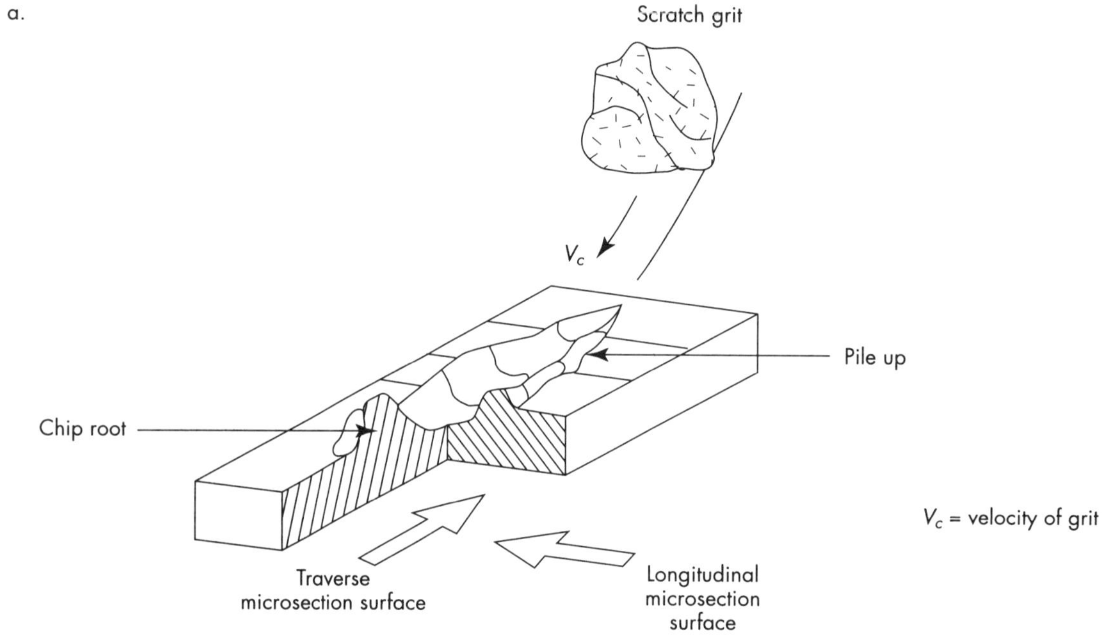

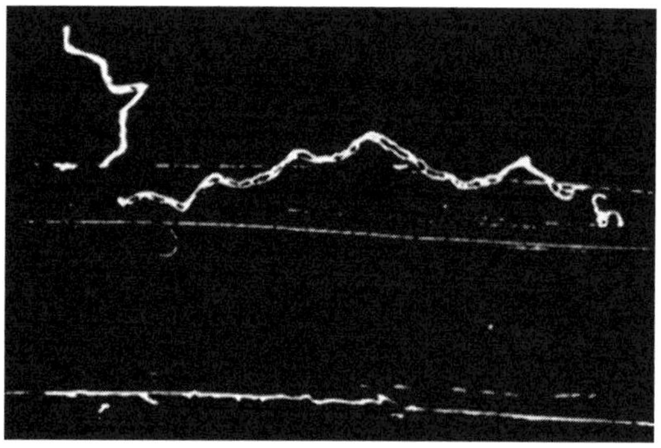

Secondary chips attached to the sides of abrasion grooves in copper
Scanning electron micrograph × 180

Figure 4-10. (*a*) Raised material at sides of scratches, on machining scallops, and on ground surfaces; (*b*) scanning electron micrograph (180×) showing secondary chips attached to the sides of abrasion grooves in copper.

4.4 OVERVIEW OF BURRS PRODUCED BY SPECIFIC OPERATIONS

There are an estimated 5,000 manufacturing processes or process variations. Providing details of burr formation and other edge conditions for each of these is clearly beyond the scope of this handbook. Most of the known literature on burr formation is concentrated on drilling, milling, and pressworking operations. This chapter will provide only a brief overview of burr formation in specific operations to allow those involved with edge finishing to understand the basic concepts and provide some direction to more extensive research and solutions for some specific material, shape, and process combinations.

The following paragraphs provide a quick overview of how the various formation mechanisms occur in specific processes. Subsequent sections in this chapter provide in-depth comments on how some of these burrs can be minimized.

4.4.1 Burrs from Turning Operations

Poisson burrs typically occur in turning operations (see Figure 4-1). A rollover burr forms when the tool passes over a groove or whenever an interrupted cut occurs (Figure 4-11). A small rollover burr also occurs when a facing cut is made. Tear burrs form when producing grooves (Figure 4-12). If sufficient rubbing occurs between the grooving tool and the walls of the groove, material will also flow from the sides of the groove as a result of the Poisson effect (Gillespie 1976b).

4.4.2 Burrs from Drilling Operations

In drilling operations, the burr that forms at the entrance of the hole can be the result of tearing, a bending action followed by clean shearing, or lateral extrusion (Poisson effect) (Figure 4-13). The Poisson effect is formed similar to the Poisson burr shown in Figure 4-2 for turning. If one thinks of the drill lip as a lathe tool entering the hole, it looks similar to Figure 4-2.

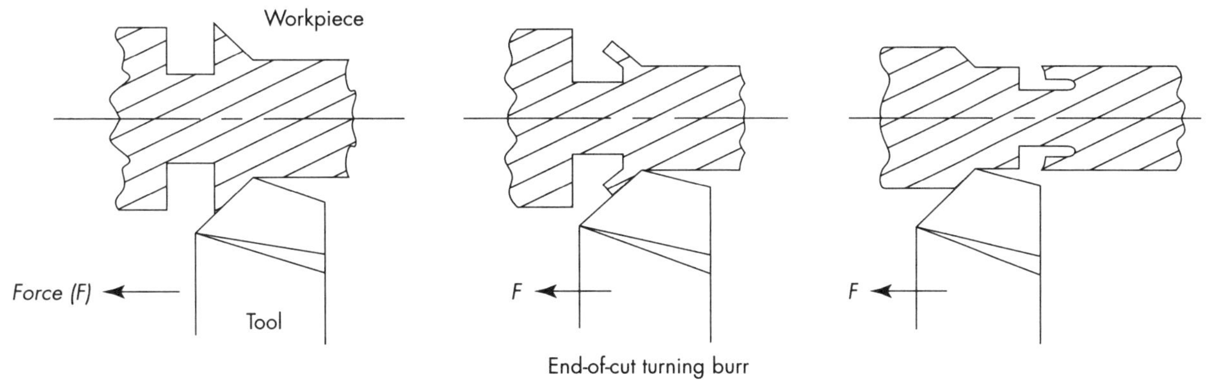

Figure 4-11. When lathe tool passes over an interrupted cut, a rollover burr forms.

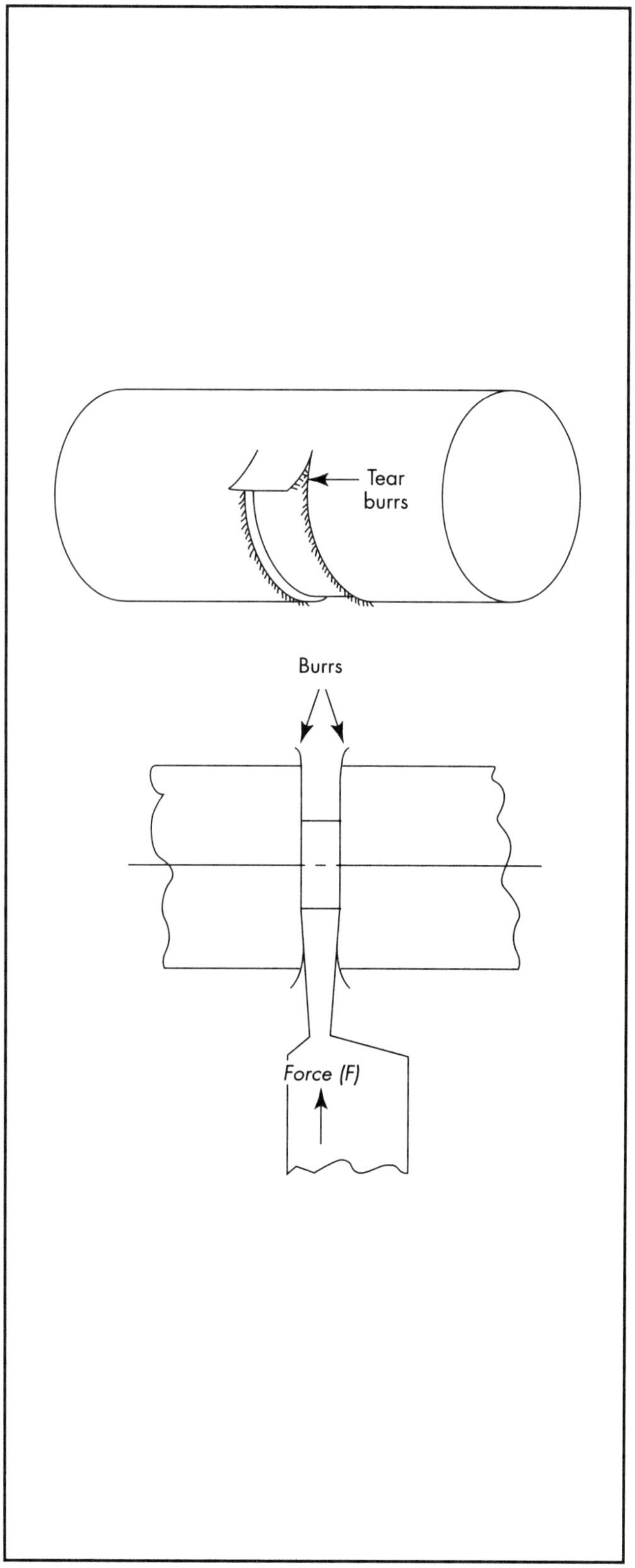

Figure 4-12. Tear burr at sides of lathe groove cut.

The burr that is formed when a very sharp drill exits from the workpiece is actually a Poisson burr resulting from rubbing at the margins of the drill. When a normal or dull drill exits from the bottom of a workpiece, the uncut chip rolls over, making a rollover burr (Figure 4-14). Rubbing against the hole walls by the drill margins can increase the thickness of the exit burr.

4.4.3 Burrs from Milling Operations

Side Milling

In a side-milling slotting operation, burrs are produced on eight edges (Figure 4-15). Tear burrs occur at edges 1, 2, 3, and 4. An entrance burr occurs at edge 8. A hybrid Poisson tear burr occurs at edges 6 and 7. A rollover burr occurs at the edges when the cutter exits from the workpiece. The height of the burr, b_h in Figure 4-16, is equal to or less than the depth of the cut *d*.

If the milling cutter has a helix angle, the properties of the burrs at each of the eight edges will be different from those on every other edge. There will be eight different sets of thickness, height, radius, and hardness for the burrs. *For precision deburring, these variations are important because they mean that each edge will respond differently to a single deburring process.* Properties of the burrs also vary along the length of all edges.

End Milling

In end-milling operations, Poisson burrs are formed on edge 1. Depending upon cutter geometry, they also form on edges 2, 4, and 10 (Figures 4-15 and 4-17). An entrance burr occurs on edge 6. Rollover burrs are produced on edges 3, 7, and 9. On half of edges 5 and 8 an entrance burr is produced, whereas a rollover burr occurs on the other halves. The height of the rollover burr equals the radial depth of cut until the cut exceeds 0.6 times the tool diameter (Figure 4-18). For deeper cuts the height tends to remain constant at 0.6 times the tool diameter, although it may vary with workpiece properties and tool forces.

4.4.4 Burrs from Grinding Operations

Grinding burrs form by two basic mechanisms—a lateral flow of material (Poisson burr) and chip rollover (rollover burr). Although the basic cutting phenomenon in grinding often has been considered to be similar to that of hundreds of minute milling cutter teeth, there are many significant differences. These differences, in turn, produce burrs that are significantly different in properties than those expected from burrs produced by minute milling cutters.

Poisson Burrs

A Poisson burr is formed on both sides of an abrasive grain as it cuts the workpiece (Figure 4-19). The formation of the burr is the result of the lateral forces that occur in any material having a Poisson ratio greater than zero. Conceptually,

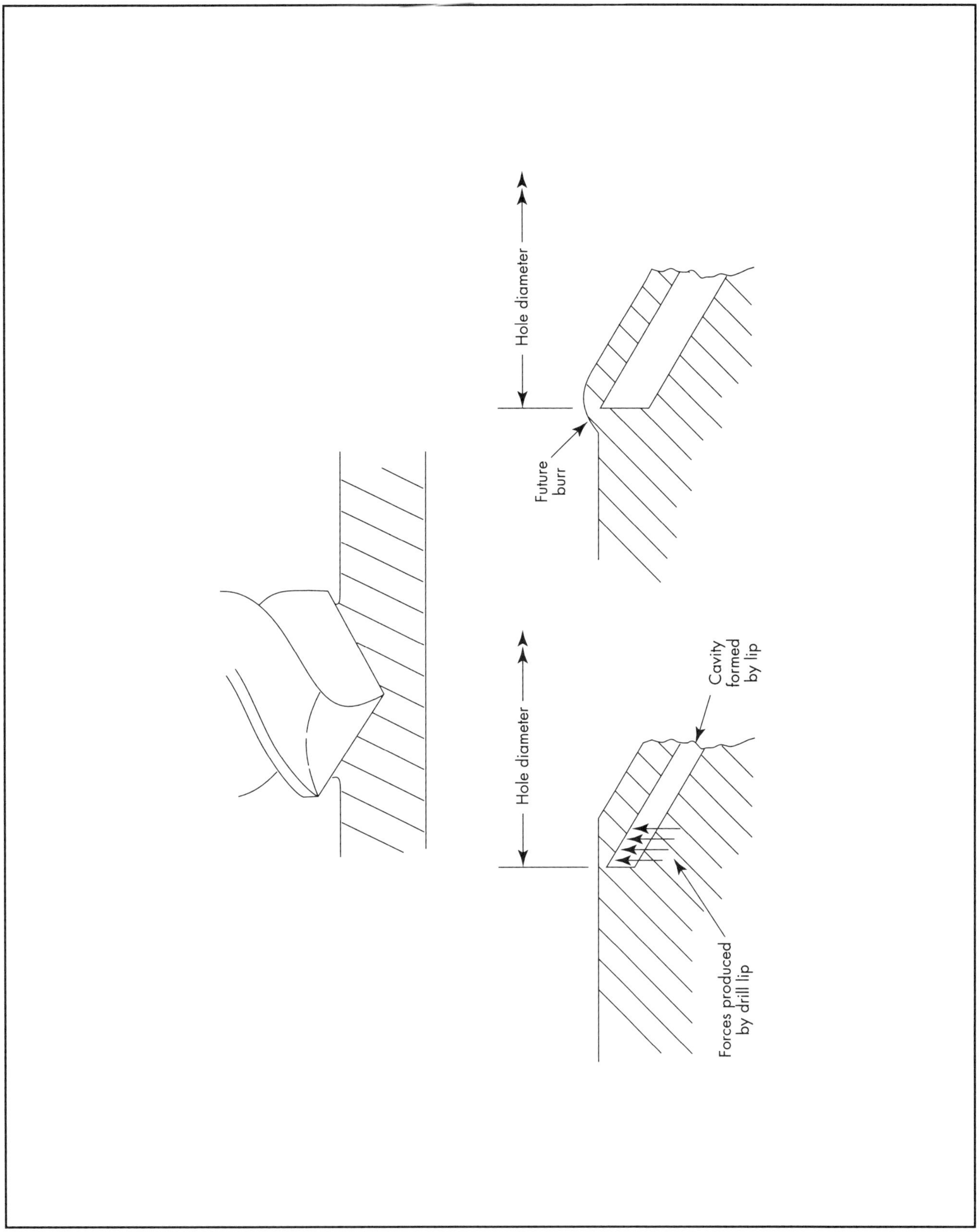

Figure 4-13. Burr formation at hole entrance.

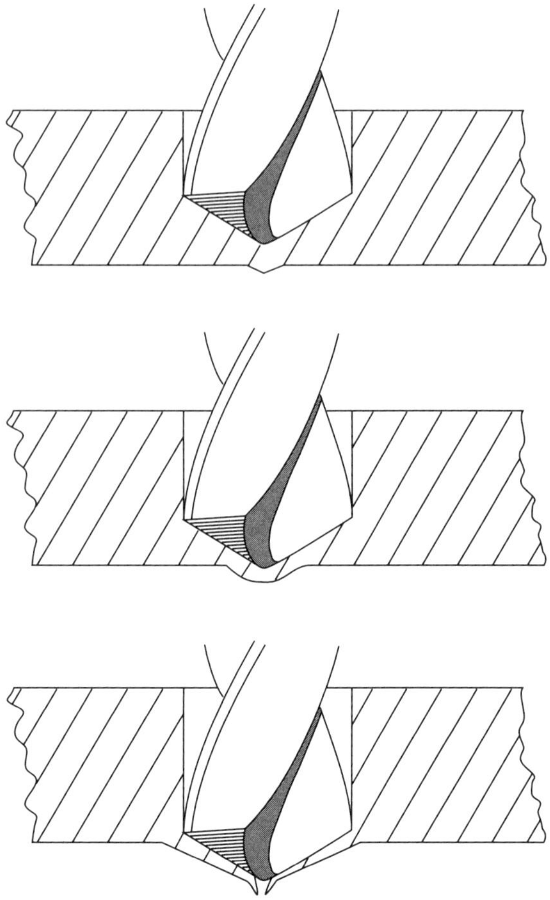

Figure 4-14. Formation of drill exit burrs.

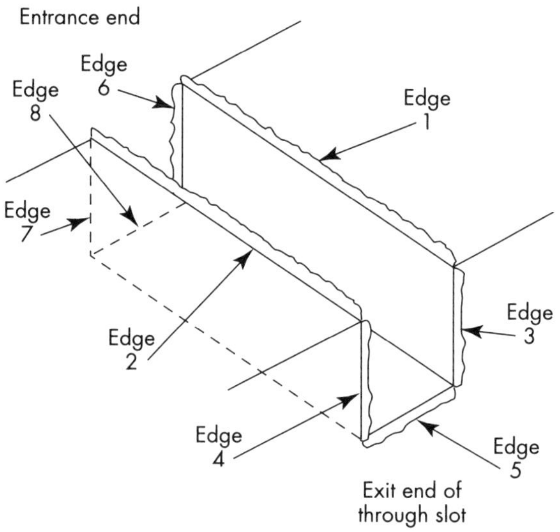

Figure 4-15. Burr locations in slots produced by side or end mills (Gillespie 1975c, 1976a).

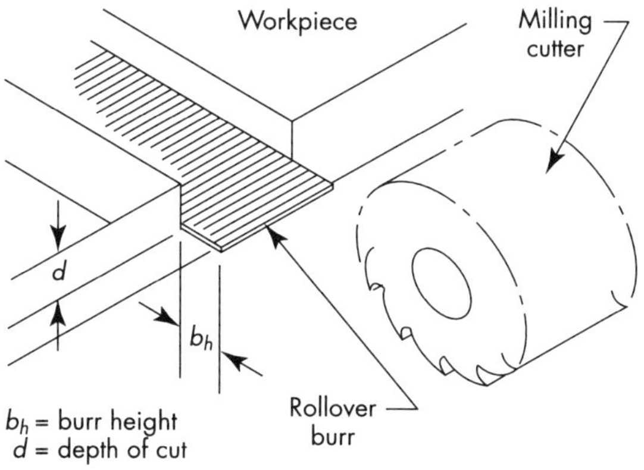

Figure 4-16. Location of rollover burr in milling a slot (Gillespie 1975c).

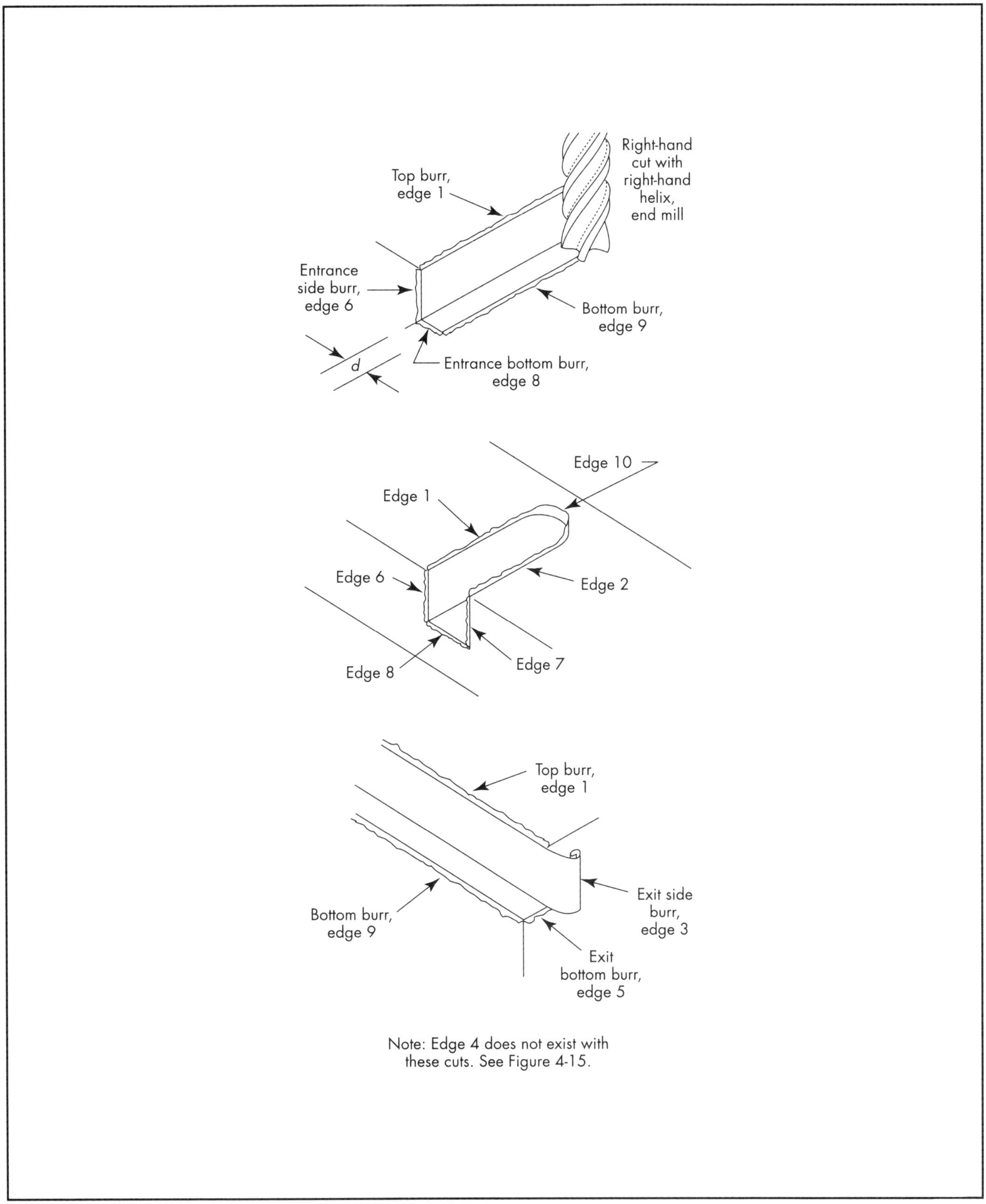

Figure 4-17. Identification of burr locations in end-milling edges of workpieces (Gillespie 1976a).

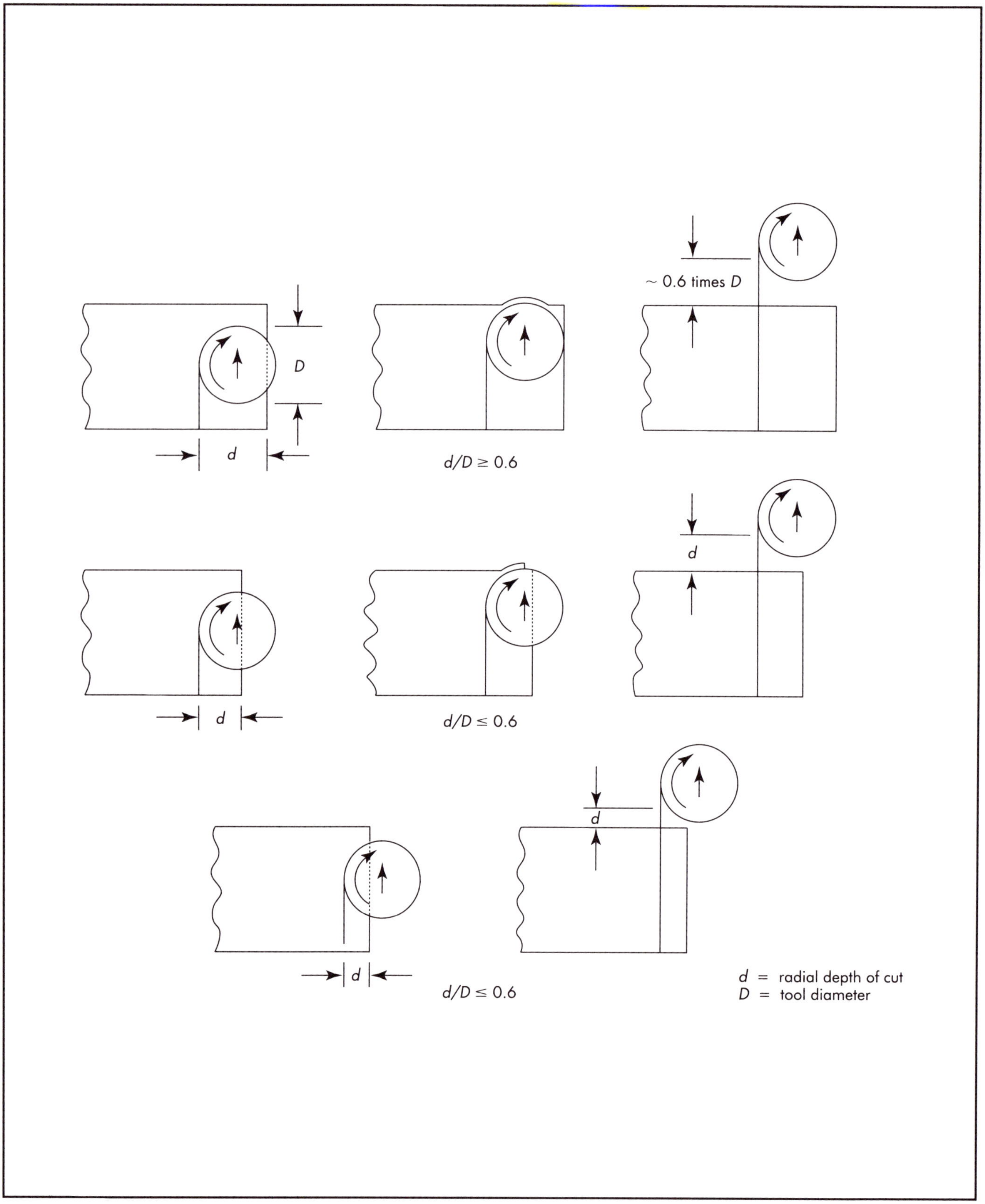

Figure 4-18. Effect of radial depth of cut on rollover burr.

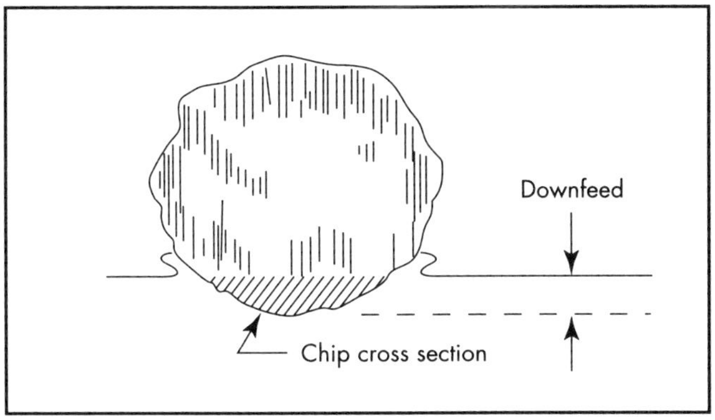

Figure 4-19. Poisson burr formed by an abrasive grain (Gillespie 1975b).

the burr formed is similar to the ridge produced when a marble is pushed through loose dirt. In the case of the abrasive grain, the grain also produces a chip as it ploughs through the workpiece.

Ignoring the effect of cross feed, Figures 4-20 and 4-21 show that a grain near either edge 1 or 2 will force material to move laterally over the edge of the workpiece. Because the wheel is turning, an individual grain is in contact with the workpiece for only a short distance.

Because of the random spacing of the grains, one grain will be positioned slightly to the left of the grain shown in Figure 4-19, and another will be positioned slightly to the right. This alternating random spacing tends to push the burr produced by the first grain farther over the edge. At some points, however, the grains will be positioned in such a manner that they will cut off the burr formed by a preceding grain. Thus, a close examination of a workpiece edge having a grinding burr should reveal a noticeable burr attached at intermittent intervals (Figure 4-22).

Materials that become noticeably work-hardened may exhibit a continuous burr attachment, rather than the intermittent attachment common to other materials. The work-hardening effect tends to swell the material to such an extent that the grains cannot entirely remove the swell.

Rollover Burrs

When abrasive grains exit over edge 3 (Figure 4-20) of the workpiece, they produce a rollover burr. This burr consists of a grinding chip that bent out of the grain's path rather than being severed from the material. Because the grain also pushes metal ahead of it, some of the rollover burr consists of swelled metal as well as unsevered chips (Figure 4-23).

With each downfeed movement, the rollover burr lengthens. As shown in Figure 4-24, an abrasive grain produces a definite chip that separates from the workpiece when the wheel approaches the end of the cut. The last few grains, however, do not produce a loose chip (see potential chips 1, 2, and 3 shown in Figure 4-24).

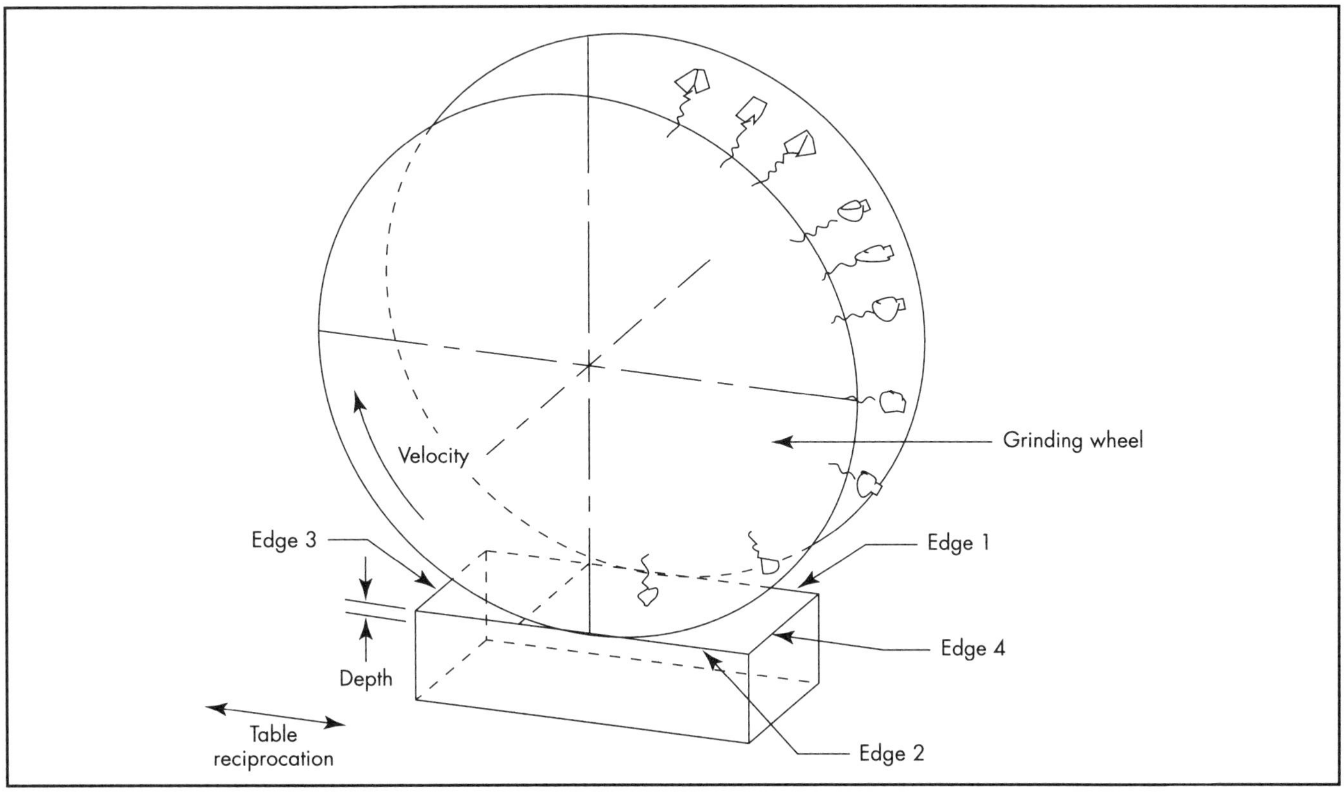

Figure 4-20. Identification of workpiece edges at which burrs will form in grinding (Gillespie 1975b).

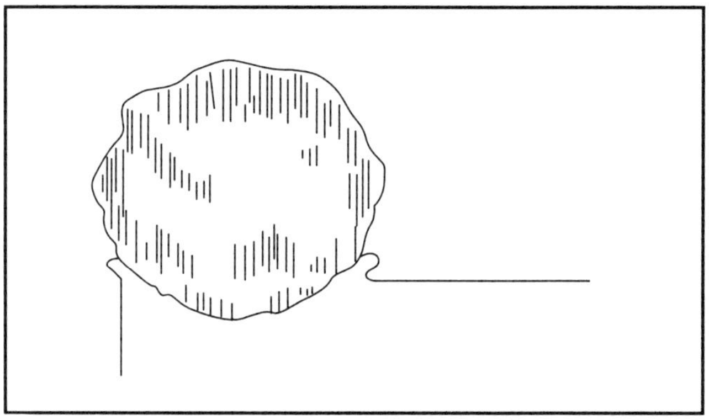

Figure 4-21. Grain cutting near edge of part (motion is away from observer into paper) (Gillespie 1975b).

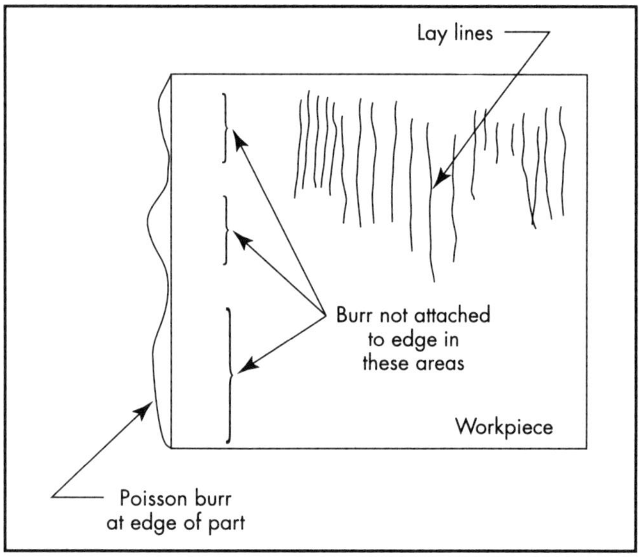

Figure 4-22. Intermittent attachment of burr to edge of part (Gillespie 1975b).

Entrance Burrs

A burr forms on the edge at which the grinding wheel enters the workpiece (edge 4 in Figure 4-20). This entrance burr consists of material that has flowed opposite to the direction of the grain motion (same concept as shown in Figure 4-2), and it is similar to the ridge that forms around the indentation made by a Brinell hardness tester. With conventional cutting tools, this burr may or may not form, depending on the shape of the tool and the properties of the workpiece material. Because of the high forces involved and the spherical shape of the abrasive grains, entrance burrs will likely form on most workpieces.

4.4.5 Burrs from Ball Broaching Operations

Ball broaching operations are used to enlarge hole sizes and produce a fine finish in a single operation. A ball-like tool, larger in size than the initial hole, is pushed down the hole, forcing metal outward and downward as it goes (Figure 4-25). There is no attempt to make chips—this is a plastic

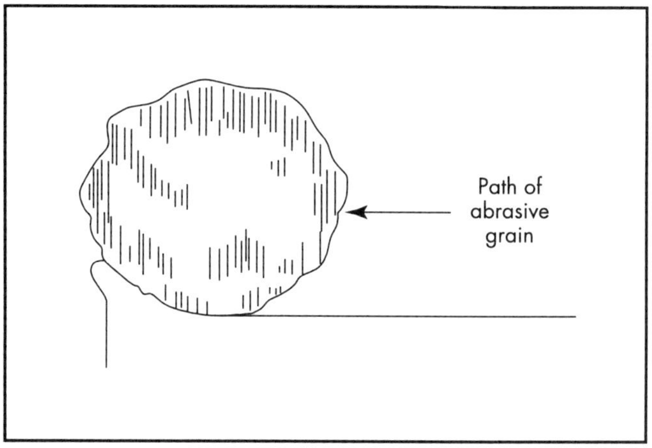

Figure 4-23. Formation of rollover burr as grain exits from back edge of workpiece (Gillespie 1975b).

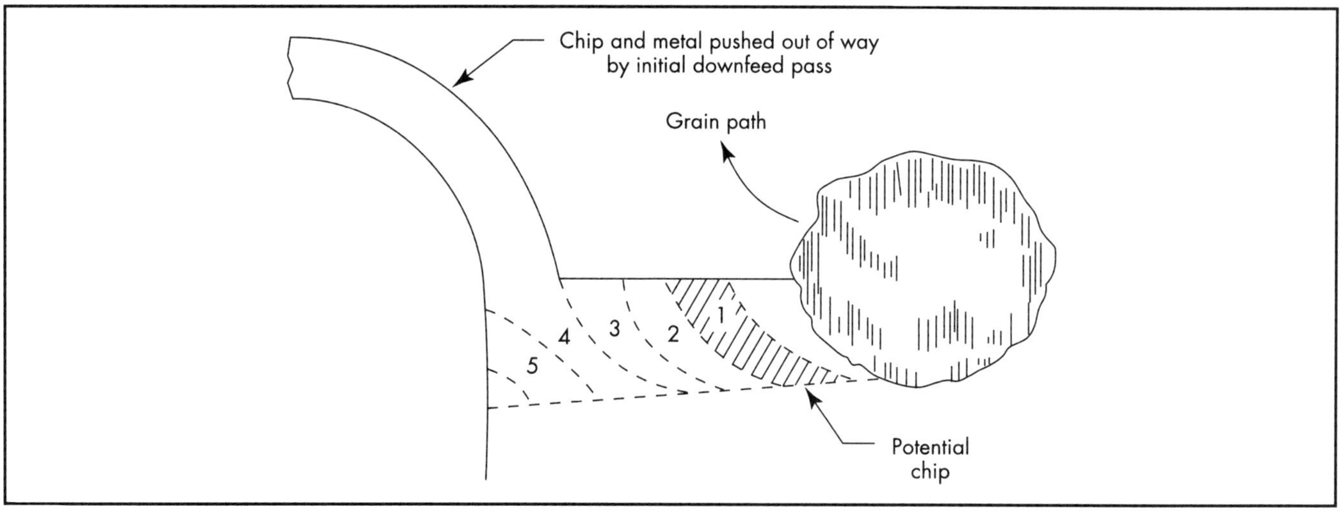

Figure 4-24. Grain path and material flow near end of cut (Gillespie 1975b).

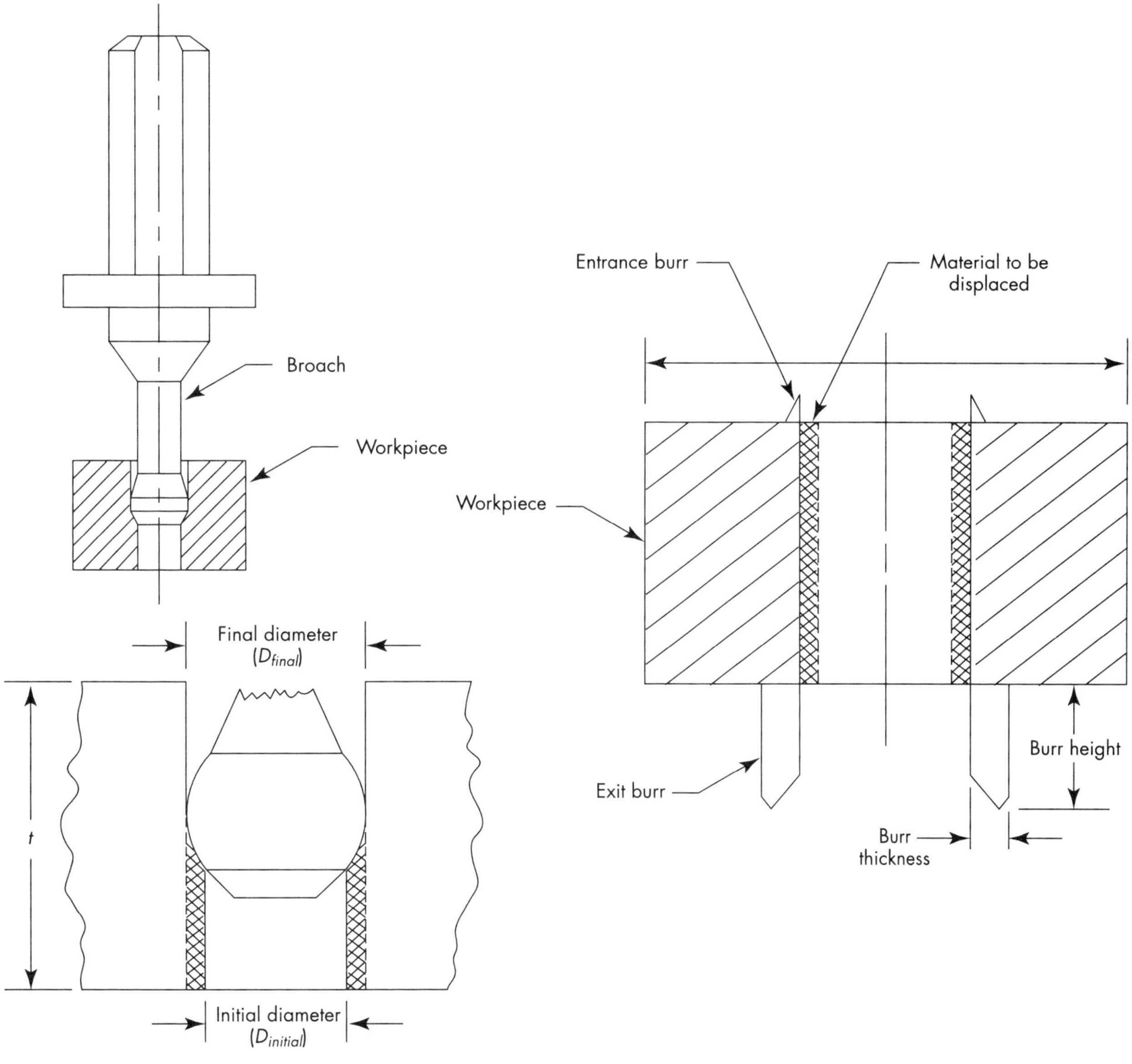

Figure 4-25. Technique of ball broaching and raised material dimensions and locations (Gillespie 1974c).

deformation process. Hole-expanding operations that involve a hydraulic radially expanding tool have some of the same characteristics as this process, though with different results.

As shown in Figure 4-26, the entrance burr is a result of Poisson ratio effects, and the exit burr has many of the same properties as rollover burrs (Figure 4-27).

4.4.6 Pressworking

Figure 4-28 illustrates the formation of a typical burr in a punching operation. As shown in the figure, a radius will always be formed at the point where the punch enters the metal. The burr on the side from which the punch exits the part is not a rollover burr like those found in machining; the stress fields are significantly different. At one time, the exit burr was called a "tensile burr" or a "tensile-compression burr," depending on what was believed in the field to be causing the stress. There is no new terminology universally accepted for this burr, but for the purposes of this handbook, it is referred to as a pressworking burr or sheet metal burr. (It is important to note that in the sheet metal industry the term "burr" can also mean a washer [used on bolt and nut] or a specially formed tab used to lock in other components. Neither of these items is the kind of burr referred to in this handbook.)

4.5 BURR PROPERTIES AND THEIR EFFECTS ON DEBURRING

The difficulty of removing, minimizing, or preventing burrs varies with each situation. The following burr properties influence the difficulty of deburring. In order of significance, from most to least problematic, these properties are:

1. Burr thickness.
2. Burr location.
3. Burr toughness.
4. Burr height (length).
5. Workpiece tolerances.
6. Burr radius and shape.

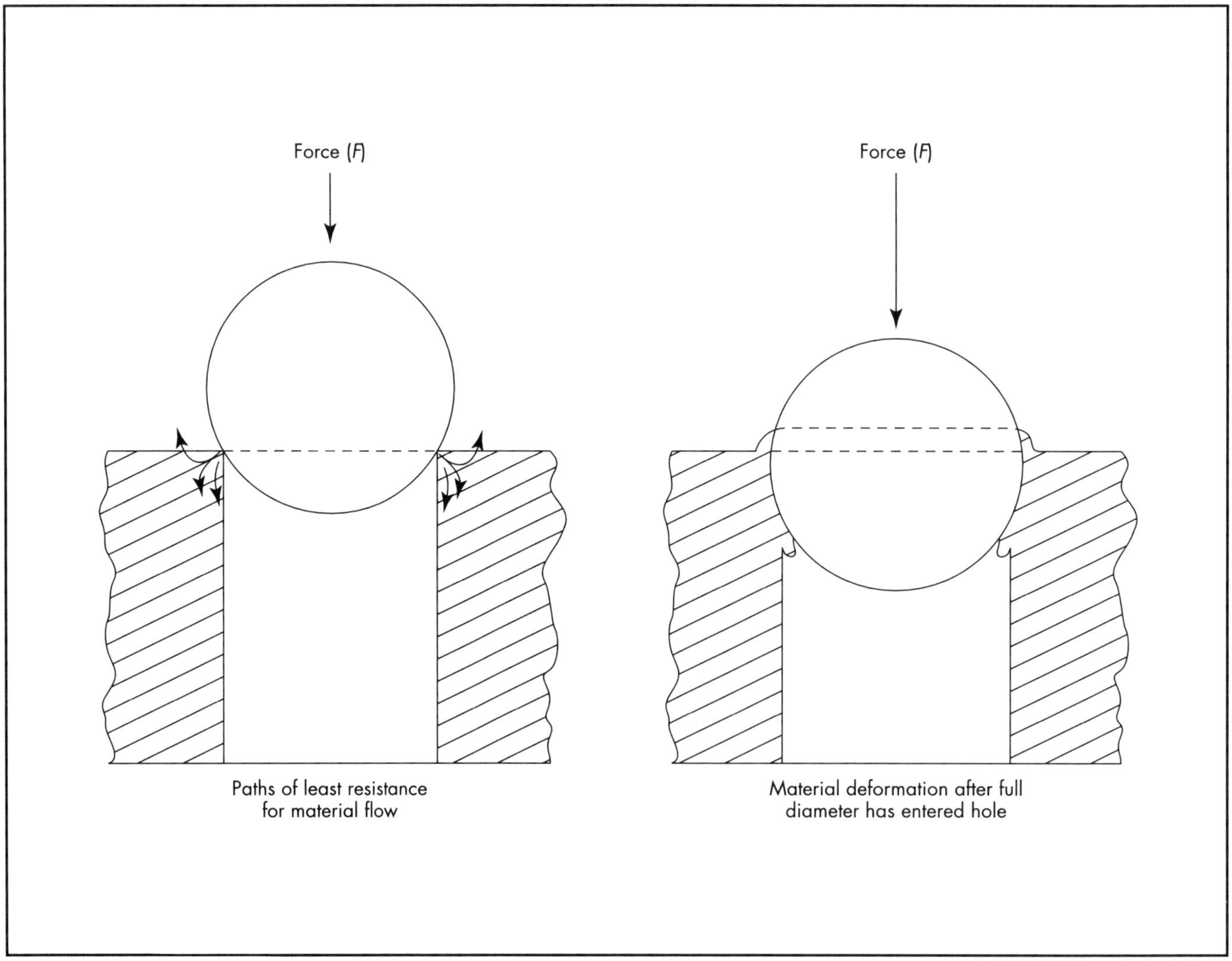

Figure 4-26. Material flow at hole entrance for ball broach (Gillespie 1974c).

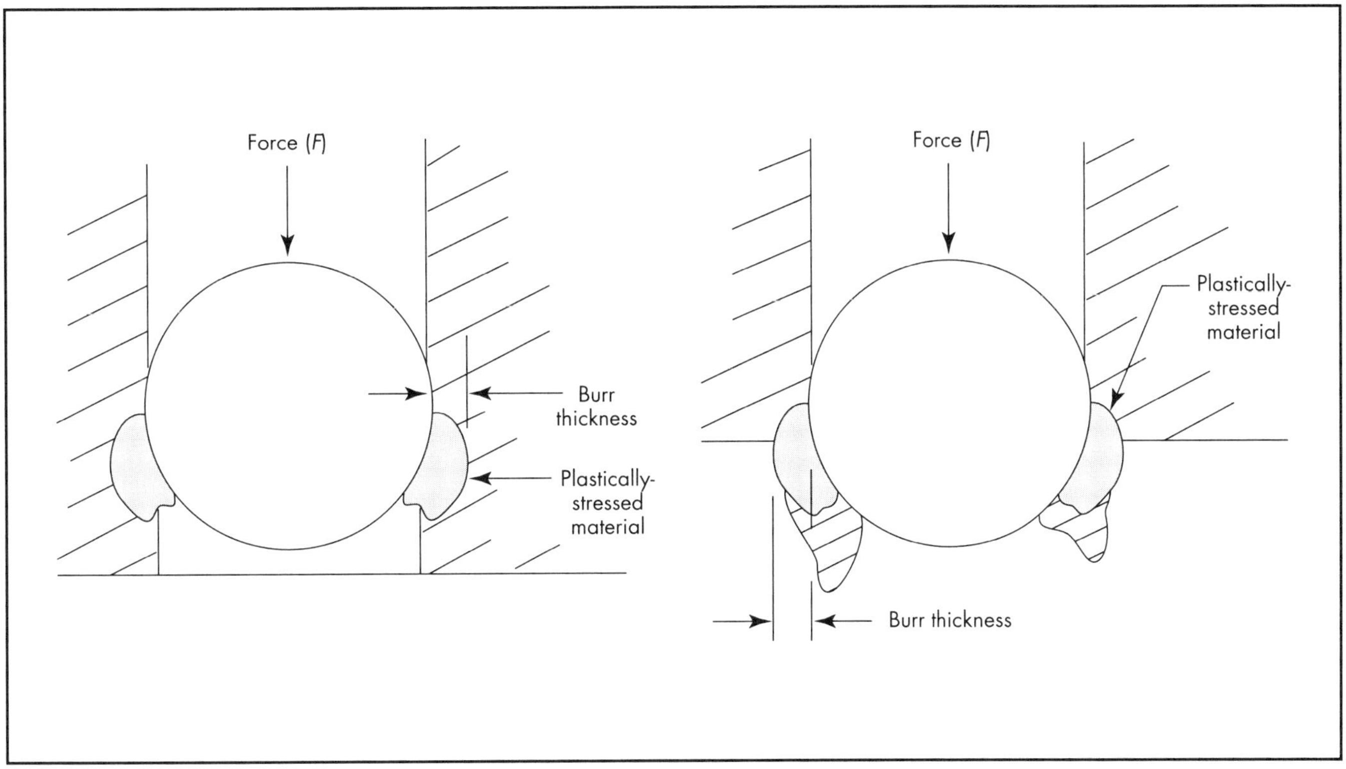

Figure 4-27. Formation of stressed material from ball broaching near exit of hole (Gillespie 1974c).

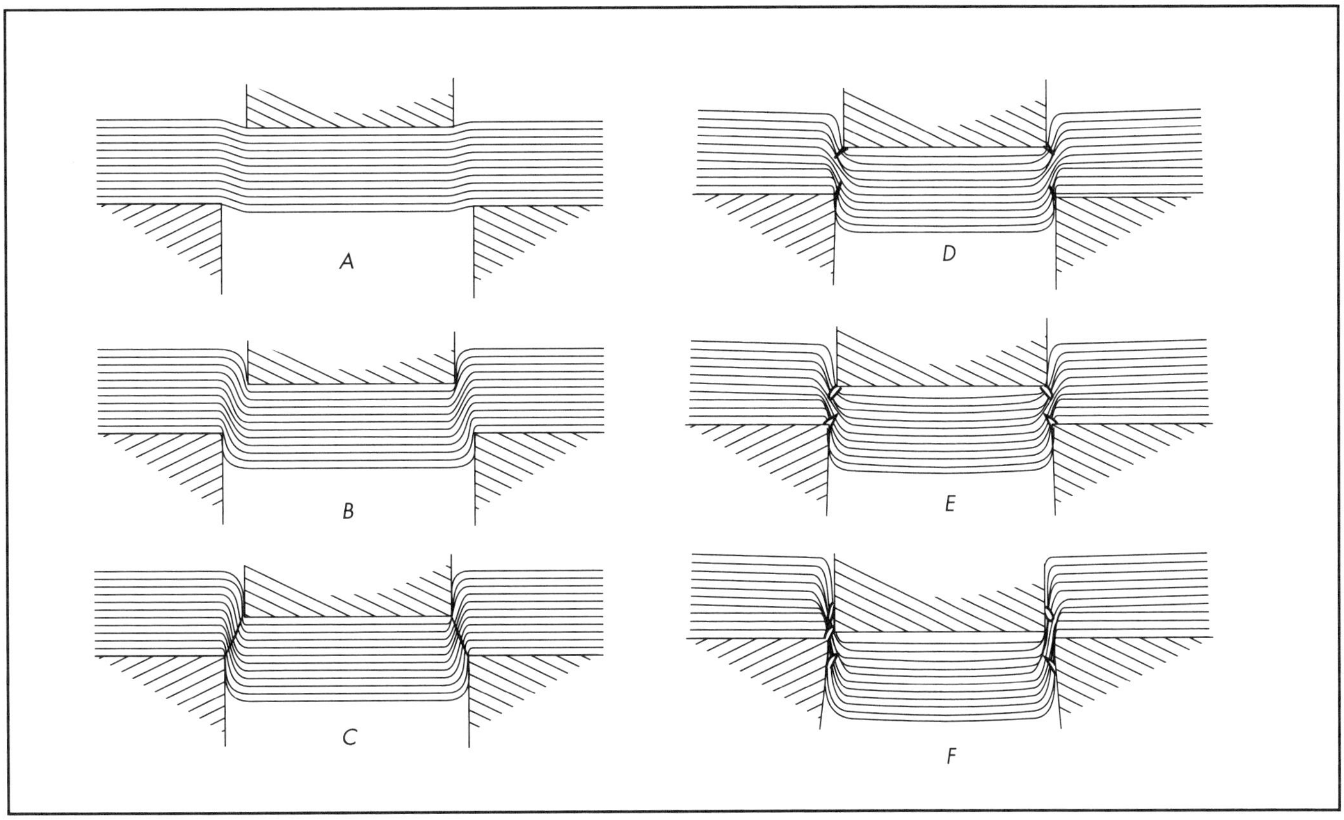

Figure 4-28. Burr formation in pressworking (*left*) typical; (*right*) desired (Crane 1939).

In the past, measuring burrs meant measuring burr height. However, *burr thickness at the root of the burr, not burr height, makes a burr difficult to remove.* In most situations, the combination of thickness and height, or thickness and location, rather than just a single measurement, is important. Burr properties depend on which metalcutting mechanism produced the burr.

From a theoretical standpoint, burr height is directly related to burr thickness, but from a practical standpoint, the relationship is not readily evident. Burrs assume many shapes. They can be triangular or rectangular in cross section, and they can be uniform or ragged (irregular) or feather-like in height. Each of these shapes indicates a different metalcutting phenomenon. Some burr configurations are easier to remove than others. For example, burrs shaped like chicken feathers with fine "spines" from the edge may come off easier than a solid sheet of burr. Burr properties vary considerably from part to part and even within the same part. As an example, an abusive face-milling operation on 303 Se stainless steel produced an average burr height of .0095 in. (0.241 mm). Measuring 100 edges, however, revealed heights varying from 0 to .020 in. (0.51 mm).

4.6 MINIMIZING BURRS FROM SPECIFIC OPERATIONS

The following sections provide detailed suggestions for minimizing burrs from specific operations. Since burrs vary with part geometry, material, and other broad parameters, these suggestions are useful, but probably will require users to perform additional research for application to specific situations.

4.6.1 Minimizing Burrs from Turning Operations

The following are recommendations that help to minimize the formation of burrs from turning operations on lathes and from other machine tools:

- *When possible, use a form tool to cut diameters.* As noted in Chapter 3, no burr can occur when the tool produces the diameter and the adjoining face at the same time. Any intermediate burr that is formed is wiped off. This does not mean the part is burr free. A burr can, and normally will, form at the edges of the form tool, but fewer edges now have burrs.
- *Break all edges with a chamfering tool.* Since tool position can generally be controlled accurately to .0002 in. (5.08 μm) on many materials, the chamfering tool can remove the normal burrs and still ensure small, final edge breaks. Chamfering tools will leave two small burrs.
- *Generate corner radii rather than machining perpendicular, sharp edges by cam design or NC program.* As discussed in Chapter 3, this method prevents burrs from being left on parts.
- *Reduce cutting forces and the weight of the workpieces to reduce the diameters of the cutoff projections.* The use of sharp, narrow cutoff tools is also helpful (Gillespie 1977).
- *Remove built-up edges, increase rake angles, use sharp cutting edges, minimize flank wear, and use shallow cuts to minimize burrs.* These factors reduce cutting forces when cutting all materials.

4.6.2 Minimizing Burrs from Drilling Operations

Burrs form on both the entrance and exit sides of drilled holes. The burrs on the entrance side are typically small, whereas those on the exit side are typically long and ragged, or short with a cap of material still attached. There are numerous combinations of helix angle, point angle, and feed rate that may produce the same burr thickness. Significant reductions in burr thickness are possible by the correct selection of these parameters. Maximum burr height on any hole is proportional to the drill diameter.

A burr thickness of .0043 in. (0.109 mm) can be expected when drilling AISI 1018 steel having a hardness of 150 Bhn, with a .25-in. (6.35-mm) diameter HSS twist drill with a 27° helix angle and 120° point angle, and using a feed rate of .006 inches per revolution (ipr) (0.15 mm/rev) (Sofronas and Taraman 1976). Increasing the helix angle to 36° reduces the burr thickness to .0026 in. (0.066 mm). Lowering the feed to .0035 ipr (0.089 mm/rev) and reducing the point angle to 98° will reduce the burr thickness to .0015 in. (0.038 mm). A burr this thin can normally be easily removed by barrel or vibratory deburring. Burr thickness under these conditions can be expressed by the following equation:

$$b_T = C_1 H^{-1.72} p^{0.84} L^{-0.36} f^{0.86} \qquad (4\text{-}4)$$

where

b_T = Burr thickness.
C_1 = 4.271 when b_T is in inches and f is in ipr (= 6.797 when b_T is in millimeters and f is in millimeters per revolution).
H = Helix angle of twist drill (degrees).
p = Point angle of twist drill (degrees).
L = Lip clearance angle of twist drill (degrees).
f = Feed rate (ipr).

Equation (4-4) shows that increasing the helix angle and lip clearance angle decreases the burr thickness. Thickness is only slightly affected by the lip clearance angle. Decreasing the point angle and feed rate also decreases the burr thickness.

Burr thickness also can be expressed in terms of the spindle speed, hardness of the workpiece material, and stiffness of the drilling system by the following equation:

$$b_T = C_2 N^{-0.78} B K^{-0.007} \quad (4\text{-}5)$$

where

b_T = Burr thickness.

C_2 = 252 when b_T is in inches and K is in pounds per inch (= 7,128 when b_T is in millimeters and K is in Newtons/meter [N/m]).

N = Spindle speed of the drilling machine in revolutions per minute.

B = Hardness of the workpiece material as Brinell hardness number (Bhn).

K = Stiffness of drilling system.

Equation (4-5) shows that increasing the spindle speed or material hardness reduces the thickness of an exit burr. Because of the low exponent over the range of values examined, the system stiffness has an insignificant effect. The effect of spindle speed and workpiece hardness when K is 27,000 lb/in. is shown in Figure 4-29.

A practical aspect of the results shown in Figure 4-29 is that for a given workpiece hardness, the burr thickness can be reduced by increasing the spindle speed. It can be seen from the figure that, for a hardness of 125 Bhn, the thickness can be reduced from .010–.007 in. (0.25–0.18 mm) by increasing the spindle speed from 800 to 1,300 rpm. Changes in material hardness of 25 points on the Brinell hardness scale cause variations of .001–0.002 in. (0.03–0.05 mm) for the higher and lower Brinell numbers, respectively. Use of high speeds for drilling aluminum, however, can cause mushroom-type burrs at the entrance sides of holes and large burrs or breakout at the exit sides.

The effect of drill lip radius on exit burr thickness when drilling 6061-T6 aluminum alloy is illustrated in Figure 4-30. The radial-lip drill (Figure 4-31), introduced to minimize burr size and increase drill life, makes more effective use of the cutting forces. A standard drill at breakthrough has a large amount of material to remove at the corners of the drill. The radiused cutting lip, however, has a much smaller amount of material to remove.

The drill thrust pushes material down, and the torque pushes material radially outward when separating the chip from the workpiece. Higher feed rates produce larger chips, which increase the burr-forming forces. At the "corners" of the radial point drill, the chip thickness is extremely small (Figure 4-32); thus, the torque and thrust at these points is also small. Consequently, the forces that push metal away from the hole are small, and the material pushed away (burr) is also small. Because the radial point cannot produce a significant thrust force at the "corners," only a very minute amount of material can be pushed down. Therefore, burr length should be very small.

Figure 4-33 illustrates the dramatic difference between stresses caused by torque from the two drills—in this instance, bulk stresses in the workpiece (causing part distortion and large burrs) rather than stresses at the cutting edge (which improve ease of cutting). As drills and reamers dull,

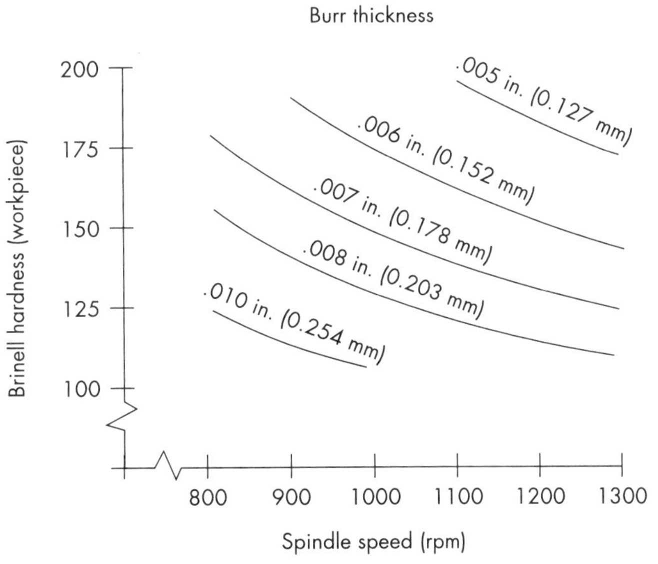

Conditions: No cutting fluid. Feed rate: 6 ipm (152 mm/min). Twist drill: 118° point angle, 27° helix angle.

Figure 4-29. Effect of spindle speed and hardness on burr thickness when drilling AISI 1018 steel with a .25-in. (6.35-mm) diameter HSS twist drill (Sofronas and Taraman 1976).

bulk stresses increase, producing larger burrs. One additional advantage of the radial-end reamer and drill is that it almost eliminates the corner disintegration that heralds the end of tool life.

Figure 4-34 shows the effect of drill point angle on exit burr thickness when drilling 1050-0 aluminum alloy. Increasing the point angle to 100° increases burr thickness. Above that angle, thickness decreases to a minimum at about 180°.

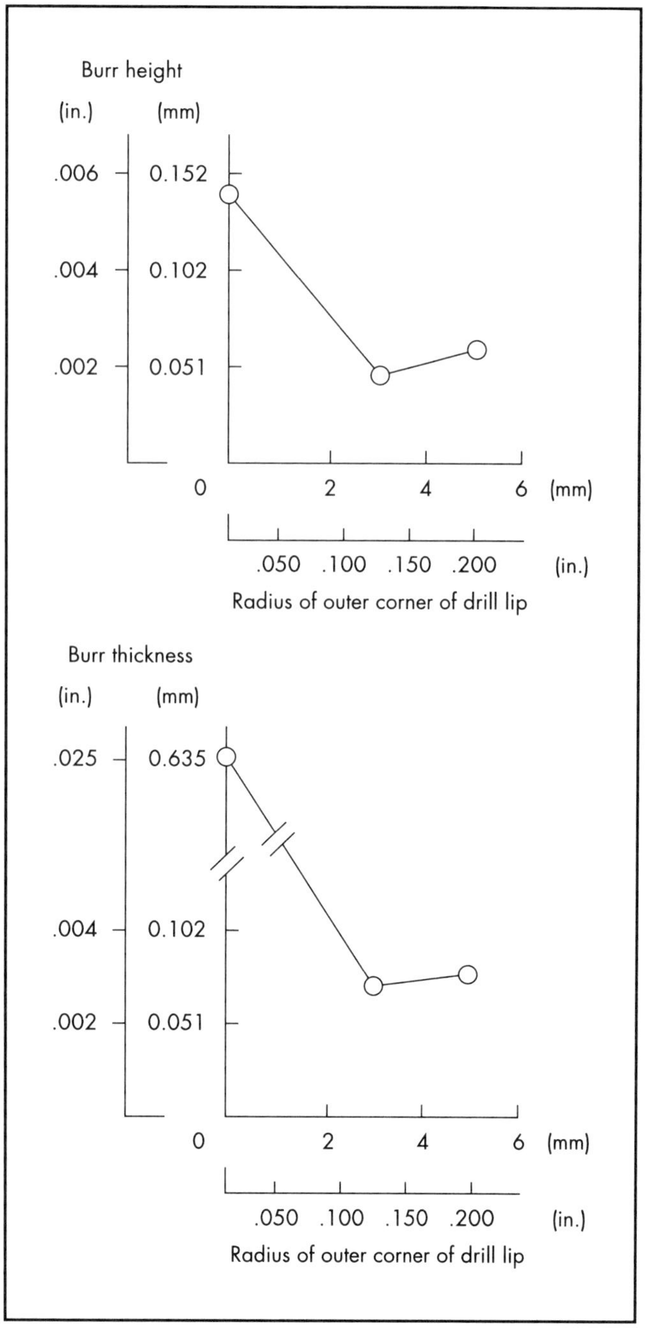

Figure 4-30. Effect of drill lip radius on burr height and thickness when drilling 6061-T6 aluminum alloy (Hasegawa, Zaima, and Yuki 1975).

The generally recommended point angle for drilling aluminum is 118–140°, but this range of angles results in comparatively large burrs.

Two approaches to drill point angles reduce burr thickness: (1) increasing the point angle decreases the angle at the outer corner, and (2) making the point angle as large as 180°. For a small point angle, the zone of plastic flow diminishes at the hole exit due to the small chip thickness. For a 180° point angle, the radial component of cutting resistance becomes small. Both methods reduce burrs. The latter is employed widely in drilling thin sheet metals using a fishtail point drill.

Drills with chamfered or dubbed corners, such as those used for drilling cast iron, are used to decrease the angle at the outer corner, as is the radial-lip drill geometry. In some cases, the radial-lip drill can reduce the chip thickness at the drill

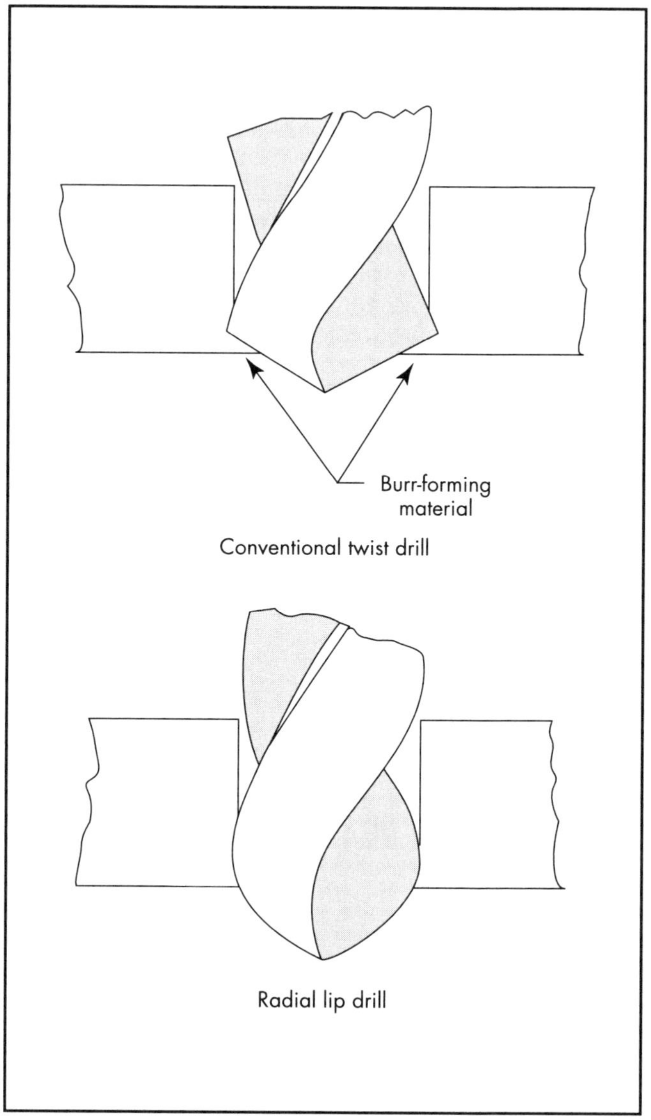

Figure 4-31. Radial lip drill has less material to remove than a conventional drill at the corners (Pond 1967).

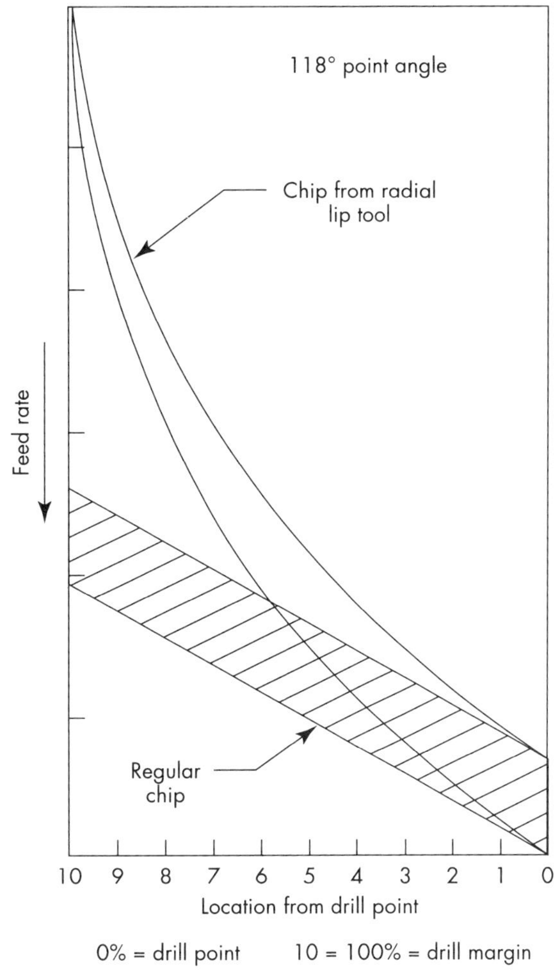

Figure 4-32. Chip size variation from radial lip and standard drills and reamers (Pond 1967).

corners to a very small value, producing the same effect as slowing the feed rate. Table 4-2 indicates the relative importance of variables on burr size for drilling operations.

Oil-hole drills having a radial-lip design with a .118 in. (3 mm) radius at the drill corners produce a smaller burr in aluminum than do conventional drills. Cutting fluids with 20% lard content produce much shorter burrs than those with lower lard content. Forced application of the cutting fluid provides the smallest burr.

Large differences occur in burr size when the feed rate employed for drilling is manually controlled. Although drilling occurs under supposedly constant conditions, the operator cannot control the breakthrough thrust. Positive-feed drilling minimizes feed-rate fluctuations. If there is flexibility between the workpieces or fixture, end play in the spindle of the drilling machine, sloppy gears, or leaky hydraulics, the resultant lurch on breakthrough—in effect, a rapid feed-rate increase—will cause large burrs, even with optimum drill points. Rigid machinery, sturdy fixtures, well-supported workpieces, and regulated feeds are necessary for best results.

Exit burr size in drilling can be reduced by minimizing the size of the clearance hole in the fixture below the hole to be drilled (Gillespie 1975a). For commercial parts, a clearance hole .002 in. (0.05 mm) larger than the drill diameter is recommended (Figure 4-35). Such a clearance should result in burrs only .001 in. (0.03 mm) thick. For precision holes, the clearance should be smaller, although tool life may be greatly shortened with this approach.

The use of a hard (R_c42) sacrificial backup material in drilling also minimizes exit burr size (Gillespie 1976c). The effect of backup material hardness on burr height and thickness when drilling Type 303 Se stainless steel is illustrated in Figure 4-36. Hole saws will typically produce a smaller burr than twist drills in ductile material, partly owing to the fact that they greatly reduce the cutting forces that create the burrs.

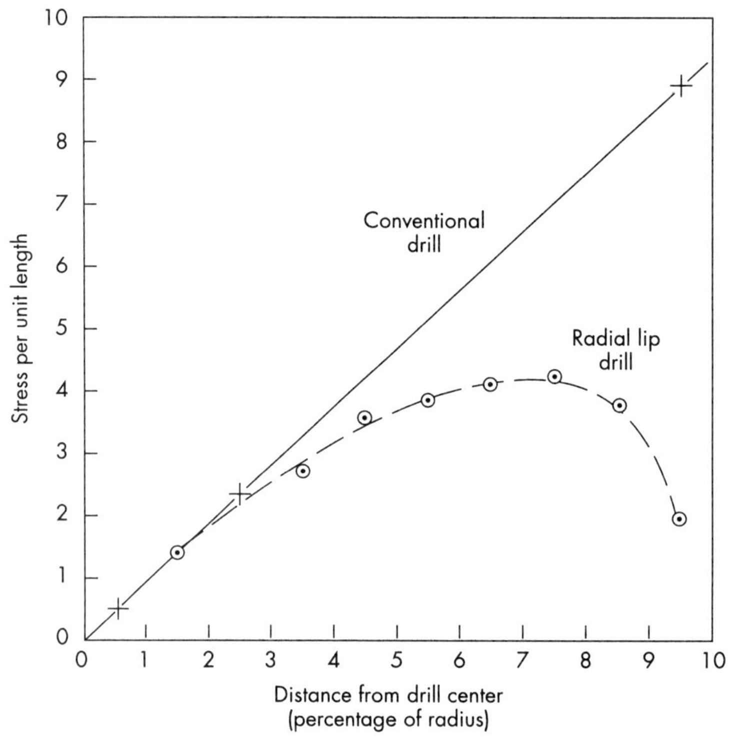

Figure 4-33. Torque-caused stress distribution along the face of the 118° point drill (Gillespie 1974a; Pond 1967).

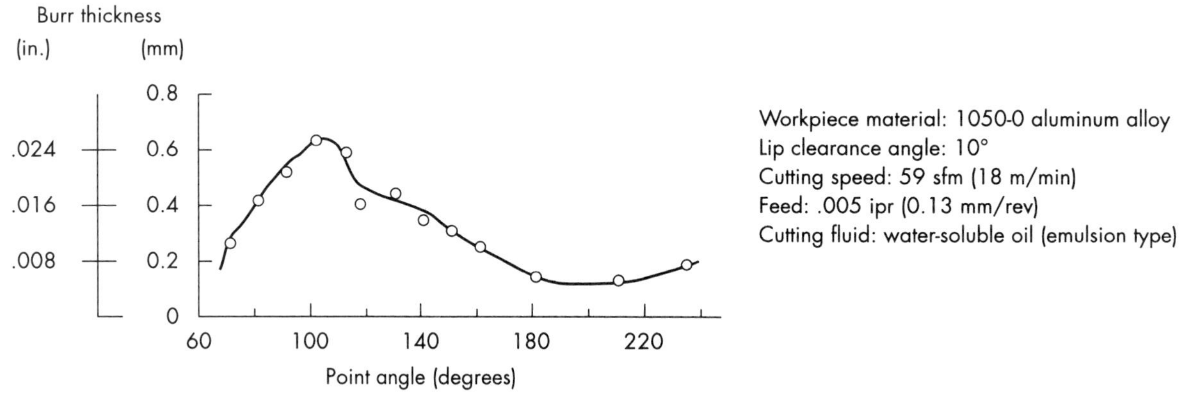

Figure 4-34. Effect of point angle of drill on burr thickness (Hasegawa, Zaima, and Yuki 1975).

Table 4-2. Variables affecting burr size in drilling 303 Se stainless steel

Variable	Entrance Burr Thickness	Entrance Burr Height	Exit Burr Thickness	Exit Burr Height	Ranked significance
Helix angle	–	–	–	–	1
Feed rate	0	0	+	+	2
Diameter	0	–	+	+	3
Surface velocity	0	–	+	x	4
Corner angles*	x	x	–	+	5

Note: – indicates that an increase in this variable reduces burr property; + indicates that an increase in this variable increases burr size; 0 indicates that this variable has no effect on burr properties; x indicates that conditions have not been studied.

* For conventional drills, the corner angle equals 180° minus half the point angle.

(Pond 1967; Gillespie 1975a)

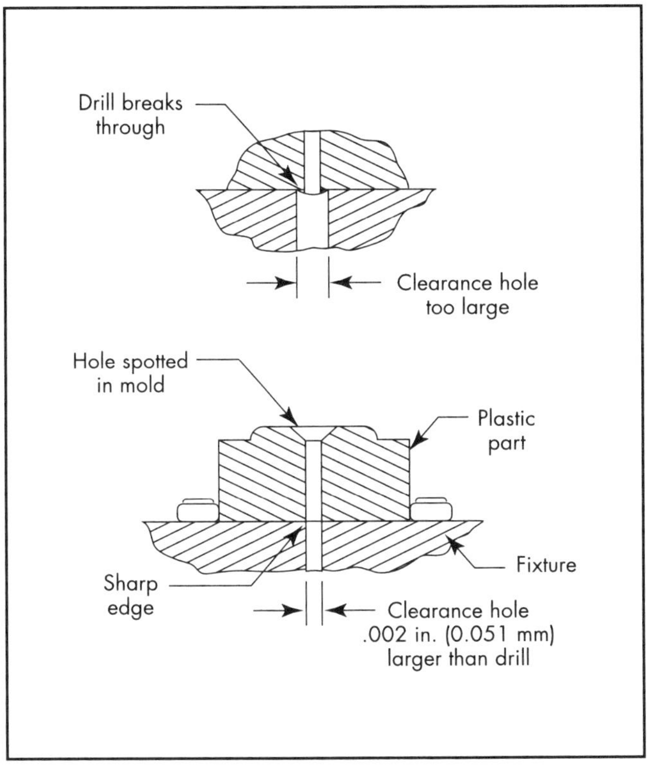

Figure 4-35. Drill clearance in fixture for burr minimization (Gillespie 1975a).

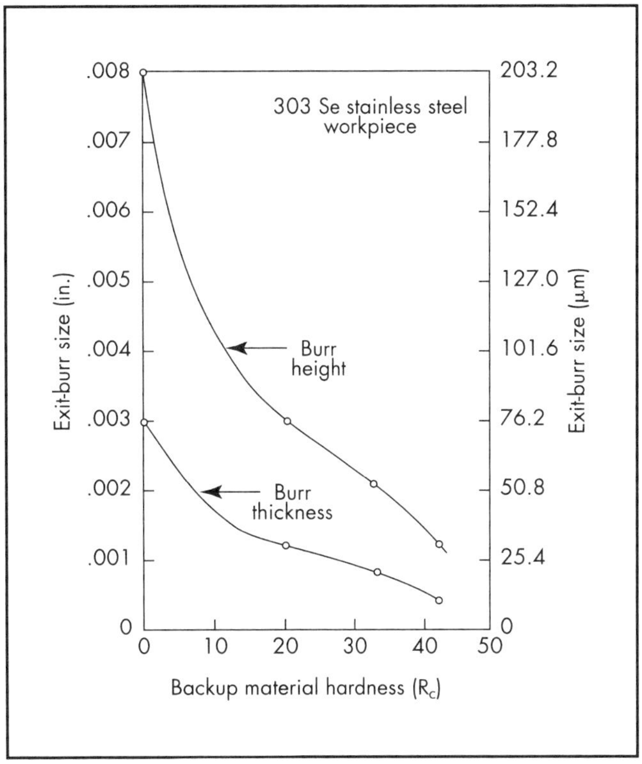

Figure 4-36. Effect of backup material hardness on burr size when drilling 303 Se stainless steel.

Some researchers have found that as silicon content is increased in aluminum, drill exit burrs become more brittle and are thus more easily removed.

Studies of drilling burrs sometimes provide conflicting results, in part because of undocumented differences in setup or operation or actual drill geometry used. Small hole drilling may provide different results than larger hole drilling. Holes drilled at angles to the main workpiece and those drilled into other holes do not have uniform properties around the intersection. The equations presented in this chapter are valid for a limited range of materials, hole sizes, drill geometries, and speeds. In contrast to the equations presented earlier for drilling low-carbon steel, data from one researcher indicates that drill burr height is a parabolic function of feed rate and of material hardness in one material (Figure 4-37) (Pande and Relekar 1986). This result actually provides some benefit to researchers, since there will be a specific feed and a specific hardness at which burr height is minimized. Note that Pande and Relekar's finding also contradicts the fact that maximum burr height for large diameter drills can be higher than for small diameter drills.

Vibratory pecking or oscillation has been suggested as one means to reduce exit burr size. It has been shown to be effective in some studies and not as effective in others.

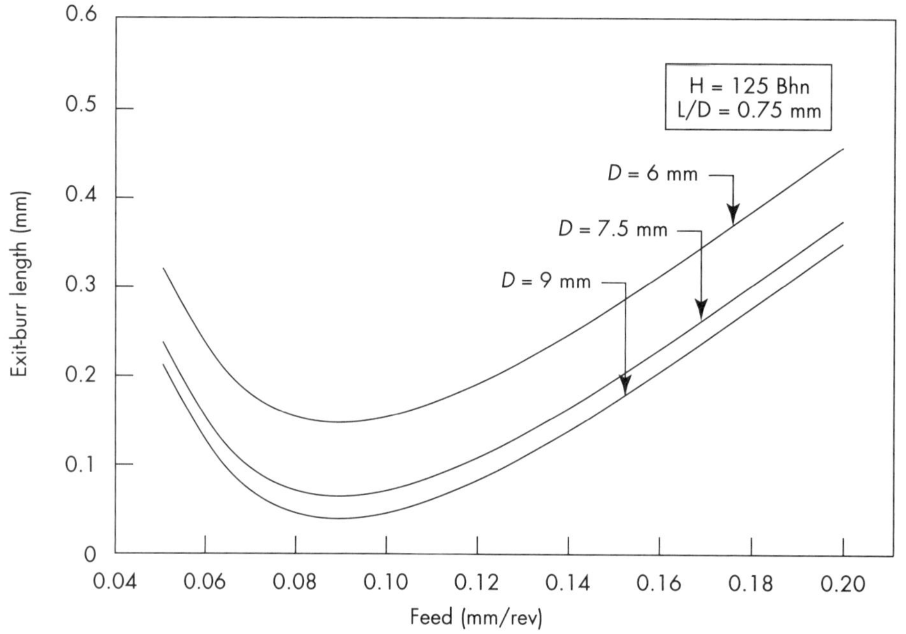

Figure 4-37. Effect of feed and drill diameter on exit burr height in stainless steel (Pande and Relekar 1986).

Effectiveness will in some part depend on whether a built-up edge occurs on the part, the actual feeds and speeds used, drill geometry, drill wear, and part material properties. It can be an effective solution in certain applications. Any technique that reduces the effective chip load should reduce burr size, except for work-hardening materials like the 300 series stainless steels. For these steels, if the drill is rubbing instead of cutting, this technique will cause severe plastic deformation and similarly large burrs. There is a range of chip loads and feeds and speeds that minimize burrs in these materials.

When drilling plastic materials, such as Teflon® or delrin, users may find it useful to freeze the material so that it hardens, reducing the ductility necessary for plastic burrs to form. This process is less effective when the drill is working in the material for a long time dissipating heat.

Detailed information on the burrs formed when drilling small holes can be found in the references and bibliography at the end of this chapter (especially Hasegawa, Zaima, and Yuki 1975; Pond 1967; Gillespie 1975a, 1976c; Pande and Relekar 1986; Stein 1977; Phillips 1976; Kent 1976; Zaima, Yuki, and Kamo 1968). Readers are particularly directed to a work by Julie M. Stein (Stein 1995), which provides details on the subject for small precision holes and holes breaking at angles into other holes.

In summary:

- High helix angles reduce burr size.
- Low point angles reduce some burrs.
- Fish-tail drills reduce burr size.
- Radial-lip drills reduce burr thickness at all feed rates.
- Radial-lip drills reduce burr height at high feed rates.
- Multifacet drill points reduce burr height at low feed rates.
- Low feed rates reduce burr size except for very low rates in work-hardening stainless steel.
- High cutting speeds reduce burr size or do not affect it all.
- Cutting fluids, such as sulfur-chlorinated fluids, reduce burr size.
- Reducing drill corner wear reduces burr size.
- Vibratory assists in the axial direction may reduce burr size in some conditions.
- Freezing plastic materials before drilling may minimize burrs.
- Reducing built-up edges should reduce burr size.

4.6.3 Minimizing Burrs from Milling Operations

End-milling Burrs

The number of different burrs produced on various edges of workpieces in end-milling operations was discussed earlier in this chapter (see Figures 4-15–4-17). The height, thickness, radius on the back side of each burr, hardness, and appearance of the groups of burrs in this category vary widely (see Table 4-3). The term *group* is used to indicate that the burr on each edge is the linear accumulation of many scallop-like cuts, each spaced one revolution apart (see Figure 4-38) (Gillespie 1976a; Hongtao, Qiping, and Lei 1994).

The large number of edges and burr properties prohibit a full discussion of machining effects on burr size, but the following observations are significant:

- Fast feed rates, .002 in. (0.05 mm) versus .0005 in. (0.013 mm) rev/tooth, reduce burr height up to 50%, but slightly increase burr thickness.
- Dull cutters significantly increase burr height and thickness.
- Changes in the helix angle increase some burr sizes while decreasing others.

Side-milling Burrs

Side-milling operations can produce burrs at ten different edges (Figures 4-15 and 4–17). For conventional side-milling cutters, burr thickness can be minimized by reducing the feed rate and using a 0° helix angle. Full-radius cutters (cutters that produce a U-shaped slot cross section) should not be used unless essential on materials that elongate noticeably. These cutters produce thicker, as well as much longer, burrs than do square-edge cutters. Table 4-3 provides some relative values of burr size along edges of one cut.

Face-milling Burrs

Although almost no data have been published on minimizing face-milling burrs, the process is similar to end milling. It is significant that burr height is a function of the angle at which the cutter teeth exit over the workpiece edge. Tests indicate that burr height can be controlled by varying the position of the cutter teeth with respect to the edge of the workpiece. Table 4-4 provides comparative data for burrs produced in milling operations.

Tool geometry, tool-to-part edge geometry, and tool sharpness, discussed in detail in Chapter 5, play major roles in determining burr size. The burr formation force is that component of the resultant cutting force that is vertical to any exit edge. As shown in Figure 4-39, any force applied into the vertical wall or into the bottom surface will cause a burr. Smaller forces applied in these directions will result in smaller burrs. Any factor that reduces burr formation force will reduce burr thickness. Such factors include slowing the feed rate, using sharp cutters, employing a cutting fluid, and using maximum rake angles.

4.6.4 Minimizing Burrs from Grinding

The following grinding factors are significant in the formation of grinding burrs:

- Abrasive grains have a large negative rake angle.
- Random placement of grains results in overlapping cuts.
- Temperatures produced while grinding are twice as high as those produced by most cutting processes.

Each abrasive grain on the surface of a grinding wheel produces a chip. Because of the roughly spherical shape of the grain, it cuts with an effectively negative rake angle (Figure 4-40). This angle, which approaches –30°, requires much greater force than does the positive rake angle of a conventional cutting tool. This up to sevenfold increase in the cutting force increases the plastic stresses at the sides of the cut, which, in turn, produces a larger burr. In other words, a microscopic cutting tool would concentrate the cutting action within a very small area; an abrasive grain, however, increases the width of the zone in which work is performed on the part. Essentially, it produces large bulk stresses, which, in turn, increase the burr size. The magnitude of the burr produced by an individual grain appears to be a function of the following variables:

- Strain hardenability of the workpiece.
- Size of the grain.
- Wheel speed.
- Table speed.
- Downfeed.
- Grain "sharpness."

Consider that for grinding a 2-in.-long (50.8 mm) specimen at a longitudinal table feed rate of 50 ft/min (0.254 m/s) and a spindle speed of 3,450 rpm, a single grain would make 138 separate cuts (Gillespie 1975b). This alone will produce a burr having constantly varying properties.

The results of milling tests indicate that burrs from climb milling are somewhat smaller than those from conventional milling. In one cycle of table reciprocation during surface grinding, one conventional pass is followed by a climb cut. The properties of the resultant burr, therefore, are caused by a combination of both climb and conventional types of cut.

Potential chips 1, 2, and 3 shown in Figure 4-20 will separate from the workpiece; a rollover chip will not form at edge 4. Because material can be pushed out of the grain's path easier than a chip can be formed, the grain will not cut. Instead, it will push material out of the way. This material, which normally would become a chip, lengthens the burr, just as peeling a banana produces a longer peel. Each succeeding downfeed thus lengthens the rollover burr by an amount approximately equal to the downfeed. By analogy, a similar phenomenon occurs when Poisson burrs form at the sides of the part.

Based on Brinell hardness results, a lip of material has been shown to form when the material has a low strain-hardening exponent (Figure 4-41). As shown in Figure 4-41b,

Table 4-3. Typical burr sizes produced in milling 303 Se stainless steel (R_c 32)

Milling operation and burr size	Burr location (see Figures 4-7–4-9)								
	1	2	3	4	5	6	7	8	9
Side milling[a]									
Burr height, in.	.0028	.0020	.0140	.0073	.0600	.0001	.0004	.0002	—
(mm)	(0.071)	(0.051)	(0.356)	(0.185)	(1.524)	(0.003)	(0.010)	(0.005)	
Burr thickness, in.	.0014	.0016	.0020	.0016	.0010	.0014	.0015	.0011	—
(mm)	(0.036)	(0.041)	(0.051)	(0.041)	(0.025)	(0.036)	(0.038)	(0.028)	
End milling[b]									
Burr height, in.	.0010	.0020	.0850	.0001	.0024	.0002	.0600	.0007	.0029
(mm)	(0.025)	(0.051)	(2.159)	(0.003)	(0.061)	(0.005)	(1.524)	(0.018)	(0.074)
Burr thickness, in.	.0007	—	.0008	—	.0009	.0003	—	.0003	—
(mm)	(0.018)		(0.020)		(0.023)	(0.008)		(0.008)	

[a] Measured using a 4 in. (102 mm) diameter, .5 in. (12.7 mm) wide HSS cutter, with 18 teeth and 0-10° helix angle, operating at 98 rev/min and feed rate of .00057–.002 in. (0.0145–0.0508 mm) rev/tooth.

[b] Using .125–.375 in. (3.18–9.53 mm) diameter, four-flute end mills, operating at 1,240 rev/min and fed at rates of .0005–.0033 in. (0.013–0.084 mm) rev/tooth.

(Gillespie 1975c)

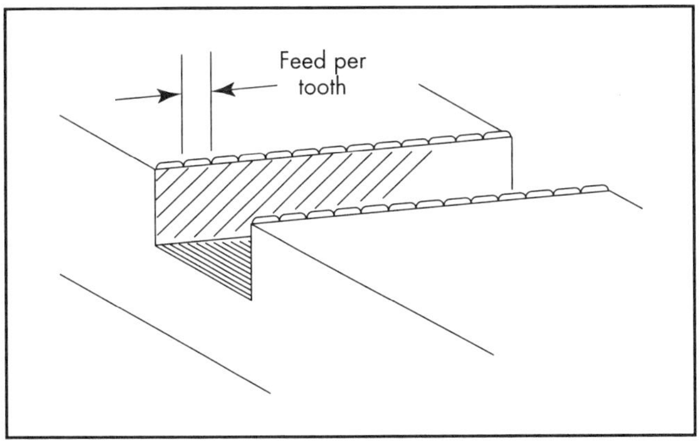

Figure 4-38. Periodic nature of milling burrs.

the flow of highly worked (nonstrain-hardenable) metals produces a "piling up" around the indenter. For annealed metals, the displacement of metal occurs a short distance from the indenter so that a "sinking in" is produced. Materials having high strain-hardening exponents therefore cause a bulge, but not a sharp burr.

Significant Variables

From the preceding discussion, it can be seen that the size of burrs produced by grinding will increase whenever the cutting forces increase without a corresponding increase in the amount of material removed. Thus, "dull" grains will increase the magnitude of the Poisson burr, the rollover burr, and the entrance burr.

High wheel speeds should minimize burr size since they increase the strain rate and thus effectively reduce the plasticity of the material.

Increasing the downfeed and the grain size increases the cutting forces. Although cutting stresses, not forces, determine burr size, the unique shape of the abrasive grains allows the use of forces as indicators of burr size. Larger grains should widen the area of plasticity-deformed material and, as a result, produce thicker burrs. Increasing the table speed also increases the strain rate and thereby minimizes burr size. (There is a critical velocity for each workpiece material that must be exceeded before this effect becomes noticeable.)

High strain-hardening exponents increase the thickness of both entrance and rollover burrs, and decrease the height of entrance burrs. Their effect on rollover burr properties is linear.

Although the preceding discussion has explained the underlying nature of burr formation, the actual mechanism is more complex. W.A. Mohun, for example, notes that the temperature of tungsten at the wheel-workpiece interface can approach 6,500 °F (3,593 °C) (Mohun 1962). For most materials, the interface temperature approaches or exceeds the melting point of the workpiece. Although these temperatures occur only in front of the wheel for a depth of only a few ten-thousandths of an inch below the surface of the workpiece, they are high enough to significantly change the mechanical properties of the workpiece in the zone of chip formation. For this reason, burrs can be found on such normally brittle materials as cast iron and tungsten.

Bond-strength variations, grain friability, the frequency of dressing, wheel loading, and the type of grain material also affect the burrs formed. A comprehensive analytical determination of grinding-burr properties would require the use of complex statistical equations to define the microprofile of the wheel, the cutting action, and the constantly changing conditions of the wheel. Such equations have not as yet been applied to this subject.

In one test of surface grinding 17-4PH stainless steel, 303 Se stainless steel, and 1018 steel with aluminum oxide wheels under gentle conditions, downfeed and the number of passes affected the burr length, but the wheel composition did not. The wheel composition, as referred to in this chapter, implies a combination of grain size and wheel hardness. The wheel-hardness values used in the study were as close to one another as was practical (Gillespie 1975b).

In general, the workpiece material did not influence the length of any of the burrs, but there were some combinations in which the workpiece material did make a difference. For example, after ten passes at .0010 in./pass (25.4 μm/pass) downfeed, the burr at edges 1, 2, and 3 were twice as long as when a single pass was made (see Figure 4-20).

Burr thickness was generally independent of all the variables studied. However, workpiece material and wheel construction did affect the thickness of burrs (Figure 4-42) at edge 3 shown in Figure 4-20.

In all cases, a .0002 in./pass (5.1 μm/pass) downfeed produced shorter or thinner burrs than did the .0010 in./pass (25.4 μm/pass) downfeed, or it resulted in no difference. Burrs became longer at all edges except edge 4 (see Figure 4-20) as the number of passes increased when a fast feed rate was used. The thickest burrs occurred on 1018 steel with the 46-grit abrasive grain (see Figure 4-42). As shown in Table 4-5, the smallest burrs occurred on the entrance side (edge 4) of the workpiece, and the longest and thickest occurred on the chip-rollover side (edge 3).

Significance of Experimental Results

Three different types of burrs are produced by grinding the top surface of flat specimens: the entrance burr; the Poisson burr; and the rollover burr. These burrs differ from each other in thickness and length. Of the three, the entrance burr is the thinnest and the shortest. Statistically, test results indicate that the entrance burr is unaffected by downfeed, number of passes, grain size, wheel hardness, or workpiece material.

The length of the Poisson burrs that formed at the sides of the workpiece was found to be proportional to the downfeed. For fast downfeeds, the length also was proportional to the number of passes taken, but it was not affected by either the wheel composition or the workpiece material. Except for one of the 48 combinations studied, the burr thickness was found to be independent of the variables.

Table 4-4. Comparative summary of burr sizes produced in milling operations

		Burr thickness, in. (mm)		Burr height, in. (mm)		Hardness, Knoop[a]	
Parent material	Condition[b]	Average	Standard deviation	Average	Standard deviation	Metal	Burr
Aluminum	Abusive face milling	.0078 (0.198)	.0011 (0.028)	.0138 (0.351)	.0045 (0.114)	115	129
	Gentle profile milling	.0023 (0.058)	.0003 (0.008)	.0015 (0.038)	not recorded not recorded		
Low-carbon steel	Abusive face milling	.0046 (0.117)	.0010 (0.025)	.0255 (0.648)	.0191 (0.485)	230	275
	Gentle profile milling	.0014 (0.036)	.0002 (0.005)	.0024 (0.061)	.0009 (0.023)		
Stainless steel, 303 Se	Abusive face milling	.0049 (0.124)	.0020 (0.051)	.0095 (0.241)	.0053 (0.135)	305	428
	Gentle profile milling	.0020 (0.051)	not recorded not recorded	.0013 (0.033)	not recorded not recorded		
Beryllium copper	Abusive face milling	.0045 (0.114)	.0008 (0.020)	.0189 (0.480)	.0017 (0.043)	249	268
	Gentle profile milling	.0007 (0.018)	.0005 (0.013)	.0016 (0.041)	.0011 (0.028)		

[a] Knoop hardness number with 3.53 oz (100 g) load (249 Knoop hardness number = R_c 24; 115 = R_B 73).

[b] Abusive face milling: dull cutter and .025 in. (0.64 mm) depth of cut. Gentle profile milling: burr produced by the bottom of a .25 in. (6.4 mm) diameter end mill rotating at 2,730 rev/min and fed at the rate of .1875 in./min (4.76 mm/min).

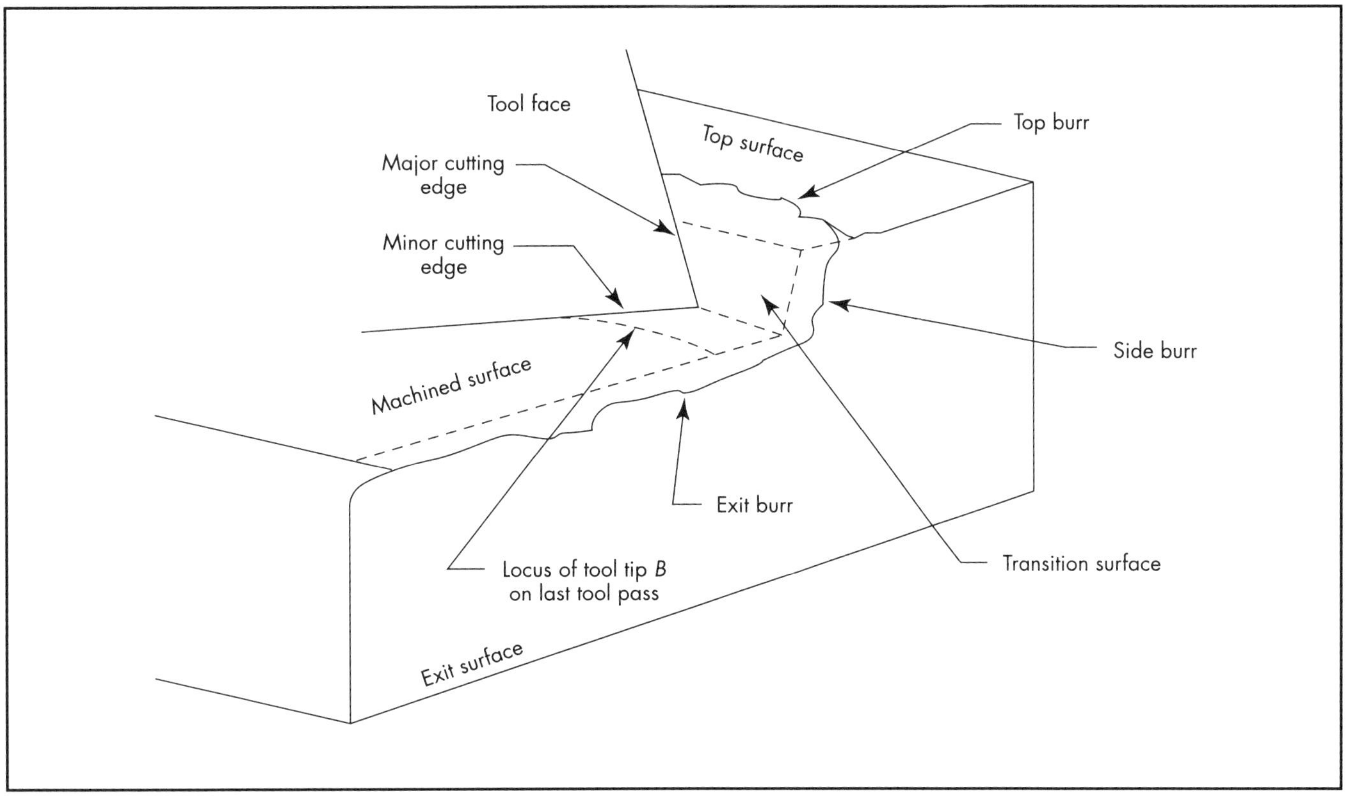

Figure 4-39. Location of burrs and cutter as it exits from the part in face milling (Hashimura, Hassamontr, and Dornfeld 1995).

The length of the rollover burr (formed where the wheel exits from the workpiece) was found to be proportional to the downfeed. In a few instances, the workpiece material and the number of passes affected the length of these burrs. The thickness of rollover burrs from 1018 steel specimens was approximately twice that of burrs from other materials. Wheels having small grains produced thinner burrs than those produced by the 46-grit wheels by a factor of two.

Experience indicates that burrs produced by grinding are relatively easy to remove by any deburring process. Previous studies indicate that any burr that is .002 in. (50.8 μm) in thickness or less can be readily removed by vibratory deburring.

4.6.5 Minimizing Burrs from Reaming

The width and length of a reaming burr are controlled mainly by the amount of stock removed by the reamer. For example, as illustrated in Figure 4-43, exit burr length almost equals the amount of stock removed per side (.0045 in. [114 μm]) for a .143 in. (3.632 mm) diameter reamer. Exit burr thickness is also controlled by the amount of stock the reamer removes, but not in so linear a ratio.

Notice the fourth column in each set of data in Figure 4-43. As shown in Figure 4-44, this dramatic reduction in burr size resulted from the use of a left-hand spiral, right-hand cut, solid carbide reamer with a starting radius. In contrast, the other reamers used in this study were right-hand spiral, high-speed steel reamers with chamfers.

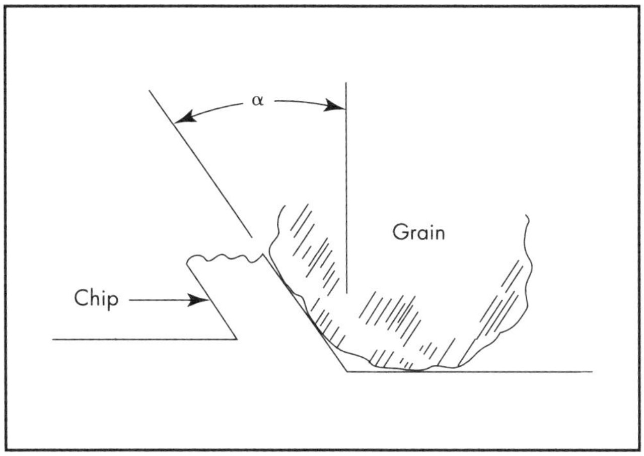

Figure 4-40. Effective rake angle of an abrasive grain (Gillespie 1974b).

Like the radial-lip drill, the starting reamer radius makes more effective use of the cutting forces: the chamfer angle or starting radius on a reamer does all the cutting. With the exception of a flat point and a difference in helix angles, these reamers were similar to the drills shown in Figure 4-31. A standard drill at breakthrough has a large amount of material to remove at the corners of the drill. The radiused cutting lip, however, has a much smaller amount of material to remove.

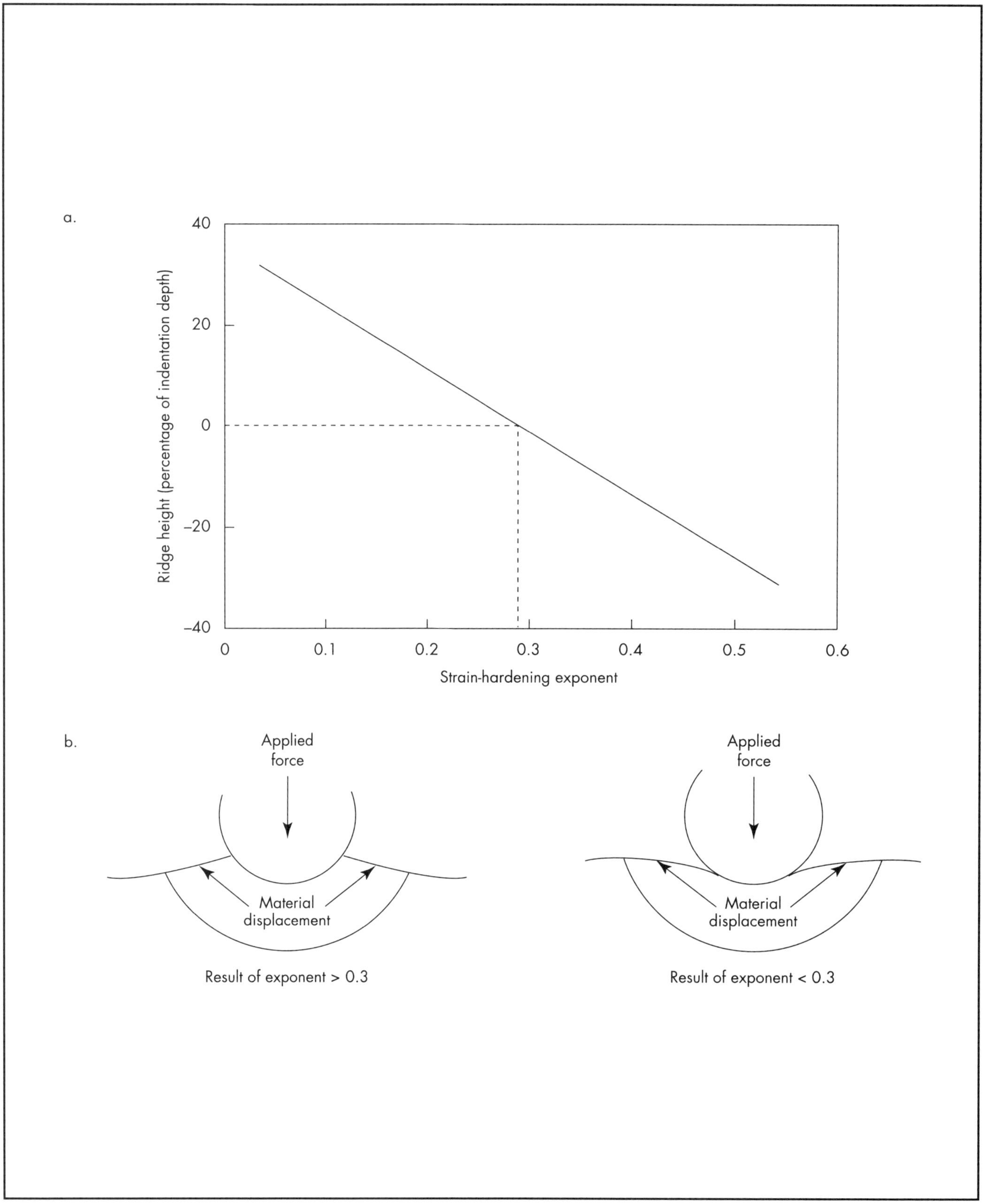

Figure 4-41. Effect of strain-hardening exponent on raised-material formation (Gillespie 1975b; O'Neill 1934).

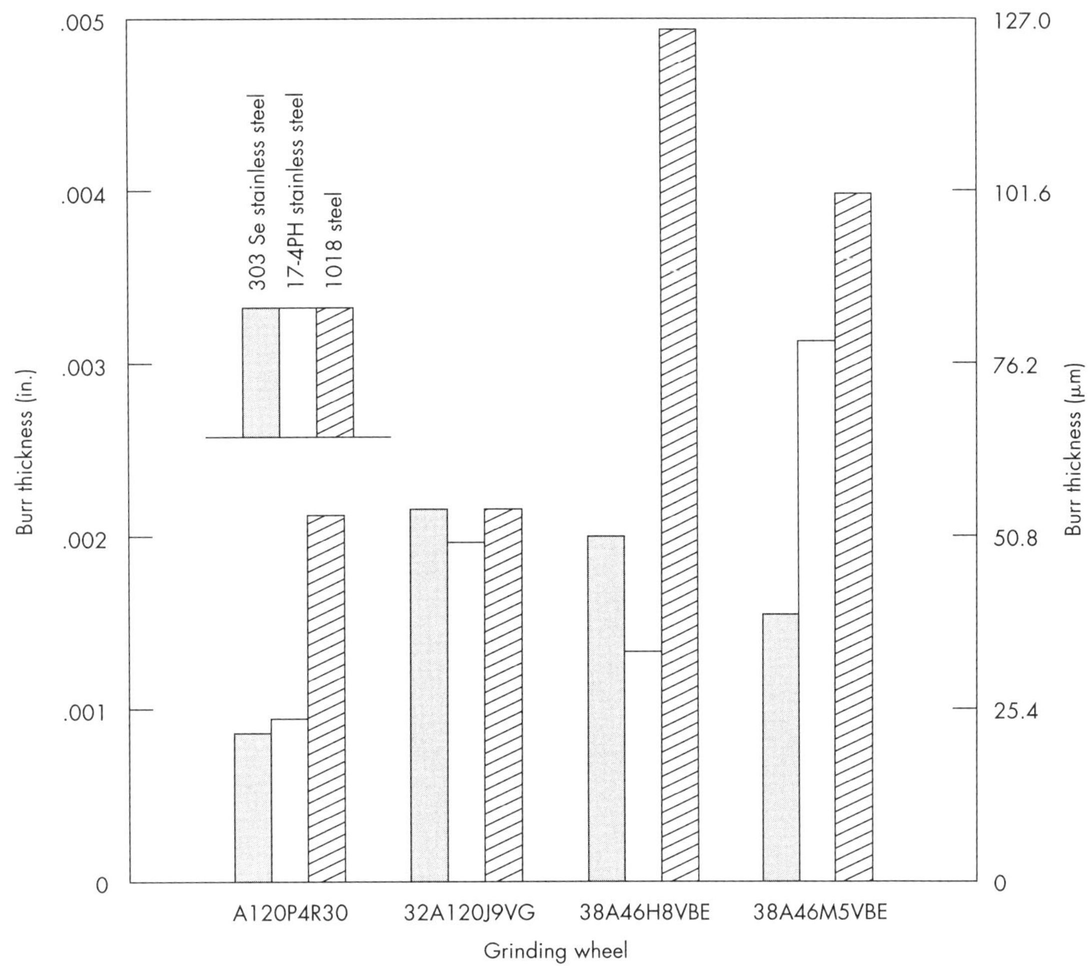

Figure 4-42. Burr thickness at location 3 as a function of type of grinding wheel used (Gillespie 1975b).

Table 4-5. Typical burr sizes from gentle grinding

Burr location (edge)	Burr property Length in. (µm)	Thickness in. (µm)
1	.0025 (63.5)	.0015 (38.1)
2	.0028 (71.1)	.0014 (35.6)
3	.0034 (86.4)	.0023 (58.4)
4	.0018 (45.7)	.0009 (22.9)

Burr height (b_h) and burr thickness (b_t) can be approximated by Equations (4-6) and (4-7). The constants C_1, C_2, and K given in Table 4-6 were obtained by a regression analysis of test data. The geometry factor shown in the following equations indicates the reduction in size caused by a starting radius as opposed to a chamfer. If the tool used has a starting chamfer, the geometry factor is 1.0.

$$b_h = C_1\,(\text{stock removed/side})\,(\text{geometry factor 1}) \tag{4-6}$$

$$b_t = C_2\,(\text{stock removed/side})^{K2}\,(\text{feed rate})\,(\text{geometry factor 2}) \tag{4-7}$$

Thus, for .0045 in. (114 µm) stock removal in brass using a reamer with a starting radius, the predicted exit burr height is

$$b_h = (0.83)(0.0045)\ (0.5) = .0019\ \text{in.}\ (48\ \mu\text{m})$$

Exit burr thickness for 25.4 mm stock removal, a feed rate of 25.4 mm/rev, using a standard reamer in 303 Se stainless steel, would be

$$b_t = (0.000031)\ (25.4)^{2.1}(25.4)\ (1) = 0.7\ \mu\text{m}$$

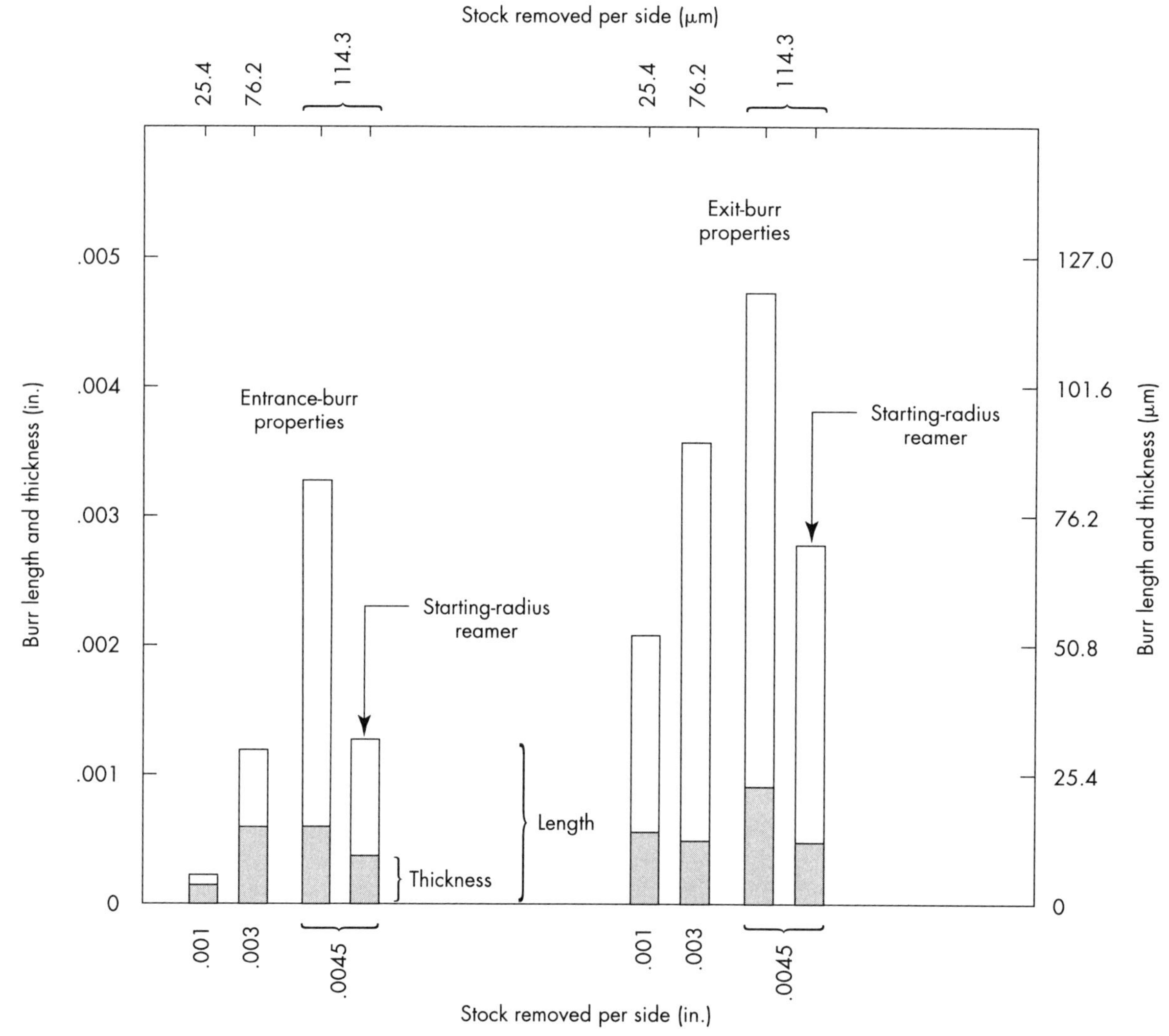

Figure 4-43. Burr thickness and length increase as stock removal increases. Using a starting radius rather than chamfer on the reamer results in smaller burrs (data are averages for aluminum, stainless steel, and brass and two-feed rates) (Gillespie 1974a).

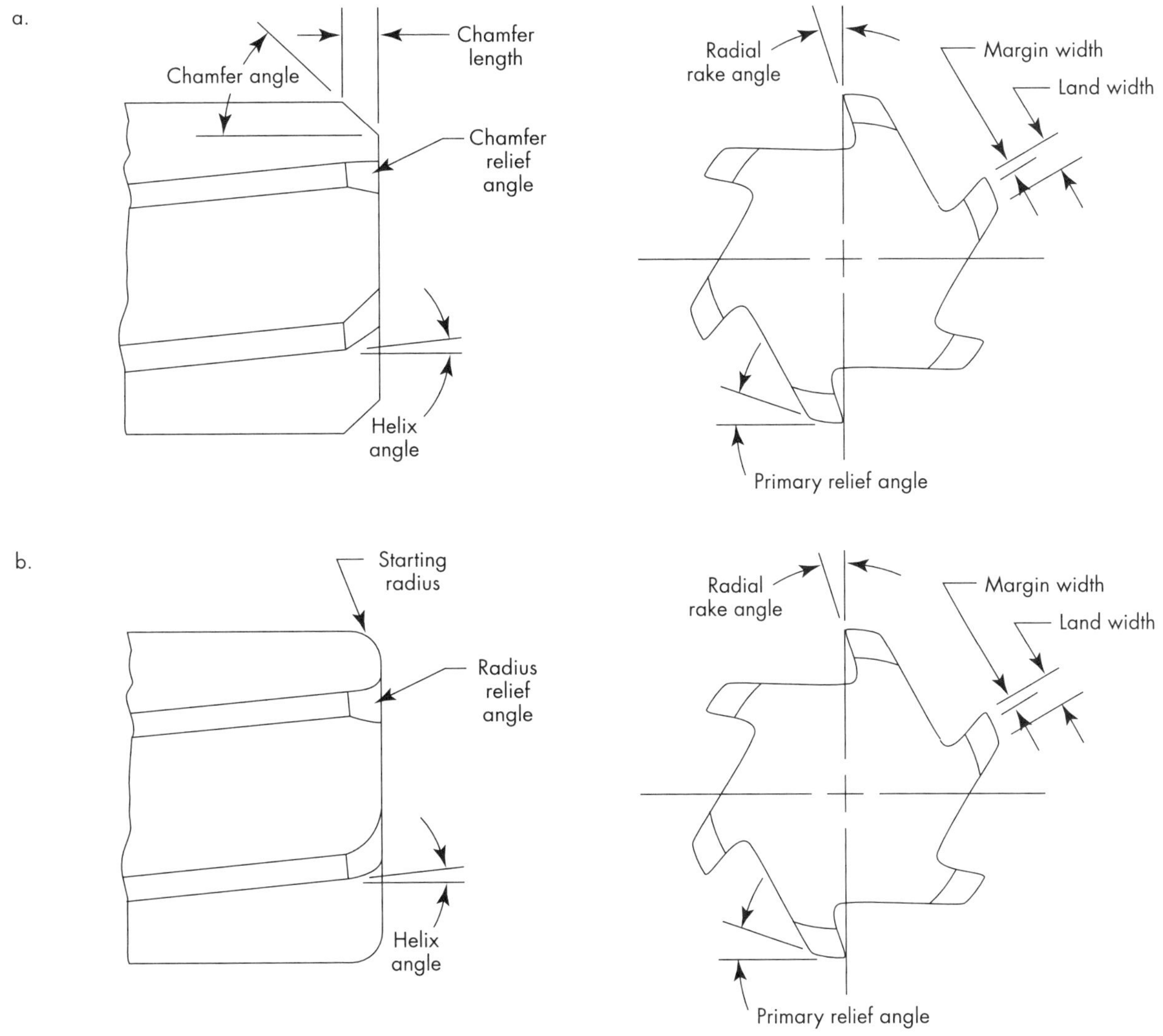

Figure 4-44. Differences from starting chamfer and starting radius (Gillespie 1974a).

Table 4-6. Burr size constants for reaming equations

Workpiece	Entrance burr constants					Exit burr constants				
	C_1	C_2	K_2	$geom_1$	$geom_2$	C_1	C_2	K_2	$geom_1$	$geom_2$
Aluminum, brass, and 303 Se stainless	0.61	202.6 (0.0470)	0.82	0.5	0.5	1.14	3.8 (0.039)	45	0.62	0.5
Aluminum	0.58	18.1 (0.0029)	0.86	1.0	1.0	1.31	0.11 (0.0333)	−0.11	0.5	0.5
Brass	0.39	5.57 (0.0170)	0.55	0.1	0.1	0.83	1.21 (0.079)	0.27	0.5	0.5
303 Se	0.89	144.0 (0.0011)	1.16	0.25	0.5	1.26	56037 (0.000031)	2.1	0.5	0.5

Data in parentheses is constant to use when feed rate is given in micron/rev.

(Gillespie 1974a)

Burr Removal

Most reaming burrs are approximately .0005 in. (12.7 μm) thick and are readily removable by light sanding or other techniques. No problems should be encountered, especially if reamers selected have starting radii rather than chamfers. Table 4-7 compares burrs produced by drilling and reaming for 303 Se stainless steel in a precision shop.

Reaming is typically used to improve hole size repeatability. It also may be advantageous to ream holes to decrease burr removal costs. On most machining operations, burr size varies considerably even when all controllable variables are held constant. Reaming burrs, however, tend to be consistent: the standard deviation of reaming burr thickness in this test (Table 4-6) was typically less than .00007 in. (1.8 μm). By comparison, the standard deviation for 303 Se stainless steel burrs using a .025 in. (0.635 mm) drill is as much as .00045 in. (11.4 μm). In a roughing cut with a face mill, the standard deviation of thickness in 303 Se stainless steel is up to .0020 in. (50.8 μm). The repeatability of burr size afforded by reaming makes burr removal efforts more repeatable also.

4.6.6 Minimizing Pressworking Burrs

The factors influencing burr formation in shearing can be estimated by considering the mechanism of burr formation. Burrs will never be generated if cracks simultaneously occur from the edges of both punch and die (Figure 4-28). However, crack formation will usually not take place from the cutting edges of both punch and die at the same time. Cracks inevitably occur at the point located slightly apart from the cutting edge, because a lower hydrostatic pressure exists there. This causes burr formation. Furthermore, the crack will occur preferentially from one or the other of the cutting edges. In the case of ductile materials, if a sheared or burnished surface is formed even after crack origination, the burr height will be further increased. In the extreme case, a crack occurring from one cutting edge will grow toward the flank of the other cutting edge in a lower hydrostatic pressure state. In this case, the burr will be noticeably larger than usual. Generally, it can be reasonably concluded that burr formation cannot be avoided in the conventional shearing process.

Table 4-7. Typical exit burr sizes produced in 303 Se stainless steel with a .143 in. (3.63 mm) reamer and standard drill

Cutting tool	Burr thickness in. (mm)	Burr height in. (mm)
Standard drill	.003 (76.2)	.020 (508)
Standard reamer	.0006 (15.2)	.003 (76.2)

(Gillespie 1974a)

Three conditions influence burr size in press operations:

1. Cutting-edge geometry.
2. Clamping condition of the material.
3. Material properties.

Clearance

In conventional blanking and shearing, the clearance between the punch and die, workpiece clamping conditions, cutting-edge sharpness, and workpiece properties are the factors that most affect burr size. *Extremely large or small clearances will cause large burrs.* Clearances of 5–15% are usually adopted to minimize burr height. *Burr height is proportional to tool dullness.*

Die Radius

Die radius affects burr height more than clearance, punch speed, die taper, die throat length, punch slide taper, or punch bottom taper. Burr height at the beginning of a run is greatly influenced by the construction of the die button and is minimized when a straight land is used. When a tapered relief is used instead of a straight land, the burr height at the beginning of a run is inversely proportional to the clearance. The burr height increases in direct proportion to the number of punch strokes made. In conventional dies, burr height will generally increase at a rate of .002 in. (0.05 mm) per 100,000 press strokes, or faster, depending on die construction and workpiece material.

Some researchers have suggested that a large die radius reduces burr size in conventional blanking, but this suggestion is not universally accepted. Burr thickness and height increase as punch-to-die clearance increases, but the relationship is not linear. For steels, a clearance of 2.5% of the stock thickness generally produces the smallest burr, but such tight fits greatly increase tool wear. Initial burr size can be reduced by improving the finish on the punch face.

If cutting edges become dull, the crack will break out at the terminal point of the radius. Consequently, the rounded portion will be transferred as the burr of the sheared blank. The condition of the material, clamping, and restriction can influence the hydrostatic pressure state by shifting the crack breakout point. This adjustment will change burr height to some extent. Generally speaking, clamping conditions that cause a lower hydrostatic pressure state may be useful for minimizing burr height.

From a materials standpoint, hard and/or brittle materials are advantageous for minimizing burrs since the crack breakout points will come near the cutting edge. However, hard materials wear down the cutting edge, which in turn increases the burr height. In addition to the material condition, the working temperature and velocity of the punch can also change burr height. High temperatures and low velocities effectively increase the ductility of the material, which, in turn, causes bigger burrs.

As noted in Chapter 3, the prevention of punching burrs has long been one of the concerns in the pressworking industries. A number of improvements have been made and touted as burr-free shearing, but most of the improvements only result in smaller burrs, not their elimination. Processes that can produce burr-free edges are described in Chapter 3. Fine blanking, finish blanking, shaving, and high-speed blanking all typically produce burrs, but they are smaller than those produced in the basic blanking processes.

Freezing materials, such as Teflon, prior to blanking has prevented some burrs and may make others much smaller. Each material has a unique temperature at which the material loses its ductility. Frozen plastics can also result in crazing and breakage during pressworking unless adequate attention is devoted to material properties. Presses that provide high-speed blanking also appear to reduce burr size in many operations.

4.7 SIGNIFICANCE OF RESEARCH ON BURR FORMATION

This chapter provides a basic starting point for understanding burr formation. The prediction of actual burr conditions today remains a demanding, challenging endeavor. In fact, although the dimensions of a burr, rather than its shape, are generally the problem in burr removal, researchers are still defining the various shapes of burrs because they reveal information about formation mechanisms and their effect on removal. Detailing the wide variety of burr shapes is beyond the scope of this handbook; readers interested in more in-depth information are referred to the reference section at the end of this chapter (see, especially, Gillespie 1996).

Software is being developed for specific applications to minimize burrs and to place them in locations that make them easy to remove (Stein et al. 1993), but the breadth of industrial need will for many years extend far beyond the few conditions studied to provide automatic solutions. Empirical studies are still needed, and professional research involves finite element analysis, plasticity theory, continuum mechanics, slip-line analysis, fracture mechanics, scanning electron microscopes, statistical experimental design (design of experiments [DOE]), regression analysis, and in some cases, stochastic analysis. Measuring burrs is as challenging as understanding their formation. Choosing the optimum deburring conditions involves considering thousands of variables. There are 50 different basic designs to chamfering and countersinking cutters, and hundreds of variations among the basic designs. When an estimated 1,500 different chamfering and countersinking cutters exist, it is clear that a full understanding of all the possibilities will require decades of work. That work has begun, and important strides have been made. Nevertheless, burr formation, minimization, and prevention still require research and close attention.

Our current knowledge of burr formation is summarized here in an attempt to help production staffs understand both the potential savings of making burrs easier to remove and the extreme challenge of making that happen for many operations and part configurations.

- Smaller burrs allow many more processes to be used and result in lower finishing costs.
- Lower cutting forces typically result in smaller burrs.
- Minimizing burrs requires a general understanding of burr formation and plant-specific test results.
- Burrs are not consistently reproducible. To provide consistent results, engineers may have to use averages of several burr measurements (Gillespie 1974b; Wang, Taraman, and Wu 1969).
- The tremendous diversity of cutting tool, design, quality, and consistency produces similar diversity in burr properties.
- Burr thickness is the principle factor in burr removal. Burr height can be a significant factor for long burrs. Successful, economical burr removal requires an understanding of the burr to be removed.
- Burr minimization requires burr measurements.
- There are few generalities of burr minimization in industrial use. There are many exceptions to every "rule" when trying to produce minimum burrs.
- Tool wear will always increase burr size. Each user must establish the tool-change frequency needed for each cutting application.

References

Gillespie, LaRoux K. 1974a. "The Effects of Reaming Variables on Burr Properties." Report BDX-613-1083 (rev.). Kansas City, MO: Bendix Corporation (available from National Technical Information Service [NTIS]).

———. 1974b. "The Measurement of Burrs." Technical paper MR74-993. Dearborn, MI: Society of Manufacturing Engineers (SME).

———. 1974c. "Properties of Burrs Produced by Ball Broaching." Report BDX-613-1084. Kansas City, MO: Bendix Corporation (available from NTIS).

———. 1975a. "The Burrs Produced by Drilling." Report BDX-613-1248. Kansas City, MO: Bendix Corporation (available from NTIS).

———. 1975b. "The Burrs Produced by Grinding." Report BDX-613-1572. Kansas City, MO: Bendix Corporation (available from NTIS).

———. 1975c. "The Burrs Produced by Side Milling Cutters." Report BDX-613-1303. Kansas City, MO: Bendix Corporation (available from NTIS).

———. 1976a. "The Burrs Produced by End Milling." Division report BDX-613-1503. Kansas City, MO: Bendix Corporation (available from NTIS).

———. 1976b. "The Burrs Produced by Turning." Report BDX-613-1748. Kansas City, MO: Bendix Corporation (available from NTIS).

———. 1976c. "The Effects of Drilling Variables on Burr Properties." Report BDX-613-1502. Kansas City, MO: Bendix Corporation (available from NTIS).

———. 1977. "Estimating the Size of Cutoff Projections." Technical paper MRR77-01. Dearborn, MI: Society of Manufacturing Engineers (SME).

———, ed. 1996. "Standard Nomenclature for Researchers in Burrs, Deburring, and Edge Finishing." WBTC-STD 2.1996 draft. Kansas City, MO: Deburring Technology International.

Hasegawa, Yoshio; Zaima, Shigeo; and Yuki, Akiyasu. 1975. "Burr in Drilling of Aluminum and Prevention of It." Technical paper MR75-480. Dearborn, MI: Society of Manufacturing Engineers (SME).

Hashimura, Masayuki and Dornfeld, David A. 1995. "Analysis of Burr Formation Mechanism in Machining Process." Technical paper MR95-178. Dearborn, MI: Society of Manufacturing Engineers (SME).

Hashimura, Masayuki; Hassamontr, Jaramporn; and Dornfeld, David A. 1995. "Effect of In-plane Exit Angle and Radial Rake Angle on Burr Height and Thickness in Face Milling Operation." *Manufacturing Science and Engineering,* ME-Vol. 2-1/MH-Vol. 3-1: 203–17.

Hongtao, Liang; Qiping, Wu; and Lei, Cao. 1994. "Research of Mechanism of Edge Burr Formation Caused by Face Milling." *Proceedings of the 3rd International Conference on Precision Finishing and Burr Technology.* Kansas City, MO: Deburring Technology International: 384–98.

Kent, Brian. 1976. "Case Histories of Burr Minimization by Radial Lip Drills." Technical paper MR76-833. Dearborn, MI: Society of Manufacturing Engineers (SME).

Mohun, W.A. 1962. "Grinding with Abrasive Disks." *Journal of Engineering for Industry (Transactions of ASME),* Vol. 84 (November): 442–82.

O'Neill, Hugh. 1934. *The Hardness of Metal and Its Measurement.* Cleveland, OH: Sherwood Press.

Pande, S., and Relekar, H. 1986. "Investigations on Reducing Burr Formation in Drilling." *International Journal of Machine Tool Design and Research,* 26 (3):339–48.

Phillips, Joseph L. 1976. Multi-layer Fastener Systems. Final Report, AFML TR-76-76, vols. 1–4, June. Seattle, WA: Boeing Commercial Airplane Company.

Pond, J.P. 1967. "Drill Points Assume New Shape." *Cutting Tool Engineering,* 19, no. 5, (May): 9-13; no. 7-8: 8-11.

Sofronas, Anthony, and Taraman, K. 1976. "Model Development for Exit Burr Thickness as a Function of Drill Geometry and Feed." Technical paper MR76-253. Dearborn, MI: Society of Manufacturing Engineers (SME).

Stein, Julie M. 1977. *The Burrs from Drilling.* Kansas City, MO: Deburring Technology International.

Stein, Julie M.; Narayanaswami, Ranga; Ho, Samuel; Lam, Anselm Y.; Park, Ilwhan; Babu, Madhu; Afza, Asim; and Dornfeld, David. 1993. "Intelligent Deburring of Precision Components." Technical paper MR93-319. Dearborn, MI: Society of Manufacturing Engineers (SME).

Wang, K.K.; Taraman, Khalil; and Wu, S.M. 1969. "An Analysis of Punching Variable by Two Level Fractional Factorial Design." Paper 69-WA/Prod-27. New York: American Society of Mechanical Engineers (ASME).

Zaima, A. Yuki, and Kamo, S. 1968. "Drilling of Aluminum Alloy Plates with Special Type Point Drill." *Journal of Japan Institute of Light Metals,* 18 (5): 269–76; 18 (6): 307–13.

Bibliography

Adachi, Katushige; Arai, Norihisa; Wakisaka, Shoichi; Harada, Susumu; and Hatutori, Hidetoshi. 1987. "A Study on Burr in Low-frequency Vibratory Drilling." *Japan Society of Precision Engineering Bulletin,* 21 (4): 258–64.

Barash, M.M. and Schoesh, W.J. 1970. *Advances in Machine Tool Design and Research.* Elmford, NY: Pergamon Press: 603-613.

Barry, Mark W., and Vanderwert, Terry L. 1995. "Five Axis Laser Machining of Complex Features." Technical paper MR95-185. Dearborn, MI: Society of Manufacturing Engineers (SME).

Bauer, Diana J. 1989–90. "Expert System for Burr Size Prediction." Research Reports 89/90, 5–8. Berkeley, CA: Laboratory for Manufacturing Automation, University of California.

———. 1990. "The Framework for a Knowledge-based System to Reduce Burr Size." Master's thesis. Berkeley, CA: Department of Mechanical Engineering, University of California.

Biegel, E., and Holmes, Robert E. 1961. "Development of a Punchability Rating Method for Electrical Steels." Research report 30, 1 March. Dearborn, MI: American Society of Tool and Manufacturing Engineers (ASTME) (now SME).

Blotter, P. Thomas; Huang, S.F.; Spear, Carl D.; and Canfield, Ronald V. 1980. "Deburring Costs Related to Cutting Speed." Technical paper MR80-661. Dearborn, MI: Society of Manufacturing Engineers (SME).

Buhler, Hans, and Pollmar, Fedor. 1971. "The Formation of Burrs in the Cutting of Thin Sheet." *Bander-Bleche-Rohre,* 12 (3): 105–111.

Chern, G.L. 1993. "Analysis of Burr Formation and Breakout in Metal Cutting." Ph.D. dissertation. Berkeley, CA: Department of Mechanical Engineering, University of California.

Chern, Gwo-Liang. 1989–90. "Burr Formation and Fracture Mechanism at the Exit Stage of Machining Metals," pt. 2. Research reports 89/90, 13–14. Berkeley, CA: Laboratory for Manufacturing Automation, University of California.

———. 1990–91. "Modeling the Burr Formation and Fracture Mechanism in Metal Cutting." Research reports 90/91, 13–14. Berkeley, CA: Laboratory for Manufacturing Automation, University of California.

———. 1991–92. "Burr/Breakout Model Development and Experimental Verification." Research reports 91/92, 6–9. Berkeley, CA: Laboratory for Manufacturing Automation, University of California.

Crane, Edward V. 1939. *Plastic Working of Metals and Power Press Operations*. New York: John Wiley & Sons.

Deck, J.D.; Thacker, J.G.; and Cushing, R.L. 1976. "Response of Pig Skin to Selected Cutting Edges." *Proceedings of the ACEMB*, 45, November. Boston, MA: ACEMB.

Dornfeld, D.A. 1992. "Intelligent Deburring of Precision Components." *Proceedings of the 1992 International Conference on Industrial Electronics, Control, Instrumentation, and Automation*. San Diego, CA: Institute of Electrical and Electronics Engineers (IEEE): vol. 2–3: 953–60.

Dornfeld, D.A. and Masaki, T. 1987. Shoureshi, R.; Youcef-Toumi, K.; and Kazerooni, H. (eds.). "Acoustic Emission Feedback for Deburring Automation." *Modeling and Control of Robot Manipulators and Manufacturing Processes*. New York: ASME: DSC-Vol.: 81–89.

Fleming, C.M. 1973. "Precision Hole Generation Methods." AFML-TR-73-135. St. Louis, MO: McDonnell Aircraft Company.

Gillespie, LaRoux K. 1974. "The Effect of Cutting Edge Radius on Poisson Burr Properties." Technical paper MR74-990. Dearborn, MI: Society of Manufacturing Engineers (SME).

———. 1975. "The Formation and Properties of Burrs." Technical paper MRR75-03. Dearborn, MI: Society of Manufacturing Engineers (SME).

———. 1976a. "Burr Formation and Properties." *Deburring Capabilities and Limitations*. Dearborn, MI: Society of Manufacturing Engineers (SME).

———. 1976b. "Machinability as Related to Precision Miniature Parts." Technical paper MRR76-09. Dearborn, MI: Society of Manufacturing Engineers (SME).

———. 1978. *Advances in Deburring*. Dearborn, MI: Society of Manufacturing Engineers (SME).

———. 1981. "Visual Thread Quality for Precision Miniature Mechanisms." Division report BDX-613-2600. Kansas City, MO: Bendix Corporation.

———. 1994. "Process Control for Burrs and Deburring." Report TR94-1. Kansas City, MO: Deburring Technology International.

Gillespie, LaRoux K., and Blotter, P. Thomas. 1976. "The Formation and Properties of Machining Burrs." *Journal of Engineering for Industry (ASME Transactions, Series B)*, Vol. 98 (February): 66–74.

Hashimura, M.; Ueda, K.; Dornfeld, D.; and Manabe, K. 1955. "Analysis of Three-dimensional Burr Formation in Oblique Cutting." *Annals of the CIRP*, 44 (1): 27–30.

Hofman, P., and Kvasnicka, I. 1996. "Study of Microdeformation on Edge of Metal Bodies." *Workshop 96 Proceedings*. Praha, Czechoslovakia: Czechoslovakia Technical University, Department of Mechanical Engineering.

Hugo, Harding R. 1971. "How to Improve Metal Stamping Die Performance." *Sheet Metal Industries,* 48 (February): 120–135.

Jameson, Elman C. 1982. "Hole Edge Condition Using Electrical Discharge Machining (EDM)." Technical paper MR82-237. Dearborn, MI: Society of Manufacturing Engineers (SME).

Kanai, M., and Kanda, Y. 1978. "Statistical Characteristics of Drill Wear and Drill Life for the Standardized Performance Tests." *Annals of the CIRP*, 27: 61-66.

Kawamura, Y., and Matsumoto, K. 1992. "Application of Piezoelectric Actuator to Burr-free Blanking." *Ferroelectrics,* 13: 121–26.

Kienzle, Otto. 1959. "Wear in the Cutting of Thin-gauge Steel Sheets." Research report 22, 1 May. Dearborn, MI: American Society of Tool and Manufacturing Engineers (ASTME) (now SME). Translation and discussion of article by Kienzle, Otto and Kienzle, D. Werner. *Stahl und Eisen,* 78 (12): 820–28, 1958.

Kitajima, Koiche; Takazawa, Koya; Miyake, Teruaki; Yamamoto, Akihiro; and Tanaya, Yukoio. 1990. "Study on Mechanism and Similarity of Burr Formation in Face Milling and Drilling." Technology report 32. Osaka, Japan: Kansai University.

Ko, Sung-Lim. 1989. "Burr Formation and Fracture Mechanism at the Exit Stage of Machining Metals." Ph.D. dissertation. Berkeley, CA: Department of Mechanical Engineering, University of California.

———. 1989–90. "Burr Formation and Fracture Mechanism at the Exit Stage of Machining Metals," part 1. Research reports 89/90, 9–12. Berkeley, CA: Laboratory for Manufacturing Automation, University of California.

Ko, Sung-Lim, and Dornfeld, David A. 1991. "A Study on Burr Formation Mechanism." *Journal of Engineering Materials and Technology,* 113: 75–87.

———. 1994. "Burr Formation and Fracture at the Exit Stage in Oblique Cutting." *Proceedings of the 3rd International Conference on Precision Finishing and Burr Technology,* Seoul, Korea: 116–25. (Available from Deburring Technology International, Kansas City, MO).

Koide, Sakae. 1981. "Press Technique, Questions and Answers." *Journal of the Japan Society of Deburring and Surface Conditioning Technique,* 1 (1):19–41.

——. 1977. "Fine Edge Blanking Press and Deburring" (in Japanese). *Kinzoku Press* (Metal Press) (November): 10–18.

Kondo, Kazuyoshi; Ishikiriyama, Tetsuo; Yoshida, Motoaki; and Takada, Kunio. 1974. "Application of Opposed Die Shearing Process." *Proceedings of the International Conference on Production Engineering.* Tokyo, Japan: pt. 1:251.

Konig, W.; Steffens, K.; and Ludewig, T. 1985. "Single Grit Tests to Reveal the Fundamental Mechanism in Grinding." *Milton C. Shaw Grinding Symposium.* New York: ASME, PED-Vol. 16: 141–54.

Kuznetsov, V.D. 1966. "The Built-up Edge in Scratching." *Metal Transfer and Build-up in Friction and Cutting.* Oxford/New York: Pergamon Press: 80–81.

Lange, K., and Liebing, H. 1990. "Blanking and Piercing Without Any Burr." *Annals of the CIRP,* 39 (1): 177-181.

Lee, G.B. 1989. "Digital Control for Burr Minimization in Drilling." Ph.D. dissertation. Berkeley, CA: University of California.

Lenz, E. 1978. "Investigation in Drilling." *Annals of the CIRP,* 27 (1): 49–53.

Link, R. 1992. "Gratbildung und Strategien zur Gratreduzierung." Ph.D. dissertation. Aachen, Germany: Rheinisch Westfalischen Technischen Hochschule.

Maeda, T. 1958. "Reciprocating Blanking Method." *Science of Machine,* 10 (1): 140.

Maeda, T., and Murakawa, M. 1977. "The Development of Burr-free Slitting." *Journal of Japan Society for Technology of Plasticity,* 18 (193): 114.

Marciniak, Mieczyslaw. 1995. "The Tribological Phenomena in Nontraditional Grinding." *Proceedings of the 1st International Machining and Grinding Conference.* Dearborn, MI: Society of Manufacturing Engineers (SME): 505–12.

McBride, James Nielson. 1969. "The Magnitude of Burrs Caused by Electrical Discharge Machining." Master's thesis. Logan, UT: Utah State University.

Murakawa, Madao; Maedo, Teizo; and Nakagawa, Takeo. 1977. "Investigation on Burr-free Shearing in Japan." Technical paper MR77-472. Dearborn, MI: Society of Manufacturing Engineers (SME).

Nakayama, K., and Arai, M. 1987. "Burr Formation in Metal Cutting." *Annals of the CIRP*, 36 (1): 33–36.

Narayanswami, Ranganarajan. 1990–91. "Burr Prediction for Part Design and Deburring Process Planning." Research reports 90/91, 6–9. Berkeley, CA: Laboratory for Manufacturing Automation, University of California.

———. 1991–92. "A Framework to Minimize Burr Formation in Machining." Research reports 91/92, 7–9. Berkeley, CA: Laboratory for Manufacturing Automation, University of California.

Ohshima, I. Kuya; Maekawa, Katsuhino; and Murata, Ryoji. 1993. "Burr Formation and Deburring in Drilling Cross Holes." *Journal of Japan Society for Precision Engineering,* 59 (1): 155–60.

Pekelharing, A.J. 1978. "The Exit Failure in Interrupted Cutting." *Annals of the CIRP,* 27 (1): 5–10.

Samuels, L.E. 1972. "Abrasive Surface Finishing Processes: Mechanisms." *New Developments in Grinding.* Shaw, Milton C. (ed). Pittsburgh, PA: Carnegie Press: 283–302.

Samuels, Leonard E. 1982. *Metallographic Polishing by Mechanical Methods,* 3rd ed. Metals Park, OH: ASM International.

Sato, Hisaya. 1969. "Burr and Burr Prevention in Grinding." *Science of Machine* (in Japanese), 21 (4): 589–93.

Shikata, H.; DeVries, M.F.; and Wu, S.M. 1980. "Experimental Investigation of Sheet Metal Drilling." *Annals of the CIRP,* 29 (1): 85.

Shin, Y.; Chen, Y.; and Kumara, S. 1991. "Framework of an Intelligent Grinding Process Advisor." *Journal of Intelligent Manufacturing,* 3: 135–48.

Sofronas, A. 1975. "The Formation and Control of Drilling Burrs." Ph.D. dissertation. Detroit, MI: University of Detroit.

———. 1976. "The Effect of System Stiffness, Workpiece Hardness, and Spindle Speed on Drilling Burr Thickness." Technical paper MR75-132. Dearborn, MI: Society of Manufacturing Engineers (SME).

———. 1976. "Model Development for Exit Burr Thickness as a Function of Drill Geometry and Feed." Technical paper MR75-253. Dearborn, MI: Society of Manufacturing Engineers (SME).

———. 1976. "Reduction of Burr Formation in Drilling." Technical paper MR75-376. Dearborn, MI: Society of Manufacturing Engineers (SME).

Spur, 0., and Masuha, I.R. 1981. "Drilling with Twist Drills of Different Cross Section Profiles." *Annals of the CIRP,* 30 (1): 31–35.

Stango, R.J. and Fitzpatrick, P.R. (eds). 1989. "Analysis and Modeling of Burr Formation and Breakout in Metal." *Mechanics of Deburring and Surface Finishing Processes.* New York: ASME: PED-Vol. 38: 79–92.

Stein, Julie M.; Narayanaswami, Ranga; Ho, Samuel; Lam, Anselm Y.; Park, Ilwhan; Babu, Madhu; Afzal, Asim; and Dornfeld, David. 1993. "Intelligent Deburring of Precision Components." Technical paper MR93-319. Dearborn, MI: Society of Manufacturing Engineers (SME).

Stein, Julie M., and Dornfeld, David A. 1995. "An Analysis of the Burrs in Drilling Precision Miniature Holes." *Manufacturing Science and Engineering*. Kannetey-Asibu, E. (ed). New York: ASME: MED-Vol 2-1/MH-Vol. 3-1: 127–48.

Strasser, Federico. 1960. "How Control of Burrs Aids Sheet Metal Stamping." *The Iron Age,* 185 (3): 90–92.

Stumpf, William R. 1983. "Drilling for a Consistent Burr in AISI 0-1 Oil Hardening Tool Steel." Technical paper MR83-676. Dearborn, MI: Society of Manufacturing Engineers (SME).

Subramanian, K. 1977. "Sensing of Drill Wear and Drill Prediction of Drill Life." Master's thesis. Cambridge, MA: Massachusetts Institute of Technology.

Sugawara, A., and Inagaki, K. 1978. "Effect of Shape of Tool Point with Dwindling of Drill Diameter on Drilling Burr in Case of 0.02% C Steel." *Journal of the Japan Society of Precision Engineering,* 44 (2): 179–84.

———. 1982. "Effect of Workpiece Structure on Burr Formation in Micro-drilling." *Journal of Japan Society of Precision Engineering,* 4 (1): 9–14.

Takazawa, Koya; Xiong, Shiyi; Kitajima, Koichi; Miyake, Teruaki; and Tanaka, Yukio. 1994. "Burr Formation Mechanism in Drilling and Its Control—Cutting Behavior and Basic Considerations on Burr Formation" (partly in English; partly in Japanese). *Proceedings of the 3rd International Conference on Precision Finishing and Burr Technology*. Kansas City, MO: Deburring Technology International: 232–45. Also available as Research report B (19): 9–20, 1995. Kanagawa, Japan: Kanagawa Institute of Technology.

Takeyania, H., et al. 1993. "Study on Oscillatory Drilling Aiming at Prevention of Burr." *Journal of Japan Society of Precision Engineering,* 59 (10): 1,719–24.

Wang, Guicheng; Wang, Shulin; and Wang, Jianmin. 1966. "Cutting Direction Burr Formation in Orthogonal Cutting." *Proceedings of the 4th International Conference on Precision Finishing and Burr Technology*. Kansas City, MO: Deburring Technology International: 58–63.

Wukusik, S., and Zeno, R.S. 1958. Improving Punchability of Silicon Steel." *Tool Engineer,* 41 (12): 63–70.

Yamada, Takashi; Kishi, Kozo; and Hasuda, Yuichi. 1994. "Burr Occurrence on Stainless Steel Thin Sheet in Round Tip Saw Cutting." *Proceedings of the 3rd International Conference on Precision Finishing and Burr Technology*. Kansas City, MO: Deburring Technology International: 209–13.

5 Part Design and Processing Considerations

Edge-finishing costs are significantly affected by both how parts are designed and the processes used in their manufacture. In this chapter, a variety of design and processing techniques are presented that have helped some companies reduce the normal costs of edge finishing.

Edge-related problems can be reduced by considering the following guidelines:

- Design parts and processes to accommodate burrs.
- Design parts and processes to prevent burrs (see Chapter 3).
- Design parts and processes to minimize burrs.
- Design parts and processes to make burrs, flash, and other protrusions more accessible.
- Design parts to allow more process choices for removing burrs.
- Design parts and processes using notes that clearly define needs.

Chapter 3 described burr prevention techniques that reduce edge-finishing problems. Chapter 4 described how speeds, feed rates, depths of cut, and other process variables affect burr size and in turn how burr size can be minimized by judicious choices of these variables. This chapter provides numerous examples of how part design (primarily geometry), process sequences, and the geometry of cutter-to-part angles can be used to reduce burr size. Part designers greatly influence finishing costs through their choice of angles, features, and specifications. Although many burrs cannot be prevented, most can be minimized, which in turn reduces finishing costs. The principle underlying each of the following suggestions is that small burrs are relatively easy to remove and allow a variety of removal processes, thus providing greater manufacturing flexibility.

This chapter is for both the product designer and the process engineer. The product designer tries to make the part easy to manufacture and wants the first attempt at doing so to be successful. The process engineer provides many ideas for improving the design and the process. When both the product and process designers share and discuss the ideas contained in this chapter with others in the plant, the company will obtain the maximum benefit from collaborative design.

The following design factors affect burrs and edges:

- Part geometry.
- Part tolerances.
- Part surface finishes.
- Part material properties.
- Part aesthetics.
- Sequence of operations.
- Tool design and tooling approaches.
- Choice of operation types.

5.1 DESIGNING PARTS TO ACCOMMODATE BURRS

Some components and assemblies can operate adequately without deburring. Sheet metal edges are often more aesthetically pleasing and trouble-free if a rolled edge is produced (Figure 5-1). In such cases, deburring is not required on the hidden edge; thus, the designer can use the geometry of the part to reduce deburring. Formed sheet metal parts can be designed so that all the burrs are located on interior surfaces and not exposed to later cutting or assembly problems. The majority of assemblies may not lend themselves to such obvious design changes; however, if deburring can be eliminated from even one part in an assembly, cost savings will result. A common example of workpiece design in which burr removal is not required is shown in Figure 5-2. Pins that are pressed into holes often do not have to be deburred (that is, made free of projections) if provisions are made in the holes to accommodate the projections.

Figure 5-3 illustrates another method of designing parts to accommodate burrs. In this instance, the designer would also provide a note on the drawing of the component stating, "No raised metal or burrs allowed on vertical surface when viewed at 10× magnification. Material in the undercut is allowed unless it prevents press fit in assembly."

If a small burr is allowable on the outside diameter (OD) of a slotted part, an optional V-groove can be placed at the bottom of the slot (Figure 5-4). This groove permits the

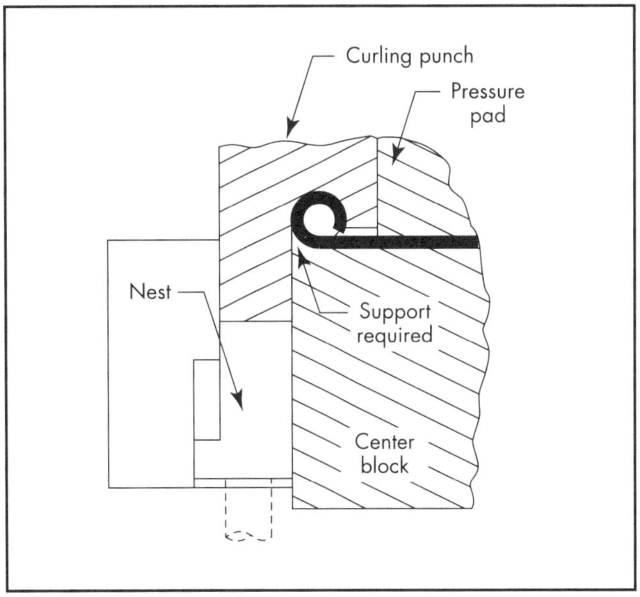

Figure 5-1. Rolled sheet metal edges eliminate need for deburring (Wilson and Harvey 1965).

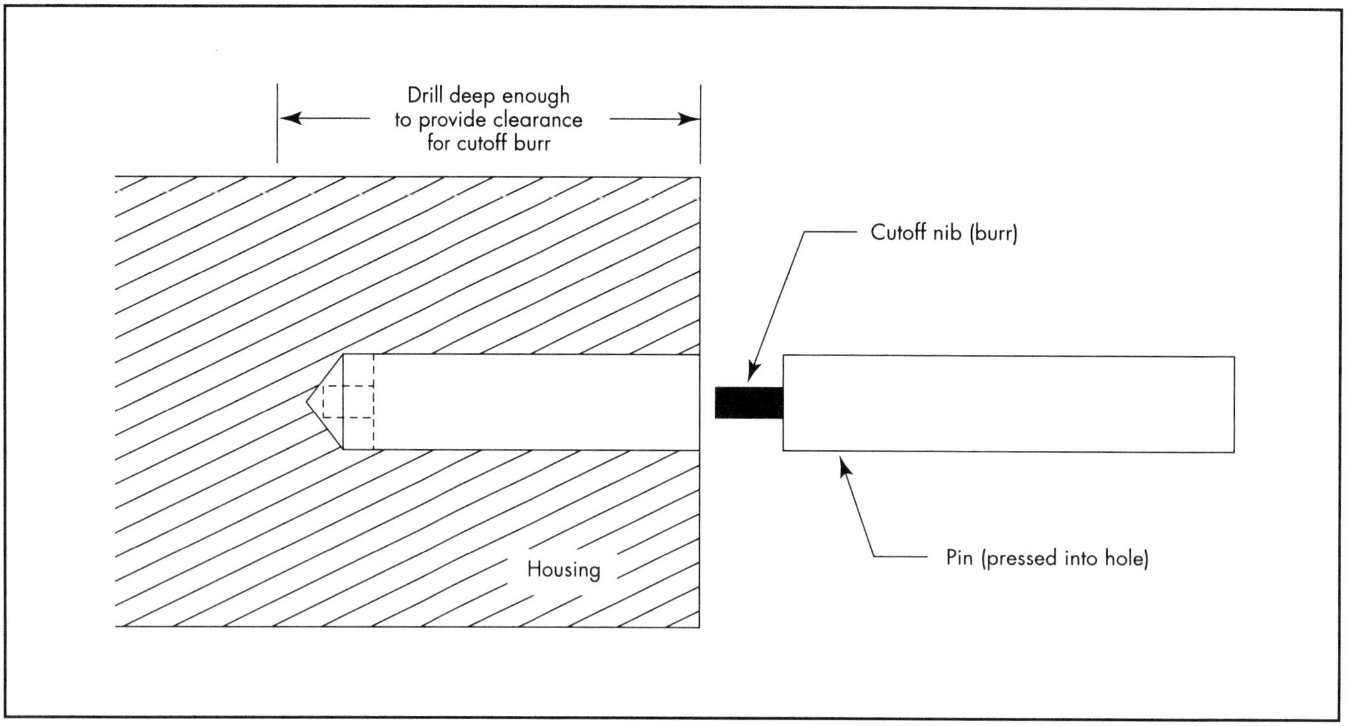

Figure 5-2. Example of workpiece design that eliminates the need for deburring.

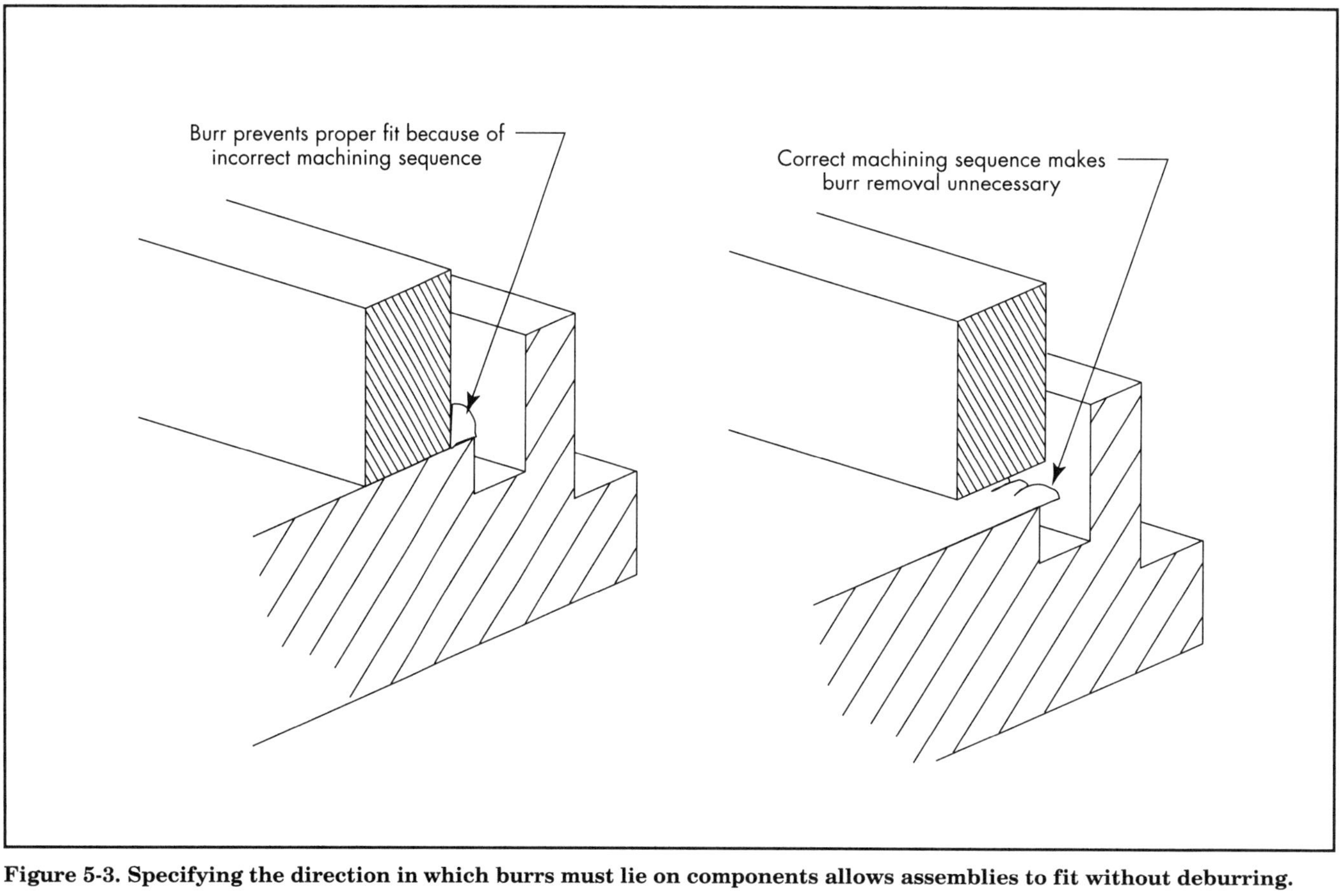

Figure 5-3. Specifying the direction in which burrs must lie on components allows assemblies to fit without deburring.

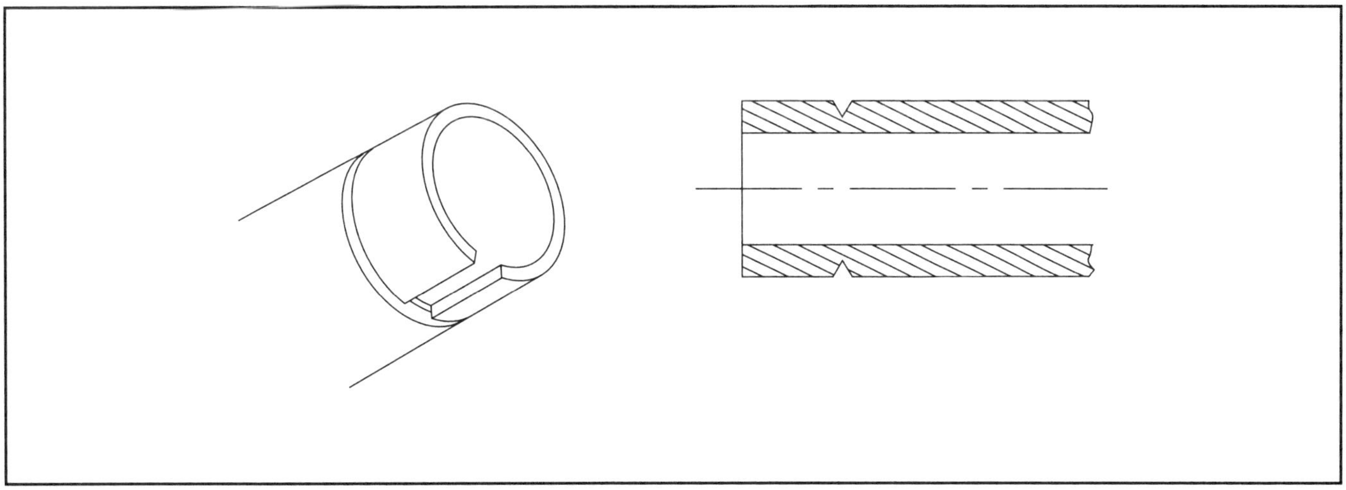

Figure 5-4. Groove at bottom of slot reduces slotting burr size.

existence of a burr at the bottom of the slot without affecting OD size. Although a small burr also forms at the sides of the slots, it may not be large enough to require removal. If it does, it is much easier to remove than the burr normally left at the bottom of the cut.

When a V-shaped groove is placed on the inside diameter (ID) of a cylindrical part so that the cutoff tool will pass through the groove (Figure 5-5), the cutoff burr will not be as objectionable as on most parts. In such cases, the improvement is the result of both the edge angle created by the groove and the fact that when properly placed, the burr will form downward into the groove rather than upward on the ID. If total burr removal is required, a vibratory operation should prove adequate. In this case, the designer may have to specify the allowable chamfer that the groove can introduce to the completed part.

The techniques just described allow parts to function without deburring, but designers still must address all the concerns outlined in Table 1-7. If users might cut hands or fingers as a result of the burrs or sharp edges, the design still has an implicit liability. Cut wires, clogged filters, and so on are all long-term concerns that also must be addressed.

5.2 DESIGNING PROCESSES TO ACCOMMODATE BURRS

The examples of "V" notches provided in Section 5.1 and the checklist in Table 1-7 are as relevant for the process engineer as for the part designer. In some instances, for example, an engineer may need a smaller burr to prevent an in-process deburring operation, so the manufacturing process is revised without changing the original part design.

Workholding jigs and fixtures should be designed keeping in mind that burrs may prevent easy removal of the workpieces. Providing clearances and corner relief (Figure 5-6) will generally eliminate some of the in-process deburring frequently required. In some cases, knockouts (such as the spring ejector shown in Figure 5-7) are required.

5.3 DESIGNING TO PREVENT BURRS

Chapter 3 presented the principles of burr prevention. Although most of the suggestions presented there involve manufacturing processes, some involve product design. They may also require a product designer's understanding of how using a particular burr prevention approach may affect salient properties, such as residual stresses, which in turn may affect product performance.

5.4 DESIGNING TO MINIMIZE BURRS

Both product designers and process engineers can minimize finishing costs by minimizing the size of burrs. The previous chapters have touched on several aspects of minimizing burrs. Edge angles and part geometry, in particular, greatly influence burr size.

The following list contains some of the ways burrs can be minimized:

- Properly designing the workpiece or changing its configuration.
- Using a more appropriate workpiece material.
- Changing the types and/or sequence of production operations (deburring can sometimes be done on NC/CNC machines, along with other operations).
- Changing machine stiffness and/or minimizing cutting forces.
- Changing operating parameters (cutting speed, feed rate, depth of cut, etc.).
- Changing the cutting tool material, configuration, and/or sharpness.
- Specifying back-up materials.
- Specifying better workholding methods.

5.4.1 Part Design

Figure 5-8 illustrates two designs of a sheet metal part. Part designs such as the one shown in Figure 5-8a lead to early die wear because of the sharp edges required. This design creates heavy burrs, particularly at the needle-like

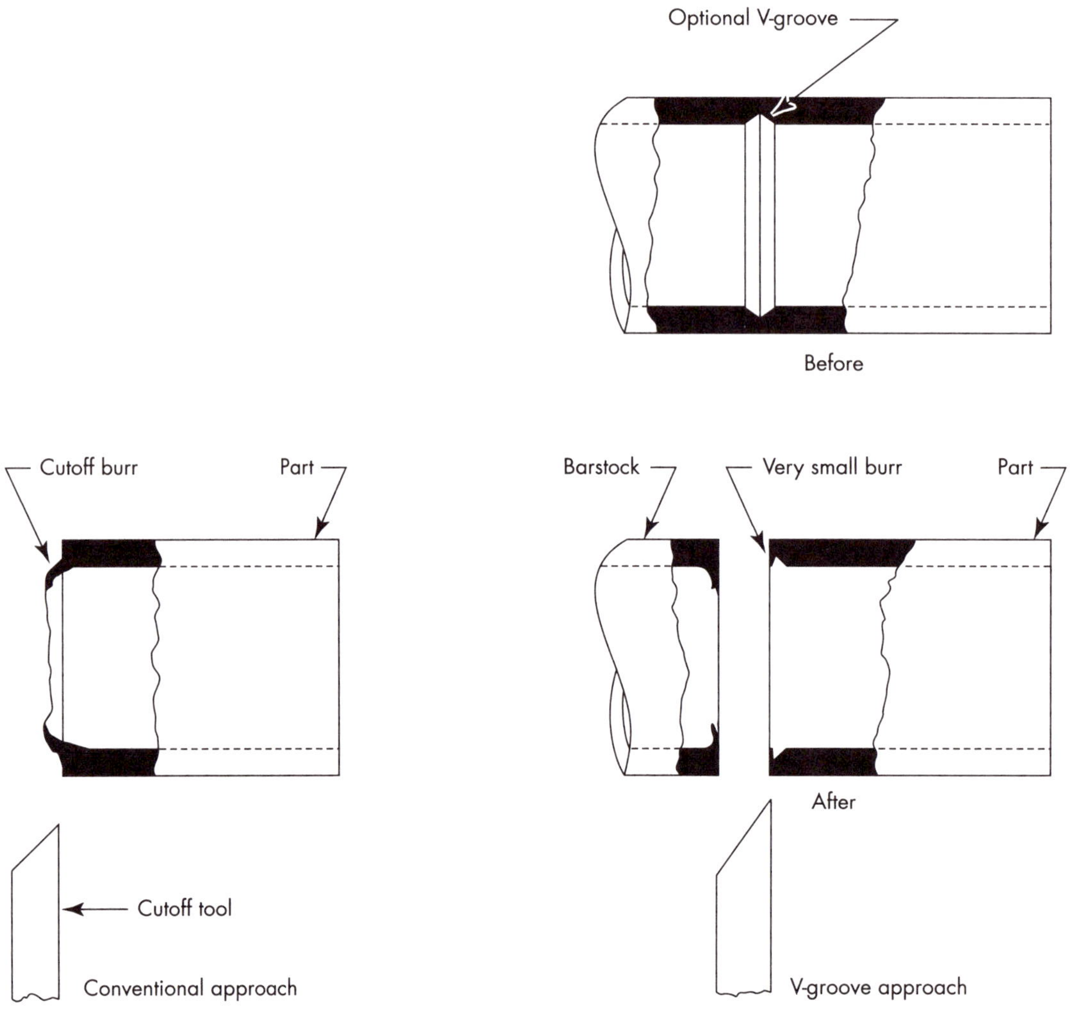

Figure 5-5. Placing a V-groove on inside diameter minimizes size of cutoff burr.

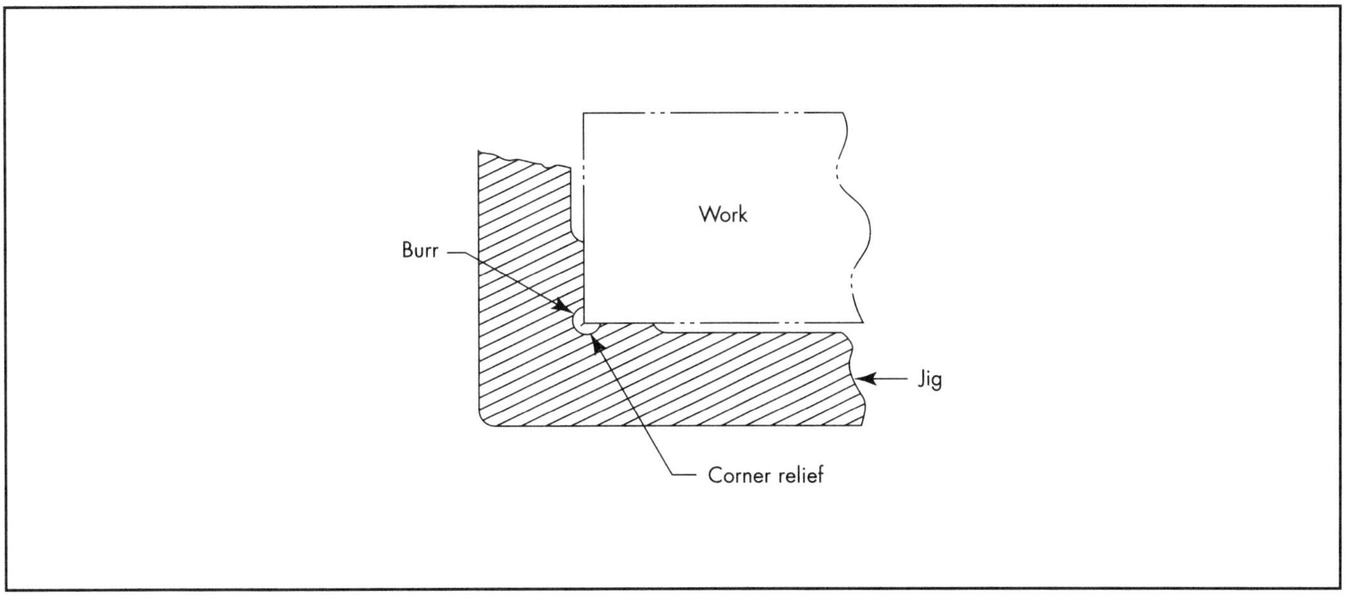

Figure 5-6. Burr clearance provided by corner relief in jig (*Tooling and Production* 1973).

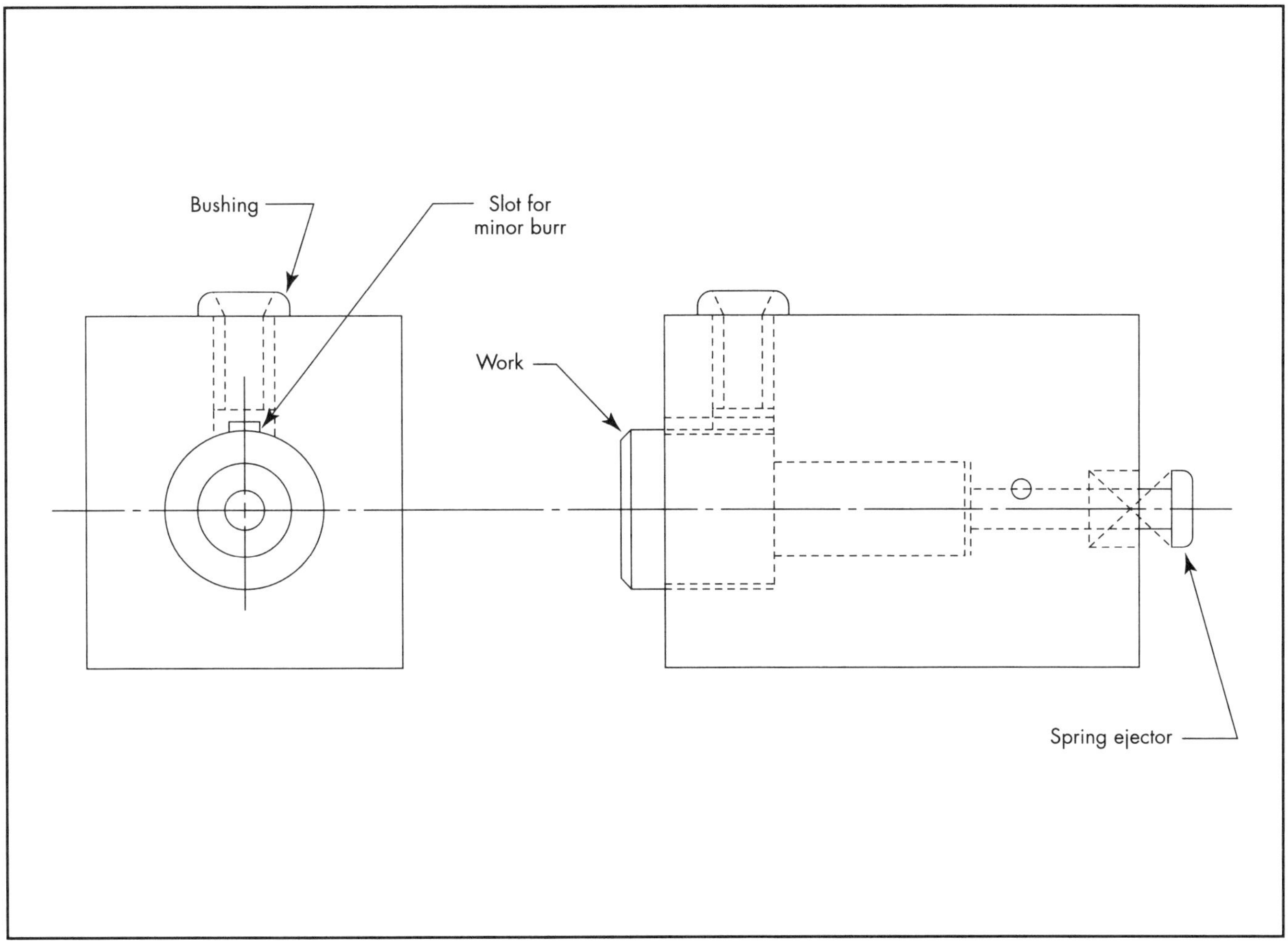

Figure 5-7. Knockout (spring ejector) for removing workpiece with drill burr.

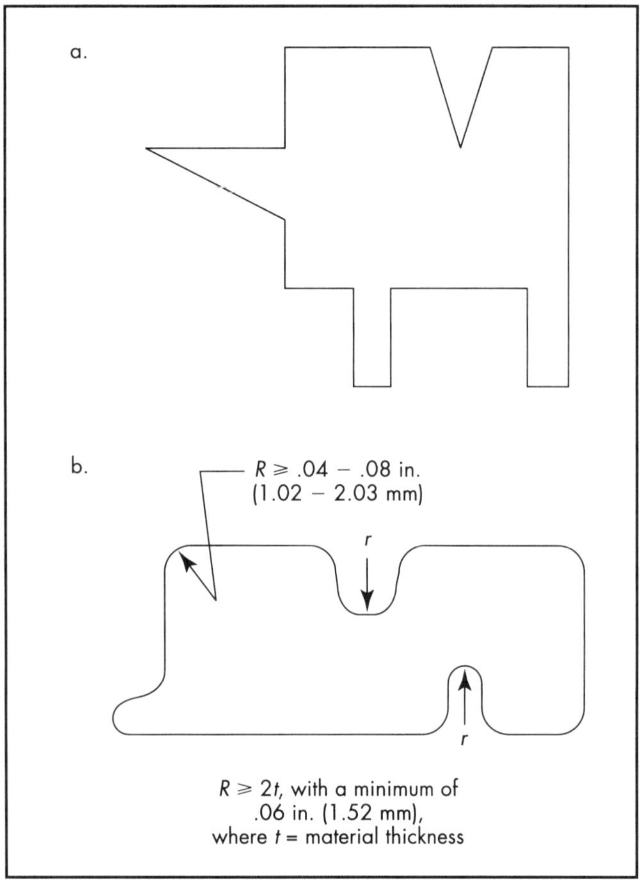

Figure 5-8. Large radii are preferred for blanked parts to minimize burr size and lengthen die life (Strasser 1960).

projections and cutouts. The design shown in Figure 5-8b extends die life and minimizes burrs significantly. When sharp corners are necessary, they can be provided by ordering more expensive progressive dies.

Burrs formed by machining through threads are extremely difficult to remove. This problem can be eliminated by turning a relief diameter that is smaller than the minor diameter of the thread (Figure 5-9). The designer must indicate that these conditions are allowable or desirable by adding drawing notes or specifying in-house standards.

Designing to minimize machining impact requires that the designer and the manufacturing engineer work together closely, since parts can be fabricated by several processes. Changing a single process operation can eliminate all the advantages just cited; similarly, modifying a single feature can invalidate the benefits just cited. Concurrent engineering and frequent interchange are the means by which benefits are gained when designing to minimize burrs.

Figure 5-10 illustrates a machined casting that originally had a large milling burr at the point where the cutter broke through the inner diameter. The large burr formed because the included edge angle ϑ at the point the cutter broke out was very small. When the casting was redesigned with a larger breakthrough angle, only small burrs formed.

Although 90° edges are a fact of life for most companies, a number of exceptions can be found daily in some large companies. Taking advantage of these exceptions requires constant vigilance in both the design and manufacturing phases of any project. For example, if a casting, such as the one shown in Figure 5-11a, has to be drilled and deburred, one can explore the possibility of revising the design to something like that shown in Figure 5-11b. Investigation should be made before the mold is made.

Figure 5-12 reveals that simple countersinking with angles greater than 90° creates an exit angle of 135° or greater. Those familiar with shop practice will note that in some materials no burr forms at the bottom intersection.

Chapter 4 described some of the material influences on burrs. When designers have a choice in materials, they should keep in mind that soft materials often create larger burrs, and strain hardening materials also result in larger entrance burrs. Choosing materials with less ductility is one key to minimizing burr size.

5.4.2 Process Design

Machining sequence can affect deburring time and needs to be taken into consideration in the design. For example, a cylindrical part having a triangular cross section for part of its length would typically be turned, and then the three flats would be milled (Figure 5-13). This sequence, however, creates two undesirable conditions. First, milling typically makes a heavy burr; second, deburring would require a separate operation. The company producing this part chose to mill the flats before turning the stem diameter. This procedure produced a smaller burr and allowed the operator to remove the burrs with a wire brush as part of the lathe operation.

One manufacturer of pumps found a technique for preventing the heavy rollover burrs that occur when turning the outer diameter of an impeller. Because of a convenient projecting hub on one side of the impeller, it was always turned so that the tool exited over a 30° edge, resulting in a long, thick burr. By reversing the part in the spindle, the tool would exit over the 150° degree edge and not produce a burr (Figure 5-14). In cases where reversing the part would add another operation, it is possible to reverse the spindle rotation and correspondingly adjust the tooling.

Figure 5-15 illustrates the angles made when a face-milling cutter passes over a round slug. The cutter leaves the part at angles very small at the bottom and very large at the top. Table 5-1 presents the size of burrs produced at various exit and entrance angles. The only difference in the results is the size of the exit angle. The longest burr was almost 100 times larger than the smallest. Although burr height is not the best measure of deburring difficulty, it does demonstrate the impact that part and process geometry have on deburring. Software is available that allows the manufacturing engineer to choose the optimum cutter position and size to minimize these problems (Stein et al. 1993).

The part shown in Figure 5-16 has five holes, four of which require countersinking or counterboring on the bottom surface. By drilling this part with the "to be countersunk" side

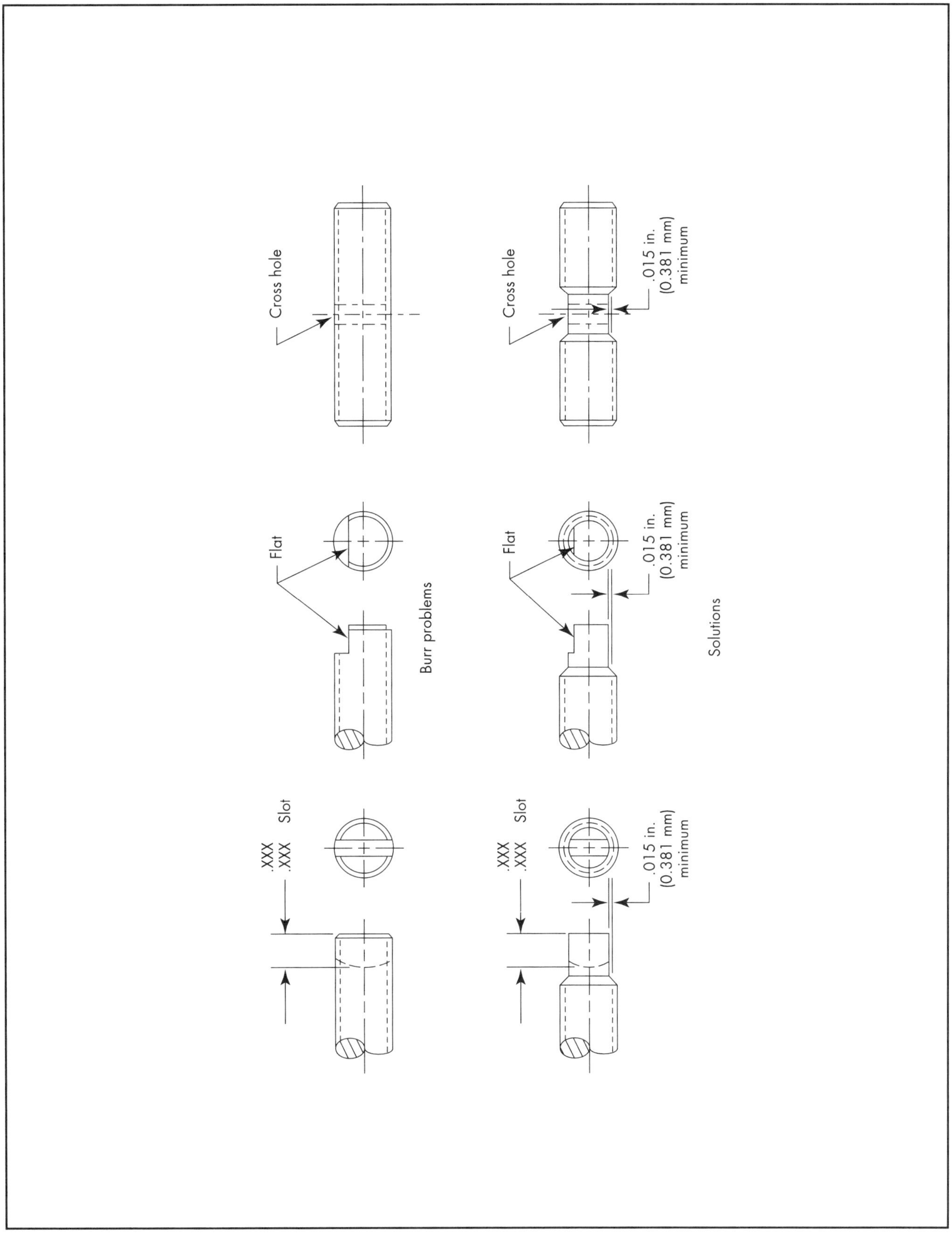

Figure 5-9. Methods of eliminating deburring problems when machining through threads (NSMPA 1983).

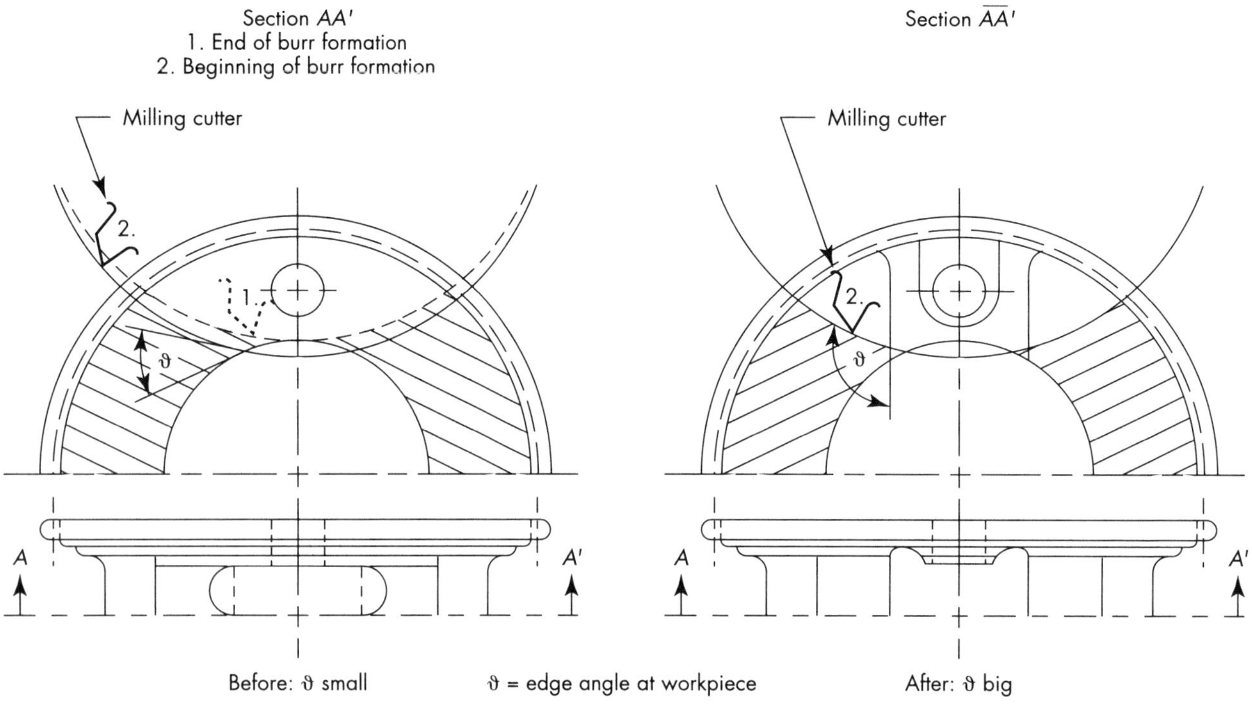

Figure 5-10. Casting before and after redesigning for burr minimization (Schafer 1975b).

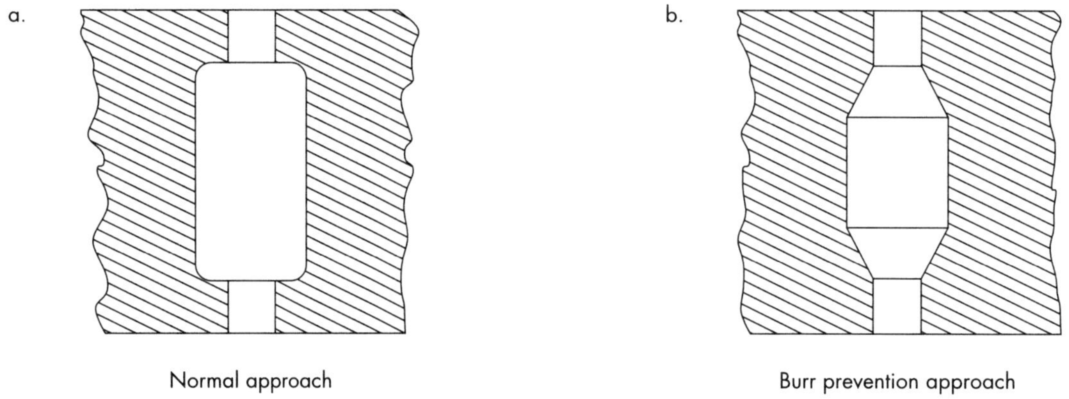

Figure 5-11. Casting design for preventing or minimizing drill burrs.

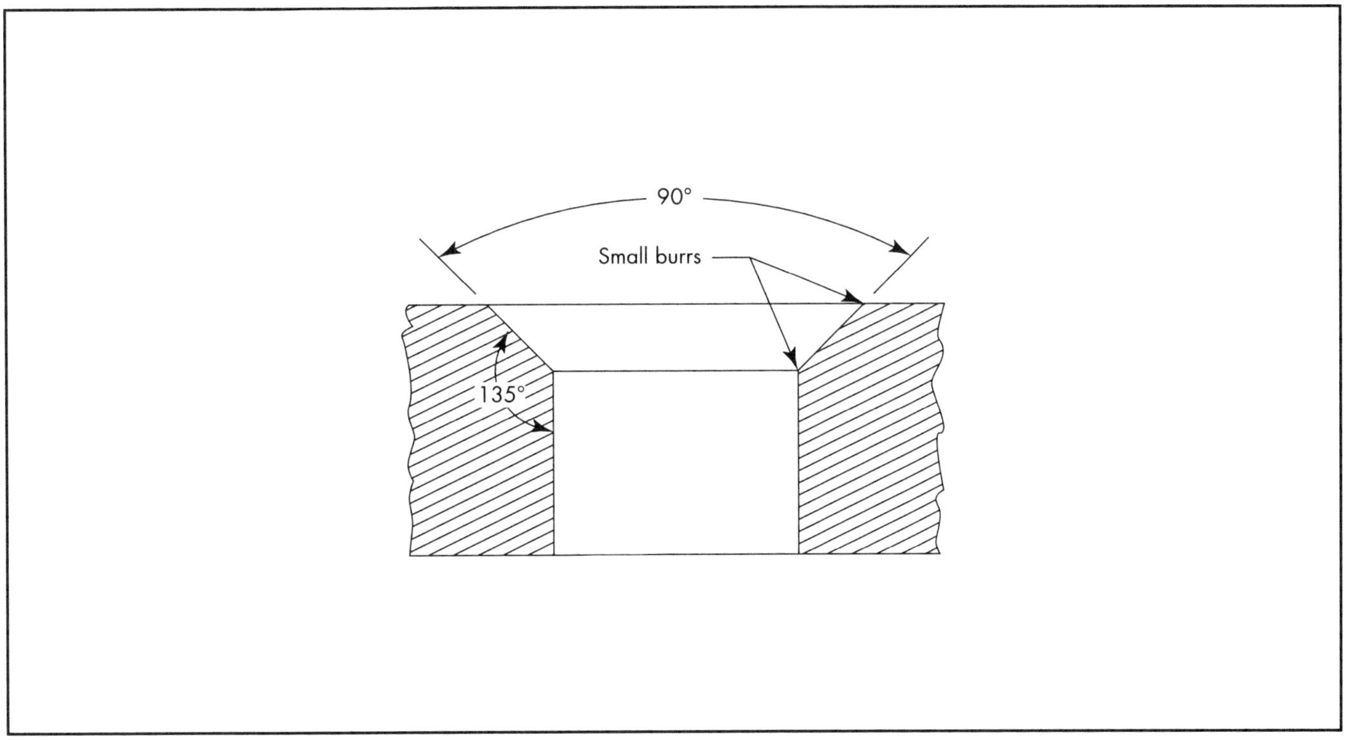

Figure 5-12. Geometry of countersunk holes.

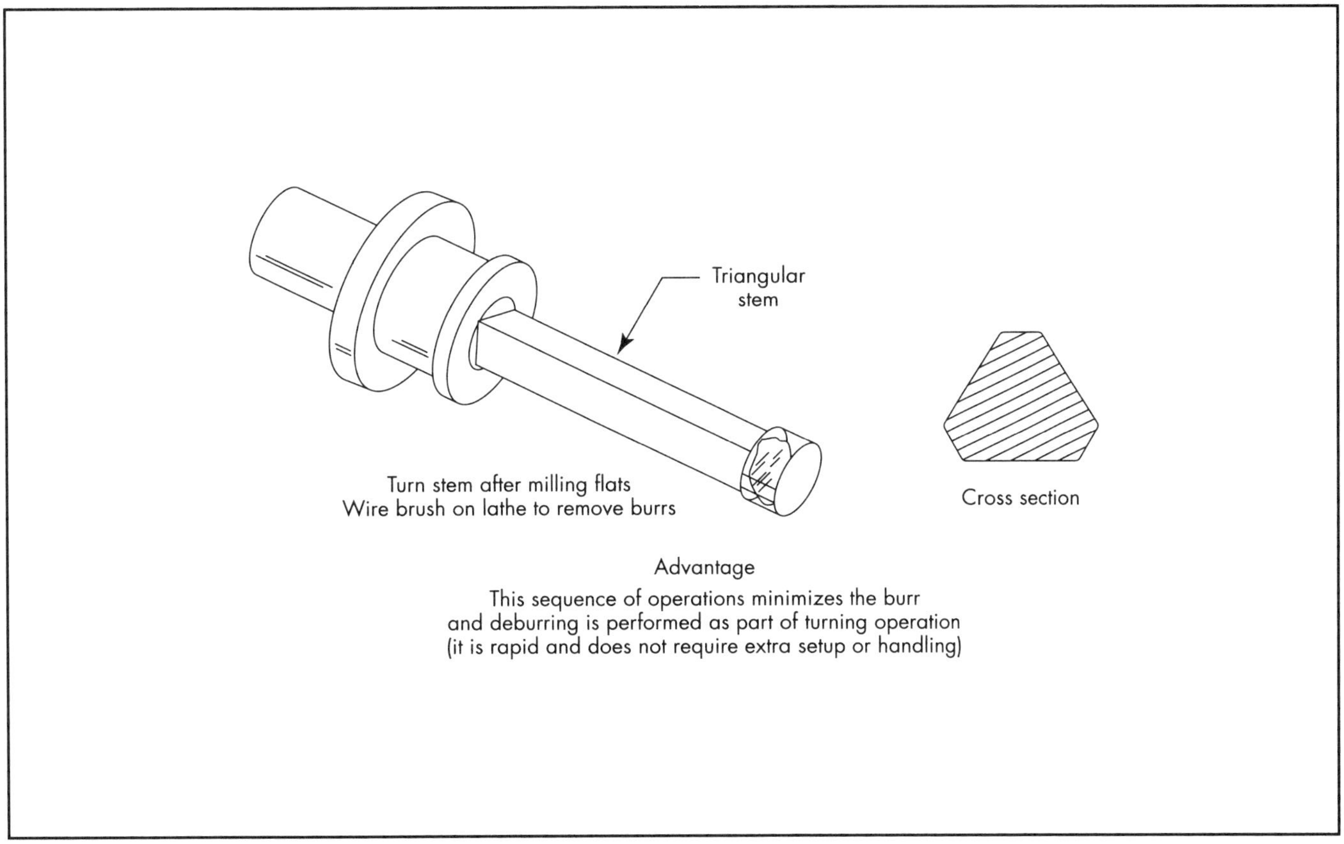

Figure 5-13. Process sequence affects deburring costs.

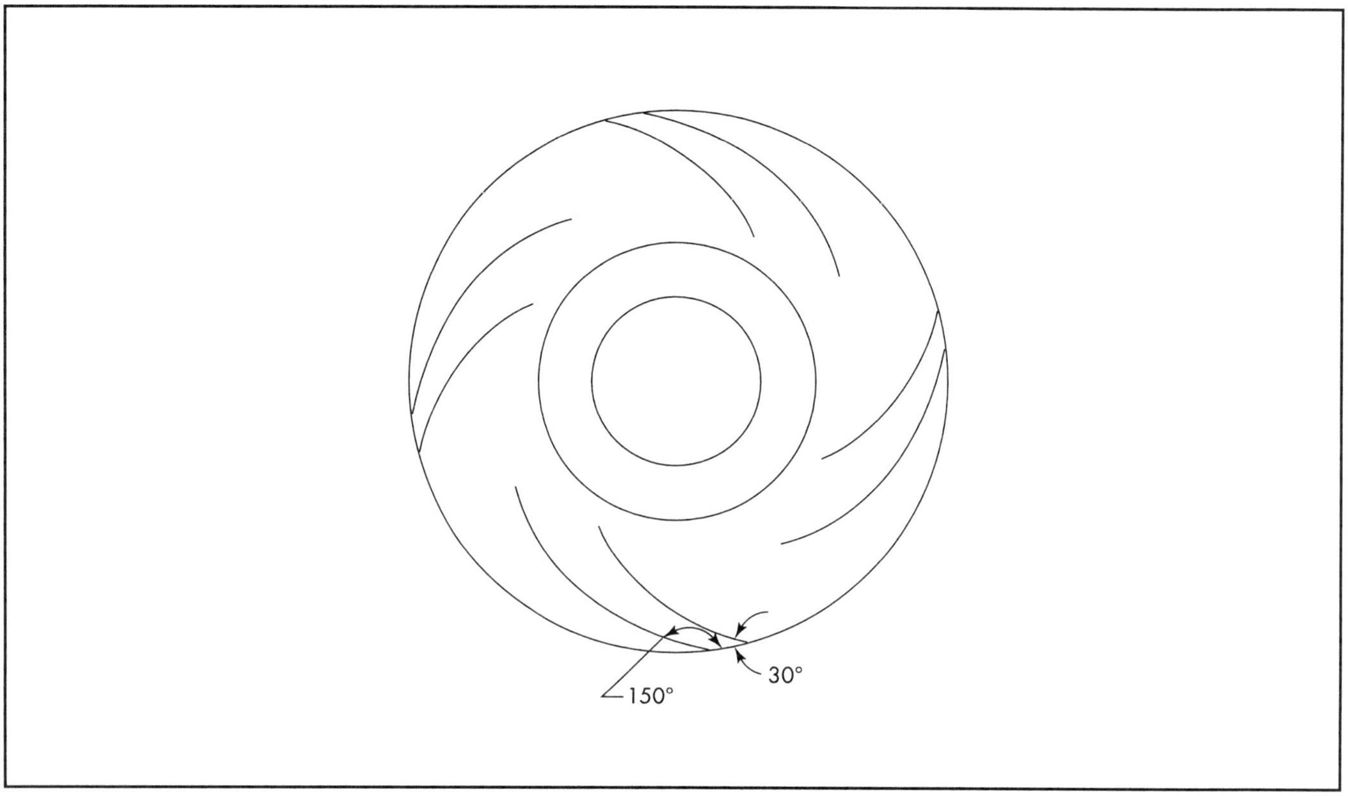

Figure 5-14. Angles on impeller blade.

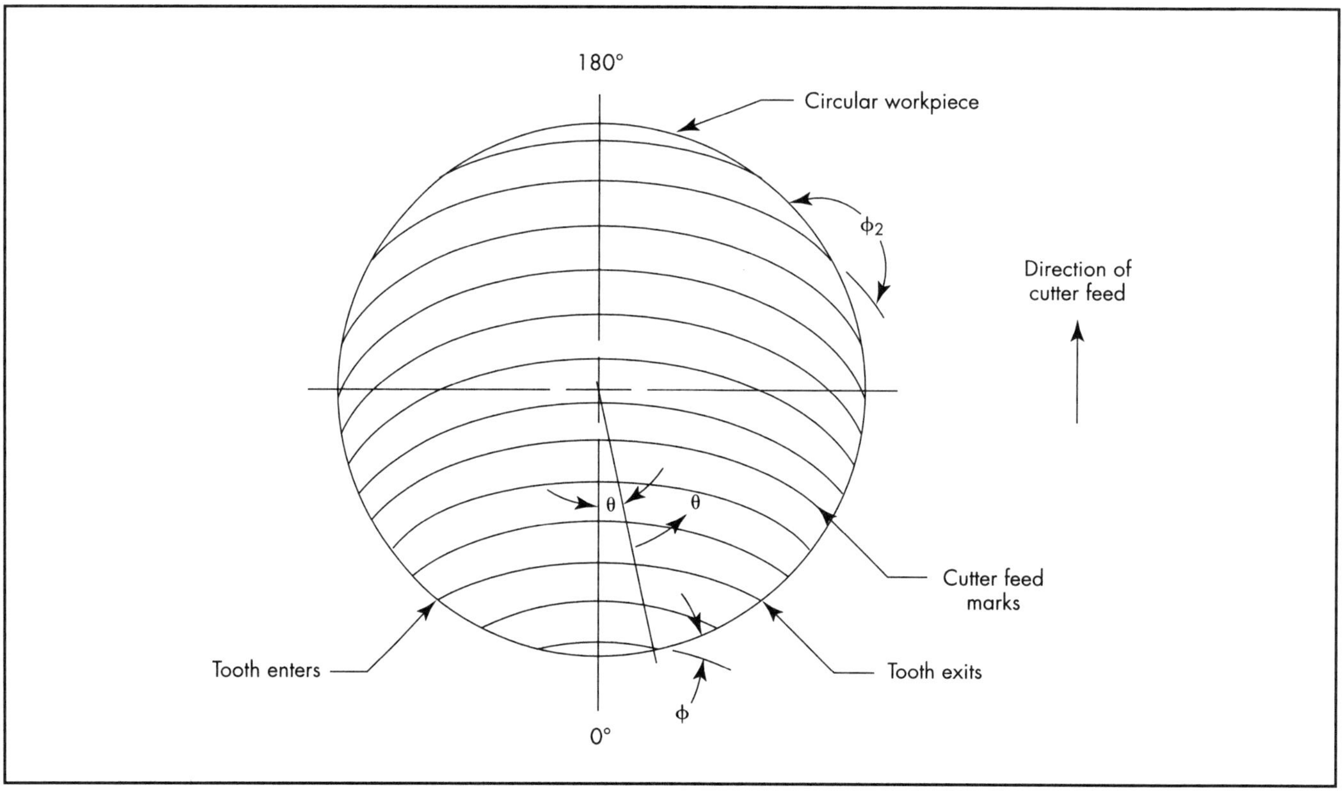

Figure 5-15. Exit angles in face milling.

Table 5-1. Effect of angle ϕ on burr height in face milling*

	Burr Height					
	303 Se (R_B 84)		15-5PH (R_C 37)		BeCu (R_B 96)	
Angle (°)	in.	μm	in.	μm	in.	μm
0	.0017	43.2	.0035	88.9	.0025	63.5
30	.0033	83.8	.0070	177.8	.0045	114.3
60	.0054	137.2	.0036	91.4	.0050	127.0
90	.0069	175.3	.0033	83.8	.0117	297.2
120	.0083	210.8	.0242	614.7	.0064	162.6
150	.0087	221.0	.0273	693.4	.0770	1955.8
180	.0372	944.9	.0215	546.1	.0953	2420.6
210	.0045	114.3	.1010	2565.4	.1328	3373.1
240	0	0	.0072	182.9	.0206	523.2
270	.0004	10.2	.0015	38.1	.0074	187.9
300	0	0	.0019	48.3	.0015	38.1
330	.0022	55.9	.0038	96.5	.0021	53.3

* See Figure 5-15 for definitions of angles—note that Θ and ϕ between 0 and 180° correspond to rollover burrs, and angles larger than 180° correspond to entrance burrs.

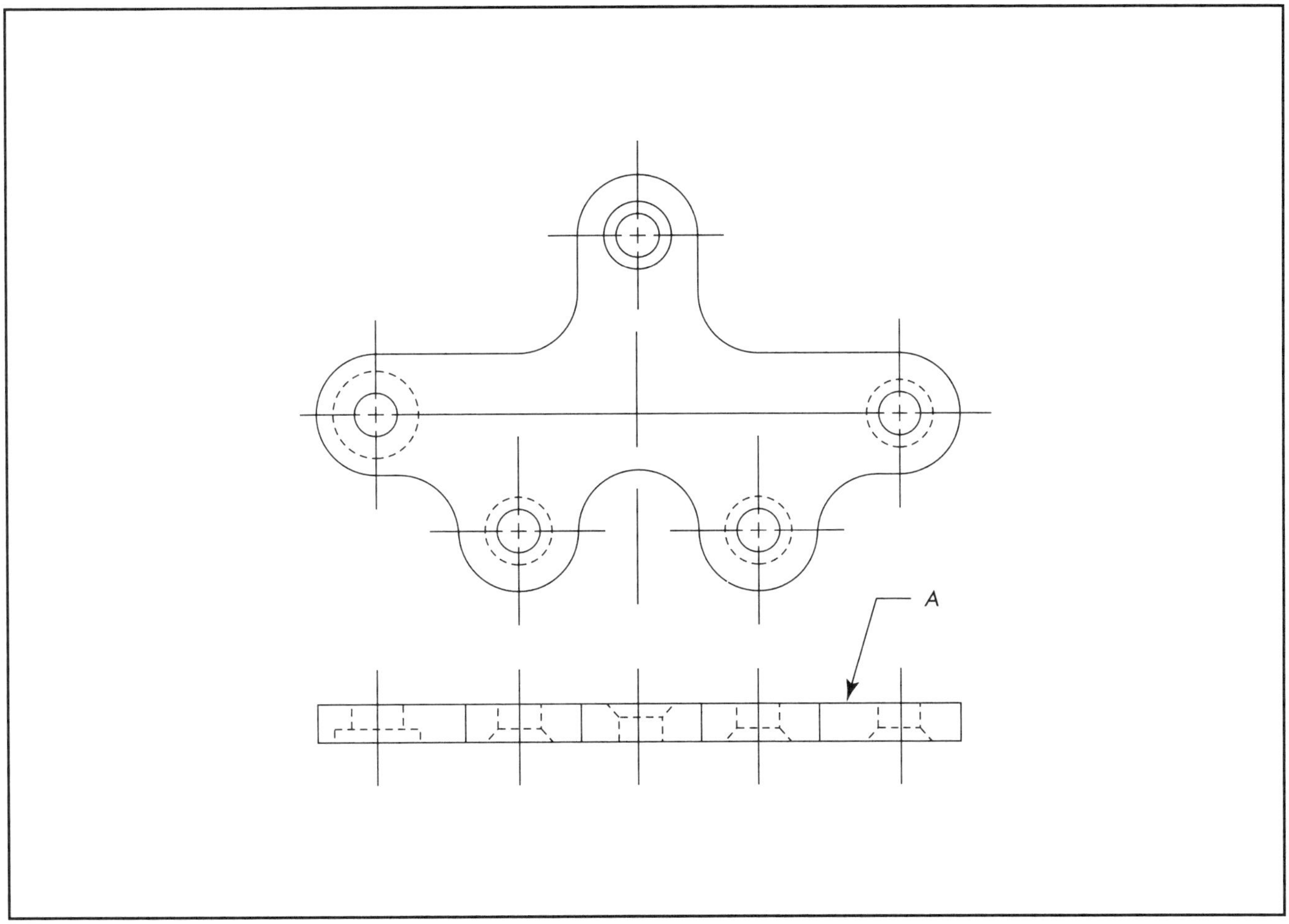

Figure 5-16. Part orientation for minimum deburring.

down, some deburring of heavy burrs can be minimized. The countersinking tool will remove these heavy burrs while countersinking. If the part is drilled with side "A" down, the heavy burr will have to be removed in a special deburring operation.

Figure 5-17 illustrates a 3 in. (76.2 mm) diameter hobbed gear having a large hub on the side facing down. The heavy exit burr is on the hub side, which violates the principles described earlier for best placement. Despite the poor design, the innovative manufacturing engineer designed a simple sanding block with a slot cutout that provides clearance for the hub. With only a few parts per month to make, this simple sanding approach, followed by brushing, eliminated hours of manually deburring the teeth.

Figure 5-18 provides an excellent example of burr placement. If the cutoff burr occurs on the large-diameter flat end of the part, burr removal by vibratory finishing is difficult. If the burr occurs on the rounded end, the burr is more exposed to the media action of vibratory deburring and can be readily removed, provided tolerances are adequate. Note that the size of the projection shown in Figure 5-18 is unrealistically large for good finishing, and some applications will require total burr removal, which may not be possible with this design. If the nose radius is an important operating feature that must be smooth and uninterrupted, then the cutoff projection should not be located on the nose.

On intersecting features, placing the burr on the most accessible edge can sometimes reduce burr costs by up to 50%. The correct placement of burrs can change a hand deburring situation to a much more economical and repeatable vibratory deburring operation. In some operations, burr placement can be controlled by a simple change in the program of an NC/CNC machine. In other cases, climb milling rather than conventional milling will provide better burr location.

When a manufacturing engineer considers the machining sequence to follow, the burr should be considered as significant a factor as the desired surface finish and tolerance. The following checklist contains the factors that need to be considered when designing manufacturing sequences:

1. Does the burr have to be removed?
2. Can the part be redesigned for less machining? for less deburring?
3. Will the burr be cut off in a subsequent machining operation? If the machining sequence were changed, would the burr be cut off?
4. Is the burr accessible?
5. Should the sequence of operations or the direction of the cut be changed?
6. Can a cutter be chosen that gives a smaller burr?
7. Will the feed rate selected form the smallest burr?
8. Will subsequent heat treating make the burr brittle?
9. Would a change in cutting fluid or method of application make the burr brittle?
10. Does the burr have to be removed now?

Several specific operating parameters for minimizing burrs are presented in Chapter 4. In summary:

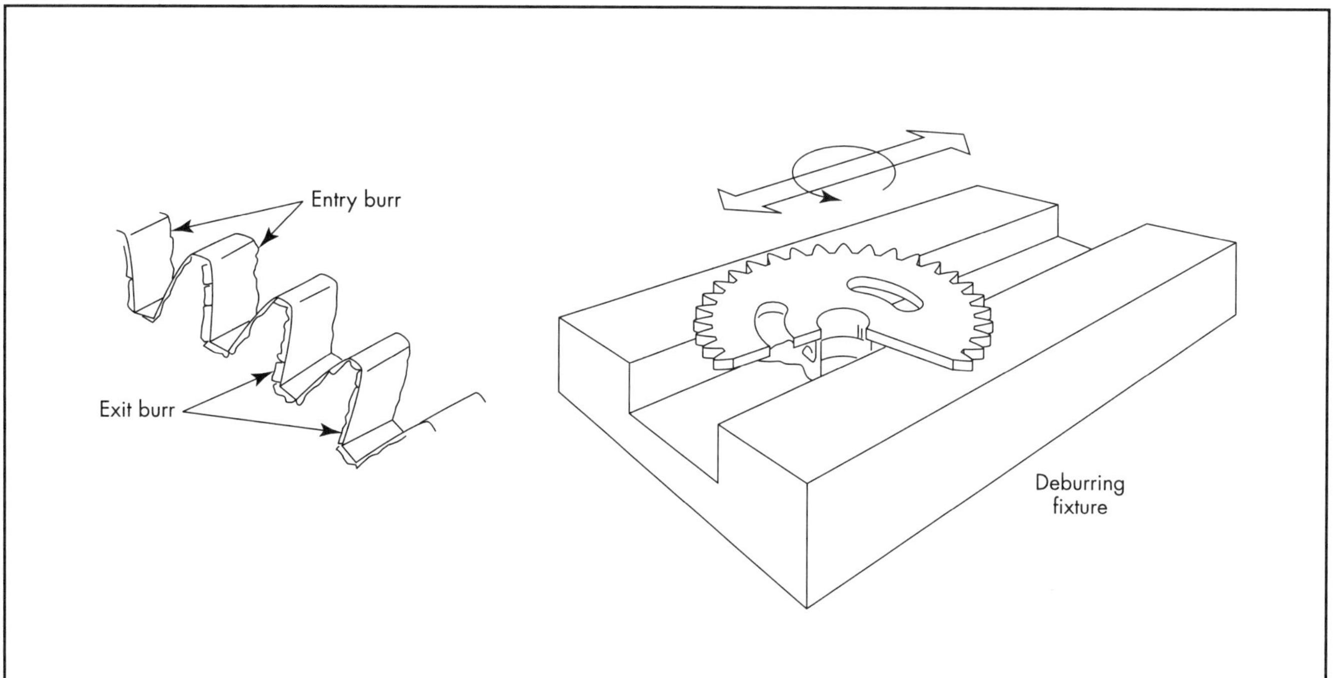

Figure 5-17. Sanding block for gears with hubs.

- Reduce cutting forces and the weight of the workpieces to reduce the diameters of the cutoff projections. Using sharp, narrow cutoff tools is also helpful.
- Remove built-up edges, increase rake angles, use sharp cutting edges, minimize flank wear, and use shallow cuts to minimize burrs, since these factors reduce cutting forces when cutting all materials.
- Simplify burr removal of blanked plastic parts by freezing the plastic stock before it is put into the press. Teflon, for example, is very hard to deburr, and it makes rather large burrs. Using cryogenic temperatures prior to blanking produces a near sharp edge, which can be rounded readily if necessary.

5.5 DESIGNING PARTS TO MAKE BURRS AND FLASH MORE ACCESSIBLE

Designing parts so that burrs, flash, or other protrusions are easily removed can result in major cost savings in finishing. Some simple approaches exist to make these design adjustments.

5.5.1 Burrs

Making a burr easy to remove involves locating the burr in the best position for easy removal. Such location should be established before the machining process and tooling are finalized.

When a rollover-type exit burr forms near a hub, removing it by mechanized means is difficult. If the rollover burr is located at the opposite end of the part (Figure 5-19), no projection will interfere with the deburring operation.

If the hobbing exit burr shown in Figure 5-20 is allowed to form on the hub side, precision deburring will be very time consuming. By fixturing the part so that the hob exits over the flat face of the gear, hobbing burrs can be removed easily by a quick hand-abrasive operation. Any small burrs remaining can then be removed and edges radiused by brushing or by one of the mass finishing processes.

Even when backup material is used in multiple-part hobbing, the cutter should exit from the back surface. Because the direction of feed or rotation of the cutter cannot be changed during the cutting of ratchet teeth and nongear shapes, burr location must be chosen before hobs are ordered. When milling L-shaped configurations (Figure 5-21), exit burrs should be located on the back and top, rather than under ledges. Figure 5-22 illustrates another example of this principle. Slotting this screw below the head causes another surface to have burrs, and their location, under the head, makes them difficult to remove.

Although tubular sieve-like parts normally are drilled or punched from the outside inward, placing the burrs on the outside may be more advantageous. Figure 5-23 illustrates a punch that pierces outward to put the heavy burrs on the exterior where they can be easily removed (by sanding). When wall thickness is adequate, locating the burrs on the inside may be desirable so that they can be removed by reaming.

Locating burrs for easy access and minimization becomes complicated when part geometry has many cutouts and pockets. Software has been developed to help simplify the design choices (Stein 1993).

5.5.2 Flash

Predicting the location and size of flash from die casting and molding processes is easier than for burrs produced in machining. This is because flash corresponds to die configuration rather than to the path of a cutting tool. As a result, designing to minimize flash has been widely practiced for a number of years.

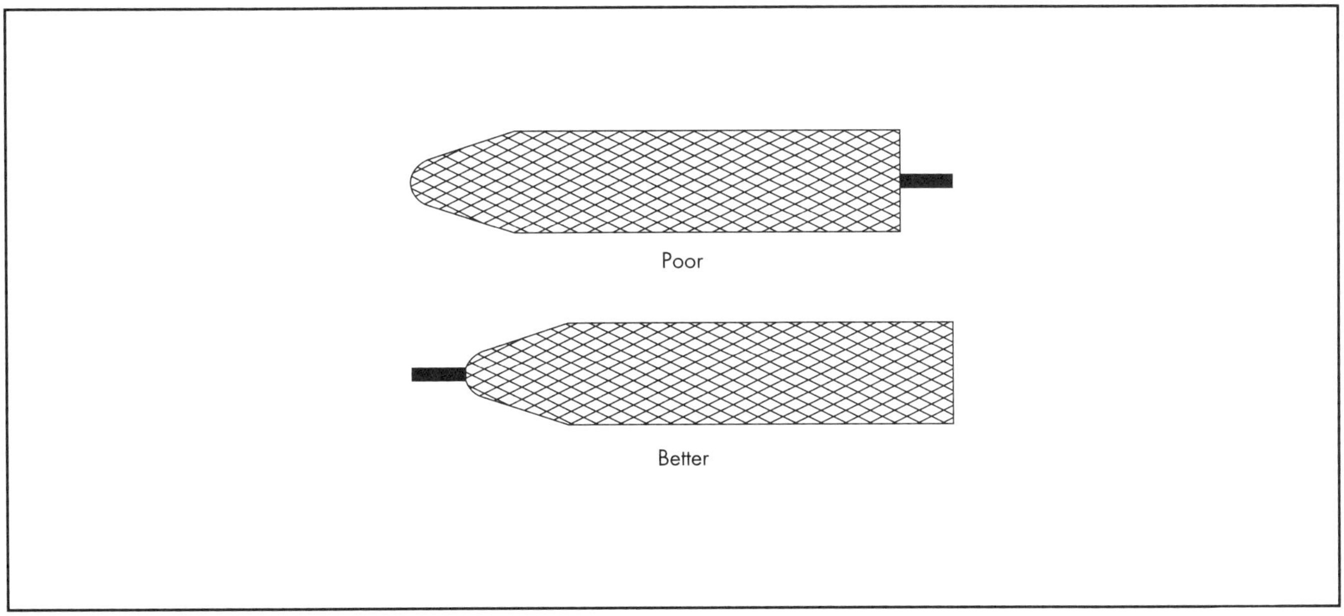

Figure 5-18. Placing the cutoff burr on the rounded end facilitates removal by vibratory finishing (Kramer 1971).

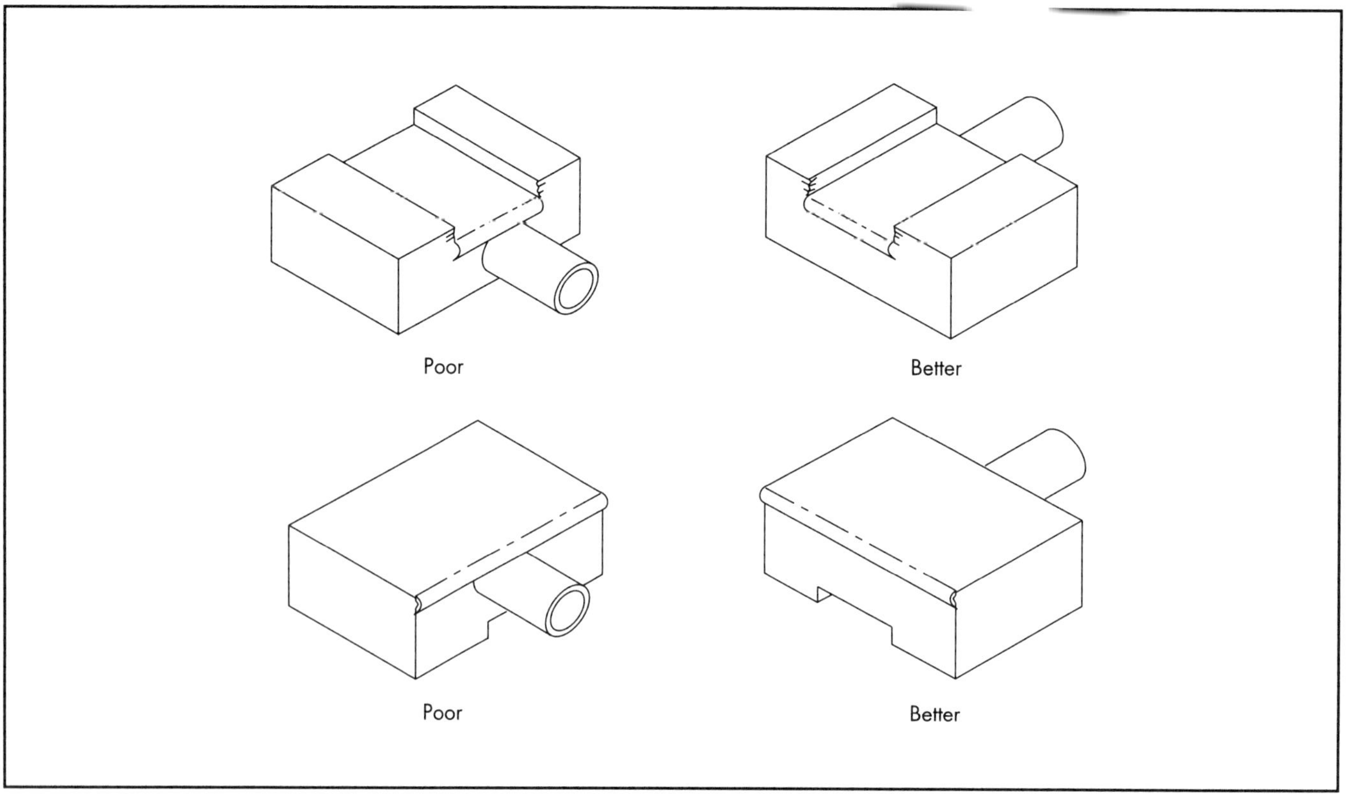

Figure 5-19. Locations of rollover-type exit burrs affect finishing costs.

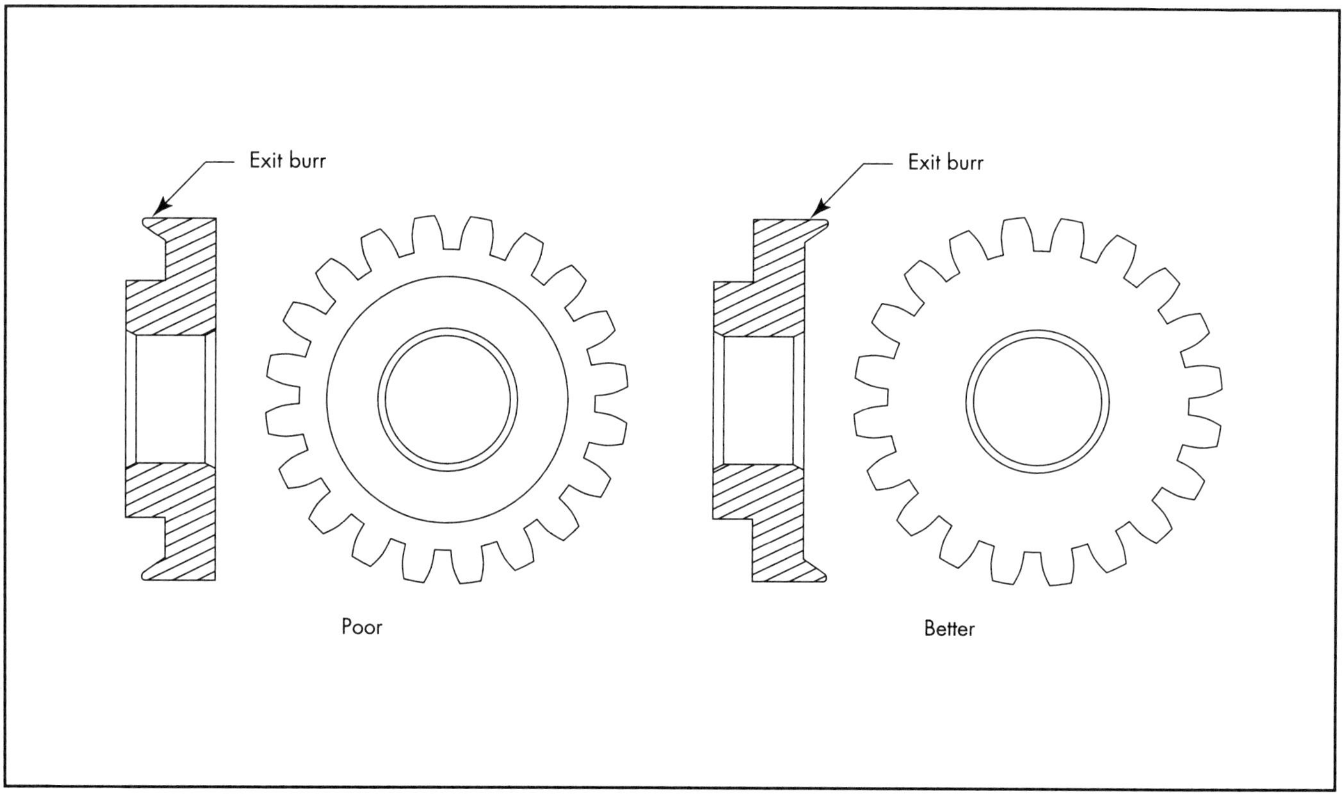

Figure 5-20. Location of exit burr from hobbing affects deburring time.

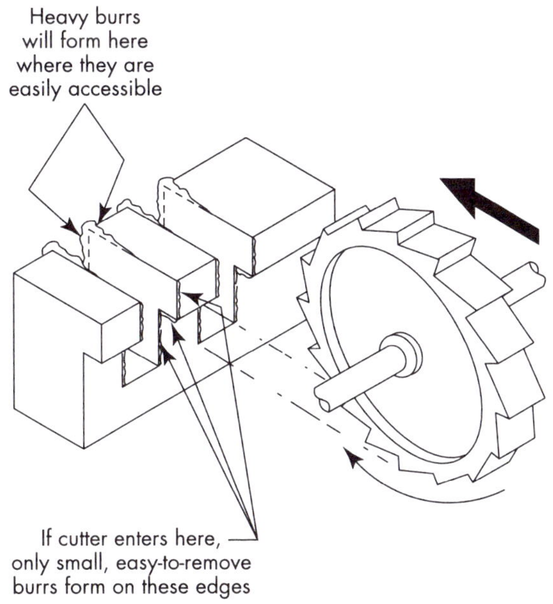

Figure 5-21. Heavy exit burrs should be located on back and top when milling L-shaped workpieces.

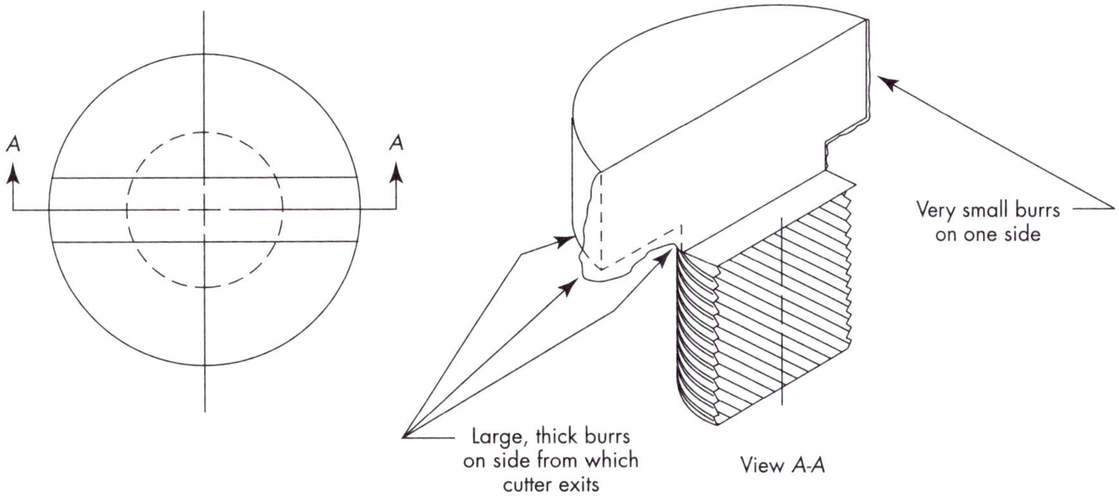

Figure 5-22. Slotting through flanges makes deburring difficult.

An aspect of molded parts that simplifies deflashing is that trimming equipment generally conforms to the part contour. Trimming attacks the flash and not the entire workpiece surface, as is the case in many deburring processes. If flash cannot be prevented at a nonworking (aesthetic) surface, it should be on a rolled surface rather than on a parting plane edge (Figure 5-24). This procedure allows trimming to be used, and any mold offset is not visible after trimming.

When surfaces must be ground to remove caps of flash around holes, any projecting features should be designed with recesses to provide an unobstructed grinding path (Figure 5-25). Although piercing dies can be used to remove these caps, their use can result in tearing of the edges.

Die and mold designs that result in rolls are depicted in Figure 5-26. The labels "poor design" and "good design" are only appropriate for aesthetic conditions. In many functional applications, the design shown in Figure 5-26a is the most applicable. Design considerations, such as rolls and unobstructed surfaces, are major factors in much of the aluminum and zinc die casting and plastics molding industries. This is because deflashing costs represent 20–35% of the total manufacturing cost. The design of gating systems, ejector pins, and parting lines must consider deflashing approaches. Overflow wells, for example, should be designed to allow the use of strong trimming punches (Figure 5-27).

Gates also should be designed to facilitate the removal of flash. In addition, gates should always have a shape that ensures that the fracture of the gate occurs at the edge of the part (Figure 5-28), and they should be as thin as possible to produce a clean fracture (Figure 5-29). Reportedly, with a shallow gate, the fracture is almost straight and follows the vertical face of the component. A wider gate breaks on an in sweeping curve, which finishes at a point a few thousandths of an inch inside the correct line. Feathered edges at the ends of threads should be avoided because they make mold fit more critical and promote flashing. This same concept applies to flash that is to be sheared from the molded part.

5.6 DESIGNING TO ALLOW MORE DEBURRING PROCESS CHOICES

Some part designs prevent the use of economical deburring. Examples of designs that allow more process choices are presented in the following sections. These approaches take advantage of simple geometry improvements and consider the impact of tolerances and surface finish on deburring choices.

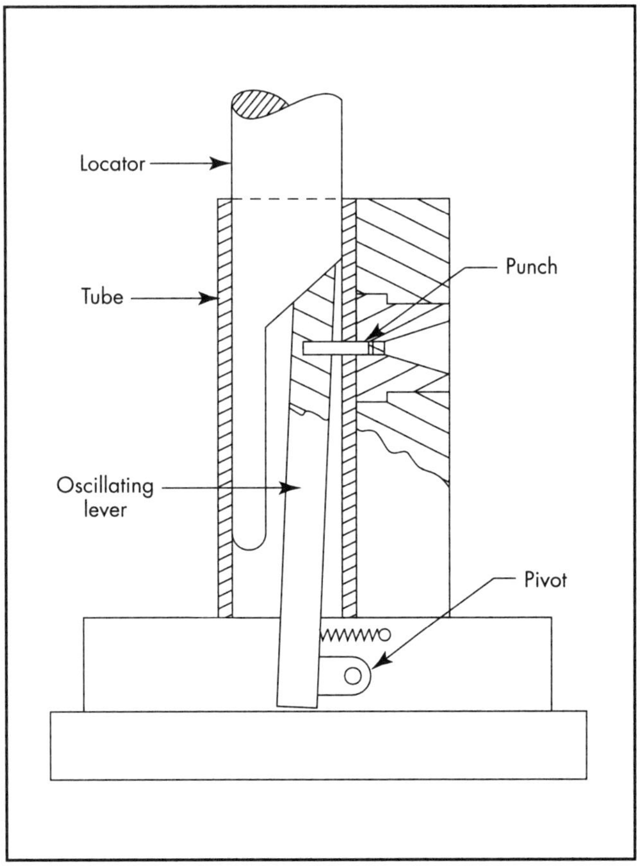

Figure 5-23. Piercing burrs placed on exterior of tubular elements.

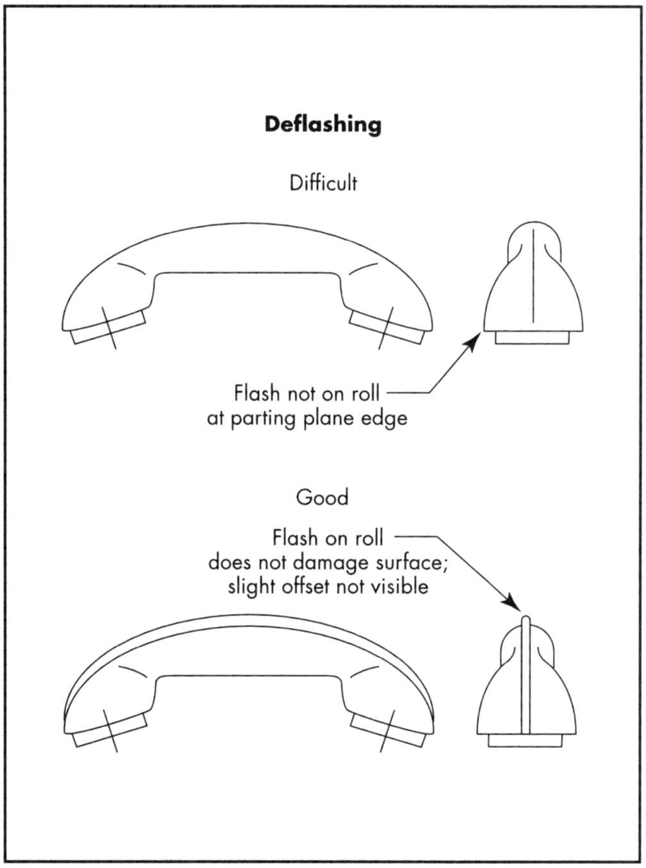

Figure 5-24. Design for aesthetic conditions and easy flash removal (Schafer 1975a).

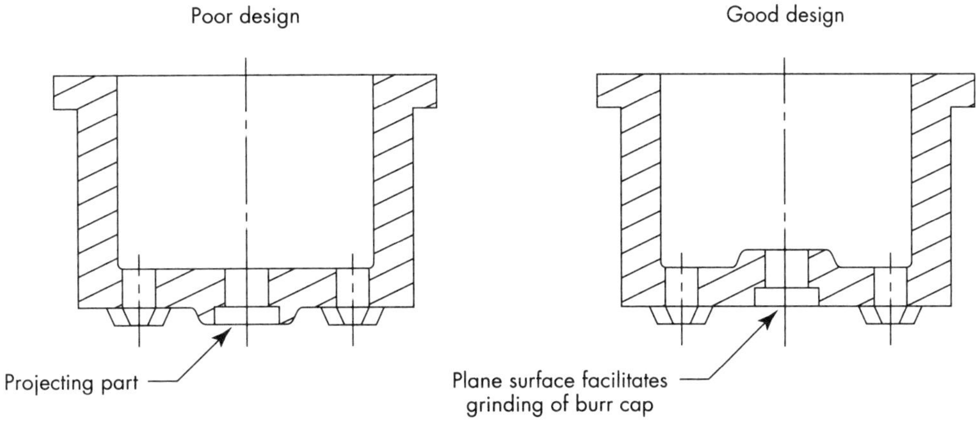

Figure 5-25. Design for ease of machining after molding (Schafer 1975b).

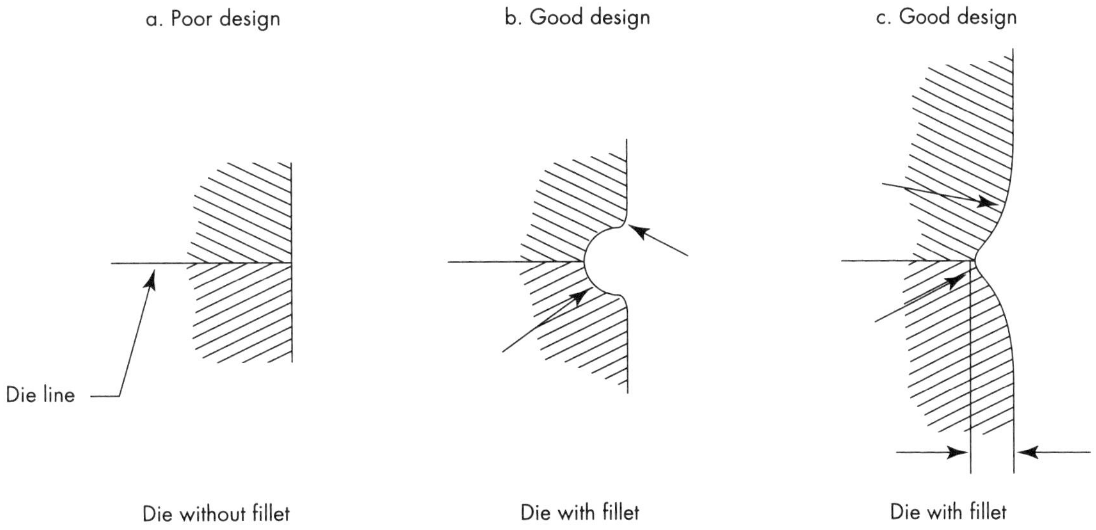

Figure 5-26. Die and mold designs that result in rolls (Schafer 1975b).

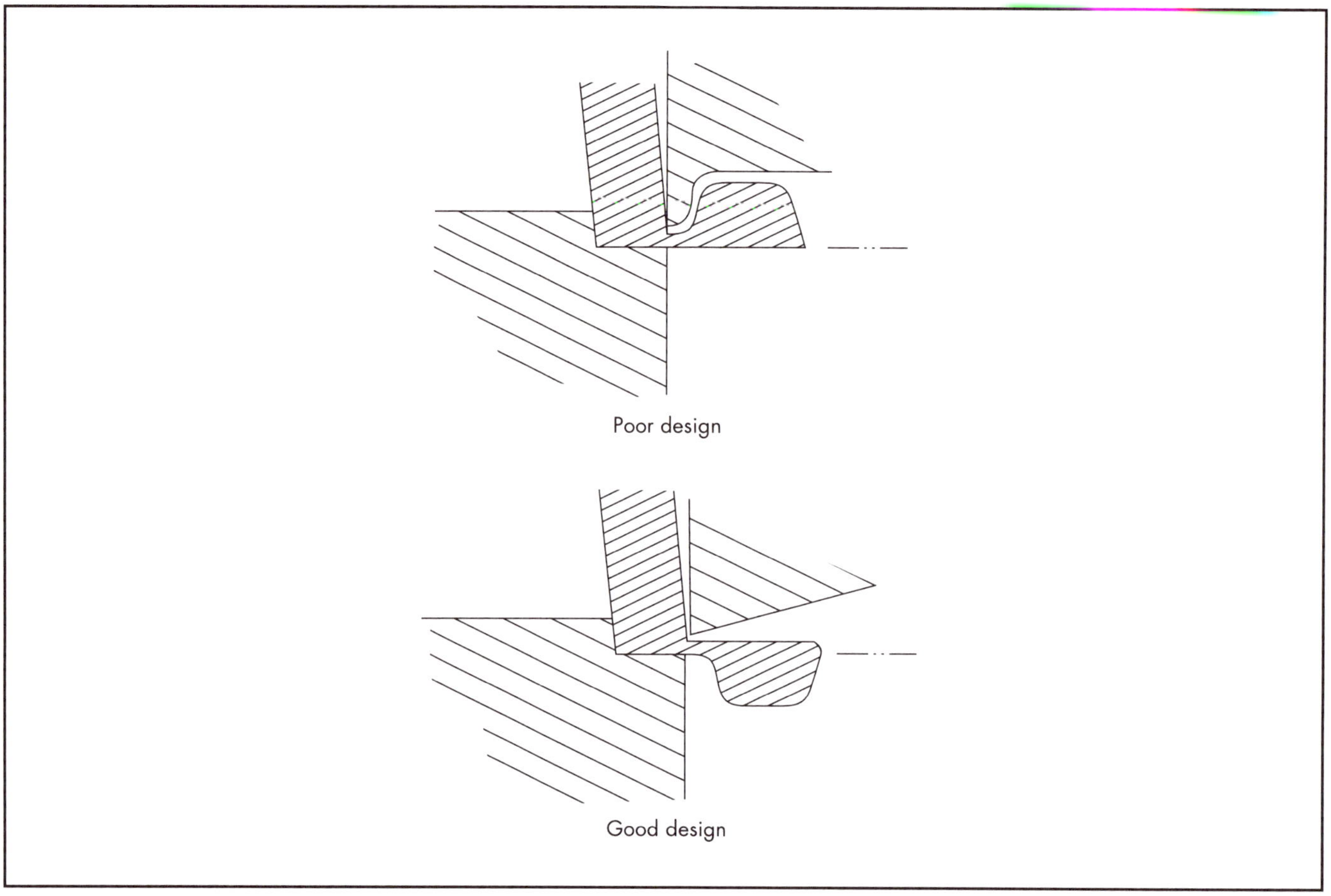

Figure 5-27. Design of overflow wells for easy trimming and long tool life (Schafer 1975b).

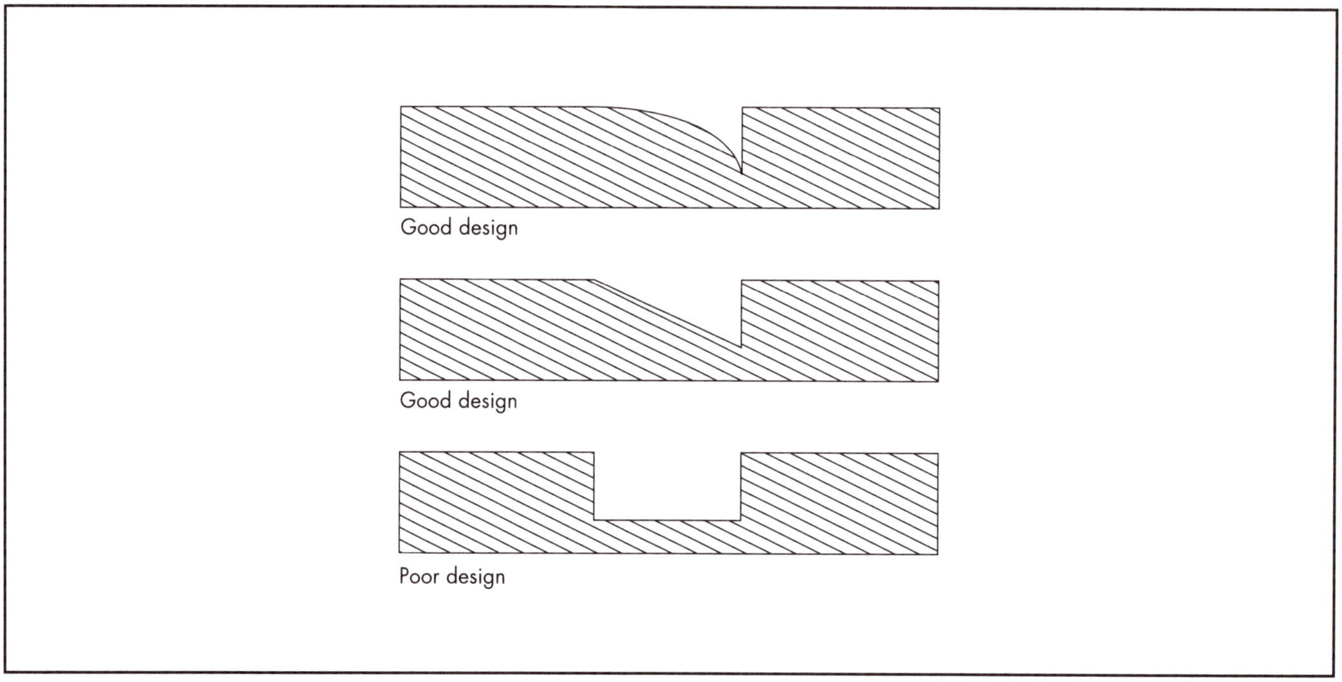

Figure 5-28. Gate design (Kramer 1971).

Figure 5-29. Effect of gate thickness on edge quality (Barton and Barton 1957).

5.6.1 Designing Parts for Mass Finishing

Significantly, at least 50% of all parts produced in the world are subjected to mass finishing processes. This is important because mass finishing costs are among the lowest of any processes. In fact, mass finishing can cost as little as a fraction of a cent per part in some circumstances, whereas hand finishing can cost up to $5 a part for hand-size parts. Mass finishing processes will be discussed in detail in Chapter 6. Because these processes all use the same basic metal removal mechanism, there are a few common considerations for workpiece design. Essential factors in all mass finishing processes include the following:

1. The tumbling media used must have access to and be able to slide over all critical surfaces and edges that require deburring.
2. Large tumbling media deburrs faster than small media, except on very small parts. (Media used for mass finishing is discussed in Chapter 6.)
3. The use of special preformed media shapes, such as triangles, cones, and cylinders, has advantages in removing burrs and flash over gravel-like media.

In Figure 5-30a, the step height *H*, is smaller than the radius R_s on the deburring media, which in this case is triangular in shape. As a result, the media will often bend the burr over rather than remove it. Complete removal can be obtained only with long cycle times or by using blends of media with smaller sizes present. If the designer specifies a larger step height, as shown in Figure 5-30b, this problem is eliminated because the media can work over the edge and against both surfaces of the workpiece. For deburring slots with triangular media, the slot width *b* (shown in Figure 5-31) must be greater than $3R_s$, and the depth must be greater than R_s.

When inner surfaces near fillet radii must be smoothed by mass finishing processes, it is essential that the fillet radii specified be equal to or exceed the major radius on the media (Figure 5-32). This is a particularly significant consideration when the feature is an extruded item, which is controlled by the die shape. Small steps in a workpiece will not receive significant deburring action, even with small media or balls, because size and geometry prohibit contact. Other media shapes, however, usually overcome these problems.

For simple parts, when the same edge radius and deburring quality is required on all edges of a part, smaller media sizes are required. This is because radiusing action is greatest where larger media can most readily contact the part. Although such differences are not critical on most parts, they are on many others. Complete deburring can occur only when the burr is totally accessible to the media.

Each of these factors should be considered while parts are in the design phase. Although smaller deburring media or different media shapes can be used, they create other problems, such as incomplete deburring, long finishing times, and media jammed in holes, slots, and other cavities. Whenever possible, workpiece drawings should be reviewed by manufacturing engineers responsible for machining and deburring before the drawings are finalized.

Figure 5-33 illustrates another significant aspect of angles. The amount of radius that can be produced economically by vibratory finishing, after removing the burrs, is a function of the angle between intersecting surfaces. Large radii can be produced relatively quickly when the included angle is large. It will be seen in Chapter 6 that finishing times are 20 times longer when the included edge angle is 30° than when it is 120° (Figure 6-2). Precision edge radius tolerances are harder to maintain, however, when large angles are present. When a component has features involving several different edge angles, edge radii will vary significantly. Designers must recognize this variation when assigning tolerances to edge radii if they want to eliminate the extra costs required to produce equal radii. With close edge tolerances and such wide variation in angles, processes more expensive than mass finishing must be employed.

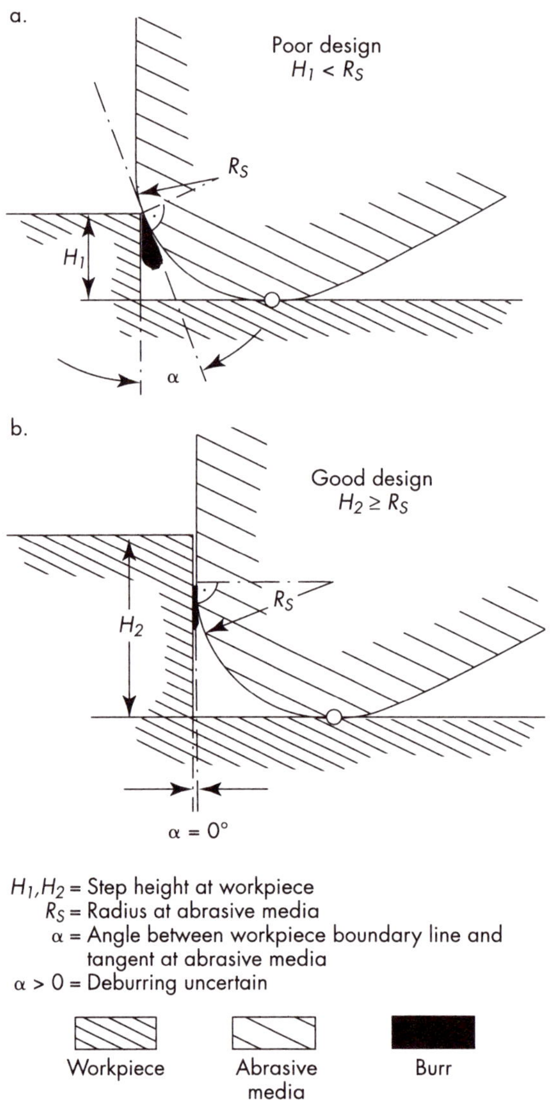

Figure 5-30. Complete deburring requires that angle α is 0° (Schafer 1975b).

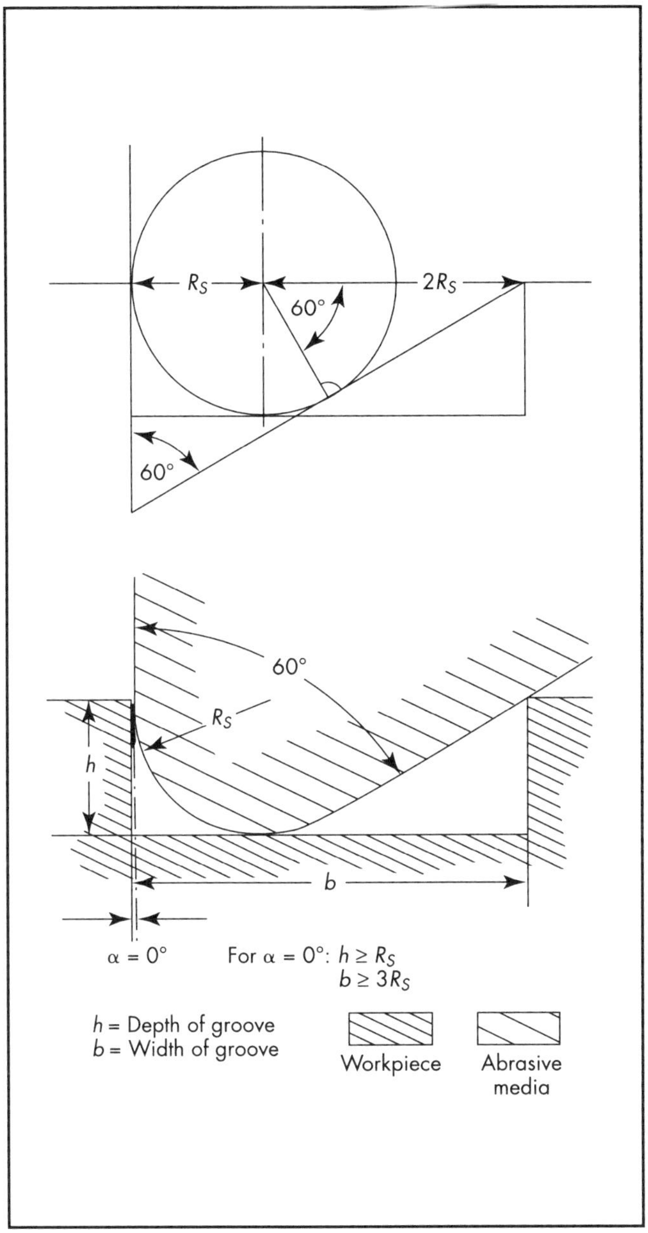

Figure 5-31. For deburring slots with triangular media, slot width, *b*, must be greater than $3R_s$, and the depth must be greater than R_s (Schafer 1975b).

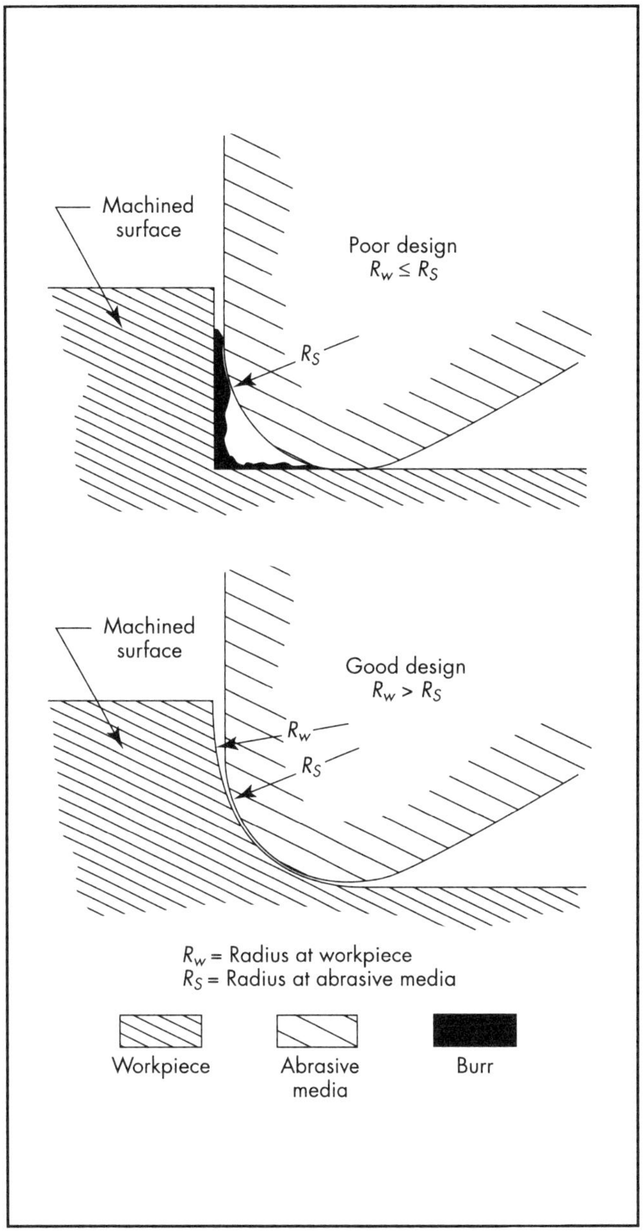

Figure 5-32. Deburring action at fillets (Schafer 1975b).

Figure 5-34 illustrates the effect of mold design on O-ring flash. Again, good designs save hundreds of hours of work and tens of thousands of dollars.

5.6.2 Designing for Tolerance, Surface Finish, and Aesthetics to Accommodate Deburring

Chapter 1 presented an overview of considerations in deburring processes, including the need to understand how precision surface finishes and tolerances on exposed edges limit the number of processes that can be used for deburring. Some burr sizes on precision parts can only be removed by manual deburring or specially designed machines. The data shown in the tables in Chapter 1 are not widely known, but are critical to process selection and part cost.

There are no precise instructions for how to best design tolerances, surface finish, or aesthetics for optimum deburring. The basic concepts discussed in Chapter 1 include the need to use the widest possible part tolerances and general surface finishes as opposed to precision finishes to allow a wider selection of deburring processes.

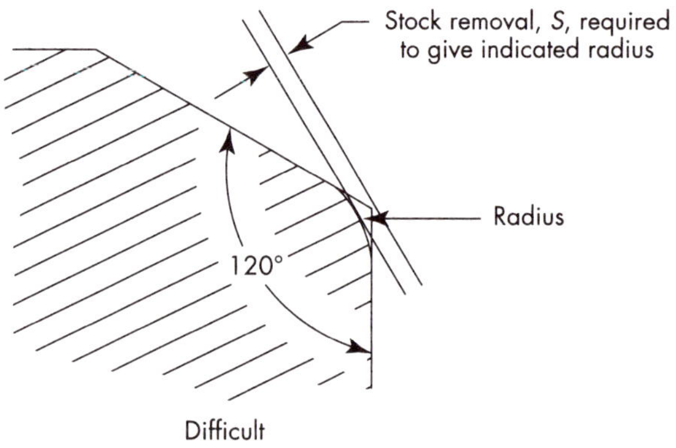

Radius in.	Radius µm	S in.	S µm
.002	50.8	.0003	7.6
.003	76.2	.0004	10.2
.005	127.0	.0007	17.8
.010	254.0	.0015	38.1
.015	381.0	.0022	55.9

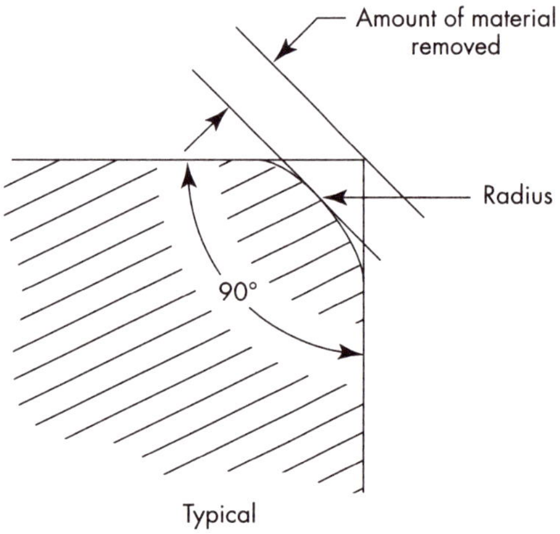

Radius in.	Radius µm	S in.	S µm
.002	50.8	.0008	20.3
.003	76.2	.0012	30.5
.005	127.0	.0021	53.3
.010	254.0	.0041	104.1
.015	381.0	.0062	157.5

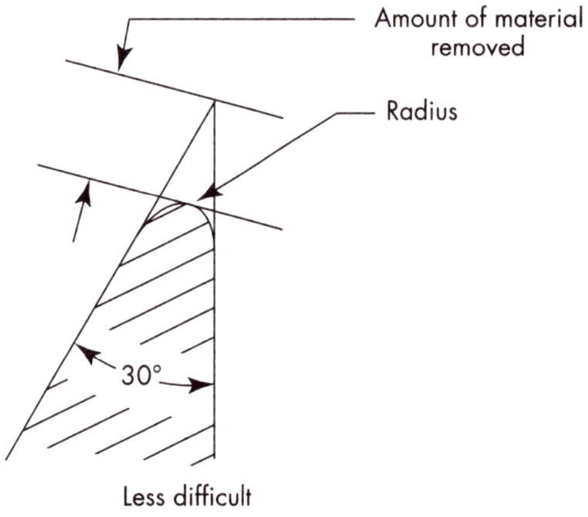

Radius in.	Radius µm	S in.	S µm
.002	50.8	.0077	195.6
.003	76.2	.0116	294.6
.005	127.0	.0194	492.8
.010	254.0	.0387	983.0
.015	381.0	.0581	1475.7

Figure 5-33. Effect of edge angle on edge radiusing.

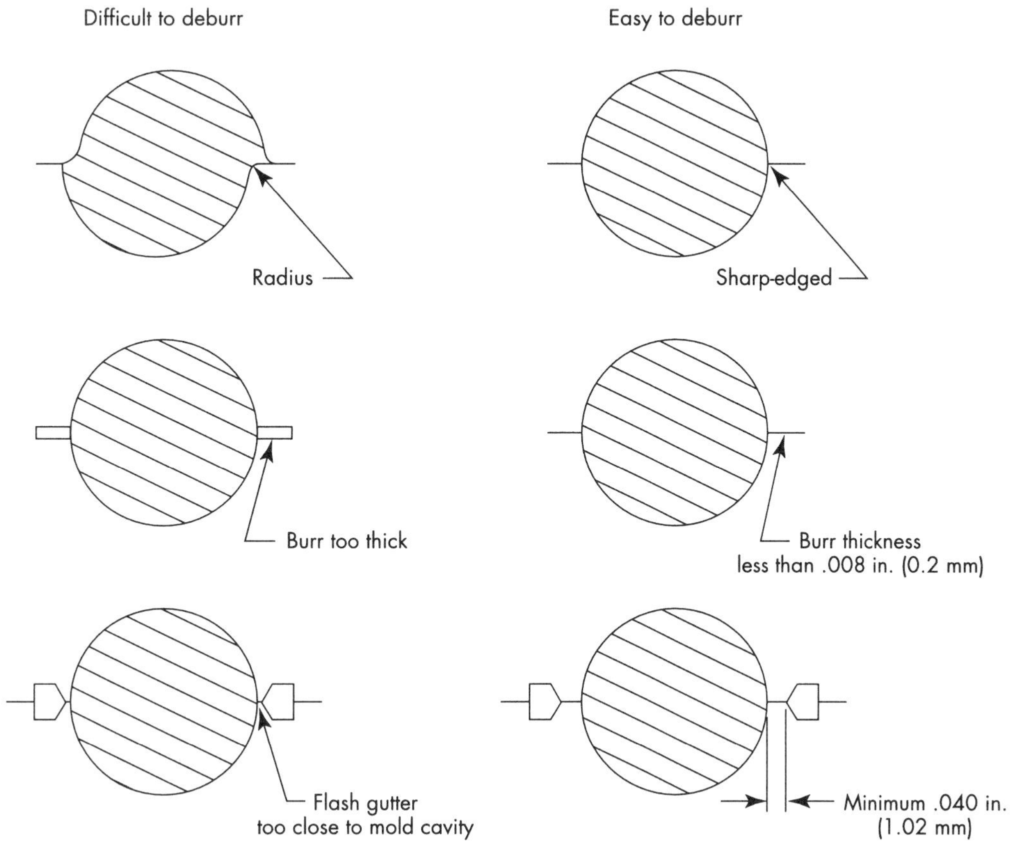

Figure 5-34. Design considerations for successful deflashing of compression-molded rubber components (Schafer 1975b).

Aesthetics are also an as yet unresearched area in deburring. In general, the deburring processes tend to smooth surfaces and round edges, which result in generally pleasing surfaces.

5.7 DESIGNING USING CLEAR NOTES OR STANDARDS FOR EDGE QUALITY

The preferred practice for specifying edge quality, with allowable burr sizes described, is illustrated in Figure 5-35. Chamfering produces a generally smaller burr than the burrs produced by other processes and may be the only deburring required. Either drawing notes or an in-plant standard should be used to indicate whether chamfering represents adequate deburring. When a smooth blend is required, the edge should be specified with a radius.

Edge breaks (chamfers) should be so specified that either a chamfered or a radiused condition is allowable. This specification permits the manufacturing engineer to determine whether a machining or a deburring process will provide the most economical edge condition. Typical corner breaks are .015 in. (0.38 mm) × 45°, or .010–.015 in. (254–381 μm) × 45°. Radii should not normally be specified larger than .010 in. (254 μm) or smaller than .003 in. (76 μm), unless required by function or stress concentrations.

The direction a burr faces is sometimes more critical than its actual size. In these cases, the orientation of the burr should be noted on the part drawing. With symmetrical threaded parts, it is helpful to the manufacturer if the designer indicates which end of the part the thread is started from, eliminating the need to deburr both ends.

Note that a burr always forms at the intersection of two holes. When a burr cannot be tolerated in one hole but can in the other, it should be noted on the part drawing (Figure 5-36). Defining where burrs can exist on formed parts may eliminate the need to deburr the sheet stock. With proper thought and communication between product designer, tool designer, and manufacturing engineer, forming dies can be designed so that burrs on the blank will be in an out-of-the-way location in the finished part.

On many parts, the only significant edge requirement is that all sharp edges be removed. In such cases, beating over burrs and dulling edges is adequate. Designers can communicate this requirement in at least two ways:

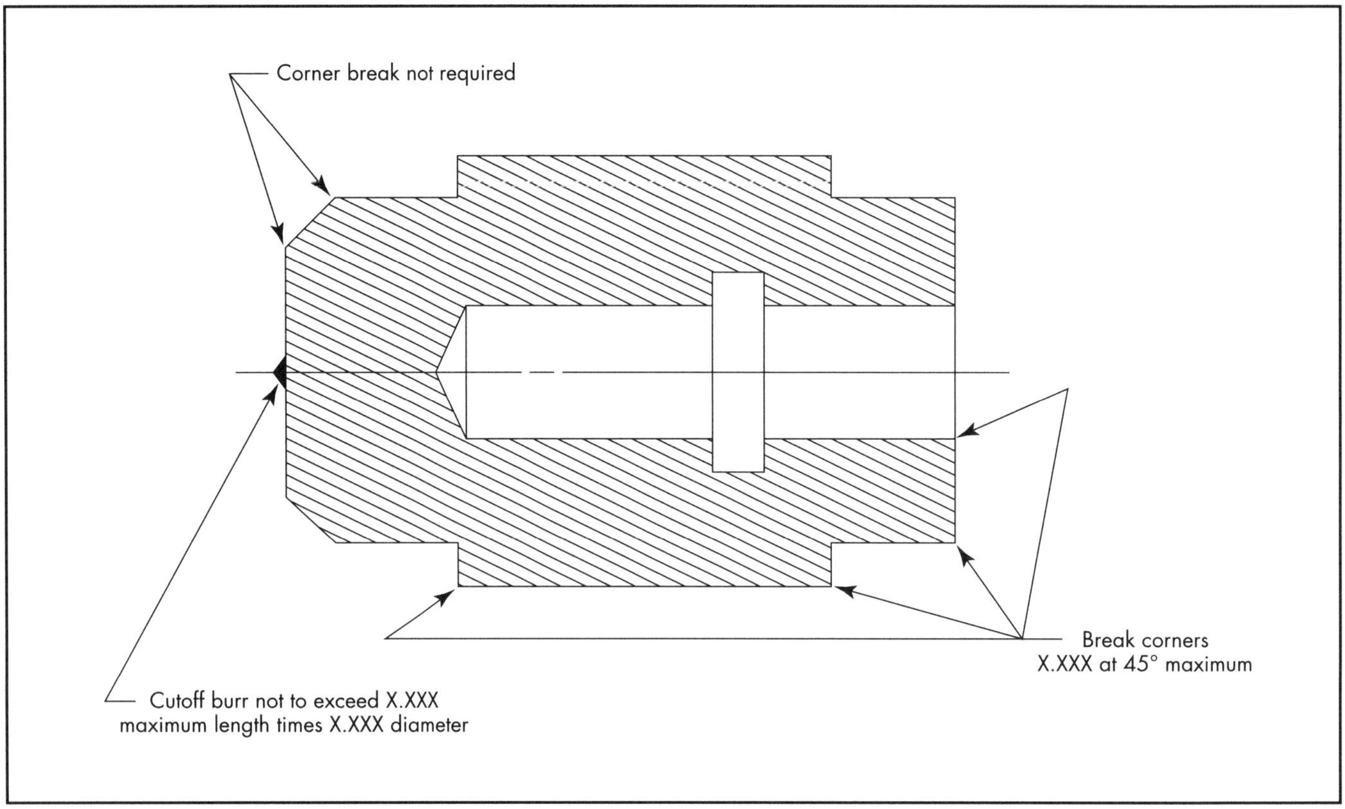

Figure 5-35. Typical burr notes for external edges.

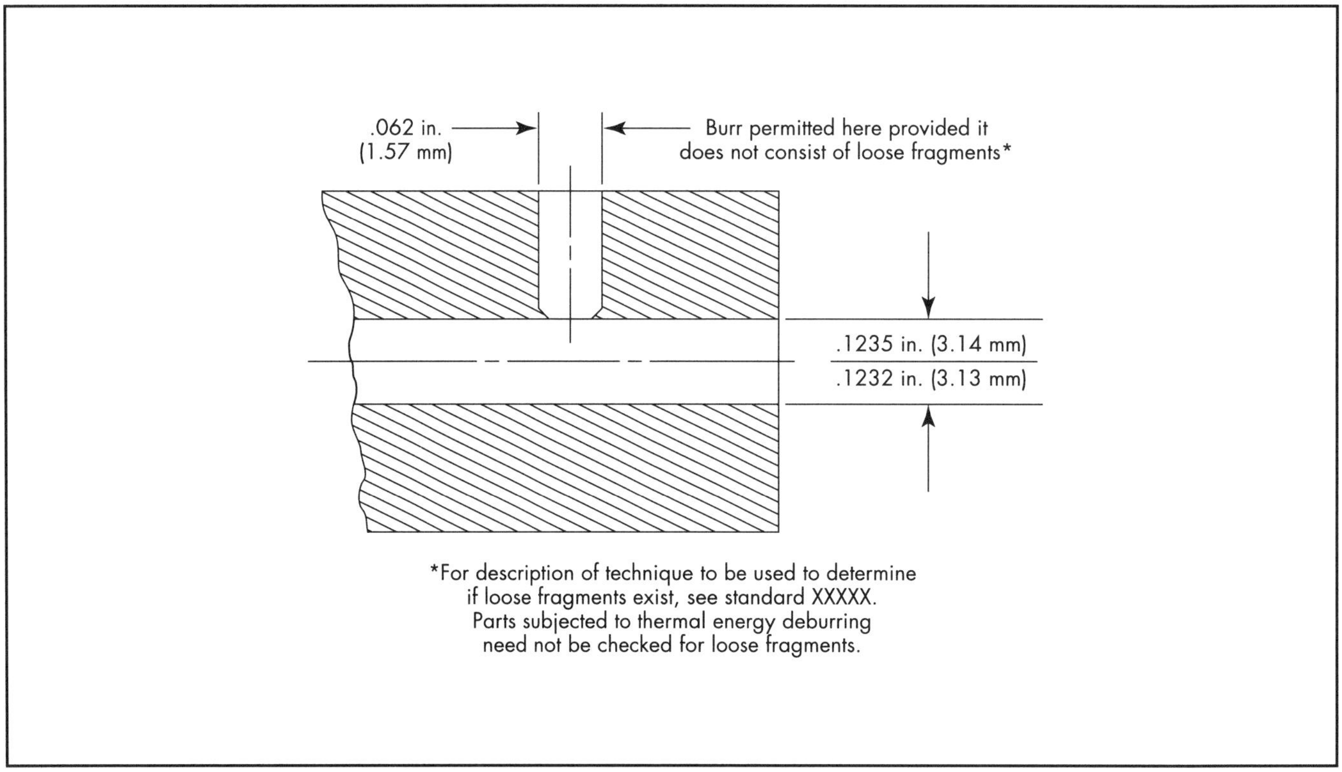

Figure 5-36. Allowable burr and its location should be noted on drawing.

1. By specifying the process that gives an acceptable edge.
2. By defining the actual edge quality needed.

For example, vibrating parts in steel media will dull the edges of most parts economically. Similarly, on many materials, the thermal energy method (TEM), discussed in Chapter 15, can be used to ensure that no loose burrs or particles will be present to jam assemblies. For parts made within the same plant, drawing specifications stating that "parts should be vibrated in steel media to dull edges" is often adequate. However, such notes are not complete enough for work that is contracted to others. Outside vendors must know what size steel ball to use, how long to run the machine, and the amplitude and frequency of the machine to be used. These requirements can be specified by developing explicit processing standards and referring to them on the part drawings. As discussed in Chapter 2, such an approach is easy and relatively problem-free when the majority of parts have similar requirements.

Although requirements should be unambiguous, they should not be inflexible. When a wide variety of parts are designed and manufactured every year, drawing notes alone may restrict the manufacturing engineer's ability to make parts by the least expensive process. For example, a requirement may specify that abrasive blasting be used. On some parts, a centrifugal barrel finisher might be more economical. Although drawing notes can be changed, the paperwork and delays involved in doing so can add unnecessary costs. Different manufacturers have different approaches to providing the same edge quality, and each uses the most cost-effective method available at the time. Specifying the actual functional requirement of the edge avoids unnecessary delays and extra costs.

References

Barton, H.K., and Barton, L.C. 1957. "The Relationship of Runner Layout to Ease of Trimming." *Machinery,* 90 (26 April): 945–55.

Kramer, Robert. 1971. "Good Die Design Makes Barrel Trimming Easy." *Precision Metal,* 29 (September): 31.

National Screw Machine Products Association (NSMPA). 1983. *Designer's Guide.* Brecksville, OH: NSMPA.

Schafer, F. 1975a. "Deburring Processes in Perspective." Technical paper MR75-482. Dearborn, MI: Society of Manufacturing Engineers (SME).

———. 1975b. "Product Design Influences on Deburring." Technical paper MR75-483. Dearborn, MI: Society of Manufacturing Engineers (SME).

Stein, Julie M.; Narayanaswami, Ranga; Ho, Samuel; Lam, Anselm Y.; Park, Ilwhan; Babu, Madhu; Afzal, Asim; and Dornfeld, David. 1993. "Intelligent Deburring of Precision Components." Technical paper MR93-319. Dearborn, MI: Society of Manufacturing Engineers (SME).

Strasser, Frederico. 1960. "How Control of Burrs Aids Sheet Metal Stamping." *The Iron Age,* 185 (3): 90–92.

Tooling and Production (anonymous). 1973. Chip Control Saves Tool Lives. *Tooling and Production,* 39 (December): 44-45.

Wilson, Frank W., and Harvey, Philip Z. 1965. *Die Design Handbook,* 2nd ed. New York: McGraw-Hill: 9–21.

Bibliography

Alting, Leo. 1981. "Integration of Design and Production for Optimum Control and Removal of Burrs." Technical paper MR81-386. Dearborn, MI: Society of Manufacturing Engineers (SME).

Anonymous. 1964. "Design of Screw Machine Products." *American Machinist,* Vol. 108 (7 December): 127.

Blotter, P. Thomas; Huang, S.F.; Spear, Carl D.; and Canfield, Ronald V. 1980. "Deburring Costs Related to Cutting Speed." Technical paper MR80-661. Dearborn, MI: Society of Manufacturing Engineers (SME).

Fickers, John J. 1982. "The Machinist's Look at Burrs." Technical paper MR82-266. Dearborn, MI: Society of Manufacturing Engineers (SME).

Gillespie, LaRoux K. 1976. *Deburring Capabilities and Limitations.* Dearborn, MI: Society of Manufacturing Engineers (SME).

———. 1978. *Advances in Deburring.* Dearborn, MI: Society of Manufacturing Engineers (SME).

———. 1981. *Deburring Technology for Improved Manufacturing.* Dearborn, MI: Society of Manufacturing Engineers (SME).

———. 1982. "Designing Parts for Economical Deburring." Division Report BDX-613-2792. Kansas City, MO: Bendix Corporation (available from National Technical Information Service [NTIS]).

Hignett, J. Bernard. 1981. "Mechanical Finishing—The Future of This New Technology." Technical paper MR81-383. Dearborn, MI: Society of Manufacturing Engineers (SME).

Miller, Emery P. 1983. "Substrate Preparation." AFP monograph FCR83. Dearborn, MI: Society of Manufacturing Engineers (SME).

Murphy, James A. 1971. *Surface Preparation and Finishes for Metals.* New York: McGraw-Hill.

Rhoades, Lawrence J. 1981. *Cost Guide for Automatic Finishing Processes.* Dearborn, MI: Society of Manufacturing Engineers (SME).

6 Mass Finishing

Mass finishing, also called loose abrasive finishing, is the term mostly commonly associated with abrasive tumbling operations. Typically, a large number of parts are placed in the deburring machine, and they are all deburred at the same time (en masse). Several different styles of machines are used, and the equipment can generally operate as a batch process or as a continuous flow process. Although the exact number of mass-finishing machines in use is not known, it probably approaches 50,000. Mass finishing is one of the largest families of edge finishing processes. These processes have the advantage of finishing most exposed surfaces as well as the edges. Because they operate on many parts at once, they typically are the lowest-cost processes for a wide range of parts and materials. Mass-finishing processes are simple to use, provide repeatable results, and are the workhorse of the industry. Furthermore, because they involve media, compounds, and generally innocuous materials, the processes are considered environmentally friendly.

The processes included in this family are:

- Barrel tumbling (barrel finishing, barreling).
- Vibratory finishing.
- Vibratory shaker tumbling.
- Roll-flow finishing (centrifugal disk finishing).
- Centrifugal barrel finishing.

Variations of these processes include cryogenic and chemical loose abrasive finishing. Spindle finishing, described in Chapter 7, uses the same media, compounds, and has many similarities to mass finishing; however, it works on only one part at a time. It is not considered a mass-finishing process, but a loose abrasive–finishing process. Other loose abrasive processes are described in Chapter 8.

The basic process for any equipment in the mass finishing family involves placing parts in tubs of rotating media. The movement of the parts against the media serves to scrub the surfaces and edges and provide a (usually) gentle polishing and edge-smoothing effect. The media can be metal forms that beat edges and surfaces or abrasive-laden materials that gently grind away surface and edge material in a manner similar to driftwood being smoothed by sand and waves at the seashore. In addition, media, water, and compounds containing fine abrasives and soap are typically added to clean and add abrasive action. Other compounds polish, brighten, or descale. Typical production rates are hundreds or thousands of parts per hour per machine, and costs range from a few cents per part to a fraction of a cent per part.

All metals and many nonmetals, as well as most sizes and shapes of workpieces, can be processed in mass-finishing equipment. Surface finishes finer than 1 μin. (0.025 μm) have been produced using special compounds, although a more common finish would be 8–32 μin. (0.2–0.8 μm). The processes provide a generally random texture to the surfaces that may be an advantage for holding lubrication.

The following process variables must be considered in all types of mass finishing:

- Machine size and design details.
- Media size, shape, composition, amount, and amount relative to part volume.
- Compound abrasive size, material, and amount.
- Water volume and amount relative to compound volume.
- Run time.

Each type of machine has additional variables that will be discussed in later sections. In addition to the process variables, the following part variables influence edge results:

- Part geometry.
- Part tolerances.
- Part's initial surface finish.
- Part material properties.
- Part aesthetics.
- Burr properties.
- Edge tolerances.

Simple modifications and "tricks of the trade" can help improve results when standard approaches fail to provide desired results. Examples of these include masking features with plugs, caps, or sealing wax, and providing dividers to keep parts from banging into each other.

In any finishing operation, keeping records of tests performed and developing an understanding of exactly what is happening and why can prevent major interruptions in production, ensure better part quality, and minimize costs.

6.1 ADVANTAGES

Mass-finishing processes not only handle parts en masse and provide low-cost, effective edge finishing, but they also generally improve surfaces. Mass finishing generally removes scale, dirt, and oils. In addition, generally undesirable tensile residual stresses on part surfaces are removed, and compressive stresses are added. In centrifugal barrel tumbling, for example, the compressive stresses can be high and provide increased service life for parts that will be placed under tensile stresses. In mass finishing, individual handling or fixturing of workpieces is not normally required, thus eliminating costs associated with manual and most other mechanized finishing processes. With proper control, consistent results can be attained from workpiece to workpiece and batch to batch. The process is so flexible that literally thousands of combinations are possible, enabling a single piece of equipment to handle a wide variety of needs. One authority documented that conservatively there are over 500 billion combinations to select from in developing a system and its details. Fortunately, the processes can be effective without mastering all these combinations.

6.2 LIMITATIONS

A limitation to the use of mass finishing is that its action generally occurs on *all* surfaces, edges, and corners of workpieces that contact the media. It is not normally possible to give preferential treatment to specific areas; however, masking specific areas has been successfully used for some applications. The action is greater on the edges of workpieces than on equally exposed surfaces. The action in holes and recesses is significantly less than on exposed areas; and in small, deep recesses it is unusual to be able to do any finishing unless the workpiece is fixtured.

A basic limitation to its use is that a portion of the industry does not consider mass finishing to be a predictable and controllable technology. However, mass-finishing methods can satisfy most requirements when adequate specifications and quality standards are established before and after finishing (discussed in Chapter 2), and when the proper process and reasonable operating parameters are selected.

Some applications are not suitable for mass finishing, for a variety of reasons. For example, the size of the burrs may be too large to be completely removed by mass finishing. In internal areas, burrs may be inaccessible to the media. Surfaces requiring smoothing may be too rough for finishing to the required degree, or critical dimensions may go out of tolerance during long cycles.

The media may become lodged in workpieces with different sized openings, but this problem can generally be avoided by selecting a different size or shape of media or by sealing the openings. Mass finishing for surface finish can produce corners and edge radii; if such radii cannot be tolerated, mass finishing should not be attempted. Generally, the smoothening of a plane surface from 20 to 5 μin. (0.51–0.13 μm) will be accompanied by the generation of radii on the order of .015 in. (0.38 mm) on exposed edges, depending on the workpiece material, size of the media, and other factors. Corner radii (intersection of three surfaces) will be two to five times larger than edge radii (intersection of two surfaces); thus, a typical corner will have up to .030–.065 in. (0.76–1.7 mm) radii. The process can produce radii as low as .003 in. (76 μm) under gentle conditions when the process begins with very small burrs, but a more typical edge radii is .010 in. (0.25 mm). Large radii also can be produced by aggressive conditions, but the entire part will also undergo aggressive cutting.

As was shown in Table 1-5, some size change will occur on exposed surfaces. Typically this amounts to .0005–.005 in. (13–127 μm) across a diameter or flat. Size change, radii size, and other conditions are a direct function of time and conditions and can be controlled.

Parts can become bent, dinged, scratched, dented, and interlocked if proper consideration is not given to preventing such damage. Parts will bang into each other in this process unless special dividers are placed in machines, but these dividers reduce some of the action and move the process from a mass-finishing process to more of a one-at-a-time process.

Parts that are subsequently electroplated will not plate well over the very fine abrasive that is beaten into the surface in these operations. Plating will not adhere in particular to aluminum oxide abrasives and media. Aluminum oxide acts as a barrier to solder, and minute aluminum oxide particles will explode in some welding operations, leaving voids in the weld. The particle sizes that cause these problems require high magnification to detect and may require laboratory analysis, such as auger spectrometry, to verify.

6.3 PROCESS SELECTION

Many variables affect the selection of the best mass-finishing process for a specific application. Major factors that should be considered prior to equipment selection include production rates, quality requirements, floor space, process variables, and costs.

Total production quantities, batch sizes, and the variety of workpieces to be finished are also critical variables. Size, weight, and configuration of the workpieces, as well as the material from which workpieces are made, must be analyzed. Any anticipated changes in production requirements and automation needs also must be considered.

Specific requirements for surface finish, edge and corner conditions, cleanliness, and uniformity must be known and considered before a process and equipment can be judiciously selected. The consistency of workpiece quality entering the finishing department and post-treatment requirements and variables are important factors in selecting the process and equipment. Burr locations, properties, and sizes are critical in this selection. Cycle times for the process selected depend primarily on the finishing requirements and the condition of the parts prior to finishing. If burr sizes will vary greatly between lots or over time, the process variables must be changed to accommodate the anticipated burr sizes. Thick burrs are incompatible with requiring small edge breaks and low stock losses in this process.

Table 6-1 provides a simple comparison among processes, and Table 6-2 provides a checklist for process selection based on the factors that equipment manufacturers will ask users. The data provided represent typical expectations. Because mass-finishing processes are so flexible, it is possible to obtain faster and cheaper conditions and even better finishes than cited, but they would require relatively unusual conditions. There are many exceptions to any capability or limitation cited. When the manufacturing engineer completes the checklist shown in Table 6-2 for specific parts, vendors and other staff will be better able to make informed discussions. Table 6-3 provides additional information on potential part-on-part damage and quality of finish. The "*X* factor" shown in the table is a measure of the relative level of aggressiveness of the process.

Table 6-1. Comparison of major loose abrasive-finishing processes

Process	Advantages	Limitations
Rotary barrel	Low initial operating and maintenance costs. Batch automation capability. Cycle times of 5–72 hr. Good for small and medium-size parts and for many plastic parts.	Slow processing. Large floor space required. No in-process inspection. Wet working area. Little to no work in part recesses. Typically not suited to very large parts.
Vibratory tub	Faster processing than rotary barrel. Can handle small to very large and very long parts. Open for in-process inspection. Cycle times of .25–3 hr. Batch or continuous operation. More work space for the same money than vibratory bowl machines. Lower initial cost than bowls.	Slower than high energy (centrifugal barrel). Noisy. External material handling generally required.
Vibratory bowl	Faster processing than rotary barrel. Can handle very small to large parts. Open for in-process inspection. Integral separation. Can be less noisy than vibratory tub. Generally better media/parts mixing than vibratory tub. Economical for general purpose work, heavy deburring, continuous processing, and continuous inspection. Full automation capability. Batch or continuous operation. Cycle times of .25–3 hr.	Cannot handle very long parts. Noisy unless noise suppressor is used. Ineffective for part recesses.
Centrifugal barrel	Fast processing. Can handle precision and fragile parts. Produces smooth finishes. Batch automation capability. Versatile—can handle fine finishing to heavy stock removal. Improves fatigue strength. Small floor space required. Cycle times of 10–30 min.	No in-process inspection. Complex. Initial costs are 8–15 times higher than vibratory equipment per ft^3 of capacity. Cannot handle very long parts. External material handling generally required.
Centrifugal disk	Fast processing. Open for in-process inspection. Batch automation capability. Versatile. Cycle times of 10–30 min.	Part sizes limited to about 1 ft (0.3 m). External material handling required.
Spindle finishing	Fast processing. No impingement of workpieces. Produces smooth finishes. Automation capability with robot reload. Cycle times of 5 s to 5 min.	Parts must be fixtured. Long, heavy parts and very small parts (< .25 in. diameter [6.4 mm]) generally not feasible. Typically limited to cylindrical parts. Less versatile than other processes. Fixtures wear in the media. Part weight typically limited to < 100 lb (45 kg).

Note: The data in this table represent normal practice. Because the processes are so flexible, many exceptions to the conditions cited here exist. Part size, shape, tolerances, materials, and other factors affect the choice of process used. Plant standards also affect the operation.

6.4 MEDIA

The term *media* refers to the abrasive or nonabrasive consumable elements used in mass-finishing processes. The main function of media is to abrade or burnish the edges and surfaces of components to the desired finish. Media also helps keep the workpieces from impinging on each other and serve as carriers for any compounds used.

6.4.1 Media Selection

Selecting the proper media for a specific application is critical for optimum deburring and surface-finishing results. Unfortunately, choosing the proper media is difficult because of the diverse requirements of numerous applications and workpieces and the wide variety of types, sizes, and shapes of media available. In many cases, optimum selection must be made on a trial-and-error basis.

Major factors to be considered when choosing media are the composition (type), size, shape, and weight of the media to be used. Other important considerations include the capability, versatility, and economy of the media. Workpiece material, size, and geometry, burr sizes and locations, finishing requirements, finishing equipment used, and subsequent operations and functions of the workpieces are also important factors. Magnetic separation of ferrous parts generally requires the addition of demagnetizing equipment. This is very important if parts are to be plated. Tables 6-4 and 6-5 contain general recommendations for media selection.

Table 6-2. Checklist of considerations for selecting processes and equipment (example data shown in 2nd column)

	Results desired				
Operation details	No burrs at 10× magnification	No sharp burrs	No burrs that cut hands	No burrs visible with naked eye and .015 in. (0.38 mm) break	Comments
Part number or name	111222-101				
Maximum part length	4.5 in. (114.3 mm)				Parts over 12 in. (305 mm) generally require a special vibratory tub
Maximum part height	3.6 in. (91.4 mm)				
Maximum part thickness	2.5 in. (63.5 mm)				
Impact concern[a]	.4 in. (10.2 mm)				All mass finishing results in some part on part impact
Pieces that must be processed per year	250,000				
Total cycle time (min) required per load	25				
Run time per year (hr)	1,750				
Pieces run per cycle	60				
Potential growth (%) (more parts/year)	30				
Other parts that will run in this machine	200				
Total need factor[b]	3.3				
Capacity required	10.7 ft^3 (0.3 m^3)				Centrifugal barrel equipment typically has smaller capacity/load than other processes
Machine capacity available on site					
Vendor's estimate of parts per hour with machine X					Be sure vendor understands part quality requirements
Total time per load based on vendor's estimate[c]					Automated or manual load/unload?
Amount of water per load					Some parts can be smoothed without water, but the standard is to use water
Type and amount of media per load					Steel media requires heavier duty machines
Expected media cost per piece					
Type and amount of compound(s) per load					
Expected cost of compound per piece					
Expected media-lodging (pieces/load)					Removal of lodged media can be expensive

Parts handling requirements, including screening					Screens must be stored
Total system energy costs per year or per load					Consider handling, finishing, drying
Parts cleaning requirements					
Parts drying requirements					
Equipment maintenance costs per year; typical maintenance failure					Machine breakdowns typically are not difficult to repair; some vibratory machines will shake themselves apart
Compound delivery requirements					
Waste removal requirements					
Container size and quantity for storing media					
Container size and quantity for storing compounds					Consider how media are taken out of and placed into container
In-process parts storage requirements					
Working space for operator (for example, table)					
Operator inspection requirements					
Miscellaneous supplies needed by operator[d]					

Note: When considering purchase of a mass-finishing machine, users will want to prepare a checklist like this one that covers the entire range of the capabilities of the product. After considering each item in the checklist, the user and the vendor can then determine which equipment meets the needs of the company. A different part, representative of the different quality expectations, can be placed in each column or a different sheet can be prepared for each part, and then each operating detail is filled in to allow for comparison among the different parts.

[a]Refer to Table 6-3.

[b]100% for part number in this column plus potential growth plus other parts divided by 100.

[c]Includes loading and unloading of parts and media, media separation, if any, rinsing, and cycle time.

[d]Includes bagging supplies, protective plugging of critical/problem features, paperwork storage, records, instructions, computer terminal, etc.

Table 6-3. Typical mass-finishing media-to-part ratios

X factor	Ratio by volume	Typical commercial application
3.0	0:1	No media. Part-on-part. Used for beating off burrs. No media for cutting.
2.5	1:1	Equal volumes of media and parts. Forgings, sand castings. Crude, very rough surfaces.
2.0	2:1	More gentle, more separation, still severe part-on-part damage.
1.6	3:1	About minimum for nonferrous parts. Considerable part-on-part contact. Fair to good for ferrous metals.
1.26	4:1	Probably "average" conditions for nonferrous parts. Fair to good surfaces. Good for ferrous metals.
1.0	5:1	Good for nonferrous metals. Minimal part-to-part contact.
0.63	6:1	Very good for nonferrous parts. Common for preplate work on zinc with plastic media.
0.5	8:1	For higher quality preplate finishes.
0.4	10:1	Used for very irregularly shaped parts. Exceptional quality.
—	No contact	Absolutely no part-on-part contact. One part per machine or compartment. Use doughnut-shaped machine or fixturing.

Note: The *X* factor is a relative measure of process efficiency. A value of "1" represents a typical industrial operation. A lower value indicates a slower process requiring more care of parts. A higher value indicates a much higher throughput than normal. Ratio by volume indicates the ratio of media used per part. A 10:1 ratio implies that typically 10 in.^{3} (163.9 cm^3) of media will be used to finish 1 in.^{3} (16.4 cm^3) of parts. A 0:1 ratio implies no media are used (parts vibrate against other parts).

(Kittredge 1993)

Table 6-4. Properties of various media

Media	Materials	Sizes	Aggressiveness	Shapes
Random natural	Limestone, river rock, granite, Turkish emery, American emery, novaculite, garnet	#00–24 size (.028–2.0 in. [0.7–51 mm]) (19 standard sizes)	—	Random
Random synthetic	Fused aluminum oxide (3 variations), crushed plastic, silicon carbide	#00–24 size (.028–2.0 in. [0.7–51 mm]) (19 standard sizes)	—	Random
Preformed ceramic	Porcelain filled with aluminum oxide grit or silicon carbide grit	—	Extra fast cut, fast cut, medium cut, moderate cut, slow cut	20 shapes, see Figure 6-1
Preformed plastic	No material grades standard among suppliers	—	At least 9 grades	18 shapes
Preformed steel and metal	Stainless steel, chrome alloy hardened, high-carbon steel, low-carbon steel, Monel®, tool steel, tungsten carbide, brass	27 ball sizes, 8 diagonal sizes, 6 pin sizes	—	6 shapes
Preformed wood	—	14 sizes	—	6 shapes
Ground corncob	—	—	—	—

Table 6-5. Media selection guide for mass finishing

Desired effect	Degree of effect	Media recommended
Deburring	Light	Steel or ceramic
	Medium to heavy	Ceramic or plastic
Radiusing	Light to heavy	Ceramic or plastic
	Irregular surfaces	Randomly shaped
Surface roughness	Reduce to lower value	Plastic or ceramic
	Preplate quality on softer alloys	Plastic or ceramic
		Plastic^2 plastic
		Plastic^2 steel
		Ceramic^2 steel
		Steel or wood
Surface reflectivity	Brighten or highlight	Steel or ceramic
	Best quality, hard alloys	Ceramic or ceramic^2 steel
	Best quality, soft alloys	Plastic^2 steel
	Best quality, plastics	Wood
Clean surfaces	All metals	Steel or ceramic
Least operating cost	Varies greatly with grade involved	Steel, ceramic, plastic, or wood
Fastest processing speed	Overall capabilities of media must be understood	Steel, ceramic, plastic, or wood

Note: Some media may be inappropriate for overall part requirements. Two-step processes are indicated by "^{2}" between media steps. A media not listed does not infer its unsuitability.

(Kittredge 1993)

6.4.2 Media Composition

The composition of the media determines whether it is for cutting or finishing. Cutting types of media contain abrasives, whereas finishing types have no abrasives or only very fine abrasives. The workpiece material and its hardness, the size of the burrs, and the surface finish requirements all govern the type of media to be used.

Media with aggressive cutting capabilities is generally best for removing large burrs. Requirements for smooth surface finishes necessitate using slower cutting media with fine or no abrasives. Harder media can generally be used in the high-energy methods of mass finishing.

The composition of the media also determines its weight, which affects the rate at which work is performed. For fixed operating parameters, heavier media exert more pressure on the workpieces than lighter media. Increased pressure can provide faster cutting and more polishing action, but may mar soft surfaces or distort fragile parts.

6.4.3 Media Size

The size of the media is important for several reasons. One function of size is to aid in keeping the workpieces separated during finishing. Small media help to keep small parts separated, whereas large media cut faster and produce a rougher finish. A mixture of media sizes is often used for processing parts.

Another important consideration in choosing media is that its size must allow for all surfaces to be reached during finishing without it becoming lodged in holes and recesses of the workpieces. The ease with which it can be separated from the workpieces also must be analyzed. With separation by screening, the media must be either larger or smaller than the workpieces. However, with magnetic separation of ferrous parts, media size in relation to workpiece size is not a factor. Smaller media can generally be used in high-energy methods of mass finishing to improve the surface finish produced without significantly increasing cycle times.

6.4.4 Media Shape

Different media shapes (see Figure 6-1) have advantages for specific applications, depending primarily on the configurations of the workpieces. Factors to consider when selecting a shape include access to all workpiece surfaces requiring deburring or finishing, media lodging in holes or recesses, and ease of separation of the media from the parts at the end of the cycle. Sometimes a mixture of media shapes is used.

Triangles

Triangular media provides uniform action and is effective in reaching corners and slotted areas. Angle-cut triangles have sharper corners and edges and provide deeper penetration

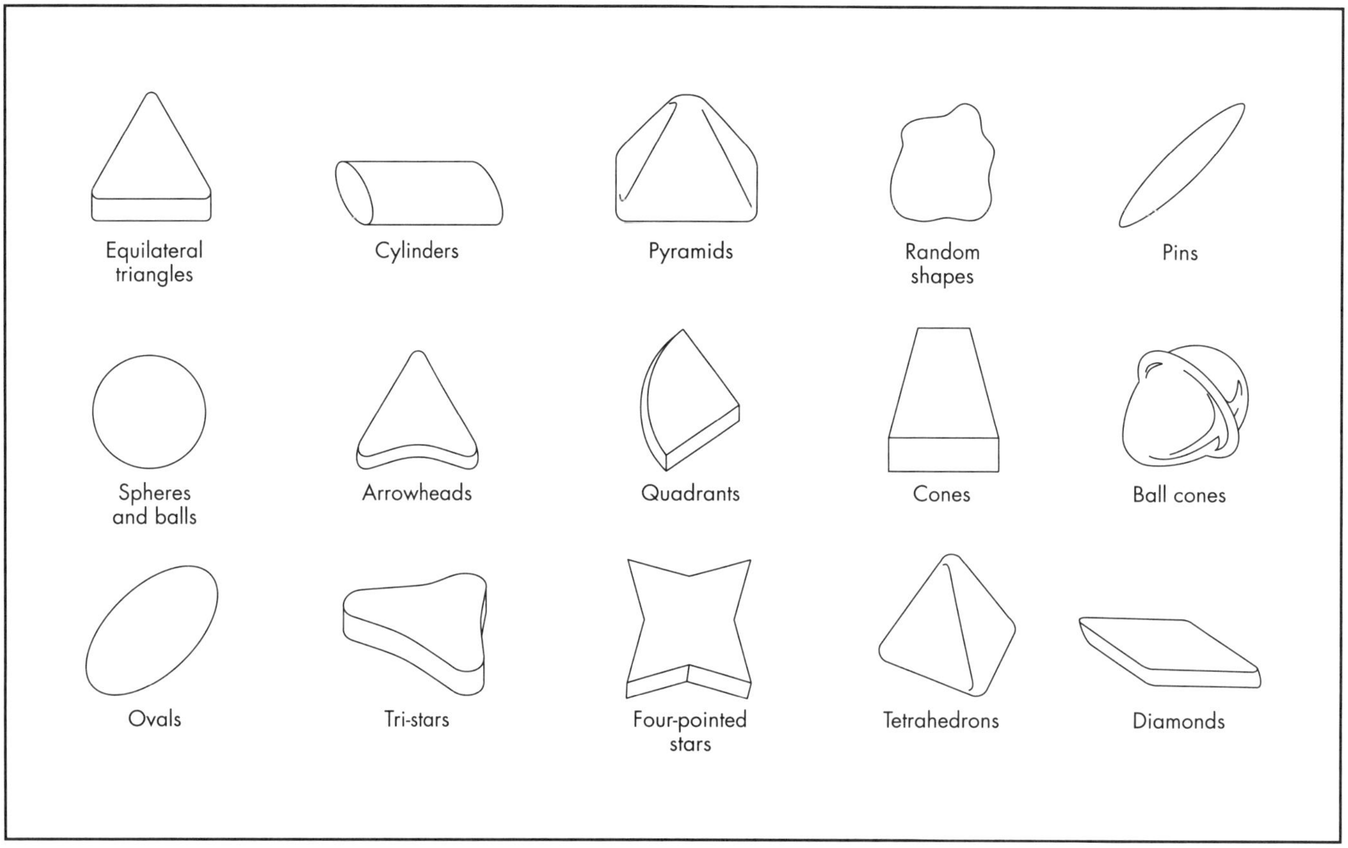

Figure 6-1. Media used for mass finishing.

into remote areas. Notched triangle (arrowhead) media provides sharper edges for reaching into slots, holes, or openings. The smallest standard triangle available has a corner radius of about .040 in. (1.02 mm). However, it is frequently desirable to use larger triangles that have larger radii, since they minimize cycle time.

Cylinders

Cylindrical media works effectively in some applications. With its ends cut at angles of 22°–60°, cylindrical media reach recesses more easily than other shapes.

Diamonds

Diamond-shaped media has sharp points that can reach into corners, slots, and recesses without lodging problems.

Cones

Cone-shaped media is very versatile. It permits partially entering holes of various diameters, without lodging problems.

Spheres

Sphere-shaped media has good flow action and surface contact. It is useful for uniformly blending and smoothing surfaces.

Wedges

Cylinders that are cut into wedge shapes have become one of the most popular shapes replacing cones, plain cylinders, and triangles in many applications. Manufacturers use several names for these shapes, including double-cut cylinders, "V"-cut cylinders, and tri-cylinders.

6.4.5 Media Capability, Versatility, Economy, and Availability

An important consideration in media selection is the ability of the media to remove burrs and produce the required corner edge and surface conditions with consistent uniformity. Cycle times required to achieve these results are also critical. Minimum media break-in requirements, wear, and reclassification needs, as well as good cushioning action, are desirable features. Other desirable features are the ability to finish a variety of workpieces in a given machine and the ability to achieve different finishes. Cost per pound of different media is of little significance in overall economic considerations. Wear characteristics of the media and cycle times to attain the required results must be considered in determining the all-important cost per finished part. The consistent availability of uniform, high-quality media from a reliable supplier is another important consideration.

6.4.6 Types of Media

The most common types of mass-finishing media are natural abrasives, agricultural products, synthetic random media, preformed ceramic and resin-bonded media, and metallic media.

Natural Abrasives

Quarried, crushed, and graded stones were the original mass-finishing medias. They have, however, been largely replaced primarily by synthetic materials because natural stones are softer, wear more rapidly, are less consistent, and occur only in random shapes. Synthetic materials are harder, provide longer life, and have more consistent cutting action, wear rate, and dimensions.

Random natural medias include natural stones of several popular types, including granite, limestone, river rock, Turkish emery, and American emery. These are low-cost media, but are not generally cost effective for deburring the majority of parts. Despite their limitations, natural abrasives still have some applications. Corundum is a natural crystalline oxide of aluminum (Al_2O_3), which can be economical for some uses. Novaculite is a fine-grained silica stone that maintains sharp particle shapes and can be of value for removing burrs from small holes and grooves. Soft limestone and hard, but more friable, granite can still be economically justified for some barrel-finishing applications. Natural abrasives are sometimes used in the manufacture of bonded preforms.

Agricultural and Wood Products

Ground corncobs, walnut or other nut shells, and hardwood sawdust are used in tumbling barrels and heated barrels for drying parts. These materials produce a good luster on some workpieces. When mixed with fine abrasives, they are suitable for fine polishing operations, particularly in the jewelry industry.

Wood shapes, such as cubes, the conventional peg, dual wedge-ended pegs, and diamond pegs, are made from hardwoods. They are being used to deburr and finish wood, plastic, ceramic, and metallic components. Sometimes they are coated with oils or waxes to transfer coatings to the workpieces or with fine abrasives for increased cutting action. Hardwood media can duplicate the effect of color buffing on plastics, wood, or soft metals.

Synthetic Random Media

Fused aluminum oxide ($A1_2O_3$), crushed and graded, has greater abrasion properties than some natural materials. Other advantages include good wear resistance and consistency of size and quality. However, the physical characteristics of fused aluminum oxide vary with different producers. Sintered aluminum oxide is the longest-lasting random-shaped abrasive material available for mass finishing. It is used extensively for fine deburring and finishing. Random synthetic bodies include at least two distinct grades of fused aluminum oxide: one designed for cutting and another for brightening parts and crushed plastic media. Users have had varying degrees of success with this media. The same lot of media has lasted for months with daily use at one plant, whereas other users indicate that it wears quickly and cuts little over an extended period of time. Lodging is a problem with any random-shaped media.

Fused silicon carbide (SiC) media is available, but sizes with a diameter larger than .25 in. (6.4 mm) fracture too easily to be well suited for mass finishing. This synthetic mineral may have to be used for components that are subsequently brazed, soldered, or welded, and when aluminum oxide cannot be used. However, preformed media is normally preferred to silicon carbide for such applications.

Preformed Media

The development of media of controlled shape, size, and abrasion and finishing characteristics has been largely responsible for making mass finishing a precision process. Preformed media is available in ceramic, resin-bonded, and metallic types, all of which are used extensively in the industry. The abrasives used in preformed ceramic and resin-bonded media are often aluminum oxide, silica, or silicon carbide. The initial reason for favoring preformed ceramics over random-shaped products, both natural and synthetic, was to eliminate lodging problems. Early preformed products wore out very rapidly, but this was soon corrected. Today, economy joins with preventing lodging as reasons to use preforms over random shapes.

Preformed Ceramic Media

Ceramic-bonded media is produced by mixing porcelain, clay, or other vitreous materials with varying amounts of abrasives (or none at all), forming the material into desired shapes (usually by extrusion and cutting), and then firing to vitrify. Fusion bonding is used to make some media. Such medias are available in a wide range of shapes and sizes. The ability to select the proper grade of abrasive, the correct proportion of abrasive to binder, and the most suitable binder make this media adaptable to a wide variety of applications and workpieces. Ceramic media with about 50% aluminum oxide abrasive of 60 grit is the fastest material for deburring, edge generation, and stock removal. Preformed ceramics, one of the two workhorse mass-finishing media, are available in a large number of fairly standard grades. Extra fast cut, fast cut, medium cut, moderate cut, and slow-to-no cut grades are typical (many manufacturers make more than five grades).

Preformed Resin-bonded Media and Preformed Plastics

Preformed resin-bonded media and preformed plastics are the other workhorse media and are available in a variety of grades or "bodies," with some manufacturers having as many as nine grades and others with as few as three. Five is typical, and other, significantly different grades are available using other binders or higher densities.

Usually preformed resin-bonded media and preformed plastics are formed by bonding abrasives (generally 40–70% by weight) into polyester or urea-formaldehyde resins by casting. These so-called plastic media are softer, have a lower density than preformed ceramic media, provide a cushioning effect, and sometimes have a shorter life. They are well suited for finishing softer metals and fragile parts, achieving preplating requirements, and producing smooth finishes. Although some users believe they assure parts free of impingement, that is not true, since all the mass-finishing processes, except for fixtures or compartmentalized operations, allow part-on-part contact. Plastic media also can be produced in a wide variety of shapes, depending on what is in favor at the time.

High-density Polyester Media

High-density polyester media is made in the same way as low-density media, except they differ in composition: a polyester resin and heavier silicate filler are used for high-density media. The additional weight of the media and sharper crystal facets improve the cutting performance compared with low-density media. Some users claim that smoother surface finishes can be produced with this media than with ceramic media.

Metallic Media

Preformed hardened steel media, free of abrasive, is available in a variety of shapes and sizes. It is used primarily for burnishing to achieve maximum luster for decorative purposes, for light deburring applications, and for heavy-duty cleaning. Advantages include uniformity of shape and size, the elimination of possible media fracturing in use, and almost no wear. The high-bulk density of steel media results in rapid peening and deburring and supports the workpieces. Preformed steel media is case-hardened, polished, and accurate in dimensions. A few through-hardened shapes are available as well. Soft steel shapes are used as carriers for fine abrasive particles.

Preformed steel media comes in a variety of shapes, including two grades of balls, ball cones, cones, diagonals, oval balls, diamonds, and serrated pins, which provide aggressive cutting action.

Media shapes preformed from zinc are occasionally used for burnishing operations as an alternative to steel. Tacks, pins, wire brads or clippings, and nails, mixed with fine abrasive, are sometimes used to remove burrs from small holes and recesses.

By being nonabrasive and nonwearing, steel media, coupled with a good compound containing cleaning properties and a built-in corrosion inhibitor, actually scrubs surfaces and interiors, resulting in economically efficient removal of both organic and inorganic soils. This method of cleaning is now commonly used to remove a wide variety of soils on many types of materials. Common carbon steels to exotics, such as silver and gold, are cleaned in vibratory equipment. Systems are able to efficiently remove soils from machining operations, stampings, and headed parts, as well as oxides from zinc, aluminum, brass, copper, and more. The unique scrubbing action and greater accessibility allow removal of metal chips from machining operations, milled grooves, threaded holes, slots, as well as soil from the parts.

Although steel media is not used to cut down heavy burrs, it is widely used for peening or hammering sharp edges to very acceptable conditions. Extremely complicated castings are deburred in large quantities due to the size control and accessibility to all part areas afforded by the media. Steel media reduce sharpness with such speed and success that the deburring cycle is usually less than 15 min.

Ball burnishing, as the process is sometimes referred to, is used to produce smooth, mirror-like finishes on parts. Metal surfaces vary greatly, from matte-type finishes to smooth, polished ones in which surface reflectivity is improved. A matte surface is nonreflective because it has numerous hills and valleys or ridges and grooves. Burnishing with steel media peens down these imperfections and produces smooth and brilliant surfaces. Vibratory equipment provides interior and exterior action, so burnished finishes can be obtained on both inside and outside surfaces.

Deflashing or removing flash from thermoset plastics has been accomplished with steel media for years by leading manufacturers of thermoset electrical components. By being nonabrasive, the steel media does not damage surfaces, and processing cycles are normally short—usually 7–12 min. Complex ceramic insulators with intricate cavities are also successfully deflashed using this media.

Surface modification reduces roughness, surface area, and improves soil and stain resistance in a workpiece. The media pounds down the surface to create uniformity and smoothness. This results in less surface area and less roughness on a part. Also, resistance to soil and stains is achieved because the microscopic peaks and valleys that entrap the soil on most workpieces are smoothed by the pounding action of the media. This application is commonly used for food-processing equipment, which requires a high degree of cleanliness.

Steel media is not abrasive by nature and does not create or absorb soils. Therefore, it is not likely to cut or scratch workpieces. Since it is not abrasive, it will not change the dimension of parts, such as investment castings, in which tolerances must be maintained. The media does not create soils; therefore, workpieces come out extremely clean and no secondary operations are required to clean parts after they are unloaded from a vibratory finishing machine. Steel media is hardened to R_C 60 minimum; therefore, it maintains its dimension and shape with continued use.

Because the density of mass steel media is approximately 300 lb/ft^3 (4,806 kg/m^3), it delivers very fast finishes in burnishing, deburring, degreasing, and deflashing operations. Cycle times are generally less than 15 min. It also has an average life of about 10 years. Steel media is too heavy for use in about 50% of the vibratory machines in operation. Users should check their machine capabilities before procuring a load of steel media.

Steel media does not wear the lining on finishing equipment, so the cost of relining machines is eliminated. Despite its mass, it can be used to process fragile parts; if small-sized media is used, it can actually cushion and support parts within the finishing mass.

The attrition rate of steel media is low, but coupled with other losses, most steel media users replace 10–25% of the charge each year.

Other Media

There are many other medias used for special mass-finishing applications. For example, pieces of leather or felt are sometimes used for light burnishing or wiping operations. Nylon media is used for cryogenic deflashing, and rubber-bonded abrasive media is used for light deburring and smoothing of soft metals.

With natural or random-sized media that has been crushed and graded to size, the best that can be achieved is one point of contact on any shape workpiece; with cylindrical-shaped media, there will always be at least one point of contact and very often line contact. With triangular-shaped media, there also will be at least one point of contact, often line contact, and sometimes flat area abrasive contact. Generally, flat-stamping work can be more efficiently processed with flat triangular-shaped media.

Screw machine work and concentric surfaces will never receive more than line contact with either the cylindrical or the triangular shape, with most of the work performed by single points of contact.

Media densities range from 20–350 lb/ft^3 (320–5,606 kg/m^3). This wide variation can be problematic, since it is tempting to purchase the media with the least cost per pound (kg). The least cost media is rarely the most economical in terms of cubic feet (cubic meters) required in the unit or in terms of media wear rate relative to stock removal. When considering lodging costs, rinse quality, wear rates, and cutting rates, some of the so-called low-cost media could cost less out in the driveway. When performing precision work, often the cost of the media is a fraction of the value of the work in the unit, and using less-expensive media often means sacrificing quality.

Wear rate versus cutting rate is an important consideration in media and compound selection. Harder media will last longer than softer media; however, softer media will probably cut much faster. All plastic, natural, and ceramic media wear easily, and as a result, eventually lose their size. They must be discarded or they will become lodged in holes and recesses, which requires manual removal and in some instances scrapping of product.

The sharpness of the abrasive grain determines the stock removal rate; thus high cutting efficiency yields a poor surface finish, and a good finish results from poor cutting efficiency. In other words, a dull 60-grit abrasive may produce a finer finish than a sharp l20-grit abrasive. Finish is not always a function of abrasive grain size, but of cutting efficiency. So stock removal cannot always be related to abrasive size (this lends some insight to the term *preconditioned media*) (Matsunaga 1971).

Users are sometimes surprised by the weight of large quantities of media, particularly steel balls. Table 6-6 provides the weights of some common media.

Table 6-6. Mass finishing media weights

Media	Weight lb/gal (kg/L)	Weight lb/ft^3 (kg/m^3)
Walnut shells	5.5 (0.7)	39 (0.11)
5/16 in. (7.9 mm) ceramic cylinders	12.8 (1.5)	90 (0.25)*
Carpet tacks	20.0 (2.4)	150 (0.42)
.5-in. (12.7-mm) steel balls	38.5 (4.6)	287 (54 balls/lb) (0.81)
120-grit aluminum oxide	15.7 (1.9)	120 (0.34)
Plastic media	13.4 (1.6)	55 (0.27)
Corncobs	3.9 (0.5)	28 (0.08)

*Ceramic media can vary from 85–125 lb/ft^3 (1,362–2,002 kg/m^3).

Cost per cubic foot (m^3) of media varies dramatically from one media to another. For example, ceramic media may be purchased for \$60–75/ft^3 (\$2,118–2,648/m^3), carbon steel balls for \$1,100–1,200/ft^3 (\$38,846–42,377/m^3), nails for \$100–150/ft^3 (\$3,531–5,297/m^3), and corncobs for \$4–5/ft^3 (\$141–177/m^3).

6.4.7 Ratio of Media to Workpieces

Determining the proper ratio of media to workload is essential for economical and successful mass finishing. High ratios reduce the chance for impingement. Any change in the optimum ratio can have a significant effect on the quality of the finish achieved. (Table 6-3 provides typical media-to-part ratios for some commercial applications of barrel and vibratory finishing.)

6.4.8 All Media are not the Same

Manufacturers often do not make the same sizes, produce identical shapes, use the same materials, or employ the same processes for preformed shapes. Media may look similar, but no standards exists to compare cutting action and media costs. Some manufacturers do provide some comparative data, but none exists from one manufacturer to another. In most instances, users pick a shape that will not lodge, yet will contact the desired edges; then they select the size needed to assure lodging does not occur. When the selection is narrowed to this extent, only a few vendors will be found that meet these criteria. Users will then purchase a sample of some product, try it, and then determine if costs and performance are acceptable. For comparison, users may run samples from several manufacturers. Although the results are generally comparable between machines, some machines are designed to be more aggressive than others, and the data developed for one machine may not be directly applicable to another, even if they are of somewhat the same type. (See Kittredge 1977 for detailed information on testing media to assure valid results.)

6.5 COMPOUND SOLUTIONS FOR MASS FINISHING

Most compounds are combinations of chemicals that dissolve in water and form solutions to maintain consistency or to modify the action of media against the workpieces in mass finishing. Compounds come in premixed liquids or powder form. Some compounds contain nonsoluble abrasives for special applications, such as part-on-part finishing. The use of too little, too much, or an improper compound can adversely affect cycle time and the ability of the media to perform as intended.

Generally, compounds include water conditioners, cleaners, wetting agents, lubricants, foam-control agents, pH buffers, rinse agents, suspension agents, and emollients. Some also contain metal brighteners and corrosion inhibitors.

There are some finishing applications where no compound solutions are used. For example, loads of workpieces are sometimes finished dry, perhaps with sawdust, ground corncobs, wood pegs, or other media for burnishing purposes.

Media for dry finishing are often treated with an oil, grease, or wax-based formulation with finely powdered abrasives to produce smooth surface finishes. Cycle times are much longer than for wet operations. There are other applications where the desired results are achieved with a suitable compound solution and no media; the workpieces themselves act as the media, tumbling against each other for the required action.

Solutions are primarily used to clean the workpieces and media and to control the process. In some applications, the solution modifies the luster or color of the workpieces. The use of corrosion-inhibiting compounds is especially important when any metal is used.

6.5.1 Cleaning Functions

All mass-finishing operations require clean machine compartments, media, and workpieces for effective and efficient results. Compound solutions remove tarnish or scale and emulsify oils and suspended solids. Solutions are also expected to maintain the cleanliness of the workpieces and media during the finishing cycle. Buildup of dirt on the media causes glazing and reduces the cutting capability. Strongly alkaline formulations are often used for cleaning iron and steel parts, whereas milder alkaline to mildly acidic compositions are generally best for nonferrous metals and alloys.

6.5.2 Process Control

Compound solutions are often required to control the hardness of the water used, the pH of the solutions, and foaming. In barrel finishing, foaming is often desirable, whereas in vibratory finishing it is not. Other control functions include wetting the surfaces of the workpieces, providing cooling and lubricity, cushioning the workpieces, and preventing corrosion. The cooling function is very important when using high-energy mass-finishing methods.

A compound solution is sometimes required to modify the action of the media used. For example, in some applications it may be necessary to change from maximum abrasion to maximum finish improvement conditions. Another important function of compound solutions is to produce effluents that meet EPA, OSHA, state, and local requirements. The development of phosphate-free, biodegradable compounds has made it easier to conform to these requirements.

6.5.3 Types of Compounds

Compounds are classified according to function: (1) abrasive, (2) descaling or bleaching, (3) cleaning, (4) deburring and edge radiusing, and (5) burnishing and coloring. Many different compounds are available for specific functions. Selection depends on many factors, including workpiece material, media type, finishing process to be used, water hardness, and cycle times. Compounds are available in liquid and powder forms. A paste compound is being used in some fine-finishing applications in which water is not used. When a finishing application requires free abrasives to influence the cutting action of the media, the compounds containing free abrasives must be powdered. Requirements for free abrasives, however, are limited. Liquid compounds are preferred for most mass-finishing applications because they are easier to handle and use with precise results.

Combination compounds are numerous and include those that (1) clean, burnish, and inhibit; (2) deburr and inhibit; (3) clean and inhibit; and (4) clean, deburr, and inhibit. Compounds are designed for various ranges of water hardness found at industrial plant sites. These can vary from below 17 to over 500 ppm (less than 1 to over 30 grain/gal).

Most compounds contain several types of ingredients, including chemicals to perform certain functions and other ingredients to facilitate finishing. Water conditioners minimize the deposit of salts on the media and workpieces by reducing the water hardness. If the available water has an exceptionally high mineral content (over 200 ppm), it may be necessary to use demineralized or preconditioned water. Detergents help to keep the workload clean. Corrosion inhibitors minimize rusting. In general, the compound used must be tailored to the workpiece material and condition and to the results required.

6.5.4 Abrasives in Compounds

In recent years, studies employing optical and electron microscopy, electron diffraction, electron probe microanalysis, and radioisotope tracer techniques have shown that abrasive particles are embedded in the part surface during both mass and individual finishing operations. The amount of abrasive embedded was extraordinarily high, and the effects of adhesions, corrosion, spotting out, and tool wear were cost considerations (Matsunaga 1971). The choice of abrasive grain structure is a cost consideration, whether the abrasive is aluminum oxide, silicon carbide, silicon dioxide, or zirconium silicate, or any combination of these.

6.5.5 Chemical Actions in Compounds

Softening Agents

Softening agents include phosphates, soluble carbonates, or the various nitriloacetic acid salts and ethylenediamine acetic acid salts. The degree of chelation that results in the softening effect will depend on the ease of reaction of the softening agent and the amount and type of salts in the water (Ramnarine 1989).

Soil Suspension

To prevent redeposition of soil or residue onto the surface and, particularly, recesses of parts, compounds should contain soil-suspending agents (Ramnarine 1989). These agents disperse and hold the soil and residue in suspension to be eventually washed out of the system. This reaction also prevents media from loading up and becoming nonfunctional by metal, soil, and abrasive fines filling and clogging its cutting surfaces. When this occurs, the media becomes like smooth stones that can no longer cut or deburr, but peen or hammer the burr away. If not corrected, finishing cycle times could be doubled, raising the cost of deburring the parts. A compound that has good soil-suspending ability keeps the media clean and parts free of smut or residue.

Soil suspension compounds are usually in the alkaline range and may include combinations of nonionic, cationic, or anionic surfactants. Silicates, phosphates, and carbonates could also enhance the capabilities for preventing redeposition. These agents could exhibit an affinity for the particular soil, reacting with and holding it in suspension until it is finally washed out of the equipment.

Wetting

Surfactants or wetting agents act as the medium for the active ingredients in the compound and are included to enable actual wetting of the soil or surface of parts. Surfactants do this by lowering the surface tension at the interface. They can also cause emulsification or dispersion of the soil, leaving a clean surface behind. This wetting effect eliminates water spotting and streaking on finished parts. Smut is caused mainly by ineffective surfactants; that is, soil is moved about the surface of the parts but not removed.

Wetting agents, depending on the substrate, can be anionic, cationic, nonionic, or a combinations of these. Saponified fatty acids, alcohols, esters, ethers, and other chemicals are added to wetting agents or surfactants. Soil/substrate specificity and the effective lowering of surface tension normally govern selection of materials.

Inhibition

The difference in velocities between parts and media in mass-finishing equipment results in a scrubbing action. This factor activates the surface of the parts, and being a wet system, the exposure to air and water makes parts susceptible to immediate corrosion. It is necessary to have an inhibitor package in the compound to passivate the metal surface during and after processing. Reversing oxidation or corrosion is difficult once it has started. Darkening of copper and brass and white powdery spots on zinc and aluminum may be a result of oxidation. Ferrous parts processed using an improper compound can cause rusting during the actual processing or flash rusting as the parts are removed from mass-finishing equipment. Hence, rinsing of parts with water is not normally recommended. A rinse-aid/inhibitor should be used to prevent corrosion later.

Compounds for ferrous metals may include nitrite as the sole ingredient or triethanolamine, amine borates, phosphates, silicates, or combinations thereof. Certain substances like triazoles, acetylenic alcohols, and fatty acid amines also could be added to compounds as both ferrous and nonferrous oxidation/corrosion inhibitors. Combinations of nitrite and amines are no longer acceptable because of formation of nitrosamines.

Lubricity

Over 75% of media wear is caused by media-to-media contact. Compounds provide lubricity, which reduces excessive media-to-media wear. Lubricity can be recognized by dipping the fingers in the recommended use-concentration of the compound and then rubbing them together. In the case of nonwearing media, lubricity may be desirable for a polishing effect. However, when using steel-burnishing media, this may not be necessary. Steel-burnishing processes require specifically designed compounds, which produce results by a combination of chemical and physical reactions on the substrate.

6.5.6 Effect of Compounds in Specific Applications

Although compounds are normally designed for a specific group of metals, it is not uncommon to successfully use one compound for several applications. However, such a compound might not always be cost effective.

Ferrous Metals

Compounds for ferrous metals and their alloys are rich in inhibiting ingredients to prevent rusting. Because of exposure to water and air, ferrous parts are notoriously susceptible to flash rusting. Normally, the pH of these compounds is higher than seven (neutral) and leans toward the alkaline side of the pH scale. Since stainless steel is more resistant to oxidation, it is possible to successfully use compounds designed for nonferrous metals for finishing stainless steel parts. This will depend on the quality of the alloy and the type of media used.

Nonferrous Metals

Being higher up in the electrochemical series, metals such as magnesium, aluminum, and zinc can be troublesome. Copper parts, although not as active, may be difficult to mass finish because of softness and the alloyed materials used in the metal. Compounds designed for these metals and their alloys exhibit near neutral pH and contain inhibiting agents specific to them. Nonferrous metals can oxidize, darken, or tarnish even more readily than ferrous parts can rust. However, ferrous parts rust dramatically and thus are more easily recognizable. The compound used should exhibit powerful soil suspending and dispersing qualities. Media fines, metal fines, and dirt easily redeposit onto the metal surface and can be hammered into the surface during scrubbing action. It is critical that soil be suspended and washed away in nonferrous applications.

Exotic Metals

Titanium is very resistant to corrosion, and other materials formed using this metal also tend to exhibit this property. Metals used in jewelry are also resistant to oxidation. Hence, the exotic metals can be processed similarly to a good grade stainless steel. However, when alloyed with aluminum, they should be processed as nonferrous metals.

Deburring

The use of abrasive media mandates a mass-finishing compound rich in suspending and inhibiting agents. A flow-through solution system that meters compound and water carries the fines or residue out of the machine. This prevents redepositing the soil and grinding it into the surface of softer metal parts. Since the abrasive action produces an active metal surface, corrosion-inhibiting agents become an integral component in the compound. When the compound does not provide adequate inhibition, flash rusting will occur on ferrous parts, and pitting, tarnishing, or similar oxidation will result on nonferrous parts. When machined parts are deburred, especially where coolants are used to machine the parts, the deburring compound should also contain emulsifiers for removing the soil. Lack of good a cleaning property will result in residues and fines smeared and imbedded into the oily surface of parts. A properly formulated compound will produce a deburred part that has a clean, normal, metal-looking surface.

Burnishing

Currently, the trend is shifting toward more steel media burnishing. Heavy steel media produces lustrous parts very quickly. The compound used in this process is specifically designed to keep steel media and parts lustrous while preventing corrosion of the steel media. This compound is normally rich in chelating agents to produce brightness and color on the parts. The burnishing compound for steel media is probably the most difficult to formulate, because it must contain ingredients that continually clean and protect the steel media from darkening and corroding, as well as from rusting when not in use. This compound determines the brightness and color of the parts, as well as the life of the steel media.

Optimum burnishing is achieved when the following conditions are met:

1. The steel media are clean and lustrous.
2. The parts to be burnished are clean.
3. The burnishing compound keeps the parts, media, and process chamber clean.
4. There is maximum contact between part and steel media.

Steel media is normally case-hardened and highly polished. If not protected completely from corrosion, the media develops pits that penetrate the case-hardened surface. At this stage, greater care should be taken to keep the media from further corrosion, or it will gradually deteriorate to a point where it will constantly corrode and hinder burnishing.

Rinse-aid/inhibitors become particularly useful in deburring processes when the surface of parts is very reactive and when it is necessary to remove residual soil from parts and inhibit oxidation. When used in dip applications, the tank containing the compound should be checked daily for strength and recharged regularly to ensure effectiveness. When excessive dragover of soil or other contamination is prevalent, the tank or reservoir should be replenished wholly rather than recharged with the compound concentrate.

Each manufacturer provides a somewhat different variation in the compounds it produces. In general, all work well, though some can cause problems, such as smut formation on the surface, lack of normal cleaning, and discoloration. Metering pump systems vary widely in quality. To be effective, they must consistently meter out the correct amount and pH and assure that blockages do not occur that would prevent compound flow. Compound metering/delivery systems can be manual, fully automatic, or semiautomatic. They can be batch systems, recirculating systems, or flow-through solution systems. Manufacturers' literature typically provides enough detail to enable reasonable first-time selection. As for media, the most cost-effective and best-quality results will require some testing. Most vendors will describe each of their products using the criteria shown in Table 6-5. In consideration of oxide contamination, pH controls, surface cleanliness, metallurgical purity, and water pollution, most modern process specifications will not permit work load contamination, microporous media impregnation, galvanic action or chemical etching, mechanical impingement, or surface crystalline changes. The cleaning agents and the water are the most important and least expensive factors. In many cases, a good, clean abrasive media will cut faster than when the best abrasive compound is added.

When using degreasers, cleaners, surfactants, burnishers, and inhibitors, only so much compound can act in a given volume. Twice as much does not necessarily double the work performed—generally over half the compound is wasted, which is a cost factor.

6.5.7 Applying Compounds

For batch operations where workpieces and requirements vary from load to load, manually adding the compounds is the simplest. A correctly weighed amount of compound is added directly to the machine compartment, together with the media, workpieces, and required amount of water. Granular or powdered compounds should be added before the workpieces. Although some authorities consider this method outmoded, it is used where high production is not a requirement.

Semi-automated systems generally require premixing the compound with water in a separate tank. This solution is then pumped into the work container of the finishing machine. Proper control of compound concentration can be a problem with this method.

For most automated mass-finishing applications, compound solutions are added by flow-through systems. With these systems, fresh compound solution is metered into the finishing machine, allowed to perform its functions, and then permitted to drain from the machine. Flow meters ensure precise control of the proportion of compound to water and the amount of solution that enters the machine, thus providing consistently uniform results.

The amount of compound added varies with the specific compound being used and the results desired. In general, .5–4 oz (14.2–113.4 g) of compound is added to each gallon (3.8 L) of water. The compound solution is generally introduced at rates ranging from .5 to 2 gal (1.9–7.6 L) per cubic foot (0.03 m^3) of workload per hour.

6.5.8 Concentration and Flow Rates

Concentration and flow rates are related to each other. In an effort to conserve, some people in manufacturing reduce the flow of either compound or solution, erroneously thinking that this is cost effective. However, this practice only reduces the effectiveness of the processes and can add hours to the cycle time. *A compound used at a concentration level below its threshold of effectiveness is like using no compound at all.*

6.6 GENERAL CONSIDERATIONS

6.6.1 Water Levels

The water levels used in mass-finishing equipment are important to cycle times and the results obtained. Barrels usually require much more water than vibratory machines. Low water levels in barrels, about 2 in. (51 mm) below the top of the mass, provide a harsher action suitable for deburring operations; high levels, about 2 in. (51 mm) above the mass, result in a gentler action and more cushioning, and are suitable for burnishing.

6.6.2 Steps in Mass Finishing

Mass finishing consists of several operational steps:

1. Load media in machine.
2. Load parts in machine.
3. Apply compounds.
4. Run.
5. Retrieve parts from machine.
6. Wash and treat parts.
7. Dry parts.
8. Inspect parts.
9. Package parts.
10. Reclassify media.
11. Store media.

Parts Handling

Parts are brought to mass-finishing equipment by hand, in tote baskets or containers, by conveyors of several types, by lift truck, and in other ways. Parts are loaded into equipment by one of several means: manual dumping or placement by the operator, or automatic dumping from a conveyor or feeder mechanism of some type. The parts may be placed in a certain area of the machine, or they may be spread around. Parts leave mass-finishing equipment by similar means.

Separation of the media from parts can be done in several ways as well: by screen deck or tie rod screen on single, double, or triple decks; by hand; magnetically; by auxiliary separation deck, either internal or external to the machine; and with various materials of construction.

Post Dip

After mass finishing, many industrial parts receive a post dip of some kind to prevent corrosion, to reduce water film thickness prior to drying, or to provide a "good" fresh-water rinse. This dip may be accomplished by spray rinse or dip tank and drag-out, either manual or automatic.

Drying

Parts may or may not be dried. They may be air dried; blow dried; spin dried; hot-air dried with gas, steam, or electricity; dried with cobmeal or wood pegs; or hand dried. Drying may be manual or automatic.

Media Handling and Storage

Media can be left in the mass-finishing machine after use, removed from and returned to the machine, sent to an overhead storage hopper, put in tote pans or steel drums, etc. These processes can be performed manually, by tilt truck, or overhead crane; automatically; or by other means. In the process, media may or may not become contaminated or corroded.

Media Classification

Separating out worn media (classifying media) is important in some applications to prevent media from lodging in holes and crevices. Worn media also may need to be removed because its cutting properties may change as its size reduces. They can be classified in one of several ways: internally (within the machine), externally (during separation of parts from media or as a separate step), or by hand. Even large shops use hand separation economically, but the volume of work required and difficulty of removal will determine whether automation is required in this process. One company found that failure to screen out undersized media on precision parts resulted in scrapping entire lots of parts because the media could not be removed without severely scratching precision surfaces. For some parts, the operator may easily dislodge media while the mass is moving in the vibrator.

6.7 MASS-FINISHING COSTS

In comparing various processes for possible use, the total cost per finished workpiece must be a major consideration. Many factors enter into the equation, including the costs of capital equipment, depreciation, maintenance, labor, and power. The cost of perishable items, such as media, compound, and water, also must be considered. Overhead costs include management, supervision, insurance, heating and lighting, floor space, and waste disposal. The possible need for subsequent cleaning or other processing can also affect costs. Scrap, rework, and inspection represent additional expenses.

Chapter 1 provides equations for approximating costs. More complex equations can be developed readily for such factors as waste removal, scrap, etc. Several examples of costs for mass-finishing processes are described elsewhere (for example, see Kittredge 1985a; Balz 1983; Johannesen 1977; Ostwald 1983, 1985, 1986; Rhoades 1981; Ioi et al. 1985). At least two software packages make evaluating alternatives much easier than manually calculating costs (Kittredge 1985a; Balz 1983). Computer-aided selection of processes and computer control of finishing is standard in the industry. At least one facility has a finishing flexible-manufacturing system (Finishing FMS) (Ioi et al. 1985). Additional cost considerations for these processes can be found elsewhere (Johannesen 1977).

In the selection of any mass-finishing equipment, the initial cost includes the invoice price, local sales tax, freight-in, installation charges (such as rigging, plumbing, electrical, gas, and oil connections), and additional contractor expense (such as welding or machining, extra labor, and overtime). Often, the initial charge of media and compounds required to start up the unit is adjusted to achieve optimum production. For example, an installation costing over $20,000 for a burnishing machine is simple enough to set up on the books. However, in addition to the costs just mentioned, for a large machine, 5,000 lb (2,268 kg) of stainless steel burnishing media (at $3–10/lb [$1.36–4.54/kg]) also must be purchased before the unit can be used. This additional $15,000–50,000 may be added to the initial cost or treated as an operating expense. (Incidentally, this high initial cost for stainless steel media has been shown to outlast three burnishing machines that became technologically obsolete. The original stainless steel media are as perfect today as they were 12 years ago.)

After equipment, supervision, and operator costs have been considered, probably the most fundamental cost consideration involves the choice of media and compound.

Tables 6-7 and 6-8 present costs from one study of an aerospace manufacturer. Costs for three plastic media and two ceramic media were compared for running under normal and other conditions. The combination of low attrition rate and high cutting power of the two ceramic bonded media clearly established them as leaders in cost effectiveness over the resin bonded media for deburring the type of part represented by the test pieces. Whereas compounds are normally used for a variety of purposes, clearly, running dry or using water only resulted in the lowest cost, and in this application caused no problems over a nine-month period (Bostock 1981).

Table 6-9 provides a worksheet for calculating total finishing costs for mass finishing.

Mass-finishing costs are characteristically low, and processes do not need to be optimized to work well. In fact, the processes are so simple and routine that it is easy to lose

Table 6-7. Media cost comparison

Media	Attrition rate lb/hr (kg/hr)	Cost of media loss/hr	Cutting efficiency parts/hr	Media cost parts/hr	Total cost, media and labor parts/hr
A	.8 (0.4)	$1.28	1.42	$0.90	Out of range
B	.8 (0.4)	$0.99	.60	$1.65	Out of range
C	2.0 (0.9)	$2.92	.56	$5.21	Out of range
D	.8 (0.4)	$1.14	2.50	$0.46	$1.10
E	.2 (0.09)	$0.28	2.25	$0.12	$0.83

Note: Media A, B, and C are resin (plastic) media containing abrasives. Media D and E are aluminum oxide ceramic media. All media were approximately 1.5 in. (38.1 mm) arrowhead, tri-stars, or triangles. Parts were milled aluminum troughs having .0005-in. (12.7-μm) thick × .0003-in. (7.6-μm) high burrs.

sight of the total costs. There is still money to be "made" by understanding daily mass-finishing operations and searching for cost-effective improvements.

One of the forgotten elements of mass finishing is the people involved. Only people make decisions. They fill the machine with media when it gets low, maintain the manufacturers' lubrication schedules, buy a "cheaper" media or compound, "save money" by reducing flow rates or classification schedules, tinker with weight or speed or other settings, etc. Experienced people can improve operations; if they are not knowledgeable, processes may be misadjusted, resulting in a more imperfect operation. Trained staff are a blessing. Untrained staff can raise havoc with an otherwise well-designed system.

6.8 DESIGNING TO ALLOW MORE PROCESS CHOICES

Chapters 1, 3, and 5 described the influence of part edge angles on burrs and deburring. Figure 6-2 illustrates another significant aspect of angles. The amount of radius that can be produced economically by vibratory finishing after removing the burrs is a function of the angle between intersecting surfaces. Large radii can be produced relatively quickly when the included angle is large. Finishing times are 20 times longer when the included angle is 30° than when it is 120° (see Figure 5-8 for an understanding of the stock loss in this operation). Precision edge-radius tolerances are harder to maintain, however, when large angles are present. When a component has features involving several different edge angles, edge radii will vary significantly. Designers must recognize this variation when assigning tolerances to edge radii if they want to eliminate the extra costs required to produce equal radii.

6.8.1 Designing for Mass Finishing

Although many designers do not know which deburring process will ultimately be used to remove burrs or flash, it is significant that at least 50% of all parts produced in the world are subjected to mass-finishing processes. Because these processes all use the same basic metal removal mechanism, there are a few common considerations that affect workpiece design. Essential factors in all mass-finishing processes include the following:

1. The media used must have access to and be able to slide over all critical surfaces and edges that require deburring.
2. Large media deburrs faster than small media, except on very small parts.
3. The use of special preformed shapes, such as triangles, cones, and cylinders, has advantages over gravel-like media.

As was shown in Figure 5-30a, the step height, H, is smaller than the radius R_s on the deburring media, which in this case is a triangle. As a result, the media will often bend the burr over rather than remove it. Complete removal can be obtained only with long cycle times or by using blends of media with smaller sizes present. If the designer can specify a larger step height, as shown in Figure 5-30b, this problem is eliminated because the media can work over the edge and against both surfaces. For deburring slots with triangular media having a standard configuration, the slot width, b (see Figure 5-31), must be greater than $3R_s$, and the depth must be greater than R_s.

When inner surfaces near fillet radii must be smoothed by mass-finishing processes, it is essential that the fillet radii specified be equal to or exceed the major radius on the media (see Figure 5-32). This is a particularly significant consideration when the feature is an extruded item, which is controlled by the die shape. Small steps in a workpiece will not receive significant deburring action, even with small media or balls, because size and geometry prohibit contact. Other media shapes, however, usually overcome these problems.

When a part requires similar edge radii and deburring quality on all edges (see Figure 6-2), smaller media sizes are required. Radiusing action is largest where larger media can most readily contact the part. Smaller media will reduce, but not eliminate, the differences shown in Figure 6-2. Although such differences are not critical on most parts, they are on many others. Complete deburring can occur only when the burr is totally accessible by the media.

Part Geometry, Tolerance, and Surface Finish

Chapter 1 presented several overview considerations for edge finishing, including the need to understand how precision surface finishes and tolerances on exposed edges limit

Table 6-8. Effect of operating parameters on media cost

Parameter	Abrasive or detergent cost/hr	Media attrition cost/hr	Cutting efficiency parts/hr, average	Materials cost/part/hr	Total materials and labor cost/part/hr
Dry	—	—	1.42	—	$1.30
Water	—	$0.36	1.86	$0.19	$1.05
Water and detergent	$1.12	$0.28	2.25	$0.62	$1.34
Water, detergent, and abrasive X	$1.52	$1.33	1.89	$1.51	$2.36
Water, detergent, and abrasive Y	$1.78	$1.06	1.13	$2.52	$3.93

Note: Data are based on media E in Table 6-7.

Table 6-9. Detailed cost worksheet for mass-finishing compounds

	Costs per load								
	Material			Labor			Machine		
	Compound (lb)	Cost ($/lb)	Extension ($)	Time (hr)	Cost ($/hr)	Extension ($)	Time (hr)	Cost ($/hr)	Extension ($)
Operation no.									
Descale									
Rinse									
Neutralize									
Rinse									
Operation no.									
Degrease									
Inhibit									
Rinse									
Operation no.									
Deburr									
Inhibit									
Rinse									
Operation no.									
Burnish									
Inhibit									
Rinse									
Unload									
Separate Manually									
Mechanically									
Screen size									
Change media									
Dry									
Load									
Process									
Unload									
Totals									

Compound cost: _________ lb $ _________/lb $_________

Labor time: _________ hr $ _________/hr $_________

Machine time: _________ hr $_________/hr $_________

Machine time: _________ hr media wear rate _________lb/hr $_________/lb $_________

Total job cost $_________

Unit cost $_________

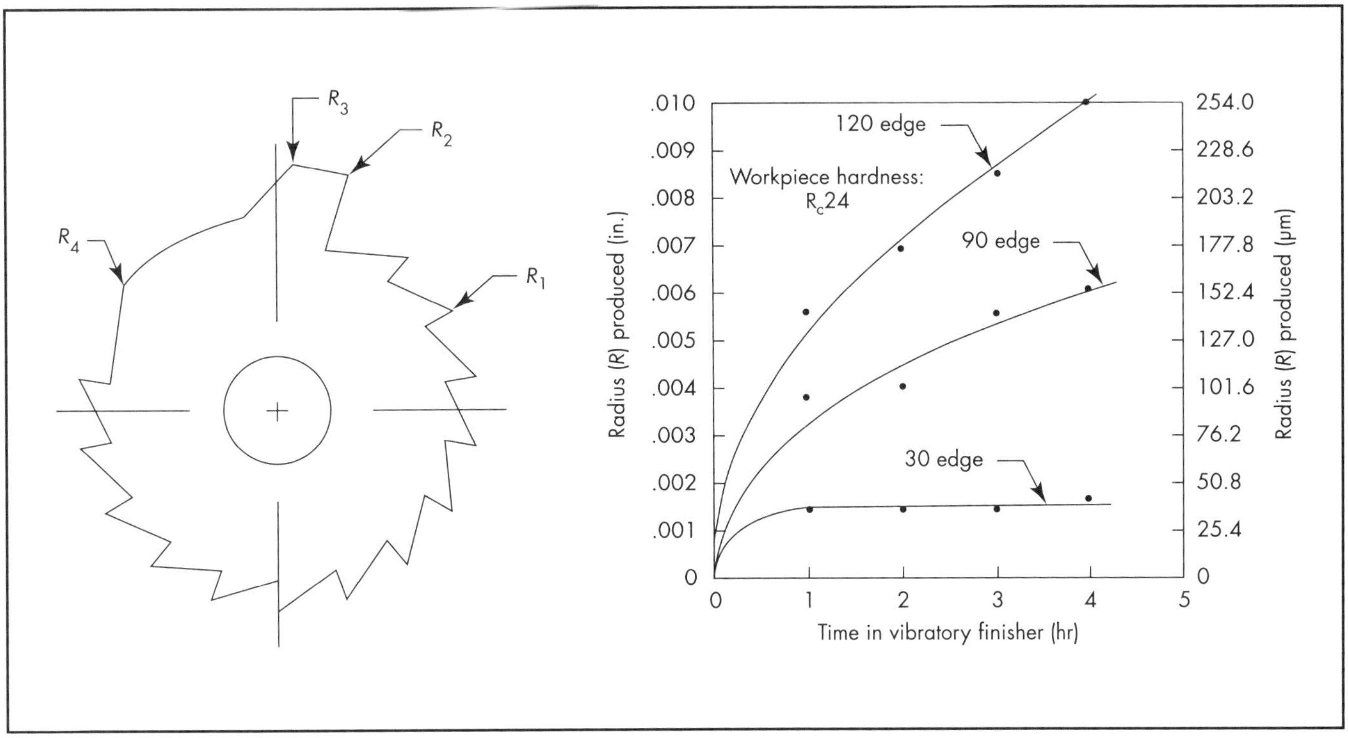

Figure 6-2. Time required to remove the burr and radius the edges of a ratchet wheel.

the number of processes that can be used for deburring. Some burr sizes on precision parts can only be removed by manual deburring or specially designed machines. The data shown in the tables in Chapter 1 are not widely known, but they are critical to process selection and part cost.

Designing for Removing Sharp Edges

On many parts, the only significant edge requirement is that all sharp edges be removed. In such cases, beating over burrs and dulling edges is adequate. Designers can handle this requirement in at least two ways:

1. By specifying the process that gives an acceptable edge.
2. By defining the actual edge quality needed.

Vibrating parts in steel media will dull the edges of most parts economically. Similarly, on many materials, the thermal energy method (TEM), discussed in Chapter 15, can be used to ensure that no loose burrs or particles will be present to jam assemblies. Specifying on the drawing that parts should be vibrated in steel media to dull edges is often done for parts made within a plant; however, such notes are not complete enough for work contracted to others. Outside vendors must know what size steel ball to use, how long to run the machine, and the amplitude and frequency of the machine to be used. These requirements can be specified by developing explicit processing standards and referring to them on the part drawings. Such an approach is easy and relatively problem-free when the majority of parts have similar requirements.

6.9 PLANT LAYOUT CONSIDERATIONS

Normally, mass finishing is performed in an area designed to accommodate water sprays and water removal. These areas, often on raised grate floors, can be as small as 15 × 15 ft (4.6 × 4.6 m). Although floors need not be wet, anytime water is handled some will end up on the floor, so planning is necessary for safety, appearance, and control of wastes.

Some vibratory deburring machines are designed to be portable and roll up to automatic screw machines, lathes, sheet metal operations, etc. This can provide immediate feedback to the machine operator that burrs may be becoming too large to remove. In addition, it provides total cell operation for faster product flow.

6.10 PLANNING FOR WASTES

The waste from mass-finishing operations generally consists of the following:

- Fine abrasive particles in a sludge.
- Metal fines or dissolved metal fines.
- Fine abrasive particles suspended in water.
- Oils (dissolved, dispersed, and emulsified).
- Plastics resins.
- Vitrified clay.

Disposal of wastes is largely controlled by the Code of Federal Regulations. EPA, OSHA, and other state and local agencies may have additional regulations. Generally, federal regulations for metal finishing will be found in 40 CFR Part 433 or 40 CFR 261.24. Regulations for plastics finishing will be found in Part 414 and others, depending on the process and material used. Since the regulations do change, users are encouraged to keep abreast of changes through local organizations, publications for the finishing industry, and technical societies. *Metal Finishing* magazine and the American Electroplaters and Surface Finishers Society are examples of such groups. There are two new amendments to the Federal Resource Conservation and Recovery Act (RCRA) that affect metal finishers using mass-finishing methods far more than did the original act.

The RCRA affects any small-quantity generator of sludge who disposes of less than 220 lb (100 kg) of sludge per month (about 26 gal [98 L] of sludge per month or a gal [3.8 L] per day). Bulk liquid and free liquid containing hazardous wastes are prohibited from being placed in any landfill. The key word in the amendment is *hazardous*.

These regulations create a situation in which users of mass-finishing machines with deburring media have three potential problems: sludge disposal, dissolved heavy metals in an effluent stream, and total suspended solids in the effluent stream. In addition, oils and plastic resins may be in some streams.

6.10.1 Sludge

A typical mass finisher uses about 1 gal (3.8 L) of water per hour per ft^3 (0.028 m^3) of vibrator capacity (Miller 1985). This water, which usually contains soap with chelating agents, washes the solids that are generated in the mass from the machine. A generally accepted rule of thumb is that the media will wear at the rate of .15–.50% by weight per hour. Thus, a 30 ft^3 (0.85 m^3) vibrator using ceramic media will have a flow of about 30 gal (114 L) per hour with about 2–6 lb (1–3 kg) of solids on a dry weight basis flowing from it per hour.

Typically, the effluent is directed into a clarifier of some kind. Many clarifiers do not have clear liquid coming from them. So a portion of the solids settles, forming the sludge, and a portion goes out the clarifier and down the drain.

The sludge that builds up is about 1–5% solids by weight and must eventually be disposed of. Strictly speaking, it is the operator's duty, under current environmental regulations, to determine whether the sludge is hazardous. The allowable limits for hazardous metals in sludge under Federal EPA regulations are shown in Table 6-10.

Some states and cities have different limits. The limits shown in Table 6-10 address only federal requirements. Users should check with local agencies, such as state health departments and county sanitation departments. In some areas, companies deburring zinc die castings will have enough zinc in their sludge for it to be classified as hazardous. High quantities of zinc may be found in some sludge where only aluminum was being deburred. Cadmium and copper has been found in sludges where the operator does not deburr cadmium or copper. The explanation for these may be that some companies occasionally reclaim copper or cadmium-plated parts by running the reject parts through their deburring operation without informing the deburring department.

Table 6-10. 1996 EPA limits for hazardous metals in sludge

Metal	Pretreatment daily maximum* oz/gal (mg/L)	Sludge TCLP limit** oz/gal (mg/L)
Zn	.00035 (2.62)	—
Cu	.00045 (3.37)	—
Pb	.00009 (0.67)	.00066 (4.94)
Cd	.00001 (0.08)	.00013 (0.97)
Cr^{+3}	.00036 (2.70)	.00066 (4.94)
Cr^{+6}	.00036 (2.70)	.00066 (4.94)
Ni	.00053 (3.97)	—

Note: TCLP = toxicity characteristic leaching procedure.
* See 40CFR433.17 (Code of Federal Register).
** See 40CFR261.24.

It is highly recommend that users have their sludge analyzed prior to disposing of it. If the analysis indicates that the sludge is nonhazardous, clarifiers can pump it into a truck and transport it to a landfill. This can cost $1,300 per 100 gal (379 L) or several times this amount. Alternately, users can filter the sludge in a filter press and produce a dry filter cake, which can be dumped into existing trash bins for disposal in conventional landfills.

However, if the sludge in the clarifier or sump is hazardous, then according to the regulations, it must be dewatered prior to disposal in a landfill certified to receive hazardous waste. There are three basic techniques for dewatering clarifier sludge: filtration; centrifugation; and heating. The most cost effective by far is filtration. Heating uses too much energy. Centrifugation uses considerable energy, has high initial capital costs, and requires routine maintenance.

Many types of filters are available, but most are not effective for dewatering deburring sludge. Bag filters and cartridge filters have the lowest initial capital costs. However, they have very small solids holding capacity, so cartridges and bags must be changed far too often for the process to be economical in high-volume use. Moving-bed filters, though excellent for easily filterable solids, such as metal turnings or chips, are not suitable for slimy sludges. As with cartridge or bag filters, the filter media of the moving-bed filter must be changed or indexed frequently, resulting in high operating costs. The continuous precoat filter is expensive. The so-called dry cake, often not dry, has a relatively high operating cost, since it must be precoated with 3–6 in. (7.6–15.2 cm) of diatomaceous earth prior to the start of each filtration cycle. Dry cake is produced by using this filtering. The pressure leaf filter is expensive, has a high operating cost, and requires considerable operator skill, time, and maintenance.

An often preferred method is a filter press. The initial cost of a press is relatively high, but operating costs, maintenance costs, and operator time are all very low. In addition, the filter press yields the driest cake from deburring sludges of all the filters. The term *filter press* is slightly misleading. It sounds as if the press presses the solids, which it does not. The dewatering action comes from the solids being pumped into a chamber or chambers at 80–90 psi (.55–.62 MPa) gage pressure. At these pressures, the solids are pushed together with such force that the liquid is driven out from among them. The liquid is driven through a filter cloth and out of the filter.

Several devices are available for dewatering sludge. Rotary vacuum filters will concentrate sludge containing 4–8% solids to 20–25% solids. Since the effluent concentration of solids is generally less than 4%, a thickener tank is often employed between the clarifier and the filter. The filtrate often contains excessive suspended solids and is recirculated to the clarifier. Centrifuges will also thicken sludges to the above range of consistency and have the advantage of using less floor space. The effluent contains excessive suspended solids and is returned to the clarifier.

Figure 6-3 shows a schematic using a small filter press. The filter chambers are formed by recessed plastic plates. The actual filtering surface is covered with polypropylene cloth. The plates are pushed together by hydraulic cylinders. Sludge from synthetic media provides the best clarity, followed by ceramic and plastic. All samples are superior to clarifiers.

An air-operated diaphragm pump is most commonly used for pumping from sump or clarifier into the filter. By setting the compressed air that drives the pump at 80 psi (.55 MPa) gage pressure, the final pressure on the solids in the filter is also controlled. When the pressure in the filter press equals the pressure put out by the pump, the pump stops. It is then time to open the filter press and dump the dewatered solids. The press is closed and the cycle is started again. There are no bags, cartridges, or rolls of paper to replace.

Costs for disposing solids can start at $85–100 per drum and go much higher. The cost for disposing of 1,000 gal (3,785 L) of hazardous sludge at 3% solids can begin at about $1,800 and go much higher. By filtering the 1,000 gal (3,785 L), it would be reduced to no more than two drums of solids costing about $200 (at the low end) for disposal, saving about $1,600 in disposal costs.

The water that comes from the filter press may be reused, sewered, or require further processing. If it contains dissolved heavy metals in amounts greater than allowed by law for discharge into the local sewer, it must be treated to remove them.

Pressure filters also may be used. In contrast to rotary filters and centrifuges, pressure filters will produce a filtrate with less than .0004 oz/gal (3 mg/L) of suspended solids so that return to the clarifier is not necessary. The filter cake contains approximately 20–25% solids. Pressure filters are usually designed for a filtration rate of .05–.06 gal/min/ft^2 (2.04–2.44 L/min/m^2) of clarifier sludge. Solids contents of 25–35% in filter cakes can be achieved with semi-continuous tank filters rated at .25–.33 gal/min/ft^2 (10.17–13.43 L/min/m^2) surface. A solids content of less than .0004 oz/gal (3 mg/L) is normally accepted for direct effluent discharge. The units require minimum floor space. Plate and frame presses produce filter cakes of 40–50% dry solids and a filtrate with less than .0007 oz/gal (5 mg/L) total suspended solids. Because automating these presses is difficult, labor costs tend to be high. The operating costs are partially offset by low capital equipment costs.

Automated tank-type pressure filters are just now finding application. The solids content of the cake can reach as high as 60%, while the filtrate may have up to .0007 oz/gal (5 mg/L) of total suspended solids. The filtration rate is approximately .05 gal/min/ft^2 (2.04 L/min/m^2) filter surface

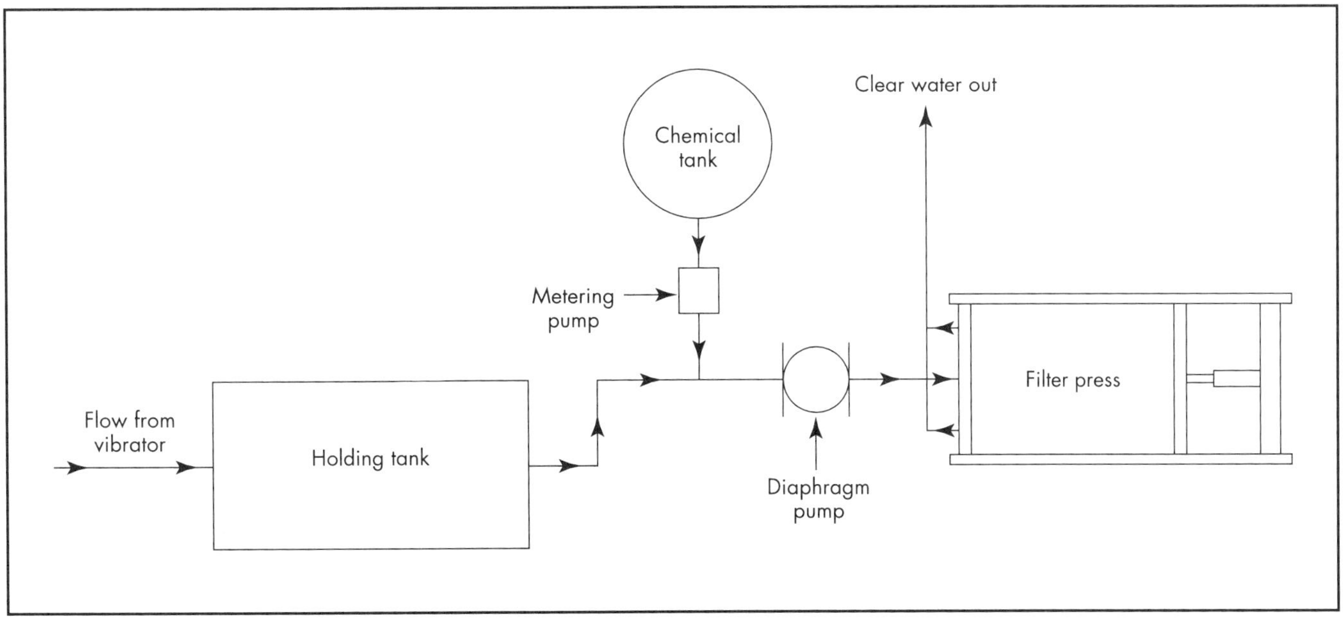

Figure 6-3. Small filter press operation.

area. Pressure filters can contain .04–.07 oz/gal (300–524 mg/L) suspended solids at a design rate of .12–.16 gal/min/ft^2 (4.88–6.51 L/min/m^2) and still maintain a low solids content in the filtrate. Filter cakes can easily be collected in solid waste containers and hauled away to landfills. There may be situations, however, where the metal in the filter cake would be redissolved if it were to come into contact with acidic water. Careful consideration should be given to where such a material is dumped (Kittredge 1977).

6.10.2 Dissolved Solids and Suspended Solids

The terms *dissolved solids* and *suspended solids* often cause confusion. A dissolved solid is truly in solution and possesses an ionic charge. It cannot be seen with the naked eye. It will not settle out nor can it be filtered or centrifuged. An example is common table salt dissolved in water. Dissolved metals must be converted back to solid form to be removed. This conversion is generally done by chemically reacting the metal with hydroxide or sulfide to form a solid particle of metal hydroxide or metal sulfide.

A suspended solid is a solid that happens to be small enough or light enough to be carried, supported, or transported by a liquid. The silt in a river is a good example of a suspended solid. Most suspended solids will settle out of a liquid if given enough time, or they can be filtered or centrifuged out. They do not have an ionic charge and will not react with hydroxides or sulfides.

If the effluent from the filter or the clarifier contains dissolved heavy metals in amounts greater than allowed for dumping into the sewer, they must be treated. The most common form of treatment is precipitation with ammonium hydroxide, sodium hydroxide, or a sulfur-containing chemical. Ion exchange, reverse osmosis, and plating out the metals are other methods currently in use. However, these latter methods are considerably more expensive than the former.

Precipitation of heavy metals with ammonium hydroxide or sodium hydroxide requires adjusting the pH of the deburring effluent to 9.5. At this pH, copper, lead, cadmium, zinc, nickel, and chromium should precipitate out. However, most soaps used in deburring contain chelating agents, which are chemicals that tend to hold metal ions in solution stoichiometrically. As a result, even at a pH of 9.5, a portion of the metals stays in solution. To prevent this effect, a pretreatment step should be included to render the chelating agents ineffective or the metals should be precipitated as sulfides. There are chemicals available containing sulfur that are more powerful than the chelating agents and will precipitate the metals in the presence of chelators. Further, these sulfur-containing compounds operate over a pH range of 3–11, so pH adjustment is unnecessary.

One approach is to use dithiocarbamate. Using deburring effluent from two zinc die cast companies, one supplier removed the solids by filtration and analyzed the remaining clear liquid for zinc. One sample contained 150 ppm, the other 50 ppm. The federal limit for zinc in effluent going to the sewer was 3.5 ppm. After using twice the theoretical amount of carbamate necessary to precipitate zinc and filtering again, each sample had less than 0.5 ppm zinc.

Another approach is to simply inject sodium dimethyl dithiocarbamate into the deburring effluent prior to the pump feeding the filter press, as shown in Figure 6-3. The carbamate precipitates the zinc and the heavy metals, which can be removed in the filter press along with the other already existing solids in the effluent. This should produce dry hazardous cake for landfill and clear nonhazardous water for reuse or the sewer.

At higher flow rates, or if the clarity of the water from the filter is not acceptable, a flocculent may be used. A flocculent is a chemical that can be added to the solution prior to filtration, causing small particles to form larger particles. A flocculent acts on suspended solids, not dissolved solids, and requires 5 min to 1 hr. If a flocculent is used, it is added to the solution in a combination holding/treatment tank or injected in line, as shown in Figure 6-4. Tank sizes vary depending on how long the flocculent takes to form and whether a continuous or batch treatment system is being used.

Many of the clarifiers used on deburring effluent do not remove enough of the suspended solids to meet regulations. If a filter press is used and the effluent is forced through the press, the effluent will generally be within allowable limits for suspended solids.

It is important to provide some treatment prior to evacuation into plant lines. Concentrations of the inert suspended solids will often cause obstructions in the plant sewer system before they reach the pretreatment area. Most of the abrasive compounds contain abrasive particles in the range of 80–100 grit sizes, maximum of .0115 in. (292 μm) to a minimum of .0025 in. (64 μm), which will further break down by normal use to colloidal size range where the particle size is 39.37–3,937 μin. (1–100 μm) (Johannesen 1981). Generally, the effluent carrying these solids is directed into the plant drainage system, which does not have high volumes of wastewater or solutions. The solids will settle in slowly moving systems and will also solidify in the standard drain gradient. The combination of these solids, the ceramic binders, and the metallic oxides is sometimes hard enough to divert a "Roto-Rooter" type of pipe cleaner out through the side of cast iron drain pipe.

Large companies may locate the deburring department near a "live drain" that also carries relatively high volumes of rinse water, cooling water, or other wastewaters destined for the plant clarifying system. In such large plants, the contribution of inert suspended solids is sometimes regarded as an ideal "binder" for other precipitants that will be dewatered by various means. The drainage distance still cannot be great, because the dense particles will settle in any quiescent area. On the other hand, the small captive shop or small job shop may actually have no drainage wastewater. The effluent from vibratory mass-finishing units, at an average rate of 2 gal per 7.5 gal (7.6 L per 28.4 L) of mass per hour, or 75 gal (284 L) discharging 20 gal (76 L) of solution per hour, and possibly 8 oz (227 g) of solids per hour is

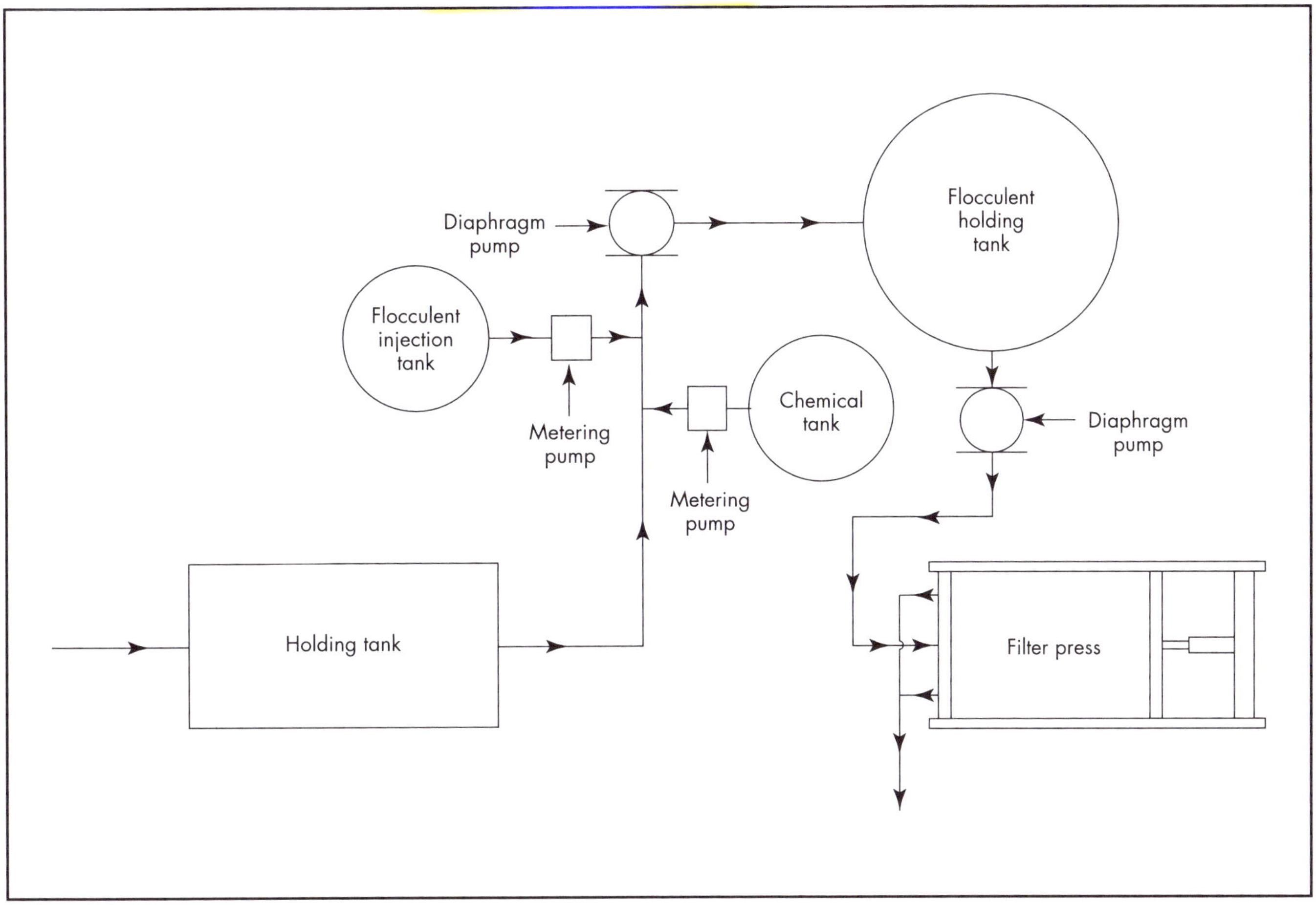

Figure 6-4. Flocculent holding tank system.

almost a "nuisance amount." It can be collected in a 10–20 gal (38–76 L) container—the overflow of which is probably within the ppm load limit.

6.10.3 The Problem of Soils

Three types of soils are encountered in metal cleaning: oily, semisolid, and solid. These types of soils are removed by one or more of the following mechanisms:

1. *Saponification:* Animal and vegetable oils may be partially removed by this reaction, but saponification is slow and usually incomplete.
2. *Emulsification:* This is a mechanism whereby the soil is broken up and suspended in the cleaner in the form of very fine liquid particles.
3. *Dispersion:* By this process, the soil is dispersed, thereby lowering the concentration of soil at the surface of the work.
4. *Flocculation or aggregation*: Once the soil has been removed and dispersed, it should reunite, away from the work, to precipitate out or be skimmed off. The ability of a cleaning solution to perform this function increases its life and results in cleaner work, because suspended soil is not redeposited on the work (Ostwald 1983).

6.10.4 The Problem of Suspended Solids

Included in the category of suspended solids are mechanically produced particles arising directly from manufacturing operations and metallic hydroxides precipitated from solution in chemical processes. The allowable limit for discharging suspended solids into a surface stream is .003 oz/gal (22.5 g/m^3) or less; when discharging to a sanitary system, the limit may be as high as .03 oz/gal (225 g/m^3) provided that no troublesome or toxic constituents are included.

Suspended solids are usually removed or reduced by coagulation and clarification. These processes involve adjusting the pH and adding chemical coagulants and/or coagulant aids to form a stable, rapid-settling floc, which is separated from the water by settling. Chemical clarification improves settling rates and provides a more consistent effluent quality than is obtainable with straight sedimentation (Miller 1985).

Removing Suspended Solids

Among the techniques for removing suspended solids is clarification. *Clarification* is the separation of suspended solids, metal hydroxide, and other settleable impurities that are heavier than water by gravitational settling. This process has been in general industrial use for many years and is the most commonly used technique for removing settleable material from wastewater. Additionally, inorganic coagulants or polyelectrolytic flocculants may be added to the waste stream following pH adjustment and prior to gravitational separation. These agents are used to enhance settling by coagulating small suspended precipitates into large particles. Common coagulants are aluminum sulfate, sodium aluminate, sodium sulfide, ferrous or ferric sulfate, and ferric chloride. Organic polyelectrolytes vary in structure, but all usually form larger floccules than coagulants used alone. Depending on the particular application, coagulants and polyelectrolytes may be used together.

6.10.5 The Problem of Sedimentation and Sludge

The major advantage of simple sedimentation (settling without the aid of coagulants) is the simplicity of the process itself—the gravitational settling of solid particulate waste in a holding tank or lagoon. It is also possible, with small sedimentation tanks, to use hydraulic sludge removal techniques, as discussed earlier. A disadvantage of simple sedimentation is that long retention times may be necessary to achieve complete settling, especially if the specific gravity of the suspended matter is close to that of water. Some materials cannot be practically removed by sedimentation alone, and chemicals must be added to achieve complete removal.

The major advantage of clarifiers is that they are effective in removing slow-settling suspended matter in a shorter time and in less space than a presedimentation system. Improved performance is obtained by adjusting the pH to nine or higher; this increases the rate of flocculation because of the pollutant solubility characteristics and, in many cases, noticeably improves effluent quality. This rapid flocculation and removal may require flushing or other maintenance to downstream equipment, such as filters and pumps. A properly operating clarification system is capable of efficient removal of suspended solids.

The clarification process has been in general use for many years and is the most commonly used technique for removing settleable material from wastewater. The process is highly reliabile, although corrosion due to caustic chemicals used for pH adjustment may cause premature failures. When clarifiers are used, the associated systems used for adding chemicals, stirring, and sludge dragout must be maintained on a regular basis. Systems external to the clarifier tank present minimal operational problems, whereas internal systems may require emptying for maintenance to be accomplished. Routine maintenance will generally consist of lubrication, checking for excessive wear, and part replacement as required. When lagoons are used, maintenance problems similar to those discussed in relation to chemical treatment arise, but the lagoon itself requires little maintenance other than periodically removing sludge. The sludge collected from clarification is usually dewatered and then landfilled, incinerated, or hauled away by a contractor.

6.10.6 The Problem of Hydraulic Loading

A common assumption in the deburring segment of the metal-finishing industry is that a vibratory deburring unit requires 1–3 gal (3.8–11.4 L) of water, premix, or solution per 7.5 gal (per ft^3) (265 L per m^3) of mass per hour of operation. This is also the effluent rate (the EPA calls this the *hydraulic loading rate*). For example, a vibratory deburring unit with 10 ft^3 (0.283 m^3) of usable volume could require 10–30 gal/hr (38–114 L/hr) of flow-through. This rate is not enough to hold solids in suspension for any length of holding tank or for any reasonable period of time.

In primary sedimentation systems, the depth of the tank is not a significant factor when the main object is to remove settleable solids (or floatable solids by a skimmer). It is the low flow rate that is important. Many tons of settleable solids have been removed from open drain trenches in the floor. Bricks, cinder blocks, steel angle iron sections, and so on serve as effective weirs in the drain trenches. Of course the drain should be wide enough to accommodate a standard square-pointed shovel, and have a standard grill or grate covering to facilitate periodic cleaning. Settleable solids have also been "trapped" in sections of 55 gal (208.2 L) barrels or any other container offering some degree of interruption to the flow, such as septic tanks.

Many settling tanks are fabricated with weirs or partitions to separate the nonturbulent areas. Each partition has a notch on the alternate end so that the effluent can flow along the longest possible routes, thus increasing the settling time. The volumes of these tanks and the number of partition/weirs can be calculated to provide the maximum nonturbulent effluent rate (hydraulic loading) anticipated. The size or sophistication of any settling tank can be negated by inadequate maintenance or cleaning to remove the solids. When the tanks or partitions are filled with solids, the system is no longer effective (that is, providing nonturbulent settling areas). Nonturbulent settling areas are what make the system ineffective.

At least one user who makes his own settling tank has a rule of thumb for sizing them: make the tank volume one and a half times the hourly flow when ceramic media are used and two times the hourly flow when plastic media are used. Equipment manufacturers reportedly make the settling tanks about five times (in gallons) the cubic foot size of the vibratory machine. Treatment costs for effluent wastes are increasing, discharge requirements are becoming more stringent, and urbanization has increasingly limited the availability of land for dumping purposes and sanitary landfills. Thus, there are many problems to be dealt with in handling the industry's wastes.

6.10.7 Equipment Selection

Several other types of effluent clarifiers are being marketed to solve effluent problems. Three segments of the industry—large captive shops, small captive shops, and job shops—now face the economics of hydraulic loads. Several reputable companies now advertise "low-cost continuous filtration to as low as 19.7 µin. (0.5 µm)"; and others claim capacities of several thousand gallons per minute. The general concept of operation in several units is to pump the unfiltered effluent onto an endless conveyor. The conveyor's sloping sides create a pool for the effluent, while the inclined conveyor or inclined discharge ramp delivers the solids into a receptacle. This simplified concept has been engineered with elaborate manual or automatic options. Many grades, sizes, and efficiencies of filter media have been developed that are directed (from rolls) onto the top of the conveyor as a thin "blanket" filter media. When the rate of liquid flowing through the filter media is appreciably slowed by the accumulating solids, the liquid level rises and activates a float control. The motor-driven conveyor or belt is advanced, eventually dumping the solid waste-laden media into a receptacle. This advance moves a section of fresh media into position.

On some models, the wet media are held securely to the conveyor by adhesion. On others, vacuum is applied on the underside of the conveyor. The vacuum draws the liquid through the media/conveyor into the filtered fluid tank. The process continues until the solids build up on the media, restricting liquid passage through the media, which also causes an increase in vacuum pressure to a preset adjustment. At this point, the vacuum is turned off, the conveyor and media advance a preset distance, and the cycle repeats automatically.

On other units, a thin layer of diatomaceous earth, textile, or synthetic fibrous material is dispersed on top of the filter media to increase the filter action. A dewatering area can be added at which the solids are dewatered by heat, vacuum, or pressure; a media cake is then delivered that has a low enough moisture content to be acceptably landfilled.

Still other manufacturers have used the principle of centrifugal force generated in a cyclonic conical chamber. These systems have been developed in sizes ranging from one small cyclone to multiple installations of large cyclones. The effluent is pumped or drawn by vacuum through the cyclonic chambers, where it is cleaned by centrifugal forces up to 7,500 gravity. These units can effectively remove particles as small as .00019 in. (5 μm) in diameter. The number of cyclonic cones can vary from one to 30 or more, depending on the hydraulic load considered. Of course, the pump output through the cyclonic cone system far exceeds the effluent input so that these units continually recycle the effluent up to 50 times the effluent flow rate. The solids clarified by the cones are discharged continuously into a sludge container or by dragout conveyor as a wet sludge, with the overflow returning to the pump area for recirculation through the cyclones.

Some manufacturers have combined the belt conveyor media principle with the centrifugal force cyclone cones whereby the cone's discharge is dispersed onto the filter media. Many other types and flow ranges of in-line filter clarifiers have been developed to further purify the effluent. Systems including chemical or carbon canisters have been developed to handle any size and type of effluent clarification required to purify the wastewater. Pairs of filter chambers can be installed with pressure drop monitors that automatically change the flow to alternate filter chambers and also signal the maintenance or operating personnel.

Filter cartridge elements are available in various porosities for the flow rate and maximum solids holding capacity. Some of the cartridges are designed to be thrown away, and others can be cleaned and reused. These cartridge elements consist of a series of fibrous layers, each progressively denser and capable of stopping finer and fewer particles as the effluent passes through them. The cartridges can be precipitated with a filter aid, making it possible to rinse and reuse, or other filter media, such as activated carbon, pleated cellulose, and porous ceramic. In both the conveyor media filter concept and the in-line filter chambers, the solids retention area in square feet (or square meters) is the measure of effective filter surface area.

6.11 MASS-FINISHING PROCESS DETAILS

The preceding material in this chapter applies in general to all the mass-finishing processes. The following sections discuss specific operational issues for barrel tumbling and vibratory finishing.

6.11.1 Barrel Finishing/Tumbling

Conventional rotary barrel tumbling was the original mass-finishing technique. Ancient Chinese and Egyptians used tumbling barrels with natural stones as media to achieve smooth finishes on weapons and jewelry. The process was known as barreling, rattling, or tubbing.

Barrel finishing is now a vastly improved process, but it is generally slow. Consequently, the more sophisticated mass-finishing processes discussed later are making the barrel obsolete for most applications. However, barrel finishing is a versatile means of edge and surface conditioning; equipment costs are generally low, operation is simple, and there are still applications where this process is the most cost effective.

Operating Principle

Barrel finishing is a low-pressure abrading process generally performed by the controlled sliding and rolling action of workpieces, media, compound, and water. Not all these ingredients are used in all applications. In a rotary or tumbling barrel (see Figure 6-5), the upper layer of the workload has a sliding movement. As the barrel rotates, the load moves upward in the barrel to a turnover point. The force of gravity overcomes the tendency of the mass to stick together, and then the upper layer slides toward the bottom of the barrel.

The barrel is normally loaded to about 60% of capacity with a mixture of workpieces, media, water, and compounds. Higher load levels (up to 80%) are desirable for some workpieces, such as large or heavy components. However, increasing the load level decreases the length of slide, thus reducing the probability of the workpieces contacting each other and the force with which they contact, resulting in much longer cycles.

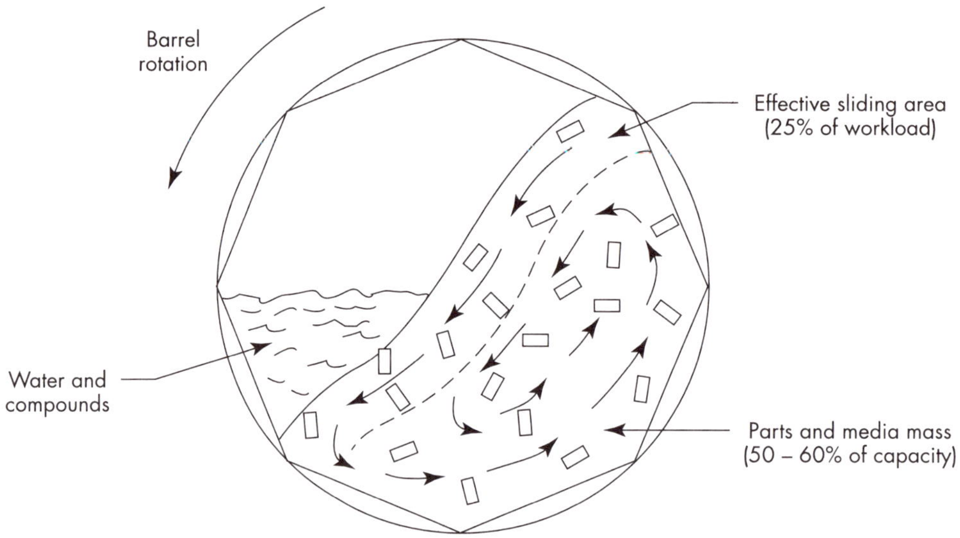

Figure 6-5. Cross section of finishing barrel showing flow of parts and media.

Although abrading action may occur as the workload rises in the barrel, about 90% of the rubbing action occurs during the slide. With horizontal barrels just over one-half filled, the most effective action occurs along the top surface of the mass. The faster the barrel rotation, the steeper the angle of slide and the faster the action. However, faster barrel rotation increases the tumbling action, increasing the likelihood of workpieces being damaged, and reducing the quality of edge and surface conditions. For some applications, small barrel diameters and low rotational speeds may be required.

With barrel finishing, large radii are produced on the workpieces in relation to the amount of material removed from flat surfaces. This process also generally produces a higher luster and, often, a smoother finish on workpieces than other mass-finishing methods. The peening action of barrel finishing can have a stress-relieving effect and can produce denser workpiece surfaces. The process also permits using the media for extended times about twice the time permissible for vibratory finishing (which is discussed next in this section), resulting in more efficient usage of the media. Not all users of these processes agree that barrels use the media more efficiently.

Barrel finishing is generally a noisy operation, and this factor should be considered in selecting the locations of the machines. Sound enclosures may be required to meet OSHA regulations. Another possible limitation is that barrel rotation must be stopped before in-process inspection can be performed when using closed barrels. Barrel finishing is also intrinsically slow, because only a small proportion of the workload is having any action applied to it at any given time. Nevertheless, the process is capable of developing surfaces of outstanding quality.

Equipment Used

Barrel-finishing machines are simple in design, relatively low in cost, and available in a wide variety of sizes and types (see Figure 6-6). Rotary barrels include the normal type of octagonal barrel with variable speed and polyurethane lining, single speed, open-end tilting or not, submerged, wood-lined, rubber-lined, and hexagonal or round barrels. Open-ended and horizontal types are the most popular. Horizontal units generally provide faster action, but open-ended units offer the advantages of inspection and making additions while they are in operation, plus they are easier to unload. Barrels are generally of multisided construction, with octagonal or hexagonal-shaped interiors being common. Most modern machines have flexible, wear-resistant, and replaceable linings in the barrels. Such linings, frequently polyurethane, extend barrel life, offer some protection to workpieces, and reduce noise levels. Machines are also available with individual compartments that allow workpieces to be processed singly.

Most machines are designed to permit changing the rotational speed of the barrel. Loading and unloading the horizontal barrels is done through wall openings that are closed during operation by watertight doors. During long processing cycles or when acidic compounds are used, heat can build up pressure within sealed horizontal barrels. This pressure must be released by a pressure-release device before opening the doors to unload the barrels.

Material handling for barrel-finishing machines is generally manual. Hoist pans and overhead hoists are normally required for servicing larger machines. Storage bins are also needed for media not in use whenever a variety of different materials, sizes, and shapes are required to produce desired

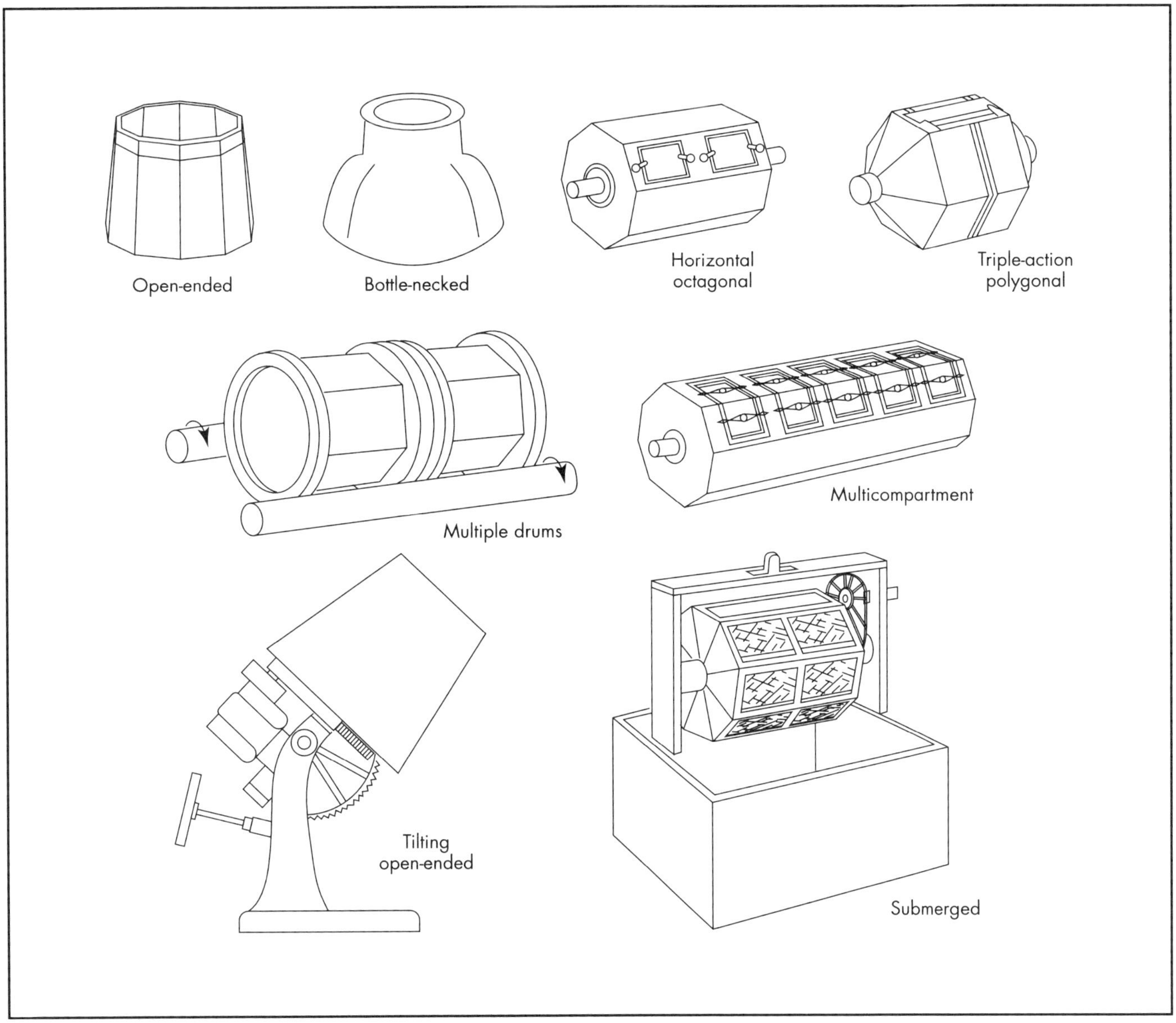

Figure 6-6. Types of barrel finishing machines. ***(Courtesy Harper Surface Finishing Systems)***

results on a wide range of workpieces. Although barrel finishing is almost always a batch processing operation, it is possible to automate the equipment by incorporating a material handling system. Feed troughs, chutes, timers, and automatic dumpers are common in such systems.

Process Variables

Process variables that must be properly selected for optimum barrel finishing include the shape and size of the barrel, the media and their proportion to the workload, the compound, the water level, rotational speed of the barrel, and processing time. Barrels are available for handling a wide variety of workpiece sizes, from small stampings or machined parts to large castings or forgings. Large or fragile workpieces are often fixtured in the barrels to eliminate part-to-part contact. Thin, long workpieces are generally not suitable for barrel finishing because of potential distortion or breakage.

Rotational speeds generally range from as low as 4 r/min for large barrels to 60 r/min for small barrels. With surface speeds usually ranging from 20–200 ft/min (6–61 m/min). For maximum productivity, barrel speed should be as high as possible without damaging the workpieces. However, the workpieces being finished and the required results may necessitate slower speeds. Burnishing is generally done at lower speeds, whereas deburring or cut-down operations are done at higher speeds.

Processing times vary from minutes to many hours, depending primarily on the workload, the media used, the desired severity of the action, and the required results. Separation of workpieces and media when unloading the barrels is generally done by dumping the contents on vibrating screens. In some cases, the parts and media are discharged into a pan and transferred to a machine designed for separation. Trench drains are generally provided under barrel-finishing machines to collect contaminants (liquids and solids). The drains terminate in sumps that retain the solids.

Vibratory barrels were an important development in mass-finishing, but they came along at the same time as vibratory machines were developed. These barrels can do work in cycles very close to those in other vibratory machines, but they have the drawbacks of other barrels: closed systems, complete operator dependence regarding quality and uniformity, they require much greater operator skill, and have dirty nonflow-through solution systems. They have not been popular in recent years.

6.11.2 Vibratory Finishing

Vibratory finishing is now the most popular type of mass finishing and, next to hand deburring, the most common surface conditioning method used in the industry. This versatile process is used for cleaning, deburring, deflashing, descaling, edge and corner radiusing, surface finishing, and stress relieving. Workpieces in a wide variety of sizes and shapes are handled, and all metals and many nonmetallic materials are processed. Large quantities of parts can be run in batch or continuous process setups without handling or fixturing, thus minimizing costs.

Equipment Basics

Vibratory equipment is made in two basic configurations: rectangular tub and round bowl types. The first tub-type vibratory finishing machine was introduced commercially in 1957, and the bowl type about five years later. Both types use an open-top work chamber containing an aggregate of media, compound, water, and the workpieces. The chamber is vibrated and the work is performed by the scrubbing or peening action of the media and compound on the workpieces. Table 6-11 provides an overview of the variety of these machines.

Tub-type Vibratory Equipment

With tub-type machines, workpieces are loaded into the open top of a container holding the media, compound, and water. The tub (container) is rectangular, generally has a shaped cross section with flat parallel ends, and is usually mounted on coil or rubber springs (see Figure 6-7). On some machines, the containers are suspended on air bags, and certain small units employ composite or metal strips for suspension. Processing containers other than U-shaped, such as keyhole shapes (see Figure 6-8), and enclosed tubular units are sometimes used to improve unrestricted mass movement. Tub liners, generally polyurethane, are used to prevent wear, reduce noise, and resist chemical attack; they also transmit energy to the media. Removable separators can be installed to divide the rectangular tub into a number of compartments, thus permitting workpieces to be processed individually or allowing the use of different media to finish various parts at the same time.

Vibration of the container is accomplished through several means, including the following:

- A vibratory motor with counterweights on its shaft is attached to the bottom of the container. This method is used on some bowl-type machines.
- A single shaft or twin shafts, with eccentric weights, driven by a standard motor.
- A system of electromagnetic vibration generators.

Action of the media against the workpieces takes place throughout the whole load. The media rubs against the workpieces while the whole load is turning over within the container. As a result, cycle times are substantially shorter than for barrel finishing. Other advantages of tub-type vibratory finishing include the capability for in-process inspection, unloading and reloading without stopping the machine, and the possibility of automating for either batch or in-line processing.

Vibratory equipment can provide more finishing action than barrels in the recesses of workpieces. Vibratory machines can also process larger parts than barrels. Tub vibrators have been built to process components, such as aircraft wing spars, 30 ft (9.1 m) or more in length. Modular-type machines mounted in a series can handle parts over 100 ft (30.5 m) long.

With long tub-type vibrators, fully automated, continuous processing of small parts is possible. Workpieces are loaded at one end of the tub and unloaded at the opposite end. At the unloading end, the media are separated from

Table 6-11. Types of vibratory deburring machines

Type of vibratory equipment	Optional configurations	Media separation
Tub, batch	Compartmentalized Fixtured Cryogenic Modular Bench top Portable	Generally none
Tub, continuous	—	With or without
Bowl, flat bottom	Compartmentalized Fixtured	With or without
Bowl with elevation	—	With or without
Bowl, big ratio*	—	With or without
Bowl, continuous	—	With
Oval	—	With or without

* Ratio is the diameter of the center section of a bowl divided by the channel diameter. Ratios larger than 4 are "big" ratio machines.

(Kittredge 1989)

the workpieces and then generally returned to the tub by a conveyor (see Figure 6-9), often through reclassification systems. Postprocessing operations, such as rinsing and drying, can be incorporated into such continuous systems.

Round Bowl Equipment

Round bowl or toroidal vibratory finishing machines have a doughnut-shaped chamber (see Figure 6-10) that permits a continuous circular flow of media and workpieces. The bowls may have either flat or spiral bottoms. As with tub-type machines, the chambers are provided with liners.

Vibration of the bowl is accomplished either by a vibratory motor or by an eccentric weight system mounted vertically in the center tube of the bowl. With either method, the amount of weight placed on the top and bottom of the eccentric system and the angular displacement between the two weights control the following:

- The finishing action (the amount of media vibration against the workpieces).
- The speed at which the mass rolls over within the bowl.
- The speed at which the mass rotates around the bowl.

Process Variables

Process variables for vibratory equipment are similar to those for barrels. The appropriate media, which must be sufficiently aggressive to remove burrs and round edges while being sufficiently gentle to achieve desired surface finishes, is the variable that generally requires the most trial-and-error testing. The size and shape of the media are also important to ensure action on all critical areas of the parts, including recessed areas of the workpieces, to prevent the media from lodging in such recesses or holes, and to facilitate simple separation of workpieces from the media. If the media can pass freely through holes in the workpieces, they will produce radii on the hole edges.

Water can have a considerable effect on the results obtained with vibratory equipment. Too much water will substantially dampen the action by flooding. For this reason, vibratory machines are designed for continuous flow-through of the water and compound solution. Compounds serve the same functions in vibrators as in barrels: they enhance abrasive action, improve the color produced, inhibit corrosion, and maintain cleanliness of the total load in the machine.

Other process variables for vibratory equipment are the amplitude of vibration, a variable provided on virtually all vibratory machines, and the frequency, which is variable on only some vibratory machines. Operational amplitudes generally range up to .375 in. (9.53 mm), and frequencies vary between 800 and 3,600 vibrations per minute. Speed of rotation equals the frequency of the transmitted vibration. Higher speeds and/or amplitudes of vibration produce faster metal removal rates but rougher surface finishes for a given media and compound. Media wear rate is also increased.

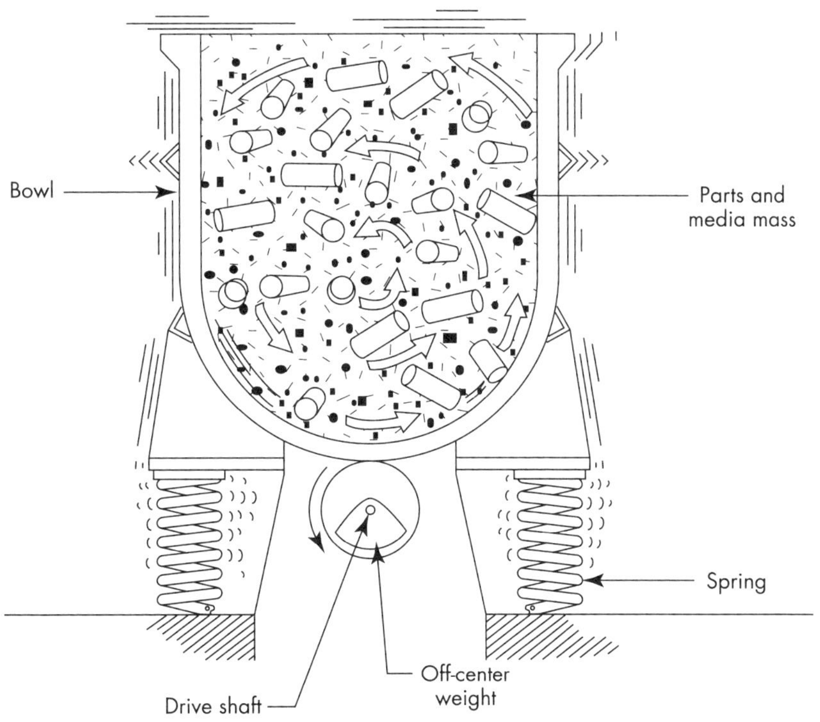

Figure 6-7. Sectional view of tub-type vibratory finishing machine.

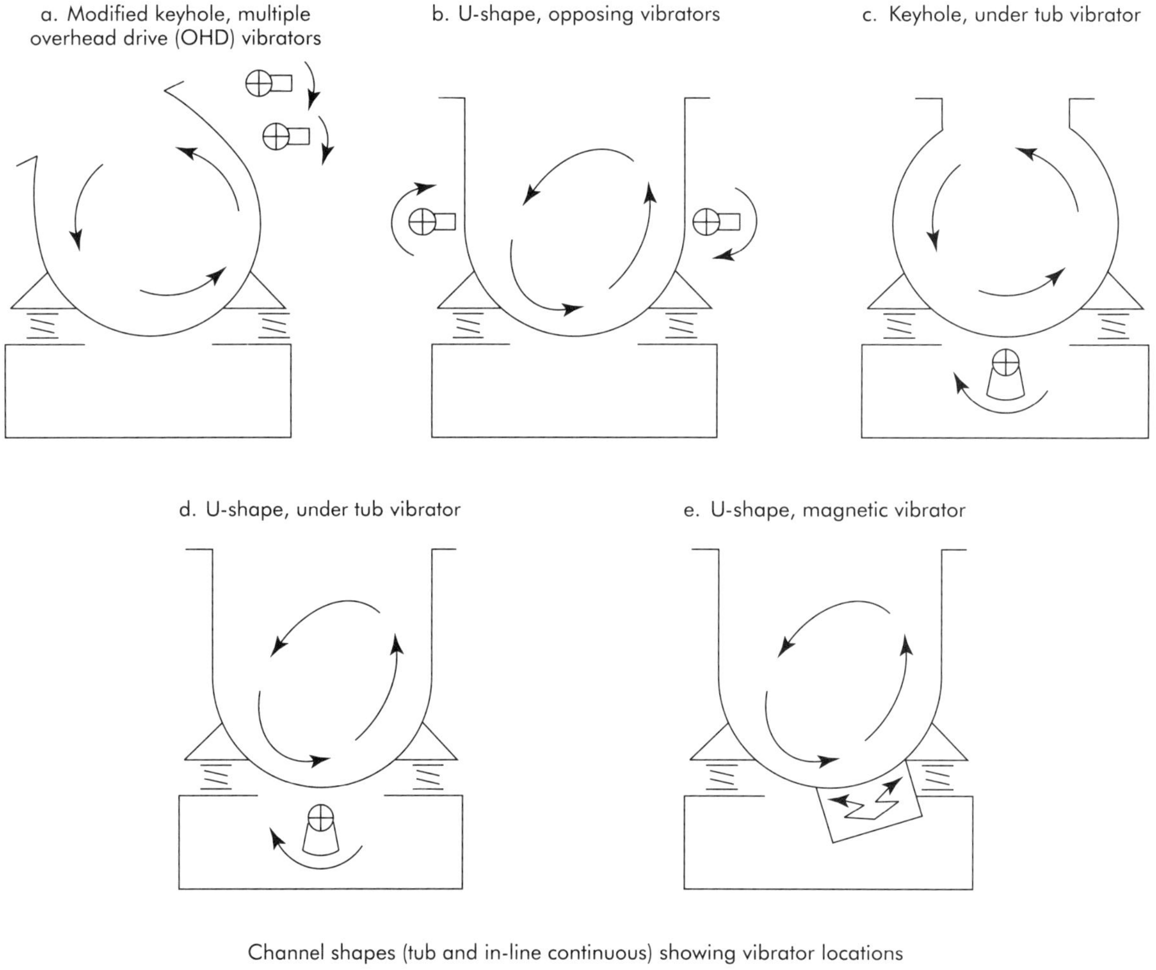

Figure 6-8. Cross-sectional view of various types of tub and in-line continuous processing containers. ***(Courtesy Ultramatic)***

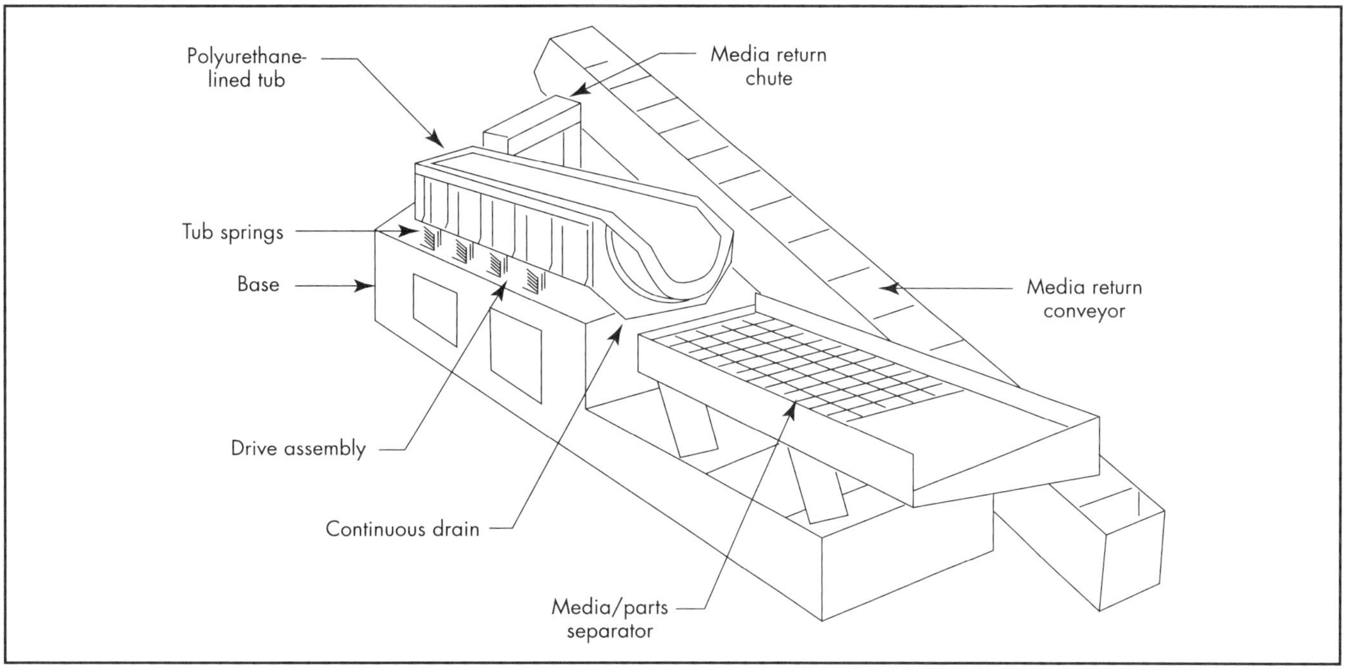

Figure 6-9. Tub-type vibrator with automatic workpiece loading conveyor and another conveyor to return media to the tub.

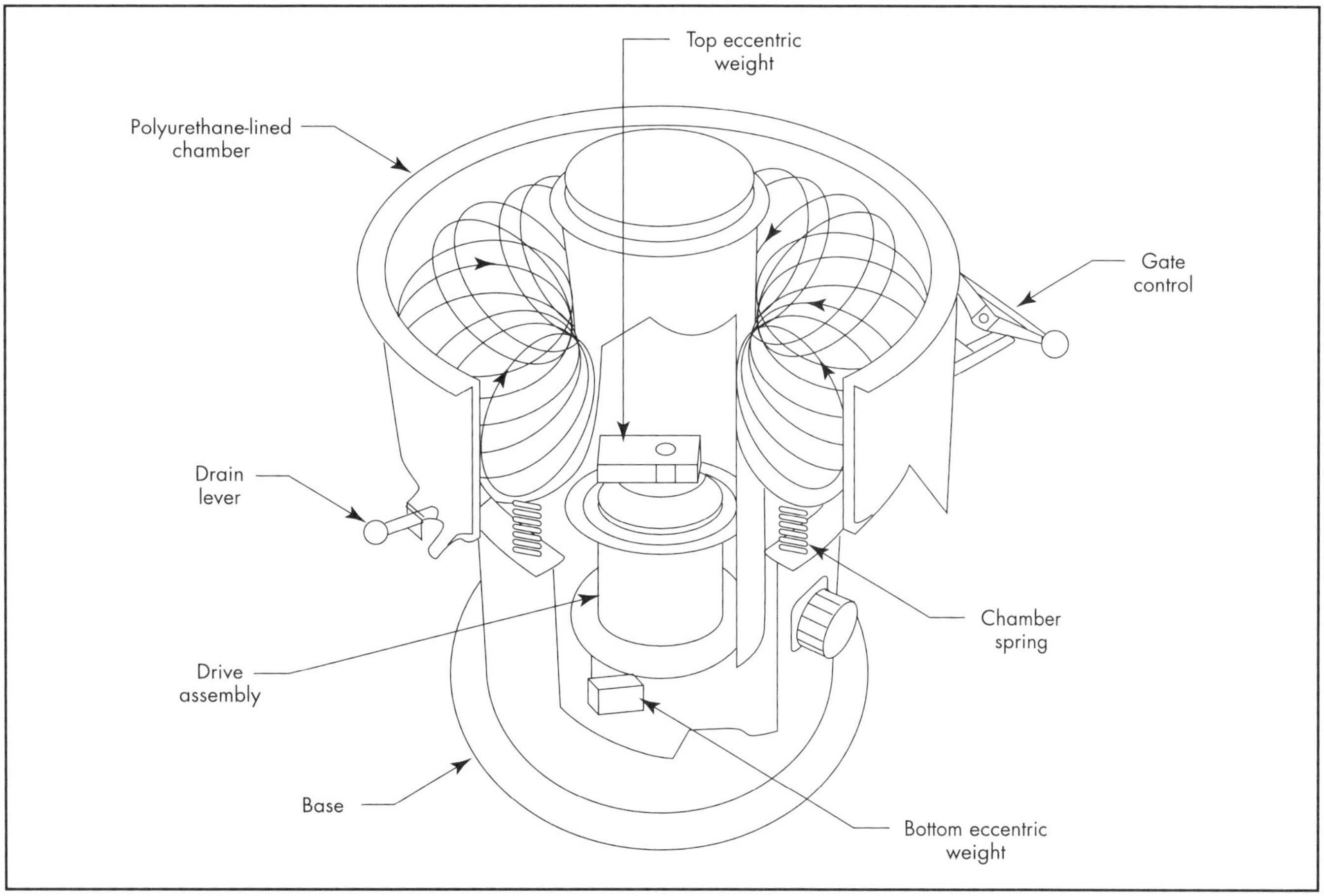

Figure 6-10. Typical configuration of a round bowl, vibratory-finishing machine. ***(Courtesy Sweco)***

Vibratory tub and bowl machines are now available with high frequencies and/or amplitudes for higher energy mass finishing than previously possible.

The following is a list of the primary factors influencing material removal rates for a vibratory machine processing with loose fill (Matsunaga 1971):

- Average number of simultaneously acting abrasive grains (media particles) in contact with the workpiece at a given time.
- Hardness of the abrasive media.
- Velocity of flow of the media.
- Pressure that is put on the workpiece through the abrasive media.
- Characteristics of the cutting edges of the abrasive media and how those cutting edges interact with the workpiece material.
- Coefficient of friction between the media and the workpiece material.
- Coefficient of friction between the media and the coating of the work bowl.

For example, with the introduction of wear-resistant polyurethane as a work bowl liner, the coefficient of friction between the media and bowl were significantly reduced, thereby raising the media wear efficiency of the vibratory machines. Presently, new materials are being investigated that will have an even better coefficient of friction than polyurethane with a simultaneous high creep strength.

By adjusting the ratio and type of raw materials, the resulting elastomeric lining can vary in hardness from about 50 Shore "D" up to over 90. Softer linings are said to make rolling of steel media in the tub easier and less harmful to delicate parts, whereas harder materials are said to last longer.

Rubber and PVC (polyvinyl chloride) are sometimes used as lining materials. Neither has the durability of polyurethane, but they are inexpensive and can be found in marginal, small equipment.

Four variables are common to all vibratory equipment: (1) single or variable speed; (2) fixed or variable adjustments for amplitude; (3) grease, recirculated oil, or sealed bearings; and (4) type, number, and size of drain patches (Kittredge 1983).

Machine Variables

There are at least six types of batch and continuous-tub vibrators: (1) U cross section, (2) inverted keyhole cross section, (3) half-U cross-section, (4) single drives, (5) dual drives, and (6) overhead drives. Batch equipment comes with or without dividers. Continuous machines may or may not have a means to control continuous process cycle time.

There are at least three types of round vibrators with no elevation in the chamber: (1) with or without dividers, (2) curved or straight outer wall, and (3) with or without a separation system.

At least two additional variables are involved for round vibrators with an elevation in the chamber and "big ratio" designs (that is, ratios larger than 4—calculated by the center section of a bowl divided by the channel diameter [Kittredge 1989]): (1) type of separation deflector and (2) its method of actuation, screen deck type, and material.

There are at least two additional variables involved with round continuous machines: (1) screen deck type and material; and (2) system, if any, for controlling continuous cycle time.

The variables associated with oval vibrators are the same as those for round vibrators with an elevation in the chamber, plus they have the potential variable of ratio of length to width.

Portable tub machines allow use at the point of machining.

Channel Cross Sections

Like the tub units, bowl channels have a curved, hemispherical bottom and, normally, straight and parallel sides (Kittredge 1989). When small parts and small media are involved, a curve to the outer wall can be used to prevent media "back-roll." Seldom is the center section wall curved (Figure 6-10).

Dryers are included in round vibratory machine designs because the simple in-and-out feed capability, which provides retention times of several minutes. Ground and graded corncob is used as a drying media and is best when heated to drive off moisture. Infrared, electric contact, and hot oil heaters are used.

Continuous machines are popular for deburring parts in relatively short cycles. In these systems, the separation deflector is fixed in the separation position so that media and parts are always deflected up onto the separation deck. Media drops through the screen where new parts are loaded continuously. Cycles are generally in the 3–30 min range, depending on media, parts, amplitude, machine design, and size. These machines hold 10–20% more media than batch machines because the area under the screen where parts are loaded is filled up.

Compartmentalized machines are, like tub vibrators, designed to essentially make a number of small machines out of one big one. The small compartments insure no part-on-part contact and are loaded and unloaded by hand. Compartments can be fixed in place in the channel, or they can be mounted on a fixture that allows them to rotate with the mass.

Some compartmentalized machines separate different media or numbers of parts. These may contain more than one part per compartment.

Oval-shaped vibratory machines are said to offer better separation of parts from media and better control of retention times as well as other advantages. They are driven by a central eccentric weight system like the round versions and, therefore, do not offer a uniform action where media and parts are close to the center versus where they are farther away. Single-shaft oval machines are no longer made, but dual shaft machines that provide better action are still produced.

Multiple-pass machines are a relatively new concept. They may permit longer retention times by making the media and parts mass travel around the machine two or more times.

Several operations may be performed on the parts in one continuous operation, such as deburring followed by drying, or two or more different types of media may be used.

Most (but not all) of the small bench-top machines are marginally designed devices that cannot compare favorably with their full-size counterparts. To hold down costs, manufacturers have removed several important items, such as the adjustable weight system and a good solution system. These two items alone can make good machines out of bad ones. There are good machines, but they cost more.

Heavy, bulky, or long parts need a tub-style machine, whereas bowls process better with parts of more equal surface area (Thompson 1981). If the machine is to be used for metal media processing, the manufacturer will supply the machine set specifically for the application. Continuous equipment is more directed in its use and usually applied to high-volume production.

Deciding whether to use tub or bowl equipment requires reviewing the part size and anticipated production schedule. The part size will dictate the channel requirements, and production and process times will determine the machine size, style, and number of units required. However, it is not unusual to find large bowl machines being purchased to process bulky parts that could be finished more quickly and efficiently in a tub unit one-fourth the capacity of the bowl. Furthermore, tubs being purchased to process larger quantities of small parts with no provision for unloading or separation is just as offending.

Regardless of the type of machine (tub, bowl, or continuous), the processing channel shape is important. Scratches, impingement, and tangling of parts occur when the channel cavity restricts freedom of mass motion. Deep channels develop forces at the bottom area that can cause distortion to delicate pieces; in shallow channels, large pieces hit the bottom and one another. It would seem that a circular channel would be the better design. Once again, however, design depends on application. A circular channel may be too forgiving and extend cycle times for deburring applications, but it may accept more parts and increase the yield per load.

A deep, narrow channel has containment, and if the part size is such that the mass motion is unrestricted, excellent results can be expected when used for deburring applications.

Shallow channels will accept long objects well, whereas bulky items will be in constant contact with the channel lining and other workpieces.

Keyhole or modified keyhole channels (Figures 6-8a and 6-8c) featuring curved walls allow for unrestricted mass movement. This design is popular where finish may be as important as deburring. Part geometry and bulk are not as critical. Just about anything that will fit in will run nicely.

Other popular channel configurations are manufactured. The preceding examples illustrate how part size, material, volume, and process are involved in the choice of configuration. Carefully review the operator work height and reach requirement. Certain units may require a platform for operator convenience.

Amplitude and Speed Adjustments

The terms speed and amplitude are often confused. *Amplitude* is the energy input to the mass. *Speed* is the frequency of using the available amplitude. The most common method of achieving amplitude is through the eccentric shaft principle. The shaft can be mounted to bearings loaded with eccentric bushings, resulting in a single-lobe crankshaft. Amplitude adjustments for increasing or decreasing generated energy forces are accomplished by adding or removing weight from the shaft.

Bowl machines have an additional adjustment. The shaft mounting of bowl equipment is located vertically in the center post or donut hole area of the machine. The counterweights are located on the top and bottom of the vibrator shaft. The amplitude is controlled by adding or taking away weight. However, since the processing channel is round, the mass travel is a combination of rotary motion and lineal extension, resulting in a spiral motion within the circular processing channel of the machine. The dual motion or mass path is controlled by adjusting the relationship of the top and bottom weight, resulting in a tight or open spiral of the mass path. The addition or removal of weight on the top or bottom flyweights governs the energy input and velocity of either the rotary motion or lineal extension of the mass and the tight or open spiral of the mass path.

Amplitude settings are the factor by which the mass path is governed and are determined by the following factors: (1) parts to media ratio, (2) size of the part, and (3) part weight or the specific gravity differential between the parts and media.

Incorrect amplitude settings can result in part impingement, poor mass motion, parts migration, parts distortion, and difficult or incomplete internal separation in bowl machines (Thompson 1981).

Speed adjustment controls the frequency of energy input and the speed of the mass. To simplify the definition between amplitude weight and speed, amplitude is the size of the hammer and speed is the frequency of its use; thus, the rapid use of a tack hammer is different from the slow deliberate blow of the sledge hammer. Many vibratory machines have no variable speed capability, whereas others do.

Since amplitude is the governing factor within the mass and is adjustable to suit needs, it is seldom changed. Where surface conditions are uncertain, the ability to change amplitude gives users flexibility. Adjustable amplitude is available on all equipment utilizing mechanical vibrators; variable speed, however, is not.

Vibrator Details

Round bowl machine vibrators are located vertically at the center of the bowl. The vibrator shaft length and vertical location in relationship to the centerline of the processing channel vary according to the manufacturer and channel configuration.

Eccentric weights are located horizontally at either end of the vertical vibrator shaft and are adjustable in their relationship to one another. When set within the allowable range established by the manufacturer, a spiral motion to the parts and media is produced. Changing their relationship allows the spiral path to be adjusted to complement the part size or configuration. Additional weight may be added to accelerate the mass velocity to move a more dense mass.

The overhead drive (OHD) vibrator is located above the centerline of the mass (Figure 6-11). The flyweight is oblong and mounted on the allowable length of the shaft between bearings. The low-profile design of this type of flyweight allows clearance for multiple vibrators. Like all other eccentric weighted vibrators, its direction is opposite that of the mass. The overhead location above the tub centerline induces a pull-up motion, which produces a uniform lift angle cascade to the mass (Thompson 1981).

The opposing vibrator method mounts the vibrator shafts horizontally on both sides of the tub (Figure 6-8b). The shafts are synchronous, and amplitude input is equal for each shaft. The theory is that by equally dividing the amplitude input at the centerline of the processing chamber, the more fluidized the mass becomes, creating a circular orbit to the tub and mass.

There are probably more machines equipped with under-tub vibrators than any other design. Tub-type vibrators introduced in the 1950s featured this method, and it continues today. The vibrator is attached beneath the tub itself, with pie-shaped flyweights fitted to the shaft ends. Vibration input is upward, producing a rise-and-fall motion to the tub; the eccentricity develops the orbital path of the mass motion opposite the vibrator rotation. Variable speed is a desirable feature to aid in controlling mass density variations.

Machines equipped with magnetic vibrators are usually limited to less than 10 ft^3 (0.3 m^3) in capacity. The vibration input is from the underside of the tub and the machine operates at a constant 3,600 cycles per minute. Voltage is controlled through a rectifier console. Increasing the voltage strengthens the magnetic field, resulting in a longer pull to the armature, thus increasing the stroke. The theory is that the variable stroke assimilates the variable speed and amplitude of a mechanical vibrator without changing weights.

Round bowl vibrators are available in a number of "diameter ratios" (the ratio of the center section diameter to the channel cross-sectional diameter). A 1:1 ratio means that the diameter of the processing channel is the same as the diameter of the center section. A 4:1 ratio means that the center section diameter is four times that of the channel diameter. Generally, the closer this ratio is to unity, the lower the cost of the machine.

Ratios of 1:1 are common for "nonelevating" chambers where no internal separation is expected. Higher ratios make internal separation easier—2:1–4:1 ratios are not uncommon for these machines when conventional ceramic and plastic media are used. Up to 7:1 ratios can be found when difficult-to-move hardened steel media are used.

Various special designs satisfy a great number of industrial requirements.

Methods of Part/Media Separation

Many methods are used to separate parts from media, and the best approach is a generous application of common sense. Separation is accomplished in three areas: (1) in the machine, (2) at the machine, and (3) remote to the machine.

In-the-machine separation involves (a) hand picking, (b) manual or automatic magnets, and (c) bowl machines equipped with an internal separator.

The most common method is to pass the parts and media over a screen, allowing one to drop through and the other to remain on top or to pass off. Should the parts have projections that can hang up in a screen hole in the pass-off method, then a bar deck would allow an uninterrupted path for the parts to pass off. Step or flip stations can provide a turnover for parts that may have dished or recessed areas that would

Figure 6-11. Overhead drive tub-type continuous finishing system. ***(Courtesy Ultramatic)***

hold media. Timing is important so that an upward-facing part holding media does not turn over, emptying its media into another upward-facing part on the lower step.

Magnetic separators work well where ferrous parts are run with a nonferrous media and the size relationship is such that it would be the faster method. Magnetic separation usually requires a demagnetizing operation.

At-the-machine external separation involves evacuating the mass to a separator, either magnetic or screen type. On completion, the media are then reintroduced into the machine by hand, using a hoist system or conveyor. External separation allows for classification of undersize media and particle fines to aid in preventing media lodging or debris collecting in part holes, corners, or crevices.

Remote-to-the-machine separation involves evacuating the mass into a container. The container is moved to an area for separation, either by hand or automatic.

Internal Bowl Separation

Flat-bottom channels generally have vertical or near vertical dams that interrupt the mass path by blocking off the channel. The continued energy input from the vibrator directs the mass against the dam, pulling the mass up on itself until it is high enough to continue over the top and across a screen deck; the screen has hole sizes large enough to pass the media back through into the processing channel, yet small enough to pass the parts off the top.

Bowl machines that feature an inclined or spiral bottom (lock washer design) use a discharge gate located at the high point of the processing channel, which, when engaged, acts as an extension of the processing incline and directs the mass to the screen deck for separation as described earlier. Various degrees of automation are easily achieved with bowl vibrators. Figure 6-12 illustrates a machine equipped with a screen for parts and media separation and a discharge chute. A gate can be raised or lowered within the bowl to act as a dam to stop the continuous flow of workpieces and media around the bowl when desired. When the gate is either raised or lowered, a screen separates the workpieces and media—the workpieces are discharged down the chute, and the media are retained in the bowl. Bowl machines with internal separation also may be fitted with a media classifier that will discharge undersized media from the machine.

Magnetic separators are sometimes used when finishing ferrous parts with nonferrous media. Machines can be equipped with magnetic parts collectors and demagnetization equipment, permitting the media to remain within the bowl.

Continuous Equipment

Continuous installations are systems oriented. The first considerations are production volume and material handling requirements, such as moving products to the machine from the loading method to the separation stations, rinsing, drying, and unloading. Other postprocessing operations may be included as required. If other than steel media are used, a media classification area may be required to automatically discard undersized or worn media to prevent them from lodging in the workpiece. Continuous equipment usually employs the in-line method, which better lends itself to in-line external automation. However, there are many applications for continuous bowl machines. Two methods of processing are available with the bowl machine. The first is the

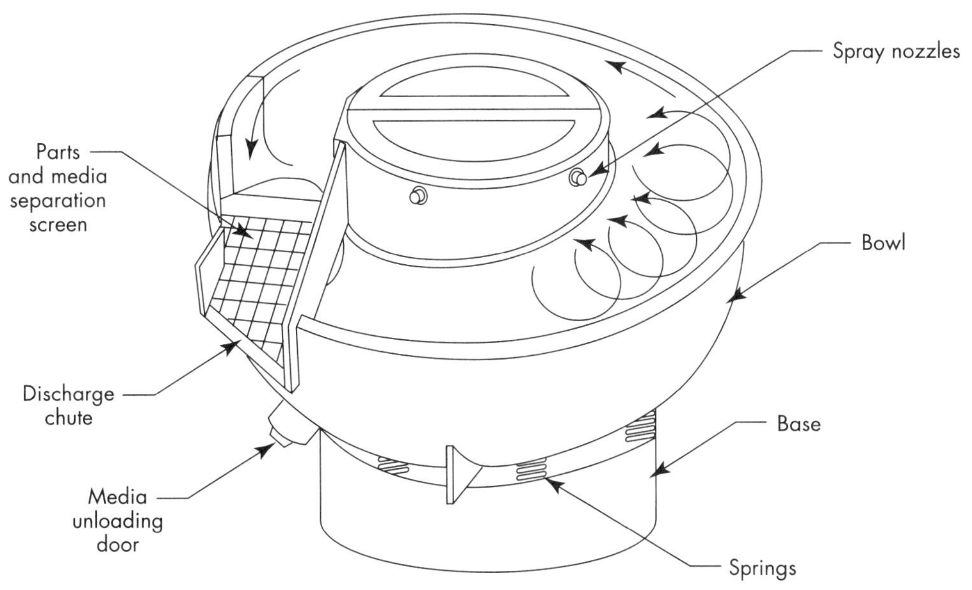

Figure 6-12. Automated continuous-feed vibratory finishing machine with screen for separating parts and media, and conveyer and chutes for returning media to the machine. ***(Courtesy Gyromatic Manufacturing Company)***

conventional once-around-and-out method, with cycle times governed by machine size and capacity, as well as part size and weight. The second method, now outdated, involves a dwell system, which is accomplished by shifting the top weight of the vibrator between preset stops, thus creating an alternating feed/dwell effect to the mass movement, resulting in a longer cycle within the processing channel.

Both bowl methods use the internal separation method in which a dam or gate interrupts the mass travel, directing it upward to be separated on a screen deck above the processing channel. The parts are sent off the top while the media pass through and are deposited back in the processing channel. Multipass machines can duplicate the time cycles and production of all but the largest in-line machines. Some multipass machines have channel lengths of over 75 ft (22.9 m).

Cycle time and production volume are important factors when considering bowl equipment. The once-around-and-out method may be too quick for deburring, but excellent for use as a parts washer or cob dryer. The dwell method may not yield sufficient production volume.

The flexibility of in-line machines is limited only by the imagination. Volume, part size, weight, and cycle time rarely present problems, nor is there the need for the machine to run with excessive amplitude to elevate the mass of bulky or heavy parts up to a separation screen, since the mass simply passes off onto a screen deck placed at a convenient level to the exit end of the processing channel (Figure 6-13). Unlike bowl machines, the separation area is independent of the processing unit, allowing for a separator speed and size compatible with the process and adequate to the task. The media are directed from the separator to a conveyor, returning it automatically into the beginning end of the processing channel. Amplitude adjustments can be made without affecting the cycle time and are accomplished by varying the speed. The primary cycle adjustment is the level of the load.

Continuous equipment is not limited to small parts only. Large parts, such as cast utility meter housings and lawn-mower decks, are examples of products that may be batch processed and are excellent candidates for continuous equipment.

In-line equipment is available in two styles: open top and tube. Both can be equipped with a plug-type door to block the discharge end. This feature gives the added flexibility of batch operation, allowing the process cycle to be extended beyond the normal continuous mode.

Open top machines are best used where a duplex continuous batch operation is planned. The open top allows manual loading/unloading of the parts during batch operation and full view of the mass during continuous cycles.

Tubular units are totally enclosed. The ends of the processing chamber feature openings engineered to the application for automated loading and discharge of the mass. The advantage of tubular equipment is that the processing chamber is enclosed and lined 360°, which usually drops the operating decibel level by 20% or more, compared with one of open-top designs.

Regardless of the type of machine selected (bowl or in line), the diameter of the processing channel is directly related to production volume and processing time.

Machines can be placed in tandem. For example, one machine can be used for heavy deburring and discharge parts into a second machine for surface refinement or burnishing and rust inhibiting, and then discharge into a third machine for drying. Parts can be automatically transferred from the dryer onto a conveyor for subsequent operations or into containers.

Fixtured and Compartment Equipment

Fixture and compartment processes are used when part-on-part contact is prohibited. Fixture processing is employed when there is a volume of one type of part, all having a common area for holding, such as a hole or threaded stud. The parts are generally attached to a rotary fixture, but non-rotating fixtures also have been used. The fixture is then lowered into an empty tub-type machine. Usually, the fixture frame is bolted to the tub. The frame mounted with bearings allows the fixture to rotate freely; media are then added. The media motion causes the fixture to rotate. Properly designed fixtures hold the parts for equal exposure to the media. Total removal of the media is usually required prior to removal of the fixture at completion of the cycle. Obviously, this method of finishing is used only when necessary. The channel shape employed is usually the straight-wall "U"-type or modified keyhole, both of which allow for ease of entry and removal of the fixture. Material-handling equipment is necessary to load and remove both the media and the fixture.

Figure 6-13. Open-top continuous system with space-saving right-angle parts and media separator. ***(Courtesy Ultramatic)***

Compartment finishing is accomplished in both tub-type and bowl-type machines. Tub-type machines are compartmentalized by bolt-in dividers (see Figure 6-14). Bowl machines use dividers that move with the lineal extension of the mass. The rotary motion is contained in the compartment with the part. Inclined or spiral-bottom bowl machines cannot be compartmentalized efficiently.

The advantage of tub machines for compartment finishing is that the compartments can be more easily moved or removed to accommodate large or small parts. The tub can be manufactured with compound spray and drains for each compartment.

The advantage of compartmentalized bowl units that cycle once-around-and-out is they have a predetermined production cycle. Compartmentalized vibratory machines make excellent parts washers for small or delicate parts that are hard to clean when nonabrasive media are used.

Media Storage

One of the automation features available is a multiple media system that uses different media for each cycle of a process, in a single vibratory machine. Overhead or container storage of media can be used after the media have been separated from the parts (see Figure 6-15). Media of another type, size, and shape can be used for the next cycle. Storage bins are available for tub and bowl machines.

Overhead storage hoppers may have a rotating carousel with four or more compartments to handle many types of media. For example, in a slow-cut plastic media process followed by a fast-cut ceramic media process, the operator would rotate the storage hopper until the proper product is over the machine, then pull the lever to dump the media into the vibrator. Several vibratory tubs can be serviced by a common separation and overhead storage system. Care must be exercised to avoid contaminating the different types of media being used.

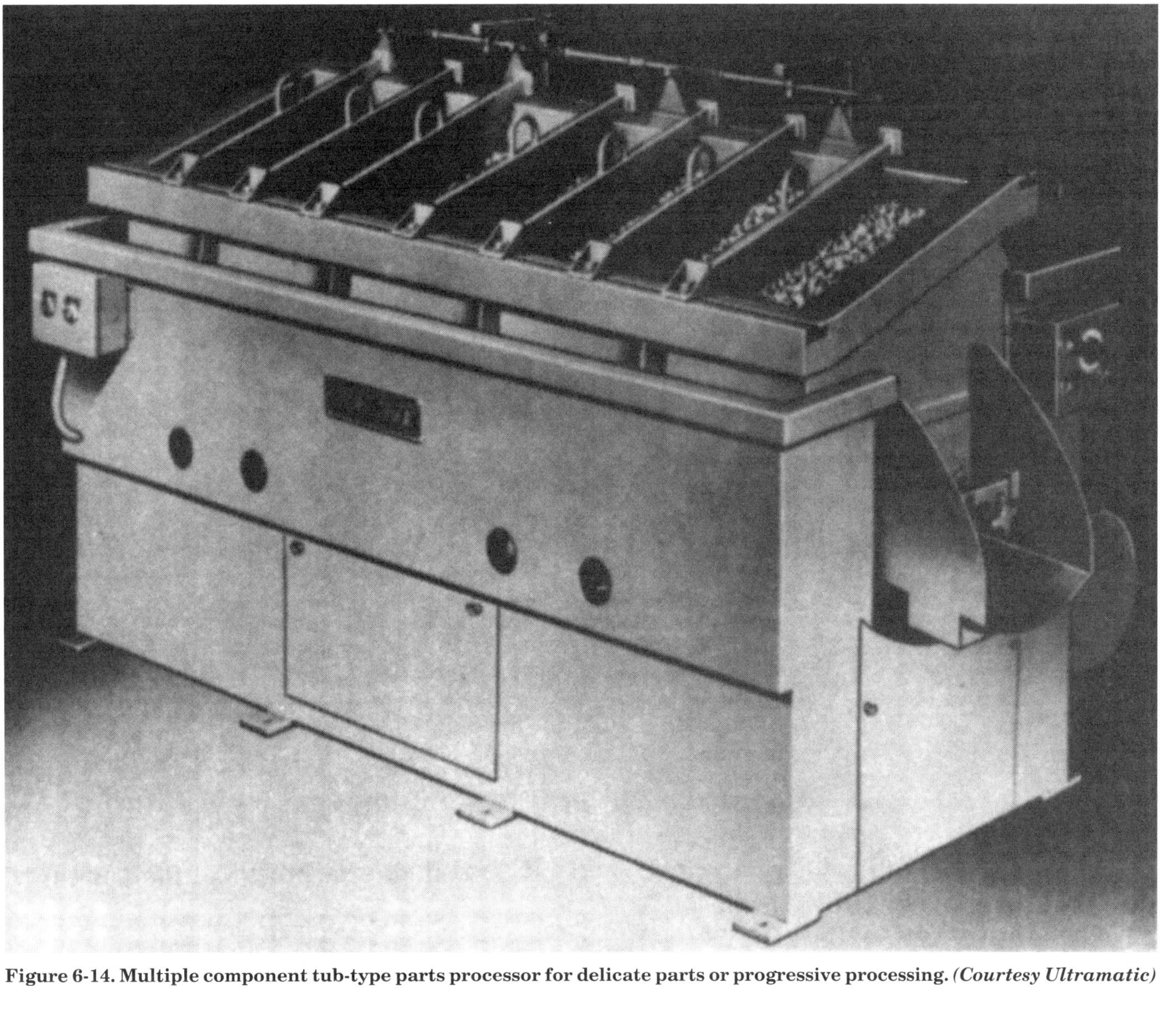

Figure 6-14. Multiple component tub-type parts processor for delicate parts or progressive processing. ***(Courtesy Ultramatic)***

Figure 6-15. Rotary media bins, incline elevator, and screen separator (four-machine capacity, shown with two machines). *(Courtesy Ultramatic)*

Fixtured Parts

Fixtures are used to hold parts. They make a big part out of a number of smaller ones. These fixtures are then placed in the vibratory tub loosely or bolted in. Figures 6-16 and 6-17 illustrate results and tooling for small gears in a large machine. The surface texture shown in Figure 6-17c is common for vibratory deburring.

Figure 6-16. Special fixture used for deburring small gears. Rubber spacers between gears permit media to circulate freely.

Because fixtures prevent any possibility of one part touching another, they allow much higher part loading in the vibratory chamber, thus increasing the efficiency of the operation and offsetting some of the cost of the fixturing process. Fixtures are used for high-quality work.

Compartments function the same as fixtures but are not quite as efficient. The tub or bowl chamber is divided by one or more bulkheads, making several smaller chambers out of one big one. Used for larger parts, one part is dropped into each compartment, insuring no part-on-part contact. Parts are normally loaded into and removed from each compartment by hand. As mentioned earlier, compartments also can be used to separate different media in the same machine.

Modular tub units are used to make very long tub vibrators from a series of shorter ones. Each unit must be separately driven, and the units must be joined in a way that maintains the same vibratory action in the connection area as elsewhere.

Load-level control is possible with small vibratory hoppers to feed media into the chamber every so often. Set up on a time basis, such devices approximate the media-use rate. Set up on a demand basis, media levels are held constant with little operator attention, except to keep the hopper full.

Getting Started

The first step is to determine which process and equipment are appropriate. Generally, this step involves discussing options with equipment vendors and other users. The checklist shown in Table 6-2 outlines the technical and economic data necessary for comparisons and decisions.

Decision-making after the equipment has been purchased involves some additional considerations, which are presented in the following list. Although these considerations are specifically for vibratory deburring, they are generally applicable to barrel, centrifugal barrel, and centrifugal disk operations (times, radii, and finishes will differ, however).

1. Determine the edge requirements for each edge.
2. Determine finish requirements for each surface.
3. Determine (estimate) what burr sizes will be produced at each edge.
4. Determine which general media characteristics are required (that is, must be ceramic for reason X) (Kittredge 1981).
5. Determine what media sizes and shapes will be needed to prevent lodging in holes or slots (see Tables 6-3, 6-4, 6-5, 6-12, 6-13, and 6-14 and Figures 6-1 and 6-18).
6. Estimate which run times, compounds, and machine settings may be needed (see Figures 6-19–6-21).
7. Determine how the conditions in the previous step affect part dimensions and features. (See Tables 6-15 and 6-16 and Figures 6-22 and 6-23. Note that though not shown in these illustrations, smaller parts in media will not deburr or finish as fast as larger parts.) (Gillespie 1975).

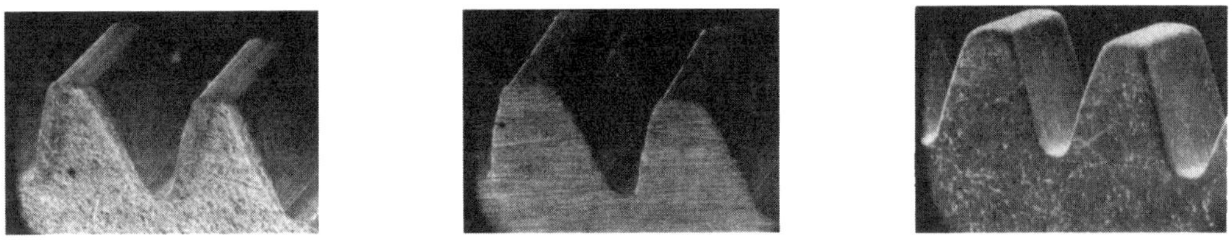

Figure 6-17. Edges of gear teeth after vibratory finishing.

Table 6-12. Basic grit sizes

Grit	Size (in.)			Size (µm)		
	Average	Maximum	Minimum	Average	Maximum	Minimum
8	.0870	.1300	.0650	2,210	3,302	1,651
10	.0730	.1050	.0550	1,854	2,667	1,397
12	.0630	.0900	.0450	1,600	2,286	1,143
14	.0530	.0750	.0370	1,346	1,905	940
16	.0430	.0650	.0310	1,092	1,651	787
20	.0370	.0530	.0260	940	1,346	660
24	.0270	.0430	.0180	686	1,092	457
30	.0220	.0320	.0140	559	813	356
36	.0190	.0300	.0120	483	762	305
46	.0140	.0220	.0095	356	559	241
54	.0120	.0195	.0080	305	495	203
60	.0100	.0160	.0065	254	406	165
70	.0080	.0130	.0050	203	330	127
80	.0065	.0115	.0040	165	292	102
90	.0057	.0095	.0035	145	241	89
100	.0048	.0080	.0025	122	203	64
120	.0040	.0065	.0020	102	165	51
150	.0035	.0055	.0015	89	140	38
180	.0030	.0045	.0010	76	114	25
220	.0025	.0040	.0008	64	102	20
240	.0020	.0033	.00099	51	84	25

This table has been prepared according to the specifications set forth in U.S. Department of Commerce Standard CS271-65 and B74.12-1974 American National Standards Institute. The values given are approximate and may vary slightly from usual commercial tolerances.

Most individuals are aware of the need to screen media frequently to eliminate worn media that causes lodging. In many companies, however, screening is only done when a problem occurs.

Table 6-13. Sizes and characteristics of large random-shaped media

Chip number	Nominal dimension in. (mm)	U.S. standard screens[a] in. (mm) First	Second	Third	Fourth	Tolerance: Maximum left on first screen (%)	Maximum left on second screen (%)	Minimum left on third screen (%)
00	1.500–2.000 (38.10–50.80)	2.125 (53.98)	2.000 (50.80)	1.500 (38.10)	1.125 (28.58)	0	5	90
0	1.125–1.500 (28.58–38.10)	1.625 (41.28)	1.500 (38.10)	1.125 (28.58)	1.000 (25.40)	0	25	85
1	.675–1.125 (17.15–28.58)	1.125 (28.58)	1.125 (28.58)	.675 (17.15)	.750 (19.05)	0	25	85
1½	.750–.675 (19.05–17.15)	1.000 (25.40)	.675 (17.15)	.750 (19.05)	.625 (15.88)	0	25	85
2	.562–.750 (14.28–19.05)	.675 (17.15)	.750 (19.05)	.562 (14.27)	.438 (11.13)	0	25	85
2½	.500–.562 (12.70–14.28)	.750 (19.05)	.562 (14.27)	.500 (12.70)	.375 (9.53)	0	25	85
3	.375–.500 (9.53–12.70)	.625 (15.88)	.500 (12.70)	.375 (9.53)	.250 (6.35)	0	25	85
3½	.250–.375 (6.35–9.53)	.438 (11.13)	.375 (9.53)	.250 (6.35)	#4	0	25	85
4	.188–.250 (4.78–6.35)	.375 (9.53)	.250 (6.35)	#4	#6	0	25	85
6	.094–.188 (2.39–4.78)	#4	#5	#6	#7	0	20	40 minimum[b]

[a] Commercial Standard CS 271-65 "Grading Abrasive Grain for Grinding Wheels" (National Bureau of Standards Publication).

[b] Minimum only. No media should pass through the fourth screen except for chip numbers four and six. A maximum of 3% of the number four chips may pass through screen four. A cumulative minimum of 65% of chip number six should remain on the third and fourth screens. No media should pass through the fifth screen except for chip number six. A maximum of 3% of chip number six may pass through the fifth screen. The fifth screen is a number eight mesh for chip number six. Other chip numbers do not use a fifth screen.

Table 6-14. Hole sizes in which media will lodge

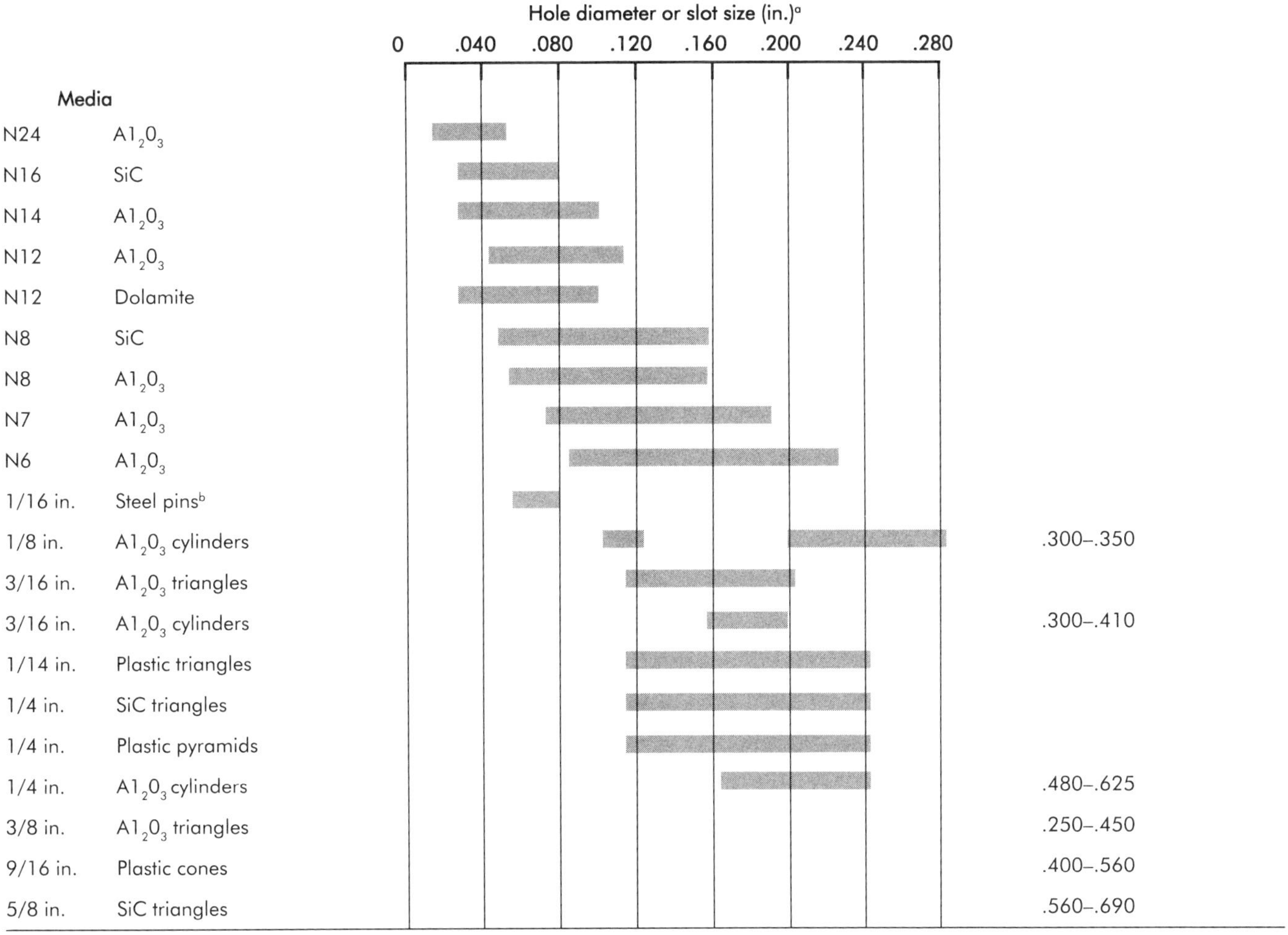

[a] .01 in. = 0.254 mm. Bars indicate size crevices in which media will lodge, allowing for approximately 20% reduction in size because of media wear.

[b] If hole is .062 in. (1.58 mm) or less deep, lodged pins can be removed readily.

8. Once conditions in steps six and seven appear to be acceptable, estimate batch size, costs, media wear (see Tables 6-2, 6-3, 6-7, 6-9 and 6-17), and cycle times.
9. Run parts, measure, and document results.
10. Develop and maintain a data file of process effects such as shown in Figures 6-18 to 6-22 based on the equipment, parts, materials, and conditions used in the plant.

Tables 6-13 and 6-14 provide evidence that media "size" is a rather elusive dimension. Random media have a range of original screen sizes they fall through. Because the media are not uniformly shaped, a single piece might not fall through a screen one time but will the next. There is one firm rule for media: they will never get bigger as they wear.

Table 6-14 indicates that .25 in. (6.35 mm) pyramids will lodge in holes or slots that are .125–.250 in. (3.18–6.35 mm) in diameter. Media lodge when a cross-corner dimension works into a hole. For a pyramid, there are three dimensions that can become lodged. Two pieces can nest together and lodge in a hole. There are at least six dimensions that can cause lodging when two pyramids are shaken together (actually, there may be many more than this). Three pieces can nest together and fill holes, and so on. For shaped media, typically a single piece is hardest to dislodge. If users select the wrong size of random media, they may have to remove particles from every hole in every part. It is possible to plug holes, but that is expensive and not practical for most operations.

Media lodged in blind holes are often impossible to remove. Any time media are removed they have the potential of scarring the feature they were removed from or distorting thin flanges.

As a general rule, select media that are smaller than the part. Ideally, when holes or slots are present, the media should be slightly larger, slightly smaller, or one-third the

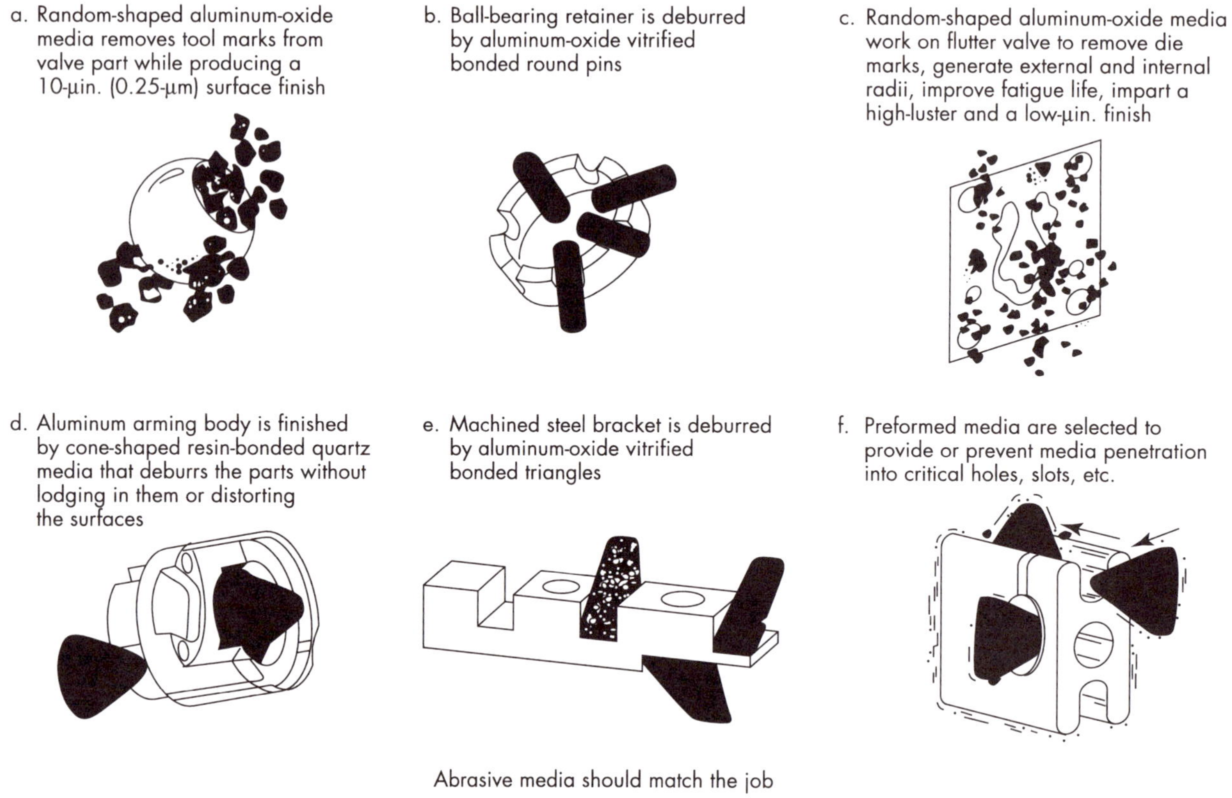

Figure 6-18. Examples of media matched to part features. ***(Courtesy American Machinist)***

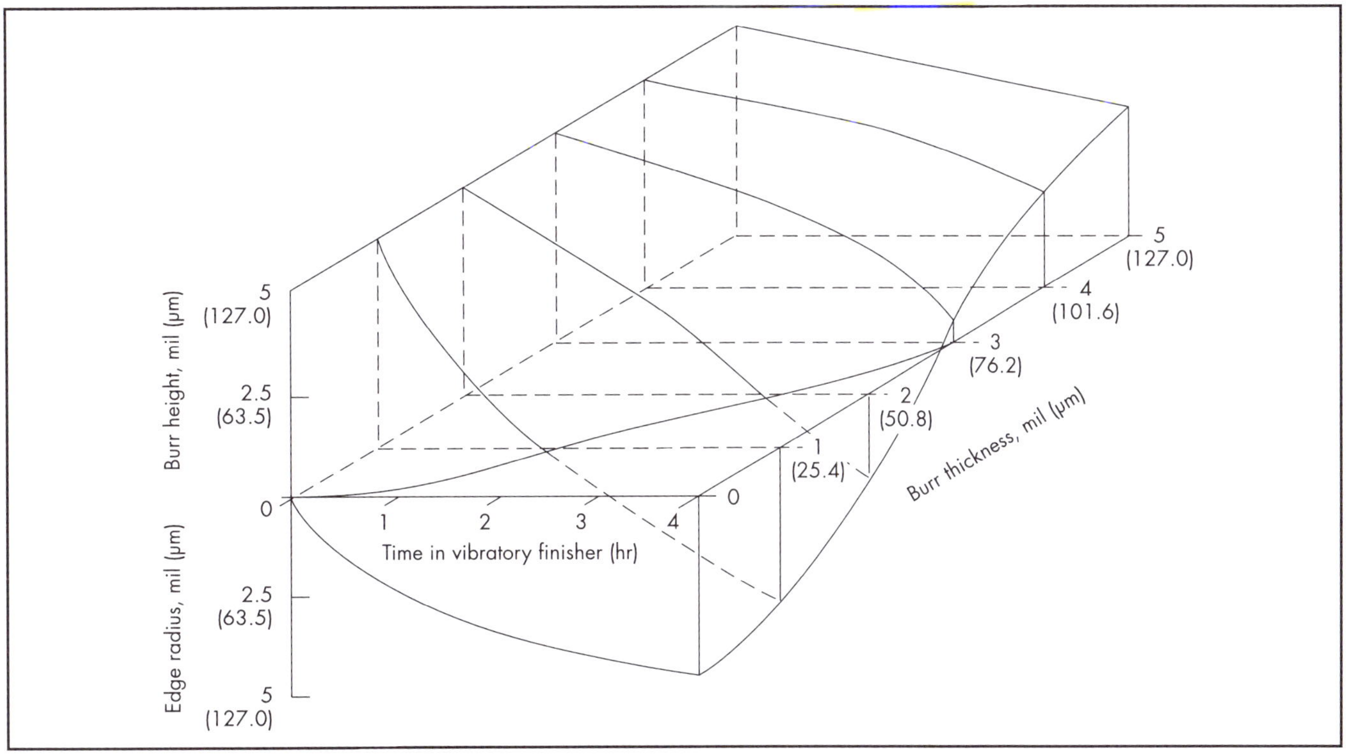

Figure 6-19. Effect of time and burr thickness on burr height and edge radii for simulated stainless steel burrs.

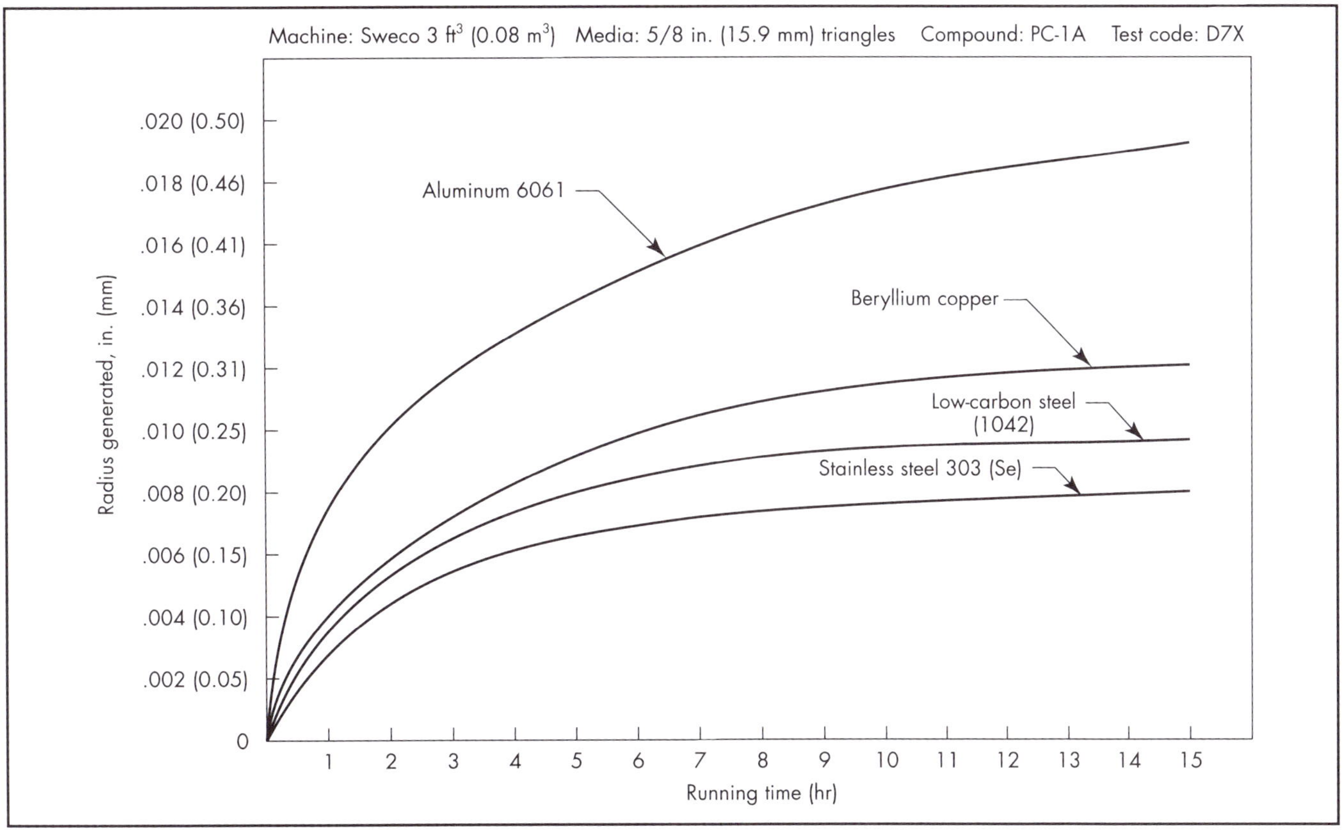

Figure 6-20. Graph of typical radius generated as a function of time in vibratory tub.

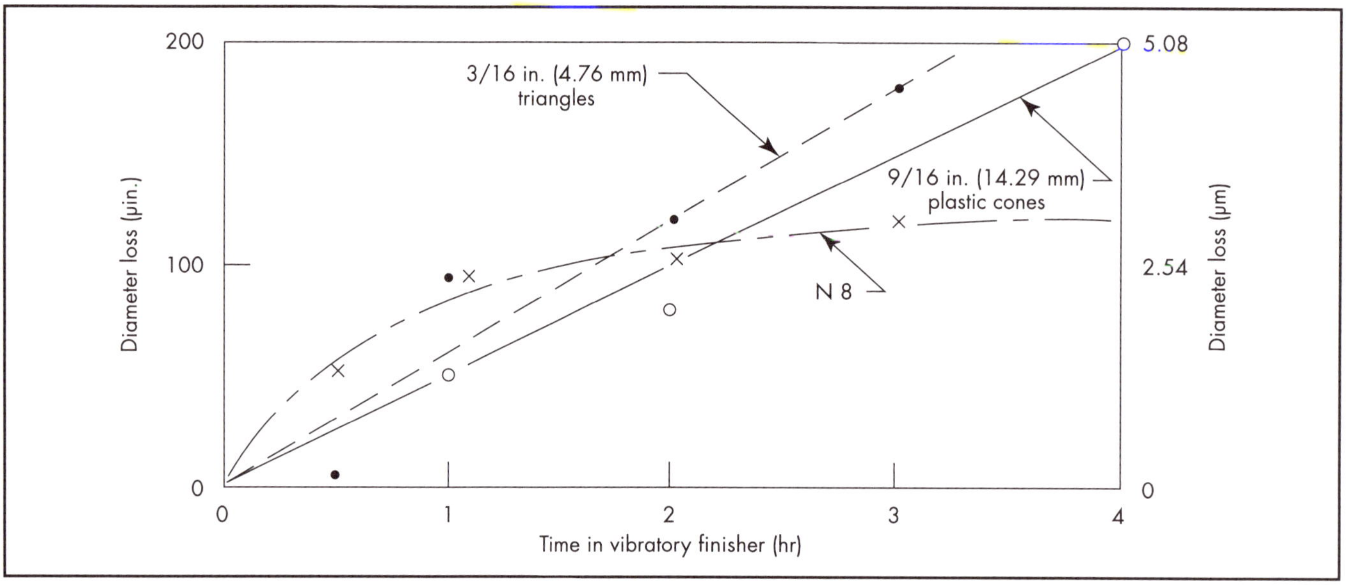

Figure 6-21. Effect of media and time on diameter change for 303 Se stainless steel pins (.490 in. [12.45 mm] diameter and .5 in. [12.7 mm] long).

Table 6-15. Measured mean change in burr height for vibratory deburring machine

Burr thickness in. (µm)	Time in vibratory finisher (hr) 1	2	3	4
	Mean reduction in burr height in. (µm)			
.001 (25.4)	.0035 (88.90)	.0043 (109.22)	.0089 (226.06)	.0121 (307.34)
.002 (50.8)	.0020 (50.80)	.0022 (55.88)	.0031 (78.74)	.0050 (127.00)
.003 (76.2)	.0018 (45.72)	.0021 (53.34)	.0031 (78.74)	.0047 (119.38)
.004 (101.6)	.0014 (35.56)	.0018 (45.72)	.0019 (48.26)	.0025 (63.50)
.005 (127.0)	.0008 (20.32)	.0011 (27.94)	.0013 (33.02)	.0018 (45.72)
.006 (152.4)	.0015 (38.10)	.0015 (38.10)	.0017 (43.18)	.0022 (55.88)

diameter of the hole. Because of media wear, the third choice is the safest with random-shaped media. For preformed shapes, the first two choices are easiest to assure.

Aluminum oxide media contributes to blow holes in welded joints (Gillespie 1976). It is desirable to use other materials when parts to be welded have to be deburred. Silicon carbide is often an acceptable, alternate material. Brazing and soldering are also affected by minute aluminum oxide particles. Solder will not adhere to a surface having such particles in or on it. The probability of plating adhesion failures increases when aluminum oxide is present on surfaces. This may not be a significant problem on copper-base alloys, since they activate easily. However, it can be a significant problem on any hard-to-activate material. The adhesion of solid film lubricants and the passivation of stainless steels do not appear to be affected by the presence of minute aluminum oxide particles.

Electrical contacts that must operate in low-resistance circuits are unaffected by dolomite, but other deburring media appears to increase circuit-loop resistance (CLR). Any dolomite film left on parts is easily removed by a 10% solution of warm acetic acid.

Burr size, allowable stock loss, and maximum allowable edge radii are the most significant factors in determining which combination of parameters will produce acceptable results. As seen in Table 6-16, it is not possible to remove burrs .005 × .005 in. (127 × 127 µm) while maintaining a .0005 in. (12.7 µm) maximum stock loss regardless of the radius allowed.

6.12 TROUBLESHOOTING

There are many variables (such as equipment, operating parameters, workpiece materials, media, compound solution, desired results) that contribute to problems in mass finish-

Table 6-16. Estimated capabilities of vibratory finishing for steel alloys with surface finish requirements of 32 μin. (0.813 μm) or better and precision low-loss conditions

Burr size thickness × length in. (μm)	Allowable thickness or diametral loss in. (μm)* .0001 (2.54)	.0005 (12.7)	.0010 (25.4)	.0050 (127)
Maximum edge radius in. (μm) allowed				
.0001 (2.54) × any length				
.002 (50.8)	X	X	X	X
.005 (127)	X	X	X	X
.010 (254)	X	X	X	X
.015 (381)	X	X	X	X
.0005 (12.7) × .0005 (12.7)				
.002 (50.8)	X	X	X	X
.005 (127)	X	X	X	X
.010 (254)	X	X	X	X
.015 (381)	X	X	X	X
.005 (127) × .003 (76.2)				
.002 (50.8)	—	—	—	—
.005 (127)	—	X	X	X
.010 (254)	—	X	X	X
.015 (381)	—	X	X	X
.005 (127) × .005 (127)				
.002 (50.8)	—	—	—	—
.005 (127)	—	—	X	X
.0010 (25.4)	—	—	X	X
.015 (381)	—	—	X	X

* X indicates process can maintain the indicated size loss and edge radius while removing burrs of the size shown. A dash indicates the process is not capable of maintaining both of these features while removing the indicated burr size.

ing, and they cannot be discussed in detail in this section. The four major problems encountered in mass-finishing processes are: (1) unacceptable results, (2) excessively long cycle times, (3) high media compound costs (discussed earlier as improper compound settings), and (4) media lodging in the workpieces (also discussed earlier) (Kittredge 1985b). Table 6-18 outlines other common problems encountered in mass finishing and their solutions.

6.12.1 Unacceptable Results

When unsatisfactory finishes, incorrect corner radii, incomplete burr removal, or other problems are encountered in mass finishing, they are probably the result of using improper equipment, media, compound solution, or workpieces. Improper operating parameters for the machines, incorrect concentrations of compounds in the solutions, and trying to finish workpieces differing in quality, all contribute to poor results.

6.12.2 Long Cycle Times

Excessively long cycle times can result for many reasons. The time can sometimes be shortened by changing the compound solution used to reduce glazing of the media and improve the cutting action. The media itself may be the problem. It may be too large and impeding separation from the workpieces. Changing to a different shaped and/or faster cutting media can often reduce cycle times. Machines should be kept in good condition and operated at proper speeds or they may lengthen cycle times.

In some cases, long cycle times are the result of over-finishing the parts. Specifications for actual requirements should be checked, and then a shorter cycle tried to see if the requirements can be met. Sometimes, the preprocessing operations performed on the parts can be changed to reduce the amount of finishing needed. Attempts should be made to increase the uniformity of quality and cleanliness of the workpieces to be finished to reduce cycle times.

6.13 HIGH-ENERGY MASS FINISHING

High-energy mass finishing encompasses processes in which the energy created within the mass in a container is greater than that obtained with standard vibratory methods. One advantage of high-energy processes is shorter processing cycles. Some high energy systems permit easy adjustment of the energy level. Maximum energy can be used for fast deburring, edge radiusing, and metal removal; reduced energy can be used for gentler action to refine edges and surfaces. With some systems there is no workpiece impingement. High-energy mass-finishing processes include centrifugal barrel, centrifugal disk, and spindle finishing, as well as chemically and electrochemically accelerated standard processes.

6.13.1 Centrifugal Disk (Roll-flow) Processes

The centrifugal disk process is the newest of the high-energy mass-finishing methods. Machines consist of an open-top bowl with stationary sidewalls and a rotary disk for a base (Figure 6-24). Media, compound, and workpieces are contained in the bowl.

Operating Principles

As the disk rotates with peripheral speeds up to 2,000 ft/min (610 m/min), the mass within the container is accelerated outward. When the media and workpieces contact the stationary sidewalls of the container, the sidewalls act as a brake, and the load starts to decelerate as it is forced upward by the mass behind it. The media and workpieces rise to the top of the load, then flow toward the center of the container, and from there downward toward the disk.

Table 6-17. Media wear for fast-cutting ceramic angle cut cylinders in a large machine

	Media in. (mm)			
	5/8 (15.9)	7/16 (11.1)	5/16 (7.9)	1/4 (6.4)
Run time (hr)	Average chip diameter in. (mm)			
27	.6123 (15.552)	.4191 (10.645)	.3253 (8.263)	.2578 (6.548)
44	.6043 (15.349)	.4077 (10.356)	.3216 (8.169)	.2457 (6.241)
89	.5726 (14.544)	.3436 (8.727)	.3004 (7.630)	.2374 (6.030)
259	.4787 (12.159)	.3436 (8.727)	.2618 (6.650)	.2061 (5.235)

Process Advantages

The action in centrifugal disk machines is substantially faster than in vibratory equipment. Centrifugal force developed is as much as 10 times the earth's gravity, which presses the media against the workpieces. Processing cycles typically vary between one-half and one-tenth those possible with vibratory finishing. One researcher found that the centrifugal disk machine could remove metal seven times faster than a vibratory machine in a 30 min period, but only twice as fast in a 3-hour run (Regan 1996). Another user notes that these newer machines were no more than four times faster at removing burrs (as opposed to removing metal from the surface). For conservative estimates, users may wish to estimate run times as 40% that of vibratory equipment.

Short process cycles result in decreased costs, reduced floor space requirements, and require less work in processing and inventory. Centrifugal disk finishing is a fairly versatile process and is easily adapted to batch automation. As with vibratory finishing equipment, workpieces can be readily inspected during the process cycle. With variable-speed equipment, it is frequently possible to combine heavy deburring and edge radiusing with a final, more gentle surface refinement operation.

Centrifugal disk equipment provides a smooth action. While the cost of centrifugal disk machines is about five times that of vibratory equipment, processing cycles are much less, thus making them economically justifiable for many applications. The cost of media and compounds per part processed is comparable to vibratory finishing. Since these machines can use lighter, less expensive, and less aggressive plastic media to do the same work, the cost of media may be less.

Process Limitations

The centrifugal disk process complements barrel, vibratory, and centrifugal barrel processes. It is capable of producing very fine finishes, but cannot yet handle as wide a variety of parts as the barrel, vibratory, and centrifugal barrel processes. Machines currently available cannot handle parts much larger than about 6 in.^{3} (98 cm^3), although parts to 12 in. (305 mm) long are being processed. Very small parts or media cannot be handled easily because there must be some gap, generally less than .010 in. (0.25 mm), between the disk and the cylinder walls. Also, centrifugal disk machines are not adaptable to continuous automation.

Centrifugal disk machines have at least six variables: speed, gap design, gap material, load height to depth ratio, contour of outer wall, and location of drains.

Equipment Available

Centrifugal disk equipment is available with capacities (workpieces and media) ranging from about 1–25 ft^3 (0.03–0.71 m^3). Machines have been developed recently for fine finishing and burnishing operations. Abrasion-resistant materials for the cylinders and disks are essential to providing a reasonable service life.

Unloading and reloading is accomplished either by doors in the sidewalls of the cylinders or by turning the entire unit over. Subsequent media separation, classification, washing, and return operations are external—similar to those used for tub-type vibratory equipment. Fully automated, computer-controlled systems are available.

Keeping the gap open and clear between the two halves of the machine was one of the early problems with this equipment and is still an area requiring some maintenance. Today, at least one company claims that maintenance is required only every 1,000 hours of operation. If dry media is used, a dust-collection system may be required.

6.13.2 Centrifugal Barrel Deburring

Centrifugal barrel finishing, like other mass-finishing processes, uses media, compound, and water to deburr and surface finish components. The process, sometimes called *mass orbital barrel finishing*, maintains a smooth rubbing action with no workpiece impingement, making it possible to produce fine finishes on precision and fragile parts. Another important advantage is the capability of imparting high compressive stresses in the surfaces of workpieces.

Operating Principles

On centrifugal barrel machines, a number of drums are mounted on the periphery of a turret (see Figure 6-25). The turret rotates in one direction, while the drums are rotated at a slower speed in the opposite direction. As illustrated in Figure 6-26, machines are available with the turret and drums rotating in a horizontal plane or in a vertical plane. Rotation in a horizontal plane is frequently used when the drums (work containers) are removable from the machine

Table 6-18. Troubleshooting problems in mass finishing

Problem	Mostly likely cause	Solution
Media fails to cut well, cycle times have to be increased	Compounds are failing to remove suspended soils	Change to better soil-suspension compound Determine what may have changed in plant processes
Parts have water spots or streaking	Surface tension too high	Use compound with more wetting action
Parts have smut on them at end of cycle	Soils are depositing on parts	Use compound with more active surfactants
Copper and brass are dark color	Oxidation	Add or increase inhibitors to compound
Zinc or aluminum parts have white powdery spots	Oxidation	Add or increase inhibitors to compound
Rust occurs when parts are removed from machine	Oxidation	Add or increase inhibitors to compound
Pitting and tarnish appear on parts; pitting appears on steel media	Oxidation	Add or increase inhibitors to compound
Residues left on parts in recessed areas	Abrasive or burnishing compound that did not dissolve in water	Thorough ultrasonic cleaning in hot soapy water Use liquid compounds in the burnishing cycle
Media lodges in holes	Poor choice of media or too rapid breakdown of media	Screen media to remove oversize or undersize particles Plug holes with rubber plugs, rubber tubing that expands when tension is released, rubber bands, or nylon line with ends expanded by heat Deburr before holes or slots are machined Acquire specially designed media Acquire specially designed screen sizes for in-house screening of media Use sizes of media that will not lodge in holes and recesses Do not use irregularly shaped media when lodging is a problem Screen the media prior to use Choose a media size/shape that can be screened or separated from the parts
Incomplete burr removal	Wrong media, compound, or water level	Use larger media Run for a longer time Increase energy or level of gravity Use a combination of large and small media Manually remove large burrs initially Use a more abrasive media compound Be sure the media shape is appropriate for part geometry If burrs are bent over, use a lighter media Place parts in a fixture Reduce the water level

Table 6-18. Troubleshooting problems in mass finishing *(continued)*

Problem	Mostly likely cause	Solution
Surface damage, impingement, or roughness	Wrong media or too long in machine	Use smaller media or well-worn media Reduce the cycle time When using two different sizes of media, use more of the smaller size Use a less abrasive compound Select a more appropriate media shape Mask sensitive areas Increase the water level Reduce the number of parts per load
Flat workpieces stick together	Surface tension	Use small random-shaped media
Parts are distorted	Large media and high energy levels bend precision, thin-wall components	Use small media, low energy levels, and short run times
Long burrs are beaten over	Media too heavy	Use plastic media Use other deburring process

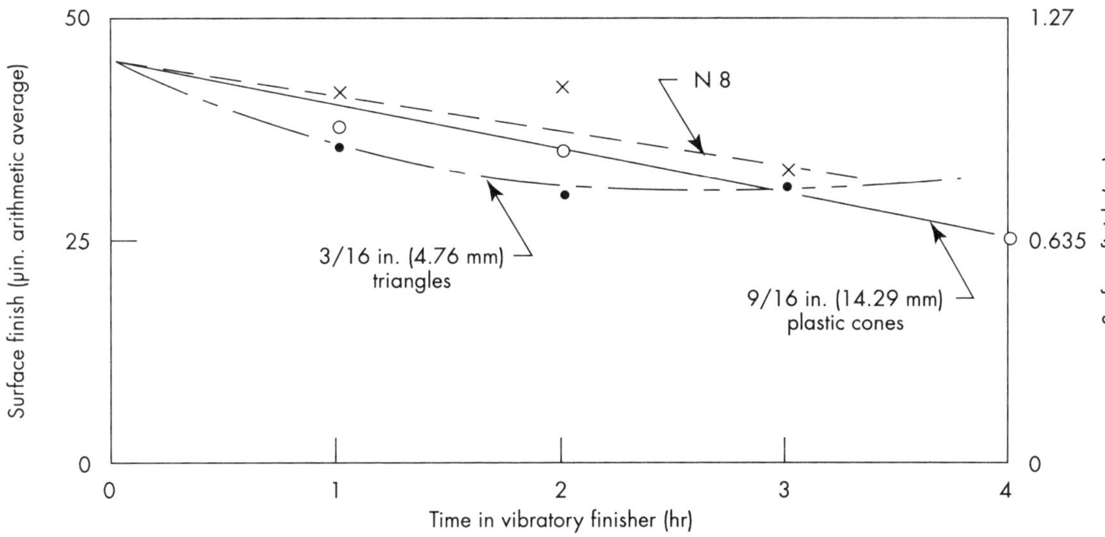

Figure 6-22. Effect of media and time on surface finish change for 303 Se stainless steel pins (.490 in. [12.45 mm] diameter and .5 in. [12.7 mm] long).

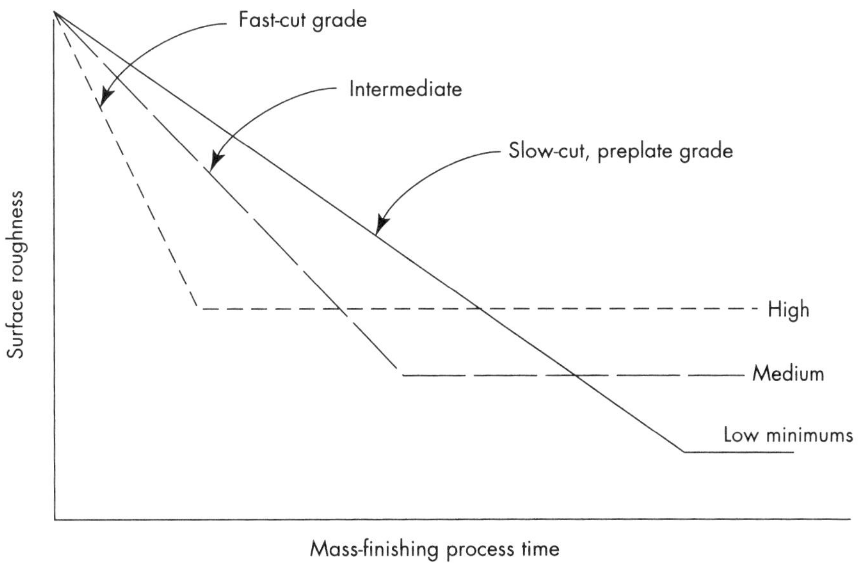

Figure 6-23. Relationship of media grade to surface finish produced (Kittredge 1981).

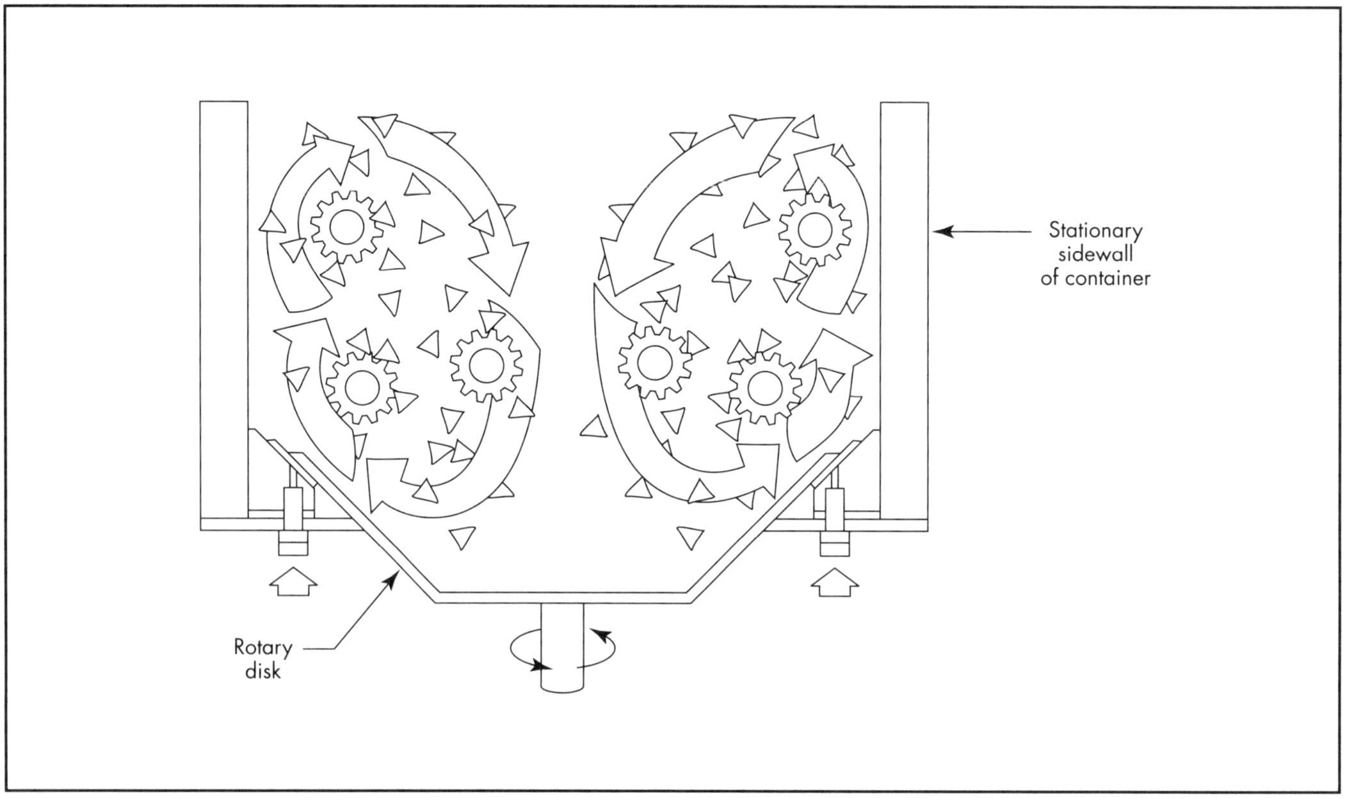

Figure 6-24. Principle of centrifugal disk finishing. ***(Courtesy Harper Finishing Systems)***

for reloading. Some research indicates that six- or eight-sided barrels are optimum for this process, whereas other studies indicate that for different conditions circular barrels are an advantage (Kittredge 1993).

Drums are loaded and unloaded in a manner similar to that for normal barrel finishing. About 60–80% of each drum's capacity is filled with a mixture of workpieces, media, compound, and water. Turret rotation creates a high centrifugal force, as much as 50 times earth's gravity. This force compacts the loads within the drums into tightly packed masses. Rotation of the drums causes the media to slide against the workpieces, removing burrs and refining surfaces.

Process Advantages

The abrading action under high centrifugal forces results in short processing cycles, 20–50 times faster than conventional vibratory finishing. Another important advantage is that counterrotation of the drums produces a smooth sliding action of media against workpieces, with no possibility of one workpiece impacting against another. The process produces consistent and reproducible results, very smooth surface finishes, and close tolerances, even with fragile and precision components.

Small-sized media can be used with centrifugal barrel finishing, thus ensuring uniformity of radii and edges and permitting greater action in holes and recesses. Small parts can be handled as well. All equipment is quiet in operation and requires no special noise-abatement measures.

Cost considerations frequently dictate the choice among conventional tumbling barrels, vibratory machines, centrifugal barrel equipment, and other finishing machines. In general, if satisfactory results can be achieved in vibratory machines with process cycles of less than 1 hour, they are usually more economical than centrifugal barrel finishing. Small centrifugal barrel machines are less convenient to operate than vibratory units because their containers are normally closed and there is no means for integral workpiece media separation. However, faster processing often offsets the higher cost and inconvenience of centrifugal barrel finishing.

Process Variables

Process variables for centrifugal barrel finishing are similar to those for other mass-finishing methods. Drums are loaded with workpieces and media, generally with a somewhat higher proportion of workpieces to media than in conventional barrel tumbling. Relative action on edges and surfaces can be controlled by the choice of media size and the amount of centrifugal force. It is not possible, however, to impart a radius on one edge while maintaining sharpness on another, equally accessible edge.

It is standard practice with centrifugal barrel finishing to operate at high speeds for maximum abrading action and then automatically switch to lower speeds, possibly with a change from cutting compound to finishing compound, to achieve finely refined surface and edge conditions. As a result, two operations are combined in a single cycle.

Applications

Originally, centrifugal barrel equipment was developed for finishing small, light components and workpieces having critical tolerances that could not be handled in other types of mass-finishing machines. With the development of large machines, up to 100 ft^3 (2.8 m^3), and automation, the process is now being used for a broader range of applications.

Automatic centrifugal barrel machines (Figure 6-27) still involve batch operation, but they provide automatic loading, unloading, and separation of workpieces from the media. The equipment may be used for heavy stock removal and edge radiusing operations for both high- and low-volume applications, as well as for precision finishing. Standard deburring and finishing operations seldom require more than 15 min.

Another important application for centrifugal barrel equipment involves imparting compressive stresses into the workpiece surfaces. This procedure improves resistance to fatigue failure and is being used for bearings, aircraft engine parts, springs, bearings, and pump components. Edge radiusing and surface finishing can be performed simultaneously, which further enhances the fatigue strength. This method generally costs less and is more effective than a combination of another finishing operation followed by shot peening.

Centrifugal barrel equipment has at least five variables: (1) turret rotation in the vertical or horizontal plane, (2) batch or continuous processing, (3) ratio of turret speed to barrel speed, (4) variable or fixed speed, (5) round or octagonal barrels.

The most significant recent development in centrifugal barrel equipment is dry finishing materials, the results of which rival hand-buffed finishes when properly applied. Media meant for dry use in centrifugal barrel or other machines do not have abrasive grains or particles embedded in them but are coated with fine abrasive materials. As a general rule, they are of much lighter density than synthetic media, which leads to generally longer cycle times than is the case with wet process materials. Their major advantage, however, is their ability to produce surface effects that are often close or equivalent to those obtained by manual buffing. A secondary consideration is that dry materials do not create the wet effluent waste-disposal problem.

Dry process media and compounds have found a growing list of applications in all of the major types of mass-finishing equipment, including conventional rotary barrels, bowl and tub vibrators, and high-energy finishing equipment, such as centrifugal barrel, centrifugal disk, and spindle or drag finishing equipment (Hignett and Gillott 1981). Dry

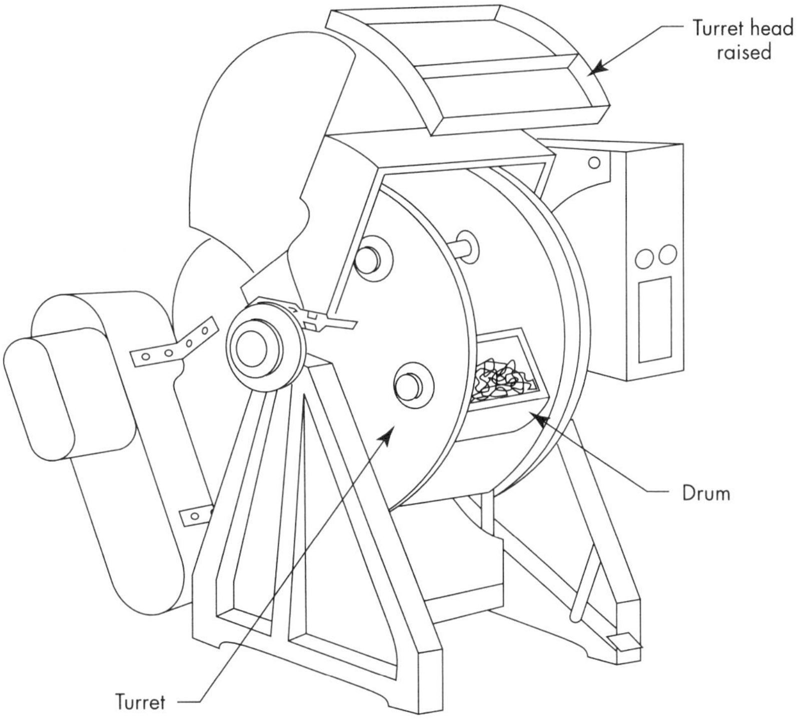

Figure 6-25. Centrifugal barrel finishing machine. ***(Courtesy Harper Finishing Systems)***

process material has been proven useful for producing both precision industrial finishes and highly reflective finishes for cosmetic or decorative applications, such as jewelry. These materials have become a key component in the surface preparation of precision parts within the aerospace, medical, electronic, ball and roller bearing, fine-blanking, plating, electroplating, and coating industries.

Barrel finishing is a positive displacement method. Most of the work is accomplished within a gravity-induced slide zone developed by the rotational motion of the barrel. In centrifugal barrel equipment, the opposing rotational forces induced by the counterrotating barrels and turret can produce pressure of media against parts that is many times the force of gravity.

The media volume requirements of each type of machine differ greatly as well, especially where dry media and compounds are concerned. Ideally, barrel equipment would be filled to 50% volume capacity to achieve the longest possible slide zone and optimum efficiency. Dry media loading requirements of centrifugal barrel equipment can take advantage of the fact that slide zone requirements are much more modest than in conventional barrel equipment, since most of the rubbing or wearing action of media against parts is caused by centrifugally induced pressures. This is a distinct advantage for centrifugal barrel equipment—fill levels of 90% or more can be used to protect fragile or delicate parts. Filling barrel compartments to only 50% of volume capacity with dry materials would in fact be ill-advised—fill levels must be higher than this to protect against part-on-part impingement.

One advantage centrifugal barrel equipment brings to the finishing engineer's arsenal is the ability to use dry media for cutting, smoothing, and polishing processes within restrictive time cycle constraints. This has been particularly useful to those firms experiencing serious wastewater disposal problems. For example, one manufacturer employed a 48-hour wet tumbling cycle to smooth its small stainless steel parts into the low surface finish root mean square (RMS) average numbers. This process used a large number of conventional tumbling barrels, which created a waste stream of effluent in excess of 10,000 gal/day (37,850 L/day). This entire system was replaced with two centrifugal barrel machines of 3 ft^3 (0.08 m^3) capacity each. A two-step process involving dry granular media was developed. Time cycles were reduced from 48 hours to 4 hours (two 2-hour steps), and effluent discharge was reduced to zero. The parts had greatly improved reflectivity, and the average surface finish on the parts was improved from the 10 RMS obtained with conventional tumbling equipment and materials to the 4–5 RMS range.

Dry process materials have been found to be useful in easing precious metal salvage problems in the jewelry industry. Many manufacturers of precious metal jewelry who use wet process media find it difficult to reclaim 100% of the precious metal fines. Dry deburring and polishing media made from

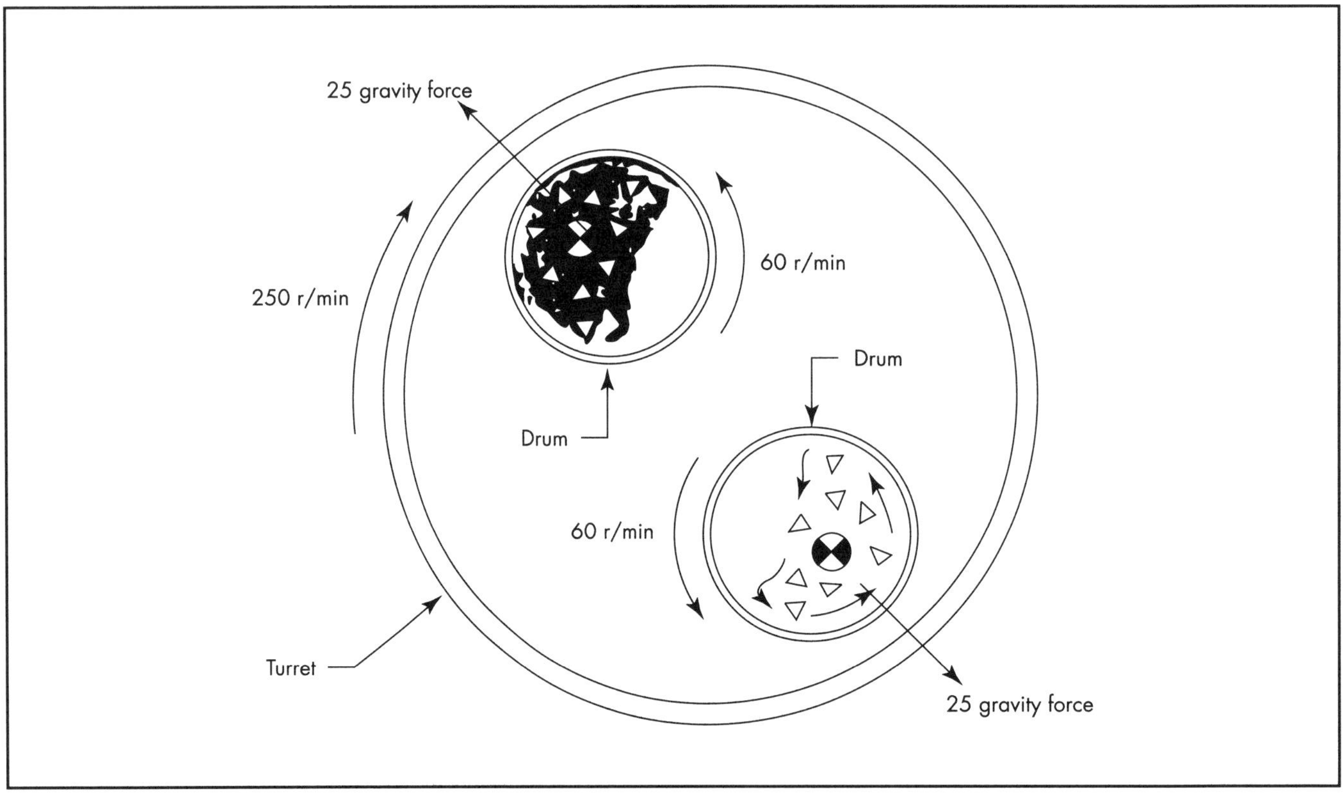

Figure 6-26. Action of turret and drums in centrifugal barrel finishing. ***(Courtesy Harper Finishing Systems)***

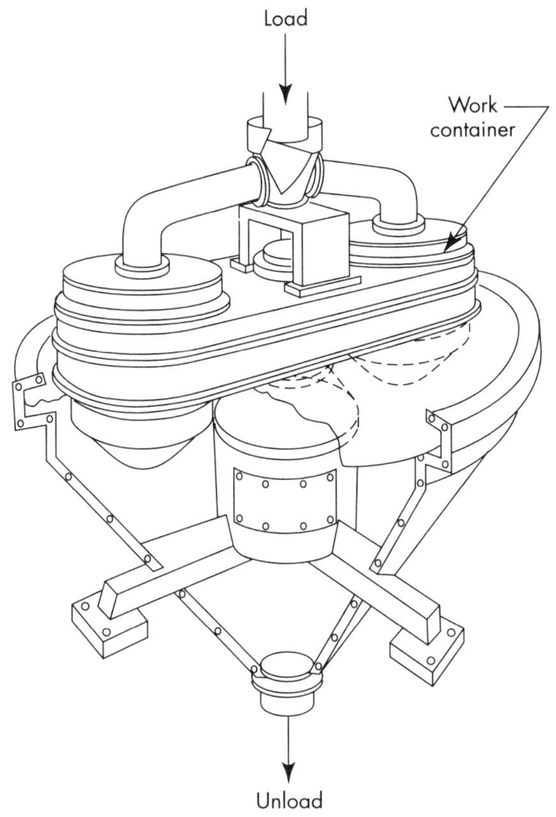

Figure 6-27. Automatic centrifugal barrel finishing machine. *(Courtesy Harper Finishing Systems)*

natural materials can ease this situation because they are easily burned and consumed in the refining process.

A follow-on technology has developed for precision industrial parts as a result of the research and development on processes to develop ultrarefined surface finishes for the jewelry industry. Similar high-energy processes have been used to produce highly refined finishes for critical components, such as aircraft engine turbine blades. Considerable documentation is now available to show just how refined these surfaces can be made.

Centrifugal barrel finishing can maintain .002 in. (50.8 μm) maximum edge radii if the burr is small. A much more desirable condition to maintain from a processing standpoint is .005 in. (127.0 μm) maximum edge radii. Allowable dimensional changes of at least .0002 in. (5.1 μm) are desirable in this process, and it is possible to maintain size within .00005 in. (1.27 μm) if burrs are small. Allowable surface finishes should be 32 μin. (0.81 μm) or rougher, although it appears possible that finer finishes can be produced or maintained. Corner radii (the intersection of three or more surfaces) will be 3–5 times larger than edge radii. When a workpiece has edges with several different requirements, the centrifugal barrel process should be based on the most tightly toleranced features. It is important to note that internal features will not receive as much action as external features. Figures 6-28–6-33 provide information on the typical process effects of centrifugal barrel finishing.

Tables 6-19–6-21 present results of a study of centrifugal barrel finishing on small pins. Using a 13-in. (330 mm) diameter barrel, 2 qt (1.9 L) of abrasive media were added in each barrel, followed by a cup of abrasive compound; parts were inserted into the barrel, and the entire mixture was covered with water. After running for 20 min, the barrel was flushed with clean water, a burnishing powder was added, the machine was cycled for an additional 5 min, and then rinsed. Parts were then separated from the media (Gillespie 1976).

In the study, solid cylinders of 1018 steel (R_B 93), 303 Se stainless steel (R_C 21), and 6061-T6 aluminum (R_B 60) were centerless ground, then subjected to various combinations of media and gravitational forces. The edge radius, change in diameter, and surface finish were recorded five times in a total deburring cycle of 60 min at a gravity force of 15. Each time the specimens were reinserted in the barrels, the barrel was flushed, and fresh abrasive compound was added. Table 6-22 indicates the three sizes of specimen used.

Media sizes used in the study are described in Table 6-19 as conditions I–VII. As shown in the table, edge radius, R, increases exponentially with time.

The large parts had larger radii than smaller specimens run under identical conditions, as shown in Figure 6-28 and Tables 6-19–6-21. From Table 6-20, after 20 min with the .25 in. (6.4 mm) plastic pyramids at a gravity force of 15, the .115-in. (2.92-mm) 1018 steel specimen had a .003 in. (76.2 μm) radius, $R = 0.00106\ (20)^{.32}$. The .5 in. (12.7 mm) specimen had a .00755 in. (192 μm) radius, $R = .00281\ (20)^{.33}$.

In the majority of cases, edge radii do not exceed .010 in. (254 μm) after 60 min. A .005 in. (127 μm) radius occurs after roughly 20 min. Edge radii as low as .002 in. (50.8 μm) can be produced after a 10-min cycle under some conditions.

Note that the coefficient in the equations estimates the radius after 1.0 min ($t = 1$). Group III has the highest coefficients for all equations, indicating that it is about twice as fast as Group I, two to three times as fast as Group II, and about four times faster than Group IV in forming the radius.

Diameter changes followed a similar pattern of size dependency. When number 14 grit aluminum oxide was used at a gravity force of 15 on 1018 stainless steel, after 60 min the diameter of the largest specimen decreased .0008 in. (20.3 μm), whereas the small specimen lost only .0004 in. (10.16 μm). These stock losses were roughly linear with time.

As a general rule, surface finishes improved from 40 μin. (1.016 μm) to 24 μin. (0.610 μm) in 60 min. Surface finish results, however, varied significantly with workpiece material and conditions used. The aggressiveness of the centrifugal barrel action was roughly proportional to the size of the media used (see Figure 6-34) and to the magnitude of the forces used (see Figure 6-35).

Extrapolating from the data, it appears that the .0013 in. thick by up to .0023 in. high (33 × 58 μm) burrs on these samples were entirely removed after roughly 2–5 min in the machine. As indicated earlier, thicker and higher burrs would require more time for complete removal. For such small burrs it is possible to assure complete burr removal with less than .0001 in. (2.54 μm) change in stock diameter. In some cases, it is possible to assure .00005 in. (1.27 μm) or less size change.

The data in the preceding study tended to be highly repeatable. Edge radii, for example, had a standard deviation of .0003–.0006 in. (7.6–15.2 μm). Diameters had a standard deviation of .00002–.0002 in. (0.5–5.1 μm). However, surface finish standard deviations varied from 3.0 to 13.0 μin. (0.0762–0.330 μm).

6.14 OTHER MASS-FINISHING PROCESSES

Chapter 8 describes some other loose abrasive and mass-finishing processes. Among them are magnetic barrel finishing, cryogenic vibratory and barrel finishing, chemically accelerated mass finishing, and electrochemically accelerated mass finishing.

6.15 ENVIRONMENTAL, HEALTH, AND SAFETY CONCERNS

Mass-finishing processes generally do not have significant health or safety issues. Noise is a concern in open-top machines of almost any kind. Operating noise levels vary depending on the part, the process, the type of equipment, and the vibrator location. Equipment manufacturers offer a variety of options tailored to the machine and application. Some may help reduce noise levels but interfere with production or maintenance operations, and others may absorb sound but retain unwanted moisture.

One death has been reported—the result of a lid blowing off a mass-finishing machine due to pressure build up. Generally centrifugal barrel machines and conventional barrel machines are covered machines. Occasionally, a drum will be thrown from a fast-spinning machine. Normal machine design should prevent this from occurring. Water hoses and splashing water are hazards of the trade, so electrical outlets need to be placed out of the way of these liquids.

Dry skin is common with operators who work with wet processes. Some acids and basic finishing compounds may cause rashes on some operators. Gloves, aprons, and other simple protective gear will prevent this problem.

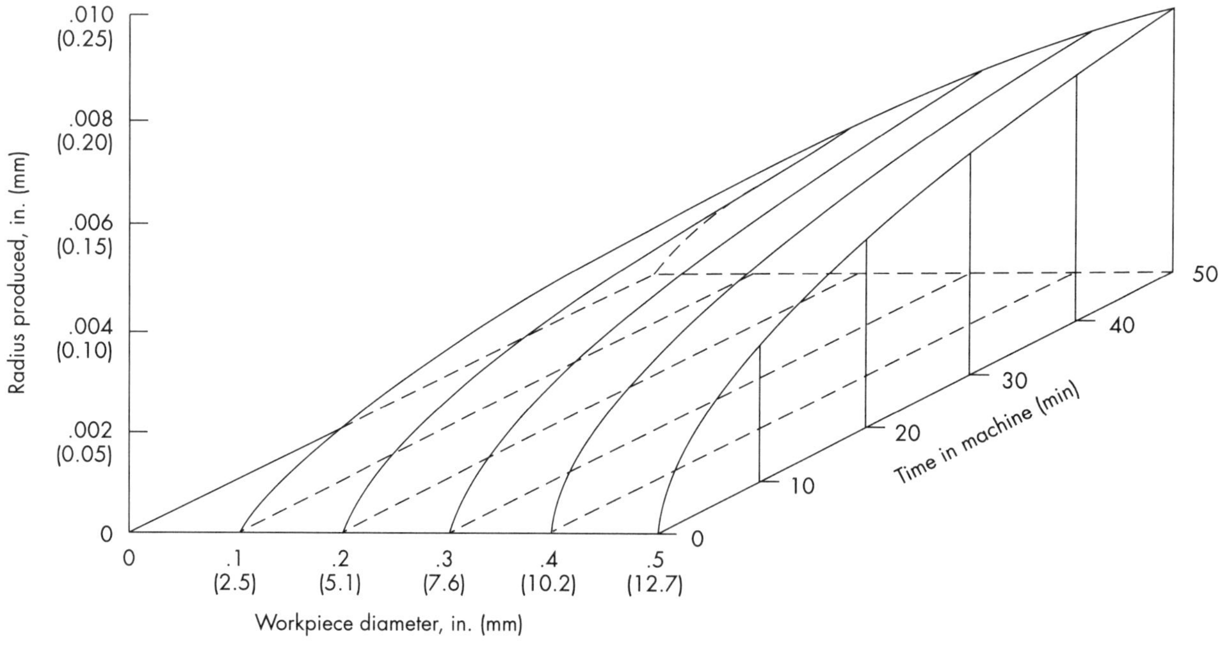

Figure 6-28. Effect of part diameter and time in centrifugal barrel. $R = 0.00459\ D^{.665}\ t^{.337}$, where: D = part diameter in inches, t = time in the machine in minutes, R = resulting radius in inches.

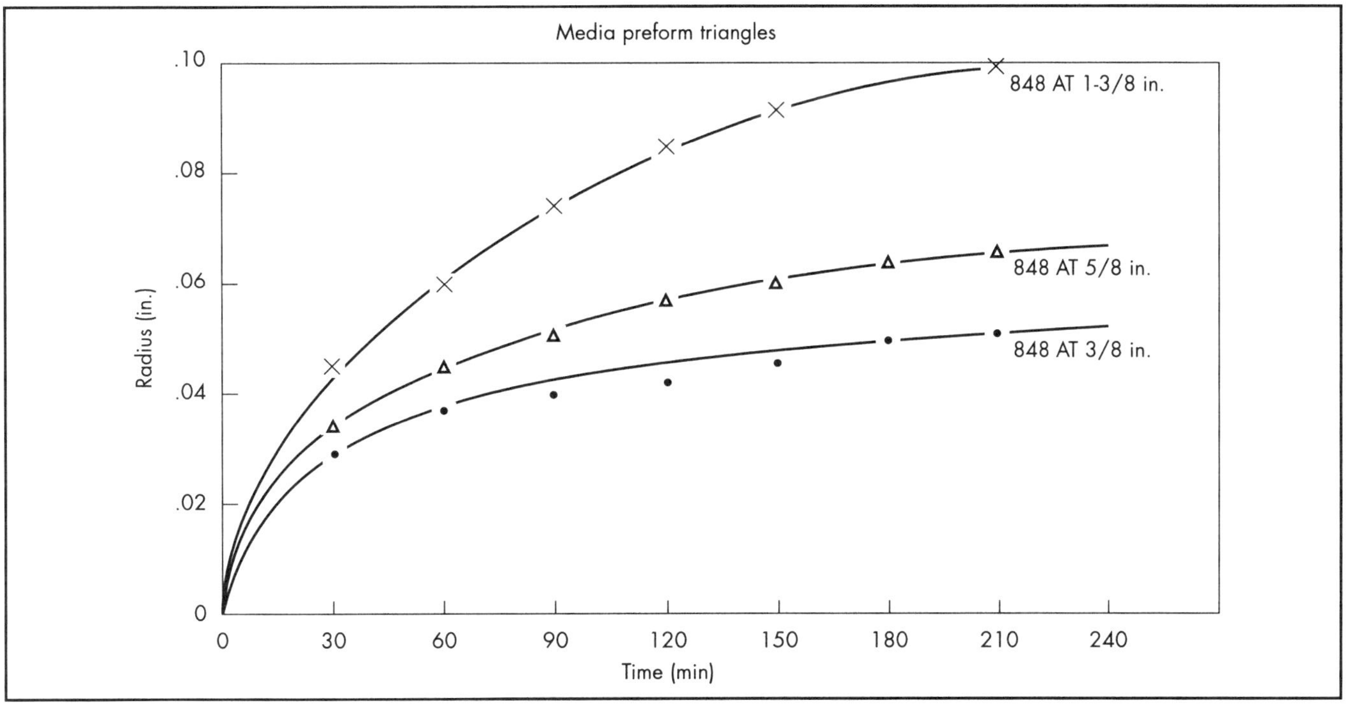

Figure 6-29. Effect of media size on edge radius in centrifugal barrel at a gravity force of 25. Data are based on using 1.5 × 1.5 in. (38 × 38 mm) cube or cylinder parts run at a gravity force of 25 in normal operation. Numbers containing the letter *R* are random-shaped aluminum oxide, and the number after the *R* represents the grit size (that is, 651R 1-1/2 represents random-sized chunks averaging 1.5 in. [38 mm]). Media numbers containing the letter *T* are plastic-bonded triangles, and those with the letters *AT* are ceramic-bonded triangles (Hignett 1976).

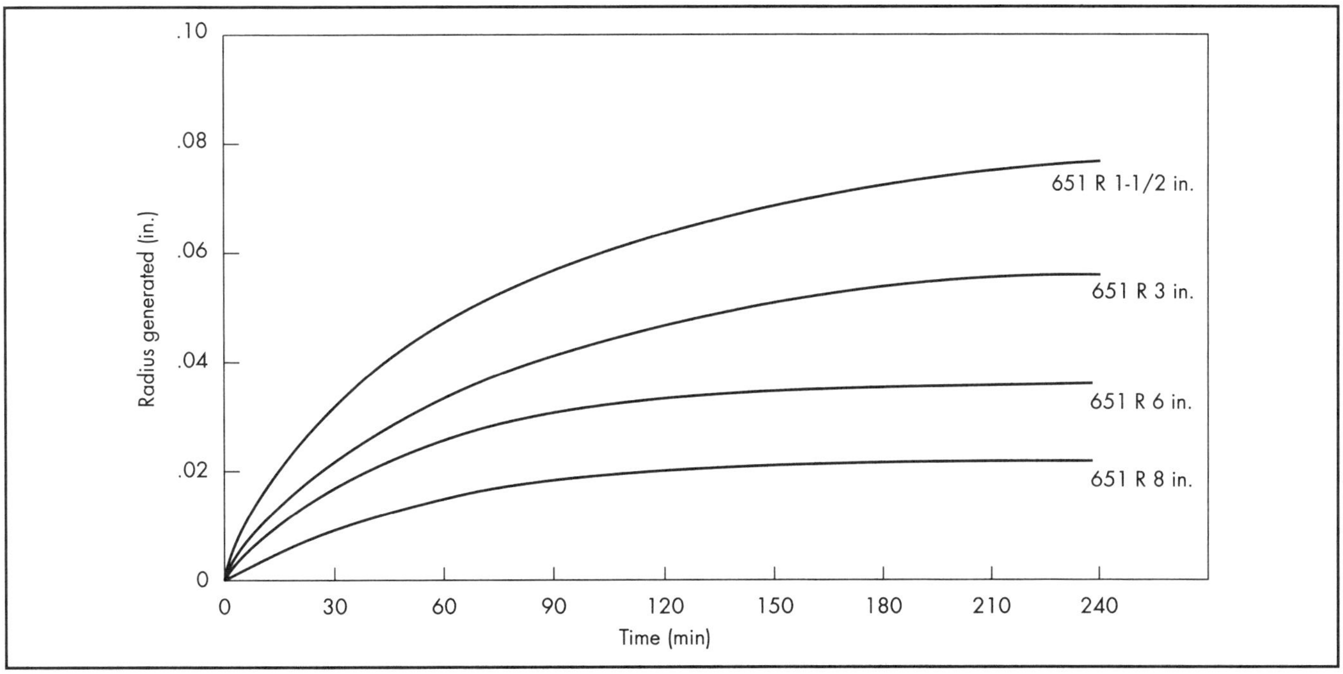

Figure 6-30. Effect of random media size on edge radius in centrifugal barrel at a gravity force of 25. Data are based on using 1.5 × 1.5 in. (38 × 38 mm) cube or cylinder parts run at a gravity force of 25 in normal operation. Numbers containing the letter *R* are random-shaped aluminum oxide, and the number after the *R* represents the grit size (that is, 651R 1-1/2 represents random-sized chunks averaging 1.5 in. [38 mm]). Media numbers containing the letter *T* are plastic-bonded triangles, and those with the letters *AT* are ceramic-bonded triangles (Hignett 1976).

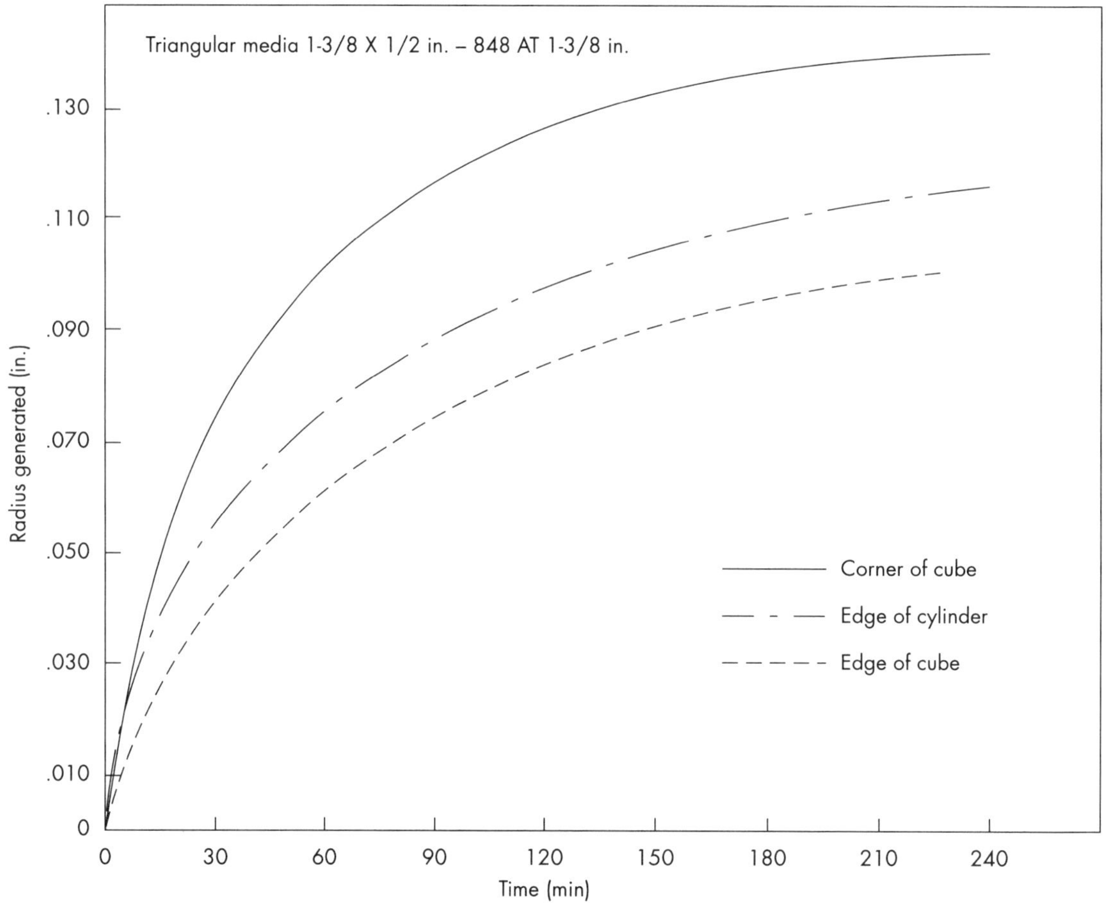

Figure 6-31. Effect of component shape on edge radius in centrifugal barrel at a gravity force of 25. Data are based on using 1.5 × 1.5 in. (38 × 38 mm) cube or cylinder parts run at a gravity force of 25 in normal operation. Numbers containing the letter *R* are random-shaped aluminum oxide, and the number after the *R* represents the grit size (that is, 651R 1-1/2 represents random-sized chunks averaging 1.5 in. [38 mm]). Media numbers containing the letter *T* are plastic-bonded triangles, and those with the letters *AT* are ceramic-bonded triangles (Hignett 1976).

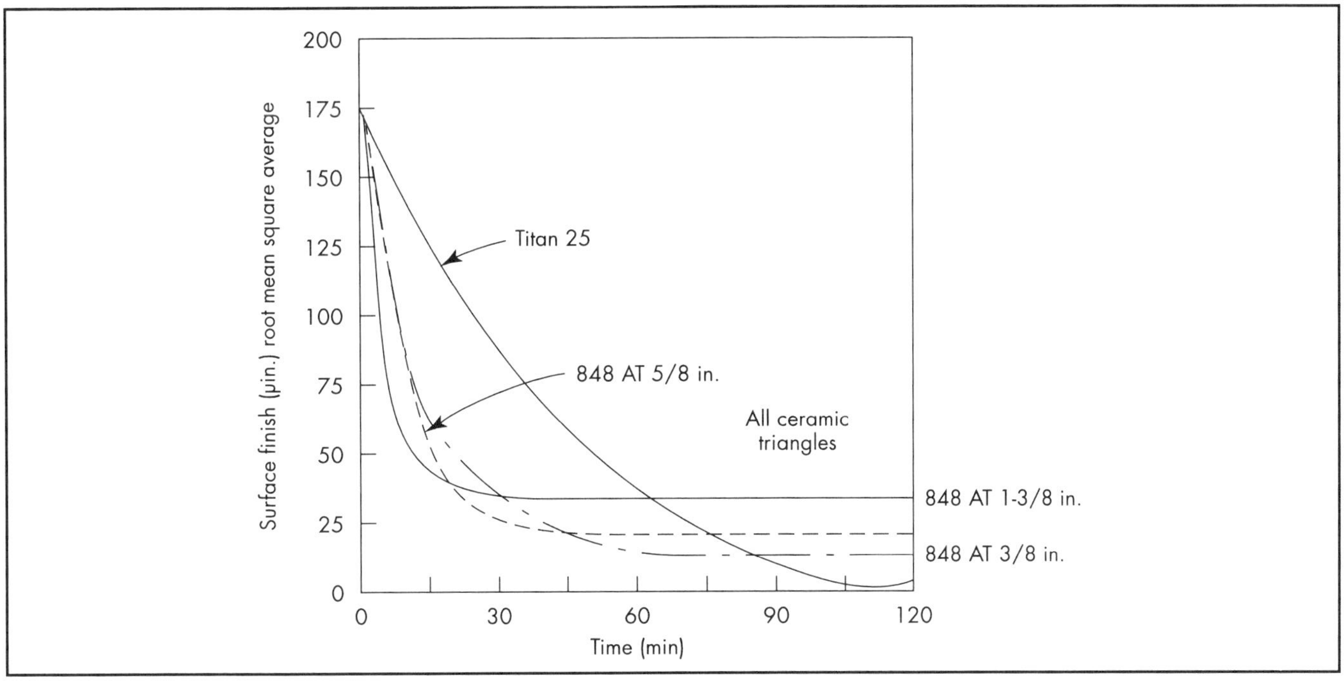

Figure 6-32. Surface finish improvement in centrifugal barrel at a gravity force of 25 using ceramic triangles. Data are based on using 1.5 × 1.5 in. (38 × 38 mm) cube or cylinder parts run at a gravity force of 25 in normal operation. Numbers containing the letter *R* are random-shaped aluminum oxide, and the number after the *R* represents the grit size (that is, 651R 1-1/2 represents random-sized chunks averaging 1.5 in. [38 mm]). Media numbers containing the letter *T* are plastic-bonded triangles, and those with the letters *AT* are ceramic-bonded triangles (Hignett 1976).

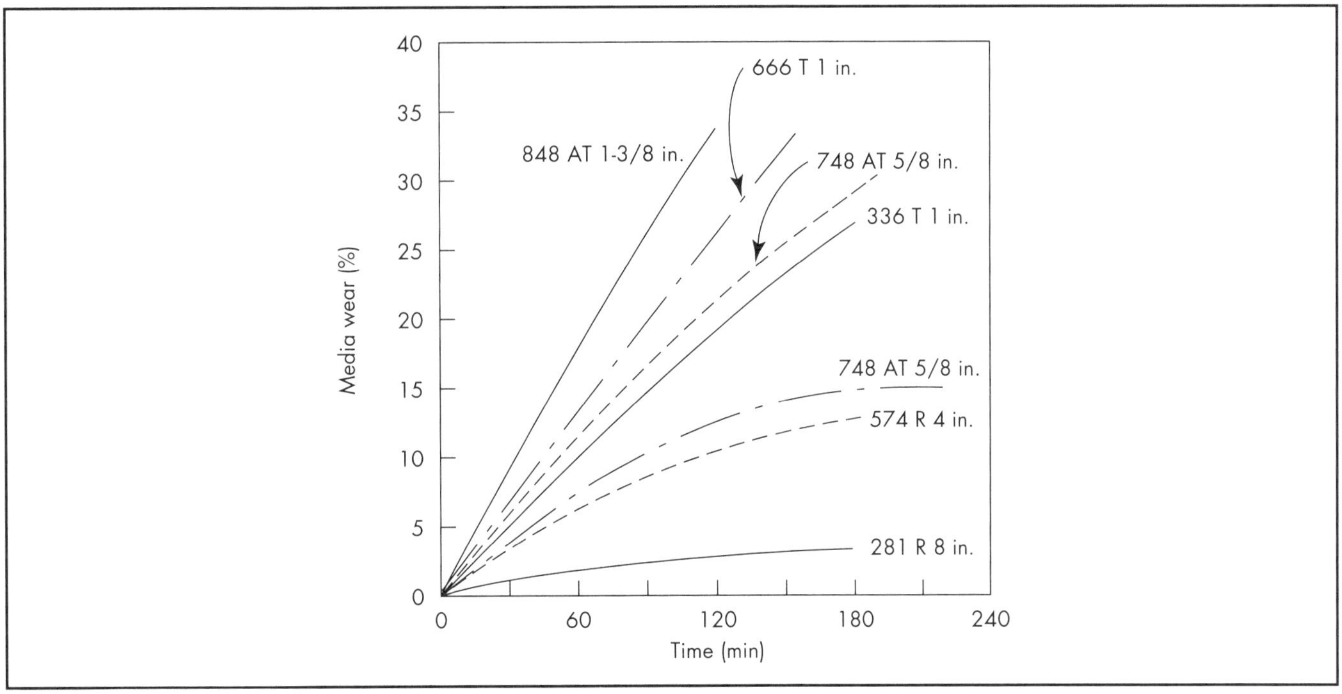

Figure 6-33. Media wear in centrifugal barrel at a gravity force of 25. Data are based on using 1.5 × 1.5 in. (38 × 38 mm) cube or cylinder parts run at a gravity force of 25 in normal operation. Numbers containing the letter *R* are random-shaped aluminum oxide, and the number after the *R* represents the grit size (that is, 651R 1-1/2 represents random-sized chunks averaging 1.5 in. [38 mm]). Media numbers containing the letter *T* are plastic-bonded triangles, and those with the letters *AT* are ceramic-bonded triangles (Hignett 1976).

Table 6-19. Equations for edge radiusing 303 Se stainless steel using centrifugal barrel finishing

Conditions	Part diameter in.*		
	.5	.25	.115
(I) $A1_{2}0_{3}$ N14 at 15 gravity	$R = .00113t^{.46}$ $\sigma_R = .00025$	$R = .00095t^{.42}$ $\sigma_R = .00016$	$R = .00063t^{.43}$ $\sigma_R = .00014$
(II) Dolamite at 15 gravity	$R = .00072t^{.48}$ $\sigma_R = .00031$	$R = .00055t^{.46}$ $\sigma_R = .00017$	$R = .00032t^{.35}$ $\sigma_R = .00012$
(III) 3/16-in. ceramic triangles at 15 gravity	$R = .00281t^{.37}$ $\sigma_R = .00029$	$R = .00241t^{.35}$ $\sigma_R = .00011$	$R = .00131t^{.38}$ $\sigma_R = .00023$
(IV) $A1_{2}0_{3}$ N24 at 15 gravity	$R = .00071t^{.48}$ $\sigma_R = .00034$	$R = .00036t^{.53}$ $\sigma_R = .00034$	$R = .00012t^{.61}$ $\sigma_R = .00011$
(V) $A1_{2}0_{3}$ N14 at 20 gravity	$R = .00106t^{.43}$ $\sigma_R = .00033$	$R = .00032t^{.68}$ $\sigma_R = .00045$	$R = .00010t^{.93}$ $\sigma_R = .00022$
(VI) $A1_{2}0_{3}$ N14 at 5 gravity	$R = .00037t^{.62}$ $\sigma_R = .00012$	$R = .00037t^{.51}$ $\sigma_R = .00025$	$R = .00017t^{.62}$ $\sigma_R = .00006$
(VII) 1/4-in. plastic pyramids at 15 gravity	$R = .00246t^{.31}$ $\sigma_R = .00053$	$R = .00179t^{.34}$ $\sigma_R = .00036$	$R = .00060t^{.40}$ $\sigma_R = .00020$

* .001 in. = 25.4 μm. Radius (*R*) is in inches, where time (*t*) is in minutes. Standard deviation (σ_R) is in inches.

Table 6-20. Equations for edge radiusing 1018 steel using centrifugal barrel finishing

Conditions	Part diameter in.*		
	.5	.25	.115
(I) $A1_{2}0_{3}$ N14 at 15 gravity	$R = .00148t^{.44}$ $\sigma_R = .00032$	$R = .00126t^{.38}$ $\sigma_R = .00035$	$R = .00076t^{.41}$ $\sigma_R = .00018$
(II) Dolamite at 15 gravity	$R = .00171t^{.38}$ $\sigma_R = .00019$	$R = .00098t^{.40}$ $\sigma_R = .00019$	$R = .00043t^{.46}$ $\sigma_R = .00009$
(III) 3/16-in. ceramic triangles at 15 gravity	$R = .00343t^{.38}$ $\sigma_R = .00013$	$R = .00251t^{.40}$ $\sigma_R = .00016$	$R = .00168t^{.35}$ $\sigma_R = .00014$
(IV) $A1_{2}0_{3}$ N24 at 15 gravity	$R = .00103t^{.45}$ $\sigma_R = .00039$	$R = .00043t^{.52}$ $\sigma_R = .00012$	$R = .00021t^{.57}$ $\sigma_R = .00005$
(V) $A1_{2}0_{3}$ N14 at 20 gravity	$R = .00143t^{.41}$ $\sigma_R = .00054$	$R = .00068t^{.52}$ $\sigma_R = .00037$	$R = .00055t^{.42}$ $\sigma_R = .00020$
(VI) $A1_{2}0_{3}$ N14 at 5 gravity	$R = .00093t^{.42}$ $\sigma_R = .00018$	$R = .00061t^{.44}$ $\sigma_R = .00013$	$R = .00028t^{.54}$ $\sigma_R = .00004$
(VII) 1/4-in. plastic pyramids at 15 gravity	$R = .00281t^{.33}$ $\sigma_R = .00081$	$R = .00192t^{.36}$ $\sigma_R = .00059$	$R = .00106t^{.32}$ $\sigma_R = .00034$

* .001 in. = 25.4 μm. Radius (*R*) is in inches, where time (*t*) is in minutes. Standard deviation (σ_R) is in inches.

Table 6-21. Equations for edge radiusing 6061-T6 aluminum using centrifugal barrel finishing

Conditions	Part diameter in.* .5	.25	.115
(I) Al_2O_3 N14 at 15 gravity	$R = .00221t^{.28}$ $\sigma_R = .00018$	$R = .00161t^{.30}$ $\sigma_R = .00012$	$R = .00108t^{.34}$ $\sigma_R = .00002$
(II) Dolamite at 15 gravity	$R = .00299t^{.36}$ $\sigma_R = .00028$	$R = .00235t^{.25}$ $\sigma_R = .00022$	$R = .00156t^{.32}$ $\sigma_R = .00013$
(III) 3/16-in. ceramic triangles at 15 gravity	$R = .00499t^{.34}$ $\sigma_R = .00039$	$R = .00289t^{.40}$ $\sigma_R = .00022$	$R = .00280t^{.31}$ $\sigma_R = .00016$
(IV) Al_2O_3 N24 at 15 gravity	$R = .00148t^{.36}$ $\sigma_R = .00034$	$R = .00085t^{.34}$ $\sigma_R = .00010$	$R = .00054t^{.36}$ $\sigma_R = .00022$
(V) Al_2O_3 N14 at 20 gravity	$R = .00084t^{.53}$ $\sigma_R = .00046$	$R = .00082t^{.47}$ $\sigma_R = .00029$	$R = .00064t^{.43}$ $\sigma_R = .00033$
(VI) Al_2O_3 N14 at 5 gravity	$R = .00069t^{.48}$ $\sigma_R = .00012$	$R = .00065t^{.49}$ $\sigma_R = .00004$	$R = .00114t^{.17}$ $\sigma_R = .00006$
(VII) 1/4-in. plastic pyramids at 15 gravity	$R = .00348t^{.35}$ $\sigma_R = .00126$	$R = .00269t^{.32}$ $\sigma_R = .00046$	$R = .00192t^{.29}$ $\sigma_R = .00073$

* .001 in. = 25.4 μm. Radius (R) is in inches, where time (t) is in minutes. Standard deviation (σ_R) is in inches.

Table 6-22. Specimen sizes used in centrifugal barrel study

Diameter in.	(mm)	Length in.	(mm)
.490	(12.446)	.500	(12.700)
.240	(6.096)	.250	(6.350)
.115	(2.921)	.125	(3.175)

The environmental issues involved in mass-finishing are related to the materials being processed and the solid wastes that result. As will be discussed in Chapter 8, when acid solutions are combined with mass-finishing processes, waste disposal is similar to that found in plating shops.

References

Balz, Gunther W. 1983. "A Computer-Simulated Mass Finishing System." Technical paper MR83-687. Dearborn, MI: Society of Manufacturing Engineers (SME).

Bostock, Ray M. 1981. "Deburring Costs—Have You Examined Yours Lately?" Technical paper MR81-385. Dearborn, MI: Society of Manufacturing Engineers (SME).

Gillespie, LaRoux K. 1975. "A Quantitative Approach to Vibratory Deburring Effectiveness." Technical paper MRR75-11. Dearborn, MI: Society of Manufacturing Engineers (SME).

———. 1976. "Deburring by Centrifugal Barrel Tumbling." Topical report BDX-613-1559. Kansas City, MO: Bendix, available from National Technical Information Sevice (NTIS).

Gillespie, LaRoux K., et al. 1976. "Effects of Deburring Contaminants on Electroplating Adhesion." Final report BDX-613-1757. Kansas City, MO: Bendix, available from NTIS.

Hignett, J. Bernard. 1976. "Centrifugal Barrel Finishing, Comparative Media Evaluation." *Metal Finishing,* 74:35-40.

Hignett, J. Bernard, and Gillott, Richard D. 1981. "The Centrifugal Barrel Process—Precision Deburring and Surface Finishing." Technical paper MR81-231. Dearborn, MI: Society of Manufacturing Engineers (SME).

Ioi, Toshihiro, Matsunaga, Masahisa, and Kobayashi, Hisamine. 1985. "Simulation Model for a Flexible Finishing System." Technical paper MR85-830. Dearborn, MI: Society of Manufacturing Engineers (SME).

Johannesen, Robert. W. 1977. "The Total Deburring Cost." Technical paper MR77-132. Dearborn, MI: Society of Manufacturing Engineers (SME).

———. 1981. "Effluent Considerations in/from the Deburring Department." Technical paper MR81-411. Dearborn, MI: Society of Manufacturing Engineers (SME).

Kittredge, John B. 1977. "The Testing of Vibratory Finishing Media." Technical paper MR77-466. Dearborn, MI: Society of Manufacturing Engineers (SME).

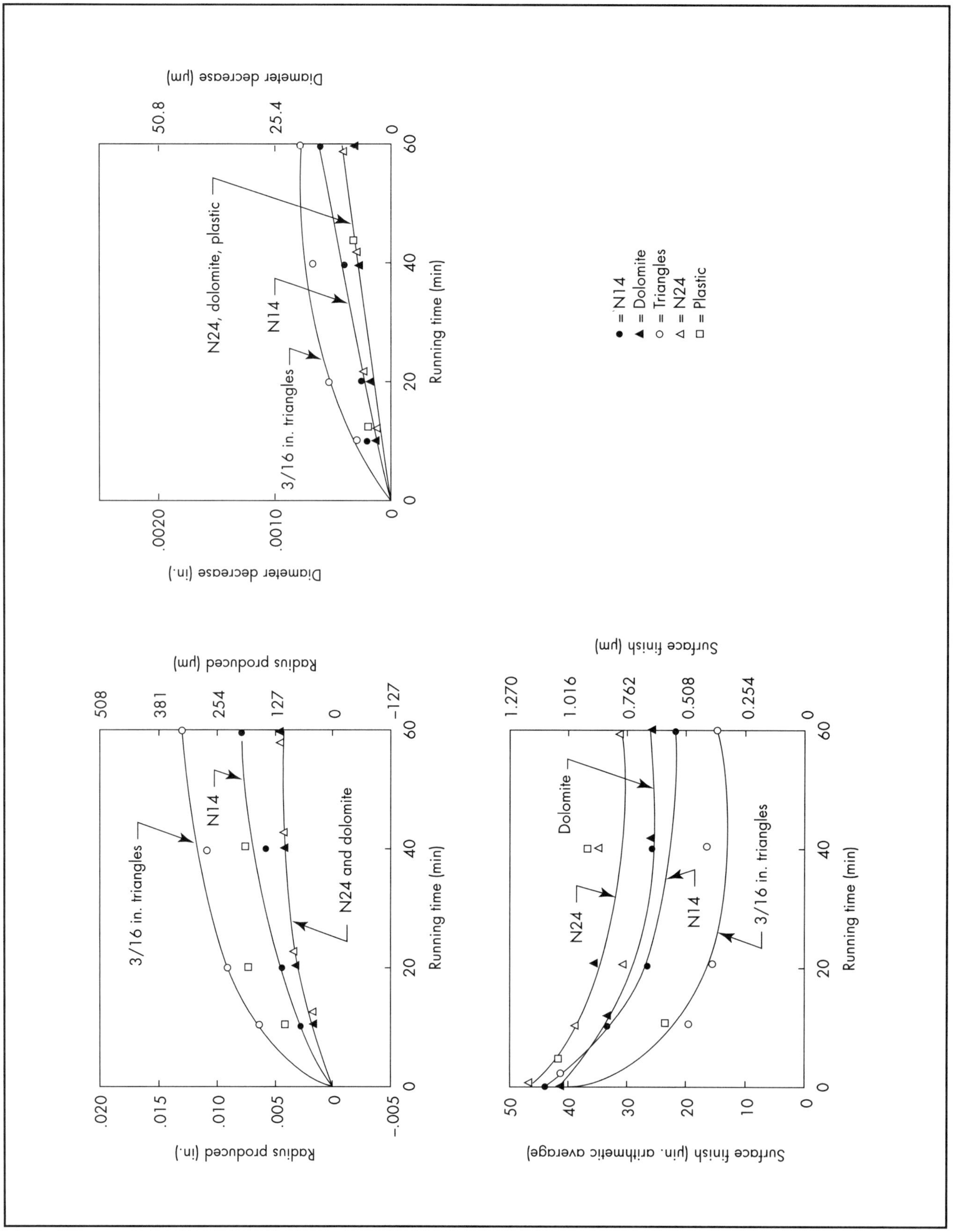

Figure 6-34. Centrifugal barrel finishing results on 303 Se stainless steel at a gravity force of 15 using various media.

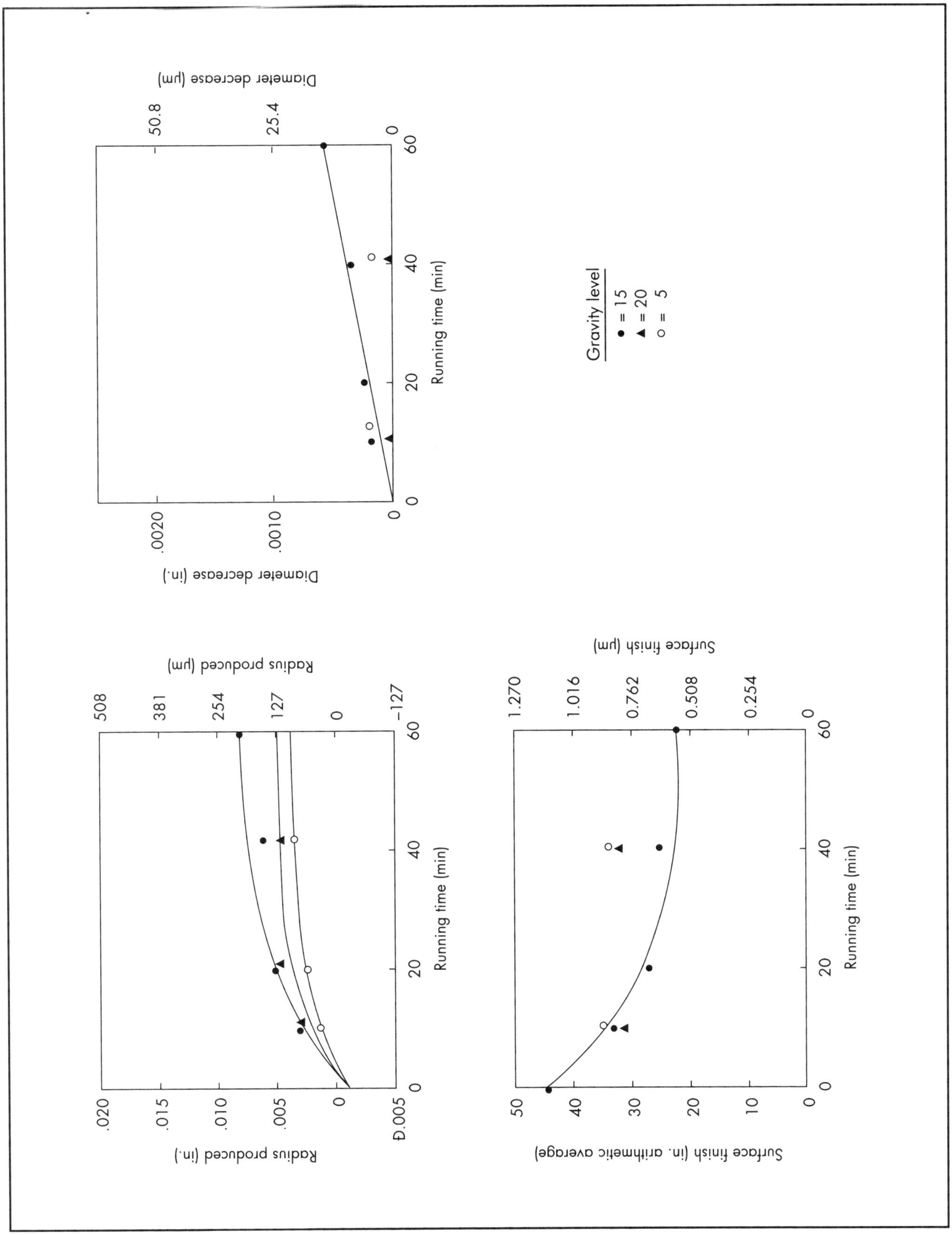

Figure 6-35. Centrifugal barrel finishing results on 303 Se stainless steel using number 14 aluminum oxide grit at various gravitational levels.

——— 1981. "Understanding Vibratory Finishing." *Products Finishing,* 45 (February–May):(2–40).

———. 1983. "Attacking the Astronomical Odds in Mass Finishing." Technical paper MR83-675. Dearborn, MI: Society of Manufacturing Engineers (SME).

———. 1985a. "Computer-aided Mass Finishing Process Development." Technical paper MR85-836 and computer disk. Dearborn, MI: Society of Manufacturing Engineers (SME).

———. 1985b. "Troubleshooting in Mass Finishing." *Products Finishing Directory*. Cincinnati, OH: Gardner Publications: 254–59.

———. 1989. "Vibratory Finishing Equipment." Technical paper MR89-149. Dearborn, MI: Society of Manufacturing Engineers (SME).

———. 1993. "Mass Finishing Media—Part II." *Products Finishing,* 57 (April):26–30.

Matsunaga, M. 1971. "Microcrystalline Abrasive Media for Mass Finishing." *Metal Finishing,* 69 (May):97.

Miller, Bob. 1985. "Deburring Effluent Clean Up." Technical paper MR85-835. Dearborn, MI: Society of Manufacturing Engineers (SME).

Ostwald, Phillip F. 1983. *American Machinist Manufacturing Cost Estimating Guide*. New York, NY: American Machinist.

———. 1985/1986. *American Machinist Cost Estimator.* New York, NY: American Machinist.

Ramnarine, Patrick C. 1989. "Mass Finishing Compound Technology." Technical paper MR89-111. Dearborn, MI: Society of Manufacturing Engineers (SME).

Regan, Dan T. 1996. "Centrifugal Disk Finishing." *Products Finishing,* 50 (May):51–56.

Rhoades, Lawrence J. 1981. *Cost Guide for Automatic Finishing Processes.* Dearborn, MI: Society of Manufacturing Engineers (SME).

Thompson, Samuel R. 1981. "Selection of Vibratory Finishing Equipment." Technical paper MR81-381. Dearborn, MI: Society of Manufacturing Engineers (SME).

Bibliography

American Society for Metals. 1989. "Heat Treating, Cleaning and Finishing." *Metals Handbook,* 9th ed. vol. 2. Materials Park, OH: American Society for Metals.

Battelle-Columbus Laboratories. 1975. "Assessment of Industrial Waste Practices." Report PB-264-349. "Electroplating and Metal Finishing Industries, Job Shops." Columbus, OH: Batelle-Columbus Laboratories.

Bennett, H., ed. 1974. EPA Ref. QD5-B-4-1974. *Concise Chemical and Technical Dictionary.* New York, NY: FAIC Chemical Publishing Company.

Davidson, David A. 1990a. "Developments in Dry Process Mass Finishing." Technical paper MR89-147. Dearborn, MI: Society of Manufacturing Engineers (SME).

———. 1990b. "High-energy Dry Process Finishing." Technical paper MR90-389. Dearborn, MI: Society of Manufacturing Engineers (SME).

Depuy, Douglas D. 1983. "Deflashing Systems for Molded Products." Technical paper MR83-681. Dearborn, MI: Society of Manufacturing Engineers (SME).

DeRenzo, D. J., ed. 1981. "Pollution Control Technology for Industrial Wastewater." *Pollution Technology Review,* no. 80. Park Ridge, NJ: Noyes Data Corporation.

Flund, Herbert I., ed. 1971. EPA TD197-142-1971. *Industrial Pollution Control Handbook.* New York, NY: McGraw Hill Book Company.

Garlinghouse, Roland E. 1985. "Heavy-duty Mass Finishing." Technical paper MR85-826. Dearborn, MI: Society of Manufacturing Engineers (SME).

Gillespie, LaRoux K. 1973. "Vibratory Deburring." Topical report BDX-613-735 (rev.). Kansas, City, MO: Bendix, available from NTIS.

———. 1976a. "The Effect of Edge Angle on the Radius Produced by Vibratory Finishing." *Deburring Capabilities and Limitations.* Dearborn, MI: Society of Manufacturing Engineers (SME).

———. 1976b. "A Quantitative Approach to Vibratory Deburring Effectiveness." *Deburring Capabilities and Limitations.* Dearborn, MI: Society of Manufacturing Engineers (SME).

———. 1981. *Deburring Technology for Improved Manufacturing.* Dearborn, MI: Society of Manufacturing Engineers (SME).

———. 1996. *Guide to Deburring, Deflashing and Trimming Equipment, Supplies and Services.* Kansas City, MO: Deburring Technology International.

Gillespie, LaRoux K., and Clay, Fred. 1975. "Deburring Effects on Plating Adhesion." *Finishing Highlights,* (September–October), vol. 7, no. 6:26–27.

Hartley, Howard S. 1981. "Considerations for Selecting a Waste Treatment System." *Finishers Management,* 25:14.

Hashimoto, F. 1996. "Modelling and Optimization of Vibratory Finishing Process." *CIRP Annals,* 45 (1):302–306.

Hignett, J. Bernard. 1982. "High Energy Mass Finishing." Technical paper MR82-264. Dearborn, MI: Society of Manufacturing Engineers (SME).

Ishida, Takao, and Cantwell, Mark. 1993. "Innovative Dry Centrifugal Disk Finishing Process for Deburring, Descaling, and Polishing." Technical paper MR93-329. Dearborn, MI: Society of Manufacturing Engineers (SME).

Johannesen, R. W. 1976. "The Deburring Job Shop." Technical paper MR76-834. Dearborn, MI: Society of Manufacturing Engineers (SME).

———. 1977. "The Total Deburring Cost." Technical paper MR77-132. Dearborn, MI: Society of Manufacturing Engineers (SME).

Kitajima, K., Sakurada, Y., Yamamoto, A., and Takazawa, K. 1996. Surface and Edge Finishing by Using Dry-type Centrifugal Barrel Finishing. *Proceedings of the 4th International Conference on Precision Finishing and Burr Technology*. Kansas City, MO: Deburring Technology International: 84–91.

Kittredge, John B. 1982. "The Vibratory Finishing Compound Solution System." Technical paper MR82-239. Dearborn, MI: Society of Manufacturing Engineers (SME).

———. 1985. "Troubleshooting in Mass Finishing." *Products Finishing Directory*. Cincinnati, OH: Gardner Publications: 254–59.

———. 1987. "Selection of Mass Finishing Equipment." Technical paper MR87-153. Dearborn, MI: Society of Manufacturing Engineers (SME).

Kobayashi H., and Matsunaga, M. 1981. "New Developments in Barrel Finishing." Technical paper MR81-402. Dearborn, MI: Society of Manufacturing Engineers (SME).

Letner, H.R. 1959. Stress Effects of Abrasive Tumbling. *Transactions of the ASM*. Materials Park, OH: American Society for Metals: 51:1,402–20.

Marino, M. 1980. "Wastewater and Hazardous Solid Waste Disposal in the Aluminum Products Industry." *Metal Finishing,* 78:21–27:40

Matsunaga, M. 1967. "Theory and Experiments on Centrifugal Barrel Finishing." *International Journal of Production Research,* 5 (4):275–287.

Matsunaga, Masahisa, and Hagiuda, Yoshiaki. 1967. "Research on Barrel Finishing." *Report of the Institute of Industrial Science,* vol. 17, no. 4, serial no. 111. Tokyo, Japan: University of Tokyo.

Miyatani, Takashi. 1981. "Low-cost Production Method by Barrel Finishing for Mechanical Parts of Engineering Ceramics." *Proceedings of the 4th International Conference on Precision Finishing and Burr Technology*. Kansas City, MO: Deburring Technology International: 101–108.

Miyatani, Takashi, Harada, Masahi, and Takazawa, Koya. 1981. "Newly Developed Deburring Machines for Precision Parts in Mass Production." Technical paper 81-382. Dearborn, MI: Society of Manufacturing Engineers (SME).

Murphy, James A. 1971. *Surface Preparation and Finishes for Metals*. Dearborn, MI: Society of Manufacturing Engineers (SME).

Neal, James W. 1981. "Finishing Jet Engine Components—New Developments." Technical paper MR81-405. Dearborn, MI: Society of Manufacturing Engineers (SME).

Noyes Data Corporation. 1978. "Electroplating and Related Metal Finishing." *Pollution Technology Review* no. 46. Park Ridge, NJ: Noyes Publications.

Ohio EPA. 1973. "In-process Pollution Abatement; Upgrading Metal Finishing Facilities to Reduce Pollution." *Technology Transfer* (July).

Okumra, Hajime, and Cantwell, Michael S. 1991. "Innovation of Centrifugal Disk Finishing." Technical paper MR91-139. Dearborn, MI: Society of Manufacturing Engineers (SME).

Punamia, Manoj, and Blotter, P.T. 1982. "Metal Removal Rate by Centrifugal Barrel Finishing." Technical paper MR82-267. Dearborn, MI: Society of Manufacturing Engineers (SME).

———. 1981. "Workpiece Kinetics in Centrifugal Barrel Finishing." Technical paper MR81-406. Dearborn, MI: Society of Manufacturing Engineers (SME).

Ramnarine, Patrick C. 1989. "Mass Finishing Compound Technology." Technical paper MR89-111. Dearborn, MI: Society of Manufacturing Engineers (SME).

Robbins, G.V. 1966. "Radius Generation in Vibratory Deburring." Topical report BDX-613-85. Kansas City, MO: Bendix, available from NTIS.

Scheetz, David R. 1995. "Rosler Keramo-Finish®." Technical paper MR95-225. Dearborn, MI: Society of Manufacturing Engineers (SME).

Shikishima Tipton Manufacturing Company, Ltd. 1975. *Barrel Finishing Manual*. Nagoya, Japan: Shikishima Tipton Manufacturing Company, Ltd.

Thompson, Samuel R. 1983. "Comparison of Rectangular and In-line Continuous Vibratory Equipment." Technical paper MR83-679. Dearborn, MI: Society of Manufacturing Engineers (SME).

Watson, Michael R., ed. 1973. "Pollution Control in Metal Finishing." *Pollution Technology Review* no. 5. Park Ridge, NJ: Noyes Data Corporation.

7 Spindle Finishing

Spindle finishing, also called gyrofinishing in some parts of the world, is a high-energy, loose abrasive finishing process that features fast and precise deburring or finishing. Good control, smooth finishes, and a high degree of reliability and uniformity are other characteristics of this process. Spindle equipment is commonly included under the mass-finishing umbrella, although it does not allow masses of parts to be finished at one time. The media variables described in Chapter 6 all apply to this process. Types of spindle finishing include single or multiple spindle, normal and high pressure, one or more parts per spindle, and conventional or fluidized bed processes.

7.1 CONVENTIONAL SPINDLE FINISHING

7.1.1 Operating Principles

Most spindle-finishing machines consist of a circular rotating tub that holds the media and one or more rotating spindles on which the workpieces are held and immersed in the media (see Figure 7-1). With the workpieces rotating in one direction and the media moving rapidly in the other, the media flows swiftly over the edges and surfaces of the workpieces. Flow-through of the compound solution using perforations in the tub is typical practice.

Processing cycles with spindle finishing rarely exceed 5 minutes (min) and are frequently less than 30 seconds (s). A limitation is that the workpieces must be individually mounted on the spindles of the machines with workholding collets or fixtures. In many applications, a cycle of 5–20 s is sufficient, although cycles of up to 5 min are not uncommon. Cycles are determined by the complexity of the parts, which can range from engine valves to complex gears. Surface finishes normally range between 10 and 5 root mean square (RMS) average, with edge radii consistently produced down to .001 in. (25.4 μm).

Because of repetitive loading and unloading due to the relatively short cycles, full-time operators are required. However, in some specific and more refined operations, mechanical loading arms and industrial robots can be incorporated to reduce operator time.

Spindles are generally rotated at speeds of 10–3,000 r/min. For some applications, however, the spindles are stationary, and in others they are oscillated vertically. Multispindle heads are used on some machines, and there can be more than one head per tub. The tubs are generally held stationary but also may be rotated and vibrated. In the fluidized bed variation, air is injected into the media to keep it loose.

Spindles are generally adjustable to vary the results obtained. For maximum action, the spindles are positioned to

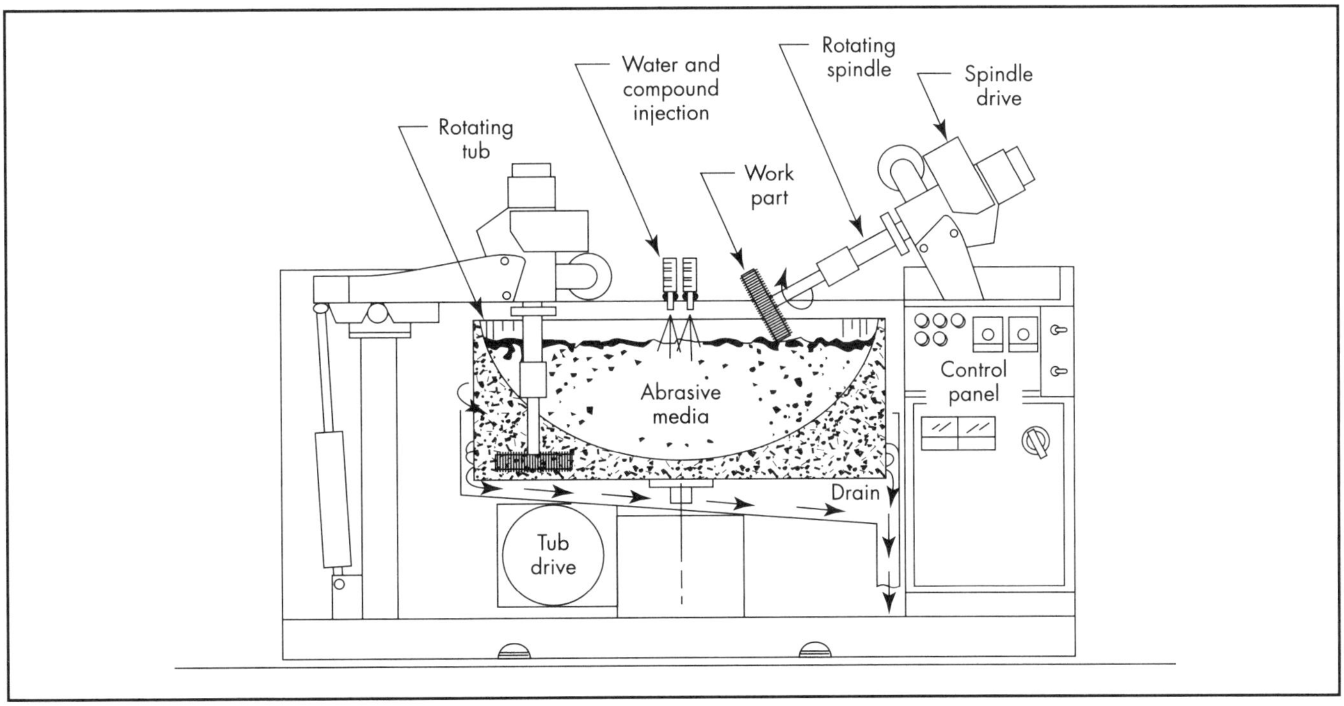

Figure 7-1. Two-spindle, spindle-finishing machine. ***(Courtesy Almco Industrial Finishing Systems)***

place the workpieces as close as possible to both the periphery and the bottom of the rotating tub. This arrangement stems from the fact that the velocity of the media is proportional to the radius from the tub's center, and hence there is an advantageous pressure effect created by compressing the media between the workpieces and the bottom and side walls of the tub. Air or hydraulic cylinders are generally used to pivot or lower the spindles into the tubs. Machines are built with one, two, four, or as many as 16 spindles per machine. Typically, parts are a few inches in diameter, but parts up to 16 in. (406.4 mm) in diameter have been finished with standard equipment.

7.1.2 High-pressure Spindle Finishing

Higher velocity and confinement of the media generate greater finishing forces. These requirements can be met by fitting a standard tub with a baffle and center dome (see Figure 7-2). The baffle tends to force the media downward toward the workpiece as the result of centrifugal force and permits the tub to be operated at a higher speed for faster processing. The center dome creates a pocket effect that does not allow the media to flow into the center area of the tub. For small workpieces, the dome allows the media to recover faster and creates a dense, solid mass. For large workpieces, the confinement generates greater finishing forces, resulting in short cycling times.

7.1.3 Continuous Spindle Finishing

To eliminate the unproductive time spent removing spindles from the tubs and unloading/reloading workpieces, continuous machines, such as the one illustrated in Figure 7-3, have been developed. With this continuous machine, a head holding six rotating spindles indexes the workpieces into and out of the media in a continuously rotating tub. Unloading and reloading the workpieces can be done manually or with a robot, as shown in Figure 7-3 to increase productivity.

One variation of spindle finishing, usually run dry for polishing of parts, uses a stationary media container so that the cluster of spindles revolves in the media.

Table 7-1 presents some possible production rates for spindle-finishing machines. To achieve these values, the operator must change parts very quickly; a continuous 5 s changeout rate is not generally realistic over a long day.

7.1.4 Typical Applications

Spindle-finishing machines are well suited for deburring and finishing uniformly shaped cylindrical components, such as gears, sprockets, and bearing cages. Such components are easily fixtured, and the action of the media is uniform over all significant surfaces. Because the parts are fixtured, there is no possibility that one part will impinge on another and

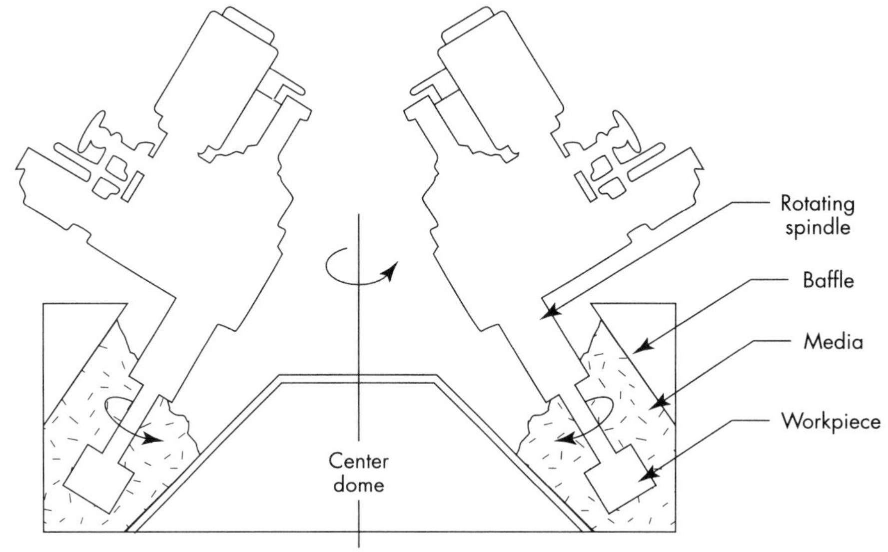

Figure 7-2. Spindle finishing tub equipped with baffle and center dome for high-pressure finishing. ***(Courtesy Almco Industrial Finishing Systems)***

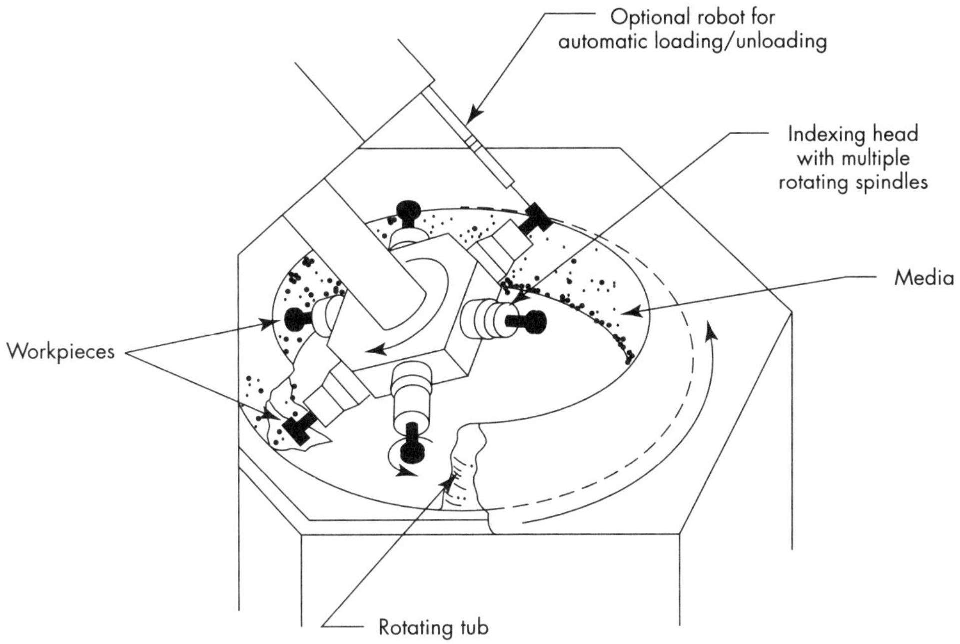

Figure 7-3. Continuous spindle finishing with indexing multispindle head and robot for loading and unloading. ***(Courtesy Almco Industrial Finishing Systems)***

Table 7-1. Production rates for spindle-finishing machines

Total cycle time (s)	Standard machine unload time (s)	Two-spindle machine maximum output (pieces/hr)	Continuous two-spindle machine output (pieces/hr)	Output increase (%)
5	5	720	1,440	100
10	5	480	720	50
15	5	360	480	33
20	5	288	360	25
30	5	205	240	17
44	5	144	160	11
60	5	55	60	9

(Olsen 1981)

cause nicks, dings, or scratches. Often, workpieces can be stacked on a spindle so that several parts can be deburred or finished simultaneously.

Variables in spindle finishing include tub speed and direction, spindle speed and direction, time, machine design, media size and material, and water level (amount). Figure 7-4 illustrates the results of theoretical work on the effects of spindle finishing. Although empirical results do not exactly match these curves, the general form of the curves are accurate and the data at higher ratios are exact.

The rotating tub is motor driven, usually by a variable speed drive to allow flexibility of operation. The spindles are also motor driven either by constant or variable speed drives to rotate the part in the abrasive mass to equally expose all surfaces for finishing. Spindles are generally pivoted or lowered into the finishing mass by means of air cylinders for processing and raised to a convenient position for loading and unloading. The majority of spindle machines used for high-production operations also include provisions for air operated fixtures for quick loading and unloading of parts.

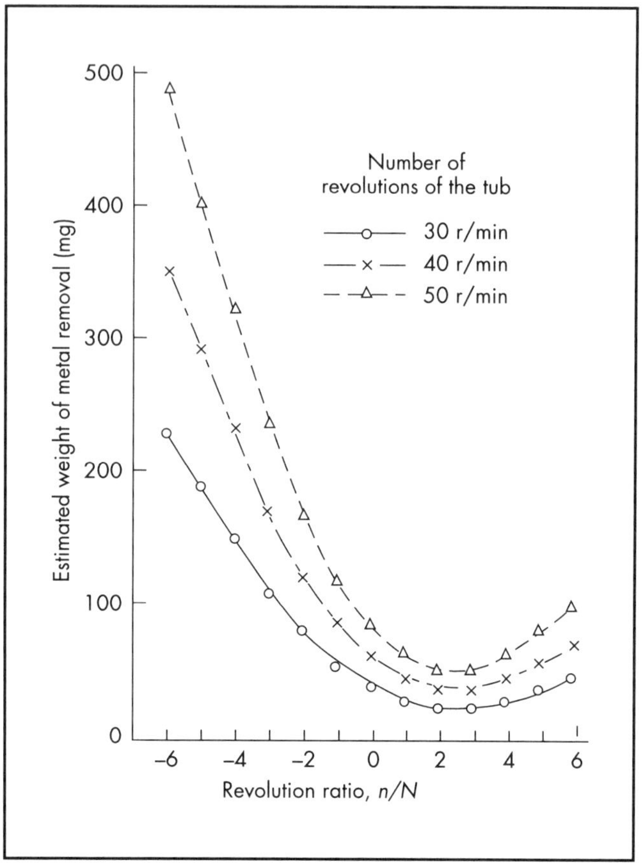

Figure 7-4. Estimated weight of metal removed as a function of spindle speed, *n*, and tub speed, *N*. Negative numbers indicate that the direction of motion of these rotating bodies was opposite each other. *(Courtesy Shiskishima Tipton)*

Machines designed for wet processing incorporate a rotating tub with perforated sections on the bottom or side wall. This design allows water-based chemical compounds to be injected automatically and continually flush away any metal chips, abrasive "fines," or sludge. This process helps to ensure efficient cutting action, continuously clean parts, and a contamination-free surface. Discharged liquids flow into a collection ring positioned around the tub and later to a common drain fitting where "spent" solutions are channeled to drains, settling tanks, or filtration and recirculation systems for reentry into the processing tub.

Accurate control of processing cycles is maintained by electronic timers. Process tub speed is controlled by direct-reading tachometers showing surface feet per minute, spindle speed is controlled by tachometers showing revolutions per minute, and liquid-flow meters measure the amount of water and/or liquid compounds introduced into the process.

To allow versatility in processing a relatively wide range of parts and configurations, most spindle equipment is adjustable; for example, spindles may be adjusted up or down or toward or away from the tub wall, and the spindle head may be pivoted or angled to allow proper positioning of parts for maximum exposure to the flowing abrasive mass.

At the end of the predetermined process cycle, the spindle is automatically raised out of the processing tub where finished parts can be manually or automatically sprayed with water or with a cleaning solution to flush away any loose abrasive particles and residue from the abrasive mass.

With the process completed, the operator can remove the finished parts, which, in most production installations, is accomplished by pressing a foot switch that actuates the fixture collet to release the finished parts. On removal of finished parts, the next process cycle begins by loading unfinished parts on the fixture and repeating the process cycle. Figure 7-5 illustrates the colleting devices used in spindle finishing.

7.1.5 Media and Compounds

Processing media includes a wide variety of abrasive materials and chemical compounds—most of the materials described in Chapter 6 can be used in this process. The most popular media are fine-grit aluminum oxides because they can effectively accomplish common deburring and edge radius operations with controlled accuracy on precision metal parts. Depending on part configuration and degree of cut or finish requirements, larger random-shaped aluminum oxide or preshaped ceramic-based media are used for more aggressive finishing action. Sometimes large-sized media are mixed with smaller-sized media to allow aggressive action on the outer edges and surface of a part. This combination deburrs or improves the surface finish of small, restricted areas and holes by allowing the smaller media to reach areas not effectively reached by the larger media.

Where fine, low microfinishes are required, plastic-based media with low abrasive content are used; and where high-luster and polishing effects are desired, such as on aluminum

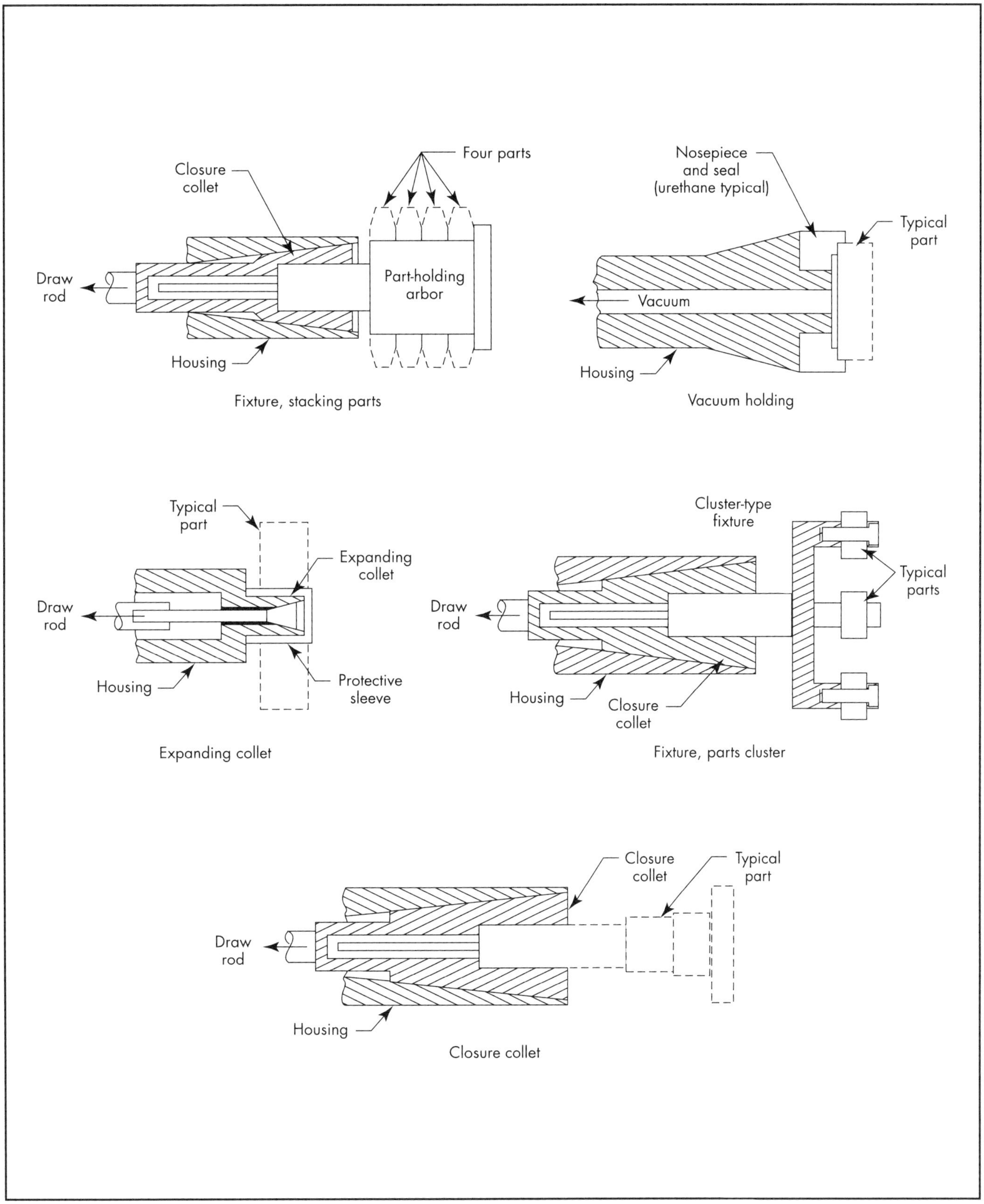

Figure 7-5. Spindle-finishing fixturing techniques. ***(Courtesy Almco Industrial Finishing Systems)***

die castings and related soft metals, dry media, such as maize or walnut shell compositions or steel balls for "burnishing," are highly effective.

Burrs in recessed areas will receive little deburring and media lodging will occur unless provisions are made to avoid these problems. While data on burr sizes, edge radii, surface finish, and stock losses are essentially nonexistent in the literature for spindle finishing, the data for vibratory, centrifugal disk, and centrifugal barrel finishing is comparable. Cycle times for spindle finishing are comparable most closely with centrifugal barrel machines.

7.2 FLUIDIZED BED SPINDLE FINISHING

Fluidized bed spindle finishing levitates the bed of dry fine abrasive using an air supply at the bottom of the container to provide a different action. The part fixtured on the end of the spindle is inserted horizontally into the machine and operates in this position. Developed in the former Soviet Union in the 1960s under the name turbo-finishing, this process has been used for a variety of aerospace rotor and similar rotational parts. Turbine disks up to 48 in. (1.2 m) in diameter have been finished with this process (Massarsky and Davidson 1995). The container or tub is stationary in this process.

Process cycle times for fluidized bed spindle finishing run from 3 to 8 min (very short, when compared with conventional barrel or vibratory finishing equipment) on most small- to medium-sized components. Very large components may take as long as 20 min. Cycle times are dependent on metallurgy as well. In comparing the cycle times required to deburr and surface finish large turbine disks, it was found in the former Soviet Union that large titanium disks could be finished in 6 min, nickel-based heat resistant alloys were finished in 12 min, and stainless steel in 20 min. Indirect labor (loading and unloading workpieces) varied from .5 to 1 min. Downtime can be considerably shortened when multiple fixtures are supplied with the machine and attaching or fixturing of parts normally occurs while the machine is in cycle. Robotics can and have been used in some applications to minimize direct manual labor costs.

The reduction in surface-texture profile generated by this equipment is significant. Both theoretical and empirical evidence indicate that the high rotational speed imparted to parts (up to 3,900–5,850 surface ft/min [1,189–1,783 m/s]) generates an abrasive particle-to-part surface contact rate of 200–500 per in.^{2}/s (0.129–0.323 per m^2/s). Surface removal rates on steel, stainless steel, titanium, and other alloys have been recorded as high as .00008–.0002 in. (2–5 μm) per minute. With appropriate media and process parameter selection, very refined finishes can be achieved. Surface reduction values from 78–200 μin. down to 7–15 μin. (2–5 μm down to 0.2–0.4 μm) are commonly achieved in the normal 3–8 min cycles for small to medium components and 6–20 min cycles for very large components.

7.3 OTHER VARIATIONS OF SPINDLE FINISHING

Spindle-finishing machines have been combined with vibratory-finishing machines to produce a hybrid machine. Although these machines are used in Russia and have been studied in Germany (Babichev 1993, 1994; Przyklenk 1985), the vibratory machines are not in production in most of the world. Chemicals can be added to conventional spindle-finishing systems to provide more polishing action. Frozen media can be used to help deflash plastic parts, but it heats rapidly as a result of the energy in the tub. Magnetic abrasive compounds also can be used. Each of these variations are more a research or laboratory curiosity than a normal production method.

7.4 ENVIRONMENTAL, HEALTH, AND SAFETY CONCERNS

The basic considerations of waste treatment that apply to mass-finishing processes also apply to spindle-finishing processes. Robot-loaded machines require normal safety considerations (such as unexpected movements injuring operators). Some parts finished by spindle finishing are large enough that they may cause ergonometric problems. For example, large jet engine rotors and automobile crankshafts may require part lifting devices to prevent operator injury.

References

Babichev, A. P. 1993. *Fundamental Vibrational Technology*. Rostov on Don, Russia. (*Vibratory Spindle Finishing and Advances in Vibratory Finishing*, in Russian: 89.)

———. 1994. *Fundamental Vibrational Technology*. Rostov on Don, Russia. (*Vibratory Spindle Finishing*, in Russian: 49.)

Massarsky, Michael, and Davidson, David A. 1995. "Turbo-abrasive Machining Theory and Application." Technical paper MR95-271. Dearborn, MI: Society of Manufacturing Engineers (SME).

Olsen, Kent R. 1981. "New Developments in High-pressure Spindle Finishing." Technical paper MR81-391. Dearborn, MI: Society of Manufacturing Engineers (SME).

Przyklenk, Klaus. 1985. "Deburring in Europe—Present State and Future Prospects." Technical paper MR85-839. Dearborn, MI: Society of Manufacturing Engineers (SME).

Bibliography

Gillespie, LaRoux K. 1981. *Deburring Technology for Improved Manufacturing*. Dearborn, MI: Society of Manufacturing Engineers (SME).

———. 1996. *Guide to Deburring, Deflashing, and Trimming Equipment, Supplies, and Service.* Kansas City, MO: Deburring Technology International.

Kobayashi, W., and Matsunaga, M. 1977. "Deburring and Finishing by Recipro-Finishing and Gyrofinishing Machines." Technical paper MR77-464. Dearborn, MI: Society of Manufacturing Engineers (SME).

Kremen, Z. 1996. "A Highly Efficient Technology for Deburring and Edge Rounding—Turbo-abrasive Machining." *Proceedings of the 4th International Conference on Precision Finishing and Burr Technology.* Kansas City, MO: Deburring Technology International: 151–59.

Miyatani, Takashi, Harada, Masahi, and Takazawa, Koya. 1981. "Newly Developed Deburring Machines for Precision Parts in Mass Production." Technical paper 81-382. Dearborn, MI: Society of Manufacturing Engineers (SME).

Shulan, Shi, and Mingzheng, Wang. 1994. "The Application of Gyrofinishing in Crankshaft Fine Process." *Proceedings of the 3rd International Conference on Precision Finishing and Burr Technology.* Kansas City, MO: Deburring Technology International.

8 Other Loose Abrasive-deburring Processes

Many mass-finishing processes have been developed that are not as widely used as those discussed in Chapters 6 and 7. Some processes, such as resonant energy, chemically accelerated, electrochemically accelerated, and cryogenic finishing, are being or have been employed and show promise for many applications. Other mass-finishing methods having only limited, if any, commercial application are discussed briefly in this chapter. Some of the processes cited are used more widely in countries of the former Soviet Union and in Japan than in the United States.

8.1 RESONANT ENERGY

Resonant energy has been used for cleaning, deburring, and radiusing of intricate internal passages and recesses. Essentially, the process involves submerging vibrating workpieces, fixtured at the ends of a resonating beam, in abrasive media for a predetermined time (see Figure 8-1) The combination of the resonant motion of the workpieces and the orbital motion of the abrasive grains cleans and finishes the surfaces (Anonymous 1976).

One or more workpieces are clamped in fixtures secured to both ends of a steel beam. High energy levels are produced by resonating the beam near its natural frequency, using a mechanical oscillator with a counterweighted shaft connected to a variable-speed drive motor. Processing is generally done at 75–100 Hz (4,500–6,000 vibrations/min), depending on the weight of the workpieces. Typical cycle times for cast iron workpieces range from 2–3 min.

Media containers are raised and lowered hydraulically, with a two-speed control to provide rapid advance and retraction of the containers, and a slower speed as the workpieces enter and leave the media. The abrasive media commonly used is randomly shaped, fused aluminum oxide. Abrasive size depends on the workpieces being processed. Water and a mild alkaline cleaning compound are sprayed on the abrasive in the containers during processing. When the containers are lowered to remove workpieces from the media after a preset time, beam vibration is continued for a short period to remove residual media from the parts and return it to the containers. Using similar principles, another machine oscillates the media container or the parts, without using a machine beam.

Although these machines enjoyed a brief flurry of popularity in the 1970s, none are known to be currently sold.

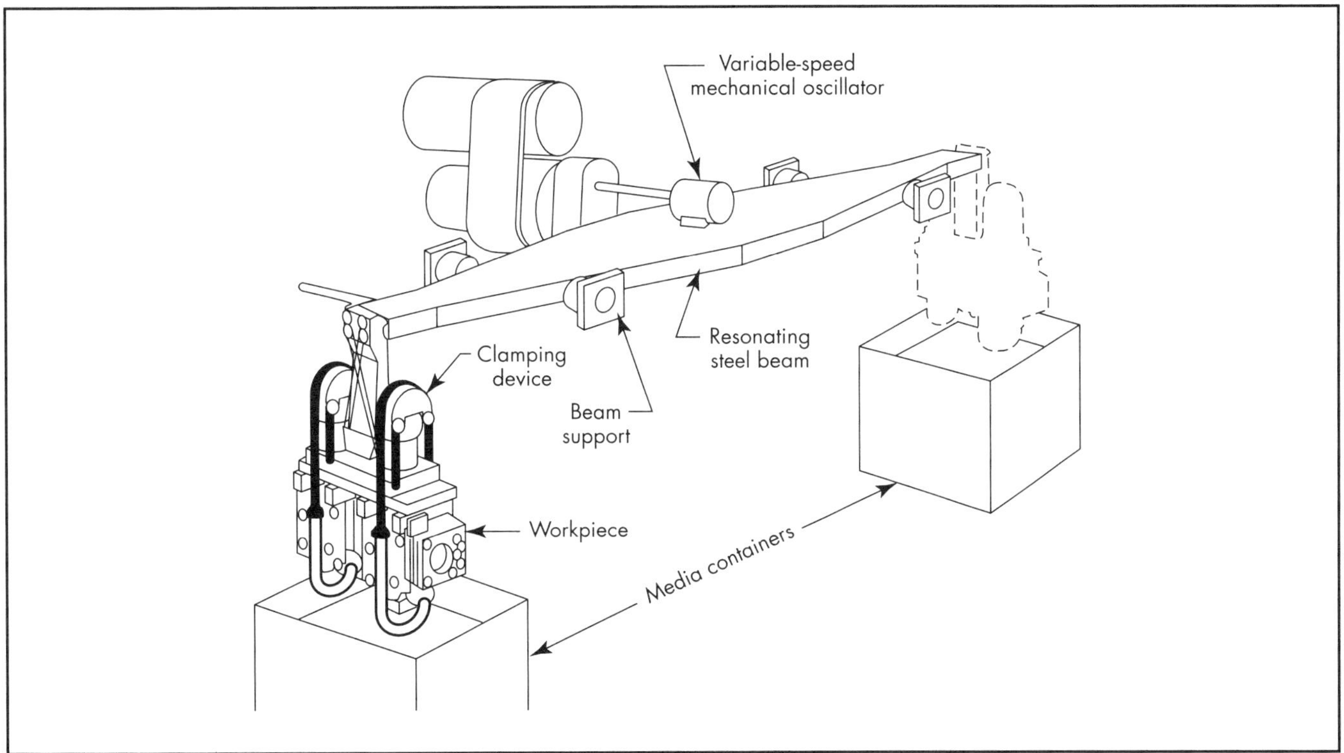

Figure 8-1. Machine for internal finishing and cleaning of precision castings, which are vibrated at high frequencies by a resonant beam. *(Courtesy U.S. Filter/Wheelabrator)*

8.2 CHEMICALLY ACCELERATED MASS FINISHING

Chemical accelerators are being used to reduce the cycle times of vibratory finishing processes. Investigations into the use of accelerators started in 1971 with a zinc alloy finishing program. Some research was done previously with the use of accelerators for barrel finishing (Dargis 1987).

When chemical accelerators are added to abrasive media in a vibratory machine, the chemicals react with the metal workpieces to remove metal. The vibrating media wipes away any passivating films that form and provides additional mechanical cutting and burnishing action. Conventional vibratory finishing media and machines can be used, although linings in the tubs or bowls may be adversely affected by the chemical solutions.

Accelerators are used for finishing steel and brass parts as well as for zinc die cast items. The chemical accelerators are most useful for vibratory deburring if the die castings have been trimmed closely to leave only small burrs. The large, jagged burrs left on many trimmed castings require the use of abrasive belts before vibratory finishing.

Aliphatic acids have been found useful as accelerators for barrel and vibratory finishing, with reductions in cycle times of up to 90%. Maleic acid appears to be the most effective aliphatic acid and the least costly. Adding sodium bisulfate to the maleic acid solution provides increased metal removal rates for deburring. However, a solution of maleic acid [.66 oz/gal (5 g/L)] with no sodium bisulfate generally produces smoother surfaces.

Chemically accelerated processes are more difficult to control than standard mass-finishing methods, and there are greater problems with effluent disposal. When using accelerators, care is required in handling the chemicals. Also, any residual chemicals left on the workpieces must be removed or neutralized. Information on chemical deburring and polishing processes is presented in Chapter 19.

The use of chemical accelerators has several benefits:

- Economics of being able to downgrade the cutting or polishing media.
- Significant reduction in cycle time to attain a given finish.
- Reduction in the problem of peened-over burrs and parting lines.
- Better removal of inclusions from prior belt grinding or polishing.
- A more active surface for subsequent finishing operations.

There are two approaches in use with chemically accelerated processes: (1) the acid solution is added initially and then allowed to dissipate with no replenishment; and (2) dilute buffered chemicals are added to the tub initially, and the concentration is maintained within production-feasible limits.

In the first approach, the pH at initial charge is in the range of 2.7–2.8. When the chemical for all practical purposes is no longer contributing to the operation, the pH will be approaching neutral (if the acid is used up, the liquid is no longer acid). A pH of 6.5–6.8 indicates effective stock removal is complete.

In the second approach, the pH of the "water and wetter" is in the pH range of 7.5–8.5. After draining and charging with the chemical deburring solution, the pH range is 2.8–3.0. Using a dilute solution and maintaining the chemistry, the pH history shows first the dilute chemistry—the pH is 3.5–4.0 as compared to 2.7–2.8 and, if adds are made periodically, the pH will rise to the 5.2–5.8 range. A simple and easy-to-use tool for measuring pH is pH paper, which is a pretreated strip of paper that when brought into contact with liquid solutions will turn different colors depending on the level of acidity or alkalinity of the solution.

Unlike open barrel chemical accelerated processes where the parts, media, and chemicals are very much in contact with each other due to the accepted linear rotation of 120–200 ft/min (37–61 m/min), a different relationship exists among these materials in tub vibrators. Although parts and media are in direct contact with each other in tub vibrators, for most of the process, parts and chemicals are not. A good part of the time the parts are "riding" in media above the solution being used. This phenomenon is directly related to the energy being generated by the process. Completely submerging parts and media in the vibratory liquid produces two undesirable results:

1. The energy is dampened, which wastes a major part of the potential energy available for deburring action.
2. Solution can be spattered within a 6 ft (1.8 m) radius of equipment making for an extremely messy operation. Without a solution-tight cover, there is one big mess.

Monitoring an "acceptable" solution in the vibratory tub, specifically circular tubs, has produced an interesting relationship—regardless of the complete cycle time in the tub, the parts are actually in the solution for only 20–25% of the time.

Figure 8-2 presents results of tests that compared different solution concentrations in different types of finishing machines. As shown in the figure, barrel and vibratory machines using higher concentrations remove material much faster than other combinations. (Values such as 10/20 refer to 10 g of organic acid and 20 g of peroxide in a 20 L tub [Kobayashi and Matsunaga 1981]). For parts plated later, an extra benefit was observed. The parts were subsequently nickel chrome plated, and from time to time, a pitting of the electroplating occurred. All pitting ceased when accelerated vibratory finishing was used. It was determined that the pitting was imbedded emery from the belt-sanding operation that was further peened into the surface of the steel. These imbedded impurities were being released during the electroplating precleaning cycle and leaving holes that were accentuated by the nickel chrome plating that followed.

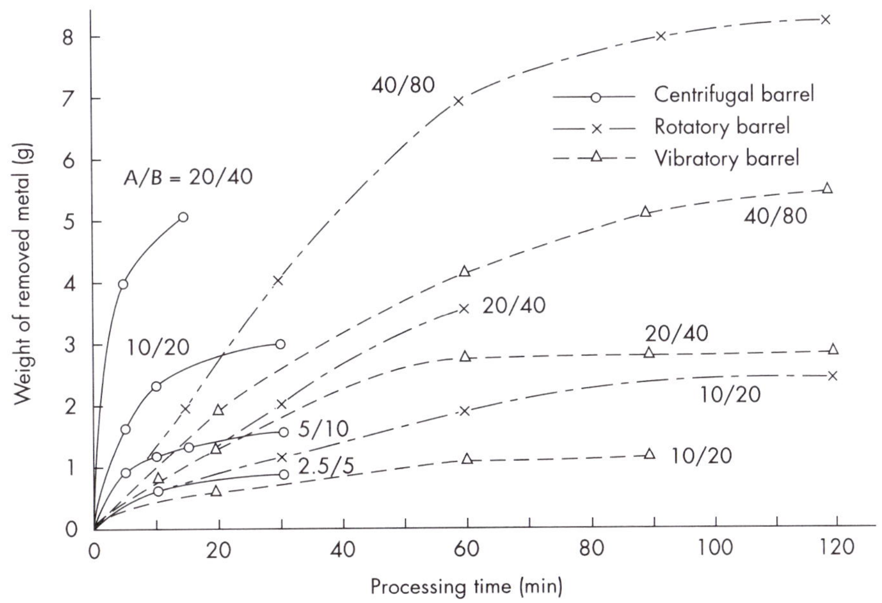

Figure 8-2. Metal removal rate for chemically accelerated mass finishing when the bright dipping solution was used. *A* is organic acid, and *B* is peroxide; mixing amount (that is, 20/40) is in grams for 20 L tub. *(Courtesy Shiskishima Tipton)*

In straight chemical polishing and deburring systems, the base metal is almost always more "active" (subject to blush rusting) and has to be rapidly rinsed and the surface neutralized to preserve the polished finish. Because of the combination of dilute chemicals and vibratory processing, the surface is still more active than that produced by normal vibratory burnishing additives; but it approaches neutral at the end of the process. A simple addition of 1–2 oz/gal (7.5–15 g/L) of sodium carbonate solution completes the neutralization process and leaves the parts with a thin, readily rinsed, protective coating. In some instances, this solution will render the chemicals in the tub acceptable for sewer disposal.

8.2.1 Environmental, Safety, and Health Concerns

Chemically accelerated mass finishing can be used with barrel tumbling, centrifugal barrel, and any mass-finishing or spindle-finishing process. Because chemicals reacting with metal can create high pressures, it is important to have proper venting and lid unloading procedures to prevent lids from being hurled at operators. Chemicals will cause skin irritations as well as more serious acid burns. Protective eye wear and outer clothing is needed. Additional EPA, OSHA, and state regulations may apply, depending on the chemicals involved.

8.3 CRYOGENIC DEFLASHING

Cryogenic deflashing of molded rubber or plastics products and aluminum or zinc die castings is being done in barrels and vibratory machines by freezing the workpieces with liquid nitrogen, carbon dioxide, or dry ice. Flash on the workpiece becomes embrittled when its temperature is reduced; mechanical means are then used to remove the flash. When using liquid nitrogen, proper ventilation is essential, and care is required in handling the cold parts and media. The cost of liquid nitrogen may be a limitation of this process.

Barrels and vibratory machines used for cryogenic deflashing require special double-walled, insulated work chambers to conserve the liquid nitrogen. The chambers are fitted with multiple nozzles for spraying the nitrogen. It is general practice to cool the media in the work chamber prior to adding the workpieces, which must then also be cooled prior to starting the machine. Temperature control during processing is essential. Temperatures as low as –320° F (–196° C) are used for some parts. Table 8-1 shows common temperatures.

Table 8-1. Embrittlement temperatures for rubbers and plastics

Plastic	Embrittlement temperature °F	°C
Acrylonitrile-butadiene-styrene (ABS)	–270	–168
Acetal	–300	–184
Delrin®	–275 to –300	–171 to –184
Nylon	–275 to –300	–171 to –184
Polylron™	–300	–184
Polypropylene	–300 to –310	–184 to –190
Teflon®	–275	–171
Rubber		
Buna-N	–225	–143
EPDM	–275 to –300	–171 to –184
Ethylene propylene	–275 to –300	–171 to –184
Hycar®	–210 to –275	–134 to –171
Natural rubber	–225 to –275	–143 to –171
Neoprene	–225 to –300	–143 to –184
Nitrile	–275 to –310	–171 to –190
Nitrile-butadiene	–250 to –270	–157 to –168
Silicone	–300	–184
Urethane	–275 to –300	–171 to –184
Viton®	–275 to –300	–171 to –184

(Meyers 1985)

Cryogenic tumbling has been used since at least 1960 to deflash plastics and rubber. Despite such long use, its characteristics are not well documented. In addition to the process variables cited in Chapter 6 for mass finishing, cryogenic deflashing in tumbling and vibrating equipment has the following variables:

1. Size, shape, and material of part.
2. Kind, size, and location of flash.
3. Cryogenic equipment.
4. Embrittlement temperature and method.

The embrittlement temperature is based on the material of the part.

8.3.1 Part Size, Shape, and Material

The size, shape, and material of the part and the kind, size, and location of flash determine the equipment (tumbling or vibrating) used in the cryogenic deflashing process. One philosophy advises that small, thin-walled parts are suited for the more gentle scrub and peen of the vibrator, and that large parts with thick walls are better suited for the more violent slide, roll, and fall of the tumbler. Size of the part can dictate use of tumbling equipment because tumblers are designed to handle large parts and a larger volume of parts. Occasionally, there are parts that lend themselves to either type of equipment. Load size and cycle time also dictate the equipment to use. Materials, to some degree, dictate the type of equipment. Polyethylene and polypropylene parts seem to need the more violent action of tumbling.

8.3.2 Flash

There are a number of kinds of flash and related protrusions: parting line, gate, knockout pin, core pin, insert, and vent. Parting line flash occurs at the line where the two halves of the mold come apart so that the part can be ejected or removed from the cavity of the mold. This is a result of wear at this area of the mold or incomplete closure of the press. Gate flash develops at the area where the material enters each cavity (the gate) if the mold has worn.

In plastic and zinc die cast molding, knockout pins are used as part of the mold to eject (knock out) the molded parts from the cavity on completion of the molding cycle. As these pins wear, flash can develop on the part in these areas.

Core pins are used in molding plastic and zinc die cast parts to establish a hole or internal passage. As these core pins wear, flash will develop in the area of the core pin.

Insert flash occurs around an insert in the plastic or rubber part. This usually occurs when the mold wears around the insert area. Misalignment of the insert in the mold will cause damage to the mold, and flash will result.

Vent flash occurs in an area where the mold is vented to permit the air to escape so that the cavity can be completely filled.

Flash can be located on the part's outside, inside, or in recesses. Flash can form at the junction of two walls, thereby forming a web. Flash in a recess is often quite difficult to remove. Flash, which is otherwise "protected" by the part, is also very difficult to remove.

Each type of flash and its location must be considered separately in the deflashing process.

Flash Thickness

As a basic rule of thumb, the thickness of the flash must be 10% or less of the adjacent wall thickness. This ratio appears to be a reasonable limit at least 70% of the time. It should be noted that very thin flash can be a problem if it tends to cling to the side of the part. The flash gains the support of the part and must be scrubbed from the wall of the part to be deflashed.

8.3.3 Burrs

Cryogenic tumbling is used routinely for removing machined burrs from plastics. Burrs may be thicker because of softening (almost melting) during cutting, or they may be smaller, making their removal even easier than flash.

8.3.4 Embrittlement Point

Embrittlement is the point at which flash (and the part) becomes stiff enough to be readily removed. Although it would be preferable to embrittle only the flash rather than the entire part (and thus save energy), it is not feasible at the present time. Each part acts as a heat sump, drawing energy from the flash to the part, until the flash becomes stiff enough to be broken off immediately adjacent to the part. To accomplish this, all flash as well as the adjacent surface of the part must be embrittled. In so doing, the entire part is essentially embrittled.

8.3.5 Cryogenic Equipment

Cryogenic deflashing is most commonly accomplished in double-walled, insulated containers, but commercial vibratory machines have been used as well for low production. The inside walls are usually stainless steel, and the outer walls are usually steel. They are insulated by polyurethane foam or similar material or by a vacuum between the two walls. Appropriate venting for the containers must be provided because liquid nitrogen expands from its liquid state to vapor by a factor of 696.5 to 1.

Cryogenic tumbling equipment is usually six- or eight-sided. The door is located on one of these sides. The diameter and length of the tumbling barrel varies with the capacity of the equipment. There are only a few manufacturers of cryogenic tumbling and vibrating equipment.

Available sizes vary from 3–35 ft^3 (0.08–0.99 m^3) containers. The key dimension in all tumblers is the diameter, which determines the length of slide, roll, and distance of fall. Increasing length is one of the easier means of adding volume.

Standard cryogenic vibrators are available in only relatively small sizes, ranging from 6 qt to 2 ft^3 (5.7 L to .06 m^3). Controls commonly used in cryogenic tumbling and vibrating include a liquid nitrogen time proportioning temperature controller, cycle timer, tachometer, and speed control.

A high-quality, time proportioning, liquid nitrogen temperature controller is an absolute requirement for successful high-production cryogenic deflashing. The temperature controller is connected to a thermocouple located inside a tumbler protected from the parts and media. It must be located in such a way that it senses a truly representative temperature in the tumbler. Temperature variations of 50° F (10° C) from one end of the tumbler to the other have been observed. This variation resulted in parts at one end breaking from being too cold and parts not deflashing at the other end because they were not embrittled. In another case, the thermocouple was so close to the large ventilation holes that it was unduly influenced by ambient air.

The temperature controller activates the liquid nitrogen solenoid valve (specially designed for use with cryogens). The solenoid controls the flow of liquid nitrogen from bulk or portable liquid nitrogen tanks. The location of the tanks and the solenoid in relation to the tumbling and vibrating equipment has a great effect on the line loss or heat leakage of liquid nitrogen. The longer the line from the liquid nitrogen tank, the greater the loss through heat leakage. Various types of lines with insulations such as vacuum, polyurethane, or multicelled foam sponge Armaflex® are used, and they vary greatly in efficiency and cost of installation. The length of the line between tank and point of use, the size of the line, and the type of insulation determine the amount of liquid nitrogen lost through heat leakage.

The time proportioning feature of the temperature controller anticipates set-point temperature and controls the liquid nitrogen supply to prevent "overshoot" (not shutting off the liquid nitrogen supply prior to set-point temperature), which creates a temperature much colder than the set point. This situation risks part breakage and wastes liquid nitrogen.

The following example illustrates line loss. One company has identical deflashing systems (vibrators and controls). One system is connected to a bulk nitrogen tank by an insulated pipe that is 22 ft (6.7 m) long. The other system is connected to a portable nitrogen tank by an uninsulated flexible transfer hose 5 ft (1.5 m) long. A comparison of the liquid nitrogen used is shown in Table 8-2.

Tachometers are used to measure tumbling barrel speed and vibratory frequency. Variable speed drive is essential. In tumbling, a slow speed provides a slow, gentle slide, roll, and minimum (if any) free fall. A rapid speed can centrifugally force parts and media to one area in the tumbler and provide very little slide, roll, or free fall. However, too slow of a speed can cause a rapid slide, roll, and violent fall. Note that the diameter of the tumbler, length of side, and speed also affect the length of slide, roll, and fall.

Speed control in cryogenic vibratory deflashing is important because it affects the "scrub and peen" of embrittled parts and media interacting with each other against the walls of the vibrator.

8.3.6 Characteristics of Liquid Nitrogen

Chemical and Physical Properties

Nitrogen is relatively inert; it does not burn or supply combustion, and it combines with some of the more active metals such as calcium, sodium, and magnesium. It is colorless, odorless, and tasteless; it is slightly soluble in water; it is a poor conductor of heat and electricity; it is slightly lighter than water, and it is nonmagnetic; and it does not produce toxic or irritating vapors. The boiling point of nitrogen is –320° F (–196° C). Its volume expansion—liquid to gas—is 696.5 to 1.

Safety Precautions

Introducing nitrogen into a confined area can dilute the oxygen content of the air and thereby cause a loss of consciousness and death if the nitrogen level is not reduced to acceptable limits. When nitrogen is in use, it is essential to keep rooms well ventilated and to have oxygen monitoring

Table 8-2. Liquid nitrogen requirements

Variable	Product		
	Rubber tubes	Pawls	Ratchet wheels
Material	Ethylene propylene 45 durometer	Zinc	Zinc
Embrittlement temperature	–285° F (–176° C)	–200° F (–129° C)	–200° F (–129° C)
Media	Carpet tacks and steel shot	None	Tacks
Time	30 min	10 min	10 min
Wall thickness	.15 in. (3.81 mm)	.043 in. (1.09 mm)	.035 in. (0.89 mm)
Flash thickness	.005 in. (0.13 mm)	.012 in. (0.31 mm)	.025 in. (0.64 mm)
Nitrogen used (lb/lb of parts)*			
Portable tank	15.6 lb/99 N	.28 lb/1.25 N	.43 lb/1.91 N
Bulk tank	25.0 lb/111 N	.44 lb/1.96 N	.69 lb/3.07 N

*15.6 lb of Nitrogen/lb of part = 99 N of Nitrogen/N of part

(Meyers 1977)

equipment. Liquid nitrogen has a temperature of –320° F (–196° C) at atmospheric pressure. Skin exposed to these temperatures will suffer painful damage that resembles burns. Care must be taken to avoid contact with liquid nitrogen and the material cooled by it. Protective eyewear, gloves, and clothing must be worn to prevent this contact.

8.3.7 Tumbling Media

Media used in cryogenic operations include carpet tacks, roofing nails, bottle caps, sheet metal screws, solid rubber balls, walnut shells, 5/16 (.31) in. (7.9 mm) ceramic cylinders, .5 in. (12.7 mm) steel balls, and golf balls. Sometimes the parts themselves act as media—that is, they self-tumble.

Steel balls are frequently used where it is desirable for media to pass through a core pin hole to remove core pin flash. Carpet tacks are commonly used in both tumbling and vibrating because the pointed end of the tack can get into many areas and "pick out" flash where flash is "protected" by the wall of the part. Ceramic media are often used on heavier, exposed flash such as parting line, gate, and knockout pin flash.

8.3.8 Relationship of Embrittlement Temperature and Material

Rubber embrittlement temperatures are affected by the amount of elements other than the base material. The durometer reading of a part within a given rubber base material appears to be a good indication of embrittlement temperature. The lower the durometer reading, the lower the embrittlement temperature. Table 8-1 shows a list of materials and temperature ranges at which molded or die cast parts have been deflashed.

As seen in Table 8-2, the amount of liquid nitrogen used varies from .28 lb per pound (1.25 N per N [metallic equivalent]) of parts to 25 lb per pound (111 N per N) of parts. Table 8-2 graphically illustrates heat leakage and why it is important to keep liquid nitrogen lines insulated and as close as possible to points of use.

Some simple operations place dry ice in a tumbling barrel and allow the natural heat transfer to chill the flash and make it brittle. This is not an economical approach to production, but may be effective for small lots and immediate needs. Some blasting operations take advantage of the same type of freezing prior to blasting to deflash parts.

Protective equipment for personnel is required for face, hands, and body to prevent freeze burns when using any cryogenic materials. In addition, providing good ventilation with any gases, even frozen gases, is necessary to prevent asphyxiation. Since these gases are typically heavier than air at low temperatures, they will fall and fill pits and low-lying passages.

8.3.9 Cryogenics as a Preventive Measure

Some burrs and flash may be prevented using cryogenics prior to part fabrication. Deburring blanked plastic parts can be simplified by freezing the plastic stock before it is put into the blanking presses. Teflon® is an example of a material that is difficult to deburr and it characteristically makes rather large burrs. Using cryogenics prior to blanking produces a near sharp edge that can be rounded readily if necessary.

8.4 ELECTROCHEMICAL MASS FINISHING

In electrochemical mass finishing, workpieces are placed in an electrically insulated drum filled with electrolyte and media, such as graphite spheres (Shimizu 1975; DeGroat 1969; Anonymous 1972; Krapivin 1972). The major advantage of this process is fast cycling. However, the equipment required is comparatively expensive, and the process is limited to workpiece materials that conduct electricity. The two basic types of electrochemical mass finishing are electrochemical barrel finishing and electrochemical vibratory deburring.

British patents (U.K. Patent 1968a, 1968b, 1970) indicate that phosphoric acid can be used as the electrolyte and that either silicon carbide or copper-coated aluminum oxide can be used as the deburring media. Electrodes connected to an external power source are placed in the drum. When the drum is vibrated or rotated, the media acts as extensions of the electrodes to remove the burrs electrolytically. The action is similar to conventional electrochemical deburring, which is discussed in Chapter 13. The media also abrades the passivating surface oxides formed on the workpieces, thus exposing fresh surfaces to the electrolytic action.

8.4.1 Electrochemical Barrel Finishing

Machining Principles

When an electric potential is applied between a metal conductor and a cathodic electrode in an electrolyte, material is dissolved from the anodic conductor into the electrolyte. The amount of material dissolved is, in accordance with Faraday's law, proportional to the quantity of electricity delivered between the electrodes.

When metal conductors of workpieces are placed between the electrodes, as shown in Figure 8-3, they will each function as a bipolar intermediate electrode and dissolution will take place at their anodic sides. In the electrochemical barrel finishing system, a multitude of conductive solid media and workpieces constitute intermediate electrodes. The conductive media is composed of a material that is dissolution-resistant; the media has an electric resistivity higher than that of the metal workpieces but lower than that of the electrolyte. Typical media is graphite balls (spheres). The conductive media acts as resistive spacers to prevent the workpieces from short circuiting, and as counter-electrodes against the adjoining workpieces to create enhanced machining current paths. Rotary or vibratory motion is imparted to a carrier drum to maintain constant random relative movement between the media and the workpieces so that the entire surface of each workpiece can evenly contact adjoining media (Shimizu 1975).

In an electrochemical machining system, contoured or protruding portions of an electrode have a higher field strength than other areas of the electrode. Accordingly, the current density is higher at the burr and contoured portions, causing those portions to be preferentially dissolved electrochemically.

Advantages of electrochemical barrel finishing include the following:

- Higher machining efficiency than conventional abrasive processes, and material removal proportional to applied current.
- Availability of tailored conductive media and special graphite balls fitted to the process.
- No need to select from a variety of media according to the workpiece material and configuration as in conventional processes.
- No significant workpiece deformation as in some conventional tumbling processes.
- No drop in material removal rate after prolonged use of media.
- Less overcut at corner or edge portions than in conventional operations.
- Absence of work hardening.
- Results are relatively independent of workpiece materials and hardnesses.
- Burrs at a hole are completely removed without moving into the interior.
- Absence of impingement marks.

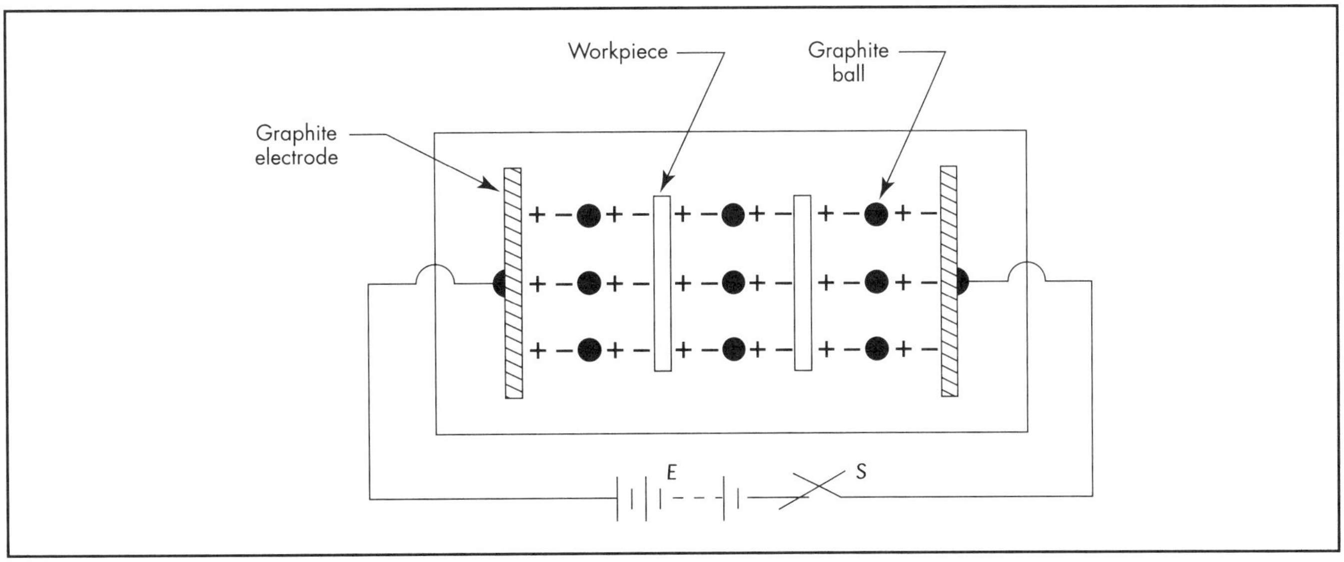

Figure 8-3. Principle of electrochemical barrel finishing (Shimizu 1975).

Disadvantages of electrochemical barrel finishing include the following:

- Appearance of a semiglossy black-brown colored finish surface on ferrous workpieces.
- Possible roughening of finish surfaces, depending on materials, type of electrolyte, and machining conditions (such as current density and electrolyte temperature).
- When mirror-face finish is essential, post-treatment, such as chemical polishing, is required.

Parameters determining the amount of material removal include (1) machining current, (2) mixing ratio of workpieces and graphite balls, (3) the agitation state of the workpieces and graphite balls, and (4) the quantity of electrolyte.

Generally, the removal rate increases in proportion to the machining current passed (the first parameter). The machining current refers to the electric current effective to machine the workpieces. Short-circuiting current and shunting current not flowing through the workpieces in no way contribute to the machining.

The second parameter, the mixing ratio of workpieces and graphite balls, is extremely important and is illustrated in Figure 8-4. In this figure, the apparent or bulk volume ratio of workpieces to graphite balls is plotted along the x-axis, and both the machining current and the amount of metal removal are plotted along the y-axis. The quantity of graphite balls was maintained constant and the quantity of workpieces varied.

When the proportion of workpieces to graphite balls is low, the metal removal is apparently reduced because the shunting current not flowing through the workpieces increases. When the proportion of workpieces is increased, a maximum amount of metal removal is achieved, with the bulk volume ratio around 1:1. When the proportion of workpieces is further increased, the opportunity for the workpieces to come in direct contact with one another and with the electrodes (fixed) increases. As a consequence of

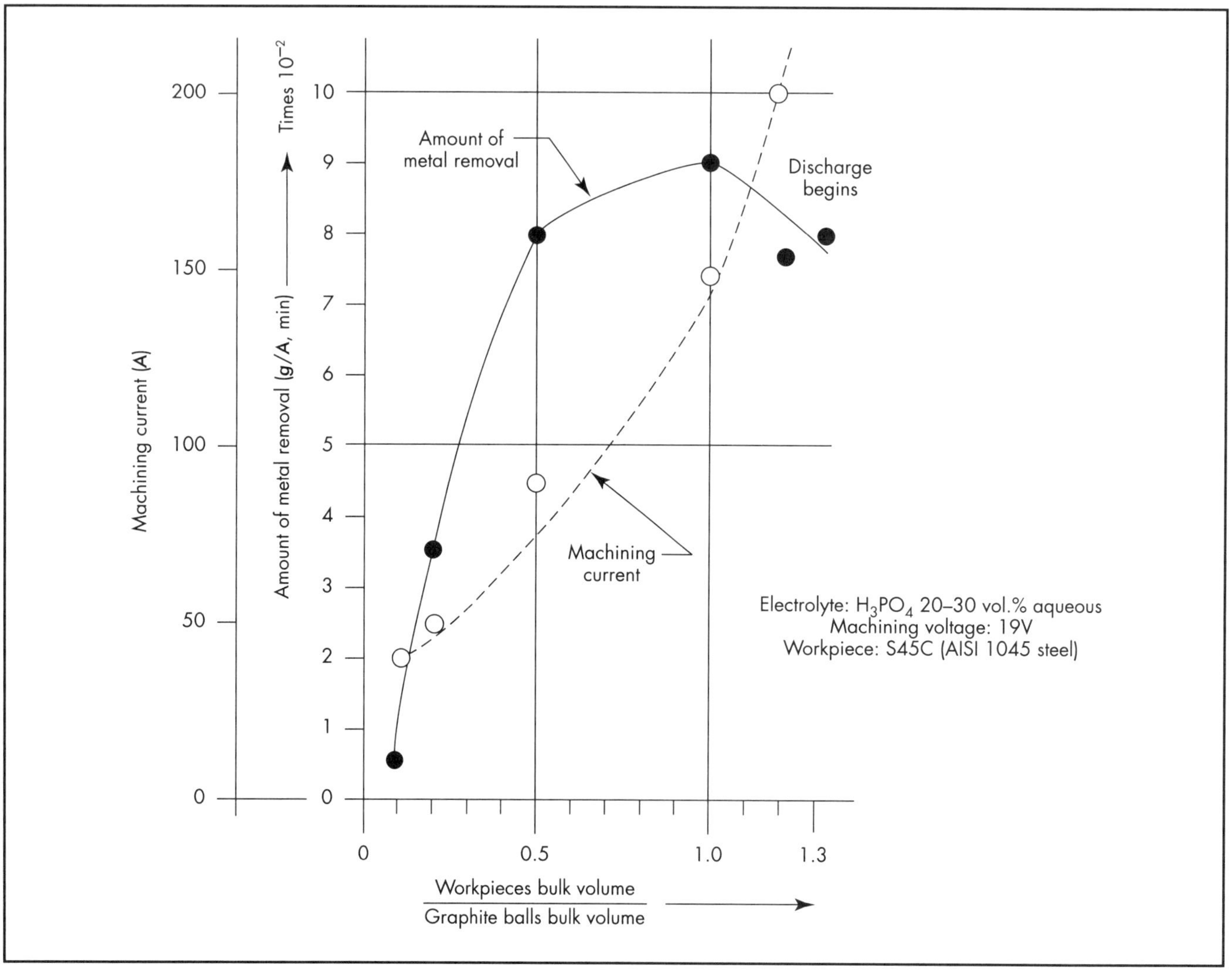

Figure 8-4. Effect of ratio of parts to graphite balls (Shimizu 1975).

the increase of short-circuiting current, machining efficiency goes down. The best results are obtained when the volume ratio ranges between 1:5 (0.2) and 1:1 (1.0). Conditions in this range prevent workpiece short-circuiting.

The agitation of workpieces and graphite balls, the third parameter, facilitates the release of gaseous products and the replenishment and repulsion of electrolyte in machining regions to enhance machining efficiency. It is imperative that the relative movement be uniform over the entire mass and continuous (without interruption).

The amount of electrolyte, the fourth parameter, must be sufficient enough to allow machining. The movement of workpieces and graphite balls and the amount of electrolyte are correlated. In rotary and vibratory systems, the state of agitation may become unsatisfactory if the amount of electrolyte is not properly monitored. In a long operation, the amount of electrolyte consumed through vaporization because of the rise in temperature may be significant and may require using an automatic electrolyte feed device with sensing and control functions.

The performance of electrochemical barrel finishing under optimum conditions can be expressed by the following formula:

$$W = K \times e \times I \times t \tag{8-1}$$

where

W = Metal removal amount (g).
K = Constant (2 to 4).
e = Electrochemical machining unit quantity (g/A – min).
I = Current (amps).
t = Machining time (min).

In the formula, K is determined by the second, third, and fourth parameters discussed earlier. The fact that K is larger than one shows that the process involves polarization effects, which distinguishes electrochemical barrel finishing from usual electrochemical machining.

Factors determining the surface roughness and accuracy include workpiece material, the kind of electrolyte, the agitation, the mixing ratio, and machining time.

The effect of materials on surface roughness is shown in Table 8-3. If a fine finish is desired, the roughness may not be satisfactory. By employing chemical polishing as a post-treatment, however, a glossy finish can be readily obtained.

The rates of electrochemical dissolution from one alloy to another can vary extremely. Since cast iron contains such alloyed components, the surface of this material becomes rough when subjected to electrochemical dissolution.

Kinds of electrolyte in customary use at the present time include aqueous solutions of H_3PO_4, $NaNO_3$, and NaCl. When a fine surface is an important requirement, an optimum electrolyte composition for the particular workpiece material should be used.

The agitation of workpieces and graphite balls is an important parameter to the accuracy of workpieces, as well as to the amount of material removal. If there is a lack of uniformity of agitation or relative random displacement, variations occur in the amount of metal removal over the entire mass of parts and over the entire surface of an individual part. This is especially the case when a workpiece is flat and long, because there may be a difference in material removal brought about between the two ends; additionally, workpieces may pile up, causing machining inaccuracy. Part thickness may vary across a part by as much as .001 in. (0.03 mm), and lengths can change by as much as .004 in. (0.1 mm) for long parts in a 20-min cycle.

Table 8-3. Materials and surface roughness from electrochemical mass finishing

Workpiece material	Machining voltage (machining current)	Roughness of treated surfaces (μR_{max})*
430 stainless steel	19 V (40–44 A)	2–2.5
1085 steel	19 V (28–32 A)	3–4
Brass (7-3)	19 V (34–40 A)	2–2.5
Zinc (die cast)	19 V (40–44 A)	5–6
Aluminum (die cast, 8% silicon)	19 V (36–40 A)	4–5
Phosphorus bronze	29 V (72 A)	1.6

Note: Electrolyte is H_3PO_4 20–30 vol. % aqueous

* 1.0 μAA = 40 μin. AA. Finishes expressed as R_{max} are not directly translatable to AA, however.

(Shimizu 1975)

When the proportion of workpieces relative to graphite balls is excessive, workpieces and electrodes tend to short-circuit. As a result, arcing marks may be produced on the workpieces. Accordingly, when shapes liable to short-circuit are used, it is necessary to limit the bulk volume ratio to 1:5. It is also necessary to use smaller graphite balls for workpieces with high volume-to-weight ratios.

The relationship between machining time and surface roughness is shown in Figure 8-5. The time required for deburring in general is 15–30 min.

The graphite balls must have a specific resistivity of 1×10^{-1} to 10^{-4} ohm-cm and be both shock and wear resistant. Graphite balls wear slightly due to electrochemical oxidation and mechanical agitation and vibration. The amount of wear for .26 gal (1 L) of graphite balls is generally .4–.53 oz/gal/hr (3–4 g/L/hr) in the case of rotary barrels and 1.34–2.0 oz/gal/hr (10–15 g/L/hr) for vibratory barrels (tubs).

Figure 8-6 illustrates several types of electrochemical mass-finishing machines. Figure 8-6a illustrates a rotary machine. The barrel is rotated relatively slowly in an electrolyte container to give uniform agitation to the workpieces and graphite balls in the barrel. A stainless steel cathode is mounted in the barrel while graphite anodes are arranged in the container. A variation of the rotary electrochemical barrel finishing system is shown in Figure 8-6b.

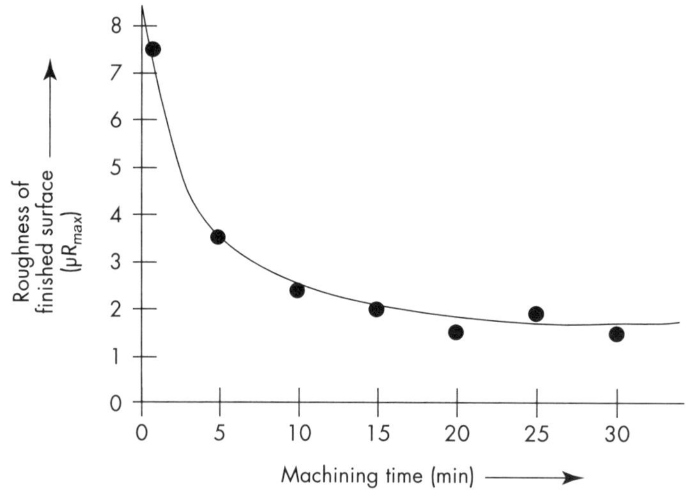

Figure 8-5. Effect of tumbling time on surface finish of brass at 19V, 40A, in a phosphoric acid concentration of 20–30% (Shimizu 1975).

Figure 8-6c shows a vibratory tub machine. A pair of graphite electrode plates is mounted inside the barrel receiving workpieces, graphite balls, and electrolyte between them. The use of excessive electrolyte in the barrel impedes the agitation, hinders uniform machining, and reduces machining efficiency. As a countermeasure to the vaporization of machining liquid due to electrically induced temperature rise, it is possible to monitor the electrolyte and automatically replenish the evaporated amount of water to the barrel as shown in Figure 8-6d.

A massive part may be finished using the method shown in Figure 8-6e. The workpiece is mounted to a holder and receives rotary and/or reciprocating movement to impart random displacement of graphite balls in the electrolyte. The graphite balls act as dynamically moving electrodes to electrochemically deburr the workpiece.

The inner surfaces of massive workpieces in a small number are effectively deburred, descaled, or finished by the method shown in Figure 8-6f. A perforated cathode is inserted into the workpiece that is poled anodic, and electrolyte is flushed from or into the cathode through the perforations. Graphite balls agitated by the electrolyte uniformly remove dissolution-resistant oxide films to achieve efficient finishing. Although not shown, centrifugal barrel units also can be made using this principle.

A comparison of performance of electrochemical barrel finishing with other mass-finishing processes indicates that vibratory electrochemical barrel finishing is competitive with centrifugal barrel finishing for some parts.

8.4.2 Electrochemical Vibratory Deburring

The electrochemical vibratory deburring process was first described in the literature in the United States in 1965 (DeGroat 1969). At that time, operating time was reportedly reduced by a factor of 10–30 over conventional processes then in use. Little information has been published on burr removal, edge-radiusing surface finish changes, or stock losses resulting from this process. One source, however, indicates that a .002-in. (0.05-mm) thick burr was removed from a brass wristwatch housing in 35 s without removing noticeable stock from the housing. This process reportedly processes millions of small precision parts of alloy steel every week. It is used in Japan, Italy, and Russia. While these processes were at one time used in high-production commercial industry in the United States, no known applications currently exist in the United States. The same principle of loose abrasive electrochemical deburring can be applied to centrifugal barrel, centrifugal disk, spindle finishing, and related loose abrasive processes.

8.4.3 Environmental, Health, and Safety Concerns

The environmental concerns related to other electrochemical deburring processes (such as handling of acids and venting), discussed in Chapter 13, apply here as well.

8.5 LIMITED APPLICATION METHODS

There are several other mass finishing methods that have only limited or no commercial application. Consequently, they are only discussed briefly in this section.

8.5.1 Loose Belt Tumbling

In loose belt tumbling, parts are tumbled on a continuously rotating slack belt while abrasive particles are dropped onto them (see Figure 8-7) (Richardson 1989). This process is similar to abrasive blasting using a rotating belt to blast

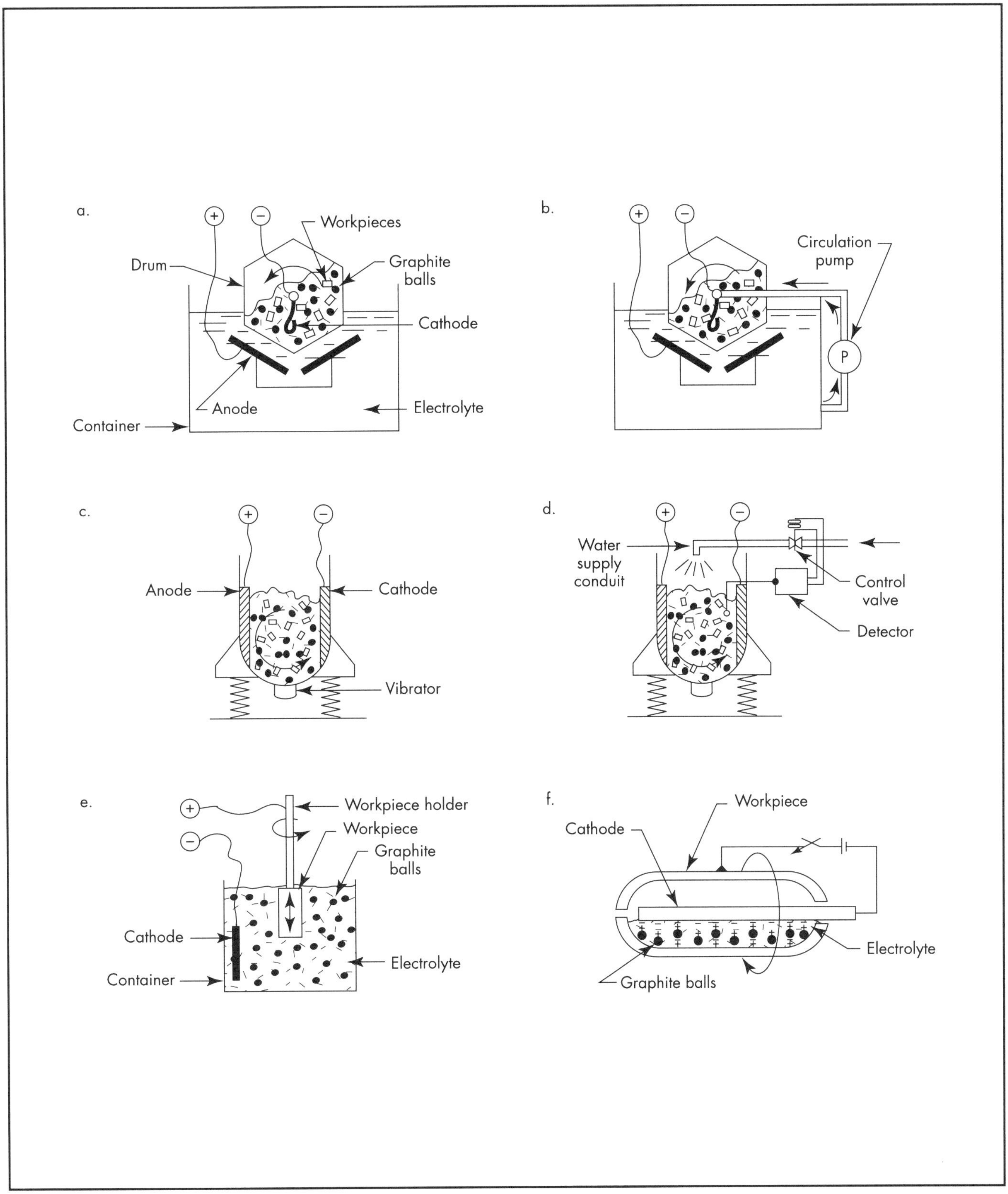

Figure 8-6. Forms of electrochemical mass finishing. (a) Rotary electrochemical barrel finishing, (b) variation of *a*, (c) vibratory tub finishing, (d) vibratory tub finishing with monitoring devices, (e) electrochemical spindle finishing, (f) electrochemical centrifugal finishing (Shimizu 1975).

off the burrs, except the action is performed by tumbling, instead of jet impingement. Although used for plastics, thin metal burrs also would be removed.

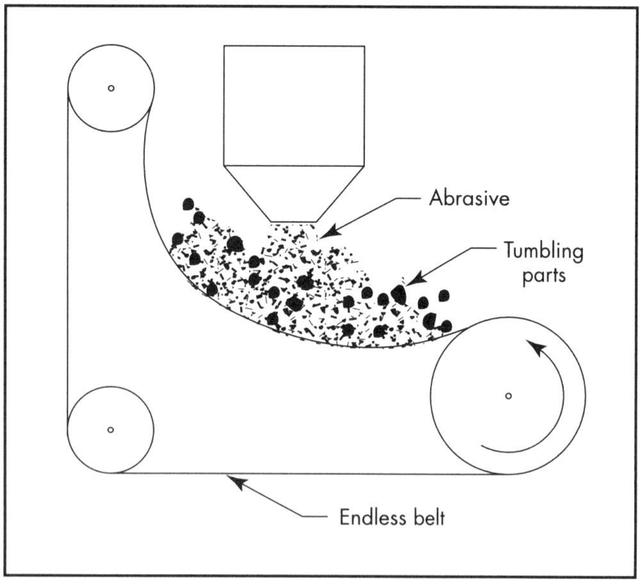

Figure 8-7. Loose belt tumbling (Richardson 1989).

8.5.2 Vibratory Rotary Barrel Finishing

Conventional barrels that vibrate as they rotate have been developed for specific applications where problems are encountered with workpieces migrating out of the media in vibratory machines.

8.5.3 Vibratory Shaker Mixer Deburring

Laboratory mixers have been used for deburring small parts. These tabletop machines generally produce a vibrating figure-eight motion in a closed container to mix powders. Small parts, including jewelry, can be finished with this process, but it is not in common usage.

8.5.4 Recipro-finishing

In recipro-finishing, workpieces are moved back and forth in a stationary tub containing media. The workpieces must be attached to a holding device. The process permits handling parts that are large or odd-shaped for spindle finishing equipment. Finishing action can be concentrated on certain areas while shielding other areas. Processing cycles are normally 5–15 min, and surface roughness changes from 120 to 44 μin. (3 to 1.1 μm). Finer finishes also can be produced (Kobayashi, Kobayashi, and Matsunaga 1977).

Recipro-finishing can be fully automated. One design incorporates a traverse unit that reciprocates on rails fixed to an upper frame, with fixture clamping arms pivoted to the traverse unit, a loading apparatus, and a transfer mechanism for loading and unloading. Several modifications of these machines are in use, including one type that employs spindle rotation for automotive wheels. These wheels are finished in 2 min with a spindle rotation of 120 r/min and 10 cycles/min reciprocation. Wavy motion machines have also been built (Kobayashi, Kobayashi, and Matsunaga 1977).

Surface Finish

Although recipro-finishing could conceivably produce finishes as fine as 2 μin. (0.05 μm) in a deburring cycle, finishes of 20–40 μin. (0.5–1.0 μm) would be more common.

The biggest apparent advantage for the recipro-finishing process is that it can deburr large parts readily and prevent part-on-part impingement. In addition, it concentrates most of the action on one side of a part.

8.5.5 Flow Finishing

In *flow finishing* (also called drag finishing) parts are inserted into a tank of moving abrasive media and compound and moved in the opposite direction of the media (Garlinghouse and Hanz 1985). The abrasive media and compound rub against the part and its edges, as in other loose abrasive processes. The action is much like a plow in the earth. In one application, a durable wire net is stretched around a pair of vertical belt wheels (see Figure 8-8). The bottom and sides of the wire net are supported by a number of bearing and guide rollers. When media is inserted into the tank, one of the vertical belt wheels drives the net in one direction, which in turn causes the abrasive media to flow in that direction; workpieces are moved in the opposite direction. Another version of the process uses a horizontal system of belts. The process can be used wet or dry (Kobayashi 1967).

Workpiece Size

Typically, flow finishing is used for hand-sized and larger parts. Since parts are fixtured, no part-on-part impingement can occur. As in all loose abrasive processes, flow finishing works best on the external edges of parts. The process does not lend itself well to deburring blind features because of media pile-up.

Cycle Time

The patent application for the flow finishing process indicates that media velocities of 230–1,640 ft/min (70–500 m/min) will allow deburring to be completed within 10 s to 15 min.

Limitations

Heavy burrs will not be removed in the flow finishing process without noticeable stock loss and long cycle times. This process will remove stock from all surfaces exposed to the media. It does not work well on edges of deep internal features. The process also impregnates minute particles of abrasive into the work surface, resulting in poorly brazed, soldered, or welded joints, and increases the probability of plating adhesion failures. Appropriate media selection can minimize these problems, however. When a workpiece contains a large number of differently sized holes, effective media that do not lodge in some of the holes may be difficult to obtain.

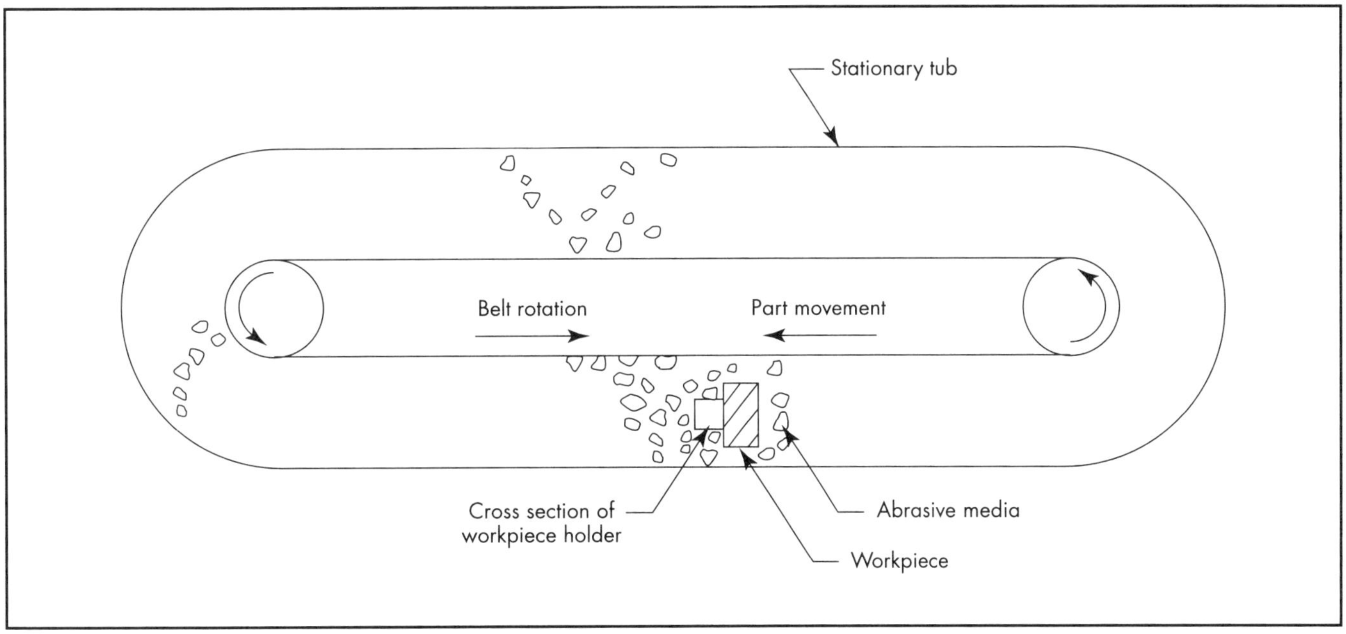

Figure 8-8. Flow finishing. ***(Courtesy Shiskishima Tipton)***

Size Change

Although data have not been published on the flow finishing process, it is reasonable to assume that .0002 in. (5.1 μm) stock loss will result from the removal of a .003 in. (76.2 μm) burr from stainless steel. Although the repeatability of this process has not been defined, it should be similar to other loose abrasive processes that have a repeatability of stock loss in the order of ±.0002 in. (±5.1 μm). It is difficult in any loose abrasive process to produce edge radii smaller than .003 in. (76.2 μm) when the initial burr is of the same thickness. Edge radii up to .030 in. (0.76 mm) should be possible. Aluminum and other soft alloys will have twice the radius found on most steels. With a consistent burr size, edge radii should be expected to have a repeatability of ±.001 in. (±25 μm).

Surface Finish

Flow finishing should produce surface finishes in the same order of magnitude as spindle finishing. Spindle finishing typically produces finishes of 10–30 μin. (0.25–0.76 μm).

The greatest apparent advantage for flow finishing is that it can be used to readily deburr large parts. It also eliminates the possibility of impingement. The fixturing used would mask one side of the part, which could be another advantage.

8.5.6 Cascading Media Deburring

One company has developed a process that fixtures valve bodies and similar housings needing *internal* passage deburring. In the cascading media deburring process, a stream of fine abrasives is fed down onto the interior edges while the fixtured part is oscillated. The oscillation moves the media through the housings and assures constant motion of the media. The free cascade reportedly removes burrs with or without use of abrasives and water. Although the process will work, the energy at internal edges is much smaller than required by most processes to assure removal of tough burrs (Meehan 1983).

8.5.7 Magnetic Abrasive Finishing

There are several variations of magnetic abrasive finishing. In the first and simplest variation, workpieces are placed in a container of magnetic loose abrasive media, and an oscillating magnetic field is produced to vibrate the media. While the workpieces can be stationary, more effective finishing is obtained by tumbling, turning, or otherwise moving the parts.

When magnetic media are used, the workpieces must be nonmagnetic to obtain relative motion. However, it is conceivable that a magnetic field could oscillate miniature magnetic parts through nonmagnetic media. Magnetic media can be coated with ceramics or other metals to minimize contamination of the workpieces. Most iron-based components will not react well to high-frequency changes in a magnetic field. When coated abrasives are used, the coating will also coat the tumbled parts and will require some cleaning after finishing.

In the most used variation of this process, cylindrical parts are vibrated, rotated, and fed axially in a magnetic field between two poles (see Figure 8-9) (Shinmura, Takazawa, and Hatano 1983, 1985; Shinmura et al. 1990). Abrasive grains are combined magnetically between the poles and the workpiece to perform the cutting action. Tests on ground and turned steel workpieces, 1.2 in. in diameter × 1.8 in. long (30 mm diameter × 46 mm long) reveal that this small burr was removed after 1 min. Finishes of 4 μin.

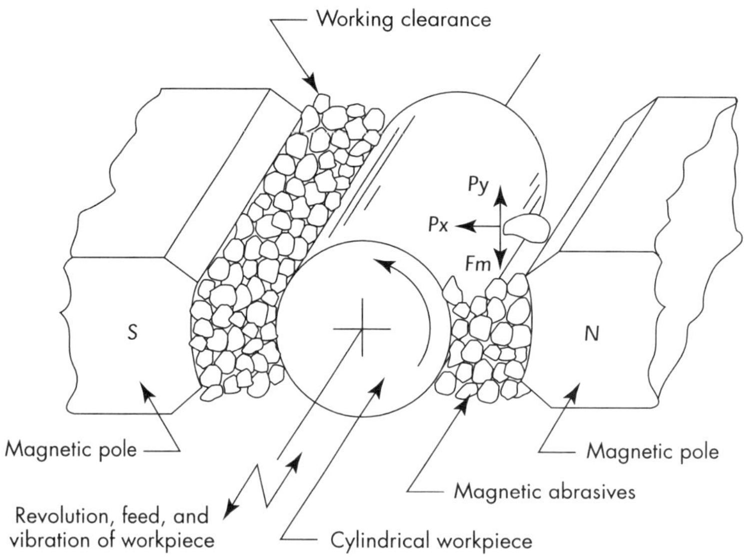

Figure 8-9. Finishing details of magnetic abrasive deburring of cylindrical workpieces (Shinmura, Takazawa, and Hatano 1983).

(0.1 μm) were reduced to .4 μin. (0.01 μm) in 4 min. This process is also used for finishing the insides of tubing, sewing machine parts, flat surfaces, balls, prismatic parts, and complex shapes. Each part design, however, will require a special head design for the magnetic field. Reportedly, 400 applications for this process are in use in Russia, and several machines are in use in Japan, but the technology is not yet used in the United States.

In one recent advance, the magnetic operation is performed in a simple CNC machining center using a rotating tool (Kumakura et al. 1977).

References

Anonymous. 1972. "Electrochemical Processing" (in German). VDI-RICHTLINIE 3401-2, January, 30. Berlin, Germany: Beuth Vertrieb GmbH.

Anonymous. 1976. "Internal Deburring with Resonant Energy." *Manufacturing Engineering,* 71 (November):38–39.

Dargis, Raymond G. 1987. "Chemical Accelerated Mass Finishing of Zinc Diecast Parts and Copper Plate—Without Cyanide." Technical paper MR87-146. Dearborn, MI: Society of Manufacturing Engineers (SME).

DeGroat, George. 1969. "ECM Deburring with Graphite." *American Machinist,* 13 (December):78–79.

Garlinghouse, Roland E., and Hanz, H.E. 1985. "Heavy-duty Mass Finishing." Technical paper MR85-826. Dearborn, MI: Society of Manufacturing Engineers (SME).

Kobayashi, Hisamine. 1967. "Beeline Flow Finishing Method and Apparatus." U.S. Patent 3,14149,871, filed 21 March.

Kobayashi, H., and Matsunaga, M. 1981. "New Developments in Barrel Finishing." Technical paper MR81-402. Dearborn, MI: Society of Manufacturing Engineers (SME).

Kobayashi, W., Kobayashi, Jr., F. and Matsunaga, M. 1977. "Deburring and Finishing by Recipro-finishing and Gyro-finishing Machines." Technical paper MR77-464. Dearborn, MI: Society of Manufacturing Engineers (SME).

Krapivin, S.P. 1972. "Vibroabrasive Electrochemical Grinding" (in Russian). *Vibrats Obrabotka Detalei mashin i priborov* (Vibratory Machining of Machine and Instrument Components). Rostov-On-Don, Russia: 80–84. (See Referativnyi Zhurnal—Tekhnolagiia Mashinostroeniia [in Russian], 1973, 4B 253).

Kumakura, Kenichi, Suzuki, Kiyoshi, Kitajima, Takayuki, Kaneko, Naoto, and Uematsu, Tetsutaro. 1977. "Now for Something Completely Different." *American Machinist,* 121 (February):53–55.

Meehan, James B. 1983. "A New Mass Finishing Process for Internal Burr Removal." Technical paper MR83-188. Dearborn, MI: Society of Manufacturing Engineers (SME).

Meyers, Richard J. 1985. "Cryogenic Deflashing of Plastic, Rubber, and Zinc Die Cast Parts." Technical paper MR85-847. Dearborn, MI: Society of Manufacturing Engineers (SME).

Richardson, Terry L. 1989. *Industrial Plastics: Theory and Application,* 2nd ed. Albany, NY: Delmar Publishing, Inc.

Shimizu, Akihiko. 1975. "Electrochemical Barrel Finishing." Technical paper MR75-837. Dearborn, MI: Society of Manufacturing Engineers (SME).

Shinmura, T., Takazawa, K., and Hatano, E. 1983. "Development of Magnetic Abrasive Finishing and Its Equipment." Technical paper MR83-678. Dearborn, MI: Society of Manufacturing Engineers (SME).

———. 1985. "Advanced Development of Magnetic-abrasive Finishing and Its Application." *Proceedings of the Deburring and Surface Conditioning Conference*, 5-7–93, Chicago, IL. Dearborn, MI: Society of Manufacturing Engineers (SME).

Shinmura, T., Takazawa, K., Hatano, E., and Matsunaga, M. 1990. "Study on Magnetic Abrasive Finishing." *Annals of the CIRP,* 39 (1):325–28.

U.K. Patent. 1968a. Patent no. 1,238,073.

———. 1968b. Patent no. 1,194,649.

———. 1970. Patent no. 5,523,834.

Bibliography

Adams, R.S. 1970. "The Economics of Line Deflashing." *Rubber World*, 172 (June):67-71.

Adams, R.S. 1969. "Deflashing Molded Rubber Parts with Liquid Nitrogen." *Cryogenics & Industrial Gases,* 4 (June):60–63.

Anonymous. 1976. "Magnetic Abrasive Deburring" (in German). *Fertigungstechnik und Betrieb*, vol. 26, no. 12:738-744.

Baron, Yuri. 1986. "Technology of Abrasive Machining in a Magnetic Field" (in Russian). Lenigrad, Russia.

"Cryogenic Deflashing of Zinc Parts." 1976. *Die Casting Engineer,* 20 (September–October):34–35. z

"Deflashing Castings by Freezing." 1974. *Metals & Materials,* (November–December):471.

"Deflashing Rubber or Plastics Molding with Carbon Dioxide." 1970. *Caucho* (Madrid), 244 (September):33–35.

"Deflashing with Nitrogen." 1971. *Plastics Rubber Weekly,* 395 (10 September):53.

Fox, M., Agrawai, K., Shinmura, T., and Komanduri, R. 1994. "Magnetic Abrasive Finishing of Rollers." *Annals of the CIRP,* 43(1)181–84.

Gillespie, LaRoux K. *Deburring Technology for Improved Manufacturing*. Dearborn, MI: Society of Manufacturing Engineers (SME).

Kobayashi, Hisamine. 1974. "Automatic Recipro-finishing Machine." U.S. Patent 3,898,767, serial no. 433,149, filed 15 January.

———. 1973. "Automatic Recipro-finishing Machine." U.S. Patent 3,885,354, serial no. 389,515, filed 20 August.

Makedonski, Boris. 1974. "Magnetic-abrasive Grinding of Parts" (in Bulgarian). *Mashinostroene,* 23 (14):59–60.

"Mechanical Rubber Deflashing." 1961. *Mechanical Engineering,* 39 (July):60.

Morozov V. A. 1972. "Changes in the Physical Mechanical Properties of Component Surfaces in Magnetic-Vibratory Machining" (in Russian). *Vibrats Obrabotka Detalei: mashin i priborov* (Vibratory Machining of Machine and Instrument Components). Rostov-On-Don, Russia: 23–29.

"Rapid Mechanical Deflashing." 1961. *Rubber World,* 144 (November):67–70.

Sachs K., and Edgers, M. 1956. "Smoothing of Mild Steel by Barrel Treatment in Oxalic Acid-Hydrogen Peroxide Solution." *Transactions of the Institute of Metal Finishing,* 33:245–69.

Shinmura, Takeo, and Toshio Aizawa. 1989. "Study on Magnetic Abrasive Finishing Process." *Bulletin of the Japan Society of Precision Engineering,* 23 (3):236–39.

Shinmura, T., K. Takazawa, and Hatano, E. 1986. "Development of Spindle-finish Type Finishing Apparatus and Its Finishing Performance Using a Magnetic Abrasive Machining Process." *Bulletin of the Japan Society of Precision Engineering,* 20 (2):79–84.

Shinmura, Takeo, Wang, Feng Hui, and Aizwa, Toshio. 1994. "Study on a New Finishing Process of Fine Ceramics by Magnetic Abrasive Machining." *International Journal of the Japan Society for Precision Engineering,* 28 (2):90–104.

Shinmura, Takeo, and Yamaguchi, Hitomi. 1994." Study on a New Internal Finishing Process of a Nonferromagnetic Tube by the Application of a Linearly Traveling Magnetic Field." *International Journal of the Japan Society for Precision Engineering,* 28 (1):29–34.

———. 1995. "Magnetic Abrasive Finishing of Inner Surface of Tubes." *Proceedings of the International Symposium for ElectroMachining*, 12–21 April. Lausanne, Switzerland: 963–76.

Shinmura, Takeo, Yamaguchi, Hitomi, and Aizawa, Toshio. 1993. "A New Internal Finishing Process of Nonferromagnetic Tubing by the Application of Magnetic Field." *International Journal of the Japan Society for Precision Engineering,* 27 (2):132–38.

Tipton, T. T. 1975. "Cryotrim—An Efficient Deflashing System for Zinc Diecastings." *Foundry Trade Journal,* 138 (3038):298–99.

Yamaguchi, Hitomi, and Shinmura, Takeo. 1994. "Development of a New Type of Magnetic Finishing Tool for Internal Finishing of Tubes Using Rotating Magnetic Field." *Transactions of the NAMRI/SME*. Dearborn, MI: Society of Manufacturing Engineers (SME): 22:135–40.

9 Abrasive Blasting

Abrasive blast deburring and deflashing are subcomponents of the broader field of blast finishing. Blast finishing is one of the most economical and widely used finishing processes and requires less labor than other methods. The equipment can be used for many different operations and a wide variety of workpieces. The media is generally reused, thus reducing costs. Most blasting processes are suitable for low- or high-volume production requirements. In addition to deburring and deflashing, blast finishing is used extensively for cleaning operations. These operations include removing sand from castings, scale from forgings and heat-treated parts, mill scale and rust, paint and other coatings, various soils, and other surface contaminants. Soils or dirt to be removed must be dry and the workpiece surfaces free of grease or oil before dry blasting.

Thousands of parts can be run per hour, and most materials can be deflashed and deburred using this method. The process does not, in general, remove all the burrs from all the edges, but it removes enough loose particles and thin attached flash and scrubs parts to an extent that parts are suitable for industrial use. The process can remove all but very heavy burrs; but removing large burrs is generally not a requirement in this process. The process is widely used for automotive and plastic parts to dull edges and remove loose materials. Cost per part can be as low as fractions of a cent per part for mass blasting operations to several cents per part for hand operations performed one at a time.

Side effects of the process may include abrasives being lodged throughout the parts, surface contamination by the blast media, and thin films of contaminants that may require cleaning (depending on application and part use). Blasting can bend thin sections, increase residual stresses, chip brittle materials, and lodge in small crevices. Blasting dust is an environmental and health issue.

Abrasive blasting processes for deburring and deflashing include:

1. Conventional dry blasting—wide fan angle blasting (including wheel thrown media) and concentrated jet blasting.
2. Conventional wet blasting—wide fan angle blasting and concentrated jet blasting.
3. Microblasting.
4. Cryogenic blasting.
5. Ice blasting.

Blasting equipment is designed to provide a concentrated stream that impacts specific edges or a series of nozzles or a broad fan-like stream to blast the entire surface of a part. Wet blasting, in contrast to dry blasting, provides a more gentle action, but cuts more slowly and has some limitations associated with water use. Shot peening is the use of round shot to break, bend, and smooth edges by impact. Abrasive blasting, in contrast to peening, cuts flash and burrs rather than knocking off material. Miniature blasters direct a very fine stream of abrasive to cut small burrs. Cryogenic blasting is widely used for removing plastic and rubber flash. Two forms of blasting with ice have been used. The advantages include clean surfaces and environmentally friendly operation.

Variables affecting the performance and physical results of the blasting process include: type of media and grain size, machine efficiency (suction or pressure blast), duration of blast, blasting air pressure or media velocity, distance from the nozzle to the part, relative angle of blast, automation techniques, workpiece material, and burr or flash size, location, and tenacity. Wet blasting has some additional variables that will be discussed later. Although not discussed in this chapter, robots are being used to perform both wet and dry blasting. In addition to all the variables cited in this chapter, the use of robots requires special protection to keep grit out of robot joints.

9.1 DRY BLASTING

There are two commonly used methods of dry blasting: air blasting and centrifugal wheel (airless) blasting. In air blasting systems, the media is propelled by a stream of pressurized air. There are two variations of this method: pressure blasting and suction. In airless systems, known as wheel blasting, the media are fed into the center of a rotating vaned wheel, which hurls the media by centrifugal force. Although the air blasting system is the most versatile, it requires a source of high-pressure air and is more energy intensive. However, air blasting systems accelerate the media to higher velocities, permitting the effective use of smaller media particles to produce smooth finishes.

Manual cabinet systems often can deburr parts as fast as the operator can manipulate the work. Automated turntable systems deflash die-cast air-conditioner compressor housings at 720 parts per hour. Basket blaster systems deburr the window-like cutouts of bearing cages at volumes of 12 ft^3/h (0.34 m^3/h). In-line deflashers can deflash thermoset, plastic-encapsulated, lead frames for the semiconductor industry at rates of 4,500–28,000 integrated circuits per hour. Hypodermic needles are batch processed in special automated equipment that produces up to 160,000 burr-free cannulas (needles) per 8 h (in some applications, 250 needles in 7 s). In this process, 400-mesh and finer glass beads are used because of their chemical purity, ability to reach into small recesses, and lack of abrasion to other parts of the needle.

Tumble-type blast units can process tons of castings or forgings in a few minutes. Some tumble-type wheel blasting machines are capable of processing 100 ft^3 (2.83 m^3) of parts in a single load within 15 or 20 min.

Unlike some other deburring processes, abrasive blasting systems are not specifically engineered for one part, which means that parts of different sizes can be deburred in one system. When not used for deburring and deflashing, the blaster can perform other finishing chores on a wide variety of parts. In many instances, blasting machines are used not for deburring, but for cleaning, peening, removing rust, and related operations. Most metalworking firms have blasting equipment because of its versatility and low cost.

Blasting is not suitable for all applications. Other techniques should be explored if burrs are heavy or fluctuate in size from batch to batch. Finishing large surfaces is more efficiently performed with such methods as wheel blasting and automatic brushing than with air blasting systems. Air blasting will not remove heavy parting lines. Some media will remove metal and alter the surface finish. Furthermore, very thin parts may be warped, and delicate parts may be chipped if improperly blasted.

Blasting works by propelling loose particles (media) against the workpiece. The particles literally bombard the surface to alter its characteristics. Burrs and flash are either eroded away or broken free from the parent material, depending on burr material and the media chosen to do the job. Materials that are hard, such as resins, can be broken off by media impingement. Cast materials fall into this category. As a general rule, the heat of processing in a mold embrittles thinner areas of many materials. The fragile cross sections are then easily broken off during bombardment. Softer, ductile materials, such as cast stainless steel, machined aluminum, and molded rubber compounds, require a different approach. Tough hang-on burrs resulting from tool smearing and edges raised from shearing are eroded away by an abrasive media. Some malleable burrs are successfully peened over and melded back into the parent material with more gentle media.

9.1.1 Media for Dry Blasting

Media ranges from soft and gentle sodium bicarbonate to hard ceramic shot; it can be angular and provide aggressive cutting action, such as aluminum oxide, or have round smooth shapes that produce a gentle effect, such as glass beads. Media can be lightweight, such as wheat starch or plastic, or relatively heavy, such as steel shot. Sizes vary from powdery dust to BB shot size. Each medium has characteristics that make it most suitable for a particular application.

The grain size of the media determines how many particles hit a given area over a period of time. Since smaller particles yield more impacts per given area, they generally finish parts faster. But when particles are so small that they have little mass (as with dust), they decelerate rapidly and lose momentum before reaching the workpiece. Large particles may lodge in crevices (in general, particles should be smaller than one-fourth the size of the cavity).

Media grain material will make a difference in some applications. Hard or abrasive media that has lodged in a cavity of a transmission casting, for instance, may come loose and lead to premature failure of the assembly if not properly blown off after deburring. Nonharmful plastic media is often used in sensitive applications like these.

Production blasting equipment usually includes a reclaimer, which extracts reusable media from inefficient dust and debris. Efficient reclaimers separate by cyclonic air movement. Heavier (reusable) particles are forced to the outside wall of a cylinder by centrifugal action; they then slide down the sidewall to be collected for reuse. Dust is extracted in a counter vortex of ventilating air, which exits through a vertical stack tube. Debris is removed from reusable media with a simple screen. Some reclaimers are tunable, rendering them better able to extract different sizes of undersized particles.

9.1.2 Surface Preparation

Parts coated with grease or oils are not easily cleaned or finished by dry blasting, and these substances can contaminate the media used in dry blasting operations. For these reasons, degreasing and drying should be performed prior to blasting. Metal surfaces prepared for painting or porcelain enameling by abrasive cleaning must have all traces of surface soil removed before blast cleaning. Any type of soil that will still adhere to the blasted surface or the shot or grit particles may produce surface defects in the subsequent phosphating, painting, or porcelain enameling operations. Normally, materials that will readily flake and shatter under the impact of blasting—such as rust, weld spatter compound, and dry dust—will be reduced to fines and removed by separation and dust-collection systems.

9.1.3 Air Blasting Systems

Compressed air is the vehicle for delivering media in air blasting systems. Air pressure is important and will be discussed later. The volume of compressed air depends directly on the inside diameter of the nozzle, the number of nozzles used, and the blast pressure. Air must be clean and dry (that is, 40° F [4° C] or lower dewpoint). The air capacity required to run a typical one-nozzle blast system is shown in Figure 9-1.

Air blasting equipment used for deburring and deflashing, whether manual or automatic, has the following main components:

- Media delivery system (suction or pressure).
- Protective enclosure.
- Dust collection and, usually, media reclamation system.
- Enhanced coverage equipment (automation).
- Material handling equipment.
- Options to improve efficiency and system life.

There are three types of air-powered dry blasting systems: direct pressure, gravity fed, and induction suction.

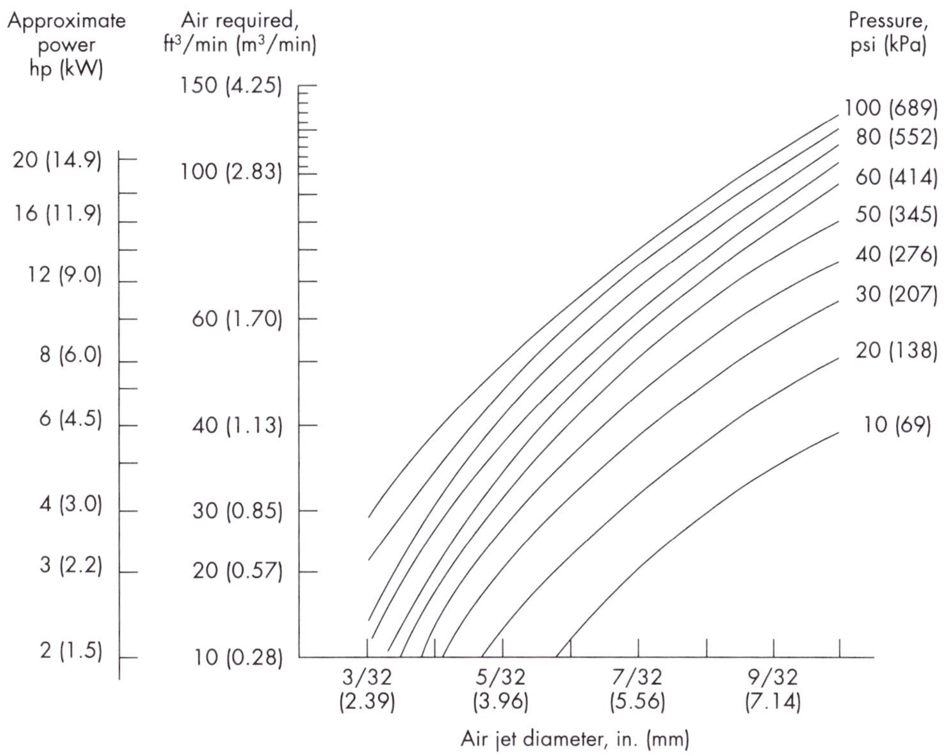

Figure 9-1. Air required for blasting with various jet diameters at different pressures. Data are based on using a single-stage compressor at sea level, where 5 ft³ (0.14 m³) of air produces 1 hp (0.75 kW), approximately.

Types of Systems

Direct pressure. In direct-pressure air blasting, the media is fed from a pressurized tank directly into the cleaning hose and discharged through a nozzle. All types of media are used. This method is the workhorse of air blasting systems and is generally used for exceptionally heavy cleaning, such as removing scale from castings and forgings. It is also ideal for cleaning interior cavities, blind holes, narrow recesses, and localized areas. This process generally provides a more uniform finish in a shorter time than other methods of air blasting. The media is ejected at higher velocities and with greater concentration than in the suction method, for example; however, nozzles and lines are subjected to greater wear, and more media is consumed. Also, the force of the blast stream may warp thin material, except when two streams are used to hit both sides simultaneously.

Pressure blasting systems force the media particle from the storage vessel to the nozzle, imparting the acceleration of the moving air to the media over a much greater distance. Particles exiting from a pressure nozzle have velocities of 15,000 ft/min (4,572 m/min) or higher. One manufacturer notes that suction guns propel media 40,800–45,000 ft/min (12,436–13,716 m/min), and pressure blasting can be up to twice that velocity. Some users believe that suction is in the high subsonic velocity range, whereas pressure machines are in the supersonic velocity range; wheel-thrown media leave at a velocity of about 15,000 ft/min (4,572 m/min). Note that impact energy is equal to 1/2 particle mass × velocity squared (Hanna 1993). When the mass of the media is doubled, the kinetic energy is doubled. But when velocity is doubled, the kinetic energy is *quadrupled*. In other words, a pressure delivery system can provide faster cutting than suction.

The effect of blasting variables on cutting ability can be modeled mathematically. Although the actual effectiveness of the process depends on many parameters in addition to the blasting variables, one researcher found the following relationships (Alwerfalli 1989):

$$B_r = 9{,}836\ PD^{.75}\ S^{.45}\ M^{.37}\ d^{.05}\ I^{-.13}\ H^{-.8} \qquad (9\text{-}1)$$

where

B_r = Burr removal rate.

P = Air pressure lbf/in.² (kgf/cm²).

D = Diameter of the nozzle in. (cm).

S = Size of abrasive particle in. (cm).

M = Mass flow rate of abrasive oz/ft (g/m).

d = Distance of nozzle tip from workpiece in. (cm).

I = impact angle of abrasive (degree).

H = hardness of workpiece material (Brinell hardness number).

Equation 9-1 shows that the removal rate is linear with pressure, increases slower than linear with nozzle diameter, is roughly proportional to the square root of abrasive size, is the one-third root of the media flow rate, rapidly decreases with distance from the nozzle, is slower for hard materials, and is affected by the angle of impact. As noted earlier, most users believe that the cutting rate increases with pressure. The point of the equation is to highlight that simple tests can provide quantitative decision-making data.

With direct pressure systems, less compressed air energy is required per given unit of work because of the increased efficiency of the process. Direct pressure offers better control at very high and very low blasting pressures, but control may not be important for many commercial applications.

Direct pressure systems will deliver a lightweight particle to the work surface with much more momentum than a suction system, making it often used with plastic, wheat starch, and agricultural media. These systems are often used for hard-to-reach areas such as cross holes.

Gravity fed. In gravity-fed air blasting systems, the media feed hopper is located above the nozzle, and media flow by gravity to the nozzle. Air entering from a separate line mixes with the media and is propelled through the nozzle. A bucket elevator, pneumatic system, or other mechanism returns the spent media to the feed hopper. A wider spray pattern results from this method. The media has less force than with direct pressure systems, resulting in slower action. However, the slower speed results in less media breakdown, which means that a larger percentage of the media can be reused.

Gravity-fed cleaning equipment is generally used when less severe cutting is required, such as to remove light scale or rust and other contaminants from castings, structural shapes, and various other parts. Maintaining unobstructed flow with traversing nozzles and blasting upward at surfaces can be difficult.

Induction suction delivery. Suction air blasting systems use the venturi principle to suck media from storage to the blast gun; the media is then blown onto the workpiece. This method is simple and inexpensive. Production times are adequate for most applications, and results are often equal in quality to pressure blasting. Unlike pressure delivery, suction systems can be left on continuously, which is important for automation. When low media velocities are sufficient, this method is the least expensive of the air blasting processes. Induction suction systems will also propel media upward, and they can blast continuously without stopping for media refills. Suction simplifies the use of multiple guns. Automatic systems with 12 guns are common; some have been built with 48 guns.

A typical nozzle produces a blast pattern the size of a half dollar. Obviously, this is insufficient when covering the entire surface of a wide part. To cover larger surfaces, the nozzle can be oscillated over the part. This technique is commonly used in manual blasting and can be automated. Oscillation replaces the need for multiple guns, thus conserving compressed air and saving running costs. Multiple nozzles are ideal for covering surfaces, with multiple facets aimed into each crevice of the part. Each additional gun speeds processing. Two guns are 100% faster than one gun; eight guns are 400% faster than two. For fast, even coverage of wide areas, multiple nozzles can be oscillated.

A rotary head is a form of multiple nozzle oscillation and spins six to nine suction guns in-place over a moving horizontal part (see Figure 9-2). Combined with a conveyor or turntable for part transport, rotary heads have high production rates on larger surfaces (up to 36 in. [0.9 m] wide for each rotary head) (Hanna 1993) and provide extremely even coverage.

Equipment

Selecting equipment to be used for blasting requires careful consideration of many factors. These factors include: the number, size, variety, and condition of the parts to be blasted; the required production rate and finish specifications; the amount of floor space and ceiling height avail-

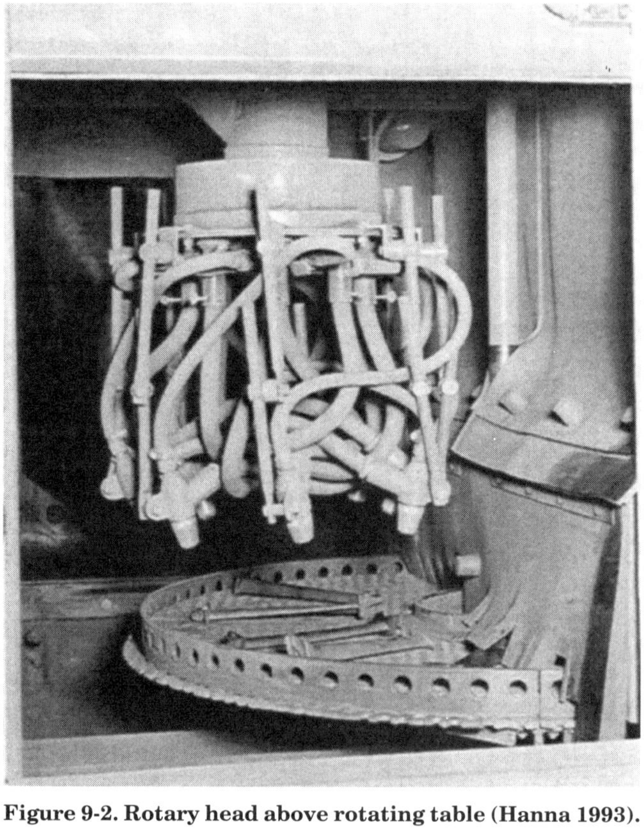

Figure 9-2. Rotary head above rotating table (Hanna 1993).

able; and environmental and safety regulations. Standard machines cost less than special machines and hence they should be used whenever suitable. However, many applications necessitate the use of specially designed machines.

The most common type of air blasting equipment requires manipulation of the nozzle by an operator, who is located outside a cabinet or inside a room or other enclosure. Small workpieces are frequently processed in self-contained hand-blast cabinets. Large workpieces that must remain stationary are blasted in rooms or enclosures with the operator wearing special apparel and manipulating a flexible air blasting nozzle.

Cabinets. Standard hand-blast cabinets are relatively low in cost, very versatile, and available in various sizes. With a typical cabinet, operators place their hands, protected by gloves, into the cabinet to manipulate a nozzle and/or the workpiece while viewing the operation through a window. Workpieces are loaded and unloaded through an access door. Cabinets are generally equipped with a media supply-and-recovery system and a means for ventilation and dust collection.

Automation cabinets may contain an in-line conveyor for production line parts feeding. A continuously rotating turntable or indexing satellite stations are usable with either a production line or work cell concept. Indexing turntable machines can be further modified with bowl feeders and pick-and-place units.

Cabinets depend on operator control and are unsuitable for large parts. However, several means are available to increase production and improve control. A simple turntable, sometimes mounted on a wheeled dolly with tracks, can be provided for handling heavy or complex parts. Additional nozzles that oscillate vertically or horizontally and are supported by adjustable brackets are sometimes used. Timers can be employed to control the duration of the blast cycle, and programmable controllers can be used for increased versatility in some applications.

When production requirements for a specific workpiece are sufficiently high, conveyorized handling is often used. In-line conveyors are generally preferable for small parts that can be automatically loaded and unloaded. Rotary conveyors, usually of the indexing type, are more frequently used for larger or more complex parts, but they are slower than in-line conveyors. Industrial robots or programmable controllers are being used when a variety of parts are to be blasted with a single machine or when consistent results are required on complex parts. Portable blasting machines are also available in a variety of sizes.

Room enclosures. Rooms or enclosures for blasting large parts provide space for positioning workpieces; they also retain the media for safety, reclamation, and recycling purposes. Manual or power-driven turntables are used for blasting heavy parts. Because of the abrasiveness of most media, the rooms should have a minimum number of moving parts and be provided with wear-resistant liners. Loose-hanging rubber sheets are effective for this purpose. Polyurethane-lined cabinets are also becoming common.

Nozzles. Nozzles for air blasting equipment are made of several different materials in a variety of shapes and sizes. Hard iron, boron carbide, silicon carbide, tungsten carbide, and ceramic are the principal wear-resistant materials used as nozzle liners. The nozzle material must be compatible with the wear-producing characteristics of the media being used. The wear of any type of nozzle is influenced by the hardness, sharpness, and velocity of the media. In general, a nozzle will last from two to four times longer when used with steel shot or grit than with sand.

Various nozzle shapes are available. Short, straight-bore nozzles are used for blasting close to the workpiece. Venturi nozzles are used for most efficient cleaning at distances of 10–24 in. (254–610 mm) from the work. Venturi nozzles produce 20–50% faster cleaning than straight nozzles. Side-shooting nozzles are available for inside-diameter cleaning of such items as pipes, tubes, and valve-body bores.

The shape of a nozzle does not influence its wear life to any great degree. However, the inlet must be well-centered in the hose and must taper smoothly from the inner diameter of the hose to the end of the nozzle. The length of the straight section at the bore should be at least five times the bore diameter. Nozzles are considered to be worn out when the outlet has worn to 1.5 times its original size. Replacing the nozzle at this point will save money by maintaining the efficiency of compressed air usage and the high terminal velocity of the media.

Compressed air. Compressed air is expensive to produce; and if excess plant air is not available, the cost of an air compressor must be considered when air blasting equipment is purchased. Consumption of compressed air by various jet diameters used in air blasting systems is shown in Figure 9-1. Excess moisture in plant air sometimes leads to problems with rust on shot mixtures or workpieces. Water and oil traps are often required, and air dryers or moisture separators may sometimes be required. If the air pressure in a plant varies with demand, care must be taken that periodic drops in pressure will not adversely affect production requirements.

Machine Operating Variables

Selecting the proper blasting pressure, impact angle, and stand-off distance of the nozzle from the workpiece are essential for efficient and economical blasting.

Pressure. When using direct pressure systems, pressures of 40–60 psi (276–414 kPa) are generally the most economical. For induction suction systems, pressures of 60–80 psi (414–552 MPa) are common. Excessive pressures accelerate breakdown of the media, with only minimal decreases in blasting times. Often the pressure used is determined by what is available in the plant. The pressure used also depends on the application and the media used. For blasting precision parts with fine, soft media, pressures as low as 10 psi (69 kPa) may be used. For heavy-duty operations, pressures of 80 psi (552 kPa) or more are often used.

Impact angle. Maximum cutting action and the production of rough surfaces are obtained with cutting media when the air blasting nozzle is held at an angle of 45° in relation to the workpiece surface. However, when smooth surface finishes are required, an impact angle of about 30° is employed. Nozzle angles are sometimes adjustable, especially on automated, multinozzle machines.

Standoff distance. The distance that the nozzle is held from the workpiece (that is, the standoff distance) depends on the specific application and the media used and is generally maintained constant during the operation. This distance generally varies from 3–24 in. (76–610 mm), with shorter distances usually employed when softer media is used. With a given media at a fixed velocity, longer standoff distances increase the diameter of the blast pattern; but with light abrasives, the impact force may be reduced.

9.1.4 Centrifugal Wheel Blasting Systems

Centrifugal wheel (airless) blasting is a mechanical method by which media is propelled by centrifugal force from a rotating vaned wheel (slinger). The media forms an elongated, cone-shaped pattern. The operator remains outside the machine during blasting. The economy of this type of equipment is apparent from the fact that a single wheel, with a 100-hp (75-kW) drive motor, can discharge up to about 2,600 lb (1,180 kg) of steel shot or grit per minute at velocities reaching 14,000 ft/min (4,267 m/min). In comparison, a single air nozzle is limited to about 100 lb (45 kg) of media per minute. Figure 9-3 illustrates this type of equipment. Media from an overhead storage hopper is fed through a spout to an impeller at the center of the wheel. From the impeller, the media enters the wheel vanes and is accelerated before being flung onto the workpiece.

Wheel blasting is used most commonly in batch and continuous equipment, especially when a large volume of media is required over a fairly large work area. The kind and size of unit will depend on the size, shape, and fragility of the metal parts to be cleaned. The mechanical wheels on single or multiple units are positioned on the equipment to expose the desired areas of the work to the proper volume of abrasive at the proper angle and velocity. The equipment must turn or roll the workpieces continuously to expose their surfaces completely. Some of the more popular wheel blasting systems use barrels, tables, hanging machines, axial-flow units, rooms, cabinets, and special-purpose machines for processing workpieces. Wheel deburring is used extensively in the automotive, machine tool, rubber, and plastics industries. It is capable of deburring machined gray iron parts; aluminum, magnesium, and zinc die castings; steel stampings; hardened machine components; and plastic and rubber molded parts. Case studies have revealed that the process is 10 times more economical than air blasting techniques for some applications.

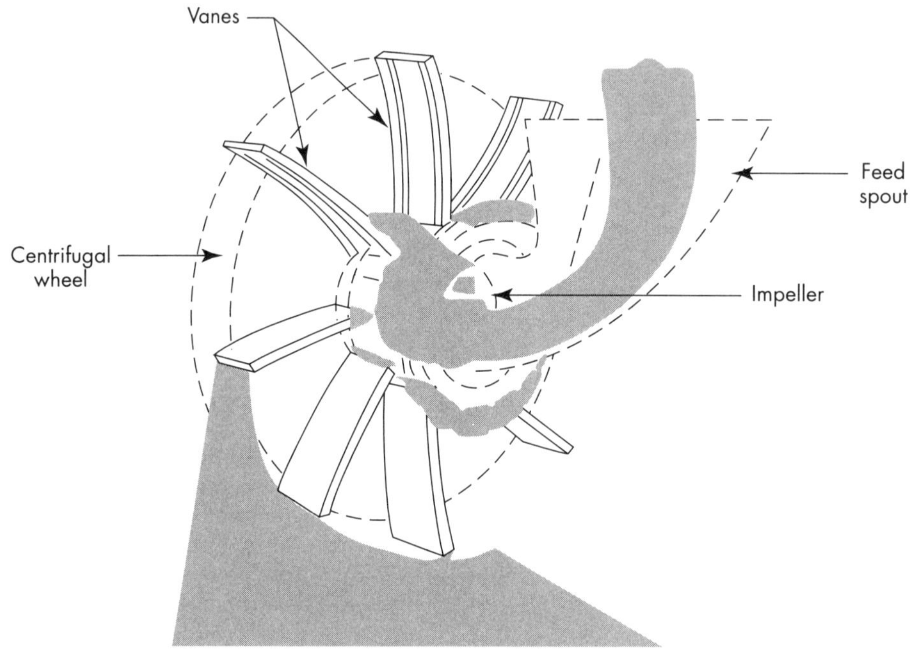

Figure 9-3. Cutaway view of centrifugal wheel blasting equipment. The impeller picks up media from the feed spout; media exits through openings in the impeller case and is deposited on the vanes of the wheel to be accelerated; the vanes then throw the media in a targeted pattern. ***(Courtesy Panghorn Company)***

Equipment

Blasting barrels. Batch-type blasting barrels are generally mechanically charged with a skip-hoist loader. Barrels continuously fed at one end and discharged at the other end are used for high-production, automated systems. Workpieces are slowly rotated or rocked under the media stream, usually by a traveling, slat-type conveyor in a closed-container system (Figure 9-4). Timing or visual inspection determines duration of the blast operation, which remains generally constant for a given operation.

The tumbling barrel normally has blast times of 3–10 min and typically uses steel shot or steel grit as the blasting media. Smaller units are widely used in the plastics industry for deflashing parts. In these instances, crushed nut shells and plastic pellets are used as blasting media. Larger units of up to 100 ft^3 (2.83 m^3) capacity are used for deburring in the automotive and machine tool industries.

The velocity of the media thrown can vary from 3,000–15,000 ft/min (914–4,572 m/min), depending on the size and speed of the wheel unit used. (As a point of reference, a shotgun blast reportedly averages 75,000 ft/min [22,860 m/min]). Lower speeds are available by varying the motor speed to control the burr-removal process.

Tumbling blast barrels are typically loaded and unloaded by automatic devices onto conveyor belts for transport to the next operation. One operator will most likely control several machines when producing high-volume deburred or deflashed parts.

Small barrels are used to remove flash from molded rubber parts, O-rings, and complex-shaped components. This type of deflashing requires a specially designed machine operating within a temperature and humidity controlled enclosure that uses small spherical steel pellets or plastic media to trim frozen organic or silicone flash (cryogenic processes are discussed later in this chapter).

Table machines. Machines with rotating tables and one or more wheels can be used to deburr parts that cannot be tumbled in barrels (Figure 9-5). The parts are positioned on the table in such a way that the edges to be deburred are exposed (repositioning may be necessary). These machines can employ a single large table or several tables mounted on each of two doors so that loading and unloading can take place simultaneously with blasting. Other units employ auxiliary tables on the main table for smaller work. Partitions between the auxiliary tables are used to prevent media from escaping from the cabinet toward the operator.

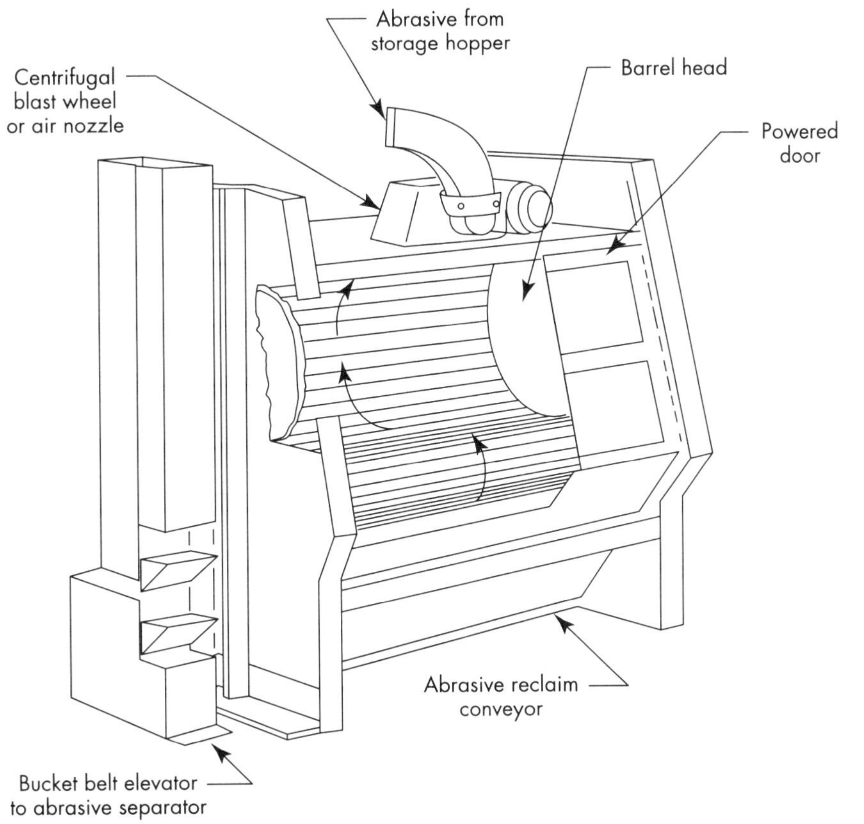

Figure 9-4. Blast tumbling machine. ***(Courtesy U.S. Filter/Wheelabrator)***

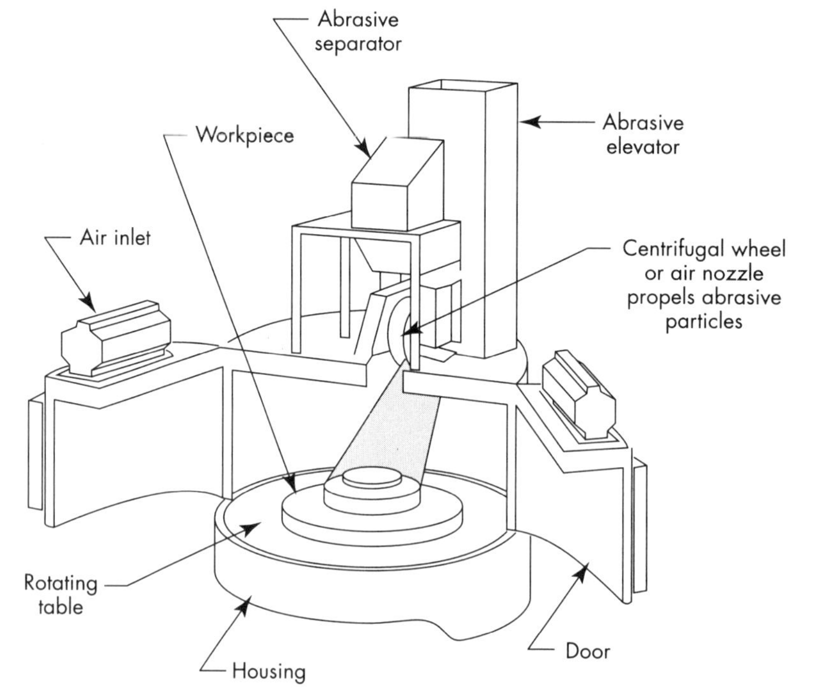

Figure 9-5. Table-type blasting machine. ***(Courtesy U.S. Filter/Wheelabrator)***

Hanging fixtures. Hanging parts on fixtures ("tree hooks") enables all surfaces to be exposed without the need for repositioning them during the operation. However, the difficulty of hanging and the number of different hanging fixtures that may be required must be taken into account.

Fixtures for hanging can be mounted on various material-handling devices for transporting into and out of the cabinet. Hand-pushed trolleys, self-powered carriers, and continuous or indexing trolley conveyors are all used. The conveyors can move in and out one side of the cabinet or pass through the cabinet (Figure 9-6). In addition to multiple blast positions and multiple wheels, a method of rotating the work while in the cabinet is usually employed.

In cleaning machines with spinner hangers, the blast cabinet is typically equipped with two or more partitioned compartments mounted on a vertical center support shaft. This permits the work to be indexed into the blast station by the movement of a controlled actuator. While one compartment is being blasted, the others are being loaded and unloaded. Each compartment has a spinner hanger for rotating the work in the blast cabinet. These types of machines usually have two or more centrifugal wheels for deburring all surfaces of large or small complex shapes.

Small parts, such as gears, fuel system components, and eyebolts, have been successfully deburred on Christmas tree-type fixtures. These fixtures are capable of handling up to 100 pieces each. Larger components, such as tractor fender stampings, have been successfully deburred to remove all sharp edges using this process. An added benefit has been surface preparation for painting. These types of spinner hanger machines are suitable for components too delicate to be tumbled or complex shapes requiring many surfaces to be deburred. Abrasive blast time for each hanger is 30 s to 3 min.

Machines with multiple tables are used for deburring parts that are delicate and have several surfaces requiring deburring. Typical applications for this type of machine include transmission parts, bearing cages, gears, and flat die castings (DeLitizia 1981). Multitable machines have a main table that rotates at an adjustable speed and carries individual auxiliary spinner work tables. As the main table carries each auxiliary table into the blast zone, the table begins to rotate so that all surfaces of the workpiece are exposed to the blast stream. As the individual tables come into the loading and unloading area of the machine, they become free-wheeling. The operator, who is protected from the blast by curtained vestibules, then flips the workpiece 180° to present these surfaces to the blast. The time needed for complete deburring of components ranges from 10 to 30 s per work table.

A wide variety of standard as well as special deburring equipment is available. In the automotive industry, special gravity roll-through equipment is capable of deburring 1,000 gears and coupling components per hour. There are,

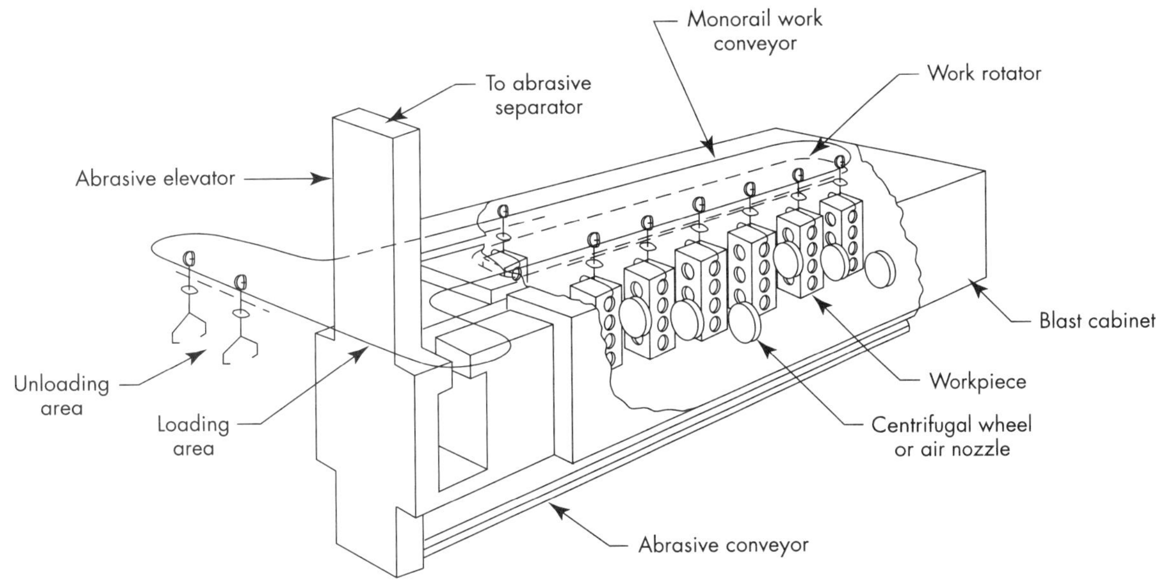

Figure 9-6. Continuous centrifugal blast machine. ***(Courtesy U.S. Filter/Wheelabrator)***

however, some applications, such as removing burrs on aircraft components, that have not been suitable for wheel blasting systems because of the potentially large size of deburring machines and because of part distortion that may result from the centrifugal blast.

Rooms and special-purpose machines. For cleaning large workpieces, such as forgings, fabrications, and railroad cars, rooms are used. These rooms have sufficient wheels to cover the part to be cleaned and material-handling devices to transport the workpieces in and out. Large self-propelled work cars with rotating tops and overhead monorails and cranes are commonly used. Rooms have been designed to be as large as 50 ft^2 (4.7 m^2) and 50 ft (15.2 m) high to accommodate workpieces weighing hundreds of tons.

Media

Dry blasting is a very versatile process because of the many types of media available, both abrasive and nonabrasive; variations in pressure used in delivering media make it even more so. Different media is used for different applications to obtain desired results and production rates, and to ensure economical media life. Properties of some blasting media are shown in Table 9-1.

The choice of media depends on the type of material to be removed, the geometry and condition of the workpieces, required production rates, media cost and wear rate, equipment wear and maintenance costs, and the intended application. Media can affect both the efficiency and the cost of the operation. Media costs can range from 30 to 75% of total cleaning or finishing costs, depending on the results required.

Blasting with shot results in a peened appearance, varying from readily visible indentations when using large shot to a very fine peened finish when using small shot. Smaller-sized media are used in applications when the flash is thinner and requires less impact. Given sufficient impact to remove a particular type of material or to provide the required finish, smaller shot will typically deburr faster, because more pellets per unit weight of shot will hit the surface of the workpiece, thus providing greater coverage. The larger the shot, however, the greater the impact.

Grit is used extensively whenever an etched finish is desired—for example, to obtain a better bond for painting or porcelain enameling. The texture of the etch varies, depending on the size of the grit. A coarse, deep etch is produced by large grit, whereas smaller grit produces a very fine etch. Nonferrous shot is used for softer metals on which grit or steel shot would be damaging. Nonmetallic abrasives are used when it is desirable to remove a minimum of material from the parts or to impart a bright finish.

Generally, air blasting systems employ a wider variety and smaller-sized media than wheel blasting systems.

Cast steel shot. Cast steel shot has a usable life that is longer than any other type of abrasive media and reduces the wear on and maintenance of equipment. Various hardness ranges are available for special peening and cleaning applications. Steel shot and grit are found in approximately three hardness ranges: 50–55 R_c, 55–60 R_c, and 60+ R_c. When a small, hard-to-obtain size, such as S-70 or S-110, is required, a metallic, low hardness grit can be substituted since it will tend to ball up during use and act like shot. Table 9-2 lists some of the sizes for cast shot as established

Table 9-1. Properties of some blasting media

	Glass bead	Ceramic shot	Stainless cut wire	Steel shot	Steel grit	Aluminum oxide	Silicon carbide	Garnet	Crushed glass	Plastic media	Agricultural shell
Finishing	Yes	Yes	Yes	Yes	Yes	Yes	Yes	Yes	Yes	No	No
Cleaning/removal	Yes	Yes	Yes	Yes	Yes	Yes	Yes	Yes	Yes	Yes	Yes
Peening	Yes	Yes	Yes	Yes	No	No	No	No	No	No	No
Surface profiling (etch)	No	No	Yes	No	Yes	Yes	Yes	Yes	Yes	Yes	Yes
Working speed	Medium	Medium	Medium	Medium	Medium -high	High	Very high	High	High	Medium -high	Low-high
Recyclability	High-low	High	High	Very high	Very high	Medium -high	Medium -low	Medium	Medium -low	Medium	Low
Probability of metal removal	Very low	Very low	Very low	Very low	Medium	Medium -high	Medium -high	Medium	Low -medium	Very low	Very low
Hardness, Mohs scale (R_c)	5.5	7 (57–63)	6–7.5 (35–55)	6–7.5 (20–66)	8–9 (40–66)	8–9	9	8	5.5	3–4	1–4.5
Bulk density, lb/ft^3 (kg/m^3)	100 (1,602)	150 (2,403)	280 (4,485)	280 (4,485)	230 (3,684)	125 (2,002)	95 (1,522)	130 (2,082)	100 (1,602)	45–60 (721–961)	40–80 (641–1,282)
Mesh size	30–440	8–46	20–62	8–200	10–325	12–325	36–220	16–325	30–400	12–80	Many
Typical blast pressure, psi (kPa)	20–55 (138–379)	20–90 (138–621)	20–90 (138–621)	20–90 (138–621)	20–90 (138–621)	20–90 (138–621)	20–90 (138–621)	30–80 (207–552)	20–50 (138–345)	20–60 (138–414)	10–40 (69–276)
Shapes: ▲ angular; ● spherical	●	●	●	●	▲	▲	▲	▲	▲	▲ or ●	▲

Notes: This information is intended as a general reference guide. Sodium bicarbonate must be treated with a flow agent. *Do not use sand silica in a hand cabinet.*

Table 9-2. Condensed SAE specification for cast shot

	None on screen		Maximum on screen			Minimum on screen			Maximum through screen		
	Screen			Screen			Screen			Screen	
SAE no.	No.	Opening, in. (mm)	%	No.	Opening, in. (mm)	%	No.	Opening, in. (mm)	%	No.	Opening, in. (mm)
780	7	.1110 (2.819)	—	—	—	85	10	.0787 (1.999)	3	12	.0661 (1.679)
660	8	.0937 (2.380)	—	—	—	85	12	.0661 (1.679)	3	14	.0555 (1.410)
550	10	.0787 (1.999)	—	—	—	85	14	.0555 (1.410)	3	16	.0469 (1.191)
460	10	.0787 (1.999)	5	12	.0661 (1.679)	85	16	.0469 (1.191)	4	18	.0394 (1.001)
390	12	.0661 (1.679)	5	14	.0555 (1.410)	85	18	.0394 (1.001)	4	20	.0331 (0.841)
330	14	.0555 (1.410)	5	16	.0469 (1.191)	85	20	.0331 (0.841)	4	25	.0280 (0.711)
280	16	.0469 (1.191)	5	18	.0394 (1.001)	85	25	.0280 (0.711)	4	30	.0232 (0.589)
230	18	.0394 (1.001)	10	20	.0331 (0.841)	85	30	.0232 (0.589)	3	35	.0197 (0.500)
170	20	.0331 (0.841)	10	25	.0280 (0.711)	85	40	.0165 (0.419)	3	45	.0138 (0.351)
110	30	.0232 (0.589)	10	35	.0197 (0.500)	80	50	.0117 (0.297)	10	80	.0070 (0.178)
70	40	.0165 (0.419)	10	45	.0138 (0.351)	80	80	.0070 (0.178)	10	120	.0049 (0.124)

by the Society of Automotive Engineers (SAE). Solid, spherical shot flattens and dimples the surfaces blasted and makes them shiny. The heavier the shot, the greater its impact on the surface. The general rule is to use the smallest shot necessary to remove the burrs, scale, sand, flash, or paint, within a normal time cycle.

Metallic grit. Cast steel grit is durable and particularly effective in applications where an etched finish is allowed. Grit is angular in shape and is made by crushing round shot. SAE specifications for cast grit are shown in Table 9-3. Grit generally dulls the surfaces blasted and removes metal. For some applications, shot and grit are mixed to produce different finishes. Materials used for metallic grit include copper, bronze, aluminum, and zinc.

Steel abrasives, though often considered suitable only for peening, cleaning, or finishing, are actually highly efficient deburring media when properly selected for the job. They are available in a wide variety of sizes, shapes, hard-

Table 9-3. Condensed SAE specifications for cast grit

	Through screen		Minimum through screen			Maximum through screen		
		Screen			Screen			Screen
Grit no.	No.	Opening, in. (mm)	%	No.	Opening, in. (mm)	%	No.	Opening, in. (mm)
G-10	7	.1110 (2.819)	80	10	.0787 (1.999)	10	12	.0661 (1.679)
G-12	8	.0937 (2.380)	80	12	.0661 (1.679)	10	14	.0555 (1.410)
G-14	10	.0787 (1.999)	80	14	.0555 (1.410)	10	16	.0469 (1.191)
G-16	12	.0661 (1.679)	75	16	.0469 (1.191)	15	18	.0394 (1.001)
G-18	14	.0555 (1.410)	75	18	.0394 (1.001)	15	25	.0280 (0.711)
G-25	16	.0469 (1.191)	70	25	.0280 (0.711)	20	40	.0165 (0.419)
G-40	18	.0394 (1.001)	70	40	.0165 (0.419)	20	50	.0117 (0.297)
G-50	25	.0280 (0.711)	65	50	.0117 (0.297)	25	80	.0070 (0.178)
G-80	40	.0165 (0.419)	65	80	.0070 (0.178)	25	120	.0049 (0.124)
G-120	50	.0117 (0.297)	60	120	.0049 (0.124)	30	200	.0029 (0.074)
G-200	80	.0070 (0.178)	55	200	.0029 (0.074)	35	325	.0017 (0.043)

nesses, and microstructures, including stainless steel. Peening shot is effective for removing burrs from stampings, springs, and hardened parts; and steel grit effectively deburrs brake housings and gray iron castings. A steel abrasive can be selected for almost any metal except for those of a ductile nature, such as nodular iron, or metals having heavy burrs. Ductile burrs bend over instead of breaking off, whereas heavy burrs require large pellets, which can mar or distort the surface of a part (DeLitizia 1981).

Steel abrasives are mainly used on cast iron and steel parts that have been hardened. For hardened burrs, deburring before heat treating has, in some cases, been replaced by deburring while shot peening after heat treatment, thus eliminating one operation.

Cutting abrasives. Cutting abrasives are used extensively for blasting or when fast cutting is a major consideration. Because of rapid wear, however, these abrasives shorten the life of equipment. Also, the overall cost of cutting abrasives may be high in light of their high initial cost and lack of reusability. For example, silica sand is still used in some outdoor blasting applications. It is initially low in cost, but is not very cost-effective because it breaks down into dust after a single use and generally cannot be recycled. It also causes considerable wear on equipment. It is subject to numerous OSHA and EPA regulations; in many areas sand is not permitted because of potential health hazards.

Crushed and graded fused *aluminum oxide* is available in a wide variety of sizes and shapes. It is fast cutting and has good wear resistance, but is comparatively expensive. It is used extensively for cutting operations and produces a desirable surface pattern for subsequent coating operations, but is rarely used with wheel blasting equipment. The physical characteristics (including durability) of aluminum oxide vary from producer to producer.

Silicon carbide is sharper than aluminum oxide and is therefore faster cutting, but larger sizes tend to fracture readily. It is used to cut very hard surfaces, such as tungsten carbide, welded or plated parts, and components that are to be subsequently brazed or welded when aluminum oxide is not suitable.

Nonmetallic media. A wide variety of nonmetallic media are available, including sawdust, crushed nut shells, fruit pits, rice hulls, corncobs, nylon, and plastic and glass beads. Nonmetallic media is used to remove light fins from aluminum castings or from intricate parts, to deflash thermoset plastic parts, and when contamination from metallic media is unacceptable.

Media made from a variety of *agricultural products*, such as crushed nut shells, corncobs, and fruit pits, is used to remove foreign material, such as carbon, oil, grease, and dirt, without etching, scratching, or marring the surface. Possible limitations include the media being dirty or dusty, lack of uniformity from batch to batch, and high usage rates because of minimum recovery. *Shellblast*, which is a soft blasting medium made from crushed walnut shells and fruit pits, is typically used for removing flash from plastic parts and light machining of burrs. It is frequently used on soft metals, such as aluminum, zinc, and magnesium, to prevent heavy etching of the surface. This material is typically inexpensive and readily available in eight sizes. Crushed shells have a severe dusting problem that is controllable by introducing either water or steam into the media mass. Its breakdown rate is 50 times higher than plastic media (De-Litizia 1981).

Media made from plastics with sharp, angular surface particles, such as polycarbonate resins, is a relatively new media for blasting. It is being used to deflash plastic parts and to clean molds, dies, electronic circuit boards, and some static-sensitive devices. There is probably a type of plastic abrasive available for nearly every application of deburring, deflashing, and cleaning, whether parts are plastic, rubber, or metal.

Nylon media, made from cut nylon wire and normally treated throughout its structure with antistatic agents, is used for applications similar to those for which agricultural-type media is used. Nylon is a low-consumption abrasive when compared with shellblast, for example, and eliminates the dust problem associated with media made from agricultural products. Using nylon media avoids the need to wash or dust off parts after blasting. It is available in three standard sizes: .06 × .06 in., .045 × .045 in., and .03 × .03 in. (1.5 × 1.5 mm, 1.14 × 1.14 mm, and 0.7 × 0.7 mm).

A new material, *polyamide nylon*, is cubical shaped to obtain optimum results from the material. This thermoplastic media is used for deflashing appearance parts (such as frying pan handles, hair clipper housings, and toaster end panels). Because it is soft, yet dense, it can deflash appearance parts without leaving marks or oil residue, which are common with walnut media, for example. Both wheel blasting and air blasting units can accommodate this type of media.

Polycarbonate spheres are in common use, as are harder *granulated polyesters*, urea, and *melamine*. Polycarbonate abrasives are similar to nylon in sizes available, consumption rate, and dust-free operation. They are more aggressive than nylon but less aggressive than crushed shells.

Media made from *cylindrical polycarbonate shot* is extruded and cut so that its length and diameter are the same size, forming a square cylinder. It is composed of polycarbonate thermoplastic, which is tough and resilient. Because of the composition of this media, it resists splintering and, during use, rounds off into a sphere. A typical size is .06 in. × .06 in. (1.5 × 1.5 mm). The two most common applications for polycarbonate shot are for deburring machined aluminum components (such as transmission parts, pistons, and gears) and deflashing thermoset plastic parts (such as circuit breakers, industrial plastic parts, and, recently, the "reactor" that goes into automatic transmissions).

Polycarbonate cryogenic grade shot is a variation of cylindrical polycarbonate shot. It is a specially blended plastic alloy, which gives it increased density and longevity when used in cryogenic deflashing machines. This is especially important because the atmosphere in a cryogenic shot blasting unit can be as cold as –350° F (–212° C) when deflashing tough elastomers such as silicone. Other plastic media will prematurely break apart due to the extreme cold caused by the freezing agent, liquid nitrogen (Globuski 1985).

Spherical polystyrene, another thermoplastic media, is a very soft, lightweight media used in deflashing sensitive, fragile parts. Most often this media is used in blast cleaning molded connectors that have gold-plated inserts; it can remove the resin adhering to the gold-plated pins without removing any of the gold plating. It differs from other plastic media because it does not have high impact strength and is not abrasive. In some cases, polystyrene is used in cryogenic shot blasting systems, usually in smaller cryogenic machines and on very small or sensitive molded-rubber parts (such as the rubber housing for hearing aids). In general, however, polystyrene is only used in air blasting machines. It is the smallest plastic media available.

Granulated plastics are available in four grades: granulated polyester, granulated urea, granulated acrylic, and granulated amino resin. Because of their jagged, angular shapes, granulated plastics clean with a unique chipping and shearing action. They also can be used in a wide range of cleaning applications because of the varied densities of each grade.

Granulated polyester is a thermoset plastic material often referred to as electronic grade. It was originally developed as a replacement for dusty walnut shell abrasives. It is most often used for deflashing electronic components (such as integrated circuits [ICs] and resistors). It can remove flash or resin bleed without distorting the plastic package or pitting the leads of components. It is most effective when used in air blasting machines.

Granulated urea, the most popular granulated media, is both harder and more dense than granulated polyester and thus is more aggressive and abrasive. Its popularity may be due to its suitability for use in both air blasting and wheel blasting machines and for a wide variety of parts and applications, including electronic components, thermoset molded parts; removing paint from aircraft; removing light flash from zinc die-cast parts; and cleaning carbon from various aluminum engine parts.

Granulated acrylic media is often used for cleaning aerospace parts.

Granulated amino resin, the hardest and most aggressive grade of the granulated plastics, is the mold cleaning or metal cleaning grade of plastic media. It is most commonly used to remove residue from chrome-plated tool steel molds (molds used for processing rubber and plastic molding compounds).

As chrome-plated molds are used, residues build up from the material being molded or from the material used to release the mold. Molds are commonly cleaned by blasting with a mineral abrasive, such as glass beads or sand; however, mineral abrasives also remove chrome plating and alter critical tolerances. Granulated amino resin, in contrast, has the ability to remove mold residues without damaging chrome plating or eroding tolerances. Granulated amino resins work in this fashion because their hardness is significantly less than tool steel and chrome, whereas mineral abrasives are roughly twice as dense and significantly harder.

Aluminum granal contains 98% pure aluminum spheroids used for deburring aluminum and magnesium parts. It is available in four sizes: small (.0017–.0331 in. [0.043–0.841 mm]), medium (.0232–.0469 in. [0.589–1.191 mm]), large (.0394–.0661 in. [1.001–1.679 mm]), and extra-large (.0469–.0937 in. [1.191–2.380 mm]).

Glass beads are a nonabrasive media; they are clear, colorless, transparent spheres generally made from a soda-lime-based, high-silicon glass. They are available in diameters ranging from a maximum of .187 in. (4.75 mm) to .0015 in. (0.038 mm) or less. Glass beads are used when ferrous contamination of the workpieces must be held to a minimum, when workpiece dimensions and tolerances must not be changed, when the small sizes available can reach areas inaccessible to other media, and when special finishes (such as bright matte to dull satin) are desirable. Four different national specifications define glass beads and their use: SAE-J1173, AMS2430J (aerospace material specification), MIL-S-13165B, and MIL-G-9954A (military specification).

Crushed glass is available in sizes similar to glass beads and has similar characteristics. However, it is more abrasive than beads because it is angular rather than spherical. It breaks down more slowly than beads because fracture lines and occlusions are obliterated in the crushing process.

Carbon dioxide pellets are one of the most unusual medias available. When the carbon dioxide pellet hits a part, it is hard enough to break flash and scrub surfaces. Since it is merely frozen gas, it evaporates once heated, so no new materials are added to the part to contaminate it or cause waste steam problems. This medium is not widely used and appears to have left the market.

All media used for blasting impregnates or transfers blasting materials to surfaces. For many applications, this is not a concern, but for others it is. Glass beads and metal shot may leave less residue than other materials, but laboratory analysis reveals that all materials leave some residue. Aluminum oxide impregnated in metals can cause poor adhesion of subsequent electroplating, holes in welds, and the inability to solder. Aluminum oxide is used as a mask to prevent solder flow.

Figure 9-7 presents comparative stock removal rates for various media.

Media mixtures. Producing uniform surfaces depends on keeping a uniform working mix of media in the machine at all times. To ensure media effectiveness, some sort of systematic procedure must be set up for replacing worn out media. If replacement is not regular, the processing unit will loose blasting capability. A screen analysis of the working mix can provide a broader range of sieve sizes than is used for as-purchased media.

A practical method of maintaining a fairly constant working mix is to add new media periodically, usually at least once every eight hours, to keep the media level in the supply

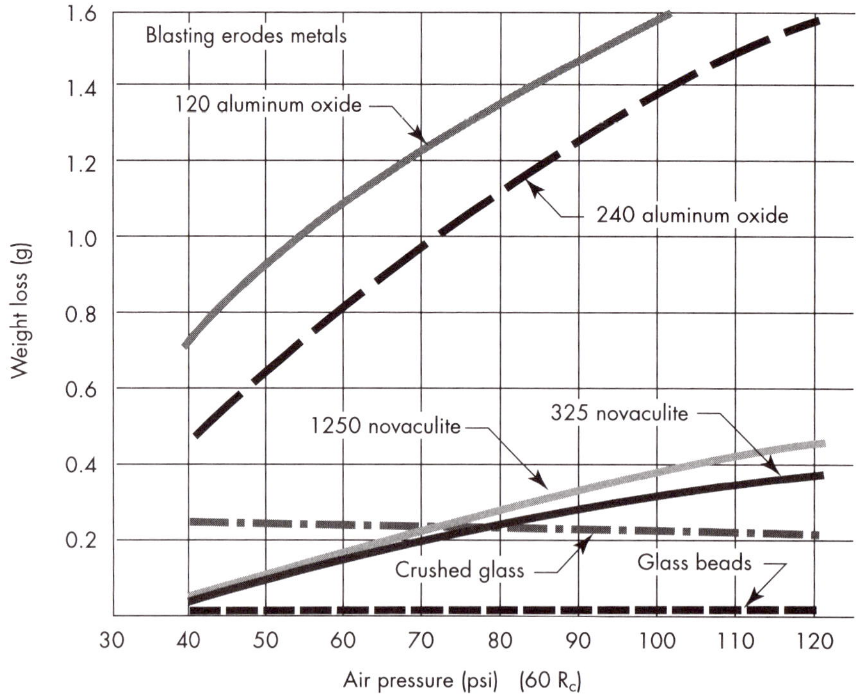

Figure 9-7. Comparison of stock removal rates for various media. Note that the curve for glass beads lies along the horizontal axis, indicating almost no loss of part weight (Woelfel 1983).

tank, bin, or hopper at a uniform level. If the required surface finish of the workpiece is critical, new media particles may have to be added as frequently as every hour. The overall amount of media required to fill the machine to operating capacity can also affect the frequency of additions. A machine with large excess capacity for storing media can run longer between additions because the quality of the working mix changes more slowly.

Media recycling systems. Reclaiming and recycling the media are important steps in controlling blasting costs. It is generally desirable to remove fines and contaminants from the media before reuse; and for some applications, it is necessary to reclassify the media so that only certain sizes are reused. Devices used for separation can be classified into the following three basic categories: air-wash separators, centrifugal systems, and screening devices. Magnetic separators are often used to remove magnetic particles from nonmagnetic contaminants.

Air-wash separators are efficient and are the most commonly used system in wheel blasting equipment in which large volumes of media are necessary. They generally create a curtain of media by pouring it down an inclined plane. Air is drawn through the curtain to deflect lighter pieces, and skimmer plates trap these particles. The size range of the working mix can be changed by adjusting the skimmer plates and the blast gates controlling the volume of air drawn through the curtain.

Centrifugal separators, often called cyclone separators, are commonly used in air blasting machines in which smaller volumes and lighter media are used. The separators combine the air-wash principle of removing fines and contaminants with the centrifugal principle of throwing heavier pieces from an air stream.

Media return systems vary in design, depending on the requirements of the blasting system with which they are used. For most wheel blasting machines, media is collected from the bottoms of the machines by one or more screw conveyors that carry the media to the sides of the machines; here, vertical bucket conveyors carry the media to the top of the separator. Stationary, rotating, or vibrating screens are usually provided along the media transfer path to remove larger contaminants or workpieces before the media passes through the air-wash separator and returns to a storage hopper. Air blasting machines generally use vacuum tubes to return the media to the separator, usually through stationary screens. The discharge end of the return vacuum line is generally an integral part of the centrifugal separator, which is also operated in a vacuum.

A *dust collector* of some type is generally used to remove dust from the separator air before it is discharged into the atmosphere or recirculated in the system.

Process Limitations

Like all deburring processes, wheel blasting has limitations. Because it is an impact process, it is not well suited to either ductile iron or heavy steel burrs. Ductile burrs bend over instead of breaking off, and heavy burrs require large pellets that can mar the surfaces of parts. As a result of blasting, all parts may have a uniform matte finish, the extent depending on the media size, impact force, and cycle length. Machined surfaces must be protected by masking to avoid etching. Parts must be tumbled, shaken, or in some way articulated to remove all abrasive media. Otherwise, costly loss of abrasive may occur. Burrs located in complex, deep internal passages may not be removed by the blast.

9.2 WET BLASTING

Wet blasting, sometimes called vapor blasting, vapor honing, or hydraulic blasting, is a process using fine media particles in slurry form. The process does not require a dust collector or ventilation equipment and, with the use of rust-inhibiting compounds, prevents immediate oxidation. Finer media with particles as small as .0001 in. (0.003 mm) can be used. Wet blasting is generally considered a precision finishing operation. It can be controlled to produce or avoid metal removal and permits maintaining dimensional tolerances to within .0001 in. (0.003 mm).

In wet blasting, media is mixed with a liquid, usually water, and then pumped to a nozzle as a slurry. At the nozzle, the slurry is introduced into an air stream and propelled against the workpiece. Additives, such as rust inhibitors, wetting agents, and anticlogging or antisettling agents, if required, may be included in the slurry. Petroleum distillates can be used in place of water; however, specially designed features are necessary in such installations.

Media particles are kept in suspension by pumping, air agitation, or mechanical means. The pumping principle has several advantages, including reducing consumption of compressed air, recirculating the slurry to keep the particles in uniform suspension, and feeding the slurry at a positive pressure into the blast gun. In machines using air or mechanical agitation, the slurry is brought to the blast gun by suction created by jet action. Unlike wheel blasting, wet blasting is ordinarily limited to use with compressed air because of the great differences in specific gravity between the fluid and the media.

Wet blasting provides a good surface for paint adhesion and is suitable for use in tool rooms and tool grinding departments. A slurry of very fine particles is used to remove feather burrs from multitooth cutting tools, such as milling cutters, hobs, broaches, and taps, thus eliminating hours of hand-honing time. Wet blasting systems set at precise angles are used to sharpen metal files. Wet blasting has also been used to deburr intersecting hole junctions in precision parts that are inaccessible by any other method. The use of glass beads for close-tolerance peening of relatively thin-walled parts has become a successful application of the wet blasting process. Other advantages of wet blasting include the ability to blast wet parts, remove oily or wet contaminants, and provide the beneficial cushioning or buffering effect of the liquid used, resulting in less wear of the equipment and longer media life. Wet systems eliminate the problem of fine dust, which poses health hazards and is abrasive to equipment surrounding the blasting cabinets. The disadvantage of wet blasting is that media reclamation is difficult.

To prevent contamination of recirculated slurry, parts may need to be precleaned; however, the decision to do this must be made on a case-by-case basis. Degreasing is recommended when contaminants, such as grease, protective coatings, and excessive and heavy oils, are present. Rust thick enough to be scraped off should be removed. In some cases, soaking in a caustic cleaner is helpful for such items as forging or forming dies and glass molds from which burnt and hardened residues of lubricating materials are to be removed. If machined parts have a light layer of oxidation or heat treatment scale, the contaminated slurry can be either discarded or, in more complex machines, separated and filtered.

The following list contains some of the operational variables involved in wet blasting:

- Metallurgy of processed parts.
- Desired work function and production rates.
- Accessibility and wedging characteristics.
- Blast nozzle size, type, and distance from part.
- Angle of attack.
- Abrasive type and size.
- Velocity of blast slurry.
- Density of blast slurry.
- Time of exposure.
- Initial and terminal surface finish specifications.

Figure 9-8 illustrates the relative effects of some of the variables.

9.2.1 Equipment

Many types and sizes of wet blasting equipment are available. Machines do not require special rooms and can be placed in production lines. Whether a machine is a simple hand-operated cabinet, just large enough to handle the product, or a complicated, multigun automated machine, a properly designed unit from an experienced manufacturer is essential. The least-expensive unit may be the most expensive in the long run if it does not produce the desired results.

Because of the specific gravity of even the finest media, media must be kept in suspension in a hopper, which is usually accomplished by air agitation, mechanical movement, or pump recirculation.

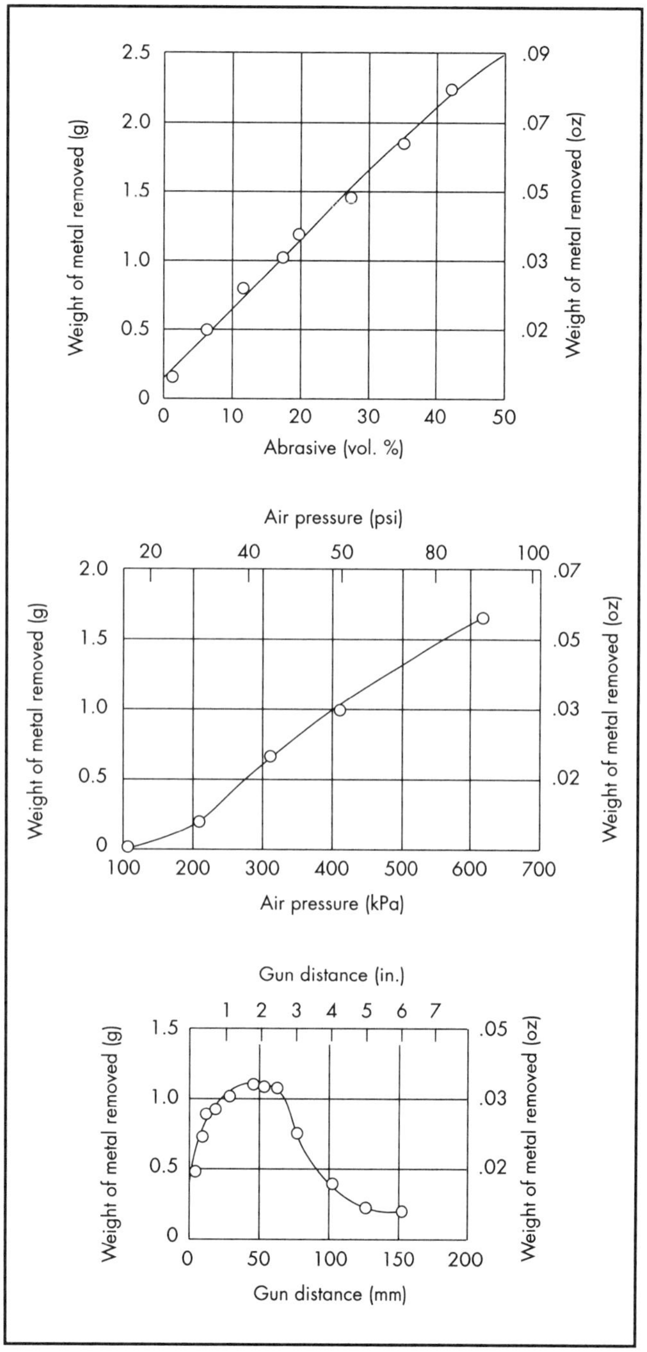

Figure 9-8. Effect of variables on the amount of metal removed during wet blasting of 1010 steel. ***(Courtesy U.S. Filter/Wheelabrator)***

Air Agitation of Media

Air agitation machines are initially less expensive than pump recirculation and eliminate the need for pump maintenance; however, nozzles that are immersed in the slurry will have to be maintained. In air agitation machines, compressed air stirs the media particles against one another with a high velocity, causing particles to wear and the hopper to erode, necessitating the use of heavy-gage steel for the hopper. This increases machine costs. Additionally, the blast gun has the advantage of serving the dual purpose of aspirating the slurry into the gun and blasting the slurry against the object to be processed.

Mechanical Movement of Media

Mechanical movement to keep media particles in suspension presents several difficulties. The moving mechanism must be powerful enough to move the total volume of slurry with sufficient force. Bearing sealing problems arise, paddles and propellers are eroded, and the blast gun must do double duty (as in air agitation).

Pump Recirculation of Media

Pump recirculation of media reduces consumption of compressed air, recirculates the slurry to keep the particles in uniform suspension, and feeds the slurry at a positive pressure into the blasting gun.

Disadvantages of pump recirculation include higher initial cost and increased maintenance. The basic requirements for effectiveness of this process are a properly designed pump with a lifetime of 1,000–4,000 working hours for the replaceable parts. The lifetime range depends on mesh sizes of the abrasives. The pump is a relatively low-maintenance item. To avoid costly maintenance, a close-tolerance direct pressure pump should be used for only the very finest meshes.

A low-velocity pump of suitable volume will not only furnish the blast gun with a positive delivery of slurry at low pressure, but also keep all particles in uniform suspension, thus preventing erosion of the hopper and excessive use of compressed air. The blasting results are uniform, and the active life of the media is increased. The form of the hopper is important to the suspension of the media and affects uniform slurry delivery in the blast gun.

Note that about 4 ft^3 (0.11 m^3) of solution at 100 psi (6.89 kPa) gage pressure are delivered per horsepower. Also, compressed air running through the inhibiting compound will reduce the pH of the solution, which may increase problems associated with corrosion.

Hopper Machines

Hopper machines are used in conjunction with a pressurized tank. The slurry is preferably pumped into the tank and, by the proper application of valving, is then pressurized. In cases when it is necessary to deliver the slurry at high pressure and velocity, more fine particles are expelled, resulting in intense blasting of confined areas, such as internal intersecting holes and similar configurations. The slurry must be correctly suspended in the feeding hopper and the pressurized slurry tank.

Tumbling-blasting Devices

Tumbling-blasting devices consist of a tumbling barrel mounted on a wheeled cart that moves on a railed stand. This device may be moved into a cabinet. The barrel is a self-contained unit with a driving mechanism that turns the barrel and oscillates the blast gun, or guns, inside the barrel. Commercial sizes range from 12–26 in. (305–660 mm) in diameter and are perforated to allow the slurry to escape and recirculate. Up to 50 lb (23 kg) of parts can be simultaneously tumbled and wet blasted.

Semiautomatic Machines

Semiautomatic wet blasting machines are available for deburring drilled, flat printed circuit boards or similar flat parts of various dimensions, compositions, and production requirements. The lateral movements of the machine are screw operated, with built-in reciprocal motion and adjustable speeds. Blast guns are vertically oscillated, with variable speeds and strokes. For higher and continuous production, conveyorized machines are used. Unitized wet blasting machines combining washing, rinsing, and drying with fluxing and soldering operations are not uncommon.

Turntable Machines

Indexing-type turntable machines are suitable for a variety of applications. A number of blasting stations may perform internal or external deburring. A separating system and a closed-circuit filtering system permit the parts to be rinsed and blown off to a high degree of cleanliness before removal from the holding fixture, thus eliminating the constant flow of rinse water and the wasteful carryout of media particles.

Nozzles

The form, shape, length, and diameter of nozzles for wet blasting vary depending on the application and size of the parts to be blasted, and the type of media used. Some commonly used types are cylindrical, special, and internal nozzles.

Cylindrical nozzles. The most common nozzle form is cylindrical and is about 1 in. (25 mm) long, with about a .5 in. (13 mm) inside diameter and roughly a .75 in. (19 mm) outlet diameter made of inexpensive mild steel. Such nozzles are used on hand-operated cabinets for most general cleaning purposes. The lifetime of these nozzles is about 40 working hours when using silicate or quartz abrasives in the 100–300 mesh range and blasting pressures from 80–90 psi (552–621 kPa).

Changing nozzles in hand-operated cabinets normally presents no problem and takes only 2–3 min. When fast-cutting abrasives, such as 80–150 mesh aluminum oxide are used, the lifetime of the nozzle may be as little as 20 working hours. However, short nozzle life does not justify using more expensive, longer lasting nozzles in a hand-operated machine, unless, of course, maintenance personnel must be called in frequently to change them. Ceramic nozzles, either plain or jacketed, are also used; although they have a longer life, they are considerably more expensive than other types of nozzles.

Special nozzles. Special nozzles, such as fan-shaped types, are usually made from good-quality cast iron, such as meehanite chilled to a high grade of hardness, or from rubber.

In mechanized machines for which surface preparation beyond cleaning is required, as in preparing aluminum for anodizing or surface conditioning for phosphatizing, higher grade nozzles made of carbides or similar alloys must be used. Cylindrical nozzles of this kind are much more expensive but may require less maintenance. Careful consideration should be given to justify their cost.

The lifetime of cylindrical carbide nozzles can be several thousand hours when using fine abrasives. After over 1,000 hours of use with 140 mesh quartz, no measurable wear can be detected. For automatically cycled machines and complex configurations, the carbide nozzle is more economical. For intermittent hand use, a good grade of steel or case-hardened nozzle is satisfactory. Inside diameters of 1/8, 3/16, and 1/4 in. (3.2, 4.8, and 6.4 mm) are not uncommon. For extremely small inside diameters, industrial diamonds are used.

Internal nozzles. Many types of internal blast nozzles are being used. The shapes and principles of nozzles having the greatest effectiveness are proprietary. Diameters and lengths are tailored to suit the customer's requirements. Nozzles of up to 30 ft (9.1 m) in length have been developed.

9.2.2 Media

Any natural abrasive in screened or ground form as well as any manufactured abrasive may be used in wet blasting. In some cases, ground fruit pits and even sodium bicarbonate compounds are used, but most agricultural media cannot be used for wet blasting. There are several forms of silicon dioxide with different physical characteristics that are widely used, such as silica sand, ground silica sand, ground quartz, and novaculite. Table 9-4 presents characteristics of and applications for some of the more popular media.

Novaculite

Novaculite compounds, ranging in size from 100 mesh to a theoretical 5,000 mesh, are very pure, screened to particle-size combinations, and have excellent cutting qualities. These compounds are successfully used for finishing, removing corrosion, and preparing surfaces for phenolic coatings, painting, and plating. They are also used for defuzzing cutting tools after grinding, blending grinding marks on tools, and magnetic-particle or fluorescent penetrant inspection. When suitable antirust compounds are added to the slurry, novaculite provides an unusual corrosion resistance.

Silica and Quartz

Silica, in mesh sizes of 50, 80, and 140 (U.S. sieve), is used for general cleaning purposes. A 400-mesh silica is used as a carrier medium for other selected abrasives. Quartz, ground and graded to 60-, 80-, 100-, and 140-mesh sizes, is a more

Table 9-4. Characteristics and applications of various wet blasting media

Media	Characteristics and applications
40–80 mesh silica	Deburring steel and cast iron. Removal of oxides from all forms of steel. Fast cutting and will remove metal. No close tolerances can be observed.
50 mesh silica	Deburring steel and cast iron; removal of oxides from all forms of steel. Fast cutting and will remove metal. No close tolerances can be observed.
80 mesh silica	Deburring steel and cast iron; roughing of surfaces; plastics bonding or rough plating; peening action. Fast cutting and will remove metal. No close tolerances can be observed.
80 mesh ground silica	Heavy burrs; light or medium scale; bad rust conditions. Use on nickel alloy steels also. Exceptionally fast cutting and will remove metal. No tolerances can be observed.
100 mesh novaculite	Cleaning of carbon from pistons and valve heads; deburring of brass, bronze, and copper. Also can be used on crankshafts. Fast cutting and will remove metal. No close tolerances can be observed.
100 mesh ground quartz	Blending in preliminary grind lines on steel, brass, and die castings. Removal of medium-hard carbon deposits. For small radius requirements. Fast cutting and will remove metal.
140 mesh silica	Removal of small burrs on steel, copper, aluminum, and die castings. Can be used on rough cleaning of dies and tools. No close tolerances can be observed. Will remove metal.
325 mesh aluminum oxide	Second stage for cleaning aluminum pistons, impellers, crankshafts, and valves. First stage in cleaning master rods and all glass. Will follow close tolerances to .0025 in. (0.064 mm). Cuts slowly.
400 mesh aluminum oxide	Excellent abrasive agent when working under oil contamination conditions. Fast cutting agent on stainless steel, zinc, and aluminum die castings. Tolerances can be held.
1,250 mesh novaculite	Second stage for crankshafts, rods, pistons, impellers, valves, gears, and bearings. For polishing all metals; also dies, tools, and die castings. Tolerances can be held.
5,000 mesh novaculite	Use on any parts where an extra-fine surface is needed.
Glass beads from 40–325 mesh and finer	Use for removal of light heat-treated scale and/or discoloration; also light oxide removal for jet engines and electronic components. No metal removal.

aggressive material than silica and has a greater resistance to breakdown. Quartz is used for blending existing grinding and machining lines on steel and brass parts, die castings, and removing semihard carbon deposits. Quartz is fast cutting and will remove metal.

Aluminum Oxide

Aluminum oxide is a manufactured abrasive that is well known for its fast cutting action and good resistance to breakdown. It is accurately graded, chemically pure, and does not contain any iron oxide or silicon dioxide, which is important in certain industries. It is available in mesh sizes ranging from 24 to 800. Ground natural ore is not as pure as manufactured aluminum oxide; it contains some iron oxide, compares favorably with aluminum oxide, and is widely used.

Glass Beads

Glass beads are available under various names and descriptions and range in size from 40 to 3,000 mesh and even finer. Some glass beads are carefully graded; others are a combination of mesh sizes. Beads are widely used to achieve a peening quality finish. The particles are globular and create hemispherical depressions rather than sharp, angular cuts. No metal is removed, and the resulting finish gives a bright, almost polished appearance.

9.2.3 Compounds

The use of water for abrasive slurries requires that rust and corrosion inhibitors, foam controllers, adequate wetting, rinse aids, and biocides be taken into consideration when wet blasting (Gillespie 1976, 1981).

Rust inhibitors are incorporated for processing ferrous parts and/or when using ferrous abrasives.

Corrosion inhibitors are used when processing nonferrous metals and/or mixed metals to avoid galvanic reactions or other undesirable reactions between the base metals and the water-media mix.

Foam controllers are used to avoid excessive foaming that may result in loss of pressure because of pump cavitation or loss of abrasive from the abrasive hopper because of froth flotation.

Adequate wetting of the abrasive, as well as the processed parts, will ensure the formation of a hydrophilic surface necessary for producing clean parts, keeping the machine clean, and avoiding loss of abrasive. Wetting can be provided by using suitable alkaline materials capable of reducing the surface tension to approximately 35 dyn/cm^2 or less. Using oil-based emulsifiable products will invariably lead to the formation of hydrophobic films, resulting in dirty equipment, dirty parts, and loss of abrasive. Using an additive with satisfactory wetting capability will also promote more ready removal of the abrasive in the flushing cycle as well as rapid drainage in the drying cycle.

Rinse aids facilitate the clean rinsing of parts.

Biocides control bacterial growth in wet systems. These compounds control smells and prevent some skin irritations.

9.2.4 Process Limitations

The following factors must be taken into account when considering wet blasting for a particular application:

- Wet blasting is not recommended for heavy work or removing heavy burrs adjacent to smooth finished areas or machined configurations on which close fits or tolerances must be held.
- It is not considered a high-production process because it requires more time than dry blasting.
- It is not as clean as dry blasting and precautions must be taken to avoid oxidation of the workpieces.
- Wet media may pack into cavities, holes, and crevices, causing removal problems. In some instances, cleaning workpieces with deep cavities is nearly impossible.
- The introduction of water when processing workpieces sets up oxidation almost immediately, causing problems in subsequent plating, coating, and other finishing operations.
- Because of the cushioning effect of water, cleaning rates are slow. Additional rinsing operations, drying to eliminate water spotting, additives or wetting agents, antibacterial compounds, and corrosive inhibitors represent additional costs and potential problem areas.
- Materials adversely affected by exposure to water include reactive metals and materials subject to moisture absorption or porosity.
- Extremely large parts may require an unduly large machine and render the process economically unattractive. Small parts that cannot be held stationary present handling problems. Extremely thin parts may be subject to distortion.
- Abrasion-resistant materials may require excessive exposure times to meet production requirements.
- Materials of high malleability may be subject to excessive plastic deformation.
- Brittle materials (unless suitably masked), such as ceramics, may be subject to unacceptable attrition.
- Areas to be finished may be inaccessible to the flow of the abrasive slurry.
- Abrasives may wedge because of part geometry. (This problem often can be overcome through a change in part design.)
- Areas of extremely low microfinish generally require shielding or parallel flow of the blast stream to avoid an increase in microfinish.
- Extremely close dimensional tolerances may be difficult to hold. (Quite often this problem can be overcome by closely controlling the angle of attack.)
- Installation costs may be higher than in dry blasting operations.
- Arrangements must be made to periodically flush residue out of the machine for disposal and to meet EPA requirements.
- Maintenance and replacement of parts (such as internal piping, valves, filters, pumps, and seals) on a slurry system are important because corrosion is a constant problem.
- Media consumption is high because of the inability of wet blasting systems to salvage good material from debris. As contaminants from workpieces being cleaned fall into the slurry and as media breaks down, the mixture becomes diluted; fresh media must be added regularly to maintain a high proportion of usable media.

Wet blasting processes pose additional problems for machine operators, such as a constant stream of water that flows over the viewing window, having their clothing and feet constantly in a wet environment, and inhaling wet air.

Despite the preceding limitations, wet blasting is a suitable finishing process for a wide variety of applications.

9.3 MICROBLASTING

Microblasting, also called abrasive-jet machining (AJM), is a specialized form of blasting used for removing material from workpieces. A high-speed stream of media particles is propelled through a nozzle by a gas, usually air. This process is used for cutting, deburring, and cleaning hard, brittle materials, such as germanium, silicon, mica, glass, ceramics, titanium, and tantalum without heating or cracking. Material is removed by the impingement of fine media particles, usually about .001 in. (0.03 mm) in diameter, entrained in a high-velocity, 500–1,000 ft/s (152–305 m/s) gas stream. By varying the diameter and configuration of the nozzle and the standoff distance (see Figure 9-9), the process can obtain a variety of results. Nozzle-tip diameters generally range from .003–.032 in. (0.08–0.81 mm). Rectangular nozzles are also used. Nozzles must be resistant to abrasive wear and are usually made from tungsten carbide or synthetic sapphire.

Because most parts are hand-held in the workstation (see Figure 9-10), changing from one part to another is possible with minimal setup time—for example, changing from cleaning a thread to deburring a cross hole may only require changing the nozzle. Also, it is possible to set up a complete blasting system on a cart when portability is required.

There is little waste of abrasives in microblasting operations. The tight stream of abrasives precisely directs particles to the work surface at localized areas as small as

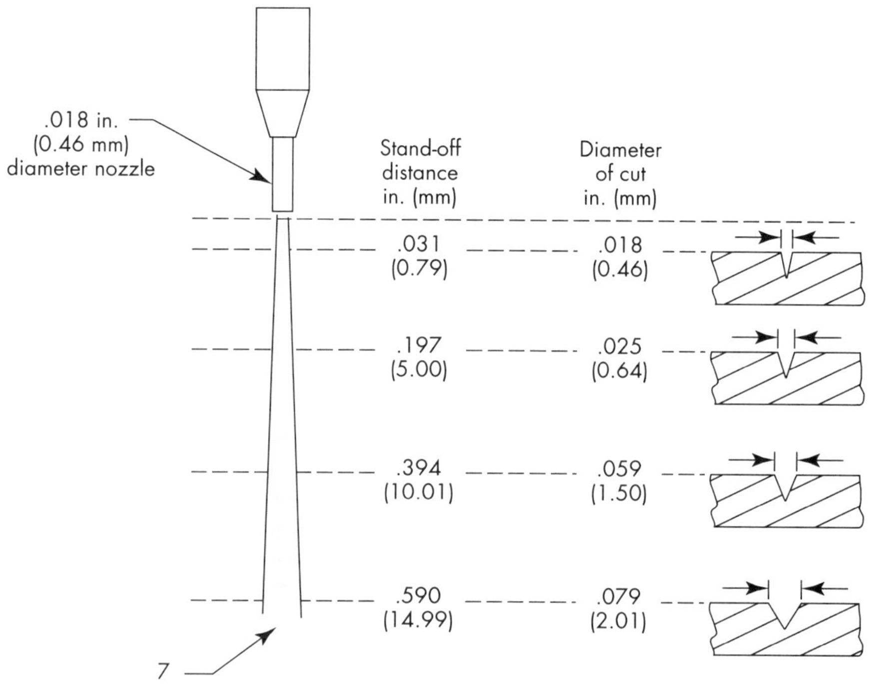

Figure 9-9. Cutting action in microabrasive blasting using various standoff distances. ***(Courtesy S.S. White Industrial Products)***

.010 in. (0.25 mm) in diameter without eroding adjacent areas. If considerable time or repeated applications are required to properly deburr a part, abrasive blasting may reduce the number and difficulty of these operations.

Microblasting tends to be most effective when cutting hard, brittle materials. Table 9-5 shows the relative cutting action on various materials, with higher index numbers indicating greater efficiency. In Table 9-5, note that the drilling efficiency of cold-rolled steel is 20 times that of neoprene. Because of this property, neoprene and other elastomers, such as polyurethane rubber, can be used as masks to confine the abrasive stream to the desired areas of the workpiece.

Variables influencing material removal rate, geometry of cut, surface roughness, and nozzle wear include the media (composition, strength, size, shape, and flow rate), gas (composition, pressure, and velocity), and nozzle (geometry, composition, and distance and inclination to the work surface).

9.3.1 Applications

Microblasting is used for numerous applications, most involving small or miniature parts, electronic components, and unusual plastic parts. Sodium bicarbonate, for example, though very soft, has been used at 40–80 psi (276–552 kPa) and .006 in. (152 µm) in diameter to deburr feather edge burrs from Celcon®, a high-quality Delrin®, and from titanium-ground needles (Weidner 1985). Cutting tools are blasted after grinding, and small slots and holes are deburred in hard-to-reach areas on stainless steels, brass, and a variety of other materials. In general, microblasting is suitable for finger-sized or smaller parts. For example, microblasting is used for deburring hypodermic needles without dulling the cutting edges. This is accomplished by using glass bead abrasive powder at approximately 175 psi (1,207 kPa) and a nozzle diameter of .018 in. (0.46 mm) on

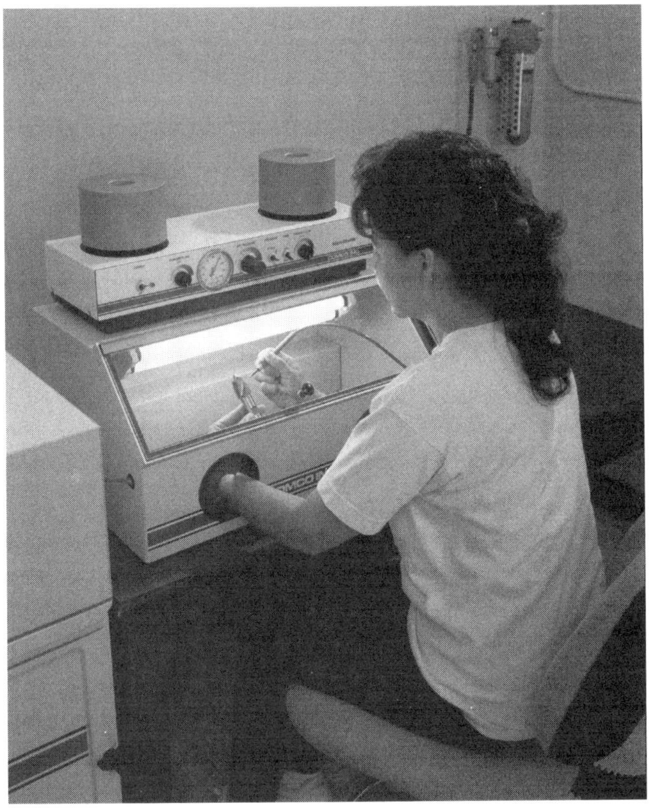

Figure 9-10. Operator using microabrasive blasting equipment. ***(Courtesy Comco)***

Table 9-5. Relative cutting action of various materials

Material	Cutting index
Glass	0.6
Aluminum alloy 2024T4	1.6
Stainless steel AISI304	0.9
Cold rolled steel	1.0
Ferrite	3.0
Neoprene	0.05
Al_2O_3	0.4

Note: Data were developed by measuring the time to drill a .019 in. (0.483 mm) hole through .025 in. (0.635 mm) stock using a .018 in. (0.457 mm) nozzle and .001 in. (25.4 μm) Al_2O_3 powder.

a .028 in. (0.71 mm) diameter opening. Media size for this application is .001 in. (25 μm). Valves are deburred with this process using aluminum oxide powder at approximately 75 psi (517 kPa).

Although microblasting is most effective on hard, brittle materials, it also can be used on other materials. For example, glass bead abrasives have been used to deflash compression-molded plastic parts—the 1-in. long × .003-in. thick (25.4 × 0.08 mm) flash was removed in approximately 1 s.

9.3.2 Media

Aluminum oxide is the most widely used media for microblasting, but silicon carbide, glass beads, and other materials are used for special applications. Media powder is generally fed from a vibrating mixing chamber into an orifice chamber where it is entrained in the gas stream and then passed through a hose to the nozzle. Good results have been obtained with media particles varying in diameter from .0006–.0016 in. (0.015–0.041 mm) and with flow rates of .07–.71 oz/min (2–20 g/min). Finer powders are used for cleaning and polishing. Increasing the flow rate tends to increase the material removal rate, but increasing the proportion of media in the jet lowers the stream velocity and tends to lower the material removal rate. Reusing the media powder is not recommended, since cutting action will be decreased and contamination may clog the orifices.

Air, nitrogen, and carbon dioxide at pressures of 30–120 psi (207–827 kPa) have been used as carrier gases. Commercial gases are generally sufficiently pure, but air should be filtered. Higher pressures increase material removal rates but also increase nozzle wear. A typical material removal rate is about .001 $in.^3$/min (0.02 cm^3/min). Masking is sometimes used to confine the cutting action. Surface finishes produced generally range from 6 to 50 μin. (0.15–1.27 μm). A dust-collection system is essential to protect operators, equipment, and workpieces.

The rate of media powder usage will vary depending on cutting pressure, abrasive type, duty cycle, and nozzle size. For example, 1 lb (0.45 kg) of 984 μin. (25 μm) aluminum oxide passing through a .025 in. (0.64 mm) tank orifice and a .030 in. (0.76 mm) nozzle at 125 psi (862 kPa) will last approximately 1 h.

The following abrasive powders are the most common commercially available media for microblasting. They are considered inert dusts by the government and by industrial hygienists. Other powders may cause operator health hazards. The cutting rate of various media in microblasting is a function of particle size and type of powder used. For the present, other parameters are neglected.

For fine cutting, 984-μin. (25-μm) size powders are a good choice. For coarser and faster cutting, 1,969 μin. (50 μm), or 220-grit (4,724–5,118 μin. [120–130 μm]) powders should be selected. Table 9-6 shows the abrasive powders that are available from one company.

Glass Beads

Glass beads are used for very light deburring, peening, and imparting a satin polish. They do practically no cutting and can only remove loose surface particles such as oxides.

Sodium Bicarbonate

Sodium bicarbonate is used for cleaning and cutting soft materials and removing coatings without damaging the underlying material. Particles are sharp but soft and water soluble, and thus easily cleaned from delicate parts.

Table 9-6. Microblasting abrasive powders available from one company

Powder	Description	Type	Application	Finish produced, µin. (µm)
No. 1	Aluminum oxide 1,063 µin. (27 µm)	Cutting	General purpose, deburring, marking, scribing	14–20 (0.36–0.51)
No. 3	Aluminum oxide 1,969 µin. (50 µm)	Cutting	Same as No. 1, but faster cutting action due to larger particle size	38–50 (0.97–1.27)
No. 4	Sodium bicarbonate	Cleaning	Electronics industry (potentiometers), wire stripping	data not available
No. 5	Aluminum oxide 394 µin. (10 µm)	Cutting	Precision cutting, polishing microdeburring	6–8 (0.15–0.20)
No. 8	Silicon carbide 1,969 µin. (50 µm)	Cutting	Heavy-duty cutting on extra-hard materials	data not available
No. 9	Glass beads	Cleaning	Cleaning, polishing; removes dirt and scale without removing parent material; hypodermic needle deburring	data not available
No. 10	Crushed glass	Cleaning	Similar to No. 9, but more abrasive	data not available

Aluminum Oxide

Aluminum oxide is used for cleaning, cutting, and deburring. It is an ideal abrasive when uniform surface finishing or etching is required. This is the most commonly used abrasive and is available in several sizes.

Silicon Carbide

Silicon carbide is used for heavy deburring and fast cutting or drilling. It is perhaps the fastest cutting of the readily available commercial abrasives. It is relatively inexpensive and very easy to use. It does not absorb moisture and flows very freely.

9.3.3 Process Limitations

In most cases, microblasting requires the constant attention of an operator; generally one operator controls one machine. Thus, the output of the abrasive station is limited by the operator's ability and handling methods. Five to ten seconds per workpiece should be considered the practical lower limit on production rate. An exception to this rule is when part volume and burr location justify using automatic handling equipment, as is the case when finishing hypodermic needles.

Microblasting is a low-energy form of grinding. Therefore, it is not a practical finishing process when large amounts of material must be removed (Weightman 1995). Abrasive particle size is limited to less than .004 in. (102 µm), and thus the abrasive action is limited. Reusing or cleaning dry, spent abrasives is not practical or economical.

Large parts or surface areas will normally be processed much more rapidly by other abrasive methods. This is because the width and volume of the stream is limited by the small size of the nozzle. However, sometimes a fine, uniform stream of abrasive is essential even when finishing large surface areas, such as when cleaning thin sheets of tantalum.

Heavy burrs do not respond readily to microblasting. When heavy burrs are present, it is more cost-effective to first deburr using a nonwoven, abrasive-filled nylon wheel, belt, or other tool to remove the bulk of the burrs. The microblaster is then used for the finishing operation.

The cost of microblasting must be weighed against the relative value of the part to be finished. If the value of the part is small, microblasting may not be cost-effective. As a general rule, if other methods of cleaning, cutting, polishing, or deburring are available and will produce an acceptable finish at a reasonable cost, they should be considered first. Microblasting is best used in problem areas when no other process will produce acceptable results.

9.4 CRYOGENIC BLASTING

Cryogenic blasting involves slowly tumbling parts on a continuous mill belt or in a barrel and simultaneously blasting parts with media propelled by a throwing wheel (wheel blast). The media is then separated from the flash and recycled for continuous use (Scott 1985). Cryogenic blasting employs some of the same processes described in Chapter 6 for cryogenic mass finishing. Cryogenically blasting (freezing) plastic or thin metal flash renders the material brittle and more susceptible to impact removal. This process is widely used in the rubber and plastic component industry. Note, however, that not all materials react to low temperatures without cracking or crazing, so care must be given to understanding part and material properties. The variables are the same in this process as for dry blasting, except for part material properties.

Cryogenic blasting machines are used to deflash thermoset rubber and plastic parts and to deburr die castings. For use on rubber parts, cryogenic blasting machines are modified to inject liquid nitrogen into the blasting chamber for flash embrittlement. Initially, the entire machine may be enclosed in a "cold room" designed to help insulate the machine and control liquid nitrogen usage. Typically, the media used is fine-mesh steel shot, which can be extremely effective for removing internal flash and for deflashing parts with complex configurations.

Cryogenic blasting machines are more costly and require more floor space than conventional blasting equipment. Wide fluctuations in temperature create moisture in the system and, in conjunction with steel shot, may cause rust and contamination problems, which usually necessitate a part-washing operation.

Systems in current use incorporate a stainless steel cabinet insulated with dense polyurethane foam. The blasting chamber with a continuous mill belt, the liquid nitrogen nozzle, and the opening for the throw wheel are located inside the cabinet. Motors and bearings are located outside the cabinet and are thus isolated from temperature variations and moisture during operation of the equipment.

To reduce rust, machines are often designed to use plastic media instead of steel shot and incorporate a sophisticated vibrating-screen, media-separation system that allows flexibility in type and size of media used. Plastic media can be used with no resulting part contamination, often eliminating the need to wash parts.

When determining the most efficient, cost-effective type of cryogenic blasting system to use for a particular application, four interrelated parameters must be taken into consideration: cycle time, wheel speed, media, and temperature.

9.4.1 Cycle Time

Cycle times range from 1.5 to 6 min. Factors that affect cycle times include number of parts per batch (which depends, in turn, on the size of parts), flash thickness, wheel speed, and type and size of media used. One way to determine optimum cycle time is to set the temperature and media, run the machine at full wheel speed, and, beginning with a short cycle time, gradually increase the cycle time until a 100% yield is produced with no part damage.

To illustrate the relationship between cycle time and media, consider a large manufacturer of molded parts with two of the first-generation cryogenic shot blast deflashers. One machine is set up to use steel shot; the other has been modified to use plastic media. Since steel shot is significantly more aggressive as an abrasive than plastic, it might be assumed that the machine using plastic would have a longer cycle time. However, both machines operate with the same cycle time. The difference between them is that the machine using steel shot runs with a wheel speed of 1,900 r/min, whereas the machine using plastic media runs at 2,400 r/min.

9.4.2 Wheel Speed

The speed of the throw wheel determines the impact force of the media particles against the part. In first-generation machines, the wheel speed was often preset at 2,400 r/min, with some machines being altered in the field to provide variable control. Second-generation machines have variable speeds up to 3,600 r/min. Although most parts are run in the 3,000–3,200 r/min range, a general rule of thumb is to use the maximum speed possible without damaging parts to conserve nitrogen and run as many parts per hour as possible.

Several companies modified their second-generation deflashers to operate up to 6,000 r/min and reported that the extra 3,000 r/min made a startling difference in the performance of the machine. In one instance, cycle time was reduced from 6 min to 2 min. Subsequently, machine manufacturers began offering units with the capacity to run at speeds of up to 7,500 r/min.

9.4.3 Media

Steel Shot

Most first-generation deflashers still use fine-mesh steel shot in the .005 in. (0.13 mm) range. Despite maintenance and contamination problems, steel shot is excellent in several applications. It is particularly effective on rubber-to-metal bonded parts, where part of the rubber flash to be removed must be cleaned from a metal surface. It is effective where the deflashing process is expected to remove any excess bonding agent from the metal. Steel shot is also used on parts, such as multihole connectors, where the media must be less than .014 in. (0.36 mm) to avoid lodging problems.

Plastic

The most common type of plastic media used in cryogenic blasting is *cylindrical polycarbonate pellets*. When the second-generation deflashers were developed, polycarbonate media was being used in ambient shot blast deflashers on thermoset plastic parts. For this media to be used with higher wheel speeds and in temperatures as low as –300° F (–184° C), it became necessary to blend polycarbonate with additives to make it more dense and, therefore, more durable. This reinforced polycarbonate can effectively withstand speeds up to 7,000 r/min with good wear life.

Polycarbonate media is available in the following sizes: .030 in., .045 in., .060 in., and .080 in. (0.76 mm, 1.14 mm, 1.52 mm, and 2.03 mm). The most common size used is .045 in. (1.14 mm). Factors determining optimum size are part size, part wall thickness, flash thickness, and lodging considerations. For example, a part with a .030 in. (0.76 mm) groove will have a lodging problem with .030 in. (0.76 mm) media. A general rule is to use the largest size of media possible without damaging parts. The larger the media particles, the longer they stay in the unit before being removed by the separation system. Part damage is usually evidenced by chipped edges. Before reducing media size, try reducing wheel speed (without increasing cycle time).

One large manufacturer has determined that the shape of the media has little effect on its deflashing action. When media strikes the part, the force shatters the embrittled flash. This is a function of the mass and speed of the media particle. It is a misconception to think of the deflashing process as a cutting or scraping away of the flash.

Another type of plastic media is *styrene-divinylbenzene copolymer beads*. This type of media is much less dense than polycarbonate and is used primarily on very delicate parts, such as hearing aid parts. This material ranges in size from .039–.006 in. (0.99–0.15 mm) and can be used on parts that require, due to small holes, media smaller than .030 in. (0.76 mm) (which is the smallest-sized polycarbonate media manufactured). To use media this small in most second-generation deflashers, it is necessary to change the separator screens.

9.4.4 Temperature

The temperature necessary to properly embrittle the flash will vary depending on the compound from which the part is molded. Compounds such as butyl, nitrile, and neoprene can normally be deflashed in the –70 to –100° F (–57 to –73° C) range. Other compounds, such as silicones, may require temperatures as low as –300° F (–184° C) (see Chapter 6).

For each compound, the important task is to determine the embrittlement point as closely as possible. After a thorough study, one molder determined that it was going colder than necessary on almost 80% of its parts. Precision on this issue can mean large savings in nitrogen consumption.

Close study of the points of heat transfer in the equipment will yield advancements in the areas of design and insulation. Research is being conducted to solve moisture and freeze-up related problems. Better methods of controlling and measuring the flow of liquid nitrogen are also being sought.

9.4.5 Machine Operation

Parts are loaded into the cryogenic deflasher in quantities of hundreds or even thousands at a time. Mixed lots of parts can be deflashed simultaneously if flash thickness and embrittlement points are compatible.

After the deflashing chamber has been loaded, the refrigerant, normally liquid nitrogen, is injected (in the form of a spray mist) into the tumbling portion of the deflashing chamber, which embrittles the flash without freezing the parts themselves. While in the deflashing chamber, the parts are gently turned over on the endless belt conveyor to expose all surfaces to the impact of media particles hurled in a controlled pattern by the centrifugal wheel. The impact of these particles fractures the frozen flash without affecting the surface condition of the unfrozen parts. Deflashing cycle times range from 3–5 min per load.

Another form of cryogenic tumble blasting is a wheel blasting system that propels abrasives—a variation on centrifugal blasting. It has a two-bladed wheel, a wheel speed of 3,600 r/min, and an abrasive flow rate of 150 lb/min (68 kg/min). Abrasive is fed onto the blades through a crescent-shaped opening in the feed spout. This results in a uniformly distributed pattern that maximizes deburring efficiency. This type of equipment is typically used in the plastics industry for deflashing and in the aluminum and zinc die-cast industry for deburring. Common applications include deflashing broiler plugs, electrical components, and plastic pulleys, and deburring fuel pump housings, light fixture brackets, and springs.

9.5 ICE BLASTING

There are two forms of ice blasting, frozen carbon dioxide (dry ice) and frozen water. The CO_2 process was discussed earlier. The newest version is small pellets of ice thrown at high speed to deflash die castings. The advantage of the process is that contaminants are reduced and media lodging is not an issue. Dimensional and surface finish changes are reportedly nonexistent with this media. In one application, the blasting time was 4 s in a transfer line moving at 5 ft/min (1.5 m/min) (Visaisouk and Fisher 1998).

9.6 PREVENTING BLASTING PROBLEMS

The basic problems in blasting are distortion, burrs being beaten over, discoloration, and part contamination. Simple considerations will prevent these problems, and constant oversight will detect them if they do occur. Table 9-7 suggests solutions to some common problems encountered in blasting operations.

9.7 RELATED BLASTING PROCESSES

Several other blasting processes are being used for various applications. These include high-volume wet blasting, waterjet deburring, abrasive-flow stream finishing, and liquid-hone deburring.

9.7.1 High-volume Wet Blasting

High-volume wet blasting differs from other wet blasting processes in that media velocity is generated by a pump rather than by compressed air; small amounts of compressed air are used only to atomize the liquid as it exits the process nozzle. The process employs high-volume, variable-speed centrifugal pumps that develop pressures to 80 psi (552 kPa). The pumps, which have no packing or seals, are commercially available to recirculate up to 1,200 gal/min (4,542 L/min) for multiple-nozzle applications. Advantages of this process include improved control of the process and the ability to provide a wide range of finishes with one type of media.

For removing small burrs and for gentle finishing operations, the pump is operated at high speeds to provide partial atomization of the spray. For removing large burrs

Table 9-7. Troubleshooting solutions for blasting operations

	Media						
Application	Glass beads	Nut shells	Crushed mineral	Steel shot	Ceramic shot	Plastic spheres	Plastic angular media
Minimize ferrous contamination	X	X	X				
Zero part size change	X	X					
Minimize damage by large shot	X	X					
Minimize contamination	X	X					
Get into fine holes, threads, and pockets	X		X				
Blast after plating	X						
Remove discoloration	X		X				
Used for fragile nonmetal parts	X	X				X	X
Used for aggressive cutting			X				
Used for steel parts	X		X	X	X	X	X

and more aggressive finishing, low pump speeds and complete atomization are employed. The scrubbing action resulting from the high volumes of slurry accomplishes simultaneous degreasing of the workpieces without the need for chemicals. Rinsing, drying, and rust-inhibiting operations can be incorporated in the same machine.

9.7.2 Waterjet Deburring

High-pressure water jets, without abrasives, are being used to rapidly and efficiently cut many nonmetals, including paper, cloth, wood, and fiberglass. This process is called hydrodynamic machining (HDM). A lower pressure version of the HDM process, called waterjet deburring or machining (WAM), is being used to deburr and deflash both metallic and nonmetallic parts. Fan-shaped blades or conical streams of water under pressures to 30,000 psi (207 MPa) impinge workpiece surfaces at high velocities, removing burrs or flash and flushing them away. The process, using no abrasives, leaves sharp corners with no radiusing. Since cycle times are longer for burrs and flash of larger size, the process is best suited for light operations. Applications include the finishing of air compressor cylinders and automatic transmission cases and components.

The waterjet deburring process is easily automated, using various designs to move the nozzles or workpieces. Production rates to 460 pieces per hour have been reported. Automation of wet blasting and waterjet deburring processes is also being accomplished with programmable industrial robots that can control the pump, pressure, and movement of the nozzle. One application of waterjet processes involves deburring laser-drilled holes in vanes and blades for turbine engines.

9.7.3 Abrasive Flow Stream Finishing

This new process relies on a stream of abrasive flow material (putty-like) being propelled at hard-to-reach areas for polishing surfaces and removing microscopic burrs. It is described in more detail in Chapter 12.

9.7.4 Liquid-hone Deburring

Liquid-hone deburring is a process in which fine abrasive particles suspended in water are forced over workpiece edges having burrs (see Figure 9-11). The steady flow abrades the burrs and generates smaller radii. It differs from wet blasting in that it does not involve the impact of a high-velocity, concentrated stream. Instead, it is a gentler process, with the media being forced over the edges instead of at them. Tooling is generally used to contain and direct the flow of media. Liquid-hone deburring has limited application and is used primarily to remove small burrs from drilled holes .062 in. (1.57 mm) or less in diameter; it requires a cycle time of

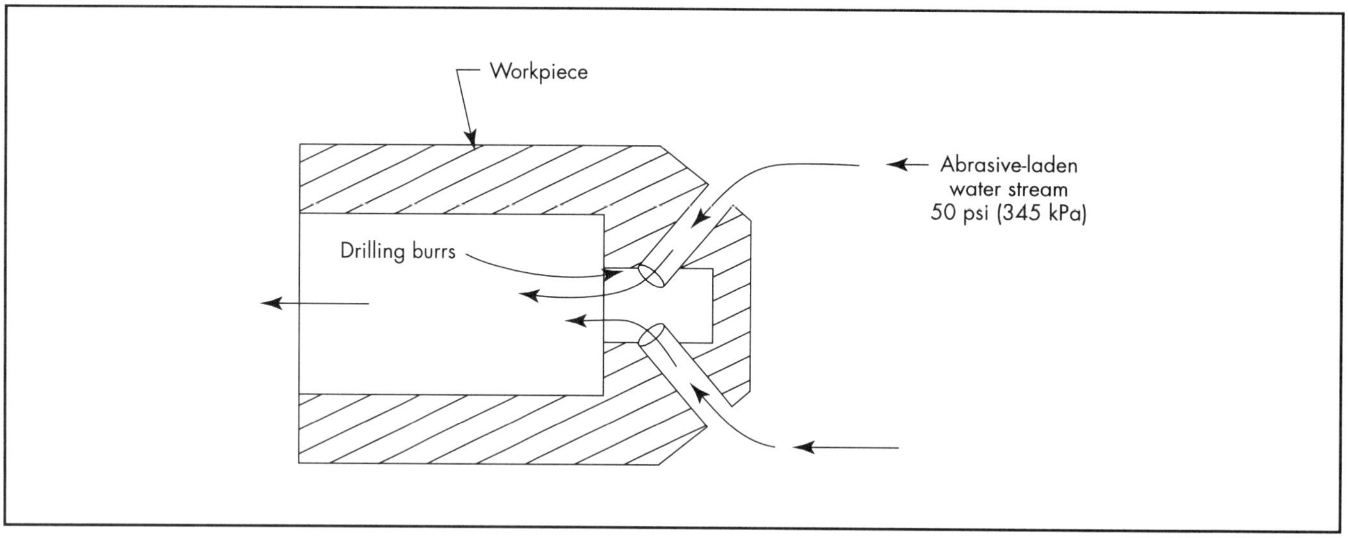

Figure 9-11. Liquid hone deburring.

1–3 min. Larger holes would require longer cycles, primarily because most of the media would be in the centers of the holes rather than at the edges. Typically, silicon carbide or aluminum oxide abrasive, 120 grit or finer, is used in a stream of water at a pressure of 50 psi (345 kPa).

9.8 ENVIRONMENTAL, HEALTH, AND SAFETY CONCERNS

Most of the environmental, health, and safety issues involved in blasting operations have been described earlier in the chapter. Briefly, they include ensuring that operators do not breath hazardous dust, that dust is maintained below explosive levels, and that carcinogenic materials are prevented from entering breathable air or are captured at point of source. As regulations continue to change, landfills may not accept wet or solid material contaminated with carcinogenic or hazardous materials. Operators should avoid directing high-pressure streams of air onto their skin because of the possibility of forcing air bubbles into the blood stream. Grit on the floor where operators walk can cause slipping and subsequent injury. Waterjet machining uses such a high-pressure, fine stream that it will cut like a knife and amputate fingers or any other item placed under it.

References

Alwerfalli, D. R. 1989. "Burr Removal Rate Model for an Abrasive Jet Deburring Process." Technical paper MR89-441. Dearborn, MI: Society of Manufacturing Engineers (SME).

DeLitizia, Alfred T. 1981. "Mechanical Deburring with Centrifugal Blast Equipment." Technical paper MR81-390. Dearborn, MI: Society of Manufacturing Engineers (SME).

Gillespie, LaRoux K. 1976. *Deburring Capabilities and Limitations*. Dearborn, MI: Society of Manufacturing Engineers (SME).

———. 1981. *Deburring Technology for Improved Manufacturing*. Dearborn, MI: Society of Manufacturing Engineers (SME).

Golubski, Michael A. 1985. "Plastic Abrasives and Their Niche in Blast Finishing." Technical paper MR85-841. Dearborn, MI: Society of Manufacturing Engineers (SME).

Hanna, Mark R. 1993. "Deburring and Deflashing with Air Blast Equipment." Technical paper MR93-132. Dearborn, MI: Society of Manufacturing Engineers (SME).

Scott, Jerry Y. 1985. "Cryogenic Shot Blast Deflashing." Technical paper MR85-828. Dearborn, MI: Society of Manufacturing Engineers (SME).

Visaisouk, Sam and Fisher, Norman. 1998. "Deburring and Cleaning by Ice Blast: A Case Study." Proceedings of the 5th International Deburring and Surface Finishing Conference, San Francisco, CA, Sept. Kansas City, MO: Deburring Technology International.

Weidner, Maureen P. 1985. "Advantages of Microabrasive Blasting." Technical paper MR85-838. Dearborn, MI: Society of Manufacturing Engineers (SME).

Weightman, Neil. 1995. "Understanding the Capabilities of Microabrasive Blasting." Technical paper MR95-221. Dearborn, MI: Society of Manufacturing Engineers (SME).

Woelfel, Michael M. 1983. "Glass Bead Blasting for Surface Conditioning and Improvement." Technical paper MR83-686. Dearborn, MI: Society of Manufacturing Engineers (SME).

Bibliography

Barber, Anthony, ed. 1989. *Pneumatic Handbook,* 7th edition. Morden, Surry, England: Trade and Technical Press.

Gibbs, Charles W., ed. 1969. "Compressed Air and Gas Data." New York, NY: Ingersoll-Rand Company.

Gillespie, LaRoux K. 1978. *Advances in Deburring*. Dearborn, MI: Society of Manufacturing Engineers (SME).

Golubski, Michael A. 1991. "Plastic Blast Cleaning Medias: Providing Cost-saving Solutions." Technical paper MR91-120. Dearborn, MI: Society of Manufacturing Engineers (SME).

Kobayashi, Shigeharu. 1987. "Deflashing Resin IC Package on the Lead Frame by Wet Blasting." Technical paper MR87-150. Dearborn, MI: Society of Manufacturing Engineers (SME).

McHugh, Brian. 1988. "Flexible Finishing with Dry Blast Deburring." Technical paper MS88-517. Dearborn, MI: Society of Manufacturing Engineers (SME).

Meyers, Robert J. 1985. "Cryogenic Deflashing of Rubber, Plastic, and Zinc Die Cast Parts." Technical paper MR85-847, Dearborn, MI: Society of Manufacturing Engineers (SME).

Plaster, H.S. 1972. *Blast Cleaning and Allied Processes,* vols. 1 and 2. Hertfordshire, U.K.: Garden City Press Limited.

Rhoades, Lawrence J. 1981. *Cost Guide for Automatic Finishing Processes*. Dearborn, MI: Society of Manufacturing Engineers (SME).

Scoville, John H. 1984. "Robotic Abrasive Process System." Technical paper MS84-361. Dearborn, MI: Society of Manufacturing Engineers (SME).

Settles, Gary S., and Geppert, Stephen T. 1997. "Redesigning Blasting Nozzles to Improve Productivity." *Abrasive Blast Cleaning News,* (Spring):5–9.

Takeuichi, Shigeo. 1994a. "The Effect of Particle Spherization on the Deburring of IC Molded Packages by Means of Botanical Abrasive Particles" (in Japanese). *Proceedings of the 3rd International Conference on Precision Finishing and Burr Technology*. Kansas City, MO: Deburring Technology International: 278–85.

———. 1994b. "Grinding Properties of Botanical Abrasives for Thermosetting Resins" (in Japanese). *Proceedings of the 3rd International Conference on Precision Finishing and Burr Technology*. Kansas City, MO: Deburring Technology International: 291–96.

———. 1994c. "Properties as Abrasives of Botanical Abrasives" (in Japanese). *Proceedings of the 3rd International Conference on Precision Finishing and Burr Technology.* Kansas City, MO: Deburring Technology International: 286–90.

Weightman, Neil S. 1993. "Why Microabrasive Blasting?" Technical paper MR93-330 Dearborn, MI: Society of Manufacturing Engineers (SME).

Wick, Charles and Veilleux, Raymond F. 1982. *Tool and Manufacturing Engineers Handbook,* vol. 3. Dearborn, MI: Society of Manufacturing Engineers (SME): 16-83–101.

Wieland, Rolf. 1991. "How to Develop and Maintain an Efficient Deflashing Operation." *Rubber World,* 193 (November):20–22, 48.

10 Bonded-abrasive Deburring

Bonded-abrasive deburring, traditionally called sanding, involves far more than sandpaper products. The process is also known as coated-abrasive finishing, and when heavy stock removal is involved, it is called coated-abrasive machining. Bonded-abrasive products are found in almost every shop in the world because of their flexibility, convenience, low cost, and wide variety of shapes and sizes. Varieties of bonded abrasives range from normal shop abrasives to highly precise tape products that produce finishes of 2 μin. (0.05 μm) or better. Bonded-abrasive products perform well in manual and automated operations and are used for deburring and surface smoothing. The variety of equipment and material available is vast—over 1,500 product variations are estimated to be in existence. Equipment is produced by over 35 manufacturers, and hand-held motors and grinders are produced by an additional 60 companies. The media used is produced or marketed by yet another 60 companies.

The following process variables are involved when using bonded abrasives:

- Abrasive material and size.
- Media type and size.
- Backing material and properties.
- Operating speed and pressure.
- Density of nonwoven materials.

Bonded-abrasive processes are sometimes used in place of loose abrasive lapping. For the purposes of this handbook, the term *lapping* will only be used to refer to loose abrasive lapping. Bonded films are used for high-precision parts, and though they are technically bonded abrasives, they have properties that are not shared with other bonded abrasives. A discussion on bonded films appears later in this chapter.

To occasional users of bonded abrasives, the possibility of providing a radius on a hole by using a flat sanding machine may not be obvious because sanding has traditionally always left sharp edges and fine burrs. However, most of today's flat-belt sanders can provide some slight edge break in holes and cutouts.

Figure 10-1 illustrates the typical elements of bonded-abrasive materials. Backing material is typically either paper or cloth, with cloth being more durable. Open cut products use less grains per surface area to allow room for swarf and chips. Closed cut products essentially fill the surface area with abrasive grain to maximize the number of cutting points. Soft materials normally necessitate the use of open cut products to prevent clogging and reduce heating. Open-cut products have 50–70% of the backing surface covered with abrasive mineral.

Figure 10-2 illustrates abrasives that are incorporated into nonwoven synthetic materials. Because the fibers in these materials are widely spaced, they are very soft and easily compressed. Their design produces an almost spring-like cutting action (see Figure 10-3).

10.1 TYPES OF ABRASIVES USED

The three most common abrasives used in bonded-abrasive deburring are aluminum oxide, silicon carbide, and zirconia compounds. Garnet, emery, crocus, and flint papers are also used, as are a few superabrasives, such as diamond and borazon. Diamond and borazon are primarily used for very hard materials and are not typically used for deburring. They are used for precision grinding of ceramics and exotic alloys. Aluminum oxide outperforms other minerals when measuring the ability to resist fracturing. Zirconia alumina breaks down with use, continually providing new cutting edges; it is preferred when heavy grinding pressures are being employed. It is not particularly effective on stainless steel, steel, or cast iron. Silicon carbide is widely used on softer metals, such as aluminum, brass, bronze, magnesium, and titanium, and on rubber, glass, and plastics. Silicon carbide cuts fast under light pres-

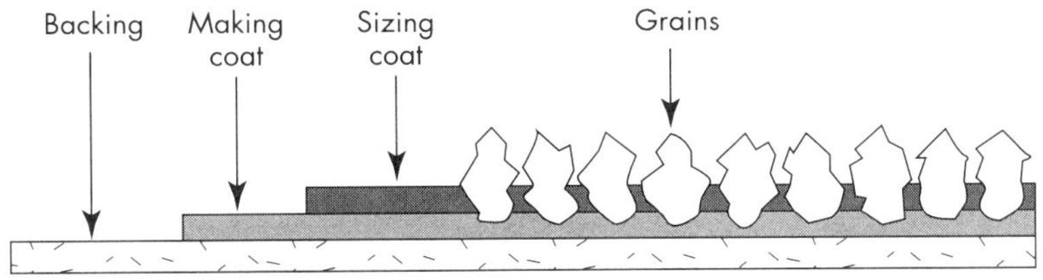

Figure 10-1. Elements of a bonded abrasive. ***(Courtesy Norton Company)***

sure. Garnet dulls too rapidly for metalworking. Crocus is used for polishing gold and other soft metals. Ceramic alumina has a submicron structure, which allows each grain to break down into more cutting edges. Ceramic materials are used for ferrous and nonferrous metals, carbon steel, and exotic alloys.

A new form of abrasive, using chunks of fine abrasives, has been developed to provide continuous cutting action. At first glance the clusters of abrasive appear to be a single grit, but as the clusters wear down, new cutting edges are exposed to the work (Gagliardi and Duwell 1989).

10.2 BACKINGS FOR BONDED ABRASIVES

Backing materials come in several grades of stiffness. One manufacturer, for example, produces eight grades. The abrasive material is attached to the backing with either glue or resin. When describing backing materials, terms such as *resin-resin* indicate that both the base coat and the top or sizing coat are resin. Zinc stearate is applied after bonding to products that need a dry lubricant. Operations requiring wet application may need to use backing materials specifically designed for that purpose. Glue-bonded material provides a softer, more uniform surface, but may not have the durability and stock removal potential of resin-bonded products. Resin over glue products provide a backing midway between pure glue and resin-resin products.

Whether the backing is paper, cloth, vulcanized fiber, or a combination of materials, it must be smooth enough for uniform adhesive coating, strong enough to withstand grinding pressures, and flexible enough to conform to contours if necessary. For economy, the least expensive backing compatible with the job requirements should be selected.

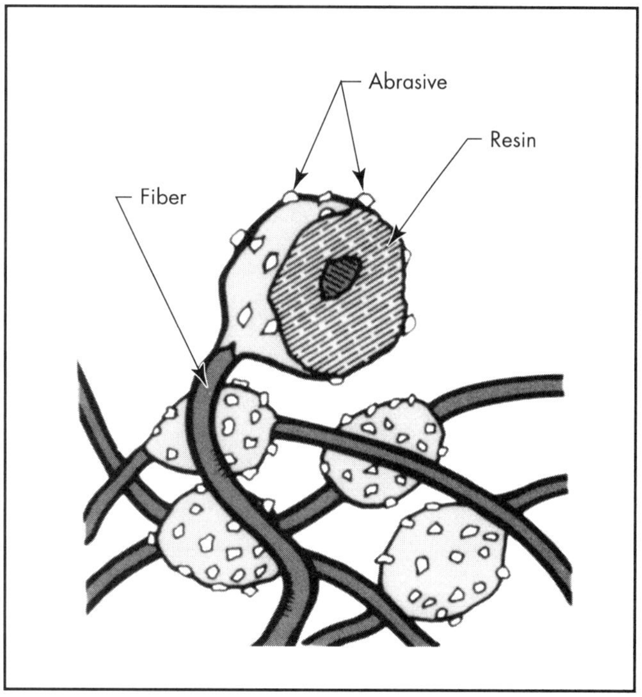

Figure 10-2. Schematic representation of nonwoven abrasive materials. *(Courtesy 3M Company)*

10.2.1 Paper Backings

Paper backings used for bonded abrasives are highly specialized technical papers made to exact specifications, thereby assuring certain essential physical properties, such as finish, strength, adhesion, flexibility, and weight. Paper weights are determined by the number of pounds in a papermaker's ream of 480 sheets, 24 × 36 in. (610 × 914 mm) in size.

The standard weights of paper used in backings for bonded abrasives are 40, 70, 90, 130, and 165 lb (18.1, 31.8, 40.8, 59.0, and 74.8 kg, respectively). These weights are coded using letters, which appear immediately after the grit size on the backing of the finished product. Briefly stated, the lighter the backing, the greater its flexibility; the heavier the backing, the greater its resistance to tearing.

A-weight (40 lb [18.1 kg])

Light and flexible, A-weight backing is used primarily for hand-finishing operations, both wet and dry. It serves as a backing for grits 80 and finer.

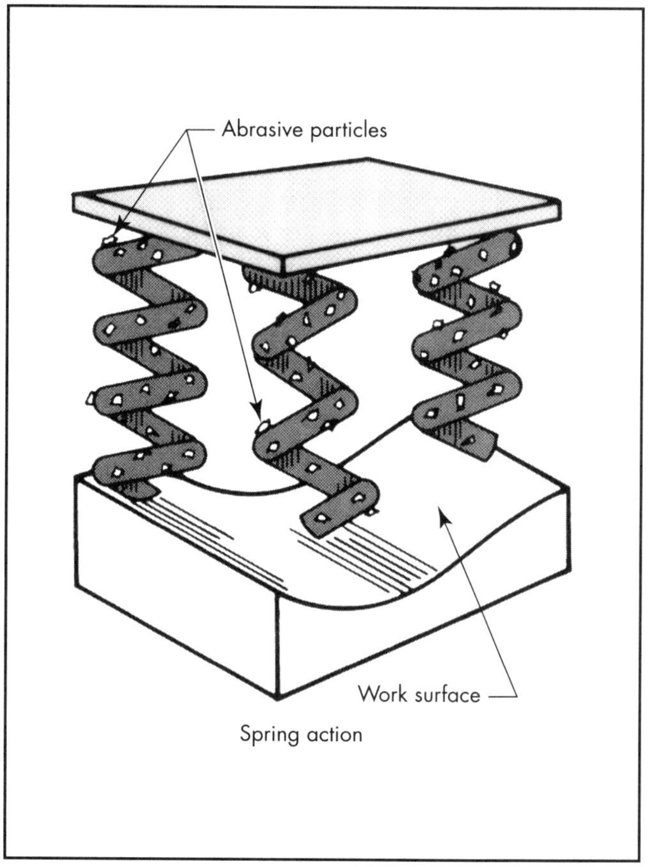

Figure 10-3. Schematic representation of controlled cutting action of nonwoven abrasive products. *(Courtesy 3M Company)*

C-weight (70 lb [31.8 kg])

Stronger and less flexible than A-weight, C-weight backing is chosen for hand sanding, dry or wet, and for use on small portable power sanders. It is used for intermediate through fine sanding and serves as a backing for grits 60–180.

D-weight (90 lb [40.8 kg])

Stronger and less flexible than C-weight, D-weight backing is chosen for hand sanding and for use on small portable power sanders. It is used for coarse through intermediate sanding and serves as a backing for grits 36–80.

E-weight (130 lb [59.0 kg])

Stronger and less flexible than D-weight, E-weight backing is used primarily on roll, belt, and disk applications where high resistance to tearing is required.

F-weight (165 lb [74.8 kg])

F-weight is the strongest, least-flexible paper backing. It is used only for crankshaft lapping rolls and tannery industry belts and rolls.

10.2.2 Cloth Backings

Although paper-backed abrasive belts have a lower initial cost than cloth-backed belts, cloth backings offer greater durability and resistance to tearing and can tolerate continual bending and flexing during use. If the severity of the operation is such that cuts, tears, and nicks may be encountered, cloth should be chosen, since such damage to a paper-backed belt would cause almost immediate breakage because of tension on the belt.

The woven cotton or synthetic yarns received from the looms are treated to make them suitable for use as bonded-abrasive backings. Desizing, shrinking, drying, stretching, filling, and calendering are all processes employed to produce the desired strength, flexibility, and coating surface. Each of the resulting backings has construction and finishing characteristics designed to make it ideally suitable for a specific product application.

Standard cloth weights used in bonded abrasives are indicated by letter codes, usually J, X, Y, H, or S, which appear immediately after the grit size on the backing of the finished product.

J-weight (Jeans)

The lightest and most flexible cloth backing, J-weight backing is used when finish and uniformity of the surface are more important than stock removal. It is ideal for finishing, blending, and when considerable flexibility and conformity are required, such as in contour work on curved surfaces.

X-weight (Drills)

Stronger and relatively stiff when compared to J-weight, X-weight is used on products designed for such applications as coarse-grit stock removal and fine-grit finishing and polishing. Consistent productivity, relatively good finishes, and long product life are characteristics of products made with X-weight backing.

Y-weight (Heavy Drills)

Stronger and more resistant to longitudinal splitting than regular drills cloth, Y-weight backing is used on products designed for severe applications, such as narrow-belt grinding of hand tools and wide-belt sanding of lumber and particle-board dimensioning.

H-weight (Heavy-duty)

The strongest cloth backing available, H-weight is used on coarse-grit products designed for extreme pressure and heavy stock removal operations. It is used exclusively with zirconia alumina abrasive for belts having a long productive life on both wood and metal applications.

S-weight (Sateen)

A special, heavy, stretch-resistant backing, S-weight cloth is used only in the manufacturing of sectional belts—belts with widths greater than 52 in. (1.32 m) for which conventional cloth backings are unsatisfactory because of the unique construction requirements of these extremely wide belts.

Several other specialized cloth constructions are available, including marquisette cloth and plain-weave mesh, which are used for backing laminates or combinations.

Cloth backings stretch under tension, which is not a consideration when using paper, fiber, or combination backings. Therefore, when cloth-backed abrasives are used, machines must be designed to allow sufficient belt-tensioning tolerances to accommodate a 3–5% stretch.

10.2.3 Fiber Backings

Fiber backings, made of multiple layers of impregnated paper, are very hard and strong, yet provide sufficient flexibility during use. They are typically used in heavy-duty portable grinder operations.

30-mil (762 μm) Fiber

With the greatest strength of any backing material, 30-mil fiber is used for bonded abrasives. It is used on resin-bonded fiber disks designed for heavy-duty portable grinder operations.

20-mil (508 μm) Fiber

Twenty-mil fiber is used primarily as a backing material for coarse grits of products designed for heavy drum-sanding operations.

10.2.4 Combination Backings

Combination backings, constructed by laminating light cloth and 130 lb (59 kg) paper, are used when resistance to tearing and breaking is a requirement. They are primarily used on products designed for very coarse drum-cover or disk sanding operations.

10.2.5 Sandscreen Backings

Sandscreen is an open mesh material that looks like window screening with fine abrasive on the surface. It is used wet or dry and does not clog. It is used for stainless steel, large parts, and other applications.

10.2.6 Cork Backings

Cork belts are constructed by bonding chunks of cork on a conventional belt, then coating the surface of the cork with abrasive and resin. This process produces large bulges on the surface of the backing, but when the abrasive contacts the part, the pressure pushes the abrasive somewhat back into the cork. This produces a cushioning effect. Although the cork is not a true backing, it does back the abrasive layer (see Figure 10-4). Cork belts are used for wet polishing of glass, ceramics, porcelain, and crystal. It is normally used after grinding with silicon carbide. Although cork backings are not normally used for deburring, they can be used on hard materials requiring fine finishes.

10.2.7 Polyester Film Backings

Polyester-based film backings offer special properties that lend themselves to the rigors of specialized applications. Precision polyester film backings provide thicknesses required for ultra-tight tolerances—a must for electronics, fiber optics, crankshaft, and some automotive disk applications.

10.3 CUTTING FLUIDS

The choice of whether to use a wet or a dry process when using a bonded abrasive is largely determined by the material to be ground and the mechanics of the operation. Under dry conditions, metals may discolor from localized heat. Thermoplastic materials may soften, distort, or load the abrasive, and some materials pose health or safety hazards if grinding swarf is not properly handled. Wet operations should not be employed with electrically operated portable tools or when chemical reactions will occur between the swarf and the coolant.

Grinding fluids are used with bonded abrasives to cool the workpiece, thus decreasing or eliminating the possibility of heat check, discoloration, melting, and warping, and to reduce friction, thus decreasing grinding temperatures, permitting higher speeds, and generally improving the surface finish. Grinding fluids also reduce the tendency to load the abrasive by inhibiting the welding or bonding of swarf to the abrasive grains. This function can sometimes be accomplished in dry operations by using bonded abrasives manufactured with an additional surface coating of zinc stearate to resist loading. A zinc stearate coating, usually whitish in color, performs as a dry lubricant for abrading soft materials. Cutting fluids also provide chip washaway from the point of grinding to improve abrasive performance and reduce potential coarse scratches associated with agglomeration of swarf.

For the best possible abrasive belt life, the belt should be run with oil sprayed on it intermittently for lubrication; a wet machine should be used with a flood of filtered, recycling water-soluble oil. The lubricant helps produce better finishes and slows abrasive breakdown. It also promotes cleaning when aluminum is being processed, since aluminum quickly loads abrasive belts, necessitating frequent cleaning to ensure economical abrasive usage.

Lubricants packaged in convenient grease stick or tube form for periodic application have been developed specifically for use on bonded-abrasive belts. They are available in a variety of formulations, and the correct selection can offer a number of advantages, including prevention of excessive heat generation at the point of contact between the work surface and the abrasive belt. Preventing excess heat is important because local surface temperatures during some polishing operations may reach 1,832° F (1,000° C). Such extreme temperatures are capable of drastically distorting or actually melting the surface layer. The frequent application of lubricant can prevent this intense heat buildup in the zone in which the friction takes place. This prevents surface galling or tearing out of minute particles on workpieces that may not be immediately visible to the unaided eye, but would show up as underlying surface imperfections following subsequent buffing, anodizing, electroplating, or painting operations.

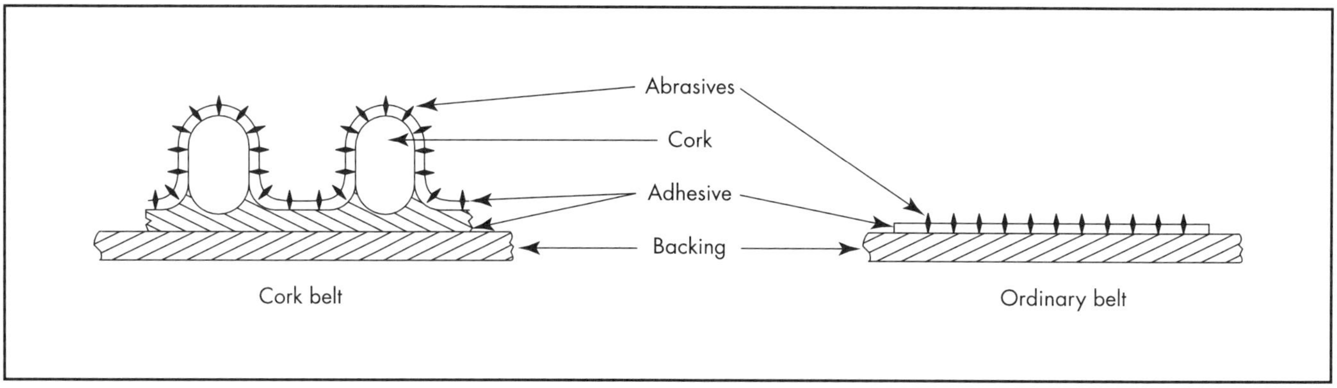

Figure 10-4. Cross section of cork belt compared with conventional bonded abrasive (Miyatani et al. 1981).

Keeping the surface heat down may also prevent light-gage metals from discoloring or warping when subjected to a polishing operation. Within limits, lubricants help to control the degree of finish obtained. For example, a heavy-bodied lubricant fills the area between the abrasive grains, which reduces the depth of cut without altering the abrasive grit size. This is sometimes referred to as "closing" or "loading" the abrasive belt and has the effect of producing a finer finish. In contrast, a light-bodied lubricant serves to keep the belt "open," which is important when rapid metal stock removal is a primary aim and the belt must be kept oily to prevent the grinding swarf from redepositing and clogging the bonded-abrasive surface.

It is possible to increase production and secure better surface finishes by using a coarse-grit belt with a lubricant, rather than a fine-grit belt without a lubricant. Obviously, a lubricant can prevent dry patches from forming on the cutting surface and provides a more uniform cutting action and, consequently, a more uniform finish.

Belt lubricants impregnated with abrasives are now in popular use. One example is the mixing of tripoli abrasive powder into the lubricant. When this mixture is used for polishing low-melting-point metals, such as aluminum, copper, or zinc, not only is lubrication provided, but also a buffing effect takes place because of the cutting action of the tripoli powder, resulting in a more lustrous finish. Using this same idea, extremely fine and bright surface finishes are obtainable on stainless steel by impregnating the lubricant with unfused alumina, an abrasive that is normally used as the cutting and coloring media in stainless steel buffing compounds.

Bonded-abrasive products are often color coded to distinguish them from each other. Abrasive grain size is based on U.S. and international standards. Many domestic bonded-abrasives manufacturers produce products that adhere to the Coated Abrasive Manufacturers' Institute (CAMI). The Federation of European Producers of Abrasives (FEPA) produces a "P" grade of abrasives. A comparison of the two systems of size grading is shown in Figure 10-5. Manufacturers also provide adhesives in spray cans or tube dispensers for adhering abrasive disks to pads.

10.4 TYPES OF BONDED ABRASIVES

Table 10-1 outlines the types of bonded abrasives currently in use.

10.4.1 Sheets

Sheets of bonded abrasives are produced in a wide variety of sizes for use in manual operations (by hand or hand block) or on machines with straight-line or orbital action. Precut sheets are also used on a number of drum machines.

10.4.2 Belts

Bonded abrasive belts are available with cloth and paper backings and contain aluminum oxide, silicon carbide, or a zirconia abrasive. Some have a waterproof material that allows longer life with wet operations. Cork-based materials are used for glass polishing as well as some metal deburring and chamfering. Manufacturers provide several grades of belts within a specific abrasive and backing type. One source provides eight different backing stiffness grades. Abrasive sizes range from 16 to 1,200 grit (grade) (see Figure 10-5). Belt sizes range from .25 to 100 in. (6.4 mm–254 cm) wide and up to 168 in. (427 cm) long. Nonwoven synthetic material provides cushioned action and consistent exposure of new, sharp grains as the belt wears. It can be used wet or dry, and its design essentially eliminates loading-up problems. At least five different end-splicing designs are in use to assure smooth action. In most cases, belt joints are at least as strong as the bonded-abrasive product itself; they run smoothly and perform as an integral part of the belt. Different machines may require different belt joints to extend life or reduce vibrations.

CAMI has endorsed a program of standard belt lengths: for a length range of 12–36 in. (305–914 mm), belt lengths are made in 3 in. (76.2 mm) increments; for a range of 42–168 in. (107–427 cm), belt lengths are made in 6 in. (152 mm) increments; and for a length range of 15–42 ft (4.6–12.8 m), belt lengths are made in 36 in. (91 cm) increments.

Zirconia alumina bonded-abrasive belts are being used for rapid and high stock removal applications in high-production abrasive machining for gate removal from castings

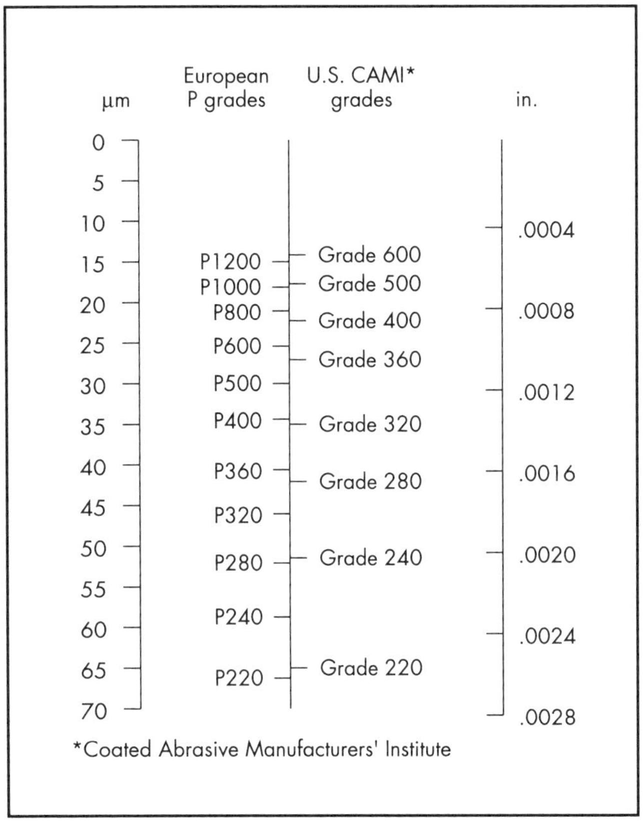

Figure 10-5. Comparative sizes and grades of fine abrasive grains.

Table 10-1. Bonded abrasive products used for deburring

Product	Variations	Abrasive media	Backing material
Belts	Conventional, narrow belt	Aluminum oxide, silicon carbide, alumina zirconia	Cloth, polyester/cloth, paper, nonwoven synthetic mesh Cork under abrasive
Film-backed tapes, sheets, strips	Conventional, brush-backed strips, sponge strips	Alumiunm oxide, chrome oxide, cerium oxide, silicon carbide, diamond, cubic boron nitrate	Cloth, paper
Pads	Conventional, sponge-backed pads	Aluminum oxide, silicon carbide	Cloth, paper
Disks	Conventional, slotted, slashed assemblies, stars, overlapping disks, raised bump disks, flap disks Fiber disks	Aluminum oxide, silicon carbide, alumina zirconia Aluminum oxide, zirconia	Cloth, paper, waterproof paper Fiber
Wheels	Conventional, disk wheels, goblet wheels	Aluminum oxide, silicon carbide	Cloth, paper, synthetic mesh
Drums, sleeves, bands	—	Aluminum oxide, silicon carbide, alumina zirconia	Cloth
Cord	—	Aluminum oxide, silicon carbide, crocus	Cloth
Sticks	—	Aluminum oxide	Paper
Cartridge rolls	—	Aluminum oxide	Paper
Spiral rolls/bands	—	Aluminum oxide	Paper
Shaped sleeves	—	Aluminum oxide	Paper
Cones	—	Aluminum oxide	Paper
Lapped points	—	Aluminum oxide	Paper
Cross pads	—	Aluminum oxide	Paper
Butterfly tools	—	Aluminum oxide, silicon carbide	Cloth, paper
Square pads	—	Aluminum oxide	Cloth
Flap wheels/mops	Conventional, brush-backed, nonwoven abrasive-backed, slashed strips (fingers), end and side-cutting flap wheels, angled flap wheels	Aluminum oxide, silicon carbide	Cloth
Abrading pads	—	Aluminum oxide	Cloth, paper
Shop rolls	—	Aluminum oxide	Cloth, paper
Miniatures	—	Aluminum oxide	Paper
Disk wheels	—	Aluminum oxide	Paper
Fingernail files	—	Aluminum oxide, diamond	Paper
Mops	—	Aluminum oxide	Paper
Abrasive sponges	—	Aluminum oxide, silicon carbide	Synthetic sponge material

and heavy flash removal from forgings. Zirconia alumina bonded-abrasive belts substantially improved deflashing operations on the upset ends of seamless tubing used in the petroleum industry. Cutoff ends of 30-ft (9.1-m) long tubes are deburred using abrasive belts to remove the outside fins and end flash. The tube ends are subsequently machined and threaded. After considerable research, several methods were developed to deflash 200–300 tube ends per hour using heavy-duty zirconia alumina abrasive belts.

10.4.3 Disks

Circular abrasive disks are produced in conventional, slotted, slashed assemblies, raised bumps, flap, and fiber types.

Conventional disks are simple round disks with adhesive backs, snap-on features, twist on/off designs, Velcro® fasteners, and integral threaded, or washer and screw attachments. Large adhesive-backed disks are used for "flat lapping"—a form of sanding in which the disk is applied to a rotating table; parts are placed on the table and held down by rollers or by hand. Disks up to 30 in. (762 mm) are standard and come in various abrasive sizes. The disk material can consist of traditional bonded materials, abrasive-coated nonwoven nylon, or other synthetic material products. The nonwoven material provides a softer pad to cushion cutting forces and is generally used for finer work, but aggressive materials are also available in this design. Precision finishes may require use of a scalloped disk whose edges have an undulating pattern rather than a constant diameter.

Slotted disks have radial cuts made in them (Figure 10-6a), allowing the disk to conform better to hole entrances and hole walls. Essentially, the slots allow the paper to fold back over itself, which results in a softer action and prevents some tearing.

Slotted overlapping disks perform heavy cutting action by providing more backing and breaking up of the cutting action as sections pass over part edges (Figure 10-6b). They are used widely for deburring holes, polishing bores, and chamfering tube ends. They are made in sets of two or four disks per assembly. The two-disk set provides cutting action only on the forward motion. The four-disk set allows deburring both on entering and exiting a hole.

Raised bump disks are hard, wheel-like fiber disks with a raised bump on their surfaces. The bump provides more of an interrupted cut.

Fiber disks are essentially thin grinding wheels designed to allow some lateral deflection. They are used for heavy stock removal, generally with disk grinders. One company provides a fiber disk cutter for easy cutting of any size disk.

Flap disks (also called mop disks), like flap wheels, consist of a series of abrasive strips laid over each other (Figure 10-7). The layers of strips look like the teeth on a face mill cutter. The disk is rotated so that the open edge trails over the surface, rather than leading into the cut as a cutter would. These products allow a cooler cut. Generally, no backing pad is required with this type of disk. It is widely

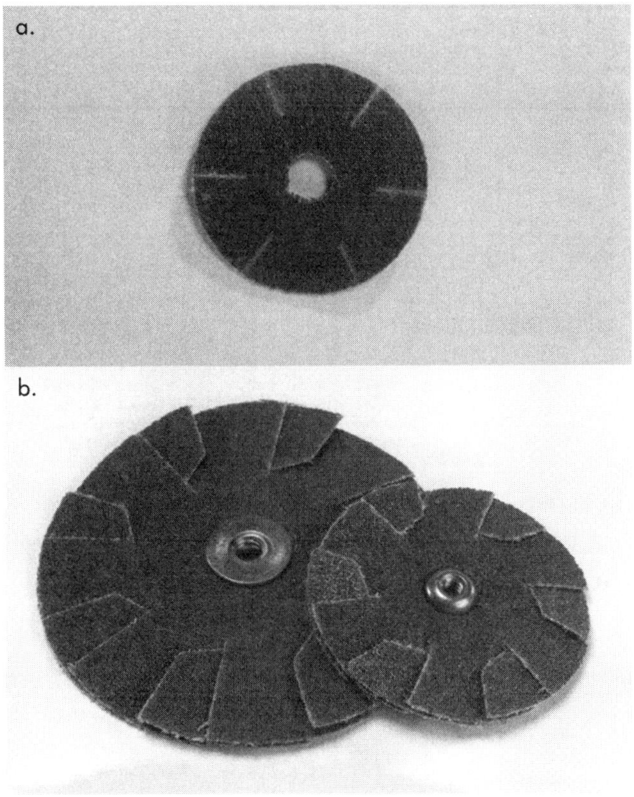

Figure 10-6. Bonded-abrasive slotted disks and overlapped slotted disks. ***(Courtesy Superior Abrasives)***

used for stainless steel, steel, aluminum, and most alloys. As the top layers wear, the layers below are exposed for additional cutting.

Stars are six-pointed disks stacked together in layers of two and offset such that a continuous face is visible. When stacks of these sets are put on a mandrel, they act like thin disks and can reach into fine grooves as well as deform into holes (see Figure 10-8).

10.4.4 Flap Wheels

Bonded-abrasive flap wheels, also called mops, deburr and polish with a slapping action. They are made in a variety of designs, sizes, and materials. The basic design consists of strips of abrasive paper bound together at the hub. Since the outer end has more space between paper strips, a gap opens that is used to perform the deburring. Materials other than abrasive paper can be placed between the paper strips. The common fill materials include soft brush strips and abrasive-coated nonwoven nylon (Figure 10-9). To further soften the action, each strip of paper can be slit into multiple thin strips (fingers) that conform better to fine detail. Flap wheels as small as 1 in. (25.4 mm) diameter and .63 in. (16 mm) wide and up to 80 in. (2 m) wide are standard.

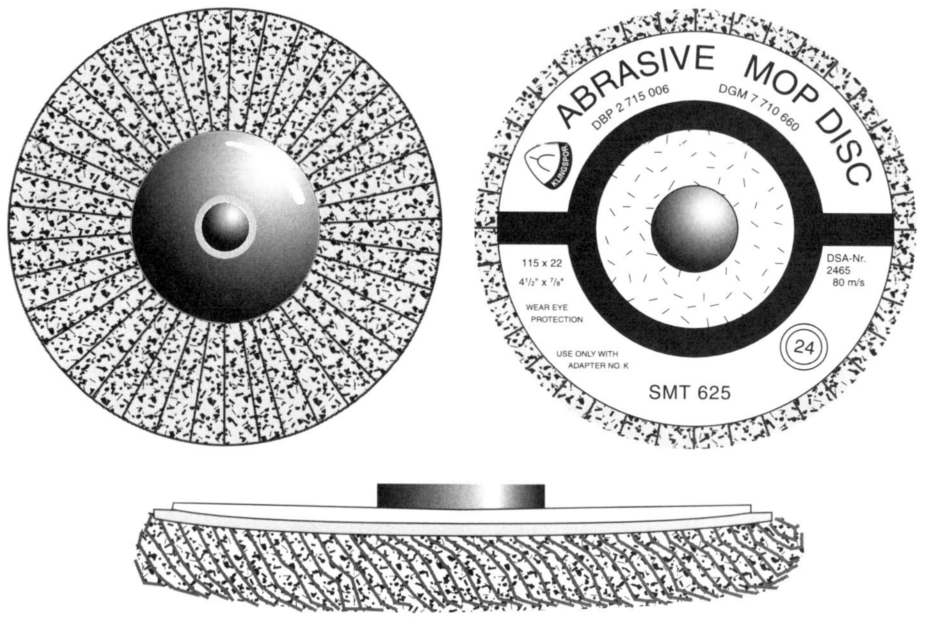

Figure 10-7. Flap disks. ***(Courtesy Klingspor Abrasives)***

10.4.5 Wheels

Wheel-style products are designed to cut on their outer diameter. Wheel diameters of up to 12 in. (305 mm) are common and can be made by wrapping adhesive-backed abrasive paper into a tight spiral and then cutting in strips.

One style, often called a unified wheel, involves compressing disks of abrasive-laden nonwoven material and bonding them together with a resin to form a solid wheel. The relatively soft wheel acts like a soft grinding wheel. These wheels are not actually bonded-abrasive wheels in the sense that grit is bonded onto the surface of the backing material; rather, the abrasive is spread throughout the wheel volume. They are often sold interchangeably with bonded-abrasive products.

Convolute wheels are formed by wrapping a nonwoven nylon strip into a tight wheel shape. The resins used for these wheels vary considerably and can provide very soft cutting or exceptionally hard, yet open cutting. Unified or unitized wheels are used for deburring metals and finishing printed circuit board edges and glass edges. They are made by compressing multiple layers of abrasive-coated, nonwoven nylon, and bonding them together to create an abrasive wheel. Low-density wheels are used for open cutting and do not develop as much heat as the harder high-density wheels. Low-density wheels can be shaped to match part configurations, and external lubricants can be used in the same manner employed for buffing wheels.

Wheel speeds typically range from 3,000–6,500 surface ft/min (914–1,981 m/min), but manufacturer's literature will indicate maximum safe speeds. Wheel-style products are sold with integral shafts or replaceable shafts.

Goblet wheels provide an almost hemispherical sanding surface and are designed for reaching into internal corners for deburring and polishing (Figure 10-10). They are widely used in the food-processing industry.

10.4.6 Shop Rolls

Shop rolls are reels of bonded abrasives; they are typically 50 yd (45.7 m) in length but vary in width depending on ultimate use. Applications for roll stock vary and include precision-slit abrasives used in crankshaft and bearing-polishing operations, wide rolls used on drum machines, hand-held operations on lathes, and a shoe-shine action performed on fixed parts. Both conventional materials and nonwoven nylon materials are available in shop rolls. Some rolls are sized to provide a continuous roll of material to cut to length for vibrating sanders.

10.4.7 Cartridge Rolls

Cartridge rolls are small, tightly wound rolls of bonded abrasives glued together to prevent unwinding. They are press-fitted on mandrels and offer fresh abrasives as each layer wears away. They are used for deburring and polishing and are commonly available in end-tapered, full-tapered, and regular (untapered) shapes. Cone-shaped bonded abrasives are press-fitted to a mandrel for grinding and finishing relatively inaccessible areas, such as curved surfaces and

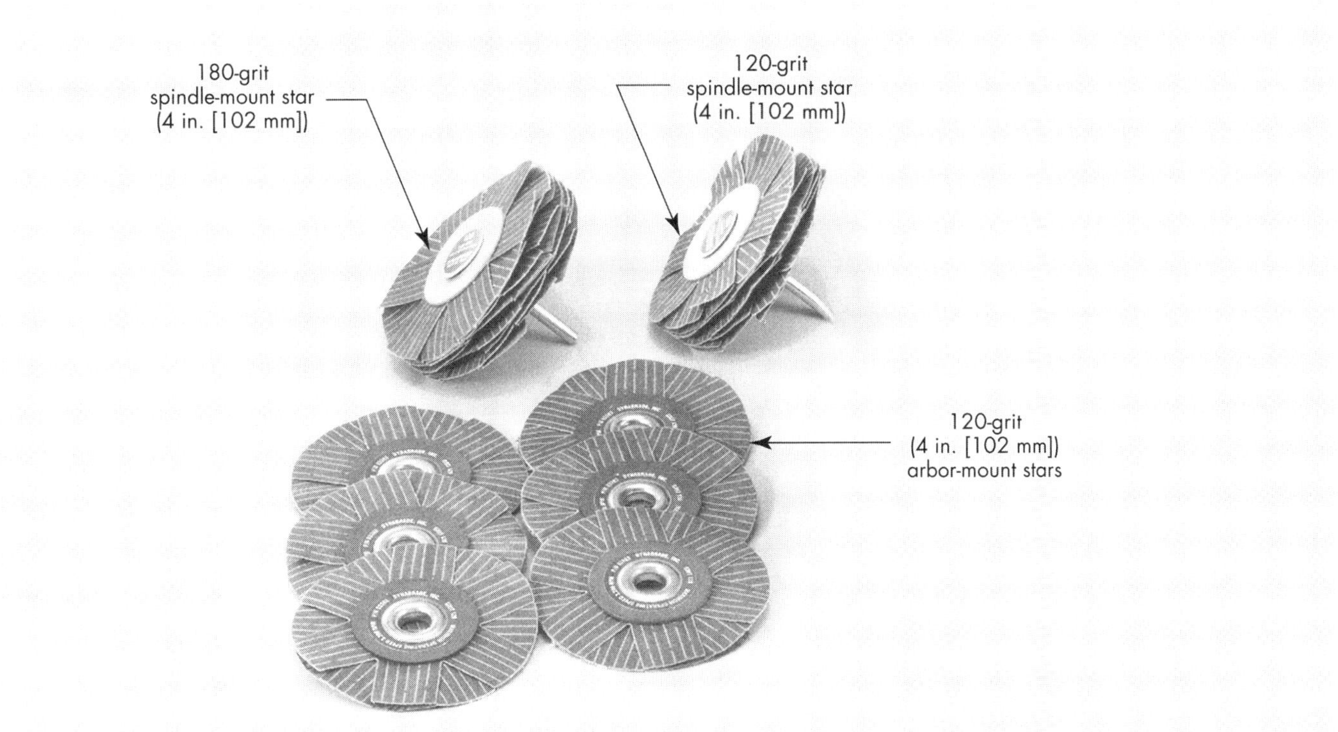

Figure 10-8. Star disks. ***(Courtesy Dynabrade)***

Figure 10-9. Flap wheel with interleaves of nonwoven abrasive-coated synthetic material. ***(Courtesy Superior Abrasives)***

Figure 10-10. Goblet wheel. ***(Courtesy Superior Abrasives)***

punched or drilled holes. These spiral-wound strips are flexible, uniform strips of abrasive wound to a specific diameter for finishing and polishing difficult-to-reach areas, such as grooves, slots, flutes, threads, holes, and other small-diameter openings.

As cartridge rolls wear, they can be unwrapped or merely allowed to wear through to the next layer of abrasive. They are available in sizes as small as .125 in. (3.2 mm). Sizes larger than 1 in. (25.4 mm) would normally be called wheels.

Cartridge rolls come in straight cylinders, tapered rolls (used for reaching into corners and small crevices), and cone or lapped points (which come almost to a sharp point for hard-to-reach areas and deep crevices). Figure 10-11 illustrates the three different taper shapes produced. Lapped points are used for scalloped surfaces, venturis, aircraft skins, and automobile parts.

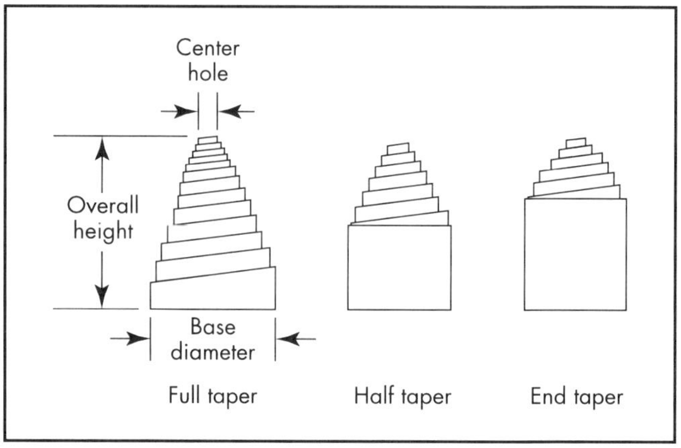

Figure 10-11. Tapered roll configurations. ***(Courtesy Norton Company)***

10.4.8 Bands, Drums, and Sleeves

Bands are bonded-abrasive materials that are spiral wound and glued to a similarly wound inner liner. They are fitted to a mandrel with a jam nut to expand the rubber mandrel and thereby tighten the band. Bands are used for deburring and grinding areas not easily reached by other abrasive tools. Most sleeves have a single ply of material, but some manufacturers produce a multiple sleeve that may wear four times longer than a single one. Diameters range from .125 to 8 in. (3.2–203 mm) and lengths can be as long as 72 in. (1.8 m).

10.4.9 Pads and Hand Pads

Hand pads are hand-sized strips of material (typically 6 × 9 in. [15.2 × 22.9 cm]); they are basically similar to sheets. They come in conventional bonded abrasive as well as abrasive nonwoven nylon materials. They are often color coded to indicate composition or aggressiveness.

10.4.10 Square Pads

Square pads are square-shaped with a hole in the middle to attach a mandrel. The intermittent strokes of the corners of the pad provide a fast action. They are used for forging dies, molds, metal patterns, filets, and corners. Typically, they are .25–.5 in. (6.4–12.7 mm) thick and range from 1 to 4 in. (25.4–101.6 mm) on a side (see Figure 10-12).

10.4.11 Cross Pads

Cross pads (see Figure 10-13) are used to deburr holes. Their shape allows the deburring motor to push them down into the holes to provide effective chamfering and deburring action. Each size will accommodate a wide variety of hole sizes. Operators can see through cross pads, so location of the tool is much easier to verify.

10.4.12 Butterfly Tools

Butterfly tools, which are referred to by a variety of other names, such as bore polishers, result when users place one

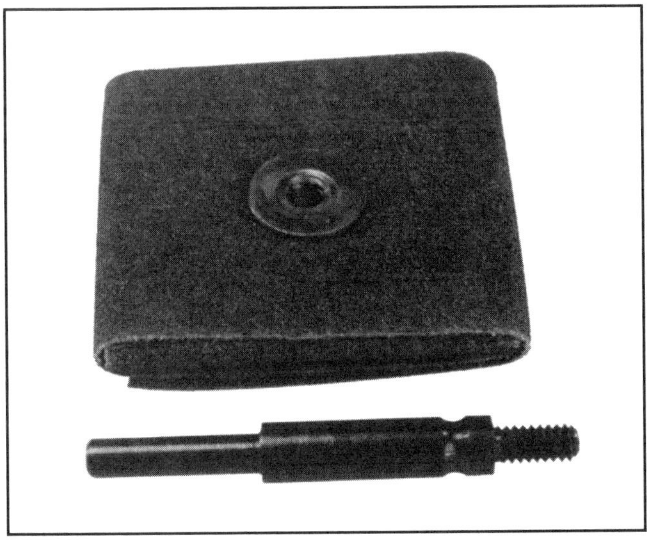

Figure 10-12. Typical design of square pads. ***(Courtesy Norton Company)***

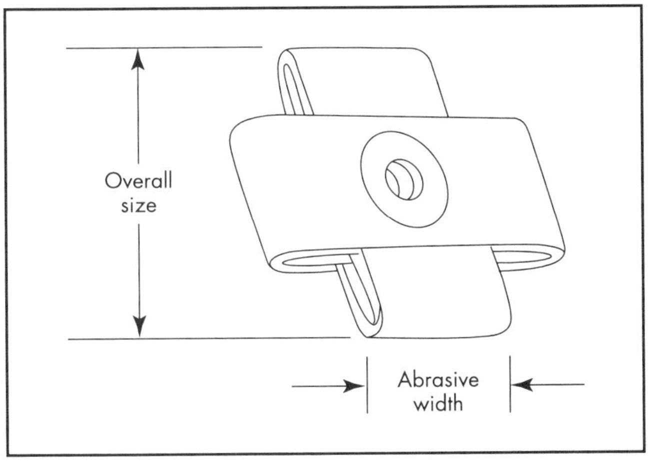

Figure 10-13. Cross pad configuration. ***(Courtesy Norton Company)***

or more strips of abrasive paper between a clothespin-like mandrel (Figure 10-14). When the mandrel is rotated, the paper spreads out and forces itself against the walls and edges of holes. Mandrels can be purchased for this purpose, but many users make their own from wood dowels or slit steel rods. Some producers provide a blunt-nose series rather than a square end. These tools can be used to reach through a hole and deburr the bottom sides, but cross pads are a better choice.

10.4.13 Chamfering Cones

Chamfering cones are small, single-thickness cones that mount onto a metal mandrel or throwaway rubber insert (Figure 10-15). They are designed to provide a 45° chamfer. A variation of the chamfering cone is the center lap designed for 60° angles. It is used to lap center holes and for fine deburring.

10.4.14 Body Cones

Body cones are designed for large parts. Diameters range from 1 to 1.5 in. (25.4–38.1 mm) and cone lengths range from 3 to 5 in. (76.2–127 mm). They are used with vertical spindle sanders on large curved-surface parts.

10.4.15 Abrasive Sponges

Abrasive sponges are coated with abrasive grain to provide a very soft cutting action. These products conform readily to part surfaces and are easy to hold and use by hand.

10.4.16 Abrasive Cords and Tapes

Cloth cord as small as .012 in. (0.31 mm) in diameter is coated with fine abrasives to provide a flexible tool that can reach into small holes and slots. Sizes up to .15 in. (3.8 mm) in diameter and .125 in. (3.18 mm) wide are available. These tools are useful when no other tools will reach into an area, but the abrasive wears quickly.

10.5 MACHINES AND EQUIPMENT

Abrasive-belt machines are available in a wide variety of types to suit specific applications, as shown in Figure 10-16. Most belt machines can be arranged for manual, semiautomatic, or fully automatic operation. Machines used include backstand, swing frame, free-belt roll, vertical, centerless, and sheet and strip grinders, as well as portable and hand-held tools. The basic design of a wide belt machine consists of a contact roll, tracking or idler roll, conveyor bed system, pinch roll assemblies, safety systems, and maintenance features.

Modern bonded-abrasive machines are available with increased power, easy and fast belt-changing devices, and improved controls. Other features include motorized belt-head positioning, rapid head retraction, automatic belt tracking, workpiece gaging, and lubrication systems. Also in use are automatic feeders and loading and unloading devices, including industrial robots (Gillespie 1996).

10.5.1 Contact Roll

The contact roll supports the abrasive belt as it comes into contact with the workpiece. The surface, hardness, and speed of the contact roll determine process characteristics. The abrasive belt normally moves in the same direction as the product being fed through the machine. This is known as climb cutting. In a typical deburring application, the contact roll is rubber covered with a neoprene or Buna-N compound that is 30–45 durometer (hardness) on the Shore A scale. The surface of the roll may be smooth or serrated with a 300 serration and .125 in. (3.18 mm) grooves (Schuster 1977).

In sizing the motor on a deburring or straight-line grain-finishing machine, the general guideline is .75 hp/in. (14.2 watts/m), the power per width of belt. For example, a 25-in. (0.635-m) wide machine belt requires approximately 18 hp (13.4 kW); therefore, a 20 hp (14.9 kW) motor is selected.

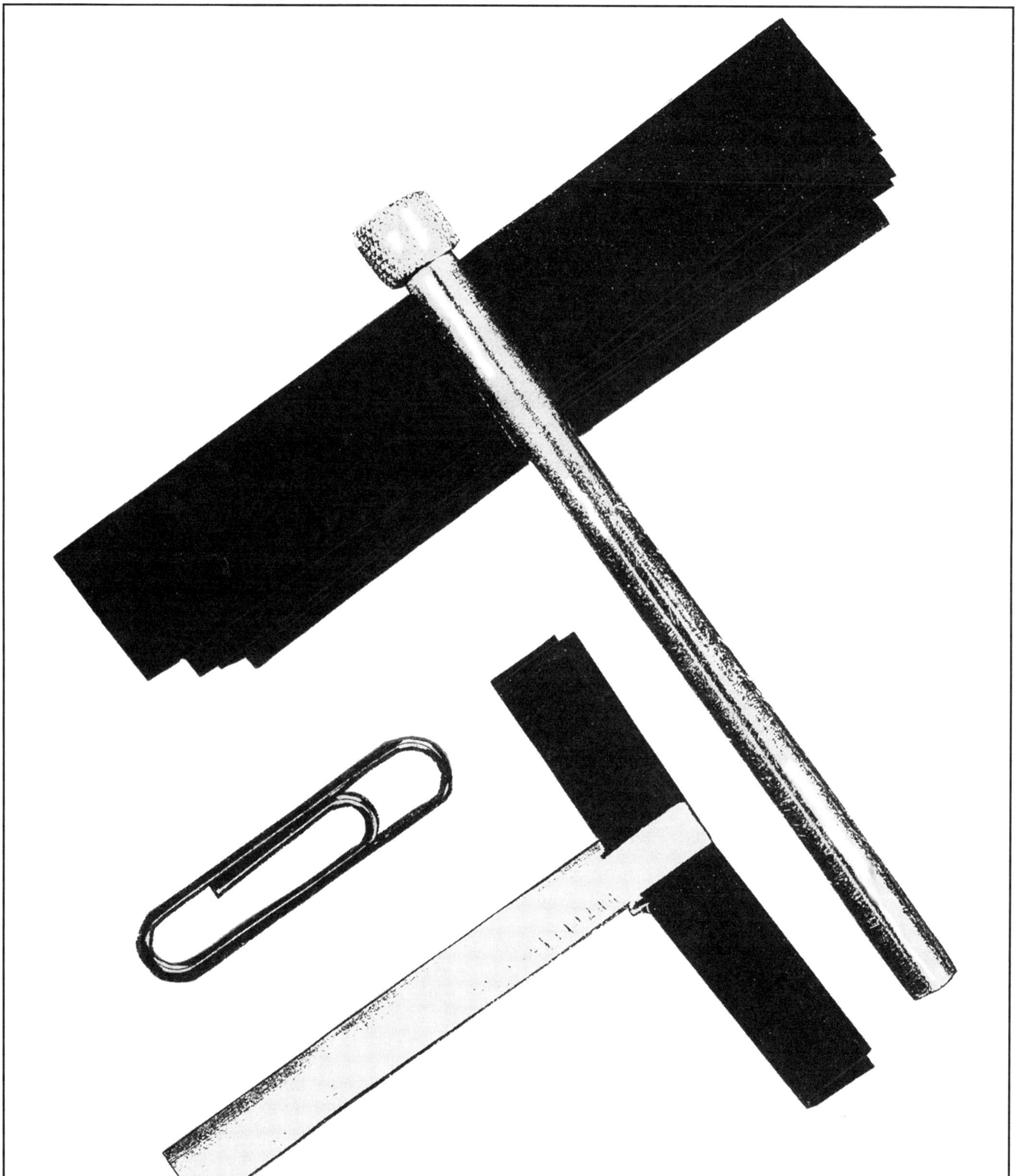

Figure 10-14. Butterfly tool.

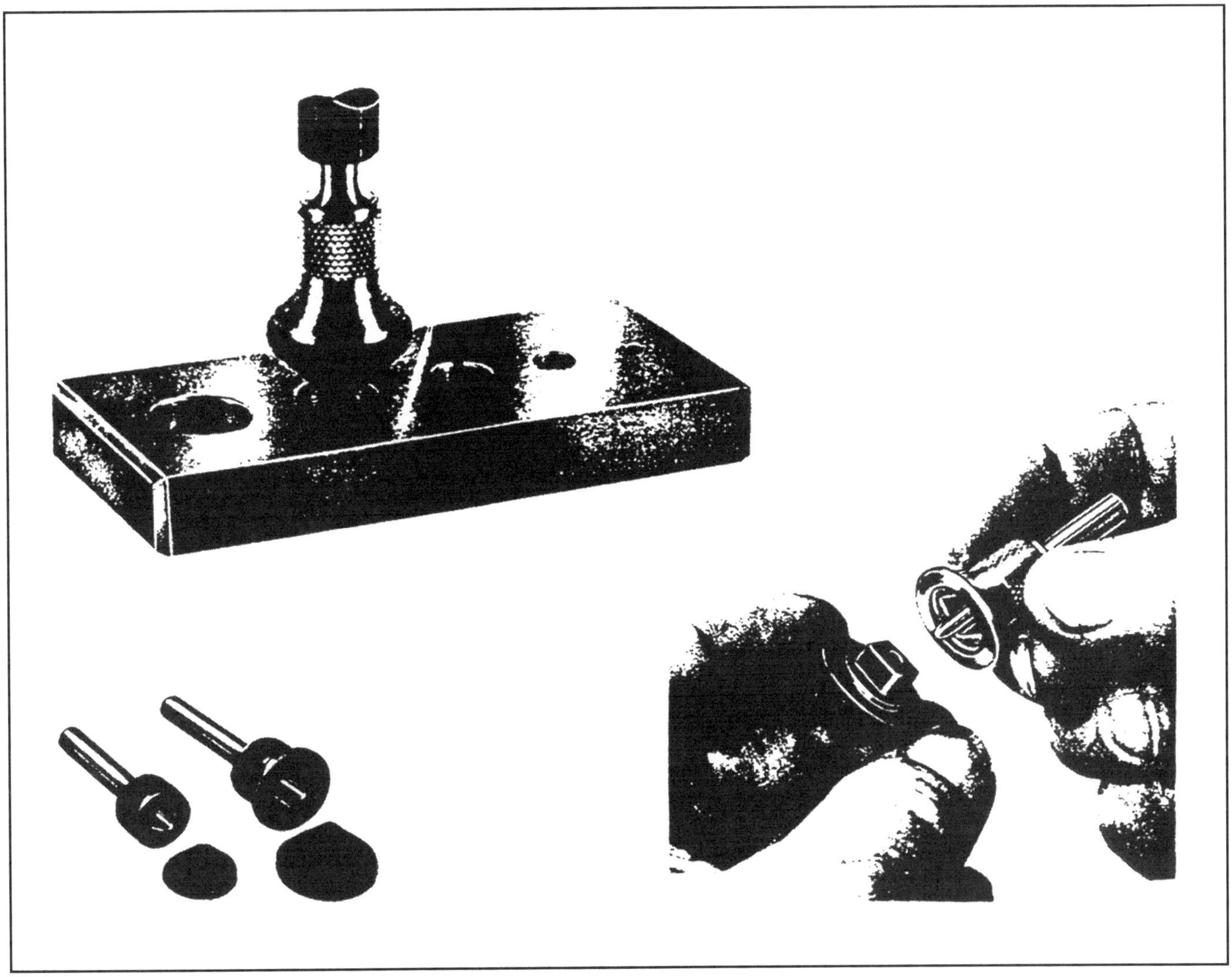

Figure 10-15. Cone shapes for chamfering holes. ***(Courtesy Superior Abrasives)***

10.5.2 Tracking or Idler Roll

The tracking, or idler, roll has two functions: to properly tension the abrasive belt and oscillate the belt in a controlled manner to provide the lowest possible abrasive cost and promote better deburring. When the abrasive belt oscillates back and forth approximately .25–.374 in. (6.4–9.5 mm), the abrasive grains can remove shear burrs from flat stock more efficiently because they approach the burrs from the side instead of head on. The side-angle approach prevents the abrasive mineral from being stripped off and thus substantially lengthens belt life.

10.5.3 Conveyor Bed System

The conveyor bed system uses a nibbed conveyor belt, which provides a high coefficient of friction and efficient feeding of small parts. The belt is frequently referred to as a Ruff-Tex® conveyor belt and can be used with various types of abrasive-belt finishing oils.

When the machine is to be used for heavy deburring, a speed of 15 or 20 ft/min (0.07 or 0.10 m/s) should be selected; for straight-line grain finishing, the normal range is 30–60 ft/min (0.15–0.30 m/s). Different metal configurations will require specific feed speeds and oils to obtain the exact texture of grain finish desired.

The conveyor belt should have a cleaning brush with a vented hood on it to help clean the grinding swarf out of the belt. This cleaning action will promote better feeding of parts into the machine and more efficient operation.

10.5.4 Pinch Roller Assemblies

Wide belt, conveyorized machines should be equipped with double in-feed and out-feed pinch rollers. These rollers are used to securely hold the workpiece during the deburring operation and must be properly adjusted to firmly grip the part. Failure to maintain the pinch rollers or to keep them adjusted for pinch can result in serious injury

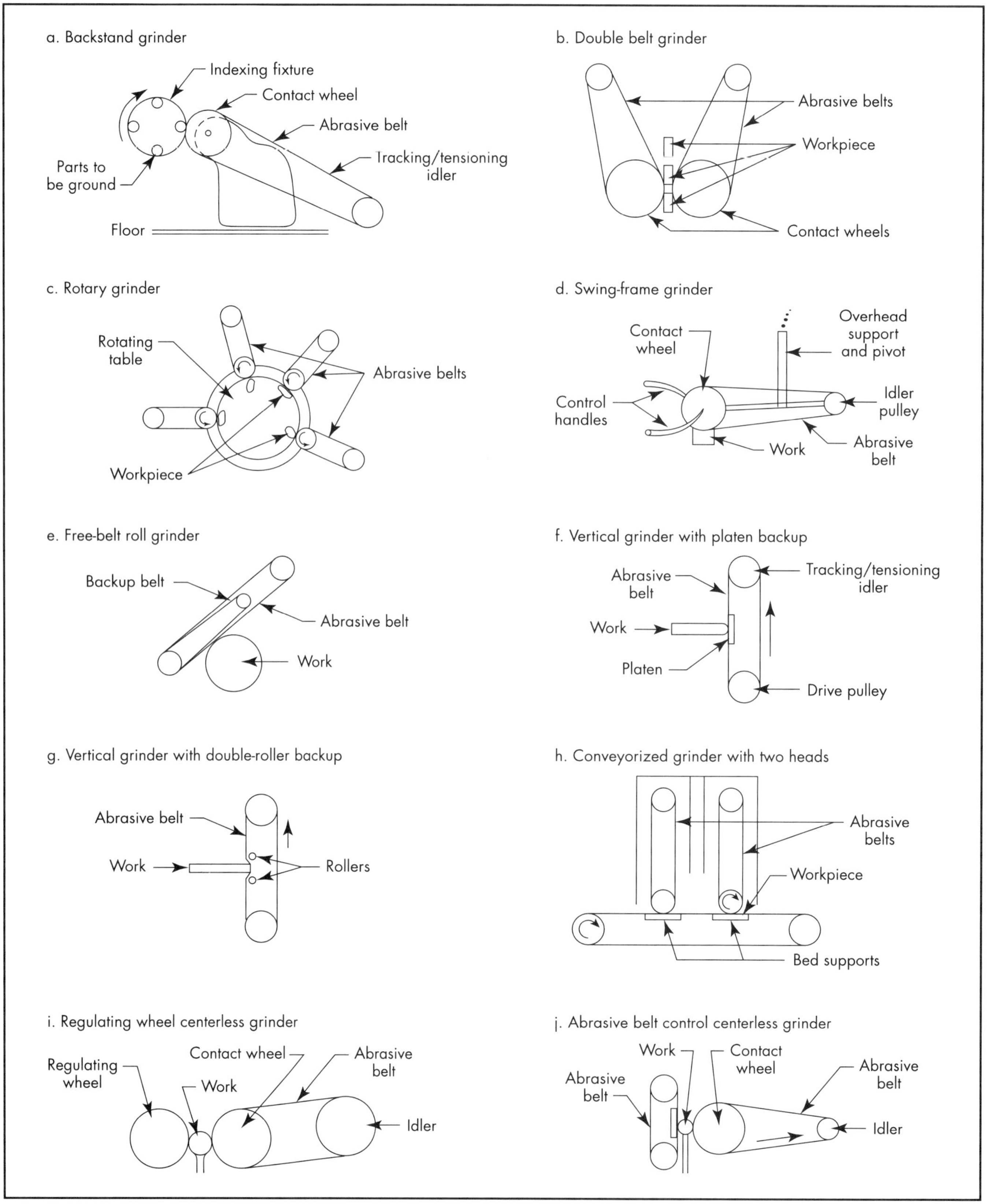

Figure 10-16. Typical configurations and applications of abrasive belt grinders.

to operators. The pinch roll should be made out of neoprene or a Buna-N compound identical to the contact roll.

10.5.5 Safety Systems

Wide belt machines should have safety provisions to (1) ensure that the machine cannot be started if the abrasive belt is not tensioned; (2) allow operators to stop the unit easily from the in-feed operating station without using their hands; (3) ensure that the machine conveyor stops with 4 in. (101.6 mm) of travel or less and the contact roll stops in 4 s or less when the emergency stop is activated; (4) ensure that all moving parts or pinch points are enclosed and protected wherever practical while assuring efficient operation; and (5) operate below OSHA 90 db sound level.

10.5.6 Backstand Grinders

Backstand grinders are some of the most simple, low cost, and versatile abrasive-belt machines (see Figure 10-16a). They are basically polishing lathes that have been converted to abrasive-belt use by adding backstand idlers. Workpieces can be held against the belt manually or with semiautomatic workholding devices, as shown in the figure, for faster production. Longer belts can be used without sacrificing floor space by using two idler pulleys, or the idler can be mounted above the contact wheel or on the wall to conserve floor space.

To permit abrasive-belt grinding operations on existing equipment, arm attachments with the belt spring loaded between a contact wheel and an idler pulley are available for installation on polishing or buffing lathes. Backstand grinders or lathes with abrasive-belt arm attachments can be arranged along work-indexing or conveying semiautomatic machines to perform any number of operations, including, through the use of different belts on successive heads, roughing and finishing operations in a single pass. Figure 10-16b shows a straight-line machine with two backstand grinders for finishing two opposed sides of workpieces. Contact wheels can be contoured or straight, depending on the surfaces required. Such machines are made in a wide variety of sizes and configurations, including straight-line, rectangular, and rotary. A semiautomatic rotary-type abrasive-belt grinder is shown in Figure 10-16c.

10.5.7 Swing-frame Grinders

Swing-frame abrasive-belt grinders (Figure 10-16d) are generally used on large and heavy parts when tolerance is not critical. They can be suspended from overhead for removing weld beads, mounted on wheels for cleaning large plates, or rigidly mounted to a carriage for grinding odd-shaped parts.

10.5.8 Free-belt Roll Grinders

In finish grinding or polishing, the work need not necessarily be applied to the belt at the wheel. It may be "strapped" against the unsupported belt or against a belt made firmer by a parallel backup or saddle belt of leather, rubber, or canvas, as shown in Figure 10-16e. Machining of this sort, which forces the belt to conform to the workpiece, requires a guide yoke or deeply flanged pulley to keep the belt tracking properly. Abrasive-belt attachments such as these are available for cylindrical grinding on center-type machines, such as roll grinders and lathes. Some are bolted to the cross slides of lathes.

With stroke-type finishing machines, workpiece contact is made by stroking the back of a moving abrasive belt with hand blocks, a hand lever with pressure pad, or an automatic traveling head. These machines usually have two or more pulleys over which the belt travels and a movable workholding table. Another type of stroke machine is an automatic gantry. The machine rides on crane rails and traverses stationary workpieces; a contact roll head strokes the belt against the work.

10.5.9 Vertical Grinders

Vertical abrasive-belt grinders are used in several ways, with workpieces generally being supported on tables. Work can be applied against a section of unsupported belt, a platen (Figure 10-16f) or contact wheel behind the belt, or a section of belt between two rollers (Figure 10-16g). The roller arrangement is useful for grinding radii on the edges of workpieces. Belts can be arranged to operate vertically, horizontally, or at an angle, and the worktable can be adjusted to various angular settings on most machines. Oscillation of the workpieces across the face of the belt is desirable for uniform belt wear and surface finish. Cutting rates can be controlled by hand, air, mechanical means, or hydraulic power.

Vertical-head contact-wheel surface grinders can be of either the reciprocating-bed (or table) type or the continuous through-feed type with a workholding conveyor. Many continuous-feed contact-wheel surface grinders are made with two or more belt heads (Figure 10-16h) and a common conveyor belt to carry workpieces under the belts. By varying the grit sizes on successive belts, parts can be rough and finish-ground in a single pass. These machines can obtain close tolerances on flatness, parallelism, and thickness.

Figure 10-17 shows three of the many types of multiple-head, conveyorized abrasive-belt machines for roughing and finishing in one pass. Figure 10-17a illustrates a continuous through-feed machine, which is often equipped with a variable-speed drive unit. Figure 10-17b shows a two-head machine, one head equipped with an abrasive belt and the other head with a wheel or brush. Figure 10-17c has air or hydraulically controlled pinch rolls for finishing coil or strip stock as well as long workpieces.

Multiple floating-type abrasive-belt heads can be used to rough and finish-grind contoured parts in one pass. Each head floats independently, with rise and fall controlled by the workpiece contour, a cam, or a template. Continuous or reciprocating conveyors can be used to move parts beneath the heads.

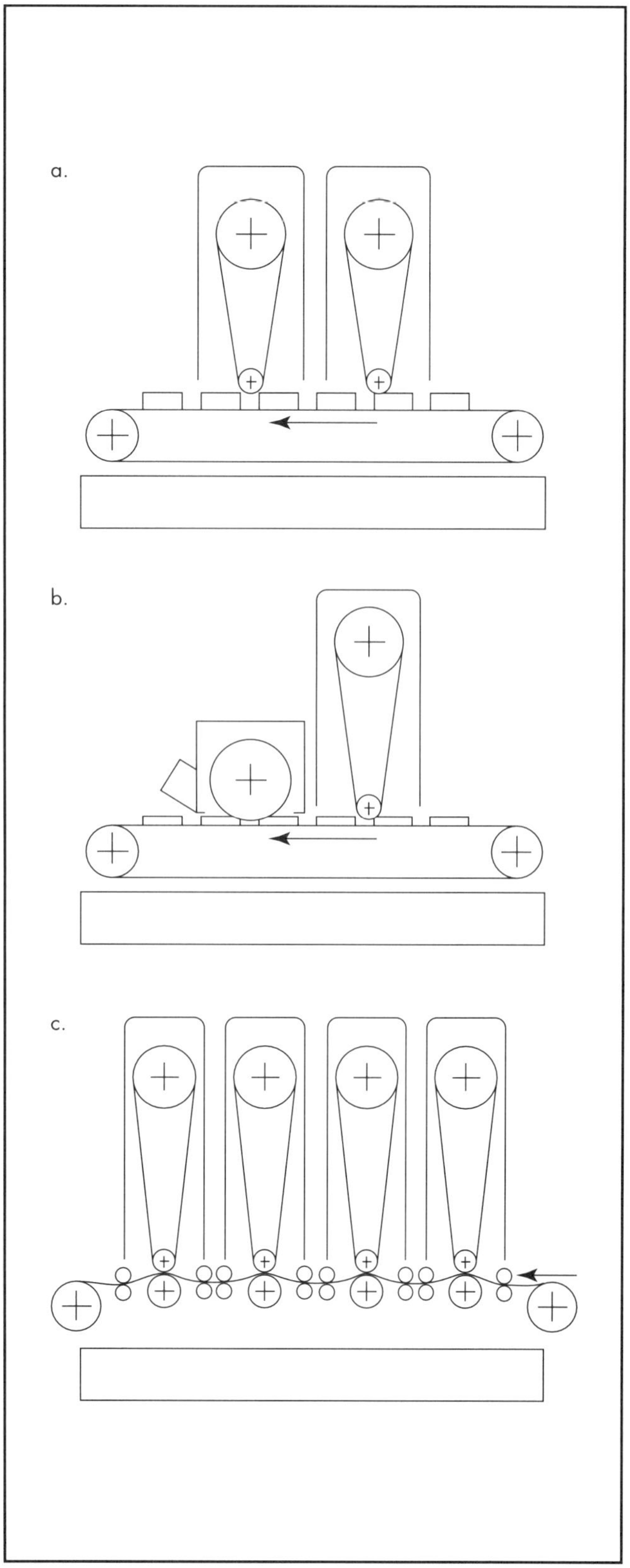

Figure 10-17. Multiple-head, conveyorized abrasive-belt machines for roughing and finishing in one pass.

10.5.10 Hand-held Grinders

The number of hand-held air, flexible cable, or electric motor tools for bonded-abrasive products is vast. As many as 300 different tools exist for use with these products, and each year more designs enter the workplace. Hand-held sanders today include units with integral vacuums, a variety of ergonomic features, low-heat dissipation devices, better operator visibility, and lighter weight. Flexible stands are available for some tools to relieve strain on the operator's body. Straight shank, angle, orbital, vibrating, and other motions are available. Hand-held grinders include disk sanders, die grinders, depressed center wheel grinders, mini belt sanders, and wet grinders (which have an internal water supply).

10.5.11 Contact Wheels, Rolls, and Platens

For most grinding operations, contact wheels, rolls, or platens are necessary as a backup support behind the abrasive belt when pressure is applied. Contact wheels are usually made of cloth or rubber, but special compositions or metal are sometimes used. Hardness or density of the contact wheel affects stock removal and the finish produced. The harder the wheel, the greater the stock removal and the coarser the finish. Wheels typically range in durometer hardness from 15 (soft) to 90 (hard). Uneven workpiece shapes need softer wheels. Large surfaces will require coarser serration, which results in a faster stock removal rate. Larger wheels provide more contact, which, in turn, results in better surface quality. Smaller wheels result in faster stock removal.

Hand-held tools utilize fixed mandrels, air inflatable rubber drums, and rubber expanding drums. Phenolic backup pads are also in use for light work. Hand-held rubber sanding blocks are often used for flat sanding of metals and wood. They hold strips of paper in a manner similar to vibrating and orbital sanders.

10.6 STOCK REMOVAL RATES

Stock removal rates are proportional to pressure applied and also depend on the type of bonded abrasive and cutting fluid used, surface finish required, and workpiece material. A stock removal rate as high as 30 in.^{3}/min per inch (323 mm^3/s per mm) of belt width is practical with standard belts. By gradually increasing the pressure, it is possible to obtain a nearly constant cutting rate for the life of the belt. When using contact wheels, grinding pressure is often expressed as the horsepower (watts) available per inch (millimeter) of belt width for driving the contact wheel. This can vary from .5 hp/in. (14.7 W/mm) for light stock removal to 30 hp/in. (881 W/mm) of belt width for heavy stock removal.

10.6.1 Belt Speeds

Belt speeds of 5,000–9,000 ft/min (25.4–45.7 m/s) are generally the most satisfactory for grinding ferrous and nonferrous metals. Toughness of the material to be ground is more critical than hardness, and some materials, such as titanium, are best ground at belt speeds in the 1,500–3,000 ft/min (7.6–15.2 m/s) range.

When using bonded abrasives, the coarser the abrasive, the faster the stock removal rate. When the surface finish produced by a coarse abrasives is too rough, a second run using a finer grit product is used. If a second run material cannot be used, then a grit size that produces an acceptable stock removal rate while yielding the desired finish is used.

10.7 OPERATING PARAMETERS

Operating parameters that must be considered in bonded-abrasive deburring include: the material, hardness, and condition of the workpiece; characteristics of the bonded-abrasive product (grit size, type, backing, bonding material, and coating method); speed of the machine using the bonded-abrasive product; workpiece feed rate; backup support; belt tension; and lubricant used. The direction of rotation of the belts in relation to workpiece feed is an often overlooked factor in improving finishes. Rotating belts in the direction of workpiece feed (climb cutting) usually improves the finish. Rotating belts *against* the direction of workpiece feed (conventional cutting) provides additional cutting action. Bonded-abrasive deburring can be effective with belts that are in planes parallel to the workpiece surfaces, but belts set at a 45° angle to the edge are more effective.

Both narrow and wide bonded-abrasive belts are being used extensively for finishing operations. Narrow belts are 6 in. (152 mm) or less in width and 42–168 in. (107–427 cm) long. Belts to 126 in. (320 cm) wide are being used to finish sheet and plate stock.

Deburring and deflashing operations are normally performed with intermediate (40–120 grit) grades of abrasive. For finishing operations, finer (150–400 grit) grades are used. For some mirror-finishing operations, abrasive grades as fine as P1200 are necessary prior to buffing to a mirror finish.

Versatility is a major advantage of using abrasive belts, since machines can be quickly converted from heavy stock removal to finishing operations, or for grinding different materials or parts, by simply changing the belts and/or contact wheels. Most materials can be ground with abrasive belts, but the harder the material, the more difficult the operation.

Flap wheels should be run counter to the direction of feed of the workpieces. For a given finish, use grit sizes one or two times larger than those used in belts. Finishes will be improved by using oils and grease; high rotation speeds will also improve finishes. Files, abrasive stones, and cutting tools can be used to shape flap wheels to particular parts.

10.7.1 Belt Speeds

Fast belt speeds generally produce smooth surface finishes, but slow speeds are often required for difficult-to-grind materials. Table 10-2 presents typical belt speeds.

Table 10-2. Typical bonded-abrasive belt speeds used for various metals

Workpiece material	Belt speed, ft/min (m/min)
Carbon steels	3,500–6,500 (1,067–1,981)
Stainless steels	3,000–5,500 (914–1,676)
Nickel chromium alloys	2,500–3,500 (762–1,067)
Inconel	3,000–5,000 (914–1,524)
Tool steels	4,000–7,000 (1,219–2,134)
Titanium	2,000–3,500 (610–1,067)
Gray cast iron	4,000–6,000 (1,219–1,829)
Aluminum alloys	4,500–7,000 (1,372–2,134)
Brass bronze alloys	5,000–7,000 (1,524–2,134)
Copper alloys	4,500–5,500 (1,372–1,676)

(Courtesy 3M Company)

10.7.2 Backup Support of Belt

Although some light polishing operations are performed by pressing the workpieces against slack-of-belt locations, most operations require support in the area of workpiece contact. Such support is provided by contact wheels and rolls or platens.

Contact Wheels and Rolls

Contact wheels and rolls are available in a variety of materials, sizes, and shapes. Cloth wheels are common for finishing operations. Rubber wheels, both serrated and plain, are also used. Wheels that are softer and larger in diameter generally produce smoother finishes.

Platens

Support with platens is provided for many types of flat surfacing operations. For precision work, the platens are often carbide faced and provided with a coolant system.

10.7.3 Belt Tension

Belt tension is generally lower in finishing operations (4–10 lb/in. [700–1,751 N/m]) than for stock removal operations (40 lb/in. [7,005 N/m] or more). For high-pressure abrasive-belt grinding operations, belt tensions approach 100 lb/in. (17,512 N/m).

10.7.4 Lubricants/Coolants

Bonded-abrasive belts are sometimes sprayed intermittently with a lubricant coolant. Many machines, including those used for finishing printed circuit boards, operate wet and produce finer finishes as a result.

10.7.5 Belt Applications

Tests have recently been performed on punched holes in steel and aluminum with a belt speed of 3,000 ft/min (15.2 m/s), a conveyor speed of 30 ft/min (0.15 m/s), using a belt lubricant, and a 120-grit aluminum oxide closed-coat abrasive belt. To produce a square edge on large holes in aluminum sheet, a hard contact roll (60–70 durometer on the A scale) was necessary. To produce a slight hollowing or chamfering of the hole, a soft contact roll (30–40 durometer on the A scale) was necessary. Regardless of the roll used in the test, the burr was cut off at its root instead of pushed down into the hole. The individual scratches produced by the hard roll were shorter and more "brush-like" than those produced by the soft drum. This is a result of the soft roll "mushing" on the surface of the panel sheet and causing a longer contact than is possible with the hard roll, which is only able to make tangent point contact (Claus 1977).

When a good production finish is pursued with abrasive sizes finer than 150 grit, a soft durometer (30–50 durometer, A scale) should be used to provide a good, clearly grained surface. If harder contact rolls are used, contact roll chatter marks (approximately .10–.375 in. [2.54–9.53 mm] apart) or "belt-splice" marks (approximately .5–1.5 in. [13–38 mm] apart) will show up on the surface of the panel perpendicular to the direction of feed and all the way across the panel.

Finishes generated with three-dimensional nonwoven abrasives varied from 120 μin. (3.05 μm) with oil to approximately 30 μin. (0.76 μm) when a belt sequence of 120–180–240-grit Scotch-brite® was used with oil. Table 10-3 compares the microfinishes generated with various abrasive sequences.

Deburring is normally done with a coarse grade of abrasive (60–120 grit); finishing is done with a fine abrasive (150–280 grit) and possibly supplemented with a final finish on a nonwoven nylon abrasive-impregnated belt or brush, such as a Scotch-brite blend and finish belt). Trying to deburr and finish with a fine-grit belt can produce a streaked panel resulting from the burrs tearing off grain from the face of the belt.

Table 10-3. Straight-line grain finishes for various abrasive sequences

Best Sequence	Aluminum Finish (μin.)*	Steel Finish (μin.)
120 to 180-grit Scotch-brite®	50–70	30–47
100 to 150-grit Scotch-brite	58–89	45–77
120-grit Scotch-brite	58–120	52–90
100-grit Scotch-brite	80–155	70–138
80-grit Scotch-brite	80–230	98–155

Note: The coarser abrasive was used in the first pass through the machine; the finer abrasive was used in the second pass. All parts were processed on a machine using a 11.5 in. (292.1 mm) diameter, 35 durometer (soft) contact roll running with the direction of product feed (climb cut) on cold-rolled 1018 steel and 3000 H14 aluminum at a belt speed of 3,000 ft/min (15.2 m/s), a conveyor speed of 30 ft/min (0.15 m/s), and a belt lubricant.

* 1 μin. = 0.0254 μm.

An abrasive belt speed of 3,000 ft/min (15.2 m/s) is normally used for deburring and straight-line grain finishing. Figure 10-18 illustrates where deburring and finishing fit into the spectrum of bonded-abrasive grinding. Deburring generally occurs in the rubbing zone and momentarily jumps into the heavy stock removal zone when actual deburring is done on the edge of the punched holes. Straight-line grain finishing occurs in both the rubbing zone and the plowing zone. The plowing zone is primarily applicable to stainless steel sheet polishing operations when the metal surface is furrowed to make the metal flow.

10.8 RESEARCH ON THREE-DIMENSIONAL ABRASIVES

This section presents the results of a study that compared three-dimensional abrasive wheels with bonded abrasives and abrasive-filled fiber brushes (Claus 1977). Table 10-4 outlines the average initial size of burrs encountered in the study.

When the proper three-dimensional abrasive product was used, burrs produced during blanking operations on both aluminum and stainless steel were easily removed. When a very fine cut-and-polish wheel was used, a radius of .002–.005 in. (50.8–127.0 μm) was put on the edge.

Three-dimensional abrasives were used on aluminum to remove burrs on drilled holes and to radius edges. In the case of a large rollover burr where the effective width is

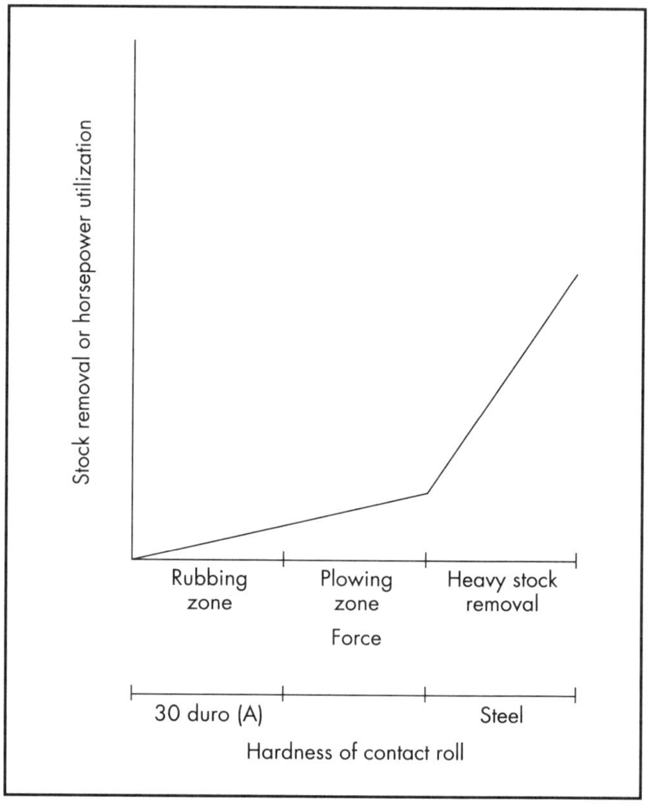

Figure 10-18. Cutting action of belts.

Table 10-4. Average burr size removed by nonwoven nylon belt on various materials

	Drilling burrs				Shearing burrs		Stamping burrs	
	Entrance		Exit		Exit		Exit	
Material	Height (in.)*	Width (in.)	Height (in.)	Width (in.)	Height (in.)	Width (in.)	Height (in.)	Width (in.)
2024-T3 aluminum	.0026	.0038	.0017	.0032	.0048	.0048	.0020	.0020
5052 aluminum	.0024	.0038	.0055	.0076	.0011	.0020	.0045	.0050
6061-T6 aluminum	.0024	.0046	.0026	.0043	—	—	—	—
304 stainless steel	.0030	.0071	.0241	.0113	.0013	.0039	.0046	.0050
Cold-rolled carbon steel	.0048	.0058	.0161	.0089	.0026	.0089	.0021	.0038

Note: Measurements include effective heights and widths when they were greater than the height and base width of the burr.

*.001 inch = 25.4 μm.

(Courtesy 3M Company)

greater than the effective height, a "crown" (slightly raised metal) was left around the hole. This phenomenon results from the conformability of three-dimensional abrasives. When the wheel cuts through the burr, it is also cutting the parent metal outside the burr at about the same rate. Since the burr was originally higher than the parent metal, after the burr is removed, the metal that was directly under the burr will be higher than the parent metal. (Figure 10-19 illustrates a folded-over burr that would accentuate this effect.)

The three-dimensional abrasive products that successfully removed burrs from drilled holes in aluminum did not have enough cutting ability to remove burrs on drilled holes in stainless and cold-rolled steel. For stainless and cold-rolled steel, a different three-dimensional abrasive product with more cutting ability proved more effective; but when the burr was removed, a smaller burr was occasionally rolled into the hole, and, in that case, the edge was not significantly radiused by this product.

Burrs resulting from blanking and drilling operations on aluminum, stainless, and cold-rolled steel were removed with 120-grit bonded-abrasive belts. In cases where the part was slightly warped so that the burr occurred below the plate where the bonded abrasive passed, the part of the burr below that plane was not touched. To remove this burr, a large amount of stock in the plane of the burr also would have to be removed.

In many flat finishing processes, the main burr on the exit side of a drilled hole is removed, but a smaller burr may be rolled into the hole. The results of the study indicate that this problem occurs quite frequently with bonded abrasives and abrasive-filled fiber brushes and only occasionally with three-dimensional abrasive products. This suggests that parts could be processed with a combination of products: bonded abrasives to remove large burrs and three-dimensional abrasives to radius the edge and pull the burr out of the hole.

It is noteworthy that inspection of the edges deburred with bonded abrasives showed that a very minute (under .0005 in. [12.7 μm]) radius, if any, was generated while removing the burrs; most of the time, the two sides of the edges formed nearly right angles. In many instances, such as in hydraulic flow systems, the ability to generate a sharp, deburred edge is highly desirable.

Abrasive-filled (80-grit and 180-grit) fiber brushes were used on aluminum burrs and produced results similar to those produced with bonded abrasives; but because brushes are conformable, burrs were not missed as they can be when using bonded abrasives. Brushes are not recommended for deburring steel because they do not have enough cutting ability; therefore, they were not tested on steel.

In many cases, parts that will be subject to stress require radiused edges. The aircraft industry, for example, has critical specifications for nearly every edge of most parts. Using three-dimensional abrasive products, a radius of .002–.005 in. (50.8–127.0 μm) can be generated and burrs can be removed in one operation. In other cases, however, radiused edges are not desirable, such as in regulatory devices for fluid flow. In these cases, burrs must be removed with something that will preserve the sharpness of the edge.

The study indicates that all three products examined (three-dimensional abrasive wheels, bonded abrasives, and abrasive-filled fiber brushes) can remove any burrs formed by aluminum operations described in Section 10.5. On steel, the three-dimensional products seemed to lack enough cut to completely remove burrs around drilled holes; bonded

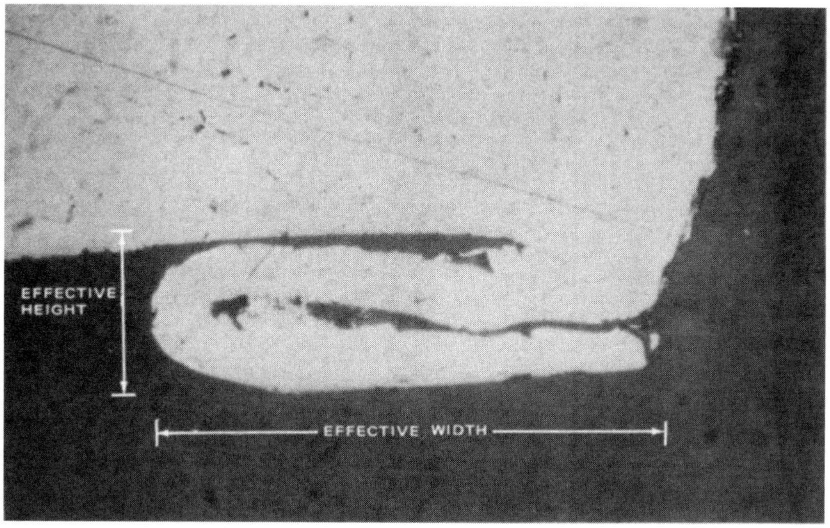

Figure 10-19. Drilled hole exit burr rolled over (1,753 × magnification). ***(Courtesy 3M Company)***

abrasives performed in much the way they did on aluminum; and abrasive-filled fiber brushes were not recommended.

The study found that abrasive-filled fiber brushes and three-dimensional abrasives are limited in their cutting and stock removal capabilities; and bonded abrasives are limited by their rigid, nonconforming construction. All three types of products roll small burrs into holes; however, this occurs more frequently with bonded abrasives and fiber-filled abrasive brushes than with three-dimensional products.

All three products can be run wet or dry; however, in most instances, bonded abrasives and abrasive-filled fiber brushes are recommended to be run dry, and three-dimensional abrasives are recommended to be run wet.

The study used belt speeds of 3,500–4,000 ft/min (17.8–20.3 m/s), feed rates of 15 ft/min (0.08 m/s), two passes per side, and .25 in. (6.4 mm) oscillation at 3.67 cycles/s. Brushes and bonded abrasives were operated dry at 30–50 ft/min (0.15–0.25 m/s).

10.9 FILM-BACKED ABRASIVES

Film-backed abrasives, which are made by electrostatically coating and resin-bonding fine abrasives to a thin, flexible backing of polyester film, are capable of producing fine finishes of 2 μin. (0.05 μm) or less. These products are available in belt, roll, sheet, and disk forms, with sheet being the least common. Aluminum oxide and silicon carbide are used as abrasives, with aluminum oxide the more common. The tough, yet flexible polyester film backing is impervious to water, which allows the use of water-based lubricants. The hardness of the film minimizes sinking of the abrasive into the backing, thereby preserving the cutting edges of the abrasive grains. Film thickness, typically .003 or .005 in. (0.08 or 0.13 mm), is more constant than with cloth or paper backings. Film has greater wear resistance than other materials and thus has a longer product life.

Rolls of film-backed bonded abrasives are available in lengths up to 450 ft (137 m). Figure 10-20 shows a typical configuration for continuously or intermittently presenting fresh abrasive to workpieces. Some machines are designed to oscillate the abrasive, which produces a crosshatch pattern on surfaces. Abrasive grain diameters of .0005–.0020 in. (13–50 μm) are commonly used for finishing operations.

10.10 ABRASIVE CHEMICAL DEBURRING

Film lapping with the addition of chemicals to provide solid-state chemical reactions has been used successfully on hard ceramics, such as silicon nitride and ceramic bearings and spindles. A diamond-coated film is held against the cylinder to be finished at a force of 75–115 N, and the part is rotated at speeds of 33–328 ft/min (10–100 m/min). The film moves at rates of .3 ft/min (91 mm/min) and oscillates axially at a frequency of 560 cycles/min. Lapping abrasives vary from 2.0–5.9 μin. (0.05–0.15 μm). This process is used for parts having surface finish roughness values of R_{max} = 45 nanometers (Suzuki et al. 1992).

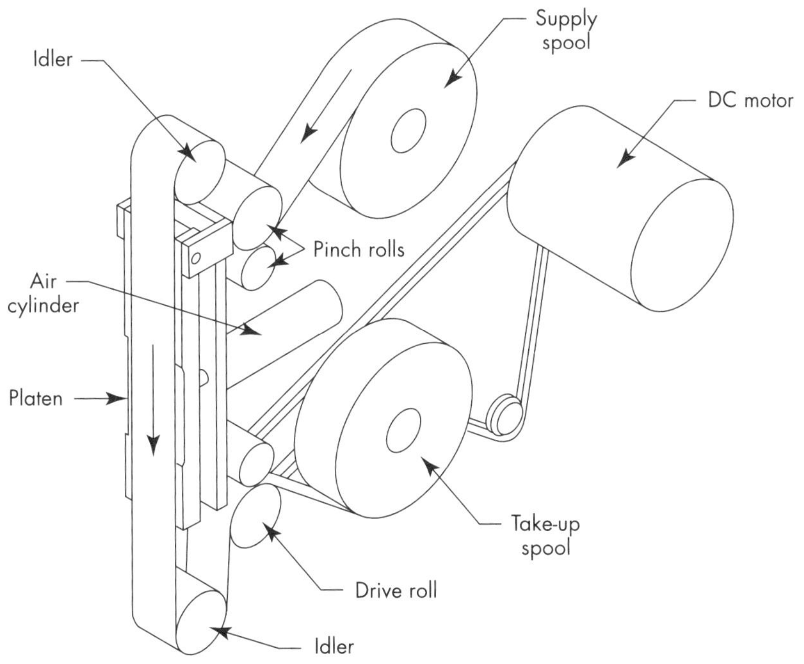

Figure 10-20. Typical mechanical arrangement for using microabrasive tapes to produce fine finishes. ***(Courtesy 3M Company)***

10.11 ENVIRONMENTAL, HEALTH, AND SAFETY CONCERNS

Bonded-abrasive finishing involves several environmental, health, and safety considerations. Dry sanding generates dust, which may include metal or plastic part particles and fines, resins, glue, paper, cloth, and minerals. Dust can be an explosion hazard as well as a health hazard. Some workers may be allergic to any of the components in the dust. Workers who sand beryllium copper often report a sharp bitter taste in their mouths. Chronic beryllium disease is a serious concern and workers must be protected from beryllium dust. Skin irritations also occur in some situations. Wet finishing resolves most of these concerns, but waste products can be an environmental hazard in the wastewater.

Rotating equipment throws particles (of mineral as well as burrs and chips), which can present an eye hazard. Parts that are hand-held against rotating equipment can be hurled through the air and become dangerous projectiles. Hands and arms can be caught between the part-carrying belt and the sanding belt, resulting in bodily injuries.

Large accumulations of belt or disk materials when hot may explode or combust and create fires or smoke. For example, after three fires occurred on one flat-lapping machine (large rotary sander) it was observed that the titanium being ground produced easy-to-ignite particles that when added to fine particles of paper, aluminum, glue, and other debris, and combined with the hot temperatures occurring during sanding operations, created a fire hazard. Cleaning out sumps or catch places is essential in dry operations. Normally, an exhaust system is used to contain dust in the work area.

Mixing aluminum particles with sparks from sanded steel parts potentially sets up the conditions necessary for a fire and an explosion, so mixing materials is discouraged unless sufficient controls are provided to prevent ignition of materials.

Machinery guards need to be kept in place to prevent machinery from grabbing clothing and pulling operators into rotating parts. Robotic loaders require special care to prevent accidents from unexpected robot movements. Any operation involving placing hand-held products against burr-laden edges can easily result in severe cuts and even amputation of fingers and hands. Operators are generally encouraged to use backing sticks (sandpaper backed by a wooden rod to provide the necessary sanding force) or other tools to keep hands away from sharp edges and rotating products.

Abrasive-bonded wheels and disks may present a safety hazard by erratically propelling broken pieces. Manufacturers frequently do not list safe operating speeds for soft materials, but many of the unified wheels have manufacturers' recommended maximum operating speeds.

References

Claus, Joseph J. 1977. "The Characterization and Removal of Burrs with Three-dimensional Abrasives." Technical paper MR77-426. Dearborn, MI: Society of Manufacturing Engineers (SME).

Gagliardi, John J., and Duwell, E.J. 1989. "Coated Abrasives Made with Abrasive Grit Clusters for Constant Performance Surface Finishing." Technical paper MR89-139. Dearborn, MI: Society of Manufacturing Engineers (SME).

Gillespie, LaRoux K. 1996. *Guide to Deburring, Deflashing, and Trimming Equipment, Supplies, and Services*. Kansas City, MO: Deburring Technology International.

Miyatani, Takashi, Harada, Masahi, and Takazawa, Koya. 1981. "Newly Developed Deburring Machines for Precision Parts in Mass Production." Technical paper MR81-382. Dearborn, MI: Society of Manufacturing Engineers (SME).

Schuster, T.G. 1977. "Use of Wide-belt Conveyorized Machining for Deburring and Surface Conditioning." Technical paper MR77-433. Dearborn, MI: Society of Manufacturing Engineers (SME).

Suzuki, K., Uematsu, T., Ohashi, H., Kitajima, K., Suga, T., and Imanaka, O. 1992. "Development of a New Mechanochemical Polishing Method with a Polishing Film for Ceramic Round Bars." *Annals of the CIRP,* 41 (1):339–42.

Bibliography

Braney, John Q. 1985. "Nonwoven Nylon and Filament Abrasive Applications." Technical paper MR85-465. Dearborn, MI: Society of Manufacturing Engineers (SME).

Carlson, Glen A. 1993. "CNC Finishing Systems Can Improve Your Buffing, Brushing, and Polish Techniques." Technical paper MR93-137. Dearborn, MI: Society of Manufacturing Engineers (SME).

Cubberly, W.H. and Bakerjian, R. 1989. *Tool and Manufacturing Engineers Handbook, Desk Edition*. Dearborn, MI: Society of Manufacturing Engineers (SME).

Drozda, T.J. and Wick, C. 1983. *Tool and Manufacturing Engineers Handbook,* vol. 1, 4th edition. Dearborn, MI: Society of Manufacturing Engineers (SME).

Gagliardi, J.J., and Duwell, E.J. 1989. "Mechanisms of Grinding Using Coated Abrasives Made with Abrasive Grit Clusters, Mechanics of Deburring and Surface Finishing Processes." *ASME Production Engineering Division Conference Proceedings,* PED-vol. 38. New York, NY: American Society of Mechanical Engineers (ASME).

Kitamura, Koichi, and Totori, Takaaki. 1994. "Application of Lapping Film for High-quality and High-performance Parts—Development of New Lapping Film and Instance of Its Application." *Proceedings of the 3rd International Conference on Precision Finishing and Burr Technology.* Kansas City, MO: Deburring Technology International: 155–61.

Murphy, Michael. 1997. *Metal Finishing Guidebook and Directory*. New York, NY: Elsevier Science.

Poferl, Mark. 1988. "Applications of Wide-belt Finishing Equipment and Abrasive Media." Technical paper MR88-181. Dearborn, MI: Society of Manufacturing Engineers (SME).

Sandberg, Paul B. 1987. "Deburring and Fine Finishing of Flat Stock Using a Combination of Abrasive Media." Technical paper MR87-161. Dearborn, MI: Society of Manufacturing Engineers (SME).

Yanagi, Kazuhisa, Tsukada, Yoichi, and Yamada, Ryuichi. 1992. "Surface Topography Assessment of High-precision Coated Abrasive Tape and Estimation of Its Machining Performance." *Annals of the CIRP*, 41 (1):377–80.

11 Brush Deburring

Power-driven brushes are used in almost every major metal industry. They deburr, clean, descale, polish, edge blend, and texturize parts. Brush deburring has many advantages, which is why it is one of the most widely used deburring processes. Brushing is fast, relatively inexpensive (just a few cents per part in some instances), and exceptionally flexible. It can be a noncontaminating process and is generally "environmentally friendly." It is readily adaptable to manual or automatic equipment and requires very little operator training or floor space. Brushing is often combined with bonded-abrasive deburring and buffing to produce final part finishes and edges. Many equipment manufacturers provide brushing machines.

The rotary action used in brushing is a major advantage of the process because it allows a tremendous variety of driving motors and fixturing to be employed. The industrial usefulness of brushing is broadened even further by the ability to change variables, such as bristle material, diameter, and length. Brushing is also safe and simple to use.

The following process variables are involved in brush deburring:

- Brush style.
- Brush design within a given style.
- Brush materials.
- Filament length.
- Face width.
- Externally applied abrasives.
- Coolants.
- Brush rotational speed.
- Brush longitudinal feed rate.
- Brush contact area (degree of interference).
- Burr size.
- Burr location.
- Part material.

With the wide variety of nontraditional materials available today, the lines of distinction between brushing, sanding, and buffing have become blurred. A single tool may include components of all three processes, and some manufacturers produce machines that incorporate cutting as well.

11.1 TYPES OF BRUSHES

There are five primary types of power deburring brushes:

1. Radial (wheel) brushes.
2. Cup and disk brushes.
3. End brushes.
4. Wide-face or cylinder brushes.
5. Tube brushes.

There are at least ten variations of these five types and for general utility, miniature brushes could be listed as yet another category.

Each of these basic types (as shown in Figure 11-1) varies in design and composition and can be modified to suit specific needs, thus making brushes extremely versatile. Each type of brush is made up of different filament materials, lengths, and crimps or knots to provide varying degrees of stiffness. The choice of filament material depends on the application. Some filaments are made from synthetic materials and others are from natural sources, and both are referred to as *brush fill material*.

Brushes constructed of metallic filaments invariably use one of the following four basic forms: crimped wire, knotted or twisted tuft, straight wire, and multiple-strand crimped filaments.

Many types of *crimped wire* are used in brushes. The best crimping results when the wire is displaced over multiple planes. Although there are a number of crimping techniques that will produce multiple-plane displacement, the end product should be similar to the crimp configuration produced by running a wire through two sets of gears perpendicular to each other. Each set of gears should have a different pitch. The number of crimps per inch is governed by the finer of the two sets of gears. The wavelength of the crimps (see Figure 11-2), sometimes called frequency, should be uniform over the entire length of wire. The displacement of the wire, usually referred to as amplitude, should be held to a tolerance of $\pm.004$ in. (±101.6 μm).

To make *knotted or twisted tuft* filaments, a bundle of straight wires of equal length is formed into hairpin shapes around a retaining member. The bundle is then twisted into a cable with a helix angle of approximately 45°. The standard twist has the tuft twisted for two-thirds of its length; however, in applications requiring maximum impact, such as extremely severe cleaning or heavy encrustation removal, the tuft is twisted for its entire length (see Figures 11-3, 11-4, and 11-5). This is sometimes referred to as cable twist. Multiple rows of twisted tuft, instead of a single row, are sometimes used to give additional stiffness for the most severe operations.

To fabricate *straight wire* brushes, rows of holes are drilled in a radial pattern on the periphery of a circular wood or plastic hub. Small bundles of straight wire are bent into a hairpin shape and then pulled into the holes and locked firmly in place with staples or retaining wire. The protruding ends of the filament wire, being relatively unsupported by other wire filaments, have a high degree of flexibility with a minimum amount of filament entanglement. This

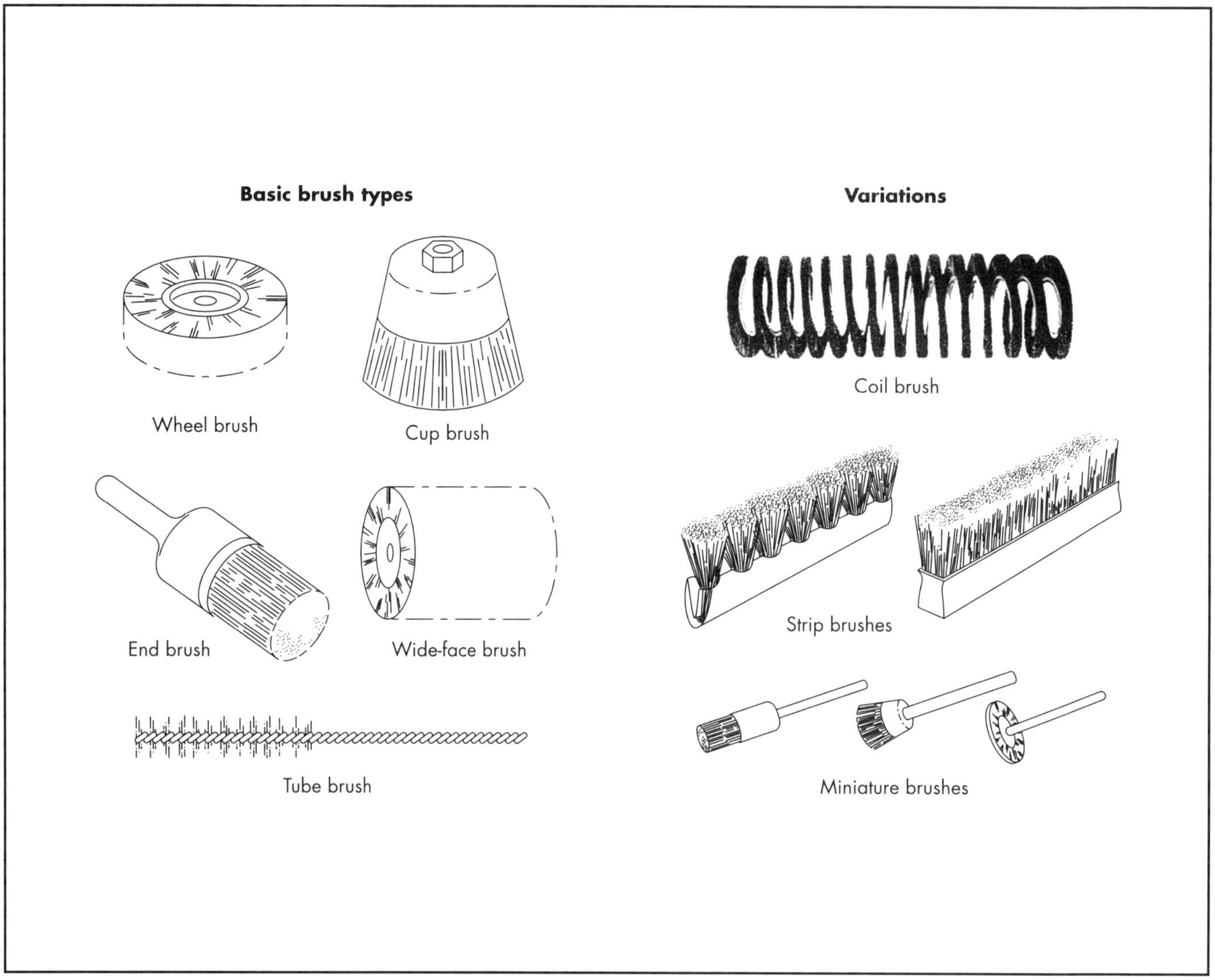

Figure 11-1. Primary types of power deburring brushes. ***(Courtesy Weiler Corporation)***

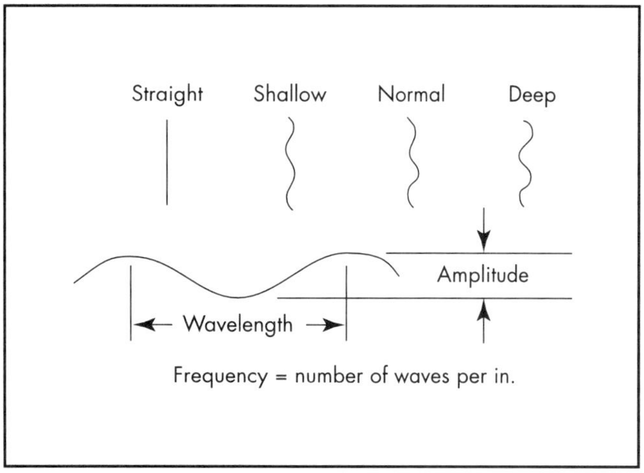

Figure 11-2. Terminology of crimped filaments. ***(Courtesy Anderson Products, Division of Wilton)***

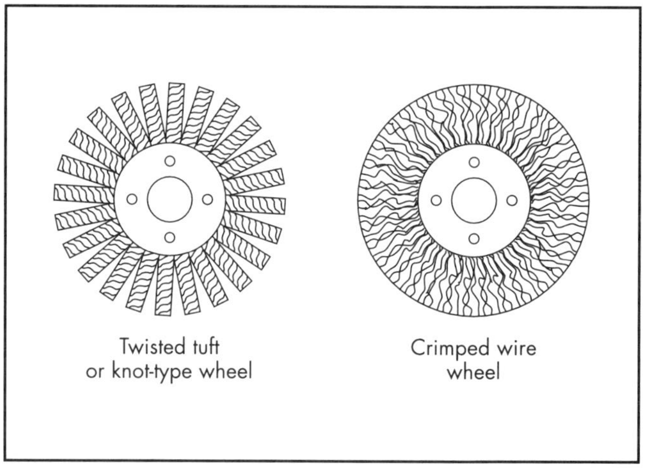

Figure 11-3. Configuration of twisted-tuft and crimped wheel filaments.

a. Tufted end

b. Multiple-row tufted end

Figure 11-4. Tufted and multiple-row tufted-end brushes. ***(Courtesy Anderson Corporation)***

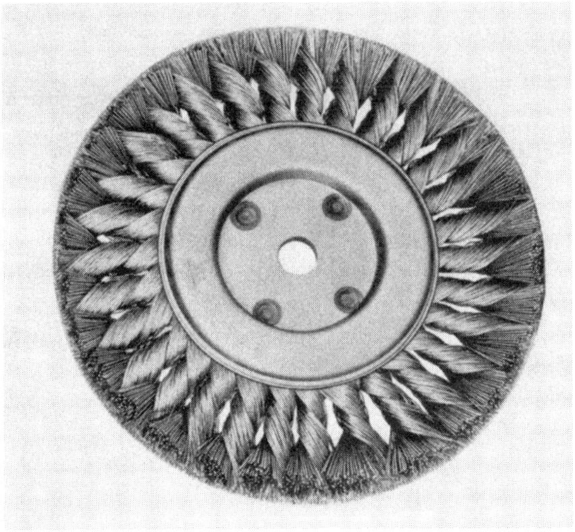

a. Knotted tuft

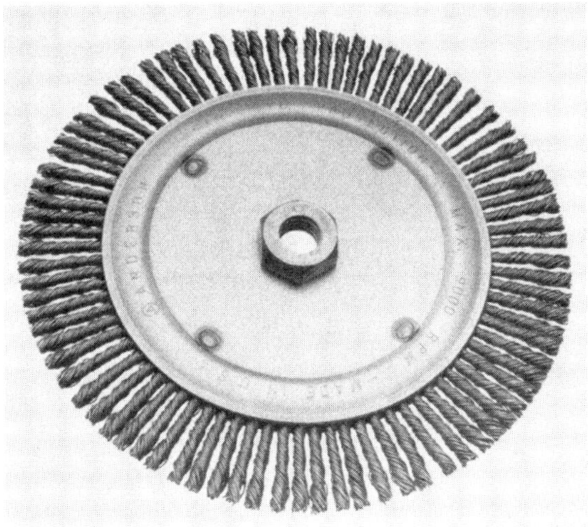

b. Twisted tuft

Figure 11-5. Knotted and twisted-tuft wheel brushes. ***(Courtesy Anderson Corporation)***

permits them to penetrate into narrow slots and other shapes better than other filament configurations.

Multiple-strand crimped filaments are made by stranding together 5–25 straight fine wires into a steel cord, usually with a diameter ranging from .020–.050 in. (0.5–1.3 mm). This cord or cable is then crimped in multiple planes as in conventional crimped wire. This type of construction is used in applications that require the stiffness of coarse wire as well as fatigue strength and multiple points of contact (Bateman 1975).

11.1.1 Radial (Wheel) Brushes

Radial (wheel) brushes are generally used for deburring readily accessible external edges. They are generally placed on fixed machines. Miniature wheels can be used in hand-held motors, and some hand-held motors can accommodate larger brushes, though they may be too heavy for prolonged use.

11.1.2 Cup and Disk Brushes

Cup-shaped brushes are typically used for open surfaces, the tops of cylinder heads, and valve bodies that have many edges. They have clearances to reach over some low bosses and features, but their main use is for reaching into short cavities to scrub the burrs away.

11.1.3 End Brushes

End brushes are used for reaching into small deep pockets and holes to remove burrs and clean counterbores.

11.1.4 Wide-face or Cylinder Brushes

Probably 90% or more of all wide-face brushes are custom-made to suit the particular requirements of the user. They are made in diameters ranging from 2–30 in. (51–762 mm) and in lengths up to 25 ft (7.6 m). Every conceivable type of fill material is used, including all the popular synthetics, hairs, natural fibers, abrasive monofilament, and metallic wires. Wide-face brushes, as well as other brushes that operate at speeds over 500 r/min, generally require dynamic balancing. Such brushes also must be designed for ventilation, especially when used for continuous and/or heavy-duty applications. Radial fin-type mountings with ends open to allow the free passage of air generally provide sufficient ventilation. They are used in flat sheet deburring, though generally not for other shapes.

11.1.5 Tube Brushes

Tube brushes are almost exclusively used for reaching into holes and, particularly, for removing burrs from intersecting holes. They can be procured with diameters as small as .040 in. (1 mm) and 6 in. (152.4 mm) or larger in diameter. Typical tube brushes have helical rows of filaments and range from .250 in. (6.4 mm) diameter to 1 in. (25.4 mm) diameter.

11.1.6 Common Variations of Primary Brush Types

Various brushes can be constructed of several radial brushes mounted together on the same spindle to provide a long continuous brush. They also can be made of *coils* that mash together or strips placed in a long spindle.

Strip brushes are readily replaceable tools that come in a variety of strip lengths, heights, channel materials, and channel sizes. Fill materials include synthetic, vegetable fiber, hair, and abrasive monofilaments, or metallic filaments. Metallic filaments are not normally used for deburring, but the concept is used for long string deburring.

Miniature brushes are small versions of end, wheel, and cup-type brushes. They are used for cleaning, deburring, and finishing delicate or miniature components. Brush diameters range from .16–1.5 in. (4–38 mm). Fill materials include steel, stainless steel, brass, bronze, and nickel-silver wire, as well as various types of animal hair and nylon filaments. The hair-filled types are sometimes used with abrasive media for deburring and polishing small and detailed parts. These brushes are generally mounted on a .09 or .125 in. (2.3 or 3.2 mm) diameter shank.

11.2 BRUSH FILL MATERIAL

Brush filaments and related materials include tempered and untempered high-tensile steel wire and tempered and untempered stainless steel wire. A number of nonferrous wires, such as brass, nickel, silver, beryllium-copper, Inconel®, and other heat and corrosion-resistant alloys, are also used. Synthetics, such as nylon, polypropylene, and styrene, are frequently used in many power brushes. Animal hairs, such as pig bristle and horsehair, are used frequently in brushing (but not for deburring), and a large number of vegetable fibers, such as tampico, habia, and bassine, are frequently used, depending on the application.

11.2.1 Carbon Steel Wire

Steel wire is used in power brushes more than all other fill materials combined, with the tempered variety predominating. The tempered steel wire used is of high quality and is similar to that used in springs. It is generally made of a .60–.75% carbon steel with tensile strengths ranging from 300,000–380,000 psi (2,068–2,620 MPa). The untempered steel wires are hard drawn to their highest tensile strength and usually have lower hardness values and tensile strengths than the tempered wire.

11.2.2 Stainless Steel Wire

Some nuclear applications and other applications involving elevated temperatures or corrosive environments prompt the use of stainless steel wire. There are two types of stainless steel wire used. The first is hand-drawn, bright-finished type 302 stainless steel that is drawn to tensile strengths of 300,000–360,000 psi (2,068–2,482 MPa). This type of wire is usually used in deburring stainless steel, aluminum, and other nonferrous materials to avoid after-rust caused by foreign deposits that occur when carbon steel wires are used (West 1975).

Type 420 tempered stainless steel, with tensile strengths ranging from 240,000–255,000 psi (1,655–1,758 MPa), is also used in deburring brushes. However, type 420 is only 80% as corrosion resistant as type 302. In spite of this deficiency, it does have some distinct advantages in many deburring applications. Type 420 is vastly superior to untempered stainless steel in its cutting capabilities, fatigue resistance, and abrasion resistance. In some instances, it has a life of seven to 10 times that of untempered stainless steel while doing substantially more cutting.

Type 420 stainless steel is not recommended for use in nuclear applications or in applications where it is desirable to deburr with brushes constructed of nonmagnetic stainless steel. Type 420 is magnetic at all times, and therefore type 302 should be used in nuclear applications. Do not make the mistake of checking the wire with a magnet. Wire from all heats of type 302, when drawn to spring temper as used in brushes, can be attracted to a magnet to varying degrees, and this amount of magnetic attraction for a given size of wire in the spring tensile range is a function of the degree of cold working that the wire has undergone. Cold work increases the tensile strength and simultaneously the magnetic permeability of the wire. Only in its fully annealed condition can type 302 stainless steel be considered essentially nonmagnetic, because in this state it exhibits permeabilities of less than 1.02.

11.2.3 Tampico

Tampico, a cellular vegetable fiber, is used more widely than any other fiber in deburring brushes and produces a flexible, resilient brushing wheel. Tampico fibers are used singly or in multiples to make a wide-face deburring brush. It has been found that tampico brushes can be treated and cured with a lacquer to improve stiffness and grit retention, and the treatment minimizes the shedding, knifing, and weaving that sometimes occur at high speeds. Treated brushes can therefore perform a much wider range of deburring and edge blending operations than the natural, untreated fiber. The stiffness and tackiness attained by treatment make it easy to apply grease stick and abrasive compounds that are used with these brushes.

One of the main reasons treated tampico brushes are used is to increase the life of the part being brushed. Over half of all failures of parts subjected to repeated cyclic stresses are fatigue failures caused by the widening of microscopic cracks resulting from stress concentrations. The radii and surface finishes in proximity to the deburred edge can almost totally eliminate the scratches, burrs, and sharp corners that cause cracks and progressive fractures.

11.2.4 Abrasive Monofilament

Nylon monofilaments that contain either aluminum oxide, silicon carbide, or pumice grit are now common deburring and edge finishing tools. Silicon carbide grit offers the best combination of hardness, sharpness, and toughness. Silicon carbide grit has harder and sharper grains than aluminum oxide, is more friable, and more effectively finishes metal. It is used for most nylon brushing jobs. Aluminum oxide grit is less likely to fracture and is generally used for finishing soft metals. Because aluminum oxide prevents carbon contamination, the aerospace industry usually prefers it for finishing jet engine blades and other parts. Other abrasive particle materials used include aluminum silicate, diamond, and boron carbide; although none of these is standard, some may be required for finishing parts made of exotic materials.

The filament contains 20–40% abrasive particles by weight, and particles are uniformly distributed throughout the filament. Grit sizes range from 80 to 800 grit, with the coarsest grit encapsulated in relatively large-diameter filaments and the finest in small-diameter filaments. Nylon monofilament can be used either wet or dry. Some users believe it provides better action when used with water, mineral oil, or soluble oil as a coolant. Grease sticks also have been used with this carrier. Nylon filaments are used for edge finishing of wood, leather, plastics, cork, rubber, ceramics, and glass, as well as metal.

The abrasive is sparsely distributed in the nylon filament, so more nylon than exposed abrasive rubs the part surface. The nylon generates heat. On poor heat conductors like stainless steel, smearing (small brush particles being embedded into the work material) occurs readily. Spraying a lubricant onto the brush surface when the tool is stopped or barely moving reduces friction between the nylon bristle and the part and prevents smearing. The brush can then accept twice the workload and forces before smearing occurs. To maximize brush life, it is necessary to maintain the temperature of the nylon filaments below 500° F (260° C). Hydrocarbons, oils, and most organic solvents have no lasting effects on the filaments.

As with wire brushes, applying less pressure to abrasive nylon filament brushes increases their life. Unlike wire brushes, however, they run at low speeds but with enough pressure to force the sides as well as the tips of the filaments to cut. Optimum cutting results when the workpiece penetrates, or "mushes," into the brush face.

Although nylon filaments are normally round, they can be made rectangular, oval, or any other shape that can be extruded. Figure 11-6 illustrates the difference in contact area when rectangular filaments are used. The diameter of

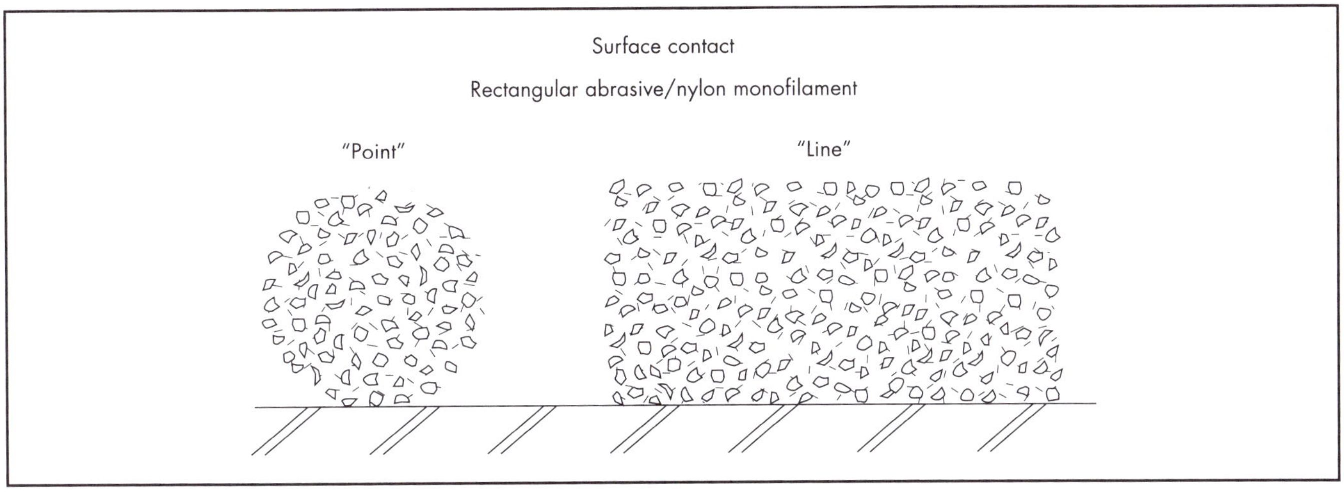

Figure 11-6. Contact area for round and rectangular nylon filaments (Scheider 1989b). (*Courtesy Osborn Brush*)

the monofilament is usually between .014 and .050 in. (0.36–1.27 mm). Figure 11-7 illustrates some commercially available filament sizes. Use rectangular filaments for more contact and impact with the workpiece in applications such as carbide edge honing, deburring injector nozzle cutouts, and finishing decorative hinges. Use round uncrimped filaments for applications such as deburring the slots in cast-magnesium computer cage backs and putting a uniform scratch-finish on aluminum chair bases.

Other synthetic materials can be used for brush filaments, but nylon has several advantages. It is resistant to most chemicals (however, strong acidic solutions will degrade and embrittle the nylon). Nylon is relatively resistant to abrasion. It also has a "memory" of its original shape, which is important when brush filaments are constantly being bent. Nylon filament stiffness decreases as temperature increases. Nylon filaments are twice as stiff at 32° F (0° C) as they are at 75° F (24° C). Stiffness drops another 50% at 140° F (60° C). Visible smearing occurs when temperatures approach nylon's melting point (410–482° F [210–250° C]). Temperature effects are minimized by controlling brush speed and pressure and by using a coolant, such as water.

11.2.5 Nonwoven Abrasives

As described in Chapter 10, abrasive grains are dispersed uniformly throughout a low-density, open-type, nonwoven web of synthetic fibers and held there with a binding adhesive (Figures 10-2 and 10-3). This uniform distribution produces a cushioned abrasive that is ideal for fine deburring. Abrasives used in nonwoven abrasives include aluminum oxide, silicon carbide, flint, and garnet. Although specific grain sizes are used (80–800 grit), nonwoven abrasives are often specified by such grades as coarse, medium, fine, very fine, super fine, and ultra fine. Other nomenclatures are also used.

Nonwoven abrasives are often substituted for bonded abrasive strips in flap wheels. They also frequently replace setup wheels, steel wool, and abrasive flap wheels. They are waterproof, chemically resistant, and can be used wet or dry. They are nonloading and highly flexible. Nonwoven

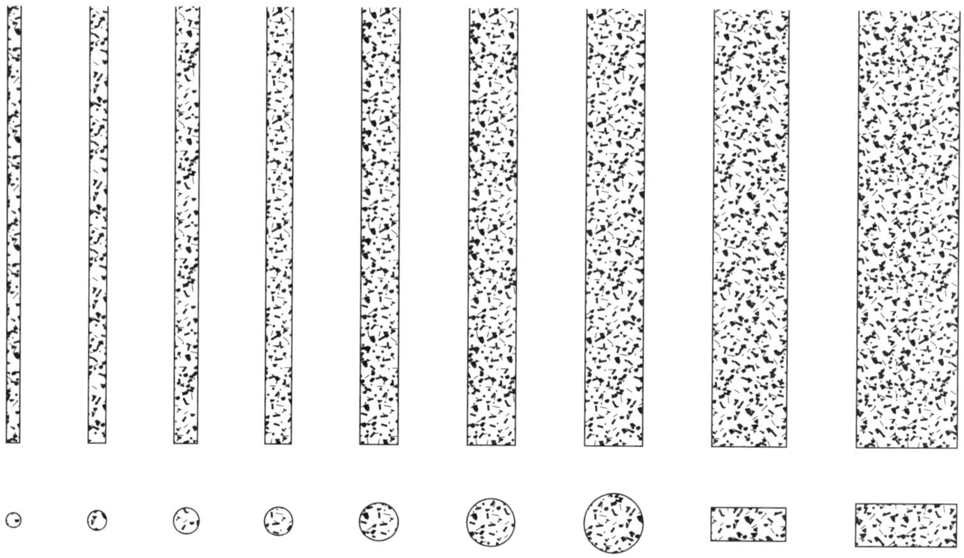

Shape									
Round, rectangle									
Size									
mil	8	12	18	22	35	40	50	30 × 70	45 × 90
mm	0.20	0.30	0.46	0.56	0.89	1.02	1.27	0.76 × 1.78	1.14 × 2.29
Grit									
Aluminum silicate	AS								
Aluminum oxide		AO		AO	AO	AO	AO	AO	AO
Silicon carbide			SC	SC	SC	SC	SC	SC	SC
Mesh									
USA	1000	600	500	320 120	180	120	80	80 120 180 320 600	80 120 180 320 400

Figure 11-7. Abrasive-filled nylon filament shapes and sizes. ***(Courtesy Osborn Brush)***

abrasives efficiently remove light burrs. They should not, however, be used on large burrs. Their scope of burr removal capabilities is substantially enhanced by their availability in a variety of densities (degrees of hardness).

Brushes made from abrasive filament and nonwoven nylon filament can be used either wet or dry. Experience seems to indicate some advantages when water, mineral oil, or water-soluble lubricants are used as cooling media. Grease sticks also have been used efficiently to provide longer life. Generally, these brushes can be used in all of the applications for which treated tampico brushes are used. Their advantage lies in the fact that they do not require the use of compounds or slurries. Some compounds, however, can be used to accelerate the cutting action of these brushes. Abrasive filament brushes also can be used in applications for which the tackiness of treated tampico brushes would be objectionable, such as in deburring wood, leather, plastics, cork, rubber, ceramics, and glass.

11.3 AGGRESSIVENESS OF BRUSHES

The following factors influence brush aggressiveness and effectiveness in removing burrs:

- Filament diameter, free length, configuration (including end of filament), texture, density, and material.
- Wheel velocity and width.
- Type of contact with part.

Brushes with large diameter, short filaments that are closely packed together provide maximum aggressiveness. Knotted or crimped filaments add stiffness, as does a wider wheel. Where the diameter of the sides of the filaments (rather than the ends) performs the deburring, a tough filament surface provides greater abrasion. Typically, cutting with the ends of the filaments is more desirable; however, with abrasive-filled nylon brushes, sliding filaments over an edge adds to the radiusing action. Faster velocities increase the effective stiffness of brushes.

It is common to think that brush filaments retain uniform diameters throughout their life, since they normally begin that way. After prolonged use, however, this may no longer be the case. For example, the ends of nylon brushes become tapered with use, and so they are less stiff and consequently cut less.

A basic economic rule in brushing is to throw away brushes when the filaments are obviously worn out. However, there are exceptions to this rule. One engineer tried to get workers who were using miniature abrasive-filled nylon wheels to throw them away when the filaments were obviously worn out. The workers, however, found that these worn brushes were ideal for use on precision ultraminiature gears. A close investigation revealed that the .014 in. (0.36 mm) diameter nylon filaments had tapered to .006 in. (152.4 μm). These fine filaments would now reach to the bottom of gear teeth on 96 diametral-pitch gears. In addition, the brush was much more flexible and as a result could maintain the required edge radius of .005 in. (127.0 μm) maximum.

Brush flexibility is the ability of the brush to conform to irregular or contoured surfaces as measured by filament deflection. This quality determines resiliency or stiffness and is measured in terms of resistance to bending of filaments. Flexibility can be accomplished in a number of ways, for example:

- The modulus of the fill material can be changed (the modulus is different for various materials such as steel, brass, hair, and nylon).
- The diameter of the fill material can be increased or decreased. For example, .020-in. (0.5-mm) steel wire is stiffer than .005-in. (127-μm) steel wire of the same analysis.
- The trim length of the fill material can be increased or decreased. For example, a 1-in. (25.4-mm) steel wire filament with a diameter of .010 in. (0.25 mm) is stiffer than a 4 in. (101.6 mm) steel wire with the same diameter.
- The speed of rotation can be increased or decreased. For example, because of the influence of centrifugal force, the faster a brush is rotated, the stiffer it becomes.
- The construction of the brush can be changed. For example, filaments and wire, twisted together as in a knotted or twisted tuft brush, are stiffer than those in a crimped wire brush. In the knotted or twisted tuft construction, the filaments being twisted together act as a family of filaments, whereas in a crimped wire brush, the filaments act as individual filaments.
- The number of brushing points per square unit of brush face can be increased. This changes the brush flexibility through wire packing and/or load division. Flexibility increases as density decreases, and less capability to conform will result as density increases.

Note that encapsulated brushes, or those otherwise treated or coated, are less flexible than untreated brushes.

The standard formula for beam deflection shows the basic relationships involved in brush stiffness (Bateman 1975):

$$D = (WL^3)/(3EI) \tag{11-1}$$

where

D = Deflection of the filament end (in. [mm]).
W = Force causing deflection (lb [kN]).
L = Length of filament (in. [mm]).
E = Modulus of filament (lb/in.^{2} [kg/cm^2]).
I = Moment of inertia of the cross section of the filament (in. [mm]).

For a round filament, Equation (11-1) becomes

$$D = WL^3/(.1473\ Ed^4) \tag{11-2}$$

where

d = Diameter of the filament (in. [mm]).

Equation (11-2) shows the following:

- The stiffness (resistance to bending) of a single round filament fixed at one end is proportional to its diameter to the fourth power.

- The stiffness of a single round filament fixed at one end is inversely proportional to its length to the third power.

In many instances, brushes are made up of tufts of filaments rather than single filaments. In the case of a tuft, Equation (11-2) becomes

$$D = WL^3/(.1473\ Ed^4N) \qquad (11\text{-}3)$$

where

N = Number of filaments per tuft.

The area of a round filament is $\pi d^2/4$, and therefore the number of filaments, N, for a given size tuft is inversely proportional to the diameter of the filament to the second power. Therefore, the following conclusions can be drawn from Equation (11-3) for tuft stiffness:

- The stiffness (resistance to bending) of a tuft of round filament is proportional to its diameter to the fourth power.
- The stiffness of a tuft of round filament is inversely proportional to its length to the third power.

From these conclusions, the following formula can be written, which relates the filament diameter and trim length of one brush to the filament diameter and trim length of another brush for equal brush stiffness:

$$L_1^3/d_1^2 = L_2^3/d_2^2 \qquad (11\text{-}4)$$

By substituting the known values into Equation (11-4), and assuming either a new trim length or diameter, the one remaining unknown can be computed. Equation (11-4) can be solved graphically using the nomograph shown in Figure 11-8.

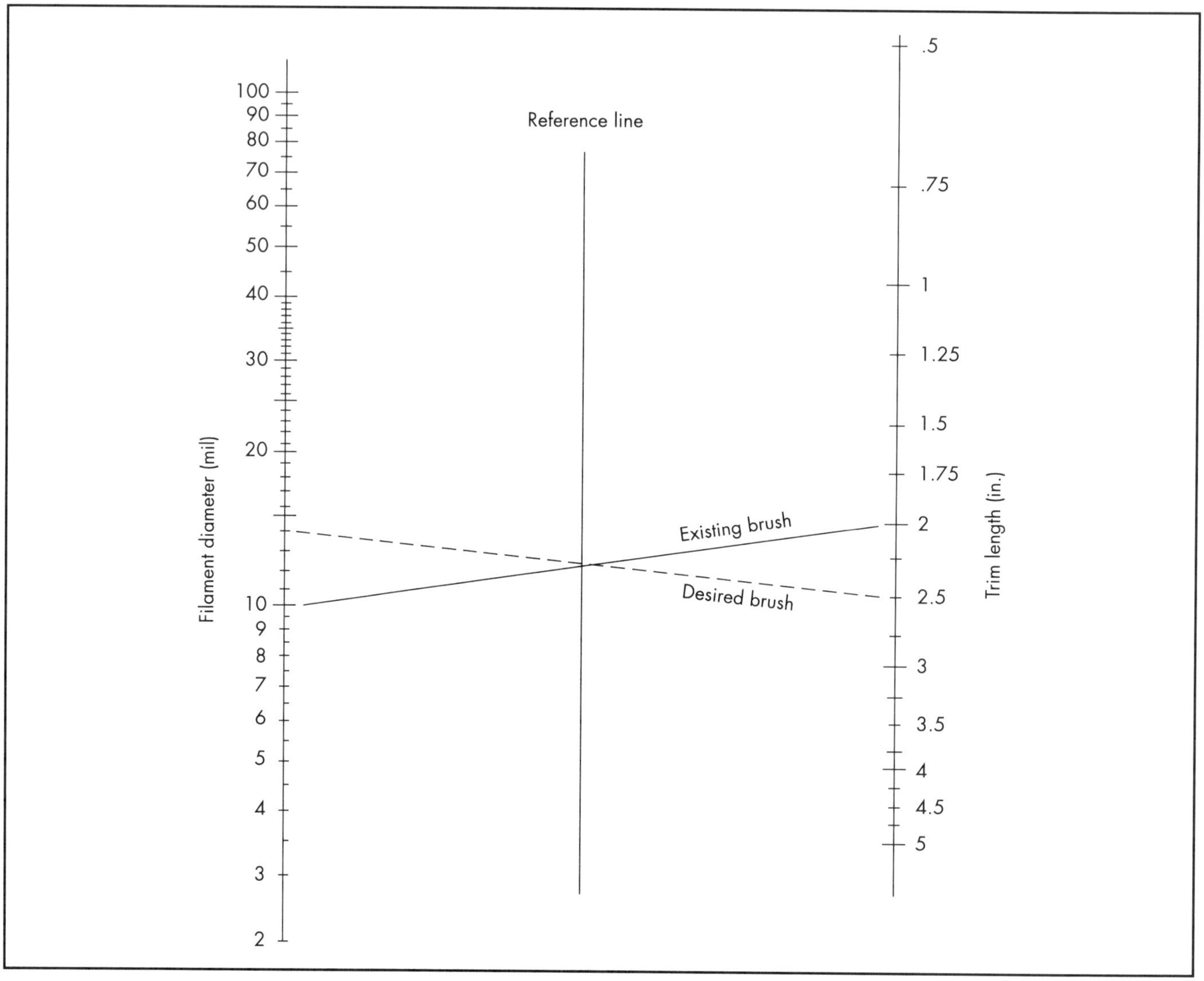

Figure 11-8. Nomograph for relating trim length and filament caliper. ***(Courtesy Anderson Corporation, a Division of Dresser Industries)***

Obviously, the nomograph shown in Figure 11-8 can be used only when brushes are of a similar construction (it would be impractical, of course, to develop a chart that includes all of the various brush constructions). In most brushes, one filament touches another and, therefore, does not function as an individual filament but as a family of filaments, and the interaction of filaments varies with tuft size, trim length, and the rigidity of the tuft in the retaining member.

11.3.1 Abrasive-filled Nylon Brushes

The effectiveness of abrasive-filled nylon brushes can be increased significantly by changing the composition and size of the filaments used. The following variables may increase tool effectiveness:

- Brush tools employing filaments with smaller diameters are often more effective than those with larger diameters. This is because the filaments can more easily adjust to the contour of the part, giving them more surface contact. For example, the difference between 50 mil and 40 mil filament is 25% in the amount of abrasive filament surface area that can contact the part, given that the tool contains the same volume of filament. A filament that is too large may hit and bounce off the part.
- Small grit can be more effective than large grit because it offers more cutting surfaces. For example, a 120-grit abrasive in the same size filament offers about 50% more grit particles on the surface of the filament than an 80-grit abrasive.
- Increasing grit loadings from 30%, which was standard for years, to 40% increases grit particles by one-third, offering increased aggressiveness.

Other variations that may influence tool effectiveness include tool speed, tool location, part position, and angle of contact.

11.3.2 Tool Wear

Tools wear out, and brush tools are no exception. As brush tools wear down, their performance characteristics will change.

Short-trim brushes wear from the tip (Przyklenk 1985). As the filament length shortens, the brush tool will become more aggressive because the filament will be shorter and stiffer. The original filament deflection, usually .125 in. (3.2 mm) with a proper pressure setting, cannot be maintained without increasing pressure. The increased pressure increases stress on the filament at the flex point, which is usually where the filament joins the tool base. This added stress can eventually cause the filaments to break.

Long-trim brushes normally wear out for a different reason. The normal wear pattern for a long-trim brush is a gradual tapering of the filament as the nylon and grit wear down. Eventually the filament diameter is reduced, and the tips begin to break off. As the diameter is decreased, the stiffness is reduced. In continuous operations, these tools are changed regularly to maintain consistent finishing effects.

Although flex fatigue can occur, brush construction methods have been developed to support the filaments better at the flex point. Filament fatigue, if it occurs, normally results from placing the part too close to the tool core, causing excessive filament flexing. Circulation of coolant at the flex point will minimize heat generation and filament fatigue.

In brushing applications that employ metal filaments, it is the sharp tips of the brush filaments that do the work. If excessive pressure is exerted, the filaments are displaced, resulting in a wiping rather than a cutting action (see Figure 11-9 for correct and incorrect techniques for wire brushes). Excessive pressure also causes excess filament displacement, which overstresses the filament wire and causes premature fatigue failure. The very nature of brushing produces forces that compress, flex, and sometimes overheat the brush filaments. These forces are conducive to fatigue when they exceed the endurance limit of the brush. When the endurance limit is exceeded, brush failure is almost immediate, and the greater the mechanical stress, the sooner the damage to the brush filaments. Although fatigue failure is possible with any size of wire, susceptibility varies directly with the diameter of the wire. The smaller the diameter of the wire, the greater its resistance to overstressing. This factor must be considered when selecting brushes so that load tolerances, cycle stress, and other fatigue-inducing characteristics can be compensated for by varying the diameter of the filament.

11.4 POWER REQUIREMENTS

Three primary factors determine the amount of horsepower required in any given operation:

1. Brushing pressure necessary to accomplish the work.
2. Resistance developed between the work and the brush.
3. Speed of the brush.

The brush configuration and the type and size of filaments used also influence power requirements. Table 11-1 shows recommended horsepower requirements for 1 in. (25.4 mm) brushes using medium brushing action. For more complex or sophisticated conditions, additional research would be necessary to determine the correct horsepower to use. Power requirements for wider brushes are directly proportional to face width.

Effective low-cost brushing is also influenced by the rotation speed of the brush. For each particular application, there is an optimum operating speed, and increasing or decreasing the speed will affect the efficiency of the brushing operation. Table 11-2 shows surface speeds that may be used for various types of brushing applications using metal brush filaments.

In automated and semiautomated applications, brushes should be operated at the highest practical speed with the lightest possible pressure. Where practical, the equipment

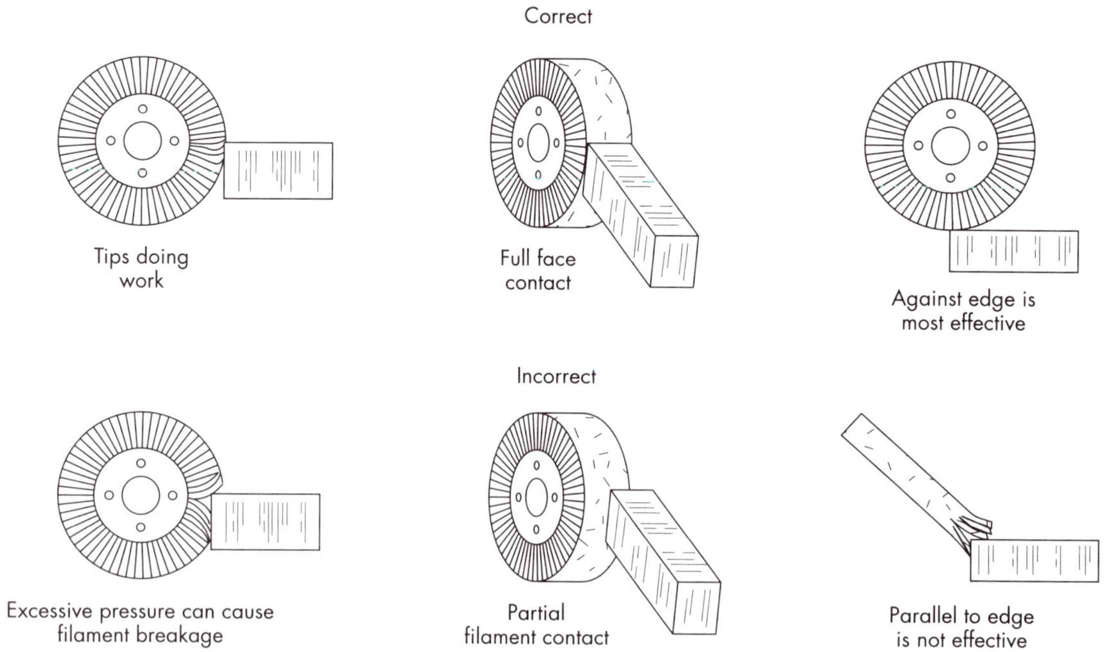

Figure 11-9. Some correct and incorrect brushing techniques for wire brushes.

Table 11-1. Horsepower ratings for a 1-in. (25.4-mm) wide brush

Brush diameter, in. (mm)	Recommended motor size, hp (kW)	Operating speed, r/min
4 (102)	.25 (0.2)	3.45
6 (152)	.50 (0.4)	3.45
8 (203)	.75 (0.6)	3.45
10 (254)	1 (0.75)	1.75
12 (305)	1 (0.75)	1.75
15 (381)	1.5 (1.1)	1.75

(Bateman 1975)

Table 11-2. Recommended surface speeds for wire brushing applications

Application	Recommended surface speed, ft/min (m/s)
Burr removal	5,500–7,500 (28–38)
Scale removal	7,500–10,000 (38–51)
Weld cleaning	7,200–9,400 (37–48)
Edge blending	4,700–7,500 (24–38)
Cleaning (dry)	4,000–5,500 (20–28)
Cleaning (wet)	1,900–4,000 (10–20)
Surface polishing	6,400–8,000 (33–41)
Surface buffing	8,000–10,000 (41–51)

should be designed with meters and aligned to measure the load on the drive motor continuously so that the amount of brushing pressure can be monitored and controlled.

In many operations, it has been found that more efficient brushing is attained if the direction of rotation of the brush is periodically reversed so that new, sharper cutting edges are in contact with the workpieces. This accelerates the cutting capabilities of the brush and reduces operator fatigue by reducing the pressure required to perform the work.

11.5 BRUSH SELECTION

The effectiveness of a given brush depends on its degree of flexibility and cutting capabilities. The most flexible brushes in any style or configuration are those made of *synthetic* or *natural fibers*, such as horsehair, tampico, and nylon. These types of brushes are used in a number of applications, including wet and dry cleaning; for applying oils, compounds, and powders; light finishing, buffing, and polishing; and edge blending. They are effective because of their high flexibility and low brushing strength, and because they can generally be used without any filament treatment or abrasive compounds.

Synthetic or natural fiber brushes can be chemically treated to make them slightly stiffer. This slightly reduces their flexibility and slightly increases the speed of their cutting action. It also permits them to be used with abrasive compounds to produce microfinishes. These medium flexible brushes are also used for edge and surface blending, removing small burrs and light scale, surface finishing, and various cleaning operations.

A variation is the previously discussed *abrasive-filled filament brush*, which is made by impregnating nylon monofilaments with aluminum oxide or silicon carbide. This combination produces a medium flexible brush capable of fast cutting action without additional abrasive compounds, since the abrasive is already imbedded in the filament. It can be used wet or dry, and the filament diameter can be varied to attain different degrees of stiffness.

If chemically treated synthetic or natural fiber brushes are still not stiff enough for a particular application, then *fine wire brushes* can be used with abrasive compounds. The wire in this type of brush is typically .006 in. (152 μm) or finer in diameter. Fine wire brushes, when used with abrasive compounds, are used for removing medium-sized burrs and light scale, edge blending, and surface finishing for appearance.

If the brushing application allows a surface finish to exceed approximately 30 μin. (0.762 μm), another group of brushes would be appropriate. The most flexible of this group is the *narrow-face, highly flexible, standard knot-type brush* (see Figure 11-3). This type of brush is used when surface impact action is necessary, such as to remove burrs, scale, and weld splatter, and for edge blending, surface roughening and finishing, and general-purpose cleaning. Knot brushes are sometimes produced in what is called *high-complement style* to decrease their flexibility. This means they have relatively short trim lengths, and the knots are close together to give increased density. This produces a brush of maximum density and medium flexibility. High-complement style brushes are used for rugged production jobs, especially in automatic or semiautomatic machine applications. They are generally used for removing heavy burrs and scale; heavy cleaning of rust, paint, rubber, and plastics; surface roughening; and edge blending.

Regular crimped wire brushes of narrow face are of medium density and are used where medium brush flexibility is required. They are used for general-purpose cleaning, surface refining and finishing for appearance, surface roughening, edge blending, and removing heat scale. These same brushes in wide face with a more dense fill provide much faster cutting action and produce fine finishes. They are usually used as general-purpose wheels for removing burrs and heat scale, general cleaning, surface cleaning and refining, and edge blending.

High-density, regular crimped wire brushes are super dense and provide low flexibility. They are used in high-production applications on automatic or semiautomatic brushing equipment for removing heavy burrs and scale, edge blending, surface finishing and refining, and general-purpose cleaning.

When maximum rigidity of wire filament is required, an *elastomer-bonded brush* is used. This type of brush has the lowest possible flexibility of filament. It is a rigid wire wheel covered with a poured plastic, making it look more like a colored grinding wheel than a brush. It has fast brushing action and a controlled brush face. This type of brush is used for confined areas, removing burrs and scale, edge blending, surface roughening, and general-purpose cleaning.

As discussed earlier, brush flexibilities can be modified by changing the diameter or density of the wire, resulting in a wide variety of possible brushing actions and finishing capabilities. Table 11-3 provides recommendations for selecting and improving brushing performance. Note that Table 11-3 has two parts—chart 1 and chart 2. Chart 1 has a left and a right side, and which side to consult depends on the surface finish required after deburring. If the surface finish must be less than 30 μin. (0.76 μm), the left-hand side should be consulted; if a surface finish must be greater than 30 μin. (0.76 μm), the right-hand side should be consulted. Brush flexibility moves from the inside of the chart to the outside edges, with the least flexible brushes listed in the middle. After a potentially suitable brush is selected based on chart 1, it should be tested to determine its effectiveness. Based on the results from these tests, brush specifications or operating conditions can be refined by consulting chart 2. Note that when large areas are to be brushed, the widest possible brush face will be the most efficient.

Because researchers have, in general, failed to record the relationships between brushing time, initial burr size, and resulting edge radius, there is little quantitative information about brush deburring. However, some tests have shown that when the intent is to remove a .0005-in. (12.7-μm) thick grinding burr and produce a minimum radius, miniature brushes can easily do this at feed rates of 1–2 in./min (0.0004–0.0008 m/s) and a speed of 1,800 surface ft/min (9.1 m/s). Table 11-4 shows some comparative data for radiusing edges using miniature brushes. As shown in the table, the nylon brushes maintained significantly smaller edge breaks than the stainless brush, which had much smaller filaments. Edge breaks as small as .003 in. (76.2 μm) have been maintained on both ferrous and aluminum parts while removing burrs. As in all deburring processes, however, to maintain small edge breaks, the initial size of burrs also must be small. Note that smaller burrs allow the use of softer wheels, which can hold finishes as low as 4 μin. (0.1 μm).

Because of the need for verifiable comparative data on brushes, many users must perform their own testing. Consequently (and unfortunately), many potentially suitable products are ruled out before they are tested. For example, one manufacturer typically removed large burrs on miniature parts by using specially designed knives. Parts were then brushed with the two nylon brushes described in Table 11-4. This process produced edge radii of the desired size and finish. Previous efforts to deburr parts using only brushes had failed because their use had resulted in too much edge breakdown. When a wider, denser brush was obtained with .024 in. (0.61 mm) diameter bristle and a trim length of .250 in. (6.35 mm) instead of .375 in (9.53 mm), engineers and workers felt that the brush was still too stiff to produce the desired result. These brushes remained in a box for several weeks until a worker, who had

Table 11-3. Recommendations for selecting brushes

Two-step Method to Select Power Brushes

Step 1: Make tentative selection from Chart 1; Step 2: Adjust first selection as necessary based on Chart 2

Step 1

Chart 1

For jobs requiring surface finish less than 30 μin. (0.76 μm)	For jobs allowing surface finish to exceed 30 μin. (0.76 μm)
Brush flexibility: high ← → low	Brush flexibility: low ← → high
Brush finish ability: low μin. ← → medium	Brush finish ability: fine ← → coarse
Brushing action strength: low ← → fast	Brushing action strength: medium fast ← → fast

Brush	Major Brush Characteristics	Major Brush Applications Listed in Order of General Usage
Synthetic filaments or natural fibers like tampico	• High brush flexibility • Low brushing action strength • Generally used without filament treatment or abrasive compound	• Cleaning wet and dry • Applicator to apply oils, coating compound, powders, etc. • Finishing, light buffing, and polishing • Edge blending
Treat filament brushes like tampico fiber	• Medium flexible fast cutting • Used with abrasive compound of a tacky composition • Produces low μin. surface finish	• Edge blending • Small burr removal • Light scale removal • Surface blending • Surface finish for appearance • Cleaning
Abrasive-filled filament brushes (Anderlon®)	• Medium flexible fast cutting • Usually used without abrasive compound since abrasive is embedded in the filament • May be used wet or dry • Available with aluminum oxide or silicon carbide grit	• Edge blending • Small burr removal • Cleaning • Light scale removal • Surface finish for appearance • Surface blending
Fine wire brushes used with abrasive compounds	• For efficient use of wire-filled brushes with compound, select brushes having high density and wire sizes .008 (0.2 mm) diameter or finer	• Medium size burr removal • Light scale removal • Edge blending • Surface finish for appearance
Elastomer-bonded brush Anderbond®	• Low flex • Rigid wheel with fast brushing action and controlled brush face	• Burr removal • Edge blending • Scale removal • Roughening • Confined areas • Cleaning
High-density regular crimped-wire or Anderlitz® brushes	• Super dense • Low flex • For high production usage on automatic or semiautomatic brushing equipment	• Heavy burr removal • Heavy scale removal • Edge blending • Surface finishing • General purpose cleaning • Surface refining
Regular crimped-wire or Anderlitz brushes	• Wide face • Dense fill • For fast cutting and fine finish • Best general purpose wheel	• Burr removal • General purpose cleaning • Surface finishing • Edge blending • Heat scale removal • Surface refining
Regular crimped-wire or Anderlitz brushes	• Narrow face • Medium dense • Used where medium brush flexibility is required	• General purpose cleaning • Surface finish for appearance • Surface roughening • Edge blending • Surface refining • Heat scale removal
High-complement knot brushes	• Maximum density • Medium flex • For rugged production jobs • Especially suited for automatic or semiautomatic machine applications	• Heavy scale removal • Heavy burr removal • Heavy cleaning rust, paint, rubber, plastics • Surface roughening • Edge blending
Standard knot brushes	• Narrow face • Highly flexible brush • Used where surface impact action is necessary	• Weld splatter and scale removal • General purpose cleaning • Surface roughening • Edge blending • Burr removal • Surface finishing

Table 11-3. Recommendations for selecting brushes *(continued)*

Step 2

After testing brush first selected based upon Chart 1,
the following refinements may be made in brush specifications or operating conditions

Chart 2

Observed results					
Brush works too slowly	Brush works too fast	Finer or lower µin. surface required	Surface finish is too smooth and lustrous	Brushing action not sufficiently uniform	Action of brush peens burr to adjacent surface
• Increase brush outside diameter • Increase brush speed • Decrease trim length, thereby increasing the brush fill density • Increase filament diameter • Change brush to heavier duty type	• Reduce trim length • Decrease filament diameter • Reduce brush outside diameter • Reduce brush speed • Select brush types having lower brushing action strength	• Decrease trim length, thereby increasing brush fill density • Decrease filament diameter • Select brush capable of producing lower µin. surface finishes • Use auxiliary abrasive compounds or oil-base lubricants with brush • Increase brush speed	• Increase trim length • Increase filament diameter • Select a brush type capable of producing coarser finishes • Reduce brush speed	• Increase trim length • Change brush type from sectional to helical construction • Select suitable brushing machinery • Oscillate brush to avoid "tracking"	• Decrease trim length • Select high-density crimped wire wheel • For ductile metals change from wire-filled brush to nonmetallic filled brushes

Brush face loads with material being removed	Brush develops excessive heat	Better control of brushing pressure	Brush wire filament ends become dull	Improve brush dynamic balance condition	Brush face wear uneven or grooving
• Increase brush trim length • Increase brush filament diameter • If wet application, apply solution through brush arbor and disperse outward through brush face • Where cleaning solutions are used with brush, modify cleaner for better solubility • Change brush fill specifications	• Use internal cooling of brush and brush arbor • Reduce brushing pressure • Select heavier duty brush • Increase brush trim length • Reduce brush peripheral speed • Use external cooling of brush	• Increase brush trim length • Install meters to measure brushing loads • When part size varies, design pressure relief in part-holding fixture • With automatic brushing equipment, allow brush to float or provide automatic pressure sensing and pressure adjusting	• Reverse rotational direction of brush to automatically recondition • Dress the brush face with an abrasive stone with the brush operating in a direction reverse of normal use • Select a heavier duty brush • Keep brushing heat as low as practical	• When wide-faced brush units are made up of thin sections, statically balance each section and stagger heavy sides • Change from sectional brush assembly to a helically formed brush • Dynamic balance brush arbor and brush assembly • Reduce brush speed • Check entire brush drive train	• If possible, have part contact brush so that the full face of brush is on the part • Consider oscillating brush or part • Provide a means for periodically dressing the brush face • Avoid excessive brushing pressure • Change from sectional brush assembly to a helically formed brush part

(Bateman 1975)

(Courtesy Anderson, Division of Dresser Industries)

Table 11-4. Comparative data for radiusing edges using miniature brushes

Brush description	Workpiece material and Brinell hardness: 3035e stainless steel (226)	Beryllium copper (187)	7075-T6 aluminum (150)	6061-T6 aluminum (95)
	Resulting edge radii, in. (mm)			
Stainless steel, .003-in. (0.08-mm) diameter filaments	.0064 (0.163)	.0085 (0.216)	.0174 (0.442)	.0186 (0.472)
Nylon, .014-in. (0.36-mm) diameter filaments with 600-grit silicon carbide	.0064 (0.163)	.0053 (0.135)	.0070 (0.178)	.0082 (0.208)
Nylon, .018-in. (0.46-mm) diameter filaments with 500-grit aluminum oxide	.0058 (0.147)	.0079 (0.201)	.0062 (0.158)	.0066 (0.168)

become tired of knife work, decided to try them. The worker's efforts using this brush produced the desired result and saved the manufacturer several thousand hours of labor time. Brushes are inexpensive, and it may be advantageous to try several before ruling out their effectiveness.

11.6 BRUSH DEBURRING EQUIPMENT

At least 15 U.S. manufacturers produce standard or specially designed machines for brush deburring. Other machines, such as bench motors, pedestal grinders, and single- and two-spindle polishing lathes, are also used for brushing. Several users have employed robots for brush deburring, although abrasive grit affects robot joints very quickly. (Robots are discussed in more detail in Chapter 16.) Because brushing machines are so inexpensive and readily available, initiating and experimenting with brush deburring is possible without adversely affecting production needs.

Before selecting automated brush deburring equipment for a specific application, the following machine features should be carefully reviewed:

- Means for measuring and relieving pressure.
- Mechanisms for adjusting, reciprocating, and conditioning the brush.
- Shaft size and ability to reverse shaft direction, which may be useful in some applications.

11.6.1 Pressure Relief

Brushing machines must have some means of relieving the pressure that builds up from misalignment, oversized parts, or improper location of the workpieces. Relieving pressure is especially important when using brushes with low flexibility (short trim and/or dense fill) to avoid damaging the brushes and the machine. For small workpieces, it is usually best to allow the parts to retract when possible. For large parts, the machine heads can have a floating design that allows the brushes to rise and fall.

11.6.2 Brush Adjustments

To compensate for brush wear, machines must be designed so that operators can adjust the pressure on the brushes; this is especially important when wide-face or cylinder brushes are being employed. Good results have been obtained by using a single adjusting screw with a fine control for both bearing slides. This design ensures that wide-face brushes remain parallel to the pass line. Adjustments can be hand-controlled or automated through power feed controls.

11.6.3 Brush Reciprocation

A means for axial reciprocation, especially of wide-faced brushes, aids in maintaining uniform action across the full face of the brush. Axial reciprocation avoids the possibility of developing grooves and prevents streaking when finishing shapes such as sheets, continuous strips, and rods. Reciprocation, however, will not eliminate a streaking pattern caused by grooves or spacing in the brush face when such grooves or spaces are continuous and in a plane perpendicular to the rotational axis of the brush. Reciprocating brushes at the rate of 1.2–2 reversals per minute is effective. When brushing multiple parts across the brush face, the reciprocation length must exceed the width of one part plus one-half the space between parts.

11.6.4 Backup Rolls

Backup rolls are often used for power brushing. They should be designed to provide proper support for the workpiece, but should not interfere with the brush. A brush cannot be used successfully as a backup roll for another brush. For thin strip or sheet material, .25 in. (6.4 mm) thick or less, two backup rolls (one on each side of the brush) are often used. For thicker materials or parts, a single backup roll is generally positioned directly opposite the brush.

11.6.5 Shaft Size

Although shaft sizes for brushing machines have become fairly standardized as the result of many years of use, it is still essential to calculate shaft diameters for special machines and operating conditions. In general, for semiautomatic machines having a shaft length requirement to provide an 8–12 in. (203–305 mm) overhang, a 1.25 in. (32 mm) diameter shaft has been found to be satisfactory. For similar service with a maximum overhang of 20 in. (508 mm), a 2 in. (51 mm) diameter shaft is generally suitable. For brushing operations, a maximum deflection of 32 in. (812.8 mm) in 6 ft (1.8 m) is the standard recommended allowance.

11.6.6 Brush Replacement

To minimize downtime, machines should be designed to allow quick and easy brush replacement. In some cases, spare heads are provided so that brushes can be replaced while the machine is in operation.

11.7 DEBURRING APPLICATIONS FOR BRUSHES

11.7.1 Removing Heavy Burrs

Most burrs with a heavy base thickness must be removed by brushes with substantial cutting action and minimal impact action. These characteristics are found in radial-type brushes because they are densely filled and have short trim. They are available in crimped wire and knotted types. The most commonly used wire sizes range from .0118 to .020 in. (0.30–0.51 mm). Knot-type brushes in diameters of 12–15 in. (305–381 mm) are extensively used in the automotive and aircraft industries for removing burrs from gears and spline bores. The deburring occurs immediately after hobbing or spline broaching, just before shaving. These short-trim brushes impart the proper radii, adequately blend the edges of gears, and remove all the minute fragments of metal that may be left behind by other deburring techniques. One advantage of this type of brush is its capacity for removing a burr from an edge without changing the dimensional characteristics of the two intersecting surfaces.

Short-trim radial brushes are used extensively for heavy-duty deburring, edge blending of stampings, metal parts, pipes, tubes, and expanded metal, and removing flash in the rubber industry. They are designed for efficient, fast, low-cost performance on automatic, semi-automatic, and manual machines. In some types of heavy burr removal imparting a circular action is necessary, such as when deburring drilled holes in condenser tube plates. Knot-type cup brush configurations lend themselves to this type of work. To avoid side flare and reduce trim length, a bridle or band is used on the brush to provide additional stiffness. This process not only deburrs the holes in the plates, but also puts a uniform radius around the periphery without any evidence of hole elongation.

11.7.2 Removing Light Burrs

Light and feather burrs are usually removed using medium- or long-trim brushes. These brushes are available in both knotted and crimped-wire construction, and each type has its own sphere of excellence. The knotted brush, which is usually used singly instead of in multiples, is a high-impact tool. Its performance is radically influenced by the speeds at which it operates. At low speeds, the brush is very flexible and soft acting. As surface speed is increased, so is stiffness, until at high speeds it becomes extremely hard and is, consequently, very fast cutting. Knotted brushes are normally used at speeds of 4,500–6,000 surface ft/min (22.9–30.5 m/s), depending on the stiffness required. Wire sizes of .0118–.023 in. (0.30–0.6 mm) are common. Deburring rubber flash from molded products can be accomplished with knotted, medium- to long-trim brushes by using substantially higher surface speeds. Speeds of up to 12,000 surface ft/min (61 m/s) are frequently used. To attain these surface speeds, brushes with diameters of 12–15 in. (305–381 mm) are used.

Burrs produced by grinding broach teeth are being deburred with 3-in. (76.2-mm) diameter wire wheels with filament diameters of .008 in. (0.2 mm). These are deburred at 17,000 r/min, moving from the back of the broach tooth to the front using hand-held motors. In this instance, the bristles rotate from the back of the tooth to the front (excessive radiusing would occur if the cutting edges of the broach teeth were hit with a brush this stiff). In some instances, vapor honing, rather than brushes, is used to remove the burrs (Yera 1990).

Crimped wire, medium- to long-trim brushes are used rather universally for many types of light deburring operations. Because these brushes are not very stiff, they can follow and conform to the contours of irregular surfaces and reach relatively inaccessible places. Crimped-wire brushes will remove feather machining and grinding burrs from edges as well as light flash found on rubber and plastic parts. Generally, smaller diameter wires are used, typically .006–.014 in. (0.15–0.36 mm). Surface speeds employed are typically 4,500–6,500 surface ft/min (22.9–33.0 m/s). Long-trim crimped wire brushes of small diameter, because of their inherent softness and flexibility, are not considered to be efficient deburring tools for heavy burrs. Light burrs are sometimes removed with brushes of approximately .005-in. (0.13-mm) diameter wire to which a deburring compound has been applied. This gives a cutting ability to the brush that it would not normally have by itself.

Removing burrs from cylindrical surfaces is sometimes accomplished with centerless brushing. The brushes used on centerless grinders normally have diameters of 20 in. (508 mm) and brush faces of 5–6 in. (127.0–152.4 mm). Fill materials for these larger radii brushes are wire (.005–.010

in. [0.13–0.25 mm] in diameter), treated tampico, abrasive impregnated plastic, or nonwoven abrasive web. When the fill is wire, these tools remove grinding burrs and produce a minimum edge of .005 in. (0.13 mm). Finishing work can be performed dry or (preferably) with normal grinding coolant. These fine wire-filled brushes, working with conventional coolant, remove feather burrs and improve surface finishes, such as on control valves for automatic transmissions. The following factors are important when using centerless brushes:

- They will not remove metal from a cylindrical surface. Parts must therefore be ground to size before brushing.
- Greatest improvement in finish (for unbrushed parts) is in the 25–30 μin. (0.64–0.76 μm) range. A part with a 25 μin. (0.64 μm) finish can be brushed rapidly to deburr and simultaneously obtain a 10–15 μin. (0.25–0.38 μm) finish; a part with a 10 μin. (0.25 μm) finish will have a 1–6 μin. (0.03–0.15 μm) finish after brushing.
- The principles involved with centerless brushing are the same as those for centerless grinding, with the following exceptions: accuracy in pressure and adjustment is not critical with brushes, and a machine is no longer needed to obtain the required finish, making it possible to locate the part below the centerline.

An old surplus grinder can often be utilized in centerless brushing operations, since the brush is flexible enough to compensate for any wear in the machine. Many manufacturers set the brushing machine in line with the parts grinding machine. The cylindrical parts can be ground to size, then fed in-line to the second machine for deburring by the same operator. Parts can be automatically moved through the machine for brushing at a rate of 2 s per piece.

Brushes filled with treated tampico fiber or cord material, used in conjunction with a grease-based polishing compound, are capable of producing extremely fine finishes in the range of 2–7 μin. (0.05–0.18 μm). Brushes filled with abrasive-impregnated nylon filaments are recommended for fast, thorough burr removal without compound and subsequent part cleaning. This type of brush is often used to deburr perforated steel tubing.

There are at least three basic approaches for applying a brush to the edges of a workpiece, as shown in Figure 11-10. In use, the lower two brushes shown in the figure are often canted or inclined to the edge of the part. One study indicated that a 25° angle of attack is optimum, but other needs may make a 45° angle more effective. Brush oscillation and the employment of multiple brushes at more than one orientation are also used to assure complete burr removal.

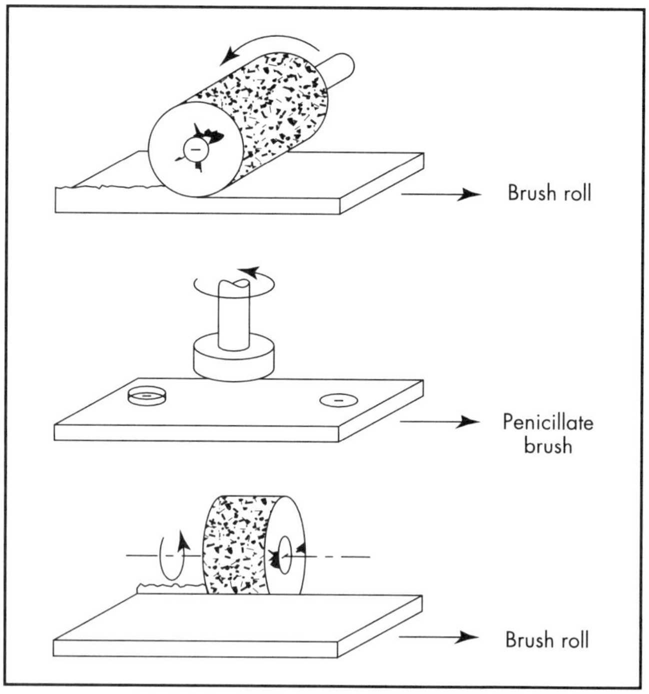

Figure 11-10. Brush arrangements for deburring (Przyklenk 1985).

11.7.3 Gear and Sprocket Deburring

High production and automated equipment used on primary gear operations cause heavier burr conditions. These burrs develop as a result of using excessive tool feed speeds and using tools beyond their most efficient cutting point.

For a great majority of gear deburring jobs, radial short-trim wire brushes (knot type) are recommended, and gears are brushed in the unhardened state. Fill wire can be any of several diameters; however, a 14 in. (356 mm) outer diameter (OD) brush with .0118 in. (0.3 mm) special analysis steel wire is most widely used. Although this type of brush develops relatively little impact action, its short trim and dense construction give the brush a high degree of cutting action. At 6,500 surface ft/min (33 m/s), this brush is extremely fast cutting.

In gear deburring, correct positioning of the brushes is very important. By spacing brushes to work on both sides of the gear face, the size of the edge break and uniformity of blend can be controlled. Brushes can produce a uniform break around the entire tooth profile or can be positioned to break the sides of the tooth more than the root. Conversely, the root can have a larger break than the sides. There are few deburring methods where this can be accomplished to result in uniformity of tooth profile on part after part.

For spur gears, brushes are spaced and contact the tooth profile on the centerline of the gear (see Figure 11-11). With this setup, brushes can be reversed, thereby keeping the wire points at their maximum cutting efficiency.

Brushing spline bores requires a different setup. Brushes must be located off-center, as shown in Figure 11-12. They generally do not have to be reversed to maintain their cutting effectiveness.

When helical gears are brushed, it is sometimes necessary to favor the acute side of the gear tooth to develop a generous radius. Such a radius is required because subsequent shaving of the gear will otherwise produce an extremely sharp edge. The correct setup for deburring helical gears is shown in Figure 11–13.

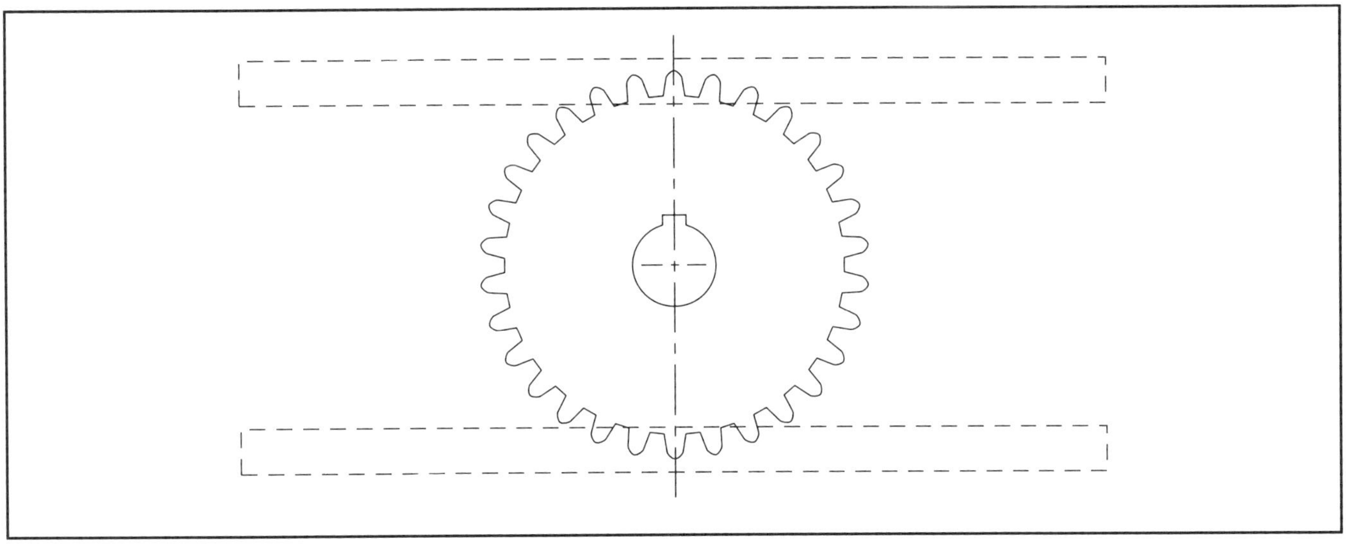

Figure 11-11. Brush position for deburring spur gears. *(Courtesy Osborn Manufacturing)*

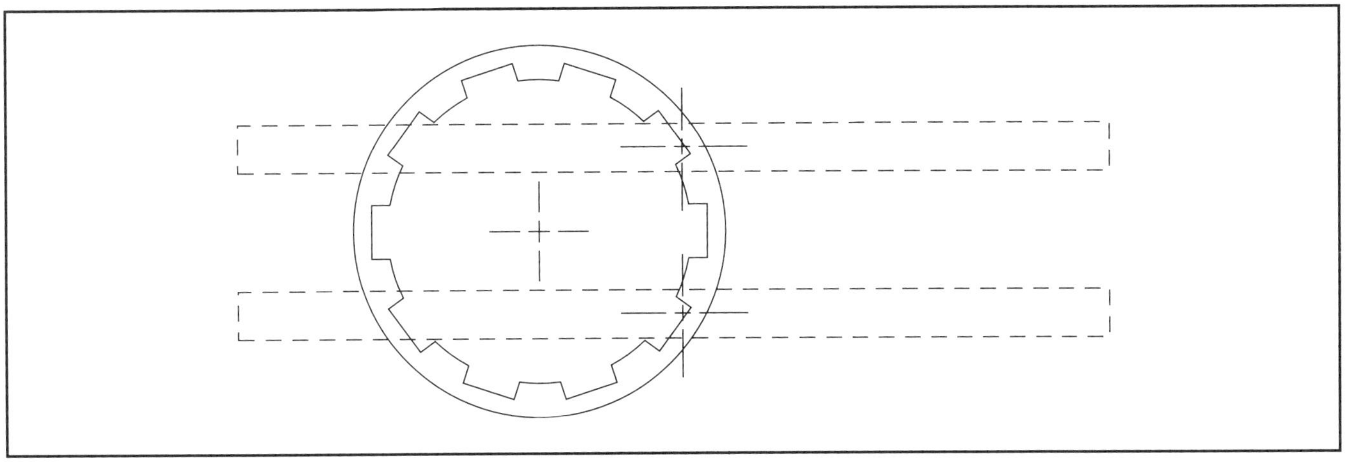

Figure 11-12. Brush position for deburring ends of splines. *(Courtesy Osborn Manufacturing)*

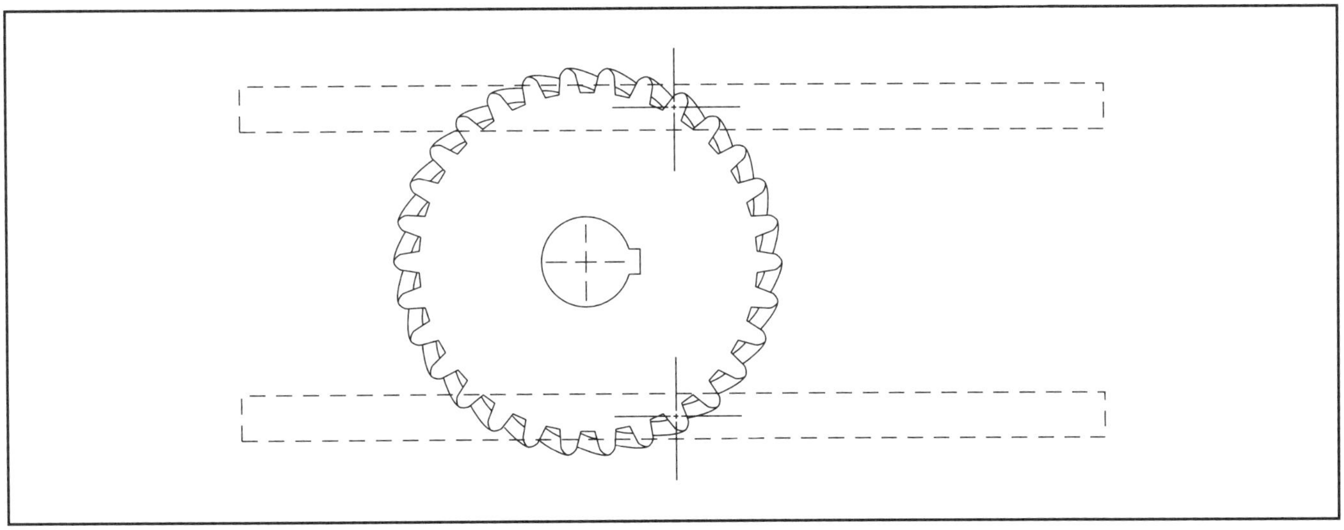

Figure 11-13. Brush position for deburring ends of helical gears. *(Courtesy Osborn Manufacturing)*

An advantage of brushing on gears is that the fillet on the gear tooth can be broken, thus relieving this high-stress concentration area. This is particularly essential on hardened gears. Many gear manufacturers experience fewer failures and replacement problems when brush finishing is used. Easier assembly and safer handling are additional benefits of this finishing method.

Elastomer-bonded wire brushes are used for deburring fine-pitch gears. These brushes have minimum flexibility on the brush face and remove burrs without leaving a secondary roll. Bonded brushes are also very effective on powdered metal and aluminum die-cast parts. At maximum efficiency, brush speeds range from 2,600–6,000 surface ft/min (13.2–30.5 m/s). Because this type of brush has such fast cutting action, surface speeds can be lower than those for conventional wire brushes.

Hardened pump gears are finished using treated tampico brushes with a grease-based deburring compound. The type of gear positioning in relation to the brushes, shown in Figure 11-11, is applied for these gears.

11.7.4 Tube, Pipe, and Rod End Deburring

Power brushes are widely used for deburring the ends of tubes, pipes, and solid rods of all sizes and shapes. A variety of methods are used, ranging from automatic machines built especially for this purpose to simple off-hand operations involving a bench motor or polishing lathe.

End burrs are removed from rough cutoff lengths of copper tubing by two heads of 10 in. (254 mm) outer diameter (OD) elastomer-bonded steel wire brushes (West 1975). The equipment utilizes a Ferris wheel arrangement to index the tubes past the brushing heads. Tubes are rotated through the brushing cycle by endless belts. The machine is adjustable for various tube lengths and diameters. For 1-in. (25.4-mm) diameter parts, the production rate is 3,300 pieces per hour.

Figure 11-14 shows an ingenious method for deburring and cleaning the ends of copper tubing after cutting to length. Three small-diameter radial brushes filled with crimped steel wire are fixed (that is, they do not revolve individually) on a metal backing plate. A wire tube-type brush is mounted in the center. The tube brush is inserted into the tubing, and the plate fixture is rotated by a portable tool. The process removes end burrs and cleans the inside and outside surfaces of the tubing. Tubes are ready for soldering or brazing. This same type of backing plate and brush assembly can be mounted on a polishing lathe or bench motor to handle long lengths of pipe or tubing. Best results can be obtained if a simple angle iron fixture is used as a steady rest to present the tubing to the brush cluster in the proper position.

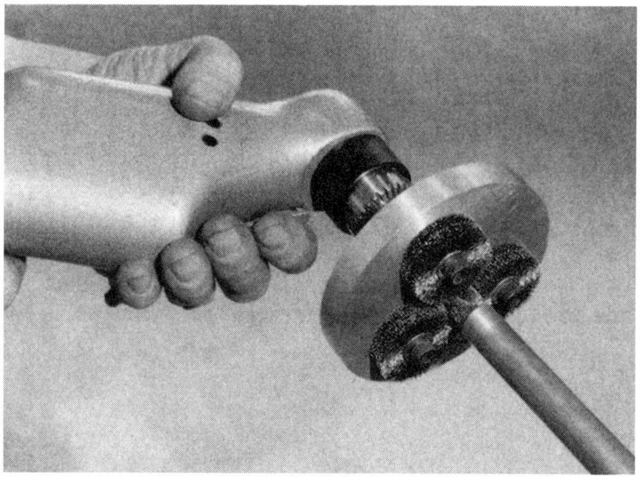

Figure 11-14. Device for deburring tube ends. ***(Courtesy Osborn Manufacturing)***

11.7.5 Deburring Internal and Threaded Areas

Small power brushes are available in a wide selection of sizes, shapes, and fill materials for deburring and cleaning recessed areas where space limitations are a factor. These brushes are designed for rotary, end, or side action and are readily adaptable for use on portable tools, drill presses, bench grinders, lathes, and special equipment (see Figure 11-15).

Small radial brushes range in diameter from 1–3 in. (25.4–76.2 mm) and are available in wire sizes from .0025–.014 in.

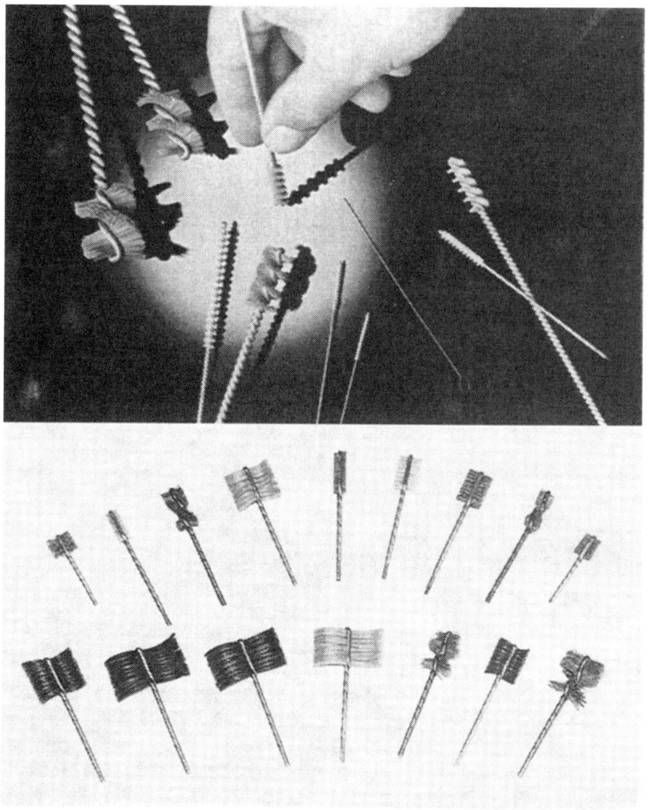

Figure 11-15. An array of cross-hole and other types of brushes used for deburring holes and edges (Scheider 1989b). ***(Courtesy Osborn Manufacturing)***

(0.06–0.36 mm). Fill materials include steel and stainless steel wire, nylon, tampico, abrasive-impregnated nylon filaments, and elastomer-bonded steel wire. A 1-in. (25.4 mm) diameter brush filled with .005 in. (0.13 mm) nylon is used to deburr small direct current (DC) armatures. The brush is driven at 8,000 r/min by a .125 hp (0.09 kW) motor and removes burrs and other foreign particles from the narrow commutator slots. S-shaped brush filaments, such as those shown in Figures 11-15 and 11-16, maximize the amount of abrasive in contact with surfaces. In some applications, the more open traditional tube brushes do better at deburring and removing contamination. For more aggressive needs, elastomer-bonded wire brushes mounted on drill presses clean internal surfaces and deburr intersecting holes in machined castings.

Side-action brushes are effective for internal deburring and for removing burrs from threaded areas. They are available in diameters ranging from .16–1.25 in. (4–32 mm) and in a broad range of styles and fill materials, both metallic and nonmetallic. Brushes with wire side tufts feather and remove sharp areas from internal threads and chips left from the threading operation. Square-trimmed brushes filled with abrasive-impregnated synthetic webbing are used to clean and deburr threaded and counterbored areas of aluminum castings in one operation. (For additional examples of applications for brushes, see Gasser 1993; Mahadev 1995; Scheider 1985; and Scheider and Carmichael 1992.)

11.7.6 Deburring Flat Shapes

Edge deburring of steel strip, sheet, or plate is effectively done by elastomer-bonded steel wire radial brushes, either off-hand using portable tools, or passed under motor-mounted brush heads. An important characteristic of this type of brush is the perpetually short trim, with each wire point firmly supported at the brush face. This produces extremely fast cutting action, maximum work efficiency with minimum pressure, and positive control over the area of brush contact.

Many manufacturers have replaced grinding wheels with abrasive-impregnated nylon filament brushes for finishing and cleaning edges of strip steel (pail stock) prior to welding. Abrasive brushes are quiet in operation (they meet OSHA noise standards), conform to slight distortions in strip, and give longer service at less cost.

Expanded metal, a form of sheet, can be thoroughly deburred by power brushing. An extremely simple setup can be assembled to convey the expanded sheet under a wide-face brush roll made up of individual knot-type steel wire radial brushes mounted in multiples on a steel shaft. The flexible brush tips reach into the uneven surfaces to deburr and blend edges, resulting in a smoothly finished product ready for decorative or other use.

Knot-type cup brushes with steel bridles can deburr thousands of drilled holes in a steam condenser support

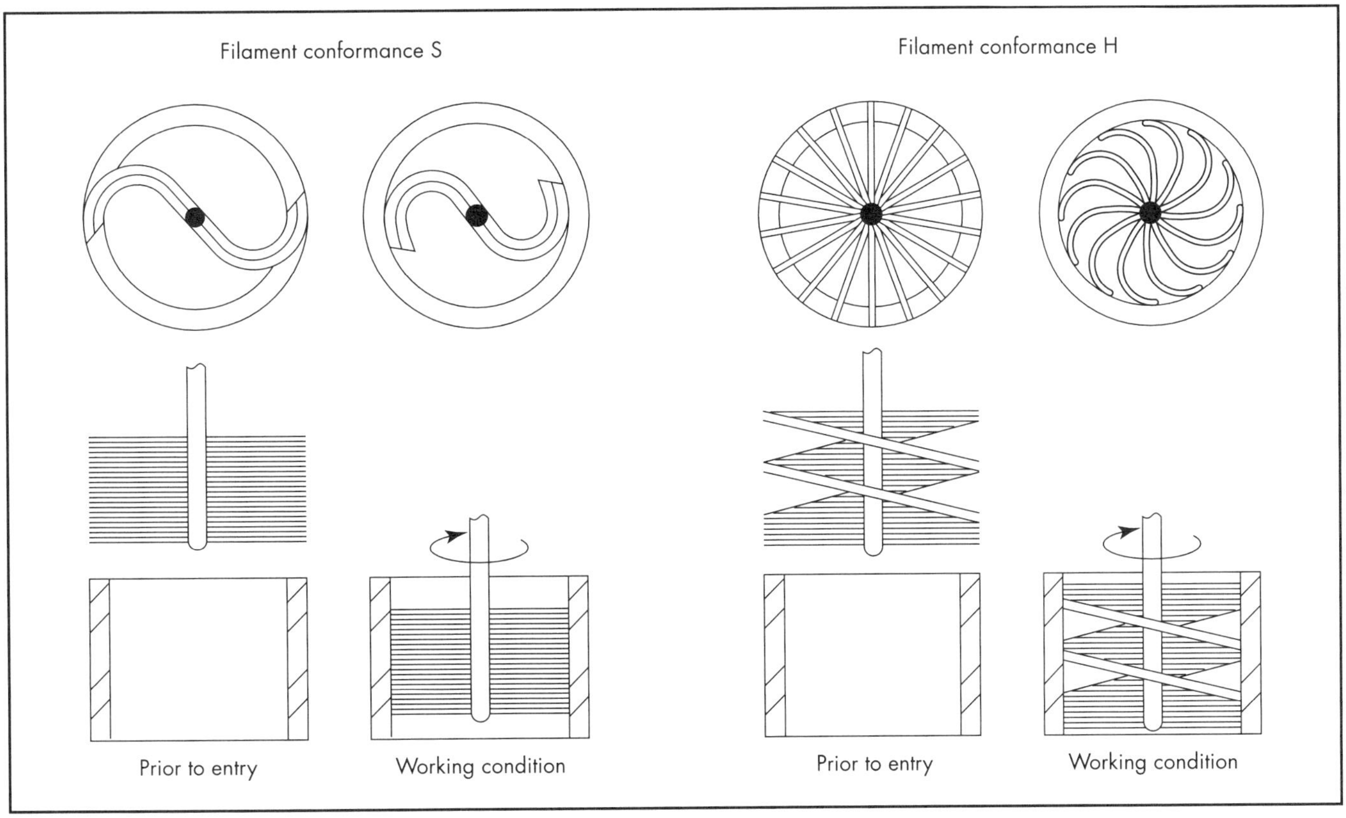

Figure 11-16. Filament bending in holes. ***(Courtesy Osborn Manufacturing)***

plate. This is accomplished by using a large boring mill that has two motor-driven brushes arranged for lateral and vertical travel positioned to deburr both sides of the steel plate at the same time. The cup brushes remove all burrs and produce a uniform radius around the periphery of each hole without elongating the hole area. For smaller plates, a radial drill press and weld positioning table with the same type of cup brush will do an excellent job.

Rounding and blending the edges of cutouts on aluminum window frames is accomplished efficiently using an interrupted-face radial brush, as shown in Figure 11-17.

Figure 11-17. Interrupted-face radial brush. ***(Courtesy Osborn Manufacturing)***

11.7.7 Deburring on the Cutting Machine

Many companies have overlooked the possibility that brushing may be more economically accomplished on the same equipment that produced the part than on secondary equipment. Even screw machines can have brushing cycles designed into cams and tooling. Putting a live spindle and brush in a lathe turret, for example, can allow one to brush the internal diameter (ID) or outer diameter (OD) without additional handling. Performing both operations on one machine has the advantage of allowing the operator to know immediately if sharper tools or different machining techniques are required. Additionally, because one operator controls both the burr and its removal, unreasonably large burrs, which often occur when cutting and deburring are separated, may be eliminated.

The abundance of computer numerical control (CNC) equipment and industrial robots provides great flexibility in on-machine deburring. CNC milling machines, for example, can machine the part, deburr it, and inspect it in a single setup. However, many companies have not taken advantage of the possibilities offered by this fast and thorough equipment. Figure 11-18 shows a tufted abrasive-filled nylon disk brush, and Figure 11-19 shows this brush placed in a tapered shank tool holder. This type of brush is used in the automotive industry for cylinder head deburring and other applications involving sharp and heavy burrs, as shown in Figure 11-20.

Figure 11-18. Tufted abrasive-filled nylon disk brush. ***(Courtesy Weiler Corporation)***

Figure 11-19. Abrasive-filled nylon tufted disk brush in computer numerical control (CNC) machine tool holder. ***(Courtesy Weiler Corporation)***

Figure 11-20. Tufted, abrasive-filled, nylon disk brush deburring aluminum valve body. ***(Courtesy Weiler Corporation)***

Even small cross-hole deburring brushes (.075 in. [1.9 mm] in diameter) are used in CNC tool holders for removing small burrs on aluminum valve parts after reaming. At 2,000 r/min, a break less than .001 in. (0.03 mm) is produced (Scheider 1990). Abrasive-filled nylon cup brushes have been used in CNC equipment routinely at least since 1988 (Scheider 1989a). These 1.5-in. (38.1-mm) diameter cups feed at rates of up to 2.5 ft/min (0.013 m/s) on aluminum parts.

11.7.8 Elastomer-bonded, Wire-filled Brushes

Elastomer-bonded, wire-filled brushes have proven to be extremely fast cutting with maximum operator safety. Because the bonding material supports the wire to the very tip, the brush provides a maximum amount of cutting action. By periodically reversing the brush, the operator has a tool that maintains maximum cutting efficiency throughout its life. Unlike a conventional nonbonded wire-filled brush, there is no loss of wire through fatigue. Bonded wire-filled brushes are made in radial, cup-style, flared-end, and straight-end designs, permitting them to be substituted for conventional wire-filled products. These brushes are effective for many deburring operations because they maintain a uniform face throughout their life and perform at maximum efficiency. Because they are so fast cutting, lower surface speeds can be used. These brushes can be used, for example, on powdered metal components, aluminum die castings, copper tube ends, and metal stampings. They also are used for edge deburring on steel strip and sheet and for removing burrs from key slots and other hard-to-reach areas. In deburring ductile materials, radii are produced without producing a secondary rolled condition. Usual wire sizes range in diameter from .0118–.014 in. (0.3–0.36 mm), and appropriate speeds range from 850–6,500 surface ft/min (4.3–33.0 m/s).

Two factors should be taken into consideration when elastomer-bonded brushes are used for deburring: First, periodic dressing of the brush face will yield a brush with a life 10 times longer than found with conventional nonbonded wheels, thus keeping end-of-service costs for a given deburring operation much lower. Second, because the cutting action of elastomer-bonded brushes is so fast, brush adjustment is critical, since the part must be held precisely if the brush is to be used efficiently.

The following are three examples for use of elastomer-bonded brushes:

1. Removing burrs from key slots and other hard-to-reach areas. The wire fill material for these brushes is generally .012 in. (0.30 mm) in diameter, and the brushes are rotated at speeds of 5,500–6,500 surface ft/min (28.0–33.0 m/s).
2. Removing stamping burrs and producing radii on ductile metals, without producing secondary rolled conditions. The wire-fill material is usually .008 in. (0.20 mm) in diameter, and the brushes rotate at speeds of 5,000–7,000 surface ft/min (25.4–35.6 m/s).
3. Removing burrs on powdered metal parts, aluminum die castings, and fine-pitch gears. The wire fill material for these brushes is usually .006–.008 in. (0.15–0.20 mm) in diameter, and the brushes are rotated at speeds of 5,000–6,500 surface ft/min (25.4–33.0 m/s).

11.7.9 Nonwoven Abrasive Brushes

Nonwoven abrasive brushes have, in many instances, replaced treated tampico, abrasive monofilament, and fine wire brushes in deburring applications. This is because they are available in a variety of densities and grades of hardness and are filled with a large assortment of grit sizes (aluminum oxide and silicon carbide). Light deburring can be accomplished with the uncompressed material. For heavy deburring applications, harder brushes are the most suitable and are used for deburring grinding lines in all types of metal components, appliances, and parts, including stampings and machined parts.

Other applications for nonwoven abrasive brushes include deburring many types of plastics, hole deburring (both wet and dry) printed circuit boards, wood finishing, removing parting lines on jet blades, and radiusing gears and machine parts. Generally, the surface speeds used with nonwoven abrasive brushes are 5,000–6,500 surface ft/min (25.4–33.0 m/s).

11.7.10 Guidelines for Brush Use

Refer to Table 11-3 for recommendations for selecting brushing and Table 11-5 for the relationship of surface speed to rotational speed for various brush diameters. Using the brush selection charts shown in Table 11-3 will be fairly simple if the basic brush technology and fundamental information given in this chapter is applied.

11.8 TROUBLESHOOTING IN BRUSHING

Table 11-3 contains some of the possible problems, as well as solutions, that may be encountered in power brushing. Since power-driven brushing tools wear out, they must

Table 11-5. Relationship of surface speed to rotational speed for various brush diameters

Rotational speed (r/min)	Brush diameter, in. (mm) 4 (101.6)	6 (152.4)	8 (203.2)	10 (254)	12 (304.8)	15 (381)
	Surface speed, ft/min (m/s)					
1,000	1,050 (5.33)	1,575 (8.00)	2,100 (10.67)	2,625 (13.33)	3,150 (16.00)	3,925 (19.94)
1,500	1,575 (8.00)	2,350 (11.94)	3,150 (16.00)	3,925 (19.94)	4,725 (24.00)	5,900 (29.97)
1,750	1,800 (9.14)	2,750 (13.97)	3,650 (18.54)	4,550 (23.11)	5,500 (27.94)	6,800 (34.54)
2,000	2,100 (10.67)	3,150 (16.00)	4,200 (21.34)	5,250 (26.67)	6,275 (31.88)	7,850 (39.88)
2,500	2,625 (13.33)	3,925 (19.94)	5,250 (26.67)	6,550 (33.27)	7,850 (39.88)	9,825 (49.91)
3,000	3,125 (15.88)	4,725 (24.00)	6,275 (31.88)	7,850 (39.88)	9,425 (47.88)	11,775 (59.82)
3,450	3,600 (18.29)	5,400 (27.43)	7,200 (36.58)	9,000 (45.72)	11,000 (55.88)	13,500 (68.58)
3,750	3,900 (19.81)	5,900 (29.97)	7,800 (39.62)	9,800 (49.78)	11,800 (59.94)	—
4,000	4,175 (21.21)	6,275 (31.88)	8,375 (42.55)	10,475 (53.21)	—	—
4,500	4,700 (23.89)	7,075 (35.94)	9,425 (47.88)	—	—	—
5,000	5,225 (26.54)	7,875 (40.00)	—	—	—	—
6,000	6,275 (31.81)	9,425 (47.88)	—	—	—	—
8,000	8,375 (42.54)	—	—	—	—	—

Note: Values containing dashes represent unsafe operating conditions for most brushes.

be replaced periodically. The primary reasons for brush failure or excessively rapid wear are abrasive wear and filament fatigue. To increase the wear life of a brush, a number of steps can be taken.

The abrasion resistance of a brush is decreased in proportion to the stiffness of its filaments, and reducing filament stiffness will prolong the life of the brush. Stiffness can be reduced by altering any of the stiffness factors discussed previously, including modulus of elasticity, trim length, and rotational speed.

Brush life can be prolonged by reducing the load on the brush, that is, by changing the relationship that exists between the brush and the work being brushed. Reducing the load also decreases the amount of filament deflection. Wear increases more rapidly than the proportional increase in load when the filaments are wiping instead of cutting.

Filament fatigue can also lead to premature brush failure. The very nature of brushing produces forces that compress, flex, and sometimes overheat brush filaments. These forces are conducive to fatigue when they reach an intensity that exceeds a certain tolerance, called the endurance limit. When the endurance limit is exceeded, brush failure is almost immediate; the greater the mechanical stress, the quicker the damage to the brush filaments. Although fatigue or failure is possible for any size of filament, susceptibility varies directly with the diameter of the filament—fine wires are capable of resisting overstressing to a much greater degree than large-diameter filaments.

In many operations, users have found that more efficient brushing is accomplished by periodically reversing the brush's direction of rotation (see Figure 11-9 for correct and incorrect brushing techniques). The filament tips of the brush become dull with use, and reversing the direction of rotation tends to resharpen the tips or at least present new cutting edges to the work. This self-sharpening action also tends to reduce operator fatigue by reducing the pressure required to do the work.

Table 11-6 provides some common solutions to problems encountered when using abrasive-filled nylon brushes.

11.9 SIDE EFFECTS OF BRUSHING

Brushing can create a number of side effects that can be beneficial or detrimental, depending on the application and results desired. The degree to which some of these effects occur depends in part on whether brushing involves entire surfaces or merely edges. Possible side effects include the following:

- Changes in surface finish, texture, color, and hardening.
- Removal of stress riser tool marks.
- Changes in residual stress.
- Changes in fatigue life.
- Contamination of part surfaces.
- Dusty environment.
- Changes in part size.
- Generation of new burrs.

Table 11-6. Suggestions for troubleshooting abrasive-filled nylon brushes

Observation	Corrective measures
Tool not aggressive enough	Use larger grit (mesh size) Use silicon carbide mineral Use rectangular-shaped filaments Increase density of filaments Use shorter trim length Increase surface speed within range Decrease feed speed—more dwell time
Tool too aggressive	Use smaller grit Use round filaments Decrease filament density Use longer trim length Reduce surface speed Increase feed rate Use aluminum oxide mineral
Filaments smearing nylon	Use coolant at point of contact Reduce dwell time on workpiece Reduce in-feed pressure on workpiece Reduce surface speed
Surface finish too rough	Use smaller grit (mesh) size Use aluminum oxide mineral Use smaller filament diameter or shape Apply honing compounds to filaments
Tool life too short	Verify consistency of burrs on incoming parts Reduce surface speed Increase density of filaments Use coolant at point of contact Use rectangular-shaped filaments Increase tool diameter

(Scheider 1991)

The fact that brushes can produce finishes as fine as 4 μin. (0.10 μm) is a distinct advantage for most companies. Other companies take advantage of the fact that brushes can leave a decorative texture on surfaces. Turbine and airframe manufacturers employ brushes, in part, because they smooth out tool marks, which are points of high stress concentration. Jet engine manufacturers, for example, have found that half the failures in their parts can be traced to stress concentrations and progressive fractures, such as scratches, sharp edges, or burrs. Eliminating these stress concentrations greatly extends fatigue life.

Wire-filled brushes can greatly affect the residual surface stresses (and harden surfaces as a result). Research over the past 100 years indicates that as a result of chemical, thermal, and mechanical interactions during surface brushing, surface residual stresses can increase dramatically. For example, one researcher noted that hardness near the brushed part of a surface was five times greater than in the center of the part. Other researchers have shown that work hardening, as opposed to brushing, can only double the hardness. If it is assumed that hardness is proportional to yield strength, then the residual stresses near the surface would be five times greater than the bulk stresses. Those high stresses occur in a layer only a few thousandths of an inch thick (Przyklenk 1985). While these typically greatly help to increase fatigue life, they can decrease life in thin materials or in materials whose inner core has high-tensile residual stresses. In addition, on thin nonsymmetrical parts, part distortion can easily occur. Another side effect of residual stresses is surface textures that resemble the texture of an orange peel. This particular condition can be corrected by reducing the frictional heat generated by the brush.

Any time a foreign material contacts a part, the part is contaminated. Thus, when a carbon steel brush is used to brush stainless steel, carbon steel is deposited on the part. This will allow rust spots to occur on a normally rustproof surface. Brass brushes used on steel will leave a yellow, brass-coated part. Nylon, if used at high speeds, will melt on parts, leaving black splotches and probably a thin invisible layer of nylon. Tampico wheels when used with grease-based "burring" compounds leave obvious mud-like deposits on the part and, in some cases, are harder to remove than burrs. Diamond lapping compounds can also be nearly impossible to remove from hard-to-reach recesses. In the case of some aluminums, anodizing will not be complete if foreign particles are embedded in the surface. Similarly, if large amounts of aluminum oxide are embedded in steel parts, welding, soldering, and brazing can be impossible to perform satisfactorily. At the present time, however, abrasive-impregnated nylon or synthetic brushes do not appear to impregnate abrasives into the workpiece in quantities sufficient to cause problems for the majority of applications. In contrast, tumbling or blasting processes produce gross surface contamination. In some situations the color of a part will change slightly as a result of brushing or brushing contamination. This can improve or diminish the aesthetics of a part.

Contamination can be carried over from one part to another. For example, if a nylon brush is used on aluminum parts and then on stainless steel, the stainless steel part will have traces of aluminum on it. Similarly, nylon brushes will pick up oil, grease, or other coatings from one part and deposit them on the next. This can be a source of concern when deburring parts by hand, which requires high levels of cleanliness.

The use of deburring compounds with tampico wheels, metal-filled wheels, buffing wheels, or felt bobs can result in large quantities of particles and clumps of compound being thrown into the atmosphere and the work area. Because of this and because the process is inherently dirty and dusty, safety precautions and regulations must be followed carefully.

By nature, brushing changes the dimensions of a part every time it is performed. Brushes are, after all, metal-removal tools. Generally, changes in part size resulting from brushing are small (typically, 50 μin. [1.27 μm], except at edges where burrs are to be removed). However, significant size changes are possible.

One advantage of using brushes instead of other products for surface finishing is that they can reach into holes while the surface is being brushed. This capacity removes some of the burrs that frequently are left by flat belt sanding. However, if an incorrect wire brush or speed is selected, the brush can generate burrs on edges that were initially burr-free.

11.10 NONTRADITIONAL BRUSHING

Although many types of brushes have been commonly used for decades or even centuries, a number of newer, nontraditional brushes are also available and being successfully employed for a variety of applications. The principles governing their success or failure are often different from those applied to the types of brushes discussed earlier in this chapter. Nontraditional brushes currently in use include miniature flap wheels, elastomer-filled radial brushes, abrasive-filled, nonwoven synthetic wheels and shapes, abrasive, ball-tipped, nylon filament tube brushes, and long-string nylon, mop-like brushes. These types of deburring tools have tremendous promise for rapid burr removal. In many cases, they can produce better surface finishes than traditional brushes.

11.10.1 Brush-backed Flap Wheels

Some flap wheels utilize sections of brushes to provide a polishing action as well as a cutting action. It is possible to combine almost any brush material with bonded abrasive products to provide longer life and different edge finishing.

11.10.2 Elastomer-bonded, Wire-filled Brushes

Elastomer-bonded, wire-filled brushes are extremely fast cutting and safe to use. Because the bonding material supports the wire to the very tip, the brush provides a maximum amount of cutting action. By periodically reversing the brush, optimum cutting efficiency can be maintained throughout its life. Compared with conventional, nonbonded, wire-filled brushes, elastomer-bonded, wire-filled brushes lose a minimal amount of wire through fatigue. They are made in radial, cup-style, flared-end, and straight-end designs, and can be used in place of conventional wire-filled products. They are effective for many operations because they maintain a uniform face throughout their life, allowing them to consistently perform at maximum efficiency.

11.10.3 Long-string Brushes

Long-string, abrasive-filled, nylon mop-like brushes are being used to remove burrs and radius edges on a wide variety of large aluminum aircraft components (see Figure 11-21) (Blanton 1983). The flexibility of the filaments, which can be 2 ft (0.6 m) long, and their tendency to wrap around an edge as they are whipped around a part, permit deburring on both sides of some parts with a single orientation of the part (see Figure 11-22). Stock removal is low (typically .001 in. maximum [0.03 mm]), and radii of .012–.060 in. (0.3–1.5 mm) are produced using 80-grit silicon carbide (Mahadev 1995). These brushes are operated at very low speeds, in the range of 500 r/min for a 22-in. (559-mm) diameter strip mop brush with 8-in. (203.2-mm) long filaments. In some applications, a robot swings the mop-like brush continuously in a wide pattern (in such applications there is little robot programming).

Figure 11-21. Mop-like, abrasive-filled, nylon brush used on aerospace parts. *(Courtesy Osborn Manufacturing)*

11.10.4 Ball-ended Bristle Brushes

Ball-ended bristle brushes have a ball-shaped cluster of fine abrasives at the end of nylon fibers (see Figure 11-23). Each abrasive globule consists of built-up laminations of abrasive materials. The globules are like pearls coated with layer upon layer of bonded abrasive. Continual wear of these little balls opens new surfaces for fresh cutting. The nylon bristles bend to allow these brushes to reach into large holes and radius cross-holes as well as entrance and exit holes. Like any tube brush, the diameter of the brush should be larger than the hole diameter to assure continuous scrubbing. In addition to deburring, these tools provide surface finishes much sought after in the automotive industry for cylinder walls. The materials recommended for cross-hole deburring are shown in Table 11-7.

Although little detailed data have been published on the selection of ball-ended bristle brushes for deburring, some excellent research has been reported describing their use for honing. In general, ball-ended, bristle brush honing has many similarities to abrasive-filled, nylon bristle brush honing and produces similar results (Lin 1992). Surface finishes of 15.7 μin. (0.4 μm) are produced with 120-grit media. Finishes of 3.9 μin. (0.1 μm) are produced with 400-grit media. These two grits produced stock removal of .0004 in. (10 μm) on 3.6 in. (91 mm) bores honed for 30 s. In general, the diameter of the ball-ended bristle brush was 3–14% larger than the diameter of the hole being brushed.

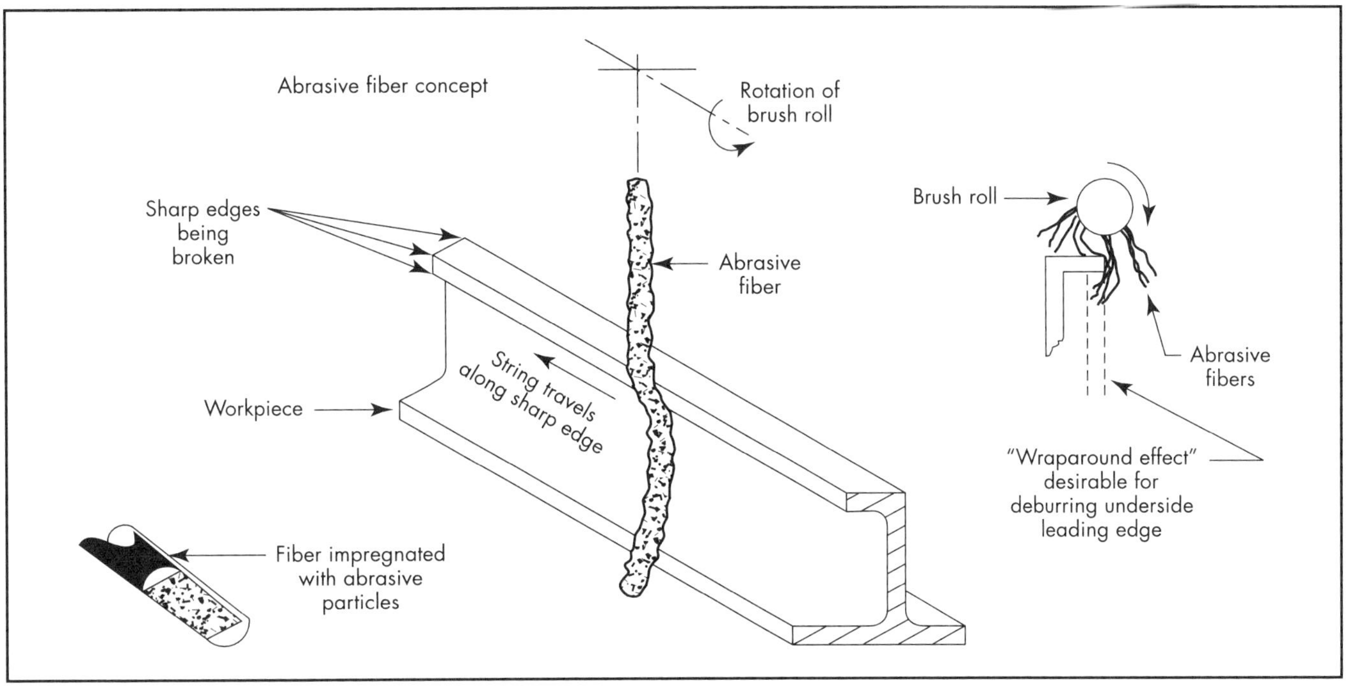

Figure 11-22. Long-string abrasive brush used for deburring large aircraft parts. ***(Courtesy Avco Aerostructures)***

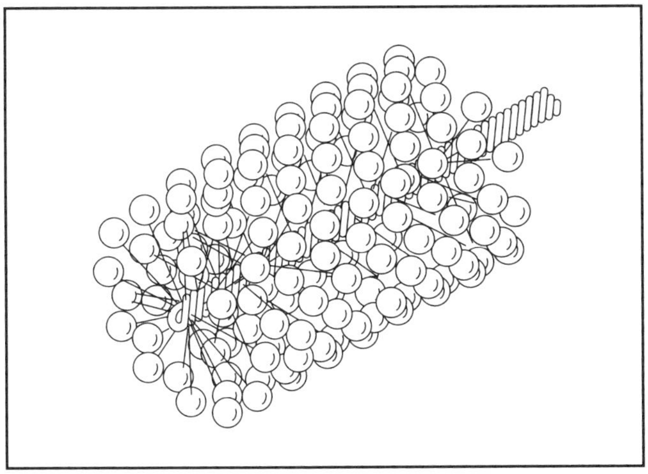

Figure 11-23. Ball-ended bristle brush.

Table 11-7. Recommended abrasive materials for deburring cross-holes using ball-ended nylon filament brushes

Part material	Ball material
Hard steel	120-grit tungsten carbide
Mild steel	80-grit zirconia alumina
Cast iron	120-grit silicon carbide
Stainless steel	120-grit silicon carbide
Aluminum	120-grit aluminum oxide
Chrome plating	120-grit aluminum oxide
Carbide	80-grit boron carbide
Brass/bronze	320-grit aluminum oxide
Nicasil®	120-grit silicon carbide
Nickel	120-grit boron carbide
Boron	120-grit boron carbide

(Miller 1993)

11.11 COMBINED-PROCESS DEBURRING

As discussed in earlier chapters, the most effective solution for many deburring problems may be to combine more than one type of process. For example, many currently available automated machines combine bonded abrasive, brushing, and buffing processes in a single machine (Frazier 1991). These machines provide much more capability than a machine using only one process, and they are fast and relatively low cost.

11.12 ENVIRONMENTAL, HEALTH, AND SAFETY CONCERNS

When properly used, power-driven brushes are safe, although inherently they have all the dangerous attributes of any rotating body. It is essential, therefore, that they be mounted, handled, and used carefully, and that all necessary precautions be taken. The operator and others in the work area must wear safety goggles or full-face shields over safety glasses; side shields; and other suitable protective equipment. A brush should only be used when it is guarded by an enclosure that protects the operator as well as other workers who are in the proximity of the brushing operation.

The machine on which a brush is being used should not be operated at an excessive speed, and the brush itself should never be operated beyond the maximum safe speed established by the manufacturer. Before operation, the machine should be jogged to ensure that it is ready for use and that the brush is securely fastened and concentric with the axis of rotation. After the machine has been started, the brush should be run at operating speed in a protected enclosure for at least one minute before applying the work; during this time, no one should stand in front of or in line with the brush.

Brushes throw fragments of wire and particles being removed from workpieces at great velocities. It is essential, therefore, that the operator and others in the area be protected by guards and other protective equipment.

In operation, workpieces should never be pressed into the face of the brush far enough to slow the drive motor. Visual evidence of excessive pressure can be observed when the sides of the filaments are in contact with the work; only the tips of the filaments should be in contact with the workpieces.

In no event should brushes larger than 8 in. (203.2 mm) in diameter ever be used on portable tools.

Wire brushes should be stored in an environmental atmosphere that precludes rusting or oxidizing and in a manner that protects the brushes from physical distortion, which might result in brush imbalance.

Safety standard requirements and excellent explanatory information are contained in the American National Standards Institute's (ANSI) Standards B165.1, "Safety Requirements for the Design, Care, and Use of Power-driven Brushing Tools," and B165.2, "Safety Requirements for the Design, Care, and Use of Power-driven Brushing Tools Constructed with Wood, Plastic, or Composition Hubs or Cores." These standards should be referred to and fully understood before setting up any new applications, and they should be periodically reviewed for existing brushing applications. The standards contain a wealth of engineering and technical information, with separate sections for users, machine tool builders, and brush manufacturers.

References

Bateman, James. 1975. "Capabilities and Limitations of Brush Deburring." Technical paper MR75-826. Dearborn, MI: Society of Manufacturing Engineers (SME).

Blanton, Albert Glenn. 1983. "Ultra-long String, Abrasive Brush Deburring." Technical paper MR83-691. Dearborn, MI: Society of Manufacturing Engineers (SME).

Frazier, Ronald. 1991. "Deburring Update: Utilizing Flexible, Radial Wheel-type Processes." Technical paper MR91-122. Dearborn, MI: Society of Manufacturing Engineers (SME).

Gasser, Joseph. 1993. "Advancements in Flexible Abrasive Finishing Tools." Technical paper MR93-135. Dearborn, MI: Society of Manufacturing Engineers (SME).

Lin, Y.T. 1992. "Honing with Abrasive Brushes." Technical paper MR92-132. Dearborn, MI: Society of Manufacturing Engineers (SME).

Mahadev, Prasad S. 1995. "Nylon Abrasive-filament (NAF) Brushes—An Alternative to Deburring, Edge Radiusing, and Surface Finishing Problems." Technical paper MRR95-02. Dearborn, MI: Society of Manufacturing Engineers (SME).

Miller, Michael L. 1993. "Flexible Honing: A Study of Cylinder Wall Microstructure." Technical paper MR93-146. Dearborn, MI: Society of Manufacturing Engineers (SME).

Przyklenk, Klaus. 1985. "Brushing—New Research Results With an Old Method." Technical paper MR85-848. Dearborn, MI: Society of Manufacturing Engineers (SME).

Scheider, Alfred F. 1985. "Deburring, Edge Contouring, and Surface Finishing Applications Using Abrasive Brushes and Buffs." Technical paper MR85-821. Dearborn, MI: Society of Manufacturing Engineers (SME).

———. 1989a. "Emerging Trends in Flexible Abrasive Tools." Technical paper MR89-113. Dearborn, MI: Society of Manufacturing Engineers (SME).

———. 1989b. "New Technology in Flexible Abrasive Tools for Nontraditional Finish Machining of Holes, Bores, and Internal Surfaces." Technical paper MR89-805. Dearborn, MI: Society of Manufacturing Engineers (SME).

———. 1990. "Advanced Flexible Abrasive Finishing Tools." Technical paper FC90-153. Dearborn, MI: Society of Manufacturing Engineers (SME).

———. 1991. "New Developments in Flexible Abrasive Finishing Tools, Buffs, and Brushes." Technical paper MR91-392. Dearborn, MI: Society of Manufacturing Engineers (SME).

Scheider, Alfred F., and Carmichael, Guy. 1992. "New Technologies in Abrasive Brush Honing." Technical paper MR92-135. Dearborn, MI: Society of Manufacturing Engineers (SME).

West, Richard. 1975. "Deburring with Power Brushes." Technical paper MR75-476. Dearborn, MI: Society of Manufacturing Engineers (SME).

Yera, Harvey J. 1990. "Proper Care and Maintenance of Broach Tools." Technical paper MS90-361. Dearborn, MI: Society of Manufacturing Engineers (SME).

Bibliography

Bateman, James. 1979. "Brush Techniques for Cleaning, Finishing, and Deburring." Technical paper MR79-757. Dearborn, MI: Society of Manufacturing Engineers (SME).

Bausch, John J., III, and Gold, Frederick M. 1995. "Design and Application of a Brush-based Active Robotic Deburring System." Technical paper MR95-224. Dearborn, MI: Society of Manufacturing Engineers (SME).

Behringer, Brian J. 1981. "Automated Deburring of Flat Sheet Metal." Technical paper MR81-387. Dearborn, MI: Society of Manufacturing Engineers (SME).

Cariapa, V., Stango, R.J., Liang, S.K., and Prasad, A. 1989. "Measurement and Analysis of Brushing Tool Performance Characteristics—Part II: Contact Zone Geometry." *Symposium on the Mechanics of Deburring and Surface Finishing Processes*, Stango, R.J. and FitzPatrick, P.R., eds. PED Vol. 38. New York: American Society of Mechanical Engineers (ASME):159–172.

Dawson, Bryan L. and Hennies, Ronald C. 1988. "Robotic, Long-string Brush Deburring System." Technical paper MR88-297. Dearborn, MI: Society of Manufacturing Engineers (SME).

Duwell, Ernest J., and Bloecher, Ulrich. 1983. "Deburring and Surface Conditioning with Brushes Made of Abrasive Loaded Nylon Fibers." Technical paper MR83-684. Dearborn, MI: Society of Manufacturing Engineers (SME).

FitzPatrick, Peter R., and Paul, Frank W. 1987. "Robotic Finishing Using Brushes—Material Removal Mechanics." Technical paper MR87-156. Dearborn, MI: Society of Manufacturing Engineers (SME).

Flores, Gerhard. 1992. "Mechanical Deburring: Process, Tools, Machines, and Applications." Technical paper MR95-272. Dearborn, MI: Society of Manufacturing Engineers (SME).

Heinrich, S.M., Stango, R.J., and Shia, C.Y. 1989. "Effect of Workpart Curvature on the Stiffness Properties of Circular Filamentary Brushes." *Symposium on the Mechanics of Deburring and Surface Finishing Processes*, Stango, R.J. and FitzPatrick, P.R., eds. PED Vol. 38. New York: ASME: 27–40.

Henderson, James A., Shia, Chih-Yuan, and Stango, Robert J. 1993. "Design, Selection, and Machining Performance of Advanced Filamentary Brushing Tools." Technical paper MR93-325. Dearborn, MI: Society of Manufacturing Engineers (SME).

Keating, Thomas V. 1983. "Today's Applications of Low-density and High-density Nonwoven Nylon Abrasives." Technical paper MR83-688. Dearborn, MI: Society of Manufacturing Engineers (SME).

Russel, W.N. 1987. "Nylon Abrasive Monofilaments—New Brush Applications." Technical paper MR87-162. Dearborn, MI: Society of Manufacturing Engineers (SME).

Scheider, Alfred F. 1991. "Precision Edge Radius Grinding of Diamond Cutting Tools and Inserts." Technical paper MR91-195. Dearborn, MI: Society of Manufacturing Engineers (SME).

———. "Production Radius Grinding of Carbide and Ceramic Cutting Tool Inserts." Technical paper MR90-532. Dearborn, MI: Society of Manufacturing Engineers (SME).

Shia, C.Y. 1988. "Analysis of Constrained Deformation of a Circular Brush System." Master's thesis. Milwaukee, WI: Marquette University.

Shia, Chih-Yuan, Stango, Robert J., and Heinrich, S.M. 1989. "Theoretical Analysis of Frictional Effect on Circular Brush Stiffness Properties." Technical paper MR89-143. Dearborn, MI: Society of Manufacturing Engineers (SME).

Spur, G., and Felling, W. 1983. "Development of a Robot-integrated Working Place for the Brushing of Rubber Metal Parts." Technical paper MS83-349. Dearborn, MI: Society of Manufacturing Engineers (SME).

Stango, Robert J. 1991. "Damage Assessment of Wire and Nylon/Abrasive Filamentary Brushing Tools." Technical paper MR91-133. Dearborn, MI: Society of Manufacturing Engineers (SME).

Stango, R.J., Cariapa, V., and Kanion, J.M. 1989. "Experimental Evaluation of Circular Brush Stiffness: Preliminary Results." Technical paper MR89-144. Dearborn, MI: Society of Manufacturing Engineers (SME).

Stango, R.J., Cariapa, V., Prasad, A., and Liang, S.K. 1989. "Measurement and Analysis of Brushing Tool Performance Characteristics—Part I: Stiffness Response." *Symposium on the Mechanics of Deburring and Surface Finishing Processes*, Stango, R.J. and FitzPatrick, P.R., eds. PED Vol. 38, 143–57. New York: ASME.

Stango, R.J., Heinrich, S.M., and Shia, C.Y. 1989. "Analysis of Constrained Filament Deformation and Stiffness Properties of Brushes." *Journal of Engineering for Industry (ASME Transactions)*, 111 (3):238–43.

Suzuki, Kiyoshi, Kitajima, Koichi, and Uematsu, Tetsutaro. 1994. "Methods for Efficient Finish Polishing with Abrasive Wheel Brush." *Proceedings of the 3rd International Conference on Precision Finishing and Burr Technology*. Kansas City, MO: Deburring Technology International: 167–72.

Tarrab, K.M.H. 1990. "Dynamic Properties of Circular Brushes and Evaluation of Filamentary Stress: Photographic Analysis." Master's thesis. Milwaukee, WI: Marquette University.

Tynex, A. 1996. *The Finishing Touch*. Technical Bulletin. Washington, WV: DuPont Filaments.

Van Sickle, Charles, and Flores, Gerhard. 1991. "Mechanical Deburring and Honing in the Automated Environment." Technical paper MR91-126. Dearborn, MI: Society of Manufacturing Engineers (SME).

Watts, James H. 1988. "Abrasive Monofilaments—Critical Factors that Affect Brush Tool Performance." Technical paper MR88-138. Dearborn, MI: Society of Manufacturing Engineers (SME).

12 Abrasive Flow Machining (AFM)

Abrasive flow machining (AFM), also called *abrasive flow deburring*, is a deburring process that involves extruding an abrasive-filled semisolid media through a workpiece passage. Deburring and edge radiusing occurs when the very viscous, dough-like media is forced over an edge. The elements required for AFM processing are the machine, workpiece fixture (tooling), and media. The process can be used for finishing holes as small as .030 in. (0.762 mm), as well as parts weighing hundreds of pounds. It employs simple, generally nonprecision tooling and has no detrimental environmental wastes. It polishes surfaces while deburring and is widely used for mold and die polishing. Most materials, from soft aluminum to tough nickel alloys and ceramics, can be abraded using AFM.

The abrasive media becomes a sort of flexible file, or "slug," that adapts to the contours of part edges and cavities. The process can be thought of as using a self-forming abrasive tool that repeatedly removes workpiece material from those areas in which media flow is intentionally restricted. In general, the media determines the kind of abrasion that occurs, the fixture determines where the abrasion occurs, and the machine settings determine how much abrasion occurs. The following process variables are involved in AFM:

- Media viscosity.
- Media rheology.
- Abrasive grit size.
- Volume of media passing over edges.
- Extrusion pressure.
- Flow speed.
- Tool design.
- Cross-sectional area of restricting passage.
- Length of restricting passage.
- Burr size.
- Burr location.

12.1 EQUIPMENT

The equipment used in AFM processes hydraulically clamps the workholding fixtures between two vertically opposed media cylinders (Figure 12-1). These cylinders extrude the media back and forth through the workpiece(s). Two cylinder strokes, one from the lower cylinder and one from the upper cylinder, make up one process cycle (Figure 12-2). Both semiautomatic machines and high-production fully automated systems are widely used.

Machines used in AFM processes primarily control extrusion pressure, which is a crucial variable that determines the amount of abrasion produced. Pressures range from 100–3,200 psi (689–22,063 kPa), with flow rates of 100 gal/min (379 L/min) and higher. The ideal media flow rate depends on the size and number of passages to be processed and the flow characteristics desired. Flow rate is affected by media viscosity and extrusion pressure and is generally not a crucial variable.

The dominant variables controlling the amount of work a specific media composition can perform are flow volume and extrusion pressure. These variables are controlled in the machine by presetting the displacement of each media cylinder stroke, the number of two-stroke cycles for the job, and the hydraulic extrusion pressure. Low-cost, low-production machines have media cylinder diameters ranging from 3–6 in. (76–152 mm). Machines for larger components or higher production rates have cylinder diameters ranging from 6–24 in. (152–610 mm).

Microprocessor controllers in the machine monitor and modify additional variables, such as media temperature, media viscosity, media wear, and flow speed. Machine accessories, such as part-cleaning stations, unloading and reloading stations, media refeed devices, and media heat exchangers, are frequently included in installations for high-production applications.

12.2 PROCESS PRINCIPLES

Two key aspects of the AFM process are the special abrasive media and its ability to precisely abrade only selected areas along its flow path. The media performs its selective abrasion by altering its viscosity (resistance to flow). One of the most used media is a "Silly Putty®-like" silicone rubber mixed with abrasive particles.

In its normal semisolid state, the media is relatively pliable and yielding. However, when its flow is drastically restricted, such as when the media mass attempts to flow through a smaller passage than previously encountered (Figure 12-3), the media temporarily becomes more viscous. This increased viscosity is due to the polymer carrier of the media reacting to the sudden force of the flow restriction. While in this more viscous state, the media in the restricted passage can be thought of as a slug. When the slug exits the restriction, its viscosity returns to normal.

Such momentary increases in viscosity explain how the media used in the AFM process can selectively abrade local areas along its flow path. A fixture can be used to direct media to specific areas in the flow path. If the area where abrasion is desired does not sufficiently restrict the flow of the abrasive, tooling can be designed to provide an insert or mandrel within this area to reduce the cross-sectional flow area and thus restrict the flow of the abrasive.

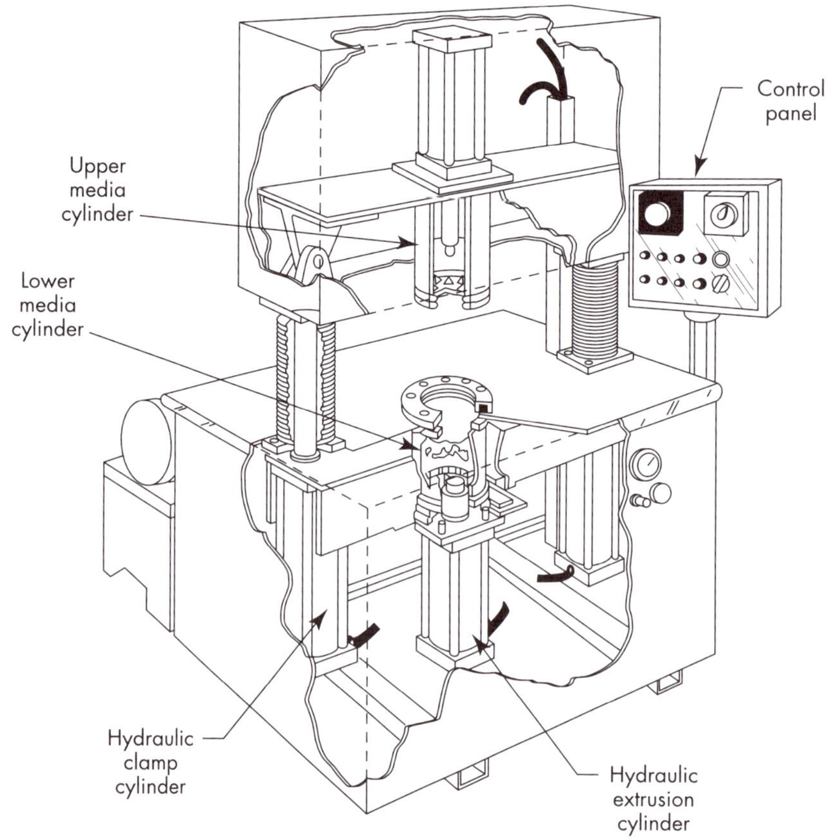

Figure 12-1. Cutaway view of high-production machine for AFM. (*Courtesy Extrude Hone Corporation)*

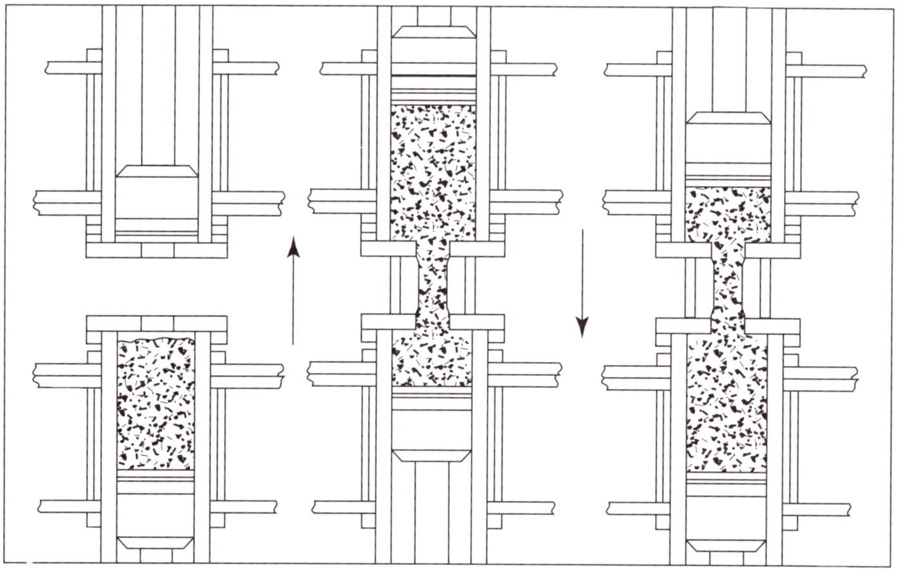

Figure 12-2. One AFM process cycle consists of two cylinder strokes that extrude the media through the workpiece twice. (*Courtesy Extrude Hone Corporation)*

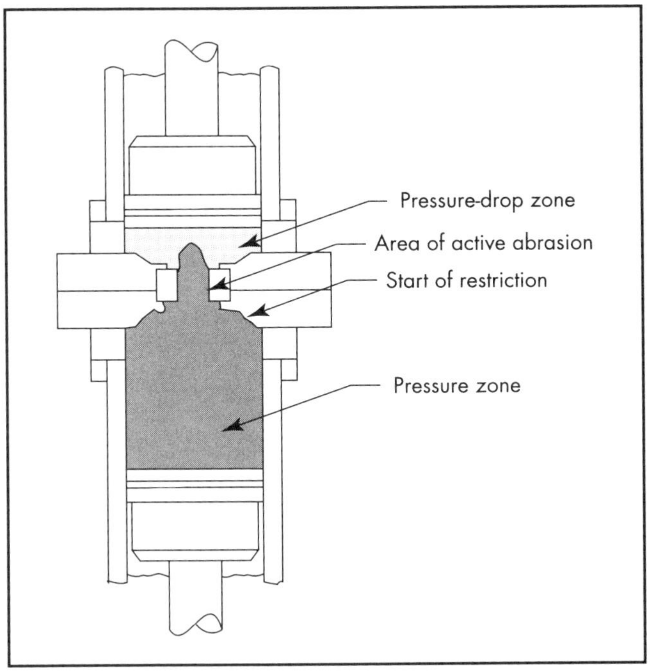

Figure 12-3. Pressure changes on media as it flows through a straight passage in a workpiece. The restricted orifice formed by the workpiece causes the media to increase in viscosity, producing abrasion in this area. (*Courtesy Extrude Hone Corporation*)

The AFM process can simultaneously abrade two, and occasionally more, successive restrictions in the same flow path, if they are equal in cross-sectional area; otherwise, only the smaller of these passages will be abraded (Figure 12-4). Any number of parallel passages can be abraded. If two parallel passages have differing cross-sectional areas, the larger of these passages will receive more flow and consequently more abrasion because it offers the least restriction to flow (Figure 12-5).

12.2.1 Controlling Abrasion

The two most important factors determining the amount of abrasion produced during AFM processes are (1) the depth of the cut made by each of the media grains in the abrasive slug and (2) the number of cuts made by the media grains during the process cycle. The cut depth depends on the size, hardness, and sharpness of the grains, as well as the machine extrusion pressure. The number of cuts depends on the length of the media slug that passes through the restriction, as well as the speed of the media that is being extruded through the workpiece.

The maximum force initially applied to the abrasive slug is equal to the machine extrusion pressure multiplied by the cross-sectional area of the restricting passage. Some of this force is lost within the media mass because of its internal resistance to flow. The remaining force is transferred through the suddenly stiffened media slug as it approaches and passes through the restricting passage and, from there,

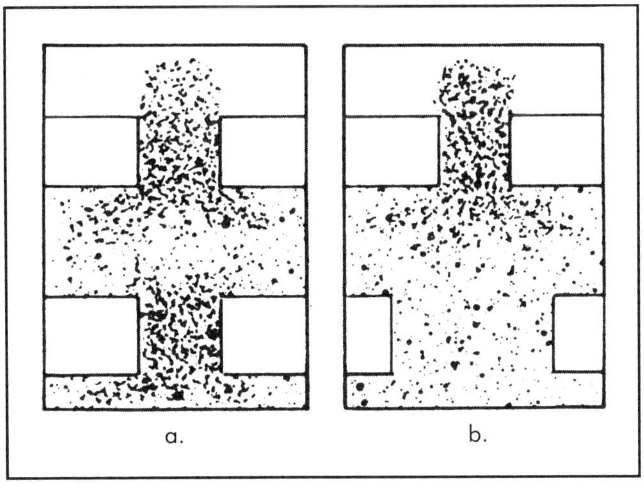

Figure 12-4. Change of media viscosity while flowing through successive restrictions. (*a*) Both restrictions will be equally abraded if they are equal in cross-sectional area; (*b*) if the successive restrictions are of different cross-sectional areas, only the smaller area will be abraded. (*Courtesy Extrude Hone Corporation*)

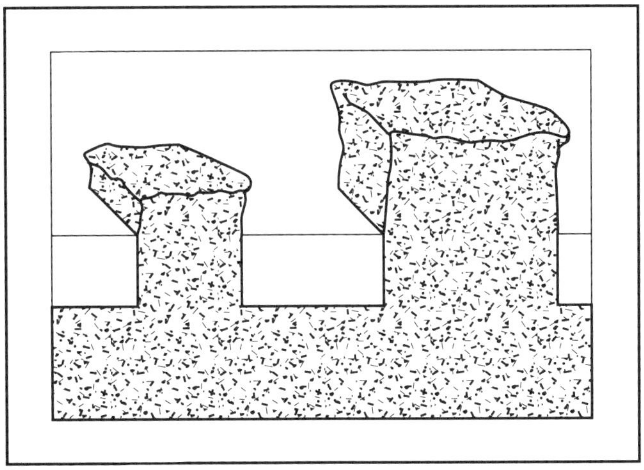

Figure 12-5. When processing parallel restrictions, the larger passage allows more media flow and, therefore, more abrasion occurs. (*Courtesy Extrude Hone Corporation*)

to the abrasive grains contacting the workpiece edges and surfaces. The force against these grains is determined by the machine extrusion pressure, media viscosity, and the type and size of the abrasive grains.

Slug flow speed determines, in part, the type of flow pattern produced (media flow patterns and their effect on abrasion are discussed in the next section). The slug flow speed is controlled by extrusion pressure and media viscosity, as well as by passage area and length.

Media flow volume is a third factor that determines the amount of abrasion produced by AFM processes. A greater volume of media flowing into the workpiece means that a greater number of abrasive grains will be in contact with the workpiece, resulting in more abrasion.

The factors controlling the amount of the abrasive contacting the workpiece determine not only the amount of abrasion produced by the media, but also the effect of the abrasive on the workpiece. Whereas flow volume primarily determines the overall amount of abrasion, slug flow speed determines the media flow pattern and the resulting type of machining action produced.

For a given size and length of restricting passage, extrusion pressure, and media viscosity, a certain flow speed will occur. Table 12-1 describes some of the effects of AFM process variables on pressure and flow speed; for example, if extrusion pressure increases, flow speed will increase.

Wall versus Edge Abrasion

The media flow pattern as it enters and extrudes through the restricting passage determines whether the greatest abrasion will be concentrated at the edges or along the walls of the passage. If a polishing or honing effect is desired (in which a uniform amount of stock must be removed from the walls of the restricting passage), the internal flow pattern of the media slug must be nearly uniform throughout the passage. Slow speeds will help the media slug to flow evenly throughout the passage (Figure 12-6a).

If the desired effect is to radius or deburr the edges of a passage, the flow speed of the abrasive can be increased, causing the edges at the passage entrance to be abraded more than the passage walls. Higher slug flow speeds accentuate this edge-radiusing effect (Figure 12-6b). Since one flow cycle normally involves media flowing in both directions, both ends of the passage will be abraded.

The following variables determine the media flow pattern: media viscosity, passage size (width and length), slug flow speed (extrusion pressure), and media rheology (the viscosity change of the polymer carrier under various stresses, due to its degree of internal bonding).

Surface Improvement

Surface improvement by AFM results from direct abrasion of the "peaks" (and to a lesser extent the "valleys") of the microsurface of a workpiece. There is no smearing of the higher material into the lower valleys, as can occur with hand polishing.

During AFM, the peaks of the microsurface are quickly abraded. As the peaks flatten into plateaus, the valleys likewise receive abrasion until a smooth and even profile is produced and no trace of the initial surface remains. The ability to remove undesirable, tensionally stressed surface material renders AFM ideal for improving surface integrity and, consequently, the service life of critical components that are machined or treated by thermal methods (such as electrical discharge, laser, and electron beam machining) or subjected to mechanical fatigue stress.

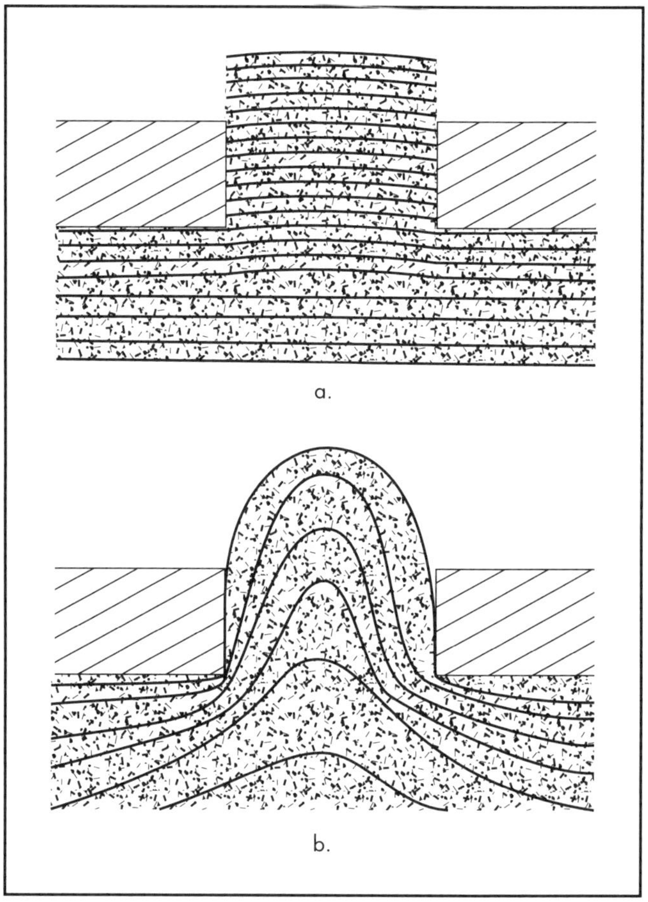

Figure 12-6. Cross-sectional views of slug flow patterns. (*a*) Uniform pattern resulting from slower flow speeds and media of higher viscosity, producing a honing or polishing effect; (*b*) uneven pattern resulting from higher flow speeds and media of lower viscosity, producing a deburring or edge-radiusing effect. (*Courtesy Extrude Hone Corporation*)

Table 12-1. Effects of AFM process variables

Variable	Effect: and Extrusion Pressure is Held Constant	Effect: and Flow Speed is Held Constant
If passage area increases	Then flow speed will increase	Then extrusion pressure will decrease
If passage length shortens	Then flow speed will increase	Then extrusion pressure will decrease
If media viscosity decreases	Then flow speed will increase	Then extrusion pressure will decrease

Courtesy Extrude Hone Corporation

A variety of effects can be achieved by manipulating process variables. One study, for example, showed the following results (Borchers 1980):

- Increasing the pressure will increase the amount of radius generated.
- Using a larger size grit will generate a larger radius.
- Using a finer grit will produce better finishes.
- Designing fixtures to maintain consistent media flow paths is essential for obtaining optimum results.

Another study showed that stock removal and surface finish results depend on the process used to produce the surface (Loveless, Williams, and Rajurkar 1993).

Selectivity and Finishing Inaccessible Areas

Abrasion occurs only in areas where media flow is restricted; other areas are unaffected. Inaccessible areas and any restricted places through which the media can be forced to flow can be finished, including complex internal passages. Shallow pockets, bottoms of blind holes, and otherwise blind features are not suitable for AFM processes. However, holes, particularly those that intersect, are excellent applications for this process.

12.2.2 Process Versatility

AFM processes are used on a wide range and varying sizes of workpieces. They can process many areas on a work-piece simultaneously. Several to hundreds of holes, slots, or edges can be deburred, radiused, and polished in one operation. Depending on the workpiece and machine size, several or even dozens of parts can be processed in one fixture load, resulting in production rates of hundreds of parts per hour.

Normal production rates for AFM are 60–300 parts/h (Rhoades 1981; Borchers 1980). Some fixtures allow multiple parts to be processed per load. For example, in some machines, up to 30,000 injector nozzle bodies a day can be deburred. In such applications, a carousel shuttles eight fixtures, loaded with 60 parts each, with a deburring cycle of 1.5–2 min/load (Rhoades 1988). When fine finishes are required, production rates will be lower. Where roundness is critical, users must perform tests to measure such changes. Increases of out-of-roundness of 24–40 μin. (0.61–1.016 μm) have been reported for 1.80-in. (46-mm) diameter holes, with corresponding hole size increases of 17–24 μin. (0.43–0.61 μm).

A number of studies have been published on surface improvements, and these are included in the reference list (Rhoades 1981; Rhoades 1988; Borchers 1980; Loveless, Williams, and Rajurkar 1993; Williams, Rajurkar, and Rhoades 1989).

Both small and large production quantities can be handled economically with the AFM process. Changing from one job to another, including replacing the tooling and media, can be normally done in minutes.

12.2.3 Process Limitations

AFM is a versatile and controllable finishing process because the workpiece is held stationary and the abrasive media is directed to and often through the passages to be finished by the tooling. However, because most workpieces require a fixture, workpieces must be loaded and unloaded from the fixture, and some parts require part orientation in the fixture (though many do not). Additionally, all of these requirements involve some expense.

After processing, the abrasive media normally remains within the interior of the part and on its exterior surfaces, making fully automated processing and handling difficult. Some fully automated systems incorporate a machine with sufficient capacity to completely process a fixtured load of workpieces in one stroke. Flowing only from the interior of the workpieces, the abrasive media exits freely and falls into a receiving cylinder or collection hopper. This approach makes tooling and automatic unloading and reloading easier because the outsides of the workpieces are relatively free of media.

AFM requires that the media be removed and retrieved and the workpiece cleaned. Media can be removed from workpieces by blowing it out with shop air. Fine particles and lubricants remaining on workpieces are normally removed in an ultrasonic cleaning operation employing commercial solvents. This type of cleaning operation is essential because fine, loose particles left after blow-off can cause plug gages or mating components to seize in the holes. Carefully choosing abrasive size and tooling will prevent abrasive particles from lodging in small holes.

The removal and retrieval of media and final cleaning of parts has been made more convenient for automated systems by using water-based solvents and vacuum. The solvents can be used as a spray or pulsed flush, without special venting, to dissolve the media carrier.

12.3 TYPICAL APPLICATIONS

AFM is being used on workpieces ranging in size from as small as .060 in. (1.52 mm) in diameter to turbine disks nearly 4 ft (1.2 m) in diameter. Passages in the workpieces have orifice diameters as small as .004 in. (0.10 mm) to splines that are 2 in. (51 mm) or larger in diameter. The process can generate edge radii ranging from less than .001 in. (0.025 mm) to more than .060 in. (1.52 mm). Table 12-2 provides general guidelines for determining the media viscosity necessary for processing passages of various sizes. By manipulating process variables, the capabilities and limitations of the media can be extended.

For a given set of deburring conditions, the edge radius produced after removing the burr is a function of burr size. Thick, tall burrs reduce the radiusing action. With standard tooling, removing burrs that are .005 in. (0.127 mm) thick by .002 in. (0.051 mm) high from stainless steel results in a radius of about .005 in. (0.127 mm). The same conditions in aluminum result in a radius of about .008 in.

Table 12-2. Media viscosity necessary for processing workpiece passages of various sizes

Media viscosity	Passage size, in. (mm)* Minimum	Maximum
Low	.02 (0.5)	.125 (3.18)
Low medium	.03 (0.8)	.25 (6.4)
Medium	.06 (1.5)	.5 (12.7)
High/medium	.125 (3.18)	1 (25.4)
High	.25 (6.4)	2 (50.8)

* Passage sizes are widths or diameters; data is based on passage lengths being twice their width or diameter.

Courtesy Extrude Hone Corporation

(0.203 mm). With standard tooling, it is impossible to simultaneously remove large burrs and produce small edge radii of .005 in. (0.127 mm) or less because of the flexibility of these processes. AFM can produce edges of less than .001 in. (0.03 mm) radii if burrs are small, that is, less than .0005 in. (0.013 mm) thick. Tooling can be designed to produce different edge radii on different holes of the same part. For example, radii could be produced in sizes ranging from .010–.030 in.(0.25–0.76 mm) during the same cycle.

Polishing with AFM can improve surface finishes of 30–300 μin. (0.76–7.62 μm) to one-tenth or less of the original finish. Material removed by the abrasion results in a dimensional change of about 25–35% more than the total roughness of the surface. For example, reducing a surface finish of 100 μin. (2.54 μm) to a finish of 10 μin. (0.25 μm) requires the removal of about .001 in. (0.03 mm) of stock. In most cases, stock removal can be held uniform throughout a passage to within 120%.

Components commonly finished by the AFM process include gears, bearing races and cages, and splined parts. Manufacturers of jet engines and turbines use the process extensively for finishing components such as fuel swirlers, combustion liners, turbine disks and blades, and compressor wheels and vanes. AFM processes are also used in the aerospace, medical, and other industries on critical components where demanding performance is essential.

12.4 TOOLING

The workholding fixtures, or tooling, for AFM are designed primarily to hold workpieces in position on the machine and to direct the media to the desired areas. If necessary, tooling also can be used to restrict passages that must be abraded but do not themselves sufficiently restrict the flow of media.

Tooling is made of aluminum, steel, nylon, urethane, or a combination of these materials. Aluminum and nylon are easily machined, lightweight materials. Steel is used for its strength and durability. Urethane is sometimes preferred because of its high resistance to abrasion and its ability to be cast in complex shapes. Tooling used as passage restrictors is most often made of urethane. For many applications, the tooling will be machined of steel, with portions made of nylon or urethane designed to fit into the more complex workpiece areas.

Tooling designs can be simple (Figure 12-7) or complex, depending on part configuration, media flow paths, and processing rates. Many applications require only simple universal fixtures. For instance, a part with a straight-through passage (such as a tube, an internal spline, or a valve plate) only needs tooling to hold the part in place between the two media cylinders, with the workpiece passage itself providing the greatest restriction in the flow path. Most extrusion die-polishing applications require no tooling. If AFM is used to process external edges or surfaces, the tooling contains the part in a flow passage, restricting the flow between the exterior of the part and the interior of the fixture (Figure 12-8).

Internal passages, surfaces, and edges will be processed as long as the media flow reaches these areas. Workpieces with complicated internal configurations may require two or more processing operations, with abrasion occurring in some areas during one operation and in other areas with different flow paths during the next. Large parts may need to be processed in sections. Counterbores, recessed areas, and even blind cavities can be processed by using a mandrel that fits inside the component and restricts the media flow at the desired areas. Tooling also may be used to protect critical workpiece areas that are adjacent to or in the flow path of a desired abrasion area.

High-production tooling is designed to facilitate part-handling operations, including loading, unloading, and cleaning. A common design for high-production processing uses two fixtures mounted on opposite ends of an index table. While the load in one fixture is being processed, the other fixture is unloaded and reloaded with parts. After the first load is completely processed, the table is indexed 180°, swinging the second fixture into position between the media cylinders. These and other types of high-production tooling arrangements are sometimes designed in conjunction with automated parts-handling systems to further increase production efficiency.

12.5 MEDIA FOR AFM

The media used for AFM consists of abrasive grains in a polymer carrier. Carriers are available in a variety of viscosities (see Table 12-2) and rheologies, each producing a different type of media flow. Abrasive grains vary in type, size, and concentration. Silicon carbide is the most commonly used abrasive; other abrasives in use are boron carbide, aluminum oxide, and diamond. Particle sizes range from 8–1,000 grit (.180–.0002 in. [4.57–0.005 mm]).

The effective lifetime of the media depends on the initial batch quantity, abrasive size and type, flow speed while processing, part configuration, and the aggressiveness of the abrasive work performed. These factors combine to wear down

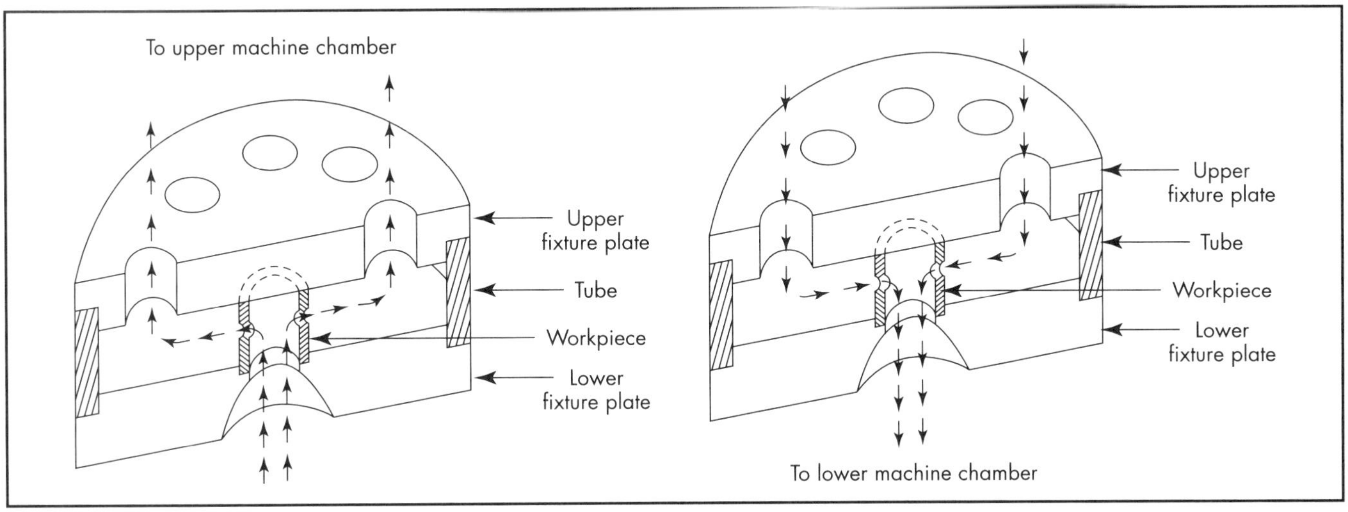

Figure 12-7. Simple sleeve tooling directs the media through the holes and over the external edges of a workpiece. (*Courtesy Dynetics Corporation*)

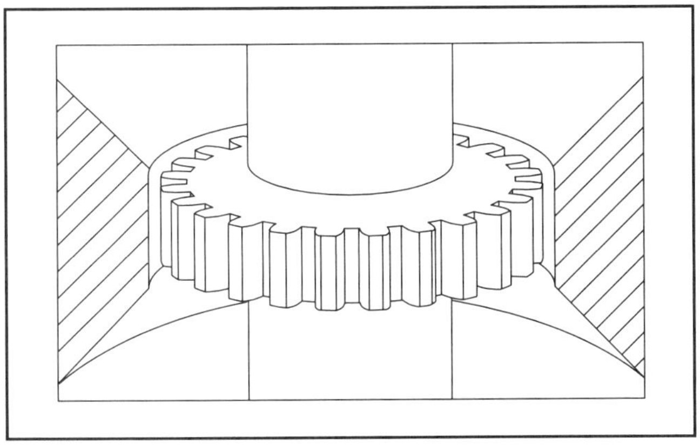

Figure 12-8. Tooling for processing the external surfaces of a gear. The media flows between the gear teeth and the bore of the fixture. (*Courtesy Extrude Hone Corporation*)

the sharp corners of the abrasive. In addition, since the media mass absorbs all the material removed from the workpiece, performance changes as the media becomes loaded. Normally, a machine load of media can be used for weeks or for thousands of parts before needing to be replaced.

12.6 VARIATIONS OF AFM

Several variations of the AFM process are in use, for example, orbital AFM, ultrasonic AFM, and abrasive stream finishing. Also, some AFM machines have been designed to push the media in only one direction or to employ multiple media cylinders to control machining and deburring when several passages exist on a part.

12.6.1 Orbital AFM

Orbital AFM oscillates a workpiece in a circular orbit of up to a few millimeters within a pressurized container filled with a special abrasive media (Figure 12-9). This approach produces mirror-like surface finishes (Figure 12-10). The process can also selectively finish edges, producing small or very large radii.

12.6.2 Ultrasonic AFM

Ultrasonic AFM employs a generic vibrating tool that is manipulated across the surface of a part (generally a cavity) in five axes. The ultrasonic vibrations excite the media flowing between the workpiece and the vibrating tool, gen-

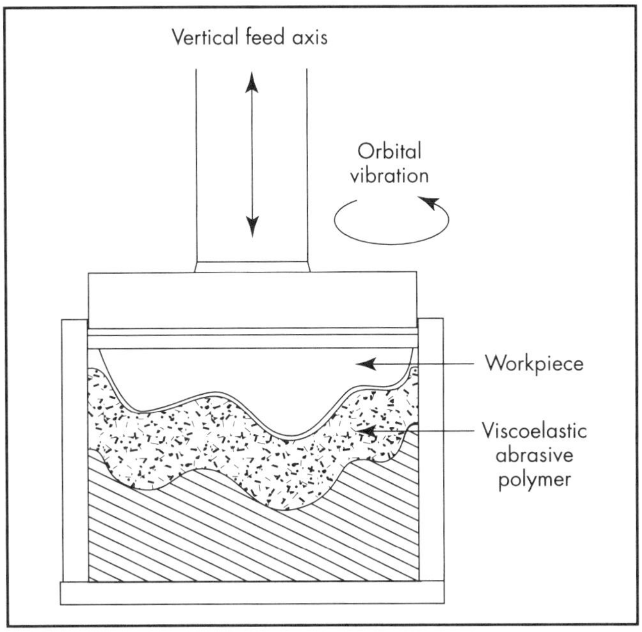

Figure 12-9. Schematic of orbital AFM. (*Courtesy Extrude Hone Corporation*)

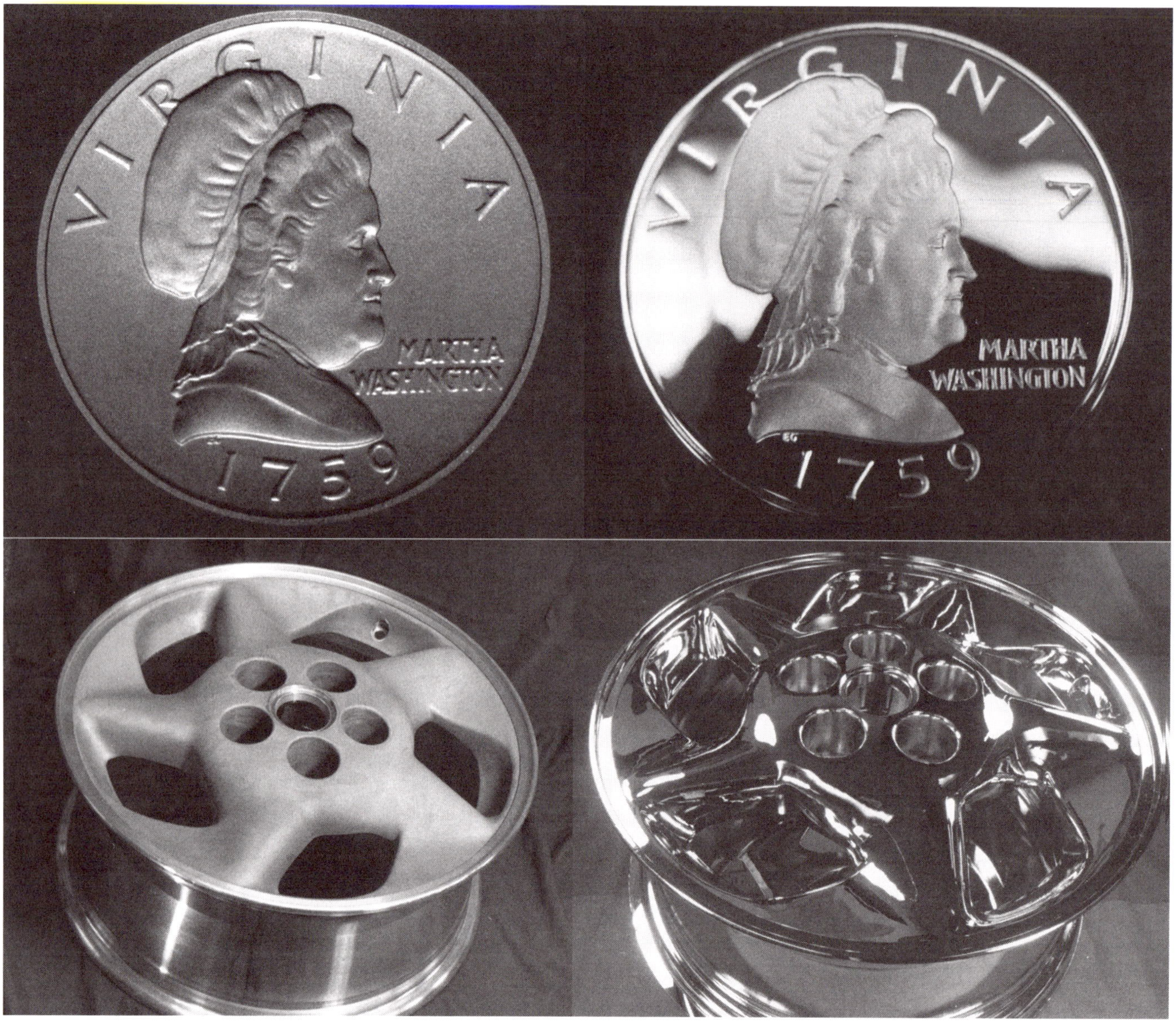

Figure 12-10. Results of orbital AFM on coins and automotive wheels. ***(Courtesy Extrude Hone Corporation)***

erating the major portion of the work. Surface roughness can often be reduced by 70–90%. The process is similar to ultrasonic machining except the action in ultrasonic AFM is a polishing or fine deburring action. The process appears most appropriate for parts having small burrs that also require high surface finishes. Figures 12-11 and 12-12 illustrate the process and equipment.

12.6.3 Abrasive Stream Machining

In abrasive stream machining (deburring), a moderate velocity stream of abrasive flow media is directed at surfaces and edges (Figure 12-13). The energy in the stream is sufficient to generate an action in which the stream engages the workpiece, producing an effect similar to that of an AFM flow restriction. The tight stream coherence provides much greater selectivity and control than abrasive blasting techniques. This approach will not remove large burrs, but it can work on edges having small burrs.

12.6.4 One-way Flow AFM

One-way flow AFM allows fast finishing, easy cleaning, and simple tool changing. Figure 12-14 illustrates the principle behind the process. It is particularly appropriate when burrs must be removed from only one side of holes.

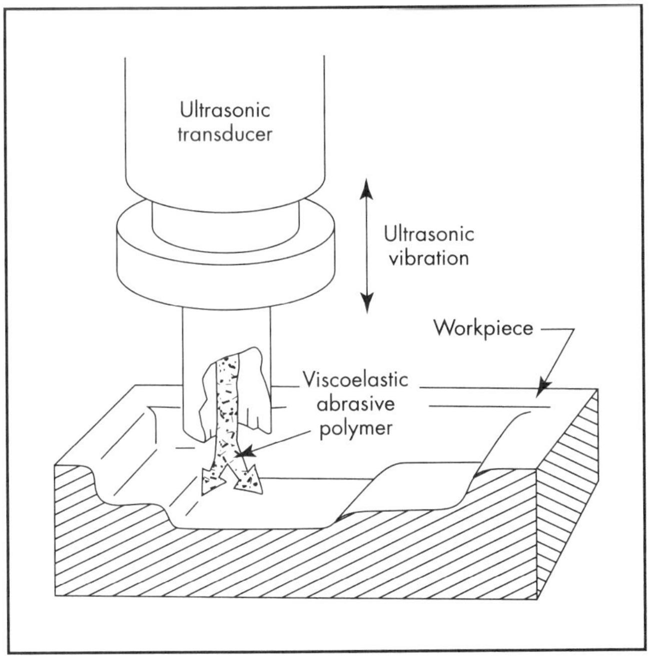

Figure 12-11. Schematic of ultrasonic AFM. (*Courtesy Extrude Hone Corporation*)

Figure 12-12. Ultrasonic AFM. (*Courtesy Extrude Hone Corporation*)

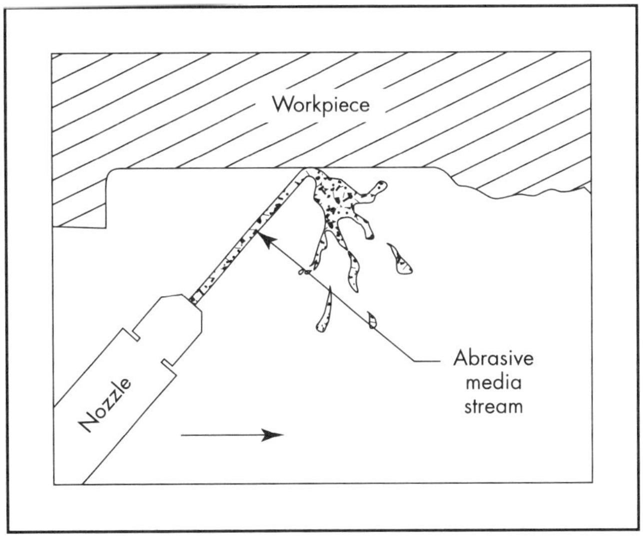

Figure 12-13. Schematic of abrasive stream machining. (*Courtesy Extrude Hone Corporation*)

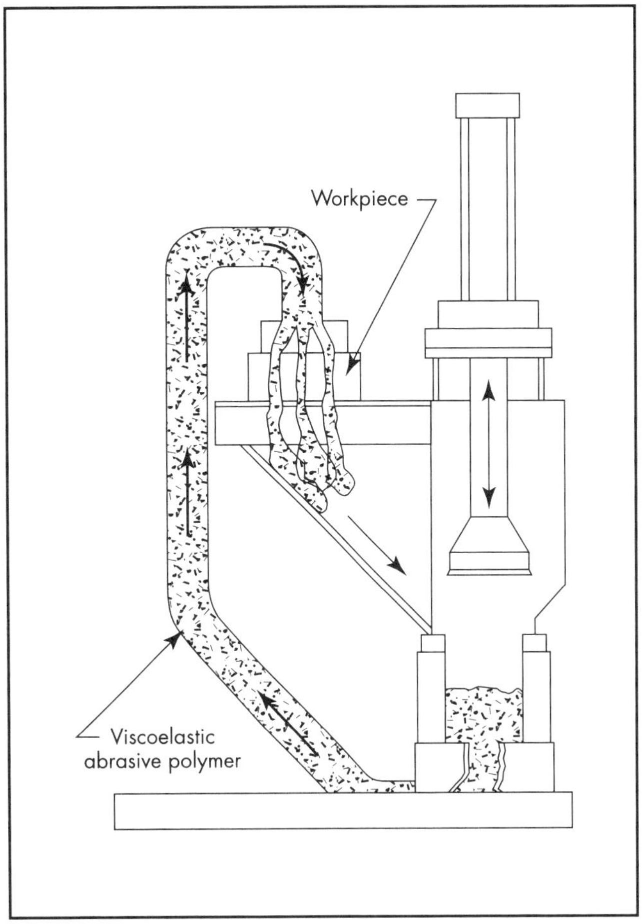

Figure 12-14. Schematic of one-way flow AFM. (*Courtesy Extrude Hone Corporation*)

12.7 ENVIRONMENTAL, HEALTH, AND SAFETY CONCERNS

There are no reported environmental or health issues for AFM. The process uses hydraulic cylinders and thus operates like many presses, so normal pressworking safety precautions should be taken. Operators may develop skin rashes from exposure to cleaning solutions used for removing media from holes.

If manual part loading is employed, operators are at risk for cuts from burrs on parts; they may also suffer from ergonomics-related injuries unless good parts-handling practices are employed. Air nozzles used for blowing media from holes and internal cavities should conform to the Occupational Safety and Health Administration's (OSHA) regulations to prevent air bubbles from being forced through the skin into the blood stream. Since air streams are used to blow off parts, a collecting box or container should be employed to prevent particles from being blown around the area and into other workers' faces. To avoid malfunctions of adjacent machines, critical surfaces of nearby equipment should be protected from abrasives.

References

Borchers, Julius. 1980. "Why Abrasive Flow Deburring?" Technical paper MR80-101. Dearborn, MI: Society of Manufacturing Engineers (SME).

Loveless, T.R., Williams, R.E., and Rajurkar, K.P. 1993. "An Investigation of the Effect of AFM on Various Machined Surfaces." Technical paper MR93-318. Dearborn, MI: Society of Manufacturing Engineers (SME).

Rhoades, Lawrence J. 1980. "Extrude Hone Abrasive Flow Machining—A Decade of Progress." Technical paper MR80-340. Dearborn, MI: Society of Manufacturing Engineers (SME).

———. 1981. *Cost Guide for Automatic Finishing Processes*. Dearborn, MI: Society of Manufacturing Engineers (SME).

———. 1988. "Abrasive Flow Finishing." Technical paper MR88-225. Dearborn, MI: Society of Manufacturing Engineers (SME).

———. 1995. "Application of Nontraditional Machining Techniques in Die/Mold Manufacturing." *Proceedings of the Net Shape Manufacturing Seminar*. Columbus, OH: Ohio State University.

Williams, R.E., Rajurkar, K.P., and Rhoades, L.J. 1989. "Performance Characteristics of Abrasive Flow Machining." Technical paper FC89-806. Dearborn, MI: Society of Manufacturing Engineers (SME).

Bibliography

Loveless, T.R., Williams, R.E., and Rajurkar, K.P. 1993. "An Investigation of the Effect of AFM on Various Machined Surfaces." Technical paper MR93-318. Dearborn, MI: Society of Manufacturing Engineers (SME).

Rhoades, Lawrence J. 1989. "Abrasive Flow Machining." Technical paper MR89-145. Dearborn, MI: Society of Manufacturing Engineers (SME).

———. 1990. "Abrasive Flow Machining of Turbine Engine Components." Technical paper MR90-392. Dearborn, MI: Society of Manufacturing Engineers (SME).

———.1991. "Abrasive Flow Machining in a New Era of Manufacturing." Technical paper MR91-523. Dearborn, MI: Society of Manufacturing Engineers (SME).

———. 1993a. "Abrasive Flow Machining—Progress in Productivity." Technical paper MR93-129. Dearborn, MI: Society of Manufacturing Engineers (SME).

———. 1993b. "New Capabilities and Applications for Abrasive Flow Machining." Technical paper MR93-321. Dearborn, MI: Society of Manufacturing Engineers (SME).

Stackhouse, John. 1993. "Abrasive Flow Machining." Technical paper MR93-148. Dearborn, MI: Society of Manufacturing Engineers (SME).

13 Electrochemical Deburring

Electrochemical deburring (ECD) involves pumping an electrolyte solution through a gap between a tool (cathode) and a metal workpiece (anode), thus completing the electrical circuit needed for the electrical current (direct current) to dissolve the burr (Figure 13-1). The burr is dissolved away by this electrochemical action. The tool and the workpiece do not contact each other, and the workpiece receives no mechanical or thermal stresses. Consequently, no changes occur in the physical or chemical properties of the metal.

Specially designed workholding fixtures and tools ensure that the conductive surfaces of the tools conform to the areas or edges of the workpieces being deburred. Proper insulation of the tools and protective shielding or masking of the surfaces of the workpiece limit the electrochemical action to the desired location.

Electrochemical deburring is ideal for selective (controlled) deburring and is especially useful for removing burrs inaccessible through most other deburring processes. For workpieces with surfaces that must not be altered or scratched, ECD is often the only possible method of deburring. The process is used extensively for deburring cross, threaded, and intersecting holes and workpieces made from metals that would work harden. Electrochemical deburring is 5–40 times faster than hand deburring, with cycle times generally ranging from 5–60 s, depending on burr size. Smooth edge breaks of .001–.020 in. (0.03–0.51 mm) are consistently produced. Large or small workpieces of various shapes and materials can be handled on the same machine. Holes with diameters as small as .034 in. (0.86 mm) can be deburred. The absence of mechanical forces permits thin sections and fragile workpieces to be deburred without distortion or damage.

The process does not affect workpiece dimensions, except at the edges being deburred. ECD can be used as a manual or fully automatic operation, and ECD stations can be incorporated in transfer machines or other automated production systems. Multiple tooling stations are commonly used for high-production applications.

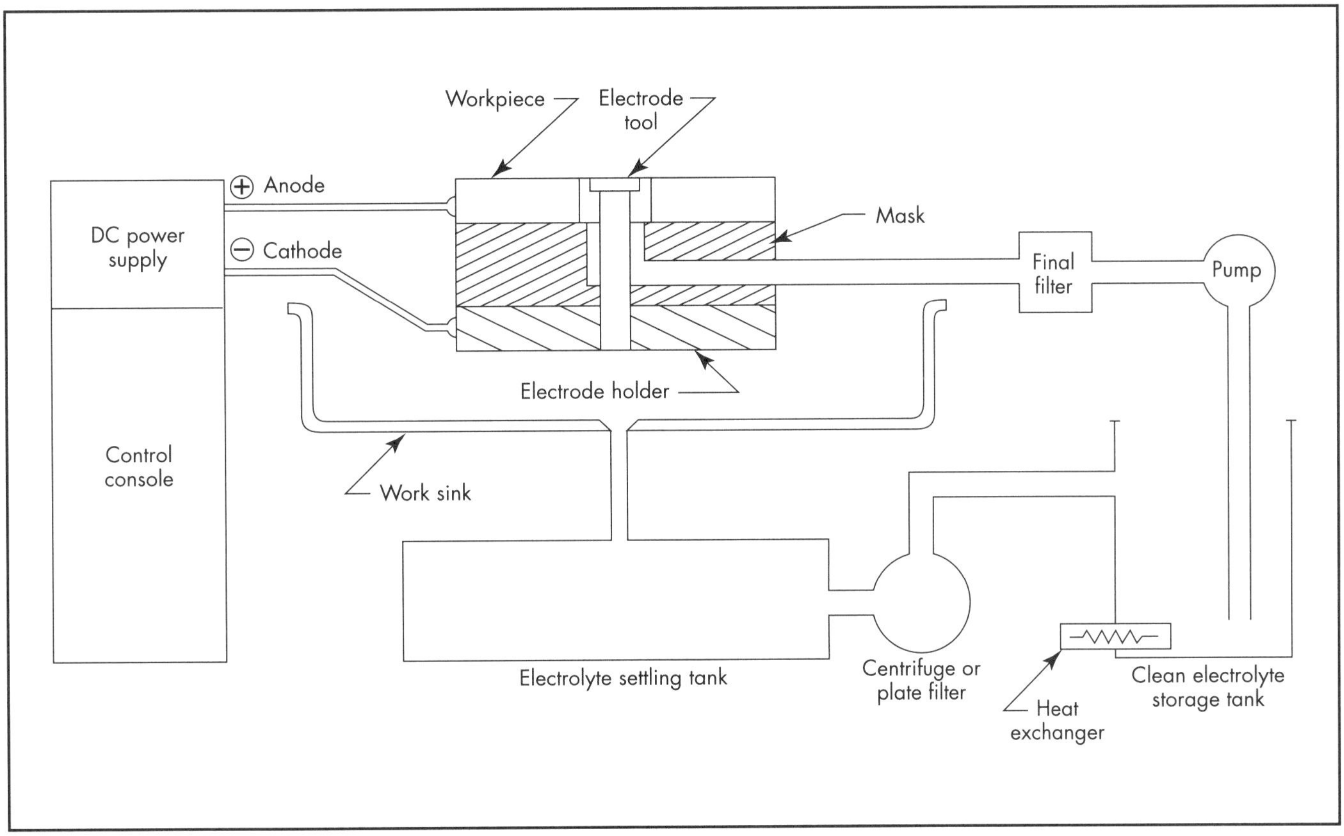

Figure 13-1. Schematic drawing of the ECD process. ***(Courtesy Dynetics)***

To produce uniform results, the following factors must be carefully controlled: tool design, current density, electrolyte, and cycle times. Variables that affect the electrolyte solution include type, concentration, pH, temperature, contaminants, pressure, and velocity.

Seven variations of ECD are currently in use:

1. Conventional ECD, which involves using a fixtured part and a salt solution, such as sodium nitrate.
2. Hand-held wand ECD, which employs a specially designed hand-held wand to reach into deep holes.
3. Rotary electrode ECD, which uses a rotating electrode to "cut away" burrs on cylinders without contact.
4. Glycol-based ECD, which employs a nonaqueous glycol-based solution.
5. Electrochemical mesh deburring, which employs a netlike mesh as an electrode.
6. Electrochemical moving-electrode deburring, which employs graphite balls in a mass-finishing machine.
7. Electrochemical brush deburring, which employs a brush as the electrode.

13.1 PROCESS PRINCIPLES

The ECD process is the reverse of electroplating; that is, metal is removed rather than added. The removal of metal accords with Faraday's law of electrolysis: the amount removed is proportional to the product of the current used and the length of time the metal is exposed to the current. Direct current at low voltage passes between a tool (the cathode) and the workpiece (the anode) through a conductive, neutral salt electrolyte.

The ECD process works in the following way. At the edges of the workpiece, electrons are removed by the current flowing through the electrolyte, which causes the metallic bonds of the molecular structure at the edges of the workpiece to be broken. Atoms removed flow into the electrolyte solution as metal ions. Because of the flow of the electrolyte through the gap (the distance between the tool and the workpiece), the ions are prevented from becoming attached to the tool. Dissolved metal, in the form of hydroxides, is carried away by the controlled flow of electrolyte, which is then filtered by various methods. According to one source, .002 µm/in. (.00008 µm/mm) of edge is a typical current density (Andrew 1982).

13.2 TYPICAL APPLICATIONS

The ECD process is used primarily for removing burrs at specific locations. It is also used extensively to remove recast (remelted) surfaces (edges) resulting from electrical discharge machining (EDM) and laser and electron beam machining (LBM and EBM).

ECD can be used on workpieces of any size or shape made from nearly all conductive metals, including most steels, stainless steels, aluminum, aluminum alloys, and exotic metals. ECD applied to aluminum alloys with high silicon content, however, will produce textured rather than smooth surfaces. Nonuniform and possibly unsatisfactory results are obtained on workpieces made from some titanium alloys unless special additives are mixed with the electrolyte. Deburring of sintered tungsten requires the use of special chemicals, and zinc die castings often present processing problems.

The automotive industry employs ECD for many applications, including deburring pistons, connecting rods, and crankshafts, wristpin holes, snap-ring grooves, and oil-return slots on pistons. Bearing lock slots, bolt holes, thrust faces, snap-ring grooves, and weight control bosses on connecting rods can all be deburred simultaneously. Deburring the edges of pump gears is another major application. Many hydraulic and pneumatic components, including valve bodies with interconnecting and cross-hole passages, are deburred with the ECD process to improve fluid and gas flow. The aircraft industry uses ECD to deburr the cooling holes in turbine blades as well as many other aircraft engine components.

13.3 PROCESS LIMITATIONS AND SIDE EFFECTS

Workpieces must be made from electrically conductive materials. To prevent short circuits from occurring during the ECD process, workpieces must be clean and free from loose chips and long, loose burrs produced during preceding operations. Relatively close tolerances are required for the workpieces because of the small clearance required between the burrs and the tool. Orientation of the burrs must be consistent from part to part to prevent short circuits and ensure uniform burr removal.

The capacity for close-tolerance edge radiusing is limited to small radii, typically not larger than .020 in. (0.51 mm). ECD may produce slight etching effects on surfaces distant from edges being deburred. Figure 13-2 shows the edge condition produced on miniature gears with one set of tooling. Figure 13-3 shows the stray etching that occurred at a distance of .125 in. (3.18 mm) from the edge on 17-4 PH (precipitation hardening) stainless steel. Discoloration (slight darkening), different from smut, generally indicates etching, which can be controlled by using insulating masks in the tooling or eliminated by bright dipping the workpieces. The amount and appearance of etching varies among materials. In some 304L stainless steel workpieces, the etching appears to have a galvanized steel large-grain pattern. Etching is not always visible and may be so slight that it is unobjectionable in many applications (Gillespie 1981).

Smut refers to the darkened surfaces left by the ECD process around deburred edges on steel components (see Figures 13-4 and 13-5). A white deposit is left on 6061-T6 aluminum when run in the same tooling, settings, and sodium nitrate solution as the stainless steel part shown in Figure 13-5.

The following practices will prevent or minimize the formation of smut:

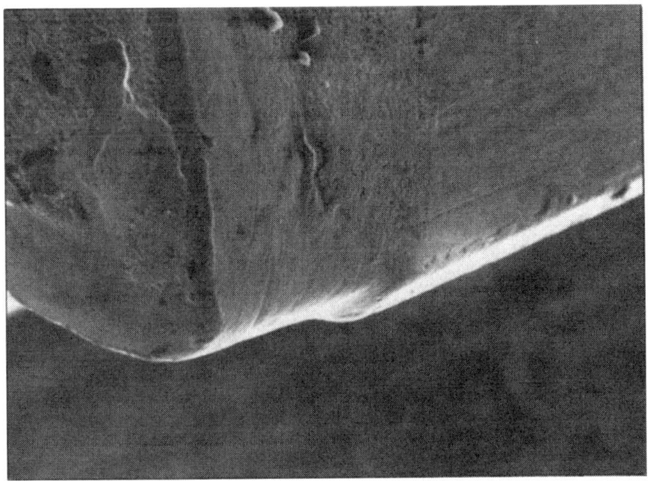

Figure 13-2. Edge condition produced by ECD (100x) (Gillespie 1976).

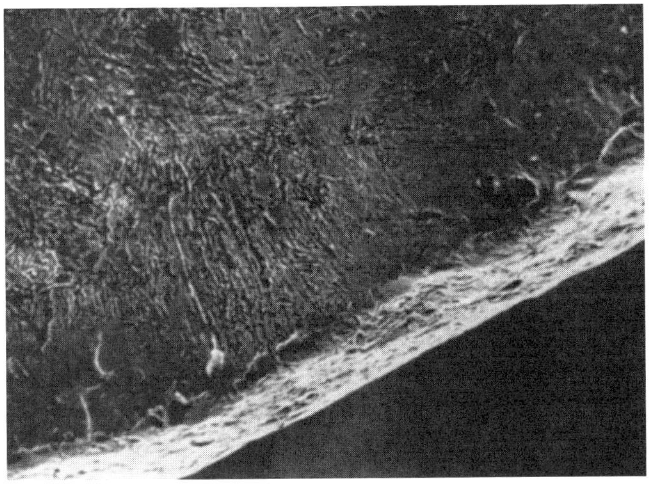

Figure 13-3. Stray etching produced by ECD that occurred away from the edge on a miniature gear tooth (Gillespie 1976).

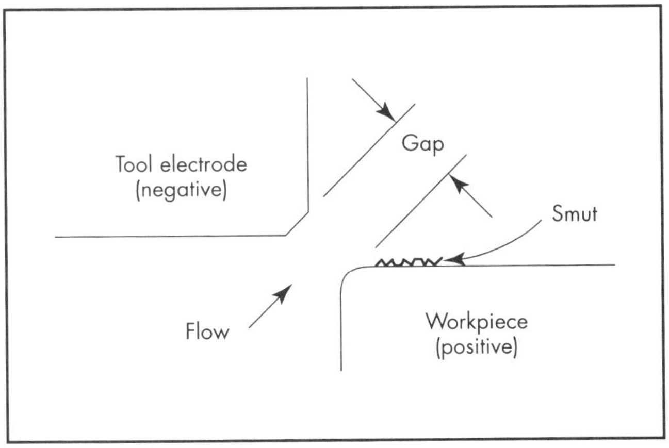

Figure 13-4. Typical location of smut resulting from ECD. ***(Courtesy Dynetics)***

Figure 13-5. Smut left by ECD on a 303 stainless steel disk processed in sodium nitrate (Gillespie 1976).

- Use tools designed to protect all surfaces except at the edge.
- Remove cutting oils prior to deburring.
- Use clean electrolyte solution.
- Maintain a pH of 6–8.
- Keep electrolyte solution at low temperatures, below 100° F (38° C).
- Select the correct voltage.
- Use pulsating DC power supplies.
- Increase the flow of electrolyte at the edges.

Note that when the electrolyte flow rate is too low, the whole area around the burr (and even the corner itself) will turn black and be covered with a thick layer of smut. As the flow rate increases, the corners start to shine and the smut begins to disappear. Too high a flow rate, however, will create turbulence, causing cavitation and irregularly deburred edges (Miller 1983).

Swirls and striations, often bright and reflective, can be found surrounding some edges after ECD. These surface effects reveal the path of the electrolyte flow. Preventing them requires minimizing turbulence in the electrolyte solution by carefully controlling the flow rate.

Each type of workpiece being deburred generally requires a specific workholding fixture and tooling. However, universal and modular-type tooling, discussed later, have been developed to permit processing with a minimum number of tooling changes.

Regular cleaning and maintenance schedules are required for effective ECD, including daily rinsing of equipment and fixturing with fresh water and maintaining the pH of the electrolyte at a consistent level. After ECD, workpieces should be rinsed immediately with warm water. Any remaining residue can be easily removed by brushing or blasting workpieces with a soft abrasive. Disposal of sludge produced by ECD processes must accord with local, state, and federal regulations.

13.4 PRETREATMENT OF WORKPIECES

Workpieces should be clean and free of metal chips prior to ECD. Cleaning is necessary because many standard machine oils used in preceding operations will contaminate the electrolyte. Some oils will affect the ECD process by insulating the workpiece, resulting in uneven deburring. Water-soluble oils may eliminate the need for precleaning. Workpieces also must be free of metal chips, which can create short circuits and damage plastic locators used in the ECD process.

13.5 EQUIPMENT REQUIREMENTS

The basic equipment required for the ECD process consists of a workstation, a DC power supply, an electrolyte system, electronic controls, and tooling.

13.5.1 Workstation

The workstation, often called the sink, consists of an insulated work and tool-mounting plate and a trough to collect the electrolyte solution. With the exception of the mounting plate, the workstation is generally made of molded fiberglass or another chemically inert material, sometimes combined with a steel weldment. The mounting plate is made from electrically conductive copper or steel. Workstations are made in various sizes, depending on the application, and multiple stations are often provided on a single machine to perform simultaneous or alternate operations.

13.5.2 Power Supply

Power supply units for ECD systems are generally solid-state rectifiers, either air or water cooled. They are available in a wide range of capacities, ranging from 2–40 V DC and 100–10,000 ampere. Generally, 10–20 ampere are needed for every inch (25 mm) of edge from which burrs are to be removed. Electrical connecting cables should be finely stranded, extra-flexible welding cable.

13.5.3 Electrolyte System

The electrolytes used for the ECD processes are neutral salt solutions. Salts used include sodium nitrate, sodium chloride, sodium chlorate, sodium bromide, and potassium chloride. Sodium nitrate is the most commonly used salt. Sometimes a mixture of salts or proprietary solutions is used.

Electrolytes

Sodium nitrate. The commercial grade of sodium nitrate is the most commonly used electrolyte. It is generally mixed with water in concentrations of 1.25–2.0 lb/gal (0.15–0.24 kg/L). This mixture produces an excellent electrolyte that is easy to handle, minimizes stray machining, and retards rust. It produces satisfactory results for most ECD applications.

Sodium chloride. Sodium chloride is common table salt. It is mixed with water in concentrations of .5–.75 lb/gal (0.06–0.09 kg/L). Sodium chloride is the least expensive electrolyte and produces the greatest amount of corrosive damage and stray etching. To reduce stray etching, inhibitors are commonly used in conjunction with it.

Pumps

Generally, centrifugal pumps are used for circulating the electrolyte solution. They are made from stainless steel, plastics, or bronze and usually have output pressure capacities ranging from 20–50 psi (138–345 kPa). Volume capacities vary with the requirements of the application.

Settling Tanks

Settling tanks generally have divider baffles to allow the electrolyte solution to clarify by settling. They can be constructed from fiberglass or stainless steel, and capacities usually range from 100–1,000 gal (379–3,785 L).

Filters

Adequate filtration of the electrolyte is essential for successful ECD. Filtration removes foreign material that tends to plug the tooling; it also reduces staining (smut formation) on the workpieces. Note that a "dirty" electrolyte solution will increase deburring cycle times.

Bag and cartridge filters. In small ECD systems, a polypropylene bag or cartridge-type filter is generally used in conjunction with the pump to remove foreign particles and metal hydroxides from the electrolyte solution. For large systems, a centrifuge or plate-press filter may be added.

Centrifuges. Centrifuge separators generally have stainless steel bowls with outer housings of reinforced fiberglass. They usually have a flow capacity in the range of 4–10 gal/min (15–38 L/min) and are suitable for medium-size systems when wet sludge disposal is not a problem. Centrifuges can be equipped for either manual or automatic sludge removal.

Plate-press filters. Plate-press filters employ a series of filter plates compressed together in a steel frame and remove extremely small (submicrometer) metal hydroxides. This method of filtration produces solid cakes that are easier to dispose of than wet sludge. It is particularly suitable for workpieces on which staining is a problem.

Canister filters. Canister filters are a feature of fully automatic filtration systems. The filters are composed of a metal canister with about 100 woven-mesh tubes made of polypropylene suspended from a flexible shaker plate. Diatomaceous earth (a filtering aid) is employed to coat the tubes, permitting particles as small as 39–79 μin. (1–2 μm) to be removed. A pneumatic cylinder automatically flexes the shaker plate to remove the diatomaceous earth and allows recoating. After several recoatings, the system is shut down, and the material is dumped into a receiving tank where a vacuum or pump pulls the electrolyte solution back into the system, leaving a solid cake of diatomaceous earth and metal hydroxides.

Electrolyte Temperature

The temperature of the electrolyte solution should be monitored and controlled to ensure consistent production conditions; consistent temperature is essential for high-production applications. Disagreement exists about the proper electrolyte temperature to use. Some experts recommend a temperature of about 55° F (13° C) for steel workpieces and 75° F (24° C) for stainless steel and aluminum workpieces. Others recommend 95–100° F (35–38° C) for all materials. Users concerned about smut buildup on parts or about overetching will need to experiment with the most appropriate temperatures and flow.

Heat exchangers. Using available plant water is an efficient method for controlling electrolyte temperature for some operations. One heat exchanger allows plant water to surround electrolyte-filled tubes made from stainless steel. The heat-exchanger unit uses a maximum of 6 gal/min (23 L/min) of electrolyte solution; only about 3 gal/min (11 L/min) are generally needed for ECD systems requiring approximately 200 amps of current. If an enclosed work chamber is used, an exhaust system is necessary to remove the gas generated by the electrochemical action on the part.

Refrigeration units. One efficient design used for controlling the temperature of the electrolyte solution is a multicoil unit, with a Freon®-like solution surrounding the electrolyte-filled stainless steel coils; the unit itself is air cooled. Multicoil units are available in various capacities to suit specific requirements.

Fittings and Hoses

Because of the moderate temperatures and pressures used in ECD processes, electrolyte supply fittings are frequently made from plastics, though stainless steel fittings are also used. For some applications, brass fittings are more economical than plastic or stainless steel and are sufficiently resistant to corrosion and require little or no maintenance.

Hoses through which the electrolyte solution flows may be made from polyethylene, polyvinyl, nylon, or similar materials; they vary in availability and cost. Hoses incorporating metal reinforcements are unsuitable because of possible problems with stray currents. Hosing connections should be made from similar materials, that is, plastic to plastic or metal to metal. However, some low-pressure tools are an exception to best practice rules, for example, nylon tubing pressed on a brass base inlet.

Storage Tanks

When deburring workpieces are made from different materials at frequent intervals, separate storage tanks should be provided so that a clean electrolyte solution can be used for each type of material. Each tank should have its own pump, valves, and filtering system. Separate work sinks are also preferable. Job shops, however, have successfully deburred stainless steel and aluminum in the same machine with no special provisions for either material.

13.5.4 Tooling

Among the important considerations in the design of tools (cathodes) for ECD are the gap between the tool and the workpiece, the flow of the electrolyte solution, and the materials from which the tools are made.

In designing fixtures for the ECD process, an attempt should be made to have the anode contact the workpiece in a dry area. Contact pressures must increase with higher amperage requirements. Round anodes should have a minimum diameter of about 7/16 in. (11 mm) to serve as a heat sink for reducing hot spots. Copper is frequently used for the anodes; but where contact is in an area where electrolyte is flowing, platinum-tipped anodes may be required.

Gap Between Tool and Work

Recent research and improvements in the ECD process have affected the positioning of the tool in relation to the burr on the workpiece. In the past, the gap between the tool and the workpiece was maintained at 0.010–.015 in. (0.25–0.38 mm). Such small gaps were problematic, however, because they led to shorts when burrs touched the tools and frequent overcutting on critical workpieces. Tool variation because of manufacturing tolerances caused some portions of the gap to be only .002–.005 in. (0.05–0.13 mm). These variations in gap size affect the velocity of the electrolyte solution, cause irregular flow, and, in some cases, create air bubbles (air acts as an insulator and thereby stops the deburring action). To address these problems in some applications, gaps of .025–.040 in. (0.64–1.02 mm) and pulsating DC power supplies are being used. Advantages of larger gaps include fewer shorting problems, longer tool life, and better finishes; cycle times, however, are generally longer.

Even flow and uniform velocity of the electrolyte solution results in the metal hydroxides being flushed away faster and reduces smutting of the workpiece surfaces. Cathode wear can be minimized by using spark-detection protection systems.

Electrolyte Flow Path

Practically all tooling for the ECD process is designed so that the tools have a corner-to-corner relationship to the burr-laden edges of the workpieces (Figure 13-6), resulting in burr removal and edge rounding. The gap (established by the x and y dimensions shown in Figure 13-6) is varied to suit the application and provide the desired corner break. The electrolyte flow path safely removes any particles of metal cut away during the ECD process.

Modifying the working gap, that is, increasing or decreasing the x and y dimensions, will affect the type of corner break produced. However, both dimensions must be sufficient to allow for the maximum variations in workpiece geometry permitted by production tolerances. Even under the most adverse conditions of dimensional tolerance stacking, a gap still must be provided to prevent electrical contact and permit sufficient electrolyte flow. The corner-to-corner relationship can be designed in different ways to accommodate workpieces of various configurations.

Through-drilled holes. Figure 13-7 shows two methods for removing exit burrs from through-drilled holes. Figure 13-7a shows a simple approach for removing the exit burr from a drilled hole. This method can only be used if the tool can enter the hole from the upper surface of the workpiece. To provide electrolyte flow in the direction necessary for washing burrs out of the gap, the electrolyte supply must be located on the side of the workpiece opposite the tool. This configuration requires a more complex fixture than if both the tool and electrolyte supply were located on the same side. Figure 13-7b shows how a workpiece can be located so that the tool and the flow path of the electrolyte are on the same side of the workpiece.

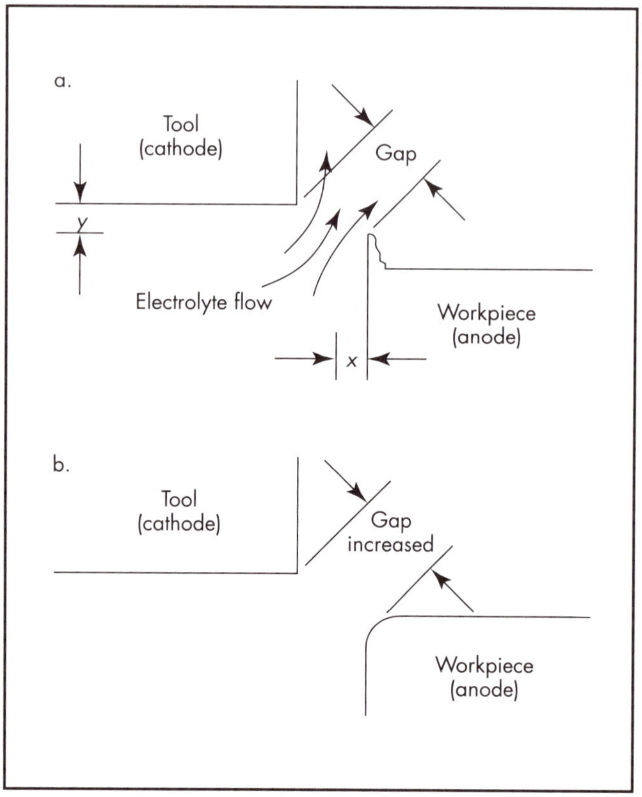

Figure 13-6. Corner-to-corner relationship for the ECD process. (*a*) Before deburring; (*b*) after deburring. *(Courtesy Chemform)*

Figure 13-8 illustrates a through-drilled hole with a radius instead of an entrance burr, a tool with two conductive corners, and a plate made from plastic or other insulating material on which the workpiece rests. This design breaks the following two rules that are generally observed in the ECD process:

1. Never have two working gaps in series in an electrolyte stream, because chips removed in the first gap may cause shorting in the second gap.
2. The direction of electrolyte flow should force chips out of the gap instead of into it.

In the case of Figure 13-8, the preceding rules can be broken because the entrance burr in most drilled holes is so minimal that no danger exists of pieces breaking loose. Also, the function of the operation shown in the figure is more for radiusing corners than for removing burrs.

To produce the most uniform results in bores and the smoothest possible radiusing of edges, a circular electrolyte flow is desirable. This type of flow pattern prevents striations on the workpiece surfaces. The centrifugal forces cause the electrolyte to follow the outer walls of cavities,

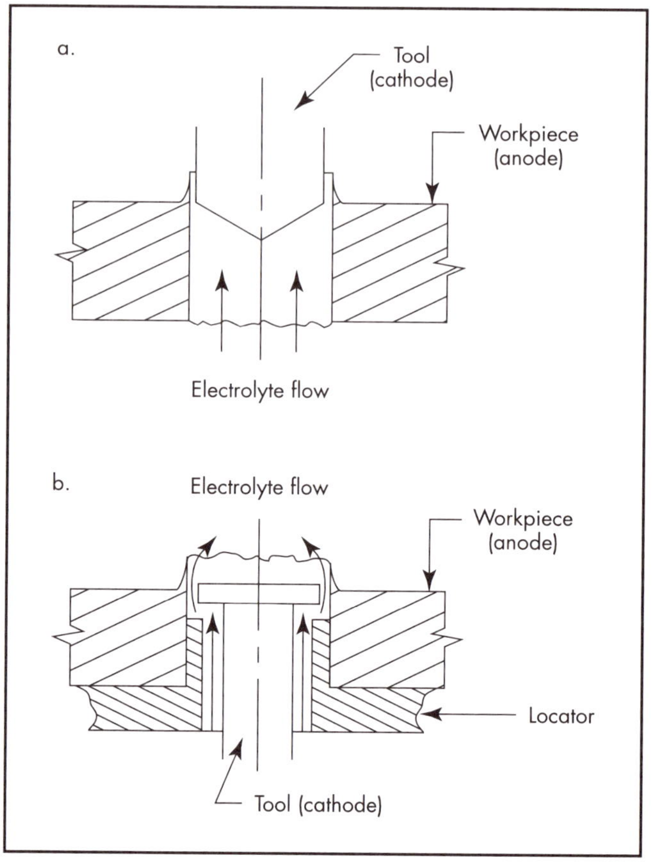

Figure 13-7. Two methods for removing the exit burrs from through-drilled holes. (*a*) The tool and the flow path of the electrolyte are on opposite sides of the workpiece; (*b*) the tool and the flow path are on the same side of the workpiece. *(Courtesy Dynetics)*

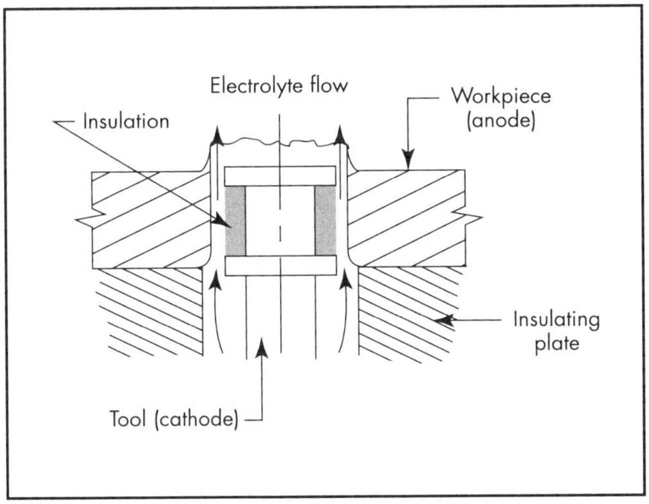

Figure 13-8. Tool with two conductive corners for ECD and edge radiusing. ***(Courtesy Dynetics)***

thus producing radii (Figure 13-9) instead of conical corner breaks. This technique is helpful where large, smooth radii are required for stress-reduction purposes.

Blind holes. Figure 13-10 shows a method for deburring the edges of blind holes. The electrolyte exits from a central hole in the tool. The exit hole must be large enough to allow the passage of chips produced during the ECD process.

Intersecting holes. Figure 13-11 shows two different approaches for deburring the intersection of a small hole (or holes) with a large center hole in a workpiece. Figure 13-11a shows a workpiece on a plastic locator, with the tool completely insulated except for the edge adjacent to the burr to be removed. This design produces smooth, uniform radii, with the size of the radii depending on the cycle time.

In Figure 13-11b, the locator protects the complete bore of the workpiece. The slot or hole machined in the locator is .020–.040 in. (0.51–1.02 mm) larger than the hole in the workpiece and at the same location. The tool is positioned inside the locator, with its end opposite the hole in the workpiece; the tool is .020–.050 in. (0.51–1.27 mm) smaller in diameter than the hole in the workpiece. This design produces a slight counterbore and a small radius. It is useful when many holes in a part must be deburred, because it provides good control of electrolyte flow for all the holes, even when they are on different levels.

Intersecting holes, with one hole at an angle to the other, can be deburred with portable ECD probes (Figure 13-12), especially in large workpieces. Figure 13-12a shows a probe placed into a straight hole. A magnetic permanent stop is mounted on the outside of the workpiece to properly locate the tool portion of the probe. A plastic diffuser on the probe forces the electrolyte solution out the angular hole. Figure 13-12b shows a probe placed in an angled hole. This method is generally used when the angular hole is drilled after the straight hole.

Figure 13-13 shows a setup for deburring five internal cross holes and two half-moon slots in one operation. Three

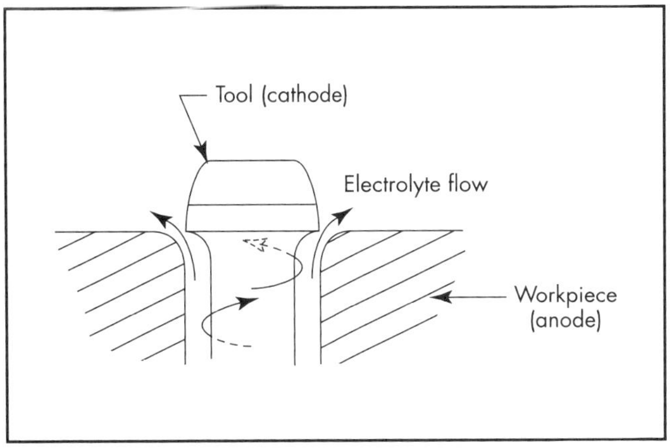

Figure 13-9. Circular flow of electrolyte can produce large, smooth radii. ***(Courtesy Chemform)***

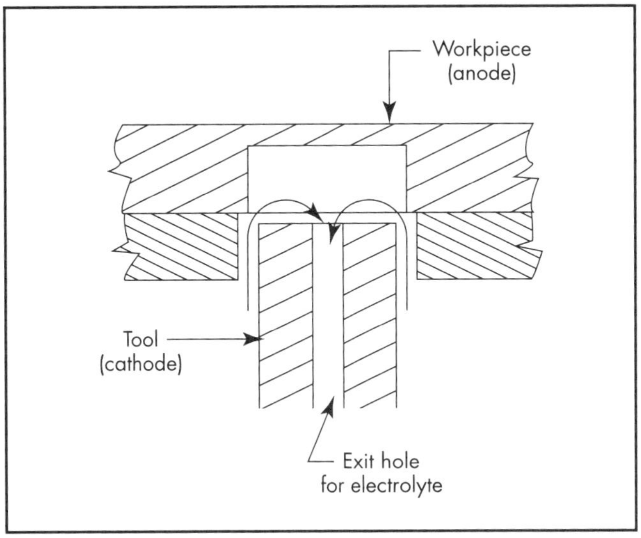

Figure 13-10. In deburring blind holes, the electrolyte exits from a center hole in the tool. ***(Courtesy Dynetics)***

air-actuated tools are used in this operation, with a programmable controller providing proper sequencing of the tools.

Gear teeth. ECD processes are commonly used to radius gear teeth and odd-shaped profiles. External gears are placed in plastic locators, with the gears positioned radially using internal gear-shaped tools. The tools are made by using a gear as an electrode. The exact opposite shape of the gear to be radiused is produced by the electrical discharge machining method.

To produce small, closely controlled radii of .001–.002 in. (0.03–0.05 mm), the electrolyte flow is generally upward through a series of small holes into a reservoir area above the tool. The electrolyte then turns 90° and flows inward against the edges of the gear teeth. To produce larger radii, the flow of electrolyte is usually in the opposite direction and then upward across the teeth. To prevent stray cutting of gear teeth, the electrolyte solution should be kept cold or adequate masking should be provided.

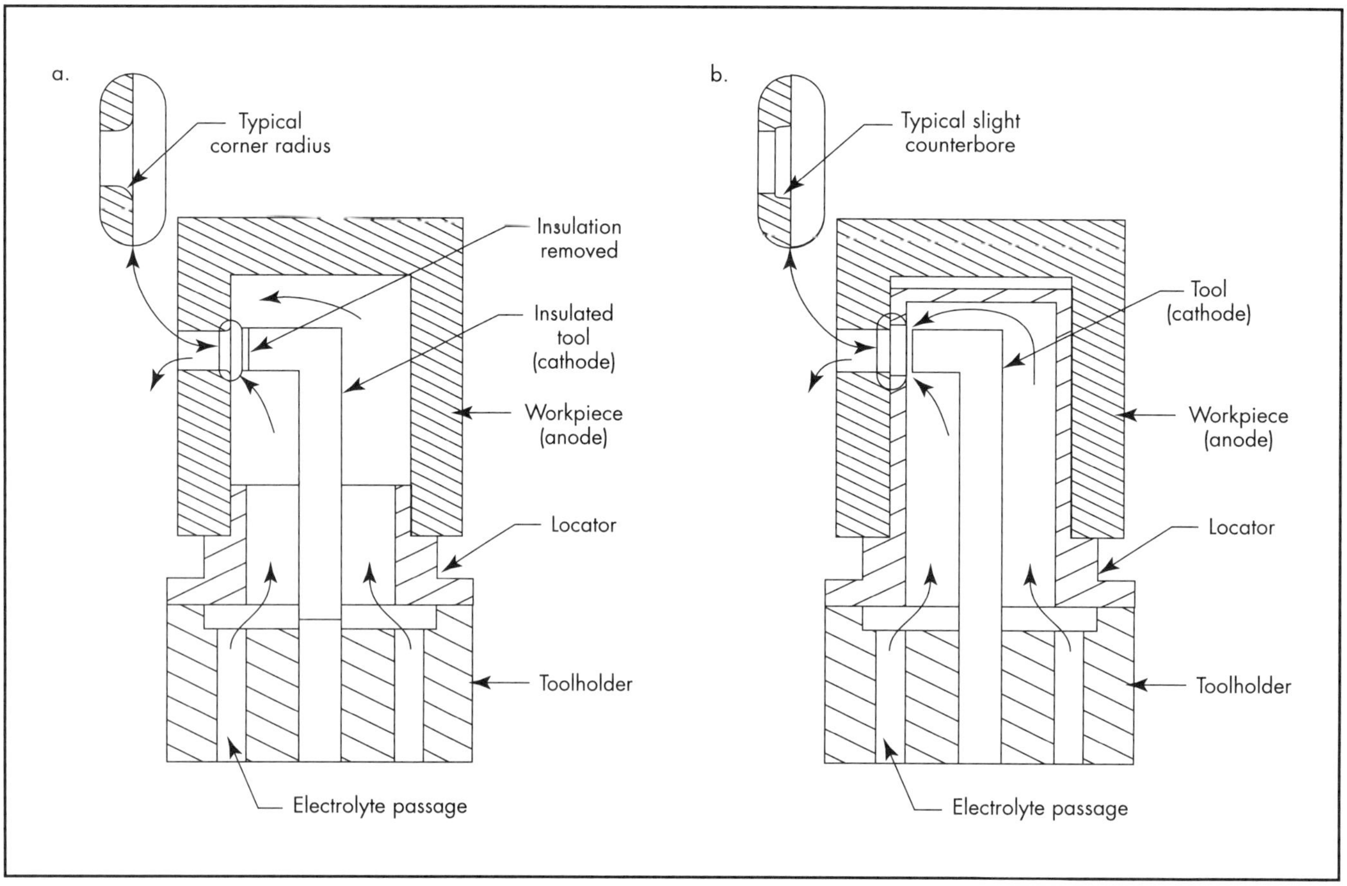

Figure 13-11. Two methods of deburring the intersection of a small hole (or holes) with a large center hole. *(Courtesy Dynetics)*

Tool Materials

Tools for the ECD process must be made from electrically conductive materials. Manufacturers employ a wide variety of materials, and the choice of material depends primarily on the specific application. Copper tungsten is recommended for the working edges of many tools because it is easily silver brazed to copper-based and ferrous metals. Mechanical mounting is generally preferable since it is changed more readily. For removing small burrs, free-machining brass is often an ideal tooling material because of its low cost and machinability.

Tool insulation. A nonconductive shield or coating is applied to ECD tools adjacent to the areas of the workpieces where deburring is required to prevent electrochemical attack on surfaces. One commonly used insulating material for both tools and fixtures is a dense, epoxy-bonded, glass-cloth laminate called *green glass* that conforms to National Electrical Manufacturers Association (NEMA) Specification G-10. This material is available in rounds, tubes, and sheets. Water absorption is low and dimensional stability is reasonably good, although warpage may occasionally occur on sections unevenly exposed to electrolyte for long periods of time. Green glass can be machined for tools with conventional equipment, but the life of cutting tools is generally short because of the abrasive fiberglass particles.

Acetyl resins (for example, Delrin® and Celcon®) are also used extensively as insulating materials. These materials machine well and produce longer tool life than green glass-laminated tools.

When using thermosetting epoxy insulation to make tools or fixture parts, an oven is required for preheating and curing. Powder-spray (flocking) guns and a spray booth with a filtered exhaust system are often used when applying insulating materials.

Fits and tolerances. For most applications, the fits and tolerances used for conventional jig and fixture designs will provide satisfactory performance in ECD tooling. Where very porous insulating materials are applied and subject to liquid saturation, allowances must be made for potential expansion.

Fixture materials. Any dielectric material that has low water absorption and satisfactory physical strength may be used for fixtures. In many cases, the choice of material is based on availability rather than the physical characteristics of the material.

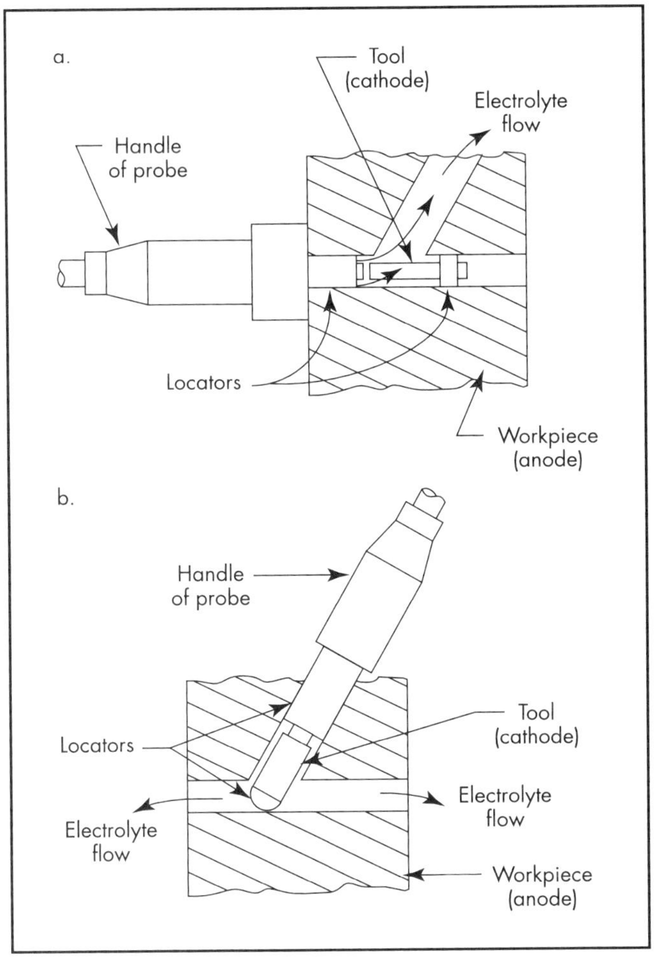

Figure 13-12. Two methods for deburring intersecting holes using portable ECD probes. (*a*) Probe placed in a straight hole; (*b*) probe placed in an angled hole. ***(Courtesy Dynetics)***

Brass is the most widely used metal for fixtures, primarily because it is easily machined and a good conductor of electricity. Stainless steels in the 300 series are also extensively used; type 316 stainless steel has the most resistance to corrosion by the electrolyte.

Copper is 10–20 times more conductive than stainless steel, depending on the alloy, and is often used for contact surfaces that provide current to the workpieces. The strength of stainless steel also can be exceeded by using beryllium copper, which provides nearly the same conductivity as pure copper.

Fixture insulation is essential because fixture components are close together, surfaces are damp, and pools of electrolyte solution may be trapped in the tool recesses. Nonconductive shields and coatings of the type used for insulating tools are suitable for insulating fixtures. Clear acrylic (Plexiglas® or Lucite®) blocks and sheets are extensively used for insulation, especially when transparency is advantageous, but these materials do not have the wear resistance or mechanical strength of green glass. Machining acrylic is fast and easy, and bonding acrylic to fixture components requires only clean, closely fitted surfaces.

Other solid plastics materials, such as Lexan® and Teflon®, are used successfully for components of ECD fixtures.

Fasteners and dowels. Typically, stainless steel bolts and cap screws are used to assemble permanently bonded fixtures. Coarse threads are preferred for tapped holes in the various dielectric materials. Socket-head cap screws permit counterboring and provide flush surfaces. Such screws also permit the use of temporary insulation to protect the fasteners, yet are easily removed for disassembly.

An ordinary silicon seal may be used to cover fasteners. For a somewhat more solid but still removable cover, a vinyl-based spackling paste may be used. In applications where dowels are necessary, stainless steel will suffice; titanium is better; and in plastics materials, glass rods, ground to diameter, are best. Solid glass rods, not laminates, should be used.

Electrical connecting cables. If any parts of the fixture are designed to move and conduct current, they should be electrically bonded to the rest of the fixture with swaged and sealed conductors. This design minimizes the gradual disintegration that occurs when a difference in voltage exists between two metal workpieces in contact with an electrolyte.

Workpiece clamping. Clamps used for positioning workpieces and conducting electrical current should require minimal operator effort to apply. For many applications, simple air cylinder clamps can be suitably adapted. They are practical, economical, and can be trouble-free if a few simple precautions are observed. Always select and install double-acting cylinders. This approach insures that any leakage across seals or joints of the cylinder will be from the inside to the outside, with no possibility of drawing in electrolyte, which can happen with a single-action, spring-return unit. If mounting the cylinder to avoid its being sprayed or soaked with electrolyte solution is impractical, a stainless steel or brass cylinder should be used.

If the tool cannot be made an immovable feature of the fixture because of the configuration of the workpiece, air cylinders may be necessary for inserting or moving the tool into position. With small round tools requiring no orientation, it is possible to use the cylinder as both an actuating and clamping device. A double-ended rod is recommended so that the rear of the rod can have an adjustable stop mounted on it. For more stringent requirements, various stainless steel slides can supplement the cylinder to properly position the tools.

Standardized Modular Tooling

Firms can realize considerable savings by using standardized modular tooling in ECD processes. Many of the components of these types of tools can be used for deburring various workpieces, eliminating the need to design and manufacture tools for each specific application. For example, Figure 13-14 shows a modular tool designed to deburr holes that break into the center holes of workpieces. In this

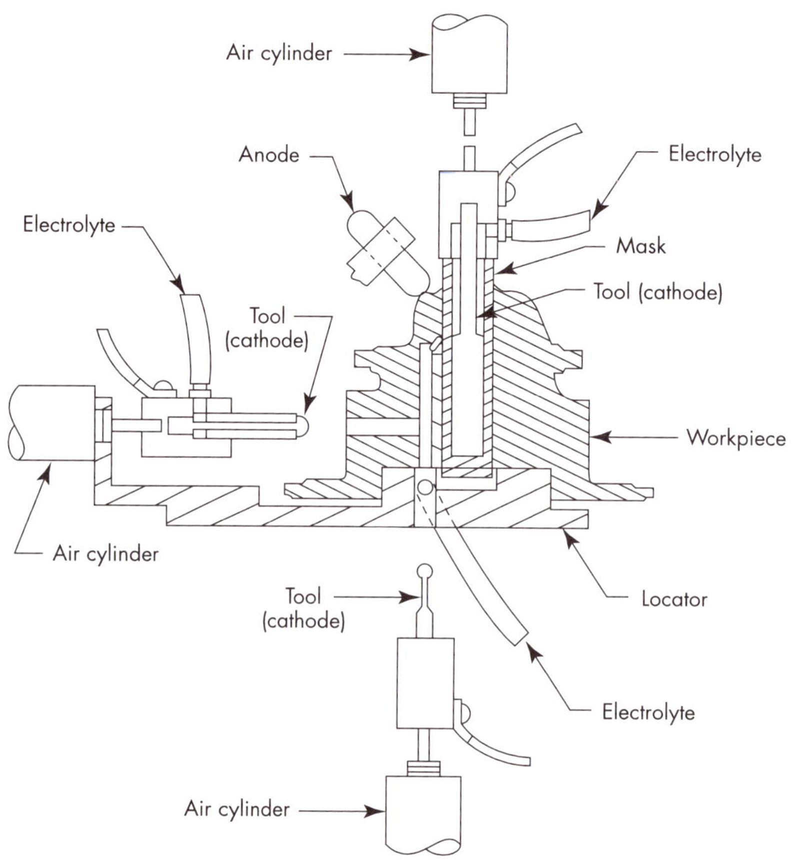

Figure 13-13. Setup for deburring five internal cross holes and two half-moon slots in the center bore of a cast-iron turbine housing in one operation. ***(Courtesy Dynetics)***

application, the mask and the electrode have been designed for a particular part, but the other components of the tool are suitable for use on many other workpieces. The mask fits over the electrode, with a protrusion on the electrode aligned with a hole in the mask. A workpiece is located on the mask, with the hole in the mask aligned with the cross hole in the workpiece. Electrolyte flows around the electrode and deburrs the edges of the cross hole. All other surfaces of the workpiece are protected by the mask.

Cleaning the Tooling

Tooling must be kept clean while in use and when stored. A regular maintenance program should include thoroughly rinsing all tooling and fixtures with fresh water immediately after shutting down the ECD machine. Some ECD machines contain flush systems as part of the unit. All brass, stainless steel, and copper-tungsten parts must be kept clean to ensure good electrical connections, otherwise slight voltage drops will occur, causing some parts of the tooling to be dissolved. For brass parts, a bright dip is available to prevent tarnishing.

13.6 OPERATING PARAMETERS

For efficient and economical deburring using the ECD process, the following operating parameters must be carefully controlled:

- Gap size.
- Current density and operating voltage.
- Cycle times and production rates.
- Electrolyte conditions (type, temperature, pressure, and velocity).

13.6.1 Gap Size

The smaller the gap between tool and workpiece, the faster the burr removal rate and the more consistent the deburring

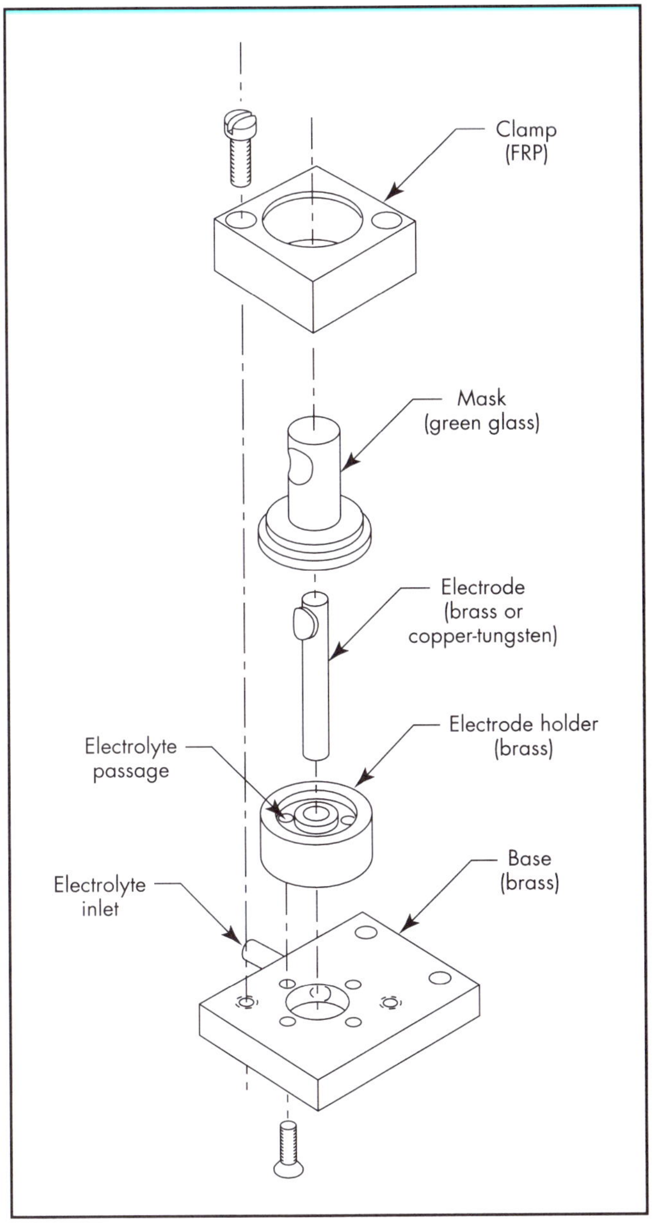

Figure 13-14. Standard modular tooling designed to deburr holes that break into the center holes of workpieces. *(Courtesy Dynetics)*

action. The size of the gap for a specific ECD application depends on the tolerances of the workpiece and tooling and on the location of the edge to be deburred. Too small a gap will cause shorts and limit electrolyte flow.

Depending on tool and workpiece tolerances, a minimum gap of .020 in. (0.51 mm) and a maximum gap of .045 in. (1.14 mm) are generally considered desirable. When production rates are low and long cycle times can be tolerated, larger gaps may be employed. For applications requiring short cycles and high production rates, proper tool design is essential (Gillespie 1981).

13.6.2 Current Density and Operating Voltage

The time required for deburring decreases as the voltage increases. For example, doubling the voltage from 10 to 20 V will reduce the time required for deburring by one-half. Figures 13-15 and 13-16 show the amount of time necessary for deburring workpieces made from iron-based alloys and aluminum, respectively, at a 10-volt setting. However, increasing the voltage can produce undesirable results, such as rough surface finishes and stray machining. Furthermore, increasing the voltage may result in hydroxide being removed faster than the electrolyte solution can carry it away. Heavy hydroxide concentration creates a condition that may act as an insulator to slow or even stop the process.

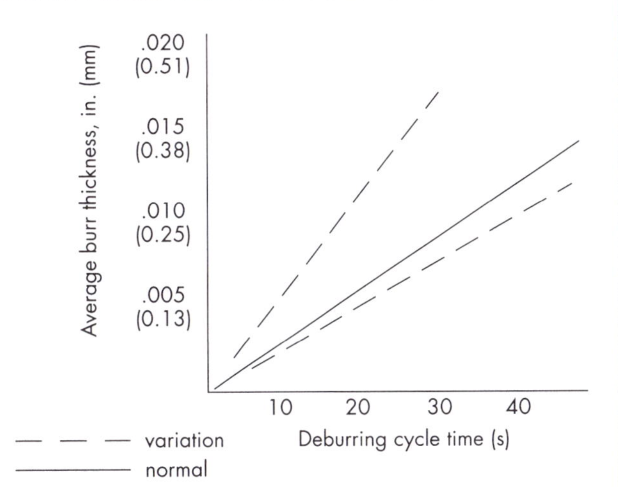

Figure 13-15. Average burr removal rate using ECD on workpieces made from iron-based alloys (Gillespie 1976).

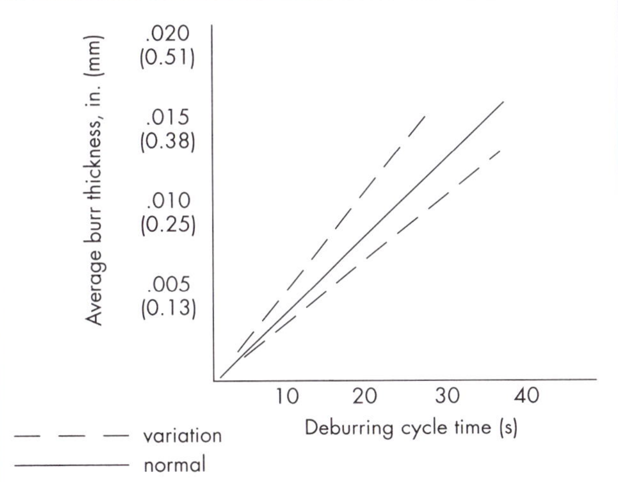

Figure 13-16. Average burr removal rate using ECD on workpieces made from aluminum alloys (Gillespie 1976).

The optimum operating voltage for a particular application must be determined by making test cuts on representative workpieces. Generally, 15 V is recommended as the voltage at which to begin testing. If stray machining is within acceptable tolerances, the voltage may be gradually increased until unacceptable conditions result. Note that higher voltages require shorter cycle times, so investigating and employing alternative means for minimizing stray machining, other than reducing voltage, may be desirable. Alternatives include adding more insulation and changing the electrolyte flow direction or velocity.

13.6.3 Cycle Times and Production Rates

Cycle times vary from a few seconds to 1 min or more depending on burr thickness, gap size, electrolyte conductivity, voltage, and tool design. To accurately determine cycle times, test cuts must be made on representative workpieces. Most close-tolerance holes can be deburred in 5–10 s. Workpieces allowing more liberal tolerances or having heavy burrs generally require cycle times of about 20 s. Workpieces allowing very loose tolerances and requiring gaps of about .030 in. (0.76 mm) may take 1 min or more to produce the desired results. Aluminum and brass workpieces can be deburred about twice as fast as steel and stainless steel workpieces.

Equations for determining removal rates and the electrochemistry of the ECD process have been presented elsewhere (for example, Jones 1981), but most removal rates appear to be straight-line functions of amperage, voltage, and time. Production rates of 240 pieces/h are common, and with multiple tool stations, rates of 1,400 workpieces/h have been achieved. Even simple semiautomatic machines are capable of processing 60–100 workpieces/h.

13.6.4 Electrolyte Conditions

Selecting the proper type of electrolyte, maintaining the electrolyte at the correct temperature, and clarifying the electrolyte by filtration were discussed in Section 13.5.3. Other important operating parameters for the electrolyte are pressure and velocity.

To prevent sludge from accumulating and plugging the flow path of the electrolyte solution, the velocity of the solution must be consistently maintained. Sufficient electrolyte velocity requires only moderate pressures, typically 20–50 psi (138–345 kPa). Low pressures minimize spray and fixture leakage and, in most cases, permit machines to be operated without enclosures (which can interfere with accessibility and decrease production rates).

13.7 POST-TREATMENT

Post-treatment operations required after ECD processes include removing salts and sludge from workpieces and disposing of electrolyte solution and sludge.

13.7.1 Removing Salt

All workpieces should be rinsed with warm water (12–140° F [−11–60° C]) after the ECD process to remove salts. The rinse tank will be contaminated with excess salt unless fresh water is supplied at a sufficient rate to prevent this condition. Standard agitating-type washing machines with heaters work well for removing salts from most workpieces. Warm water is normally better for removing salts than many degreasing systems and special cleaning solutions. Workpieces subject to rust should be rinsed in a water-soluble rust inhibitor after rinsing.

13.7.2 Removing Smut

Smut remaining on workpieces after completion of ECD may or may not need to be removed, depending on the workpiece materials and later finishing operations. On aluminum workpieces, for example, anodizing, painting, and finishing can be performed over smut without affecting quality. Brass workpieces can be plated or chromated and steel workpieces can be plated and painted without first removing smut. Carbon smut on steel workpieces serves as a lubricant, which on some workpieces is beneficial. Stainless steel workpieces can be passivated, electropolished, and painted without first removing smut. Smut *must* be removed, however, from instrument components that will later be used in ultraclean environments.

The procedures for removing smut, when it is required, depend on the workpiece materials. For aluminum, brass, or copper workpieces, either acidic or alkaline cleaning solutions can be used without affecting workpiece tolerances. For steel workpieces, different types of acidic solutions are used. Although several types of acids have been employed for removing smut from stainless steel workpieces, generally electropolishing is the most effective method. For some steel and stainless steel workpieces, brushing, tumbling, and vibratory finishing processes are employed.

13.8 ECD PROCESS VARIATIONS

The conventional ECD approach is used in probably 95% of ECD applications. Several variations exist that allow wider capability. These are described in the following sections.

13.8.1 Hand-held Wand ECD

For deburring hard-to-reach holes on workpieces that are produced only occasionally, a hand-held wand can be employed instead of having fixtures specially designed for the purpose (Tuneski and Hignett 1981).

13.8.2 Rotary Electrode ECD

In rotary electrode ECD, an electrode shaped like a rotary cookie cutter is slowly lowered over the burr-laden edges of machined rings (Vishnitsky 1989). Cutouts in the electrode cause enough turbulence to keep fresh electrolyte in the electric field area. The tool "cuts" the burr away as it descends and rotates around the part. The fixturing is simple and may eliminate some of the problems caused by long loose burrs shorting out stationary electrodes. The process is not limited to deburring machined rings, but workpieces must be cylindrical.

13.8.3 Glycol-based ECD

Glycol-based electrolyte solutions are a recent innovation in ECD (Koroskenyi 1995), and in several countries they are a standard commercial product. Glycol-based solutions are near neutral and tend to concentrate the action of the ECD process at the desired edge more closely than aqueous solutions. When using glycol-based solutions, tools do not need to follow the contour of the edges being deburred as closely; and, therefore, tool design can be simpler because the action of the electrolyte is more concentrated. Voltages range from 24–48 V (in contrast to the 10–15 V used with aqueous solutions). Deburring times range from 30 s to 2 min, and surface finishes of 4–8 μin. (0.102–0.203 μm) have been produced. A pressurized spray rinse may be required to completely flush away hydroxide particles stuck to workpiece surfaces after ECD.

Glycol-based ECD does not work well on titanium or on precious metals, such as gold or silver. For some applications involving workpiece materials made from cast iron and aluminum with high silicon content, the cosmetic appearance after ECD may not be acceptable. Hardened steel workpieces that have been ground and are bright may emerge from the process with a dull gray color.

13.8.4 Electrochemical Mesh Deburring

Conventional ECD tooling is generally impractical for deburring workpieces with hundreds of drilled holes. For this type of application, one researcher used a mesh metal screen as the electrode and placed a porous cloth between the workpiece and the screen (Fromson 1976). The workpiece, screen, and cloth were completely submerged in the electrolyte solution, and fresh solution was continually pumped through the tank. If burrs were initially small (such as the burrs that occur after sanding a flat plate), the process provided a slight edge break. A pattern of the weave of the cloth remained on the workpiece after processing. This side effect is problematic for some applications.

13.8.5 Electrochemical Moving-electrode Deburring

Electrochemical moving-electrode deburring, sometimes called *moving-electrode ECD*, employs graphite balls in a mass-finishing machine (Shimizu 1975; DeGroat 1970; Gillespie 1976). The balls provide continuous contact with the metal workpieces as well as some abrasion for removing smut and other by-products. The process requires no special tooling because the balls serve as the electrodes. Production machines are in use in Japan and Italy. There was one application for this process in the United States in the 1970s.

13.8.6 Electrochemical Brush Deburring

Electrochemical brush deburring employs a rotating conductive brush instead of a shaped electrode. Researchers claim a 3,000% increase in productivity using this method instead of conventional methods (Wicht and Kurze 1973). The belt-driven brush spindle is connected to the negative side of the circuit. As in conventional brush deburring (discussed in Chapter 11), a variety of brush styles may be used in this process.

The advantages of the process include speed, minimal tooling costs, stock removal concentrated at edge, edge radii capabilities ranging from .001–.030 in. (25.4–762 μm), and some localized surface improvement. Like electrochemical grinding, the percentage of deburring action produced by mechanical brushing could be varied from 0–100%. Deburring time does not depend on burr thickness and size (in most cases, loading and unloading time exceed deburring time). The disadvantage of this process is stray etching.

13.9 ENVIRONMENTAL, HEALTH, AND SAFETY CONCERNS

ECD has few environmental, health, and safety concerns. Voltages at the point of contact with workpieces are below life-threatening safety levels, but machine operators must be protected from power coming into transformers from the machine's power source. Since the ECD process is wet, nonslip work areas must be provided. Air cylinder clamping could create some safety pinch points for fingers. Although simple salt solutions are generally considered nontoxic, some operators may develop chemical sensitivities after prolonged contact with electrolyte solutions. This can be minimized by wearing rubber gloves.

The ECD process is normally used on workpieces made from common, nontoxic metals, but it may be used on some toxic or carcinogenic metals. Specific safety procedures for handling and processing toxic materials are beyond the scope of this handbook. Users should consult with appropriate professionals and government regulators regarding these materials, and regulations should be strictly enforced.

Many ECD operations require that air exhaust systems be installed over processing equipment to keep gases generated during the process away from operators, keep airborne salt away from adjacent equipment, and keep hydrogen from collecting in pockets above the machine. Although some shop personnel may fear that salt will accelerate corrosion on nearby machines (ECD machines are sometimes referred to as "old salty"), years of experience with low-production rates has shown that this is not the case.

Disposal of metals and pollutants filtered from the electrolyte solution, as well as disposal of the solution itself, are strictly regulated by local, state, and federal codes. Users must be aware of and comply with these codes. Furthermore, overall finishing costs must include the cost of waste disposal.

Electrolyte solutions can sometimes be reused after filtration or dumped at sanitary waste facilities, depending on the material and on waste disposal regulations. Testing for certain chemicals may be necessary before disposal. The pH level must be 7 (neutral). (Sodium nitrate, the most common electrolyte used in ECD processes, is used extensively for treating industrial and domestic waste.)

Sodium metabisulfite may need to be added to electrolyte solutions used on workpieces made from stainless steel or other metals containing high chromium contents. This is to reduce any carcinogenic hexavalent chromium to its trivalent form, enabling the solution and sludge having this chrome to be dumped in waste systems. Sludge remaining after filtration of electrolytes can either be stored or discarded in landfills, if local, state, and federal regulations permit.

Regulations for waste disposal change continuously; local regulators as well as equipment manufacturers must be consulted regularly to ensure that local, state, and federal codes are being met.

References

Andrew, Thomas C. 1982. "Advanced Technology in Surface Finishing." *Manufacturing Engineering,* December:47.

DeGroat, George. 1969. "ECM Deburring with Graphite." *American Machinist,* (December 1): 78–79.

———. 1970. "Conductive Abrasive Speeds Deburring." *Metalworking Production,* 114 (9):46–47.

Fromson, R.E. 1976. "A Conforming Electrode for the Electrochemical Deburring Process." Technical paper MR76-707. Dearborn, MI: Society of Manufacturing Engineers (SME).

Gillespie, LaRoux K. 1976. *Deburring Capabilities and Limitations.* Dearborn, MI: Society of Manufacturing Engineers (SME).

———. 1981. "Electrochemical Deburring." Division report BDX-613-2594. Kansas City, MO: Bendix Corporation (available from NTIS).

Jones, William F. 1981. "Today's Electrochemical Deburring Processes." Technical paper MR81-388. Dearborn, MI: Society of Manufacturing Engineers (SME).

Koroskenyi, James B. 1995. "Burlytic Deburring—A New Approach in Electrochemical Deburring." Technical paper MR95-228. Dearborn, MI: Society of Manufacturing Engineers (SME).

Miller, Jerome F. 1983. "Electrochemical Deburring of High-technology Machine Parts." Technical paper MR83-197. Dearborn, MI: Society of Manufacturing Engineers (SME).

Shimizu, Akihiko. 1975. "Electrochemical Barrel Finishing." Technical paper MR75-837. Dearborn, MI: Society of Manufacturing Engineers (SME).

Tuneski, Glenn, and Hignett, J. Bernard. 1981. "Some Recent Developments in Electrochemical Deburring." Technical paper MR81-964. Dearborn, MI: Society of Manufacturing Engineers (SME).

Vishnitsky, Victor. 1989. "Rotolytic Finishing." Technical paper EE89-804. Dearborn, MI: Society of Manufacturing Engineers (SME).

Wicht, Hans, and Kurze, Peter. 1973. "Deburring Still a Difficult Problem?" (in German). *ZIS Mitt,* 15 (8):930–35.

Bibliography

Fisher, Thomas. 1993. "Cationic Deburring Processes." Technical paper MR93-133. Dearborn, MI: Society of Manufacturing Engineers (SME).

Gillespie, LaRoux K., and Moore, W.M. 1982. "An Analysis of Electrochemical Deburring." Division report BDX-613-2818. Kansas City, MO: Bendix Corporation (available from NTIS).

Kissel, Horst. 1985. "Electrochemical Deburring as an Automated Process." Technical paper MR85-477. Dearborn, MI: Society of Manufacturing Engineers (SME).

Rhoades, Lawrence J. 1981. *Cost Guide for Automatic Finishing Processes.* Dearborn, MI: Society of Manufacturing Engineers (SME).

Schein, Henry. 1987. "Automatic Electrochemical Deburring." Technical paper MR87-154. Dearborn, MI: Society of Manufacturing Engineers (SME).

Tonshoff, H.K., Egger, R., and Klocke, F. 1996. "Environmental and Safety Aspects of Electrophysical and Electrochemical Processes." *CIRP Annals,* 45 (2):553–68.

14 Electropolish Deburring

Electropolish deburring involves removing metal from a workpiece surface using electrical current directed through an electrolyte solution. Under the proper conditions, the process takes advantage of the fact that raised portions of the workpiece surface serve as "lightning rods" for the electrical current, attracting more energy than other surfaces and resulting in more material being removed in these locations. After electropolish deburring, workpiece surfaces are smooth and bright, which contributes to the popularity of the process. Electropolish deburring will remove burrs as well as stock on all exposed surfaces unless they are insulated or masked. Unlike electrochemical deburring, no special tooling is required with electropolishing. The parts are connected to the anodic side of a circuit and cathode rods hanging in the solution complete the circuits, as shown in Figure 14-1.

Electropolish deburring involves the following process variables:

- Electrolyte solution.
- Solution temperature.
- Cycle time.
- Racking (electrical contact).
- Current density.
- Burr location.
- Burr thickness.

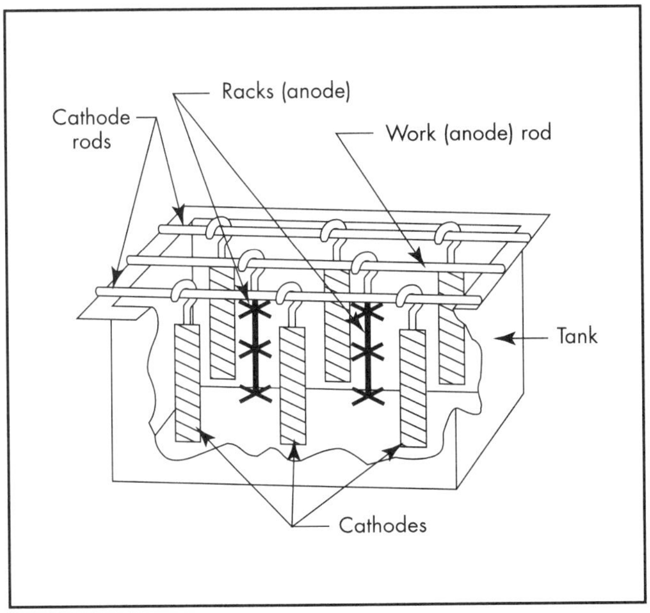

Figure 14-1. Typical electropolish deburring tank with a single work rod. ***(Courtesy Electro Glo Company)***

Throwing power is a plating term describing the conductive nature of the solution. It is an element of the solution and it changes as the tank begins to fill with metal ions and sludge.

14.1 PROCESS PRINCIPLES

When an electrical current is applied to metal, a polarized film forms on the surface of the metal workpiece. The metal part surface becomes metal ions that are converted to metallic salts and must diffuse through this film. The strength and viscosity of the polarized film affect both the brightening and leveling actions of the process. The film is thinner over the projections and thicker over the depressions of the metal, allowing the projections to be more exposed to electrolytic action and to have lower electrical resistance than the depressions. Where the film is thinner, such as over projections, surface material is more rapidly dissolved than where it is thicker, such as in depressions.

Atoms of the metal being removed pass through the polarized anodic film as metal salts and enter the electrolyte solution to be either dissolved, deposited on the cathode, or precipitated as sludge. Electropolish deburring solutions are therefore considered full sludging, semisludging, or nonsludging.

Burrs located in deep holes or masked because of the configuration of the workpiece may not receive the full "throwing power" of the electrolyte or electrical action that exposed burrs receive, and thus will not be removed unless auxiliary cathodes are employed to direct more energy to these areas, as indicated in Figure 14-2. Pitting can occur when improper conditions are employed.

14.2 APPLICATIONS

Electropolish deburring is generally used in high-production facilities for removing small burrs on precision parts. Thousands of workpieces can be deburred in an hour with proper system design. The process can be used on any conductive metal, including carbon steel, alloy steels, 300 and 400 series stainless steel, high nickel alloys, nickel silver, aluminum, brass, copper, and zinc. Surface finishes of 1–2 μin. (0.025–0.051 μm) can be produced at edges and over the entire workpiece. The edges produced by electropolish deburring remain essentially sharp after deburring, although some slight rounding may occur. Generally, the process is quick and requires no special tooling. Although stock is preferentially removed from edges and raised elements, the process will not remove extremely thick burrs without also removing large amounts of stock from other surfaces. For example, to remove grinding burrs that are .0005 in. (0.013 mm) thick, an overall stock loss of .0002–.0003 in. (0.005–

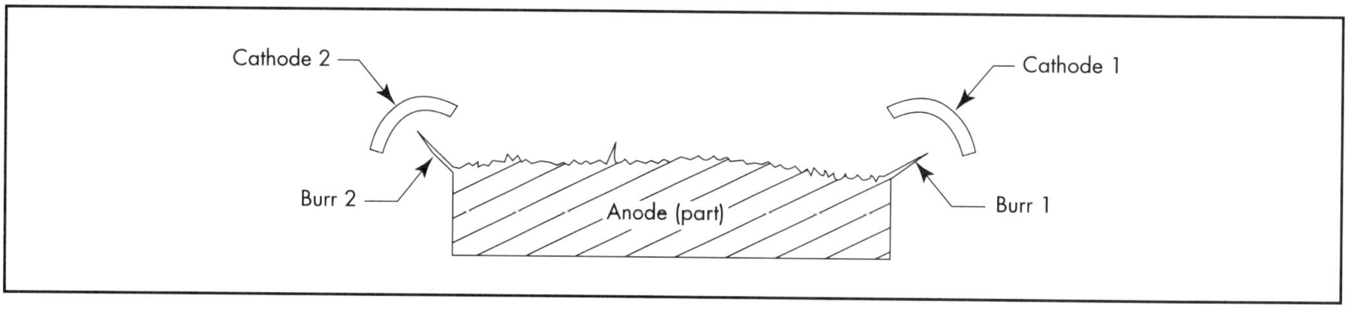

Figure 14-2. Schematic of an electropolish deburring system employing auxiliary cathodes. ***(Courtesy Madwed Manufacturing Company)***

0.0076 mm) may result on all exposed surfaces. (Diameter would change about .0005 in. [0.013 mm]). In general, the length of the burr plays no significant role, but burr thickness is pivotal. The process will remove any of the so-called microburrs that form in valleys on ground surfaces. After being deburred using other deburring methods, many stainless steel workpieces must undergo a special passivation operation to remove traces of free iron, but electropolish deburring automatically passivates the surface.

Electropolish deburring is particularly suitable for finishing workpieces with small burrs—typically .0005–.001 in. (12.7–25.4 μm) thick or less—especially those produced as a result of grinding, honing, lapping, and similar processes.

The process is also used for finishing hydraulic steering housings and valves, many punched parts, ball bearing cages, perforated Hasteloy® and Nimonic® combustion chambers, guide plates, ducts for jet engines, ground gears, pinions, notched shafts, knitting-machine cylinders, frames, watchband clasps, cutting tools, spoons, forks, knives, fine-toothed combs for the fiber industry, and needle bearings. Ground parts used in the spinning industry must have absolutely burr-free edges and surfaces and are commonly deburred using this method. Suture and hypodermic needles are also deburred using this method. Suture needles are deburred at a rate of 1,500 per hour, with one operator supporting the effort, and hypodermic needles can be processed at 8,000 per hour. A unique magnetic racking system is used for suture needles.

Another application involves deburring filter plates for paper-making machines. These 323 ft^2 (30 m^2) stainless steel plates have 300,000 conical .035 in. (0.89 mm) holes. After large drill burrs have been removed by surface grinding, the plates are electropolished at a rate of almost 4,000,000 holes per hour (Siegel 1975), with every hole receiving the same edge finish (see Figure 14-3).

14.3 EQUIPMENT

The equipment is simple—a series of tanks, power supplies, timers, and part transportation (if automated).

The deburring tank is typically rectangular and may be made of either austenitic stainless steel or mild steel lined with polyvinyl chloride, polyethylene, polypropylene, fiberglass, or special types of rubber. For some applications, either lead-lined or plain steel tanks are used. Steel tanks may be problematic in some applications since they are electrically conductive and draw away electrical action from the solution. The appropriate tank material depends primarily on the type and temperature of the electropolish deburring solution being used. The tank is fitted with tank rods or bars, and workpieces are hung on work rods (the anodes) that carry positive current. On a tank with a single work rod (Figure 14-1), the rod is centered on the width of the tank and is parallel to the length. The cathode rods are parallel to and equidistant from the anode or work rod and carry negative current. In some cases, tanks are equipped with two work rods, a configuration requiring three cathode rods (see Figure 14-4) that are arranged as they are in a tank with a single work rod.

Automatic temperature control devices are included with most electropolish deburring systems. Heating is achieved either with electric immersion heaters or with coils through which hot water or steam passes. For cooling, cold water is passed through the coils (this water can be later reused for rinsing). Coils are made from lead, stainless steel, titanium, or Teflon®.

Low-voltage direct current is supplied to the electropolish deburring tank by a rectifier similar to those used in electroplating. The rectifier must have voltage controls.

Rinse tanks are constructed of the same materials used for the deburring tanks, except when temperatures differ between operations. Less expensive materials can be used whenever possible. Rinse tanks are fitted with a means for draining and filling and may be arranged for counter-flowing to reduce effluents and water consumption.

Workpieces are usually racked in the tank. The rack not only holds the pieces in the most advantageous position for electropolish deburring, but also conducts the current to the pieces, through either one or several contact points. Racks are designed to hold any number of workpieces, depending on the size and volume of the tank. Typically, several racks are placed in the tank simultaneously.

Some workpieces may not require racking and can be hung on J-hooks from the conductive bars at the top of the tank. However, this may be less productive if the full depth of the solution is not being used.

Small pieces that will not tangle or nest can be processed in bulk, with production rates of up to 8,000 workpieces per hour if special equipment is employed. Bulk processing

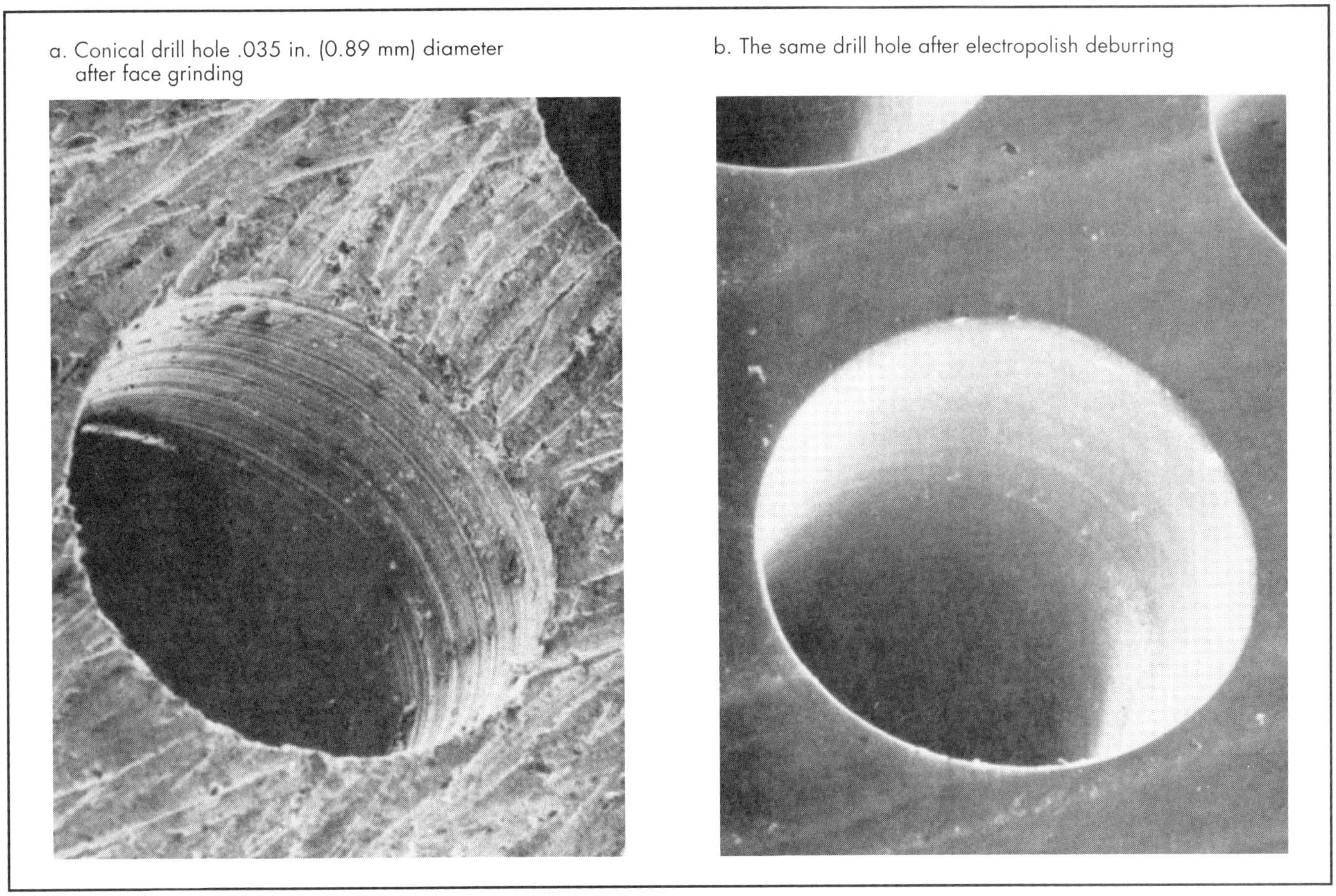

Figure 14-3. Effect of electropolish deburring on drill hole. (*a*) Conical drill hole (.035 in. [0.89 mm] in diameter) after face grinding; (*b*) the same hole after electropolish deburring. *(Courtesy Polygrat Corporation)*

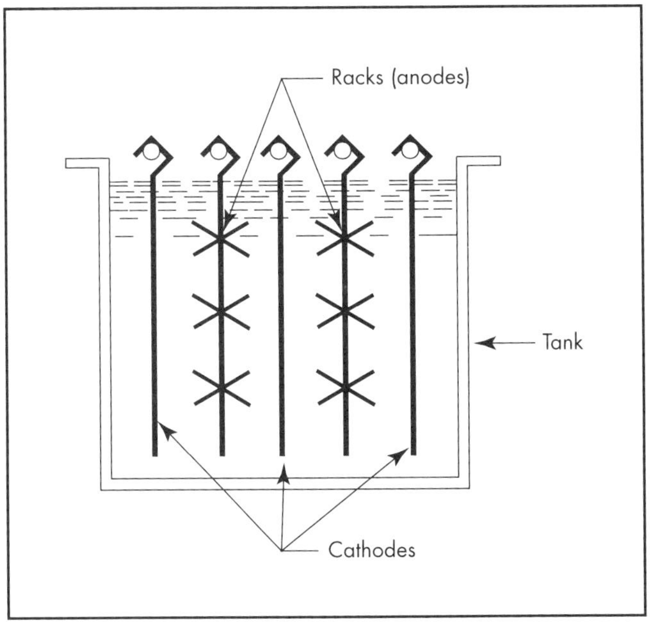

Figure 14-4. Electropolish deburring tank with two work rods. *(Courtesy Electro Glo Company)*

avoids much of the labor normally associated with hand racking. Equipment is also available for reel-to-reel or continuous electropolish deburring of wire and strip stock. Figure 14-5 illustrates several approaches for high-volume electropolish deburring of small parts.

14.4 ELECTROPOLISHING SOLUTIONS

There is no universal solution that will successfully deburr all metals and alloys well, and therefore many types of electropolishing solutions exist. Most solutions are acid-based, but alkaline, cyanide, and metal electrolytes are also used. Electrolyte solutions differ not only in composition but also in operating temperature and current density. Operating temperatures range from ambient to near their boiling points, 70–200° F (21–93° C) and current densities range from 25–250 ampere/ft^2 (2.3–23.2 ampere/m^2).

For many years, the most widely used electrolyte solutions were concentrated acids (see Table 14-1), such as acetic, citric, chromic, hydrofluoric, phosphoric, sulfuric, and tartaric acids. The disadvantages of using acid-type electrolytes include poor throwing power, high operating temperatures, narrow operating range, excessive current densities, and short useful life. Furthermore, some of these acids are

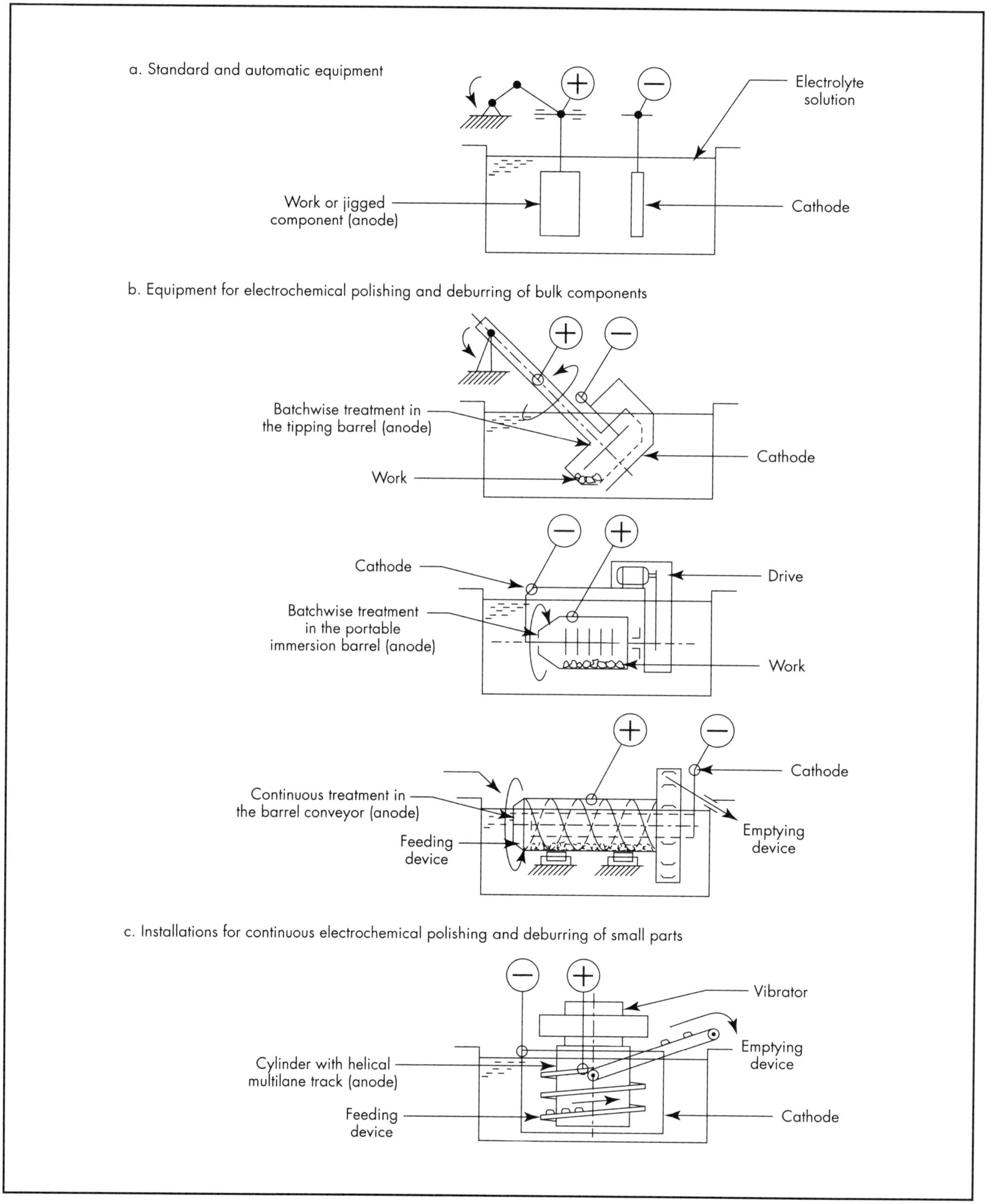

Figure 14-5. Electropolish deburring units for high-volume production of small parts. ***(Courtesy Polygrat Corporation)***

now considered toxic or carcinogenic and are seldom used. Phosphoric acid is the major constituent in many of the acid-type solutions. Hexavalent chrome is a carcinogen. Cyanides and halogens can be toxic.

Currently, organic acids and addition agents are being added to acid-type solutions to mitigate many of the disadvantages of straight acid solutions.

The majority of electropolish deburring solutions in use today are proprietary solutions and are similar to those used in the electroplating industry. An important advantage of using proprietary solutions is that considerable research and development goes into developing the product, and technical and laboratory assistance are available to users when necessary.

It is commonly believed in the industry that some electropolish solutions improve with age (actually with a slight buildup of sludge at the bottom of tanks).

Electrolyte solutions containing cyanides, chromium compounds, halogens, or other noxious materials require exhausting with hoods or lateral exhaust systems. When using solutions containing concentrated phosphoric acid, a slight spray is liberated during processing that rises only several inches from the surface of the solution. This spray is not harmful to humans, but it is corrosive to metals. The gases produced by phosphoric acid-type solutions are hydrogen and oxygen, and adequate air circulation should be provided while they are being used to prevent dangerous hydrogen pockets. Effluents resulting from rinsing operations

Table 14-1. Components of typical electropolish deburring solutions used for processing various materials

Material	Component	% of solution volume, by weight	Temperature, ° F (° C)	Current density, ampere/ft^2 (ampere/m^2)
Steel	Phosphoric acid[a]	55–85		
	Trialkali metal phosphate	1–15	160–180 (71–82)	400–800 (37.2–74.3)
	Alkali metal sulfate	0.5 minimum		
Stainless steel	Phosphoric acid[b]	35	200 minimum (93)	25 minimum (2.3)
	Glycerin	55 minimum		
	Sulfuric acid	41		
	Phosphoric acid	45	170–230 (77–110)	200–350 (18.6–32.5)
	Water	14		
	Phosphoric acid	56		
	Chromic acid	12	80–175 (27–79)	100–1,000 (9.3–93)
	Water	32		
Aluminum	Sulfuric acid[c]	1–60	140 (60)	100 (9.3)
	Hydrofluoric acid	0.2–1.5		
	Sulfuric acid[d]	4–45		
	Phosphoric acid	40–80	160–200 (71–93)	25–950 (2.3–88.3)
	Chromic acid	0.2–9		
	Trivalent metals	6		
	Sodium carbonate (anhydrous)	(pH 10.5)	176–180 (80–82)	20–30 (1.9–2.8)
	Trisodium phosphate			
Copper and alloys	Sodium tripolyphosphate[e]	(pH 7–7.5)	125–135 (52–57)	100 minimum (9.3)
	Boric acid			
	Sulfuric acid	14		
	Phosphoric acid	59	60–170 (16–77)	100–1,000 (9.3–93)
	Chromic acid	0.5		
	Water	36.5		
Nickel and alloys	Sulfuric acid	70	Room	50–200 (4.6–18.6)
	Water	30		

[a]Patented by Hammond, Edgeworth, and Bowman.
[b]Patented by H. Uhlig.
[c]Patented by Mason and Tosterud.
[d]Patent 2,550.544.1951 by C. Faust.
[e]Patented by H. Wiesner.

and tank sludge containing metallic salts must be disposed of in accordance with local, state, and federal regulations.

14.5 OPERATING CONSIDERATIONS

The surface condition of the workpiece, regardless of workpiece material, has the greatest effect on the final surface finish produced by the process. Surface conditions contributing to unsatisfactory results include large grain size, nonuniform structure, nonmetallic inclusions, directional roll marks, salt or scale contaminations, over-pickling, and insufficient or excessive cold reduction.

Most electropolish deburring is performed using low-voltage direct current, typically 5–18 V. Current densities vary, depending on the type of solution and the application. Each type of electropolish deburring solution has an ideal current density range.

The production rate of the process is determined by the current density (the number of amperes of current applied to a given area) and the amount of time the current is applied. When the distance between the workpiece (the anode) and the electrode (the cathode) is small, the voltage required to produce the desired amperage and results is low. As the distance increases, the voltage must be increased to overcome the additional resistance and maintain the same current density. Figures 14-6 and 14-7 show typical stock-removal rates at various voltages and current densities.

Throwing power refers to the amount of energy directed through the electrolyte solution toward a workpiece. Since electrical current takes the path of least resistance, areas that are more distant or obscured from the electrical power source will respond differently to the electrolyte solution than areas receiving high current densities. Figure 14-2 shows how auxiliary cathodes can be used to accelerate deburring in specific areas and thus decrease cycle time.

The following variables affect the throwing power of the electrolyte:

- Size and shape of the workpiece.
- Size and shape of the processing tank.
- Size, share, number, and location of electrodes.
- Type and composition of electrolyte solution.
- Processing time.
- Temperature.
- Current density.

Due to the nature of the electrolyte, workpiece protrusions, such as corners, edges, and burrs, naturally attract higher current densities than recessed areas, such as holes and inside corners, and thus receive more of the throwing power from the electrolyte.

Figures 14-6 and 14-7 show typical deburring time when burr size is known. For example, in Figure 14-7, if a 304 stainless steel workpiece has a burr with a thickness of .002 in. (50.8 μm) and 1,858 ampere/ft^2 (173 ampere/m^2) of current is available in the solution, the workpiece can be deburred in less than 1 min. Note that since the burr is exposed to the electrolyte from both sides of its thickness, the material removed per side is .00098 in. (25 μm). The intersection of the 304 stainless steel curve and 1,858 ampere/ft^2 (173 ampere/m^2) occurs at about .00118 in./min (30 μm/min).

Electropolish deburring will not generate large radii. Typically edges range from sharp to .002 in. (50.8 μm) radii. The use of special electrodes and long run times may, in exceptional cases, extend these limits.

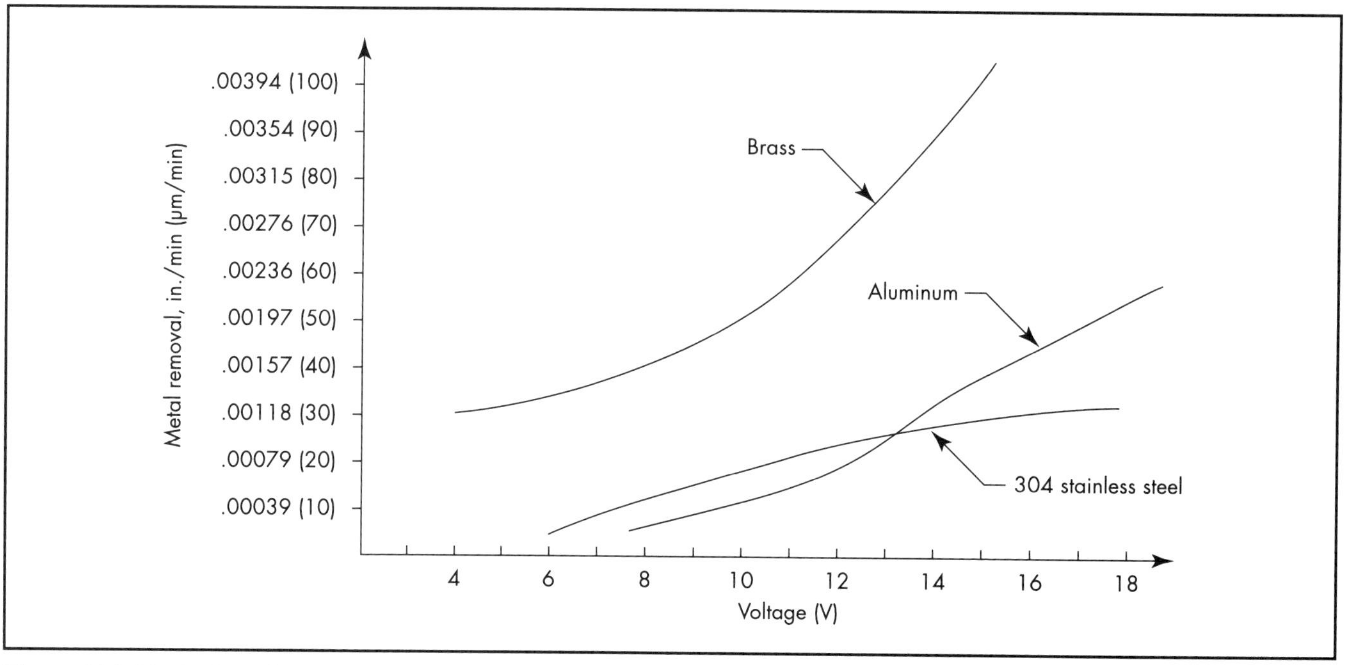

Figure 14-6. Typical stock removal rates as a function of voltage and material (Madwed 1976). ***(Courtesy Madwed Manufacturing Company)***

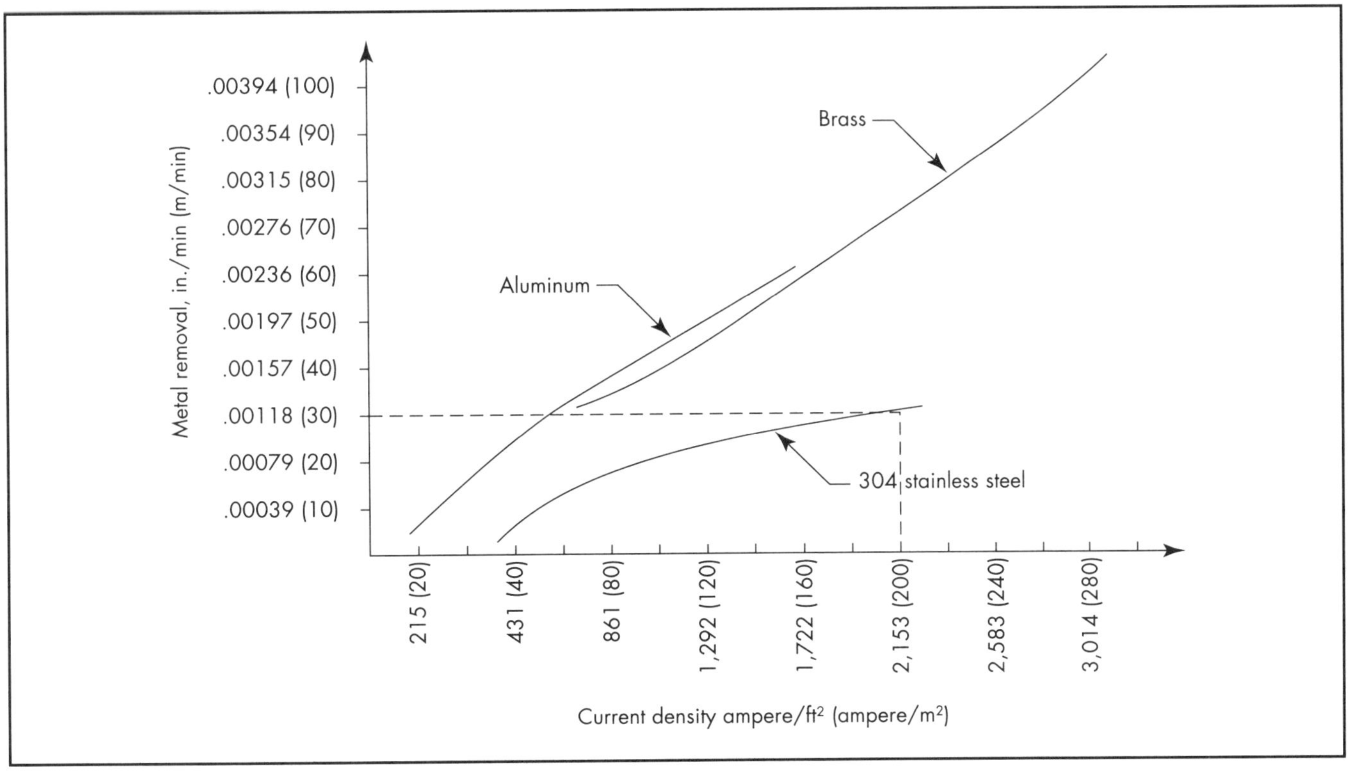

Figure 14-7. Typical stock removal rates as a function of current density and material (Madwed 1976). ***(Courtesy Madwed Manufacturing Company)***

14.6 POST-TREATMENT

Nearly all electropolish deburring solutions are aqueous systems, and water is used for rinsing. Several rinses are usually required after electropolish deburring, depending on the workpiece material and the application. Workpieces too large for rinsing by immersion must be hosed off. Acid or alkaline neutralizing rinses may be necessary between water rinses. Overflow rinsing tanks are being replaced by more efficient, counterflow rinsing tanks, and spray rinsing is increasingly being used. When static or overflow rinsing is used, agitation (either air or mechanical) is beneficial because it provides some scrubbing action.

Staining and spotting are usually caused by the final rinse water, due to either contamination carried over from previous rinses or water hardness. If the final rinse water is hot (for example, 130° F [54° C]), evaporation will further concentrate contaminants, resulting in increased staining or spotting. This problem can be alleviated by frequently changing the final rinse water or by adding suitable wetting agents to the rinse water. Wetting agents reduce surface tension and eliminate water beading. A thin water film, with fewer deposits, remains. The ideal final rinse, although more costly, employs deionized or distilled water. Freon was popular for specialized applications for water displacement and rapid drying, but is not widely used now because of environmental concerns. Workpieces may be further processed by operations such as plating, anodizing, and conversion and organic coating.

14.7 EQUIPMENT FOR SPECIAL NEEDS

14.7.1 Bench-top Operations

Excellent results, as seen in Figure 14-8, can be obtained using simple bench-top tanks, illustrated in Figure 14-9. The instrument gear is hobbed from 303 stainless steel and has normal hobbing burrs on crests and ends. In this low-volume operation (few parts/day) depicted in Figures 14-9 and 14-10, the workholding rack or clip is made from titanium, and a spring alligator clip is used. The current draw for this operation is .4–1.5 ampere/in.^{2} (0.037–0.139 ampere/m^2) at 6–12 V. Temperatures were held to 34° F (1° C). An oven dry was used after the final rinse.

14.7.2 Out-of-tank Operations

Several methods exist for electropolish deburring when workpieces are too large to be submerged in a tank or when parts have been installed before the need for electropolishing is recognized.

In Situ Electropolish Deburring

In situ electropolish deburring involves making the part or the workpiece into the "tank." The process involves closing off outlets and openings in the workpiece or part with suitable materials, placing a conforming cathode inside the part, and pumping in preheated electropolish deburring solution. To allow gases to escape, an opening must be left

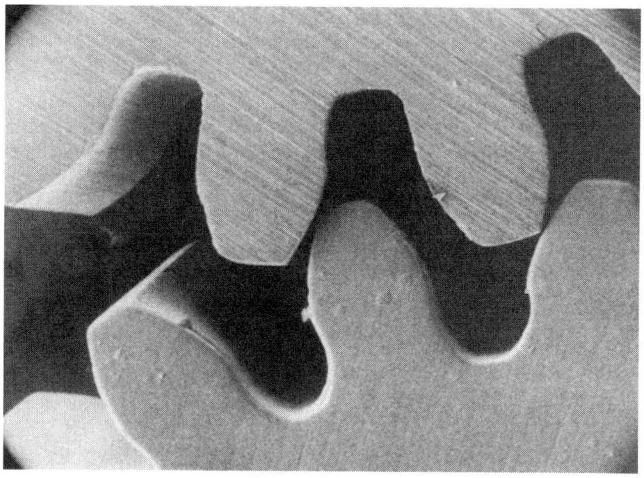

Figure 14-8. Before and after electropolish deburring on 175 diametral-pitch gear teeth processed for 12 sec at 130° F (54° C). ***(Courtesy Bendix Corporation)***

in the workpiece. A collar is usually attached to this opening to allow the solution to expand and prevent any foam that may occur from coming into contact with nearby surfaces and thus dulling them. The procedure may need to be periodically interrupted to allow gases and foam to subside.

Wand Electropolish Deburring

Wand electropolish deburring is another method for processing workpieces outside of a tank (Cutchin 1991). The wand, resembling a household iron, serves as a mini-reservoir for the electrolyte solution and is constructed of a material able to withstand heat and corrosion, usually stainless steel. The solution is pumped into an inlet in one end of the wand and then exits through many small holes, wetting and flooding the area being electropolished. A copper shoe or shield with holes that align with the holes in the wand is attached to the portion of the wand that will be in contact with the workpiece. Electrical current is supplied through a cable that runs from a DC power source to the shoe (the cathode). The workpiece is the anode. To prevent the wand and shoe assembly from coming into direct contact with the workpiece, the entire assembly is covered with a synthetic fiber wadding and a synthetic fabric "glove." A particularly effective wadding material is poly fill, and a durable outer covering is polyester fabric. A handle attached to the wand enables the operator to continuously move the wand across the area being processed to reach all burrs. The flow rate of the solution must be periodically adjusted to avoid generating excessive heat that will evaporate the solution. The process resembles putting a sock over a hollow lightning rod.

The shape and configuration of the wand depend on the application. For deburring large flat areas of large-diameter vessels, a surface contact area of 6 × 4 in. (15.2 × 10.2 cm) is effective. For processing the inside of large-diameter pipe and tubing, a curved shoe 4 × 1.6 in. (10.2 × 4.1 cm) is effective. For processing the inside of small pipe with diameters of .5–1.6 in. (1.3–4.1 cm), a short section of copper tubing covered with wadding and a fabric glove is effective and easy to assemble.

14.8 ENVIRONMENTAL, HEALTH, AND SAFETY CONCERNS

Electropolish deburring employs acids, sometimes hot solutions, and electrical currents, and appropriate safety precautions must be instituted. Face shields and protective

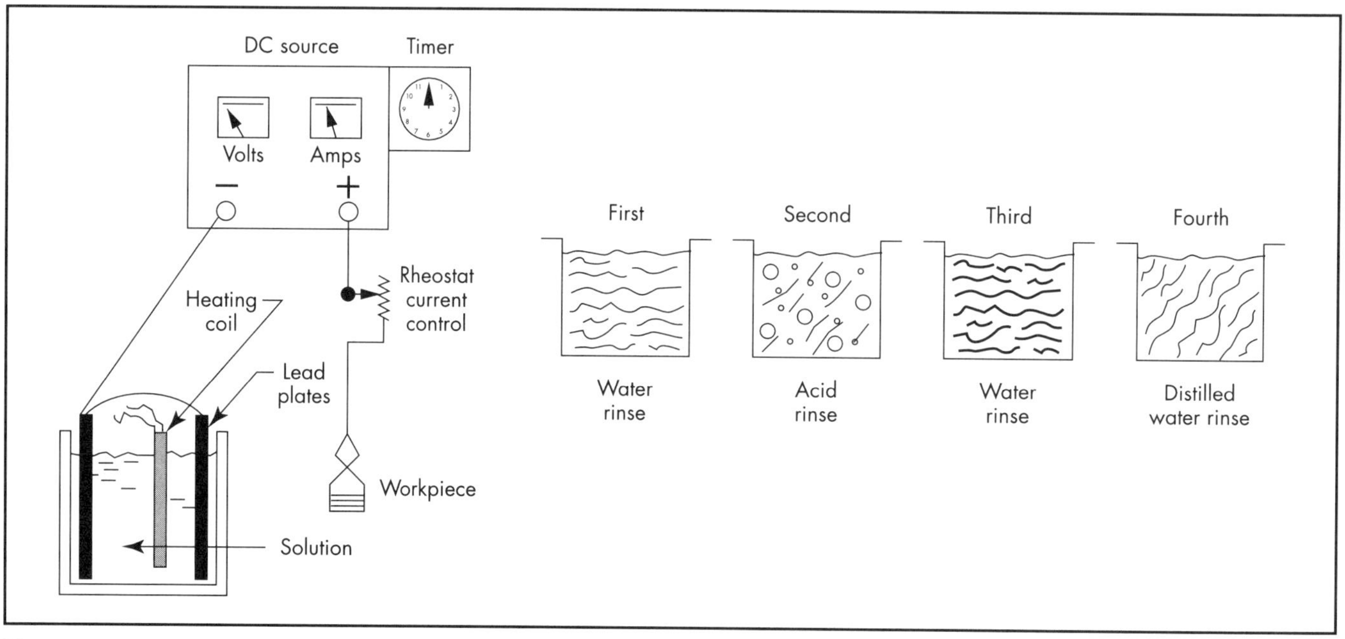

Figure 14-9. Schematic of bench-top electropolish deburring system. ***(Courtesy Bendix Corporation)***

Figure 14-10. Simple bench-top electropolish deburring operation. ***(Courtesy Bendix Corporation)***

clothing for hands and body are necessary. Voltages at the tank are not generally life threatening, but rectifiers employ industrial voltages and currents. Automated equipment can grab clothing and act unexpectedly and, therefore, provisions should be made to avoid pinch-point accidents caused by passing parts and workpieces. The process may be used for processing large panels carried by overhead cranes, necessitating training in crane safety and control.

Acid solutions are typically neutralized before being discharged as waste. Since stainless steels have high quantities of chromium, and hexavalent chrome is a health hazard, procedures for disposal of sludges and solutions containing this material must be properly carried out. All waste must be disposed of in accordance with local, state, and federal regulations. (For further information, see the discussions on waste disposal in Chapters 6 and 13.)

References

Cutchin, John H. 1991. "Practical Electropolishing from a Job Shop Point of View." Technical paper MR91-130. Dearborn, MI: Society of Manufacturing Engineers (SME).

Madwed, Albert. 1976. "Some Fresh Approaches to Electrochemical Deburring." Technical paper MR76-133. Dearborn, MI: Society of Manufacturing Engineers (SME).

Siegel, Bruno R. 1975. "How to Deburr 1,000,000 Drill Holes within 15 Minutes." Technical paper MR75-486. Dearborn, MI: Society of Manufacturing Engineers (SME).

Bibliography

Zerweck, Klaus. 1977. "Optimizing Process Parameters in Electropolish Deburring." Technical paper MR77-450. Dearborn, MI: Society of Manufacturing Engineers (SME).

15 Thermal Energy Deburring

Thermal energy method (TEM) deburring is used primarily for deflashing and removing thin burrs on workpieces. Workpieces are placed in a gas-filled chamber (pressure vessel) of a TEM machine. There the gas is ignited. The intense heat, which lasts for only microseconds (ms), ignites thin metal sections vaporizing burrs and flash on both the internal and external surfaces of the workpiece (Figure 15-1). Under proper conditions, a small radius is left at the edge. The process is particularly appropriate for high-production applications when thin burrs need to be removed or when a commercial-quality edge is required. It can also be used to rapidly deburr a large number of intersecting holes and hard-to-reach areas. Workpieces are typically deburred en masse, though larger workpieces may be deburred singly. Production rates of 200 workpieces per hour are not uncommon, and small workpieces can be deburred en masse at much higher rates.

15.1 PROCESS PRINCIPLES

The thermal shock is applied to workpieces by burning a mixture of oxygen and a combustible gas (generally, natural gas, though methane and hydrogen are also in use). The ratio of oxygen to gas is 2.5:1 up to 9:1 (Kelley and Schwarz 1991). To provide the necessary energy, the gas mixture is compressed before it enters the sealed and pressurized work chamber. It distributes itself evenly throughout the chamber and into all holes and recesses of the workpiece (Figure 15-2). The more the mixture is compressed both before it enters the work chamber and once it is within the chamber, the more deburring action it provides.

The combustible gas mixture is ignited with a 30,000 V spark, creating a temperature of 4,530–6,330° F (2,499–3,499° C). Within 2–3 ms, the fuel is burned and the fire extinguished; however, because of the intense heat created during the combustion of the gas, the burrs or flash burst into flame and burn in the excess oxygen (Figure 15-3), causing them to vaporize. Burrs and flash continue to vaporize until the heat dissipates into the body of the workpiece and the flame is eventually extinguished. Burrs, but not the rest of the workpiece, burn because they have a higher ratio of surface area to mass, allowing heat to enter the burr quicker than it enters the workpiece. Heat hitting the burr cannot flow fast enough (and thus be dissipated) into the main workpiece before the burr reaches its flash point or ignition temperature. Under normal conditions, the workpiece itself is heated but not enough to be ignited. The principle is similar to a candle wick burning but not igniting the surrounding wax.

In the process of being vaporized, the burrs or flash become oxides of the workpiece material (if the material is metal)—for example, the burrs on steel workpieces become iron oxide, and on aluminum workpieces they become aluminum oxide. These oxides settle on the workpieces as a loose, powdery residue. Although this residue may discolor the workpiece, the workpiece itself has not been oxidized.

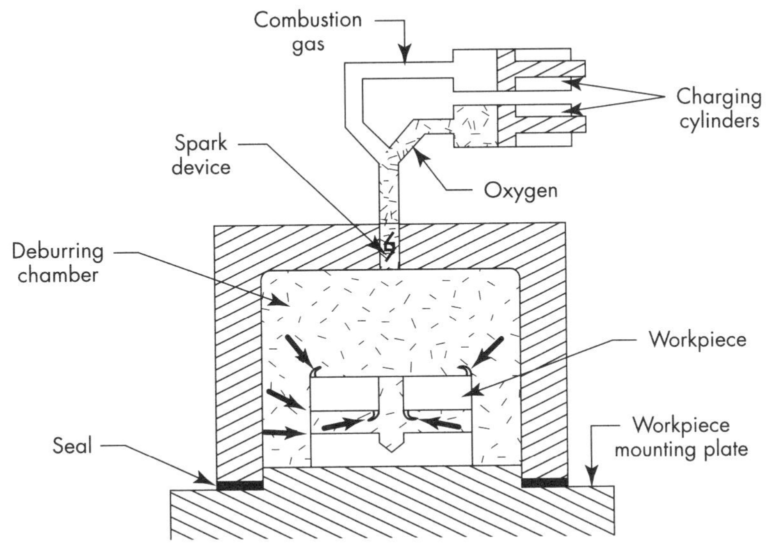

Figure 15-1. Schematic of TEM (burrs are indicated by heavy arrows). ***(Courtesy SurfTran)***

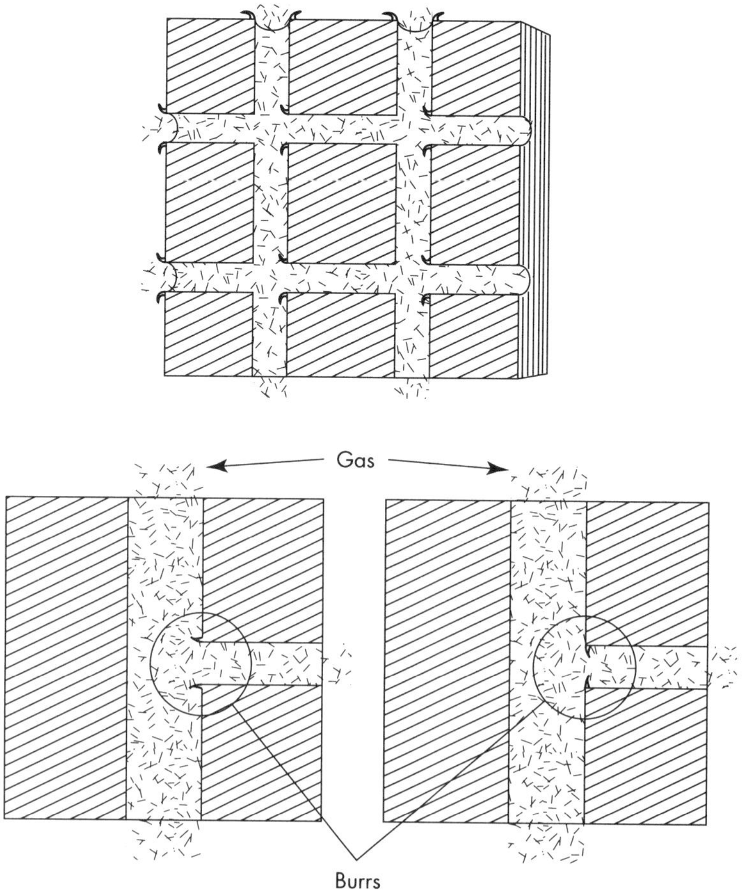

Figure 15-2. Conceptual view of TEM gas surrounding drill exit burrs and filling intersecting holes of a workpiece. ***(Courtesy SurfTran)***

On aluminum or zinc workpieces, the residue can be left undisturbed, since it does not hinder further processing, such as enameling or plating. When further processing requires that the residue be removed, it can be removed with a suitable cleaner. Chemical post-treatment of ferrous workpieces often eliminates the need for cleaning.

For effective processing using TEM, observe the following three rules:

1. Keep the burrs in the gas. Oil or water on parts will keep gas away. Some blind passages will have air, not gas around them, producing a bubble effect.
2. To save thin features of a workpiece, make them at least ten times the thickness of the burr. This process will burn thin sections. Therefore, make thin sections thicker or put heat absorbing blocks in front of them.
3. Restrain the workpiece if desired. The ignition will blow loose pieces around the chamber causing nicks, dings, and scratches on the parts.

15.2 APPLICATIONS

TEM is used for deburring and deflashing a wide variety of workpieces, especially those with thin burrs on internal intersections or threads and those that cannot be processed satisfactorily or cost effectively by vibratory deburring, shot blasting, tumbling electropolishing, electrolytic deburring, brushing, or more conventional methods. Most, though not all, applications are those requiring high production rates, such as 50 workpieces per load, with cycle times of 25 s per load (in this case, the effective production rate is 0.5 s per workpiece). Costs, production rates, and suitability of the process for a particular workpiece depend on the specific application. All metals can be treated by TEM, though certain metals are more difficult than others to process. The oxidation resistance of stainless steels, for example, makes them less adaptable to TEM. Excellent results, however, are produced on metals with low thermal conductivity, such as steels and zinc alloys.

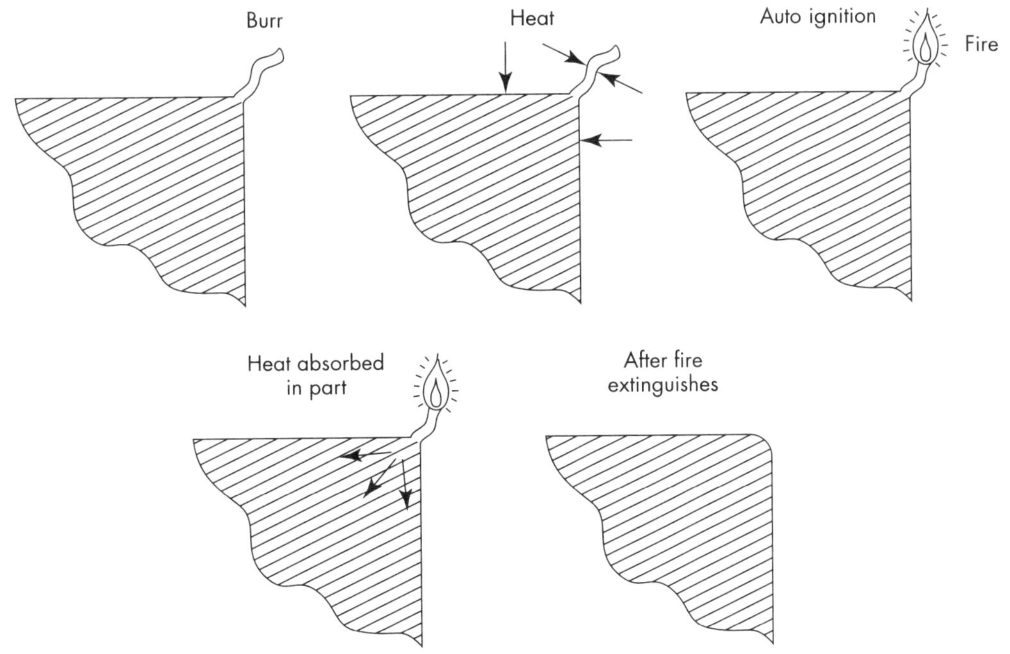

Figure 15-3. Burr on workpiece (before TEM processing, during ignition, and after oxidizing). ***(Courtesy SurfTran)***

Typical applications include automotive parts, die castings, electronic parts, bearing cages, pump bodies, internal combustion parts, aircraft engine components, pneumatic valves and gages, gears and splines, oil drilling parts, and screw machine parts.

15.2.1 Iron and Steel Castings

Deburring and deflashing iron and steel castings are major applications of TEM. The process is particularly suitable for and offers the most substantial savings on hydraulic and pneumatic valve bodies and castings with internal intersecting recesses and difficult-to-reach bores.

15.2.2 Zinc Die Castings

A wide variety of zinc die castings are deburred, deflashed, and cleaned using TEM. Carburetor components, automotive-lock cylinders, and many other automotive parts are being finished in this way. Figure 15-4 illustrates the effectiveness of TEM on zinc die castings (note that some flash remains near the body because it was thicker in this location, but leaving such small amounts of flash is often perfectly acceptable). A major benefit is the consistently high quality obtained. Substantial savings also result because no tooling is necessary and no change in setup is required to handle workpieces of different shapes or sizes.

Figure 15-4. Zinc die-cast parts before and after TEM. ***(Courtesy SurfTran)***

15.2.3 Machined Steel Components

Stainless steel valve spools and steel gears are just two of the many machined components finished using TEM. Ring grooves and slots on the valve spools are deburred at the rate of 720 spools per hour, with six spools handled per cycle. For the gears, burrs are removed from the teeth and from tapped and through holes at the rate of 120 gears per hour.

15.2.4 Plastics

TEM can be used to deflash and clean plastics workpieces. The process was first used commercially to remove "webs" in nonwoven plastic mesh. Machines used for processing plastic workpieces operate at lower pressures and temperatures than those used for metal, and gases are monitored more precisely. The chamber is generally evacuated of air before gases are introduced to maintain precise mixtures and to minimize contamination. Temperatures are held to about 2,000° F (1,093° C); and because of the short duration of the heat, workpiece temperatures rise only about 300° F (149° C), with a maximum increase of 1,000° F (538° C). Because of the low melting points of plastics, and the relatively thick flash, it is not widely used for these materials.

15.2.5 Aluminum

Heat sinks used to dissipate heat from electronic modules are one of many aluminum components being deburred with TEM. Cycle time has been reduced from 15 min to 1 min per unit, and the quality of the heat sinks has been improved.

15.3 ADVANTAGES AND LIMITATIONS

Although the equipment required for TEM is generally more expensive than that for some traditional deburring processes, TEM is currently the quickest method for deburring most workpieces. The process typically takes less than 20 ms. The usual cycle time for a high-production machine, including loading and unloading, is about 25 s. If incoming workpieces and burrs are of similar shapes or sizes, consistent results can be achieved.

The process is ideal for finishing metal workpieces through which fluids or gases must flow. It has the unique ability to deburr blind and intersecting holes simultaneously with the outside surfaces of workpieces. Since the process does not use any abrasive media, the surfaces adjoining the burrs experience no changes in dimensions. The bodies of most workpieces processed rarely exceed a temperature of 300° F (149° C). With a water-cooled chamber and fast cycling times, only thin workpieces, such as some stampings, reach temperatures hot enough to affect the parent metal. A recast layer that forms at the workpiece-burr interface is usually of negligible thickness ~.00008–.001 in. (0.0020–0.0254 mm); however, this layer may cause cracks in previously hardened steels.

With TEM, burrs or flash of uniform thickness but limited size are removed to their roots. With parting lines, however, which generally have thicker roots than do burrs, a rounded rise is left on the surface of the workpiece. When designing workpieces to be thermally deburred, the thinnest sections with useful function should be at least 10–15 times thicker than the burrs. This design limitation is beneficial because TEM will only slightly round the extreme edges of threads, not remove them. For most stamped workpieces, however, which typically have triangular-shaped burrs, the process removes only the tips of the burrs.

Burrs do not have to be facing in a particular direction to be removed by TEM because the workpiece, once inside the chamber, is surrounded by the gas mixture. This flexibility simplifies fixturing and generally eliminates the need for workpiece-holding devices other than baskets.

Workpieces to be deburred by TEM must be free of oil. If oil remains on a workpiece, the oil will be burned into the surfaces, leaving a carbon smut that is difficult to remove. Blind and tapped holes must be free of compacted chips. Chips that remain cannot be surrounded by the gas mixture and will not be removed by TEM. Achieving specific radii on individual edges is not always possible with TEM. Some edge radiusing is possible on steel and cast iron workpieces, but is seldom achieved on aluminum or stainless steel workpieces. In many cases, removing burrs but keeping edges almost sharp is possible.

Although TEM cannot normally produce specific radii, radii of .002–.060 in. (0.05–1.52 mm) have been produced on steel and .002–.010 in. (0.05–0.25 mm) on aluminum. The major determinants of edge radii are material and burr size. When burrs are large, larger radii will result; when they are small, almost no radii will result.

If burrs on exterior surfaces of workpieces touch fixturing or the surfaces of other workpieces during processing, they will be heat sunk by these fixtures or surfaces, resulting in little or no deburring action at the points of contact. Folded-over burrs behave as main body material and do not burn off because they, too, become heat sunk material.

Large burrs can be removed if the rest of the workpiece is relatively large and thus allows relatively high heat to be applied.

TEM is unsuitable for deburring thick burrs on thin workpiece features. It is also not suited for all materials. For example, steel alloy workpieces can be effectively processed only when hardness is less than R_c 40 . The intense temperatures involved in TEM can change the molecular or metallurgical structure of some materials, causing the workpieces to not function as designed. Workpiece materials containing at least 0.1% titanium are susceptible to embrittlement effects.

Martensitic stainless steel alloys exposed to the carbon that results from the high processing temperatures and combustion effects of TEM will react adversely, forming carbides in grain boundaries and at the surface and eliminating the corrosion resistance of these materials. High nickel-chromium alloys, such as cobalt, waspaloy, Hastelloy®, and Haynes alloys, have low resistivity values and high thermal conduction factors, thus heat is poorly distributed, resulting in little or no deburring or edge radiusing (Kelley and Schwarz 1991).

Exposing steels to hydrogen gas may cause otherwise ductile materials to become brittle and accelerate crack growth rates. These characteristics, as well as intergranular cracking resulting from hydrogen and temper embrittlement, have been the subject of recent research (for example, see Sonego 1993).

15.4 EQUIPMENT

Machines for TEM are available with C-frames (Figure 15-5) or three posts to hold the water-cooled chamber halves together. A toggle mechanism is used for clamping and locking the chamber. Container cylinders for TEM machines are available with diameters of up to 18 in. (457 mm) and lengths of up to 36 in. (914 mm). Typically, machines have sound enclosures, seals designed for long life (5–30,000 detonations), regular maintenance requirements, and designs that operate in conformance with local, state, and federal regulations.

A dial-type indexing table permits unloading and reloading a lower enclosure (bottom half of the chamber) while workpieces in another closure are being processed. The number of closure stations on the table depends on the size of workpieces and production requirements.

Gas pressures are adjustable to provide different levels of energy and accommodate different gases and workpiece materials. Some machines employ two work cylinders; this setup eliminates or reduces slag or oxide beads associated with some iron materials. Typically, the first shot is run at fuel rich and lower pressure, and the second shot operates at a stoichiometric mix and higher pressure. Low-pressure machines are generally used for nonferrous alloys.

15.4.1 Workholding Methods

Several methods are used to hold workpieces in the chamber. Large and heavy workpieces are generally placed directly on the closure plates. Small workpieces that can be handled in bulk are loaded into pots or baskets that are placed on the closure plates. A tightly wound coil spring is used as an inexpensive holding device for many applications.

Certain workpieces must be inserted into perforated plates, mounted on pins, or restrained in a fixture. Restraining fixtures are used for workpieces with external threads (Figure 15-6) or sealing surfaces that must remain nick and scratch-free, and for thin die castings that have little ribbing support to prevent distortion.

The deburring action on some workpieces must be confined to specific areas while other areas remain protected. Masking and/or placing heat sinks in strategic locations can increase processing capabilities. As a rule, the burr being removed should be smaller than the workpiece feature on which it is located. Heat sinks can be used to provide protection and prevent potential damage to workpiece features that are the same size (or nearly) as the burr.

Fixtures are fabricated from various alloy steels. For durability, stainless steel is the most commonly used material.

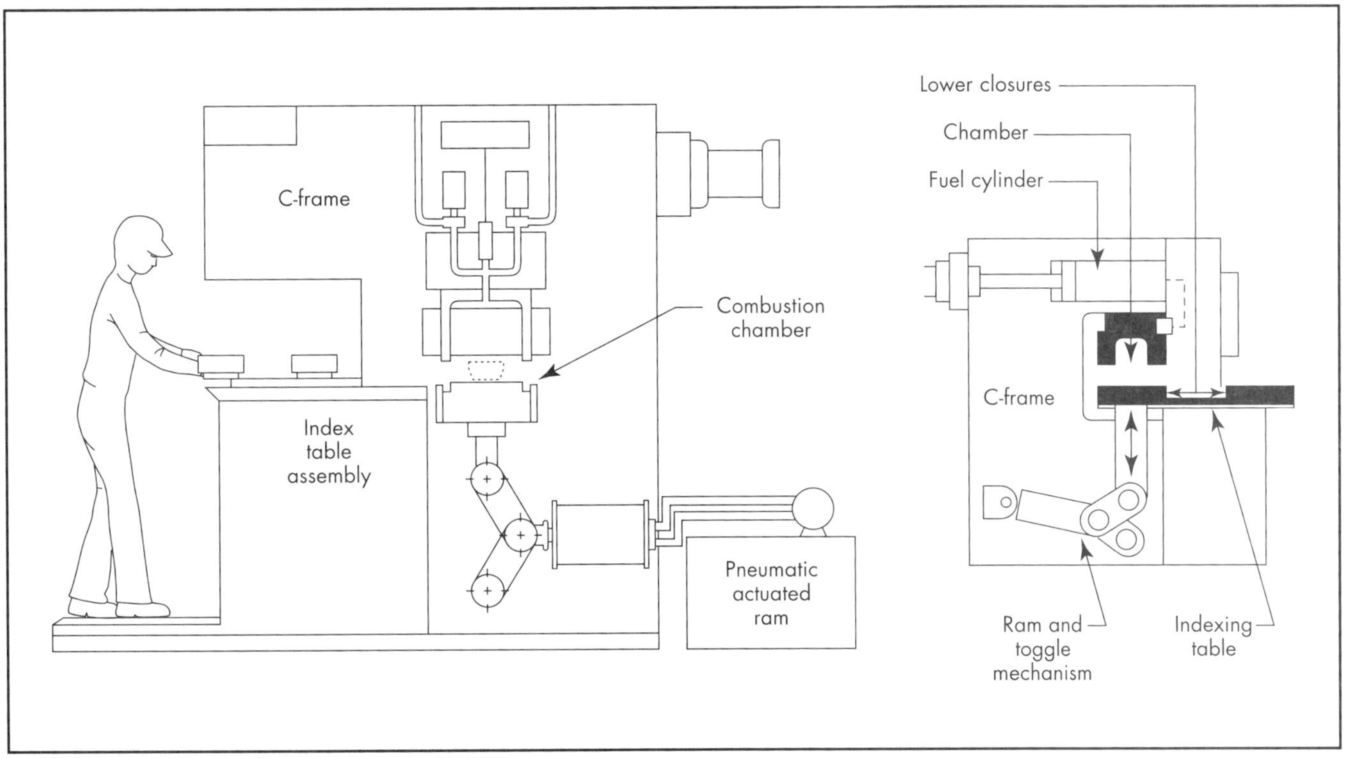

Figure 15-5. TEM machine with C-frame. ***(Courtesy SurfTran)***

However, Inconel® or other high nickel/chromium-based alloys provide excellent wear resistance against continual high temperature combustion and increase fixture life.

Figure 15-6 depicts a workpiece-holding fixture that has two functions. First, it contains workpieces, preventing them from flying about the chamber during combustion of the gases and protecting them from being dinged and damaged. Second, it elevates the workpieces in the chamber. Varying the height of the center post allows different deburring results to be achieved. A higher elevation may prove beneficial on one workpiece configuration, whereas a differently shaped or configured workpiece may be optimized at a middle or lower elevation. Varying the height is just one method of influencing the end result. By combining variations in height with changes in pressures or gas ratios, an unacceptable deburring result can be turned into an acceptable one.

Figure 15-7 shows a fixture that may be used for improving or changing deburring results on portions of a workpiece. The area to be deburred may be a fine or small hole or perhaps an excessively large or difficult burr. By "channeling" the transient heat wave, the deburring action is intensified. Although empirical evidence indicates that channeling gases improves results, at this writing there is no rule or scientific law supporting this. Deburring results may be affected by varying the diameter of the channel hole, height, distance between channel plate and workpiece, workpiece configuration, and gas pressures and ratios.

Figure 15-8 shows tooling that was created to suit a particular workpiece configuration that originally was difficult to deburr to the customer's satisfaction. Diverters such as this often result in improving TEM performance. It should be clear from Figures 15-6, 15-7, and 15-8 that the effectiveness of TEM can be improved by designing and employing imaginative and innovative tooling.

15.4.2 Gas Supply

Gas is supplied to a compressor that fills the fuel cylinder. Oxygen is generally supplied to its charging cylinder from pressurized bottles (Figure 15-1). The amount of gas charged into the cylinder is controlled by a system consisting of a series of spacers that can be added or removed to change the strokes of the cylinders. Gages are provided to monitor the fill and chamber pressures. Pressures in the chambers can be adjusted; normally they range from 50–300 psi (345–2,068 kPa). Lower pressures are used for low-energy ignition, for example, when removing small burrs; higher pressures are used for high-energy ignition, for example, when removing large burrs or burrs in hard-to-reach areas.

15.5 ENVIRONMENTAL, HEALTH, AND SAFETY CONCERNS

TEM machines have been in use for many years. Safety issues generally involve pressurized vessels, hydraulic clamping press features, explosive gases, noise levels, and hot or warm workpieces.

Safety is an important concern when working with gases in a combustion environment. Pneumatic pumps and mechanical clamps are used to toggle a device that clamps and seals the chamber to keep the ignition in the chamber. Toggle mechanisms, chamber seals, valve regulation, and metering devices should be regularly maintained to ensure proper operation. Additional safety features, such as detection devices for gas leaks, air exhaust flow, and water flow, are available to enhance the standard safety features of the TEM machine.

For increased operator safety, hearing protection is recommended. An option is to soundproof the pneumatic pumping mechanism. By varying gas mixtures and increasing

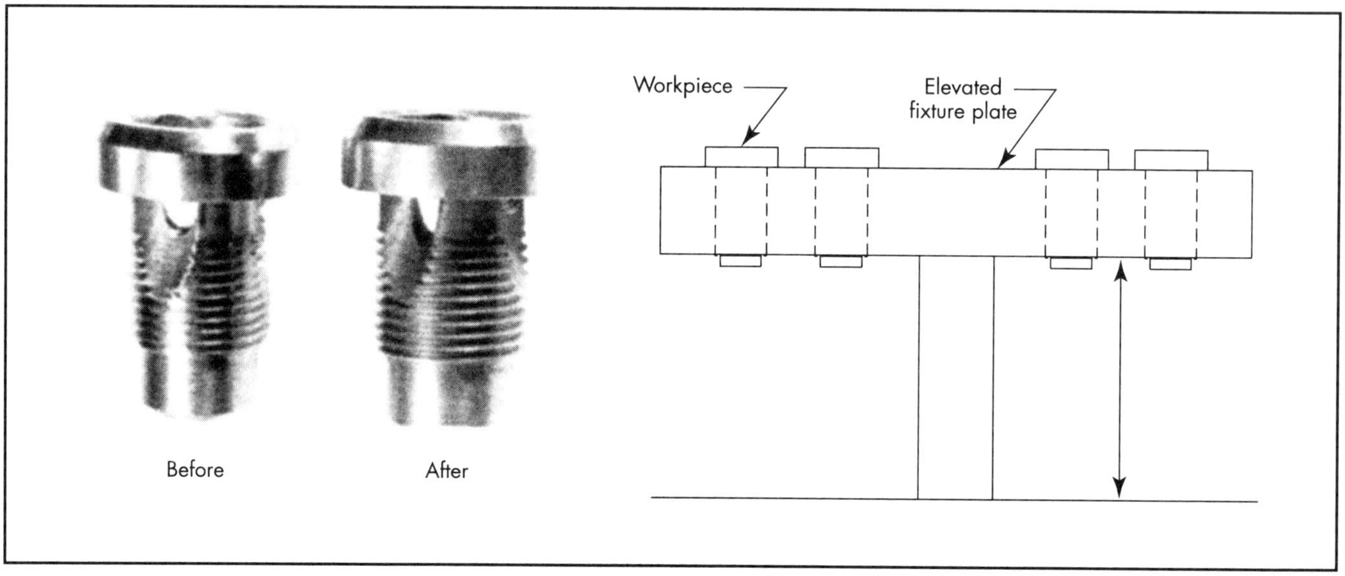

Figure 15-6. Fixtured part before and after TEM and schematic of workholding fixture. ***(Courtesy SurfTran)***

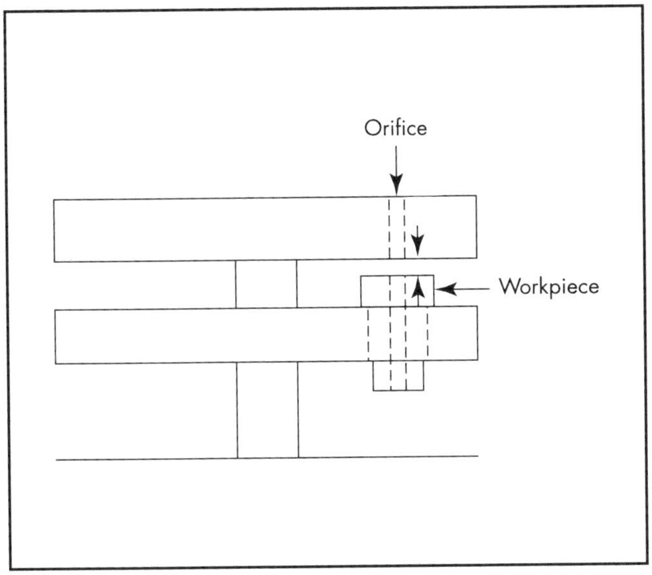

Figure 15-7. Tooling concept for directing or channeling detonated gases. ***(Courtesy SurfTran)***

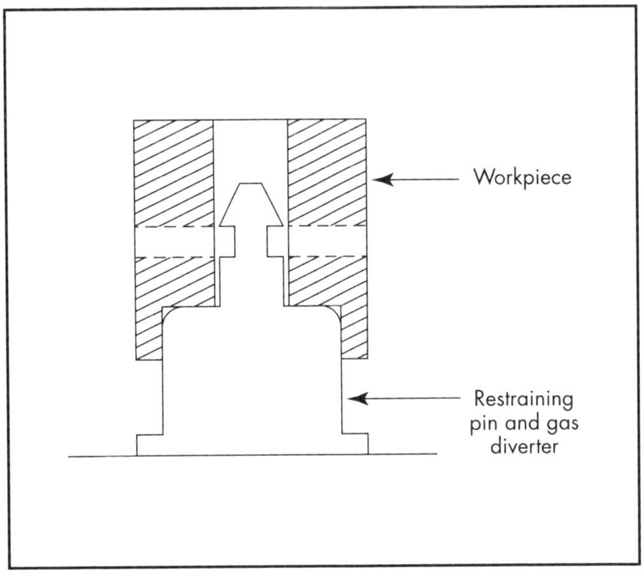

Figure 15-8. Restraining pin approach used for directing appropriate energy to burrs. ***(Courtesy SurfTran)***

chamber pressures, the sound level is similar to that of a .22 caliber shot. Typical TEM systems operate in excess of 87 decibels; but by incorporating sound enclosures to reduce hydraulic and machine noise, sound levels of 65–77 decibels can be achieved.

Cleaning solutions may irritate the skin of some operators; protective garments can be worn to avoid this problem. Most operators wear a smock or apron to keep the oxide and deposited material from smearing on clothes. Exhaust systems are required to protect operators from the by-products of TEM, such as hot gases, oxides, unburnt gas, and steam. Some materials combusted in TEM may contain hazardous materials; for example, nickel is a suspected carcinogen, and stainless steel and nickel workpieces may generate airborne nickel. Refer to local regulations for any restraints on emissions. The total volume of combustion products and exhaust is relatively low.

References

Kelley, Donald G., and Schwarz, Klaus H. 1991. "Thermal Energy Deburring." Technical paper MR91-136. Dearborn, MI: Society of Manufacturing Engineers (SME).

Sonego, Ronald. 1993. "Advances in TEM Technology." Technical paper MR93-322. Dearborn, MI: Society of Manufacturing Engineers (SME).

Bibliography

Beier, Hans-Michael. 1990. "Industrielles Entgraten." Berlin, Germany: Verlag Technik GmbH.

"Deflashing Plastic Implants." *Manufacturing Engineering*, 94 (April):23.

Koster, David L. 1985. "Improving Quality and Reducing Costs with TEM Deburring." Technical paper MR85-476. Dearborn, MI: Society of Manufacturing Engineers (SME).

Przyklenk, Klaus, and Schlatter, M. 1987. *Entgraten von Werkstucken aus Aluminum*. Dusseldorf, Germany: Aluminum-Verlag: 84–89.

Rhoades, Lawrence J. 1981. *Cost Guide for Automatic Finishing Processes*. Dearborn, MI: Society of Manufacturing Engineers (SME).

Schäfer, Friedrich. 1975. *Entgraten: theorie, Ver fahren, Anlagen*. Mainz, Germany: Krausskopf-Verlag.

16 Mechanized Cutting and Robotic Deburring

Several manufacturers produce standard off-the-shelf machines for removing burrs with cutting tools. Some companies produce machines designed for a specific part using cutters to remove the burrs. Specialized machines are being used extensively to remove burrs and finish parts when production rates are high or families of similar parts are processed. Various machines available are manually, seimautomatically, or automatically operated, depending upon production needs.

In addition to cutting tools, some of these machines can employ grinding wheels, brushes, buffs and honing stones. Typical parts processed on such machines include stampings, castings, forgings, and extrusions; sheet and plate stock; and many different types of workpieces, tubes and bars, gears, and cutting tool inserts.

While most of the machines described in this chapter use cutting tools, the approaches used are broad-based and, as noted in the following, several different types of removal mechanisms are involved. The following standard mechanical (cutting) deburring machines are discussed in this chapter:

- Workpiece edgers.
- End-finishing machines.
- Gear deburring machines.
- Edge honers.
- Sheet-metal trimming presses.
- Reciprocating files.

In addition to standard metal deburring machines, the following machine types also are used for deburring and are discussed here:

- NC/CNC machining centers.
- Manual metal-cutting machines.
- Robots.

The use of hand-held deburring devices and motors is described in Chapter 17, and equipment specifically dedicated to sheet metal parts is described in Chapter 18.

The decision to use automated machines, such as robots and other high-technology equipment, involves careful study and consideration before making an investment. (For more detailed information, see Gillespie 1987; McCoy and Hsie 1984).

16.1 WORKPIECE EDGERS

Two basic approaches are used for general-purpose workpiece edging. The first approach involves placing rectangular workpieces in an angle-iron trough and hand feeding them over a small grinding wheel, providing a small, adjustable chamfer. The second approach involves using a plane or piloted chamfering tool that extends upward through the flat table of a machine (Figure 16-1a). Workpieces are moved into contact with the pilot on the cutter and then fed across the cutter, using the pilot as a guide or stop. Chamfer depth is controlled by adjusting the cutter upward or downward. A permanent stop (Figure 16-1b) or a magnetic device can be installed and used with a standard nonpiloted chamfering tool. Contoured edges are deburred with the piloted cutter and straight edges are chamfered with the adjustable stop. Figure 16-2 illustrates this approach in use.

16.2 END-FINISHING MACHINES

Equipment for finishing the ends of tubes and bars is available with various levels of sophistication and automation. A manually operated machine capable of finishing up to 1,500 ends per hour is shown in Figure 16-3. Simultaneous inside-outside deburring, chamfering, and facing are possible on this type of machine. The operator places the workpiece through the self-centering chuck jaws, against an adjustable stop, and pulls the starting lever. This action closes the chuck jaws, pivots the stop clear, and feeds the work to the rotating cutters. As shown in the upper portion of Figure 16-3, one cutter scrapes the outer edge, one scrapes the inner edge, and one scrapes against the flat end to remove any burrs raised by the other cutters. Figure 16-4 illustrates an end-finishing machine in use.

On some end-finishing machines, the rotating cutter is mounted on the same spindle as an abrasive cutoff saw. In operation, the workpiece is sawn to length and then fed into the cutters. Double-end machines are available to deburr, chamfer, and face both ends of tubes, rods, or pipes at rates of up to 2,000 workpieces (4,000 ends) per hour. Some machines are integrated into transfer lines, using magazine or cradle loading, unloading, and storage devices.

16.3 SINGLE-PURPOSE MACHINES

Single-purpose machines are available in a wide variety of types for deburring, brushing, grinding, polishing, and buffing. They are designed for manual, semi-automatic, or automatic operation. Many machines are designed to handle specific workpieces, but different workpieces can be accommodated by changing workholding fixtures or finishing heads. A number of single-purpose machines are discussed in other chapters in this handbook that deal specifically with brushing, polishing, and buffing.

One type of single-purpose machine uses grinding wheels to deflash four different faces of cast engine blocks. A number of machines have been built to deflash, grind flat surfaces, and polish and buff hand tools to specific

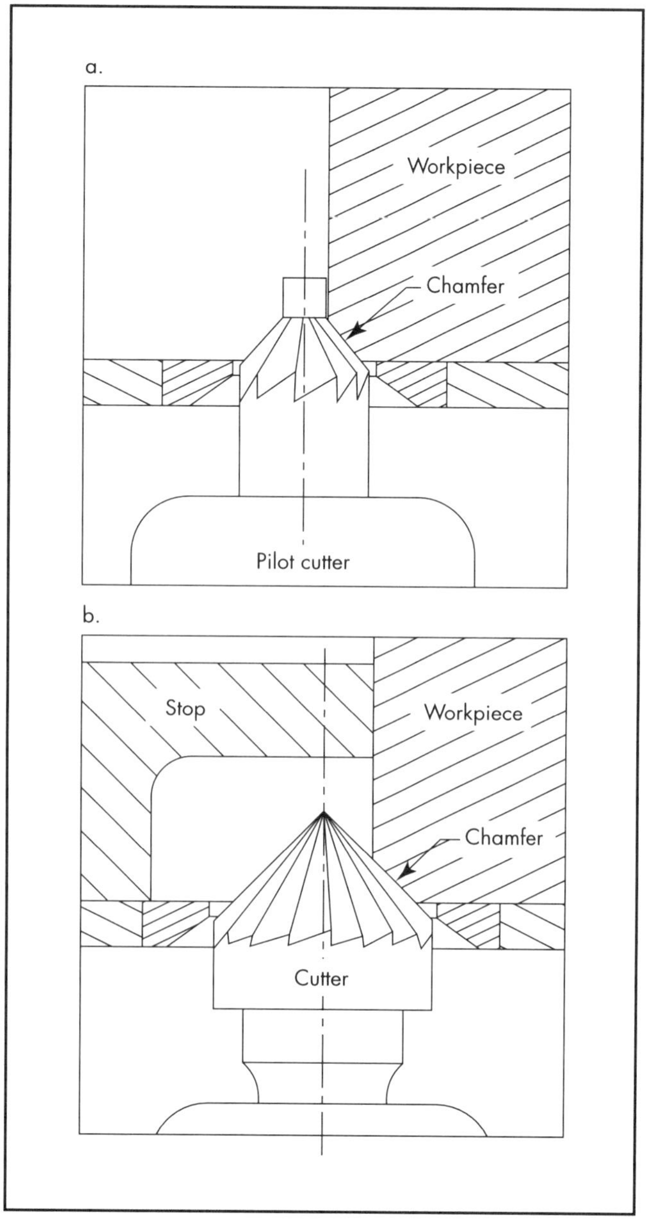

Figure 16-1. Two concepts for locating edges on a general-purpose workpiece edger: (a) piloted chamfering tool, and (b) permanent stop. ***(Courtesy Senn Company/Reishauer Corporation)***

requirements. Such machines frequently employ rotary or straight-line automatic transfer of the workpieces, with up to 30 stations per machine. Multistation machines have production rates of up to 3,000 workpieces per hour. Some machines rotate workpieces both clockwise and counter-clockwise to reach recessed corner areas.

Special contouring machines are employed for deburring and finishing compressor and turbine blades and other workpieces with complex contours. In one design, the deburring head has three-dimensional motion, fully controlled by three plate cams mounted on the same shaft. The head

Figure 16-2. Operator chamfering a chuck jaw using a piloted countersink tool. ***(Courtesy Senn Company/Reishauer Corporation)***

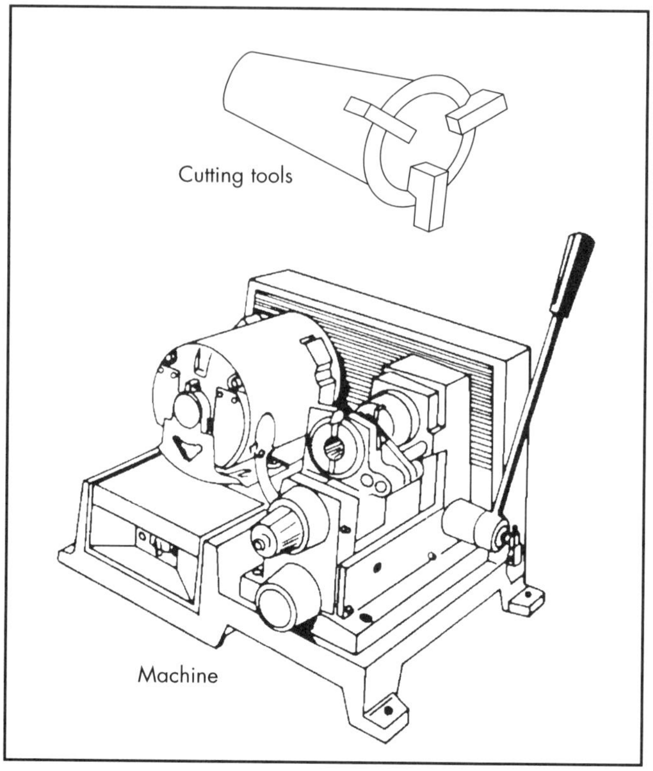

Figure 16-3. Tube- and rod-end finishing machine. ***(Courtesy Pines Manufacturing, Inc.)***

can be equipped with a high-speed spindle for handling rotary burs or milling cutters. A fourth motion can be incorporated to maintain the cutting edge of a slowly rotating, single-point tool perpendicular to the workpiece surface. Rotational speed of the camshaft can be varied to change the cycle time, usually 5–18 s per workpiece, depending on the complexity of the operation.

Figure 16-4. Tube-end deburring machine in use. ***(Courtesy Pines Manufacturing, Inc.)***

The automotive industry particularly has incorporated single-purpose deburring machines into dedicated equipment and transfer lines.

16.4 GEAR-DEBURRING MACHINES

In gear hobbing or shaping, burrs form on the end faces of the gear teeth. Such burrs can seriously impair smooth and reliable gear operation. A wide variety of methods are used to deburr and chamfer gear teeth, including using a rotating tool or small grinding wheel to traverse all tooth edges (Figure 16-5), brushes, a skiving tool and subsequent brushing, an abrasive belt grinder (sometimes followed by brushing), and blasting with steel shot.

Heavy burrs are removed by combining a milling operation with tooth chamfering (Figure 16-6). Many gears, particularly those used by the watch industry, are being deburred on gear hobbing machines, thus eliminating a secondary operation. Small pinions are deburred by automatically holding a lathe tool against the hob exit side of the workpiece, with the tool shearing the burrs. Some machines retract the burr-cutting tools several rotations prior to the end of the hobbing cycle. This action allows the hob to remove the remaining burrs between the gear teeth produced by the tool. Rotating milling cutters are used on larger gears.

Several specially designed machines are available for deburring, chamfering, or rounding gear teeth. One type uses a rotary deburring tool (a driving gear and two cutters) with the workpiece driven by the tool. The workpiece and tool are in a parallel-axis relationship, and the tool automatically swings into engagement with the clamped workpiece. Stock is removed along the edges of the workpiece teeth by overlapping cuts of the edges on the rotary

Figure 16-5. Small grinding wheel used for chamfering gear teeth. ***(Courtesy Grat-O-Mat)***

Figure 16-6. Heavy burrs are removed by milling and then chamfering. ***(Courtesy Grat-O-Mat)***

tool; simultaneously, stationary tools engage lateral surfaces of the workpiece to remove any projecting burrs. When the cycle is completed, the rotary tool swings back, and the workpiece is released for unloading. Clusters of up to four gears of different configurations can be chamfered and deburred simultaneously on these machines.

A cam-operated machine for deburring, chamfering, or rounding teeth can handle spur and helical gears as well as splines, both internal and external. The horizontal cutter spindle of this machine is advanced in a timed relationship with the rotation of the gear teeth for continuous cutting. The cutter spindle is mounted in an eccentric housing that can be offset to either side of the work spindle for operations on helical gears. One machine chamfers both ends of external gear teeth simultaneously at a rate of up to 400 teeth per minute.

16.5 GEAR-BURNISHING MACHINES

Machines are available for removing nicks and burrs from the tooth surfaces of gears. A centerless-type machine uses three hardened or abrasive-impregnated burnishing gears to finish the workpiece with adjustable burnishing pressure. Gears are automatically fed into the workstation where they are engaged by the burnishing gears. The workpiece gear is constrained by pads supported by a series of leaf springs that are driven by an eccentric shaft. This motion causes the workpiece gear to oscillate as the burnishing gears revolve. Production rates are as high as 900 gears per hour.

16.6 MACHINES FOR EDGE HONING CUTTING-TOOL INSERTS

Machines are available for deburring and honing the edges of carbide or ceramic inserts for cutting tools. These types of machines are also employed for deburring flat and cylindrical components and for producing sharp corners with no burrs.

On one such machine, the inserts are finished between two rotating, resilient, abrasive disks. The disks consist of silicon carbide abrasive in a rubber-like bond with a pressure-sensitive adhesive backing for applying them to the wheels. Each insert is held and guided in a separate opening of a carrier plate (Figure 16-7), which is part of an epicyclic sprocket holder. The machine has three independent drives, one each for the upper and lower wheels and one for the carrier plate (Indge 1991).

The required pressure is applied to the top wheel through an adjustable pneumatic device. Pressures and processing times vary, depending on the size and shape of the inserts, the materials from which they are made, and the required finish. Typical pressures are 42 psi (289.6 kPa). Depending on requirements, insert edges can be rounded on one or both sides. For one side, the top disk is replaced with a pressure pad having a wear-resistant surface. The form of the honed edge is a multiradius compound curve with an average radius of about .0017 in. (0.043 mm).

Figure 16-7. Carrier plates for edge-rounding inserts.

Figure 16-8 illustrates the elastic disk deformed over the workpiece edge. The carrier is approximately .80 in. (20.3 mm) thinner than the workpieces and the pockets in the carrier that contain the parts have about .03 in. (0.8 mm) clearance. Parts must be spaced far enough apart to allow "mushing" over edges (Figure 16-9). Burrs can be removed from cylindrical parts in the same manner (Figure 16-10).

16.7 TRIMMING PRESSES

Mechanical, hydraulic, and pneumatic presses are used extensively to remove excess material from stampings and drawn workpieces and to remove flash from forgings, die castings, and moldings. This method requires the use of cutting dies, such as pinch-trim, Brehm® (shimmy), or notching dies. Shaving dies or broach tools are employed for more precise requirements.

Die-casting trim presses require less force (tonnage rating) than for pressworking steel. In many cases, a four-corner approach is used to guide the ram, although some C-frame, open-gap, or arch units are available. Some employ a rapid-feed platen that automatically slows as the tool contacts the workpieces. Machine designs range from simple models to fully automated systems with a capacity of 100 tons (90,720 kg) of trimming force. Mechanical presses provide 90–300 strokes per minute. Part output is about 600 workpieces per hour for most machines, but if four workpieces are held together by gates and runners, then up to 2,400 workpieces can be processed per hour. Trim accuracies of .005 in. (127 μm) are possible. Several presses can be equipped with rotary load-unload tables for rapid handling of small workpieces. Small bench-top presses are also available for trimming workpieces that require 4 tons (3,629 kg) of force or less. Kick presses are used for low production runs on medium-sized workpieces. Broaches are used when conventional trim dies are not adequate.

Some plastic workpieces require heated dies and punches for trimming to prevent part delamination or tearing.

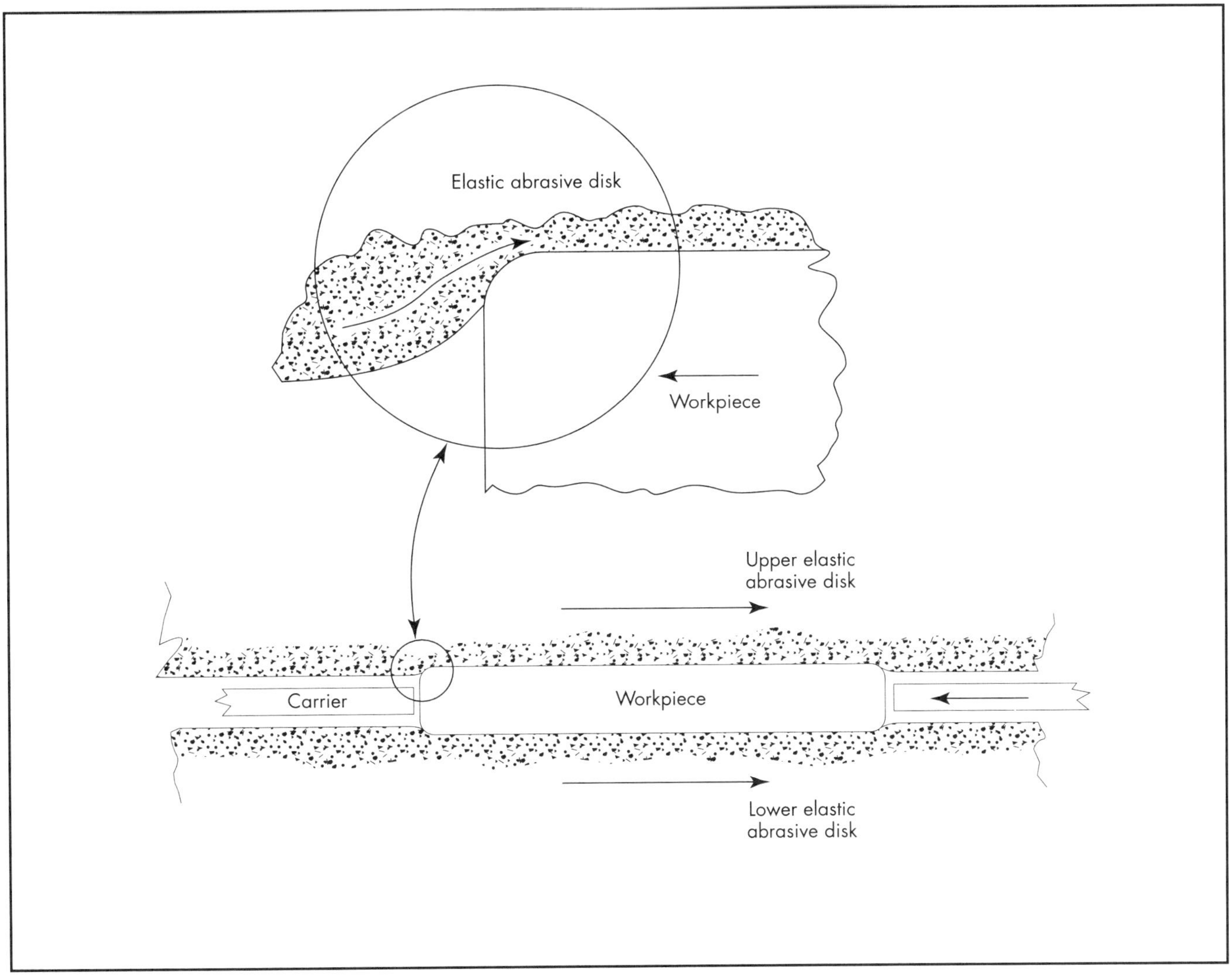

Figure 16-8. Cross section showing elastic disk "mushing" over workpiece edges.

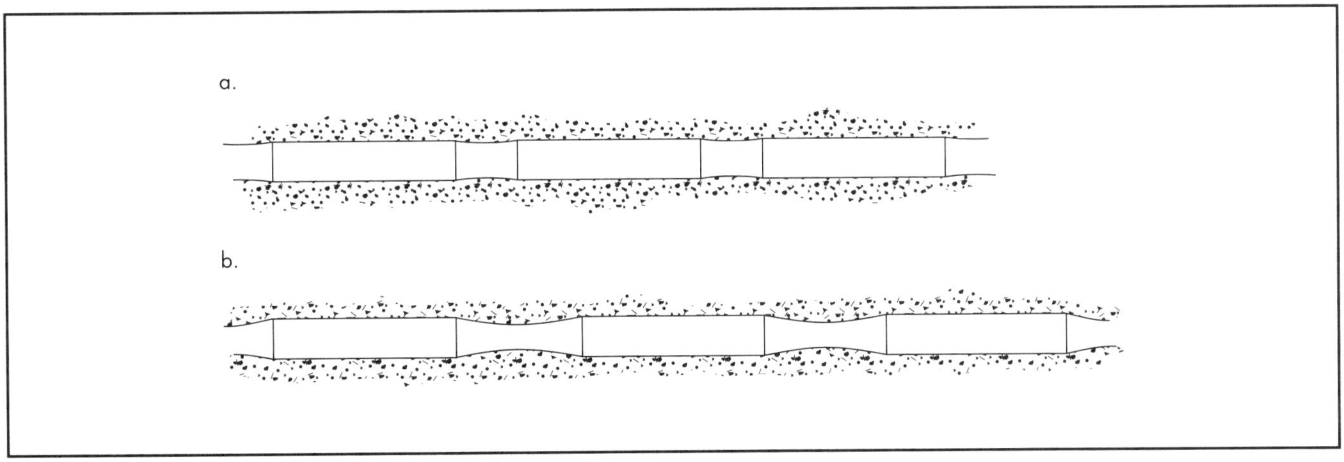

Figure 16-9. Effect of part spacing on "mushing" of elastic disk. (a) Flat parts are too close together to allow elastic abrasive to produce an adequate radius; (b) flat parts at proper spacing to allow elastic abrasive to return to its original surface before engaging another workpiece.

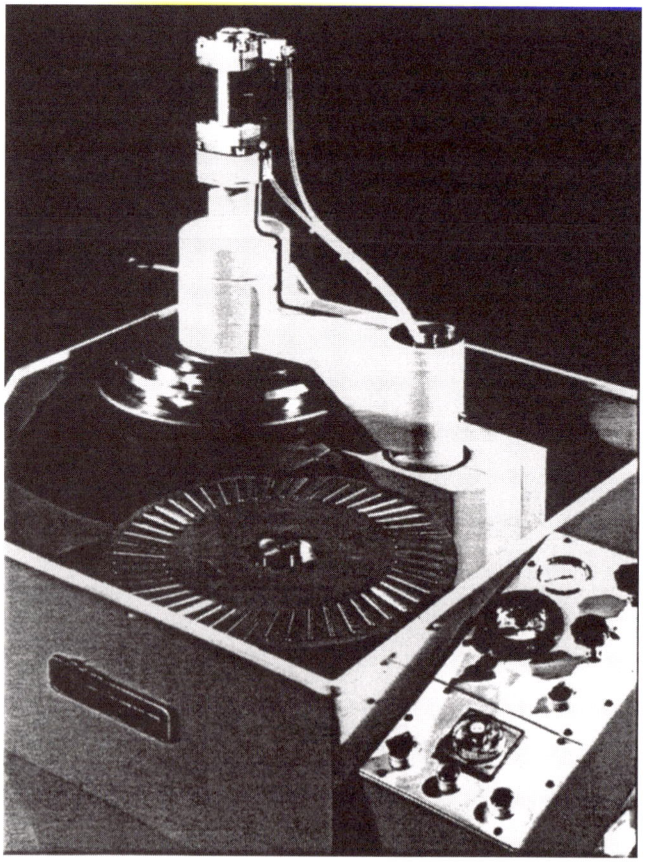

Figure 16-10. Cylindrical parts are deburred and edge finished with elastic abrasive-bonded disk.

16.8 NC/CNC MACHINES

NC/CNC machines that produce workpieces can often be effectively used for deburring or finishing the workpieces or many features of them. These machines ensure the accuracy of features produced in the machining operations and can result in considerable savings by eliminating the need for separate finishing operations. However, they may not produce satisfactory results on cast or forged surfaces because of more liberal tolerances. When wall locations vary from workpiece to workpiece, edges may be produced with too much or too little chamfering or rounding.

Deburring and finishing tools can often be stored in the magazines of NC/CNC machining centers that have automatic tool-changing capabilities. Macros are available for some CNC units to remove burrs by automatically chamfering.

Several CNC machines are specifically designed for removing sprues, risers, gates, and runners from castings. In the casting industry, this process is called fettling and cleaning. Rigid, single-tool machines are used, and touch sensors are often employed to define surface orientations and then rotate the workpiece to provide a straight-line path for the cutter. CNC machines are used on simple contours for non-precision edge requirements.

Multi-axis machining centers have been used for years for deburring and chamfering holes as part of the machining program. Lathes can routinely generate a radius on every turned-edge workpiece when programmed to do so, whereas simply facing and turning the workpiece will leave a turning burr. Edge rounding on lathes takes longer, but it prevents secondary deburring effects and may be the most economical method to use, depending on thc application. Many operators routinely hold a sandpaper-covered broom handle or paint stirrer to the edge of the workpiece while it is being rotated by the lathe; this is a simple deburring method, but it has safety hazards. Multi-axis lathe machining centers provide live spindles in which small brushes as well as milling cutters can be used. Although some areas of the workpiece may not be totally deburred to final requirements, major improvements can be made to part quality and consistency using this approach.

Examples of deburring integrated with machining include the following. Many barstock and tube cutoff machines come packaged with deburring spindles. (Gear deburring is another arena in which deburring operations have been integrated in a single machine for decades.) At least one CNC jet, rotor-blade grinder can remove burrs produced by its own grinding. Although there are few standard machines that come equipped with deburring spindles, in theory they can be used on many machines. CNC machining centers can brush workpieces by simply putting a brush and arbor in a toolholder. When appropriate, the machine changes tooling and begins a brushing path instead of a cutting path. Brushes are low cost and are available in hundreds of combinations of aggressiveness, brush material, configuration, and size. CNC machines are currently using miniature cross hole brushes as well as large aggressive brushes.

On-machine deburring has one undervalued benefit. It can take advantage of over 10,000 standard cutting tools and use them in machines for which they were designed, thus providing great flexibility. When considering NC/CNC machines for precision deburring, the following factors must be considered:

- Rotary burs are designed to be used at high speeds and low forces. Many of the spindles on NC/CNC machines will not approach the recommended cutting speeds for these tools.
- When deburring and machining are performed on the same setup, the operator has better control and visibility, thus enabling appropriate tool changing when needed. If the tools cause too large a burr, the burrs will not be removed by chamfering.
- Abrasive stones or filled rubber tools generate a large number of fine abrasive particles that can become lodged in collets, holding fixtures, and machine ways.
- The rubber in some abrasive-filled rubber products reacts with coolants, producing sticky machine-way surfaces.
- Thin flanges may distort during machining to the extent that chamfering tools may not even touch the

edges. This can occur when flanges are less than .030 in. (0.762 mm) thick and the maximum chamfers required are .003 in. (0.076 mm).

A five-axis, high-speed machining center has been developed to replace manual deburring. In contrast to CNC machining centers that can also do deburring, this machine is a CNC deburring machine that performs machining as well. With spindle speeds of 15,000–75,000 r/min, it is well-suited for use with small rotary burs and files.

Even manually operated machines can provide effective deburring. Brush manufacturers report that their products are being used on vertical mills, horizontal jig borers, and manual mills. Most companies have one or more mills or machining centers that languish in the corner, and imaginative processing can add extra brushing heads to these manual or CNC machines to remove some or all of the burrs in the final phase of the machining cycle. The advantages of using underutilized simple machine tools are repeatability in edge quality, reduced handling, shorter queue time, reduced floor space required to store large workpieces while waiting for deburring, and the reduced cost of fixtures for positioning workpieces for precise deburring.

The prevention of repetitive motion injuries to workers is also a source of major cost savings from using mechanized equipment. Lost time from work-related injuries, workers' loss of personal manual efficiency resulting from such injuries, and the high costs of workers compensation and insurance can be mitigated by employing well-thought-out approaches to deburring. In some of today's large companies, the real costs of manual deburring at low wages versus machine deburring with higher wages is not much different given all the factors involved.

Figure 16-11 illustrates a simple sanding disk tool that provides a constant force to an edge. It is used on both robotic and CNC machines. Companies have reported that this device on a CNC machine will reduce deburring time by up to 95% (Graf 1990) while improving part quality. It is designed to use nonwoven deburring disks with a twist-to-lock holding system. It can be used with the disk held flat to the surface, or it can be tipped up on edge to remove larger burrs or generate larger radii. When holding the disk flat to the surface, each edge is hit from a number of directions, providing uniform deburring and radiusing. On aluminum, radii of .006–.012 in. (0.15–0.31 mm) can be generated while removing less than .00001 in. (0.0003 mm) of material from the surface. These devices are available in force ranges of .3–6.5 lb (0.1–3.0 kg). Figure 11-19 shows a brush in a CNC toolholder. In this instance, constant force or compliance merely involves programming the brush tips to be .020–.250 in. (0.51–6.35 mm) below the top surface of the part. The bristles provide their own compliance. This allows simple programming to reach below edges and wipe over them.

One innovation involving machining centers allows the tool spindle to pick up a movable water jet nozzle and traverse it around the machined workpiece edges. Water jet blasts can be focused for deburring edges right on the

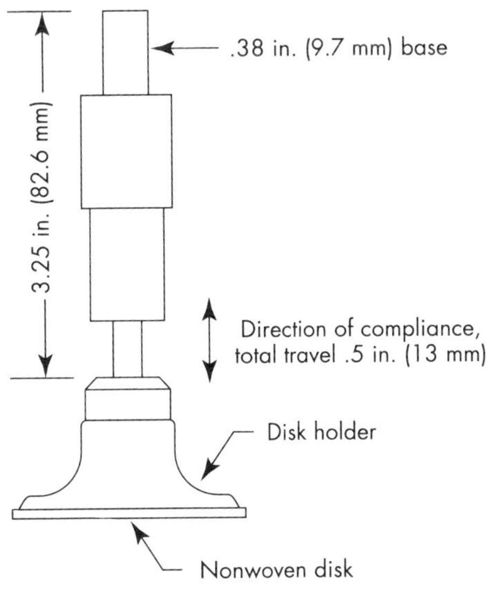

Figure 16-11. Constant-force device for disk sanding on CNC and robotic machines. ***(Courtesy 3M)***

machine, or to save time they can be spread in a fan shape. The process is particularly applicable to those workpieces requiring reasonable but not complete burr removal.

Robot technology brings with it tools applicable to CNC centers, and many of these tools are just entering the production field. As machine tool controllers increase capability, force control, data acquisition, better compliant tooling, vision, burr inspection, and other aspects of robotics will be available to common machine tools and thus allow even more effective deburring on the machining center.

16.9 MANUAL DEBURRING

Many companies dedicate a manually operated machine to cutting or brushing off burrs. Flash is generally removed from holes with deflashing or trimming dies, but for short runs or low production applications, flash is removed by drilling, reaming, and manually punching.

Workpieces produced on a lathe are often deburred with a file or abrasive paper while still on the lathe. Some manually operated machines have been especially adapted to accommodate unique combinations of live spindles and cutters to finish edges.

There are a few special compliant cutting tools that can be used in manually operated machines, on NC machines, or in robots. Among these are a rubber-backed countersink that "gives" somewhat when a heavy cut suddenly occurs. Some users also consider back-chamfering tools, clothespin deburring tools, and similar force-limited (spring-loaded) opening-closing tools to be compliant (Figure 16-12), since the exact location of the tool is not critical for most deburring. A tool known as an "ellipti-burr" is used to deburr some holes machined at angles into other holes. These elliptical intersections are not readily machined by most tools.

16.10 ROBOTIC DEBURRING, FETTLING, AND FINISHING

For many companies, the use of industrial robots for deburring, *fettling* (removal of gates, runners, and sprues from castings), and finishing has considerably reduced costs and improved quality. Robots can operate three shifts a day, virtually untended; they reproduce the same motions accurately; they can process workpieces faster than humans; they can use heavier, higher-powered tools for faster finishing; and they can work in dusty, noisy environments that are unsuitable, hazardous, or ergonomically difficult for humans. Robotic deburring, fettling, and finishing are not actually processes in themselves, but are simply automation added to mechanical cutting, grinding, brushing, abrasive blasting, and other operations.

Robotic systems are typically used for workpieces that have production rates falling between job shop quantities and automotive industry requirements. Robotic methods are not normally economical for one-time runs of very low quantities because of the amount of engineering and programming time required. Unlike many off-hand processes, the use of robotic systems requires considerable planning to ensure optimum results.

16.10.1 Typical Applications

Industrial robots are being used for many different deburring and finishing operations. Most applications involve large workpieces and are dedicated to a specific workpiece or a closely related geometric family of workpieces. Although some precise and intricate workpieces are being deburred using robots, these applications are the exception rather than the rule, because robots generally are not yet capable of performing precision deburring.

Robot applications fall into three general areas: (1) simple-shape deburring and chamfering, (2) contouring, and (3) sensor-controlled contouring. Using robotics for processing simple shapes requires little planning. For example, one company uses a robot to deburr the ends of aluminum and stainless steel cut tubing. The robot picks up a tube, pushes one end into an end-finishing machine (such as the one shown in Figure 16-3), turns the tube around, and puts the other end into the machine (Grace 1984). This is an excellent example of simple robotic deburring—it requires no sensors or path programming, produces consistent results, and is cost effective. The developers of the process in this particular company thoroughly understood the application before they began the process of converting to robotic approaches. Simple brushing routines, as opposed to using cutters that require more accurate positioning, are often the most suitable choices for robotic, CNC, and other automated deburring machines.

The more widely used applications employ sophisticated robots that use five-axis compliant tools and remove most, but not all, of the burrs. These applications generally allow generous edge chamfers. Some workpieces may be acceptable as they come off the machine, whereas others may require some final hand finishing.

The following applications have been run on robot systems: connecting rods, automotive trim, truck rims, stamped metal workpieces, track-drive sprockets, golf club heads, disk-drive housings, air-tool handles, transmission hous-

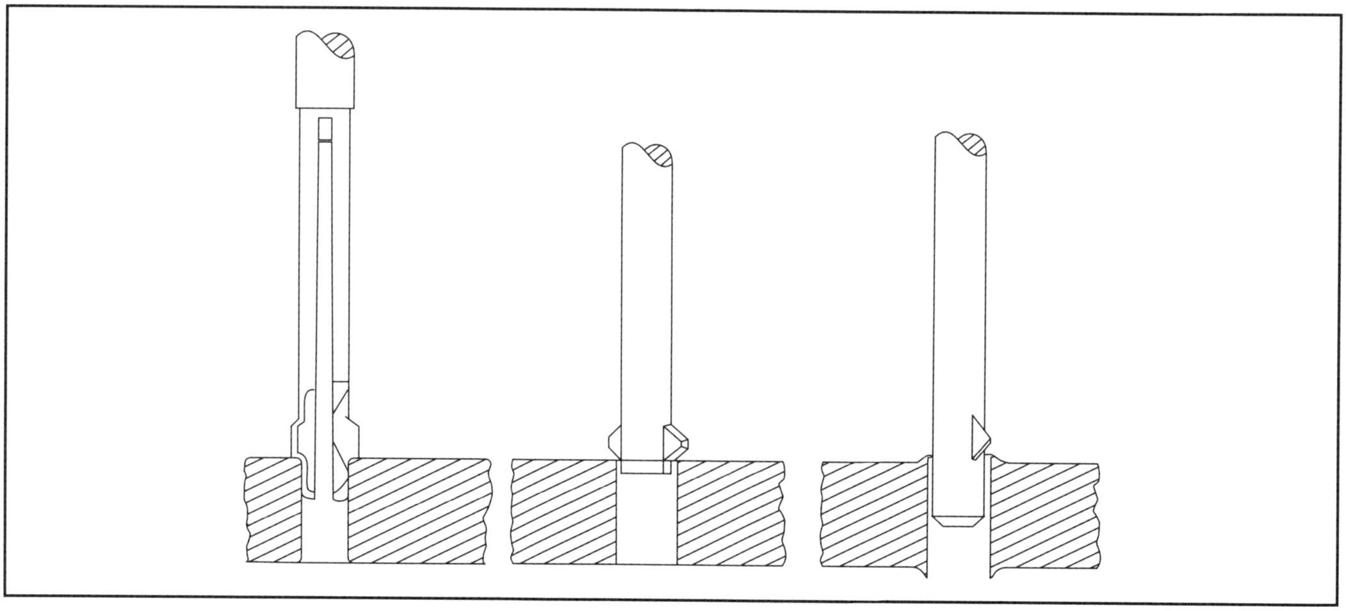

Figure 16-12. Spring-loaded tools used on manual, NC, and robotic machines for deburring holes. ***(Courtesy ABB)***

ings, helicopter rotor hubs, jet engine casings, missile bodies, crankshafts, transmission gears, engine blades, rear tail-light assemblies, car windshield edges, composite rocket tank shells, and car doors.

The foundry industry is a major user of robots. Robots can handle the castings or the tools, depending on the size, weight, and configuration of the casting; on the material from which the casting is made; on the amount of stock to be removed; on the type of operation; and on production requirements. Operations performed include deburring and chamfering internal edges and removing sprues, flash, risers, gates, parting lines, and runners (Schraft et al. 1983).

Because fettling involves cutting through thick sections of cast metal and smoothing wide sections of flash, large forces must be employed and therefore high-powered tools are required, necessitating powerful robots. Whereas a human may use a 4 hp (3 kW) power tool in fettling, robots can handle 45 hp (34 kW) tools weighing up to 50 lb (23 kg) and employing cutting forces of 100 lbf (445 N) or more. In contrast, deburring operators will typically use .33 hp (0.25 kW) tools that involve forces of 1–10 lbf (4.4–44.5 N). Other robotic operations performed in many industries include brushing, polishing, buffing, and grinding. In the automotive industry, parts being deburred and finished with robots include transmission and steering knuckle housings, connecting rods, and plastic moldings. Oil holes in crankshafts are being deburred with robot-held tools. Laser-drilled holes, which have resolidified molten-metal burrs on their edges, are deburred by wet abrasive blasting (the blasting nozzle is attached to the arm of the robot).

16.10.2 Robot Requirements

The following characteristics are essential for successful deburring, fettling, and finishing operations that use robotic systems: a high degree of accuracy and repeatability, continuous-path capability, easy and quick tool and spindle changes, rigidity, low inertia, easy off-line programming, circular interpolation, and the ability to translate movements to similar features on the same workpiece. Other desirable features are a large-reach envelope and a small wrist-arm configuration, no backlash, and calibration. Load capacity of the robot is generally specified as three times the expected working force. Five-axis capability is generally the minimum requirement; six axes are necessary when nonrotational, nonsymmetrical tools will be used.

The specifications and costs of a robotic system can vary greatly, depending on the accuracy requirements for a specific application; the closer the required tolerances, the greater the cost of the system. For applications involving workpieces allowing wide tolerances, lower cost robots with less accuracy but good repeatability often can be used. Such applications may use compliance tooling or fixturing.

Some robots may be limited by their design, and these limitations are often subtle. For example, some radial or polar-motion robots cannot move in a straight line, although at least one radial-motion robot has a wrist that does allow it to move in a straight line. However, attempts to approximate straight lines require large amounts of computer memory. Vertical travel is also limited in some robots.

16.10.3 Robot Accuracy

Some manufacturers of industrial robots fail to specify the accuracy of their products. Many robots cannot move to a predefined point or location, such as $x = 1$ mm, $y = 2$ mm, and $z = 0.00$ mm. Because of this limitation, the traditional NC machine concept of programming to a point and then measuring how close the machine came to that point is not directly transferable to robots. Even when robot controls allow this predefinition of destination, precision robots may miss their targets. This limitation is not acceptable for deburring workpieces when tolerance requirements are ±.005 in. (±0.13 mm). Although compliance may overcome some accuracy limitations, having a robot with the capacity for a high degree of accuracy is desirable. Note that fettling operations require less accuracy than deburring.

Some robot manufacturers specify "accuracy to a point in space." In this case, the programmer can move to a point in space by using visual judgment or sensory devices that identify the point. That point, however, is not associated within the machine as a dimension; rather, it is identified as a resolver count. An emerging robot technology that employs visual or electrical signals is able to achieve accuracies of .002 in. (0.05 mm) to a "point" in space. Accuracies to edge contour (continuous-path rather than isolated points) typically are ±.025 in. (±0.64 mm). Some continuous-path machines use hydraulics rather than servos, losing the inherent accuracies of servo resolver systems. Electric robots currently available provide continuous-path capability, high accuracies, and eliminate the noise and other undesirable aspects of hydraulic units.

The accuracy of an industrial robot depends on the following factors:

- Accuracy and positioning repeatability of the robot mechanism.
- Ability to consistently position the edges of the workpiece to a reference surface (clamping forces can significantly deflect some workpieces).
- Nature of the edge before deburring (burr uniformity, thickness, and location consistency).
- Ability to consistently position the deburring tool in the robot end-effector or gripper.
- Repeatability and accuracy of the cutting tool geometry. Each tool can be ground to a different geometry; there is a natural variation.

16.10.4 Control Methods

Some precision robots are point-to-point machines, with no capacity for circular interpolation, which effectively prohibits their use for deburring. For example, if a machined casting has 100 in. (2,540 mm) of edge to deburr, and points are identified every .003 in. (0.08 mm) to ensure that a .010 in. (0.25 mm) maximum break requirement is not exceeded, then more than 30,000 positions have to be stored in the robot's computer memory. This requirement may exceed the memory capacity of some older robots. For complex workpieces, the only workable solution is to select a continuous-path robot.

Continuous-path Robots

Continuous-path robots are generally programmed by the operator, who physically grasps the robot arm, or a similar teaching arm, and leads it through the desired path in the exact manner and velocity desired for the robot's motions. During this programming, the position of each axis is recorded into the robot's memory on a continuous-time basis, thus generating a continuous-time history of each axis position. The sampling frequency is typically in the range of 60–80 Hz.

Controlled-path Robots

Controlled-path robots are operator-programmed to manipulate the robot arm to the desired tool center-point location and tool orientation. Instead of moving each axis individually, computer control allows coordination of all robot axes when "teaching" the program. The position of the robot's arm is recorded in the controller's memory at each desired program-point location. During program replay, the robot's computer generates a mathematically definable path between the program points; it also controls the velocity along the path and acceleration-deceleration to and from the start-stop points.

With the controlled-path motion control, several modes of generating the robot's path are possible, depending on the type of motion-interpolation employed. The types of interpolation are joint, linear, and curvilinear. Robots are available with one, two, or three types of interpolation systems in a single unit. The type of system depends on the specific requirements of the application. Joint interpolation results in controlled-path motions in which the computer calculates the individual robot axes motions so that each arrives at the program-point location at the same time. Linear interpolation results in straight-line motion of the robot's tool center point. Curvilinear interpolation results in tool center-point motions that follow a desired curved path.

On-line Programming

Many robots use a teaching pendant to program path moves. The operator stands near the robot and uses a joy stick or directional buttons to cause the robot to traverse the path in a learning mode. The learning mode involves trying, correcting, trying, correcting, and so on. It is a convenient approach for simple workpieces because the results, and the needed corrections, are immediately apparent. Because of robot inaccuracies and servo lag, making final adjustments to the program through on-line programming is essential in many situations.

On-line programming has the following limitations:

- Production is halted until the program is developed (however, for continuous production of the same part, the downtime incurred for teaching may not be significant).
- Making precision adjustments may be difficult because the point of contact between the tool and the workpiece is masked by the deburring motor and cutter.
- The operator must be near the robot arm, which poses a safety hazard if the arm should suddenly fail.
- Obtaining precision edge results on complex workpieces may take considerable time.

Off-line Programming

When cutter paths generated for NC machining exist, *off-line programming*, that is, the insertion of a separately written program into a robot, has several advantages. The robot program can, at least theoretically, be generated quickly, and functions without having to face the problems of an operator seeing the cutter-workpiece interface, which may be masked by the tool and the workpiece. The deburring path can be generated and debugged without stopping production. The operator does not have to be in proximity to the robot arm for as long a time as for on-line programming; this increases safety. In addition, computer-generated programs can produce smoother and more precise transitions around complex geometries. Some off-line programming systems require the development of databases of tooling, tool-change frequency, tool types, and path macros (DeMotte 1990).

16.10.5 Approaches to Robotic Deburring

There are three different philosophical approaches to ensuring robot accuracy, obtaining precise results, and accommodating workpiece variations. These approaches are referred to as compliant, sensor-driven, and force feedback. A fourth approach, the fine-tuned robot approach, is described here but is not a practical way to perform most deburring. Basically, in this fourth approach, the tool deflects when it touches the part, so it does not have to be accurately located.

Compliant Approach

The *compliant approach* can be used with any robot, but for deburring, finishing, and grinding operations, robots with continuous-path capability are generally used. Five- or six-axis robots may be required.

Although every robot has some inherent compliance, the compliant approach adds extra compliance in one or more directions by using specially designed toolholders. Normally, single-axis compliance is necessary for achieving smooth edges, and sometimes two-direction axes compliance is necessary. For some buffing applications, three-directional axes compliance is advisable.

If designed properly and programmed correctly, and if proper cutting tools are used, the compliant approach produces consistent edge breaks of .010 in. (0.25 mm) or less, even though the locations of edges may vary. Since a consistent volume of material is removed per unit of time and the burr volume is normally much less than the volume of the parent material, consistent edge breaks can be maintained even if the relative burr sizes vary slightly.

A major difference between NC machining centers and robots is the variance in their dynamic rigidity. The tool needs to move around a corner quickly, but it must move so that the surface finish on the workpiece meets requirements without producing chatter marks. Compliance is often needed in the direction perpendicular to the part surface, but it must be absolutely rigid in the feed direction to prevent chatter and skipping. As shown in Figure 16-13, a robot is "blind" when the force measurements alone must determine whether any change in force is the result of a change in burr thickness, contour variation, or encountering an unexpected surface.

Figure 16-14 shows a simple sanding disk, nonwoven wheel, or brush compliant tool. This device is designed for applications where the robot handles the part. The sanding disk is enclosed to keep out dust and protect the inner components. To provide accurate force control and position feedback, any type of drive motor and deburring-finishing product can be attached to the slide table. Figure 16-15 shows a force-limiting or compliant device. This device is floor standing and contains the force control, drive motor, and position feedback. The direction of force control is vertical, but the head can be rotated 180° on the angled section of the base to provide horizontal force control.

Rigidity is the key to the success of a robot. If the tool can twist in any direction, it will climb, push away, dig into the surface, or chatter, and it will not traverse the corner as accurately as desired. An ideal robot would have servos quick enough to compensate for the system, but ideal robots do not exist.

Sensor-driven Approach

Sensors detect edges and tell the robot to adjust its actions. Numerous sensors have been developed for robots to detect edges, burr sizes, and deburring quality; and many sensors are still in development. Sensors have been designed to detect torque or motor current (Matsunaga et al. 1985), spindle speed, vision, triangulated vision, structured lighting (Loucks, Selleck, and Thunborg 1989), laser and light-emit-

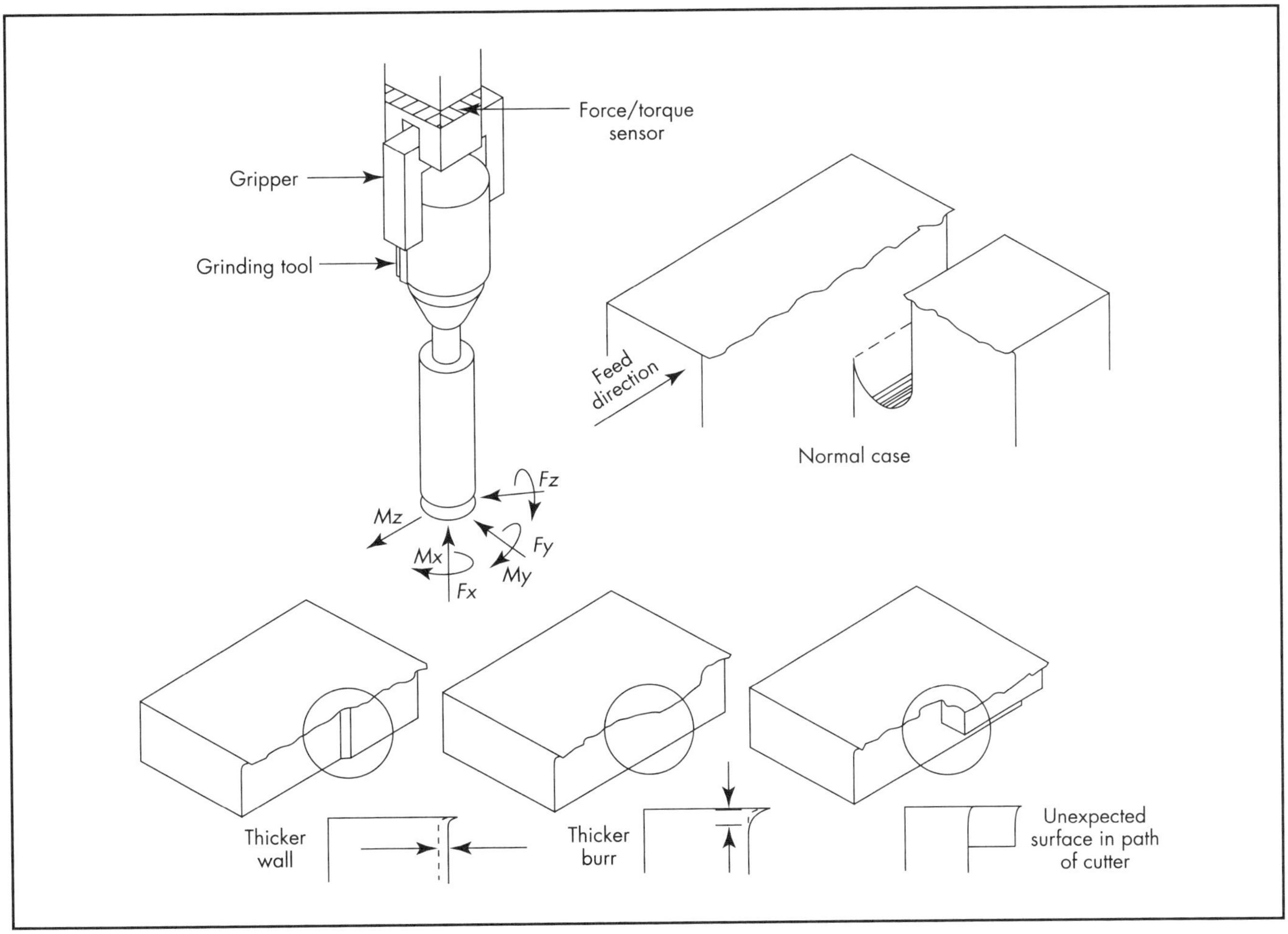

Figure 16-13. Conditions requiring compliance or force sensing.

ting diode (LED) reflection (Nicoletti 1988), acoustic emission (Dornfeld and Erickson 1989), color, and force feedback.

Force-feedback Approach

The force feedback approach for most deburring applications involves using a five- or six-axis robot with continuous-path capability. The six-axis, force-sensing unit interacts with a computer to tell the robot what to do next. For example, when a heavy burr is detected, the computer may tell the robot to slow the feed rate to prevent undesired tool and arm deflection. If necessary, the tool passes over an edge several times to achieve the desired force profile. A number of programming approaches are used, but most follow some portion of the following force control concepts (Gillespie 1987; Loucks, Selleck, and Thunborg 1989).

Compliance of the robot is a significant factor in defining control logic. When the robot deflects as a result of compliance, the system never knows the real cutting or interference force. Compliance may vary as a function of wrist position in space.

In maneuvering the robot arm on any complex workpiece, the reaction moments are not organized along the robot axis in any particular fashion. Determining what is going on at the cutting tool when a force sensor is placed on the wrist is a difficult problem. On complex workpieces, the sensing system may need another independent feedback system to close the loop around the force sensor compliance limitations. At fast feeds, the controller may have to accept data at megabaud rates to adequately provide the control for precision and complex workpieces.

Force sensing is being used in some simple, straight-line systems for fettling operations. These systems work well as long as the burrs, flash, and projections have the same dimensions at every location on the workpieces. Fettling operations become more complicated if the burrs are not in a straight line or if their dimensions vary. In such cases, the

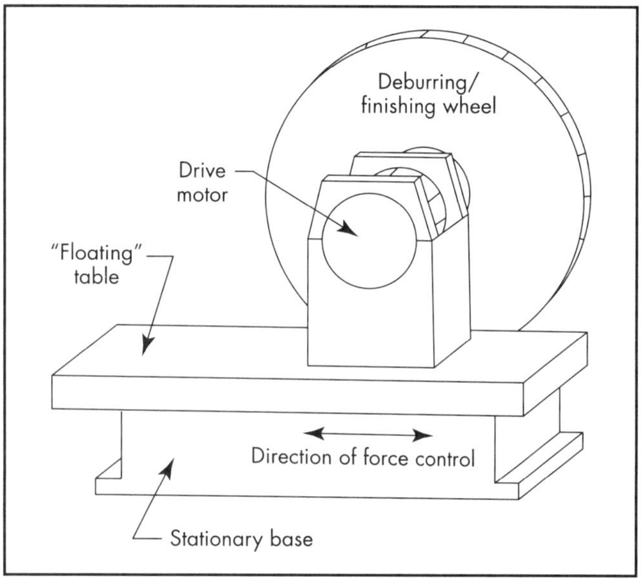

Figure 16-14. Bench-mounted, constant-force device for use when the robot moves the workpiece to the wheel. ***(Courtesy 3M)***

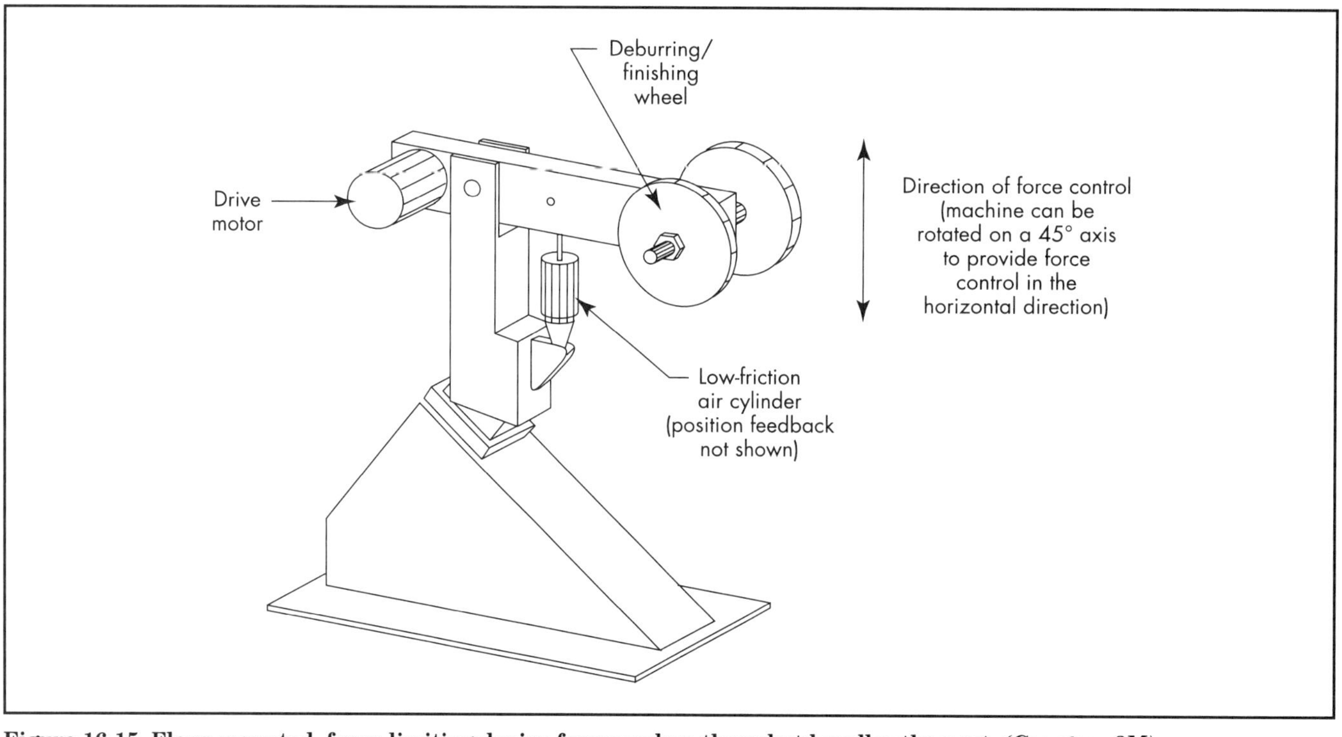

Figure 16-15. Floor-mounted, force-limiting device for use when the robot handles the part. ***(Courtesy 3M)***

reaction force of the robot cannot be used to shift the program in the z-axis, since it is not known whether the change in reaction force is caused by changes in burr dimensions or the contour of the workpiece.

Some encoders are located behind the robotic arm transmission train. As a result of losses through the train, the system cannot account for tool weight or process-induced deflections. A deburring force of 1 lbf (4.45 N), for example, fed through transducers, is overridden by system noise when the system also must be able to accommodate large cutting forces.

For a given workpiece material and set of operating conditions, the cutting force F is defined by the following equation:

$$F = C \times B_w \times B_h \times R_f \quad (16\text{-}1)$$

where

C = constant
B_w = burr width, in. (mm)
B_h = burr height, in. (mm)
R_f = feed rate, in./min (mm/min)

Thus, if the robot detects an increase in cutting force, then either the width or the height of the burr increased. Because the burr volume increased, the cutting force increases, which causes greater robot deflections. If the robot keeps feeding the cutter at the same rate, the cutter will be deflected more from the part surfaces, thus leaving an area of incompletely removed material. By slowing the feed rate, the deflecting force can be maintained constant. As a result, the finished edge or surface should not have small islands of incompletely removed material. If work-piece geometries vary significantly, then simple control logic is no longer sufficient. Increased forces may be the result of contacting a wall sooner than expected. Dull or chipped tools are similarly not accounted for by simple control logic.

For simple fettling operations and a spring-force servo, the normal stiffness of an electric robot can be adjusted to reduce the contact force values as required by the application. By using a proportional speed controller, the forces on the arm will be proportional to the position error. By using outputs from the robot, the robot stiffness can be adjusted as required in the program. The spring-force servo can be used to compensate for small variations in wheel size and small tolerance variations in the castings.

Force-sensing systems also can be used with a search function to produce a transformation of the program. In this approach, the robot intermittently compensates for wear of a wheel; it detects the edges of fins and the surface of castings to compensate for wheel wear. The robot moves the wheel to a measuring hole and moves sideways in a search routine until a force results between it and the hole walls (see Figure 16-16). An offset is calculated that accounts for the wheel wear, and then the offset is fed into the robot to define the new normal centerline position. A simpler approach, however, is to automatically offset a dimension after a certain number of given workpieces. It is desirable to have a robot that has tool center point registers that change the offset automatically. In one approach, users probe an edge before deburring to determine actual edge conditions and then deburr and measure again (McCoy and Hsie 1984).

Researchers are investigating using dynamic controls to improve robot accuracy. These controls have the potential for significantly improving deburring capabilities. Software programs, originally developed to design robots, can calculate the deflection and compliance of the robot system. If the forces are known, corrections can be made. If tool changes occur during the cycle, the system must account for the difference in spindle or tool weights to ensure accuracy.

One unique and beneficial approach using force feedback employs a milling cutter hidden within a small cylinder. A window is cut open at one position such that a fixed amount of cutter diameter will project beyond the window (Loucks, Selleck, and Thunborg 1989). When the robot pushes this tool into the part, it can only cut a specific radial distance before the part hits the cylindrical guide.

An NC program made for one robot will not work as accurately in another robot of the same model because of robot manufacturing tolerance and control-system tuning differences. Robots designed for high levels of accuracy, without sensory system programs, must have their programs generated on the robot that will use the program. It is possible to acquire matched robots that are able to use the same tape, however, by specifying such needs on purchase orders or by having them built at the same time.

Fine-tuned Approach

By fine-tuning servos and modifying resolvers, it is possible to improve robot accuracy by a factor of two or three. This fine-tuning, however, slows the system slightly, and the servo may be pushed to an overload condition when the tool tries to go to an exact position. Theoretically, this approach could eliminate the need for compliance. In practice, however, the fine-tuned approach is not normally recommended because of workpiece variations, the need to minimize chatter, and the need for maximum reliability.

Every time mechanical maintenance is performed on a compliantless, fine-tuned robot, every point in every program must again be retaught, as is the case when accidents occur. Motions are defined from workpiece drawings (off-line programming). The fine-tuned approach requires a six-axis, continuous-path robot and depends on the small amount of inherent compliance.

16.10.6 Robot Capabilities

Robot capabilities must be understood before specific robots are selected for deburring. With the radial robot arm shown in Figure 16-17, whenever the end of the arm is moved in an x or y direction, the tool remains in the same angular orientation to the arm but changes significantly with respect to the workpiece. As a result, every movement

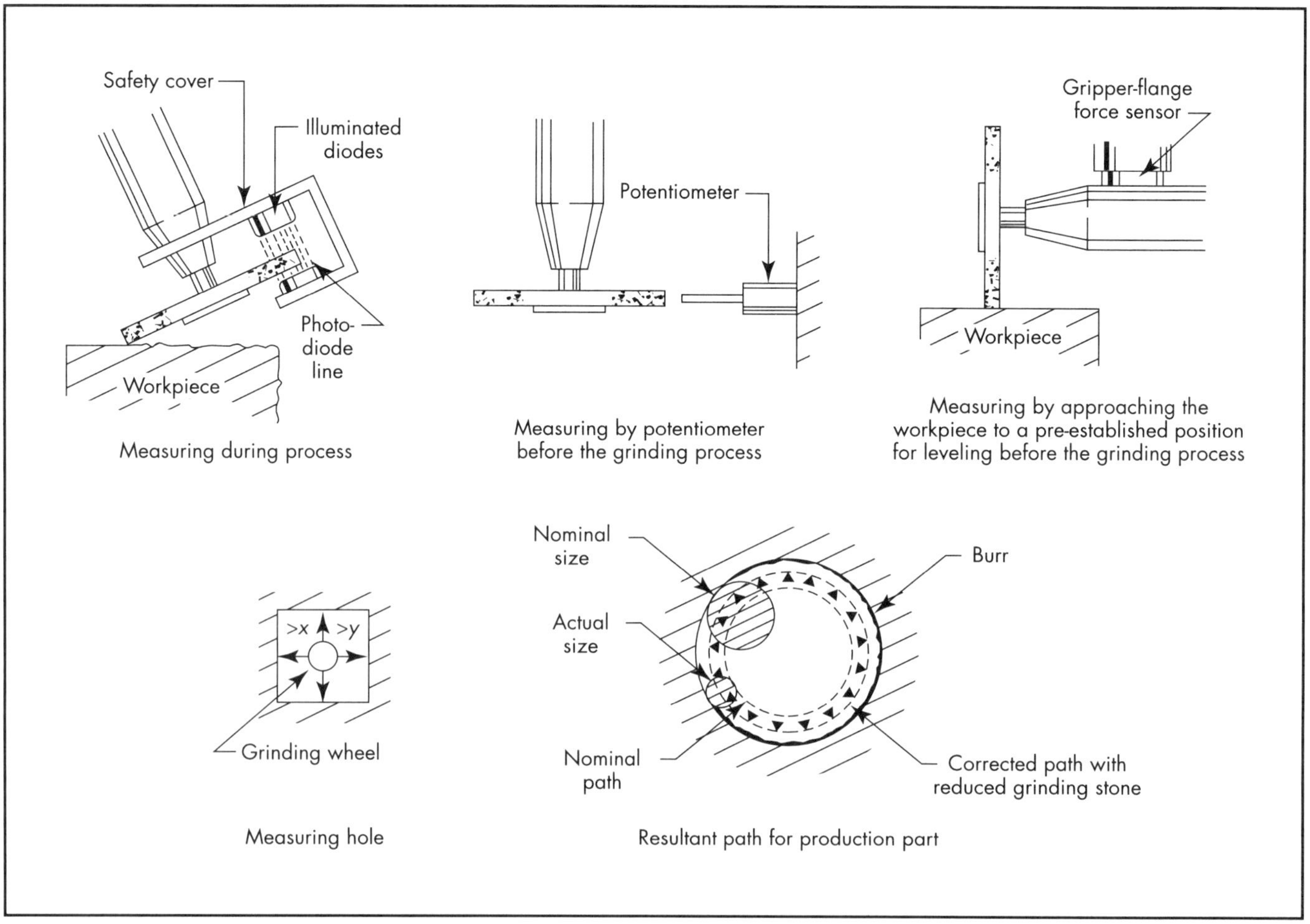

Figure 16-16. Measurement compensation for wear by moving grinding wheel inside a known hole size until wheel contacts hole walls. ***(Courtesy IPA)***

(other than a radial motion into the horizontal plane) requires a reorientation of the tool to provide perpendicularity to the workpiece edges.

Most deburring robots currently rely on an operator to manually "teach" the robot how to move. Such programming is the most convenient approach for many workpieces. For complex workpieces that already have an NC database, however, the ability to use off-line programming is desirable.

Off-line programming has the advantage of minimizing machine downtime for teaching. In theory, with a pre-existing CNC database, off-line programming should be faster and more accurate than manual teaching. Because current robots are not precise, however, off-line programming with existing robots typically requires some manual final adjustment.

In practice, deburring compliance requires offsetting the tool from the edge to be deburred (Figure 16-18) by as much as .24 in. (6.1 mm). The tool is normally presented at a 45° angle to the workpiece edges for contouring operations, as shown in Figure 16-15. This method has at least two

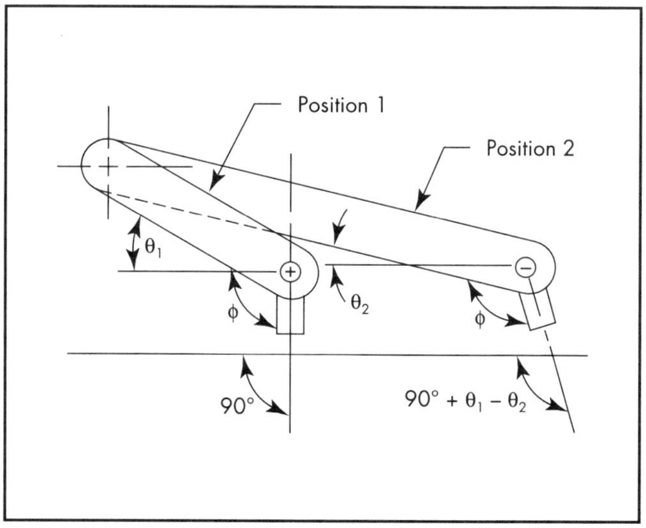

Figure 16-17. When the robot arm is moved from position 1 to position 2, the tool is no longer perpendicular to the surface of the workpiece.

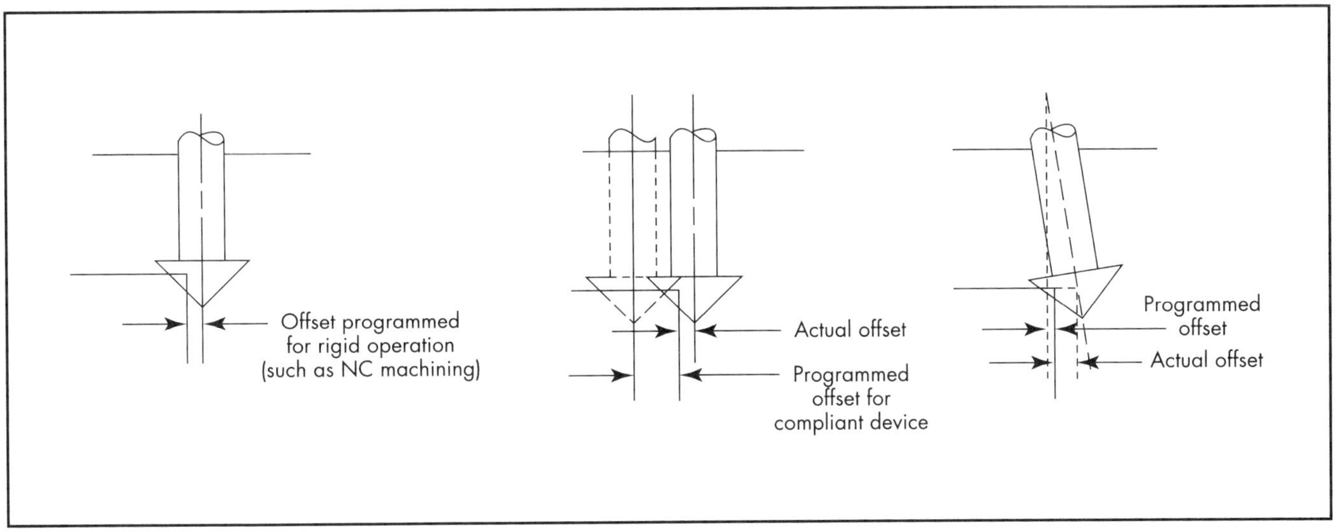

Figure 16-18. Offsets for deburring. ***(Courtesy ASEA Robotics)***

benefits. First, a cylindrical cutter can be used rather than a specially shaped cutter. Second, programming is easier because all rotation is performed around a single axis (such as "A" in Figure 16-19). This approach generally provides a smoother finish when the tool must pass through a fillet radius transition to another wall.

Offsets are also used in fettling, grinding, and finishing operations when wheels or abrasive belts are used. The offsets serve multiple purposes, such as compensating for wheel wear, permitting large variations in workpieces and locations, and reducing programming time. The degree of offset depends on the specific application and the compliance device used.

For tools that have rotational symmetry and are rigidly flange mounted to the robot, a five-axis capability usually allows any desired tool orientation within the working area of the robot. However, as soon as the working force on the tool is unidirectional (for example, when nonsymmetrical tools such as chisel hammers or grinding belts are used), six-axis capability is often required.

The usable working space of robots is far less than the volume that can be reached by the robot. The area swept by a robot's arm does not account for the fact that the robot cannot provide the desired tool orientation in all positions. In addition, many robots have singularity points within the working volume. When one attempts to drive them through these points, they lock up or perform undesirable movements. In addition, the presence of the workpiece itself forces the robot to crook its arm, wrist, and tool into difficult positions to reach all surfaces. The net result is that a robot that can sweep a volume of 10–20 ft^3 (0.28–0.57 m^3) may actually have a usable volume of only 2–9 ft^3 (0.06–0.25 m^3). Deburring requirements limit the amount of usable volume.

Edge chamfers of .005 in. (0.13 mm) can be produced in deburring operations, and contours of .010 in. (0.25 mm) are possible in some fettling applications.

Influence of Workpiece Geometry

Both the number of edges and their lineal lengths on flat workpieces affect deburring quality, cost, and programming difficulty. Deburring is difficult in shallow areas, such as counterbores and ledges with depths or heights of less than .080 in. (2.03 mm). The points of chamfering tools tend to scrape, resulting in tool marks that prevent the production of smooth chamfers. Also, the teeth of conical cutters wear rapidly near their points because of their small diameters.

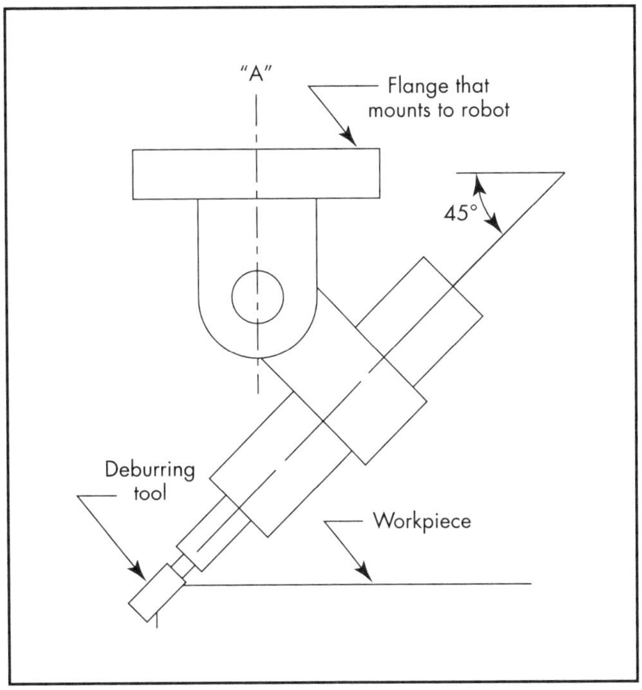

Figure 16-19. Deburring tool presented at 45° angle to edge of workpiece. ***(Courtesy ASEA Robotics)***

The liberal tolerances for most castings make robotic deburring difficult for these types of workpieces. Variations in cast wall locations are especially troublesome. Edges that flare out into adjacent areas cannot be totally finished with robots without the walls being gouged. Small, internal edge radii are difficult to deburr and require small tools and precision programming. Thin walls may distort under deburring forces. Holes close to walls may be impossible to finish because of tool motor, cutter, or spindle interference.

Burr Size

Burr sizes and locations determine the parameters of the deburring process and influence the edge breaks obtainable. Burrs vary considerably along each edge, among edges, and among different workpieces.

Deburring Standards and Requirements

Incomplete or improper specification of requirements can increase deburring costs, as discussed in Chapter 2, and may limit the use of robots. Developing and using consistent standards is recommended for deburring with robots. In general, the more stringent the deburring requirements, the less likely that robotic deburring can be easily and successfully implemented.

Cost Effectiveness

To determine whether robotic deburring is cost-effective for a particular application, it must be compared with other methods, including hand deburring, for both simple and complex workpieces. Savings generally result when robots are used because a robot is faster than a human and an operator is only required part of the time. However, the cost of programming the robot must be taken into consideration, and cutting tool or brushing costs may be higher because they receive more continuous use. Energy and space costs also can be significant.

Tool and Workpiece Handling

A robot can carry deburring tools to the workpiece or carry the workpiece to the deburring tools. Having the robot manipulate the deburring or fettling tools is the most frequently used procedure when workpieces are heavy or when only two or three deburring spindles are used. The robot manipulates workpieces when several spindles are required, when work-handling devices need to be minimized, when fixturing costs need to be kept low, or when workpieces are light, typically less than 25 lb (11 kg).

Tool handling. When the robot manipulates the deburring tool, workpieces must generally be brought to the robot. This procedure typically requires a rotary table and fixtures to orient workpieces. For dedicated applications, fixture costs are relatively low, but two to six sets of each fixture may be required to keep the robot operating at a constant rate. Figure 16-20 illustrates a robot deburring a gearbox case presented to the robot on a conveyor. Other applications involve placing workpieces on tombstone faces and using long string brushes to reach edges without actually programming to specific edges. The robot merely brushes to approximate positions (Koelsch 1991).

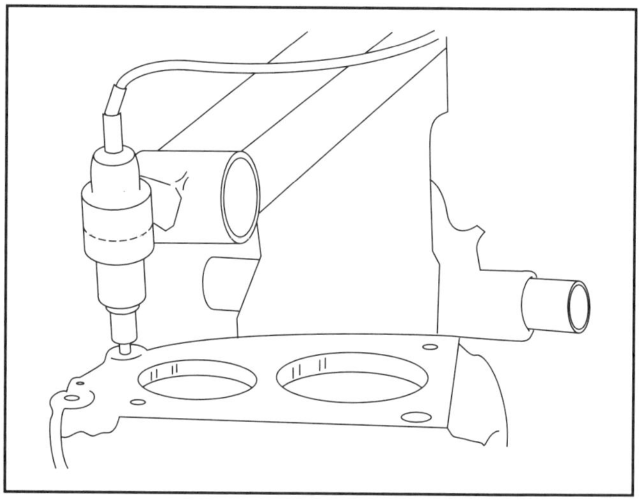

Figure 16-20. Robotic arm with live spindle for deburring gearbox case. ***(Courtesy ASEA Robotics)***

Tool-holding robots may be limited by having to handle power umbilical cords (unless power inputs can be routed through removable wrist connections). This limitation is normally only a concern when several spindles must be changed during the operation.

Deburring forces increase as the depth and width of cut increases. In practice, for a vibration-free chamfer of .020–.080 in. (0.51–2.03 mm), forces of 3–6 lbf (13.3–26.7 N) occur.

Workpiece handling. When workpieces are light, they can be lifted from a magazine or conveyor and moved by the robot between two or more fixed tools. They also can be manipulated in such a way that the contours are passed over the tools. At the end of the cycle, the workpieces are placed on a conveyor or are palletized, and the robot grasps another part. Cycle times of as low as 40 s can be achieved.

For universal applications, the robot can be surrounded by multiple pedestal-mounted grinders, files, rotary burrs, or brushes. The robot traverses the workpiece edges with the tools appropriate for the part. This approach provides flexibility for handling many part geometries with no increase in tool changing time.

With appropriate gripper design, the workpiece can be manipulated to expose most workpiece surfaces. If necessary, an intermediate regrip fixture can be used to allow access to all part edges. In contrast, when the robot holds the tools, the fixturing shields at least a portion of one part surface. With a family of similar part geometries, one gripper may accommodate many workpieces.

Tool considerations. Tool changing capability is an important consideration when using robots for deburring. The following three approaches are commonly used for tool changing:

- Use multiple motors on the end effector (at 90° to each other) to eliminate the need for changing tools (Figure 16-21).
- Change preset spindles.
- Change grippers and tools.

In many cases, the most appropriate method for tool changing is to exchange the motor that contains the tool for another motor that contains the appropriate cutting unit. Systems that sense when cutters are dull, clogged, broken, or worn excessively are desirable but not generally available. Consequently, tools must be changed at predetermined intervals.

When installing a robot, it is important that the robot not be limited by the speed at which the operator is able to load workpieces and tools or start and stop the machine. For maximum cost savings, workpieces must always be available to the robot to be finished once the robot is turned on. Therefore, the material-handling portion as well as the deburring portion of the operation must be automated. Otherwise, deburring time is limited by the speed at which the operator is able to provide workpieces to the robot.

16.10.7 Cutting and Deburring Tools

Most of the tools used for other deburring methods or mechanical cutting operations can be used for robotic deburring, such as files, rotary burs, countersinks, brushes, abrasive-filled rubber and cotton products, abrasive paper products, grinding disks and wheels, and mounted points.

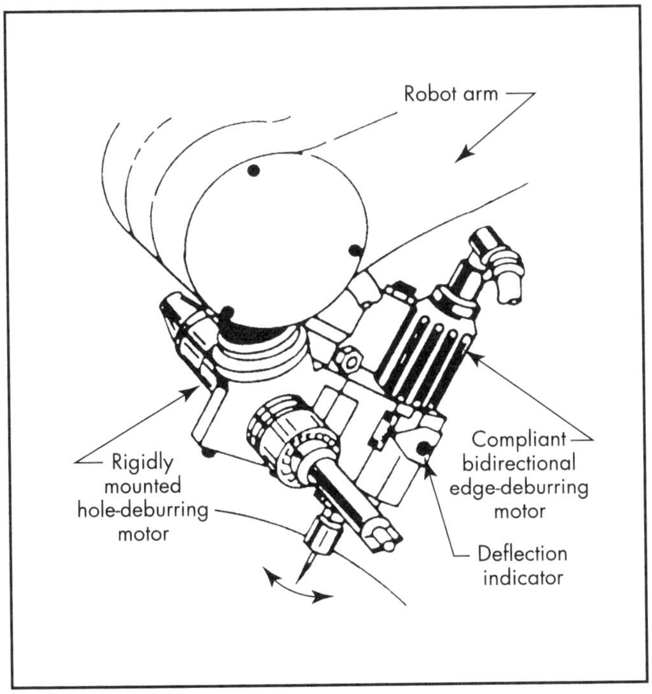

Figure 16-21. Multispindle tools. ***(Courtesy ASEA Robotics)***

Although most deburring is done using tools with tapered cutting edges, effective deburring is best accomplished using a variety of tool configurations, tool axes, and motions, including deburring with the flat ends of some tools, with the cylindrical portion in either of two axes, or with brushes (Figure 16-22). Rotary burs have a life of 30–300 h, depending on application. It is important to note that most carbide cutters generate two smaller secondary burrs. The size of these burrs can be minimized by selecting the appropriate tool geometry.

Specific motions can be used to deburr some features. The top surface of a bur can be used to deburr the bottom side of a hard-to-reach hole (Figure 16-23). In this case, the tool is used not as a countersink tool, but as a profiling tool for the hole edges. Double-sided chamfering tools are available to chamfer both the top and bottom of rectangular stock or related features (Figure 16-23); the smallest standard cutters of this design have diameters of .76 in. (19.3 mm).

The following factors are involved in the optimal use of cutters:

- Correct contact point (errors often lead to vibrations).
- Correct contact angle (to avoid or minimize secondary burrs).
- Correct path direction (to minimize vibrations).
- Correct path velocity (depends, for example, on type of burr and desired edge quality).
- Correct resilient mounting (compliance).

Rotary Files

Rotary files made of high-speed steel or carbide are the most common robotic deburring tools. A carbide cutter rigidly held by a robot can produce excellent surface finishes unattainable with hand deburring. Because chip-cutting tools blunt rather than undergo significant geometrical wear, wear compensation is not possible. A robotic system will not know whether increased forces are the result of tool wear or changed geometry, and so cutters must be periodically changed.

Reciprocating files are suitable for applications in which only minor burrs are to be removed. The speed and stroke of the file can be adjusted, and files with different cuts can be used to vary the amount of material that the file removes. These controls simplify the programming necessary. Motors can be mounted in compliant holders to compensate for variations in tools and workpieces. Oscillating files are particularly suitable for hard or brittle materials. Note that soft materials, such as aluminum, tend to accumulate between the teeth of files. Reciprocating files are a good choice for reaching into sharp corners.

Abrasive-belt Grinders

Edges easily accessible from the outside are suitable for deburring, fettling, or finishing with an abrasive-belt grinder, which removes material rapidly and leaves even surfaces. However, chamfer size can be difficult to control with this tool. The grinding belt should be spring or air-loaded,

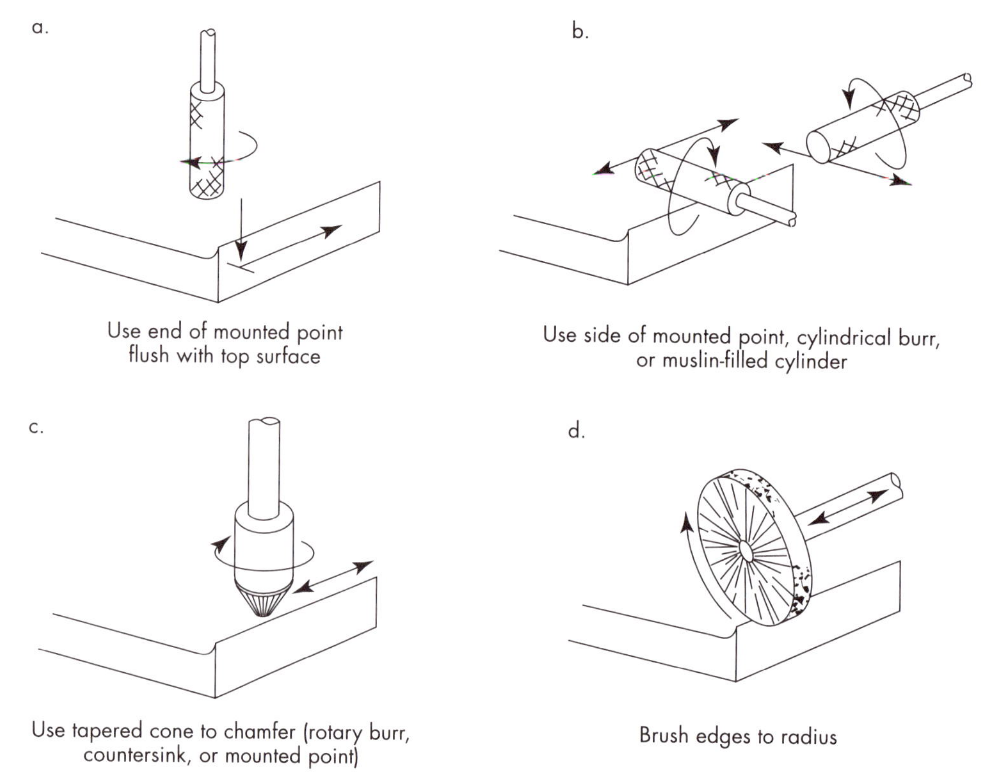

Figure 16-22. Tool orientation for robotic deburring.

or the driving equipment should be suspended in a compliant mount to provide flexibility. A longer belt can be used if the grinder is pedestal mounted. Long belts on pedestal-mounted grinders decrease the frequency of belt changes and allow the use of more powerful drive units for higher stock removal rates.

Brushes

For many commercial applications, a brush is used rather than a cutting tool. When aggressive brushing is permitted, all normal burrs can be successfully removed. However, brushes that will ensure a smooth blend not exceeding .010 in. (0.25 mm) will frequently beat small burrs flush against the surface. In precision applications, workpieces may be rejected because all burrs have not been removed. Nevertheless, a judicious choice of brushing applications and approaches may eliminate the need for cutters.

Robotic brush deburring is suitable for all external surfaces when light abrasion on areas adjacent to the brushed edges is acceptable. Internal edges can be deburred if they are located so that the brush can reach the burrs at an angle of 25–50°. Programming for brush applications is simple because of brush flexibility and because the deburring effect does not depend greatly on the applied force. Brush wear can be compensated for by automatically offsetting the path. (For more specific information about brush deburring, see Chapter 11.)

Grinding Disks

Figure 16-24 shows the working power zone for robots that use motorized grinding wheels or disks to remove flash or burrs (DeMotte 1990). In Figure 16-24, "A" indicates the feed rate at which grinding burnout occurs; "B" indicates a robot limit on feed rate under acceptable control; "C" indicates the lower limit at which the grinding force will adequately remove material, and "D" is the limit of robot grinder power. Each combination of robot, grinding motor material, and orientation will have a work space similar to this chart. Figure 16-25 shows a grinder and compliance device for use with a robot.

Multiple Tools Used for Each Edge

The best deburring method is the one that provides the necessary edge quality at the most economical rate. When high-quality edges are required, following one procedure with another may be necessary; for example, cutters may be used first, followed by brushes, bonded

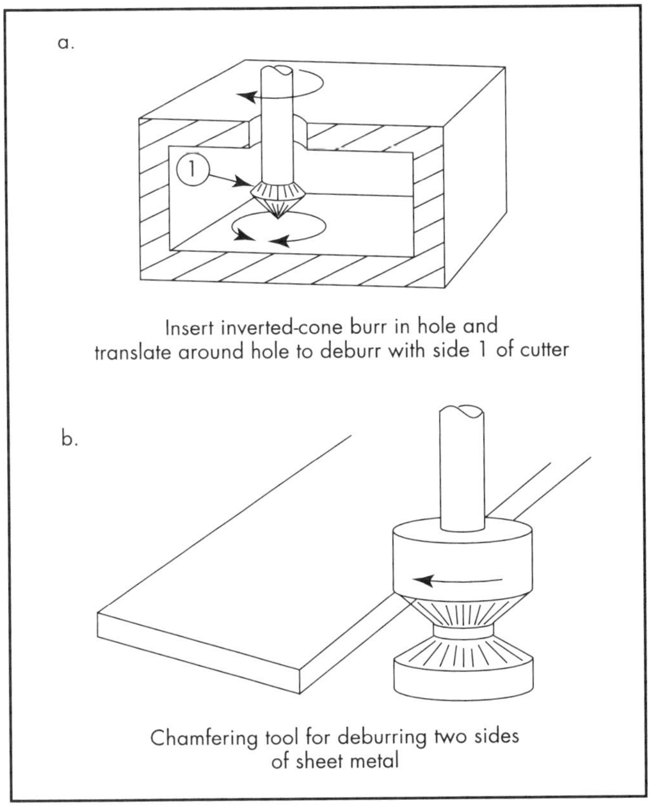

Figure 16-23. Specialized tools and deburring movements.

abrasive tools, or abrasive-filled rubber tools. Table 16-1 outlines common follow-up processing methods for producing high-quality edges.

16.10.8 Spindle Holders

One-way Spring-loaded Holders

Spindle holders are in most cases designed to give compliance in just one direction, the direction perpendicular to

Table 16-1. Combinations appropriate for robotic deburring

	Subsequent finishing approaches	
Primary approach	Abrasive rubber	Brushing
Bur balls	X	
Chamfer tools	X	
Grinding wheels		X
Reverse radius cutters		X
Abrasive rubber		X
Abrasive-filled cotton		X
Brushing	—	—
Grinding belts	—	—
Reciprocating files	—	—

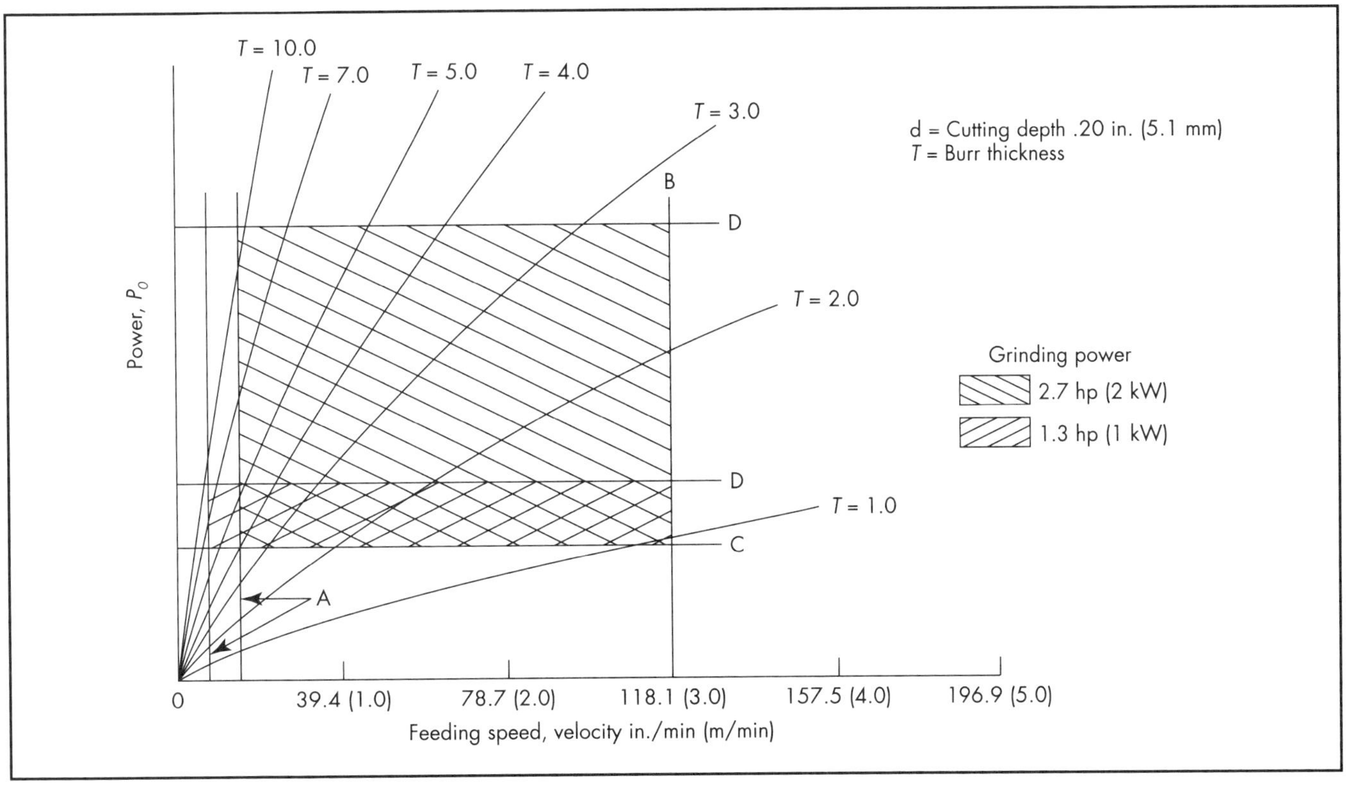

Figure 16-24. Working zone for disk grinding. ***(Courtesy Dr. Toshihiro Ioi)***

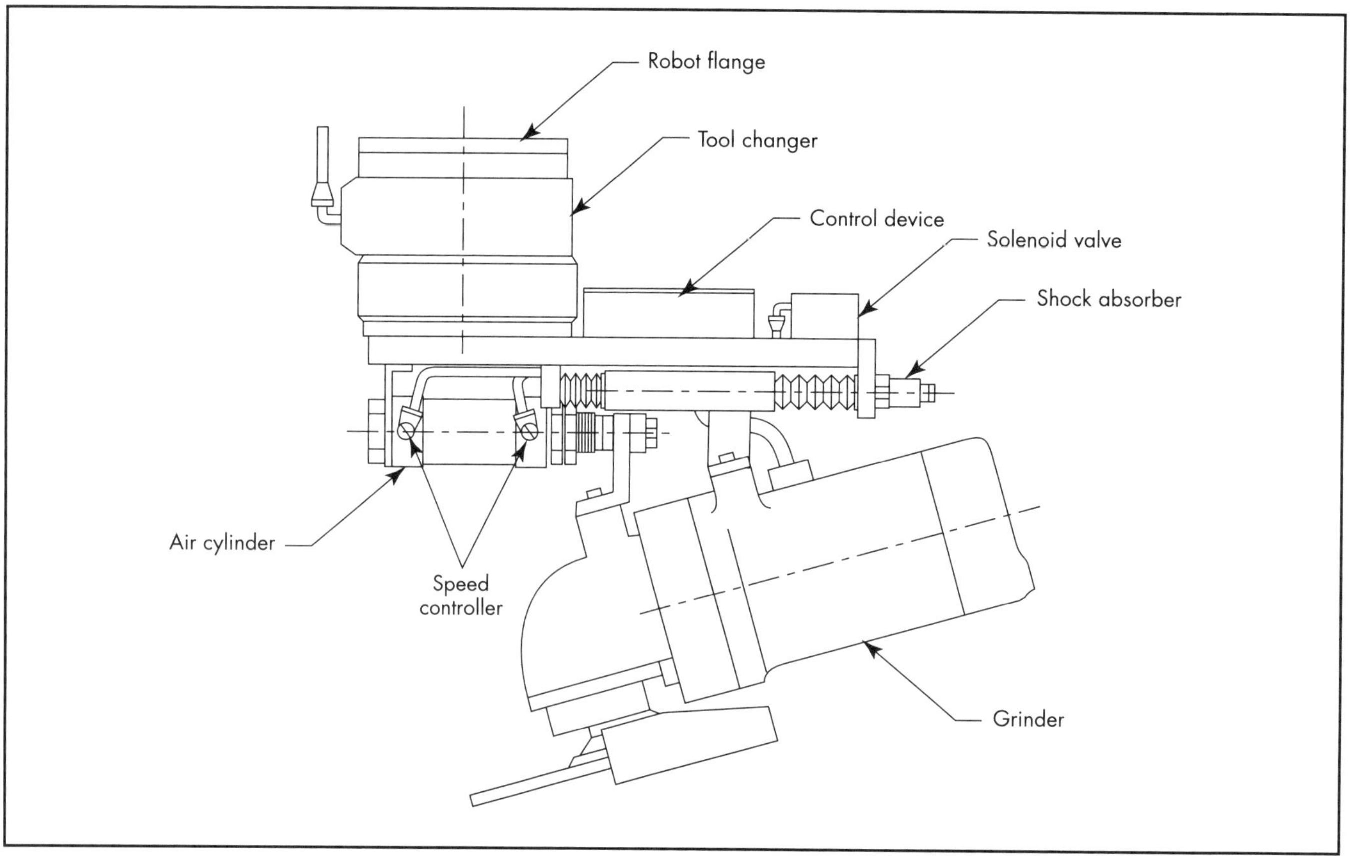

Figure 16-25. Robot-held grinder with compliance device; two small air cylinders at the back of the unit provide adjustable compliance. ***(Courtesy Dr. Toshihiro Ioi)***

the direction of feed. Mounted on the robot's fifth axis (wrist), the holder is continuously turned by the robot as the part contour changes. This action maintains the direction of deflection perpendicular to the feed direction. The spring force is adjustable, and compliance can be used with this tool. The major advantages of this type of device are that vibration is minimized and the tool follows complex contours with a minimum amount of programming.

Two-way, Spring-loaded Holders

Two-way, spring-loaded holders (Figure 16-26) can deflect in two directions (in the same plane) from a neutral position. To ensure that the tool center point is accurately located, two pneumatic cylinders are activated on either side of the pivot point. By activating the cylinder opposite the desired direction of compliance, the tool will always be firmly located, although not on a hard end stop.

The working range can be extended when using this type of holder. The holder can be offset from the robot wrist and does not have to turn 180° as the one-way holder does to change the direction of compliance. This action minimizes robot wrist motions and overall cycle time.

To further increase the accessibility of the tool, the direction of compliance can be turned continuously and integrated with the robot's control system. The tool is positioned with the robot's fifth axis, and the direction of compliance is turned to a right angle relative to the deburred surface using the robot's sixth axis.

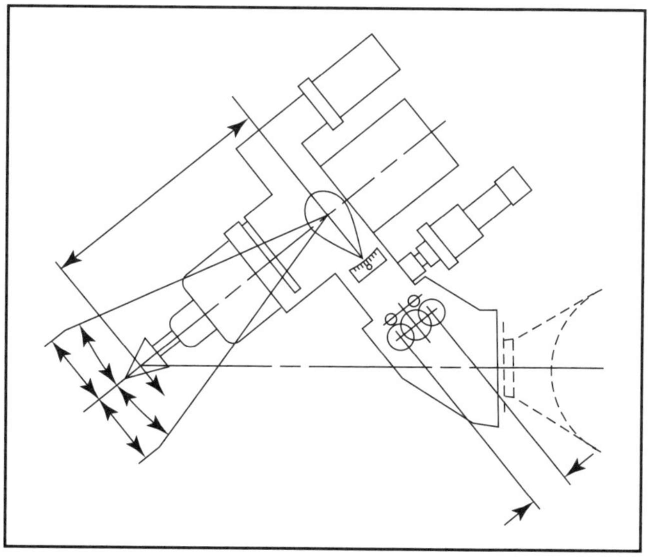

Figure 16-26. Spindle holder for robotic deburring with dual-directional compliance. ***(Courtesy ASEA Robotics)***

Grippers can be designed to be multifunctional. They can hold two separate motors, provide air blow-off, or clean while deburring. As a result, they can significantly increase output when multiple, simultaneous operations are allowed. If necessary, they can stamp workpieces, verify the existence of hole patterns, perform related tasks, or work with equipment that can perform these operations.

Spindles

For fettling operations, the spindle on the robot must have a stall force greater than the normal fettling forces. For example, a .75 hp (0.56 kW) motor may have a stall force of 11 lbf (48.9 N), but grinding or abrasive-belt finishing may require 50 lbf (222.4 N) of force to finish in a reasonable cycle time. Therefore, users should consider using 3–5 hp (2.2–3.7 kW) motors for rapid metal removal.

Deburring and fettling spindles must have a good power-to-weight ratio, such as that found in high-frequency electric tools. Pneumatic motors, however, can be overloaded without damage. High performance can be achieved with hydraulic drives, but only within certain speed ranges.

16.10.9 Operating Parameters

Most robot manufacturers recommend fast traverse rates that far exceed those normally achievable by humans. For applications in which high rates are possible, robotic deburring can greatly increase production as well as minimize costs. For most nondeburring robotic applications, the robot works at about the same pace as manual operators.

For all deburring applications, the robot must move continuously at a constant feed rate. In most applications, climb (down) cutting provides a smoother and more even surface finish than conventional (up) cutting. At optimum feed rates, when moving around sharp corners, it may be necessary to start turning the holder before it reaches the corner. The change of compliant direction is then completed as the tool moves in the new direction.

When using rotating files for deburring, the cutting speed should be about 4,200 ft/s (1,280 m/s), and the recommended feed rate is 300 in./min (7,620 mm/min). For deburring holes, the cutting speeds and feed rates used are about the same as for drilling or thread cutting. Low feed rates of 1–2 in./min (25–51 mm/min) are often used.

Procedures for programming with a rigid toolholder are different from those for programming with compliant holders. Rigid toolholders require accurate positioning of the robot and many instructions. It is important that an open surface (minimum contact area) be used as the starting point. If more than 30° of the tool's radius is in contact with the surface, vibrations and chattering may occur. The rigid toolholder limits the maximum programmable feed rate because of servo lag created as the speed of the robot increases.

Compliant toolholders decrease the number of instructions needed because they do not have to adjust for small changes in workpiece geometry. They also have a relatively high feed rate when moving around complex contours, and the positioning instructions do not have to be exactly located along surfaces.

16.11 LIMITATIONS OF MECHANIZED CUTTING

Mechanized cutting works well when workpiece edges are readily accessible to the cutters. The following workpiece features make mechanized cutting difficult or impossible:

- Shallow counterbores and steps (Figure 16-27).
- Edges flaring into walls (Figure 16-28).
- Features near other walls (Figure 16-29).
- Small edge breaks (chamfers) at the intersection of internal walls (Figure 16-30).
- Very small features (such as holes with diameters of .010 in. [0.25 mm]).
- Cross holes.
- Holes breaking into internal corners.
- Features requiring micro-level precision, extreme surface finishes, and near-sharp edges.

16.12 ENVIRONMENTAL, HEALTH, AND SAFETY CONCERNS

There are seven major safety hazards in robotic deburring (many of these are also applicable to NC and manual operations).

1. Robot runaway.
2. Inadvertent human contact with the robot.
3. Sudden release of workpiece or tooling by the robot.
4. High feed rates.
5. High spindle speeds.
6. Noise levels.
7. Flying debris, such as broken wheels, burrs, flash, wire from brushes, and sparks from grinding.

The first four items are common concerns in all robotic applications. If a robot loses its gripping power in a runaway condition, the workpiece in its gripper or the tooling on the wrist could become a flying projectile. This situation creates a hazard of some concern because of the weight of the workpiece or tooling and the potentially high travel speeds.

Feed rates of over 100 in./min (2,540 mm/min) are considered high in most metal-cutting operations but not in robotic deburring. Workers in the immediate vicinity of robots that operate at these rates must be protected, since they may not be able to react quickly enough to get out of the way if something goes awry. High spindle speeds are used in both mechanized and manual operations, but these speeds still require some care to ensure that chips do not fly wildly and that appropriate tools are inserted and correctly chucked. Pinch points are a concern in any mechanized cutting operation.

In fettling operations, high-power tools may produce noise levels that exceed allowable OSHA limits. Consequently, these operations may require walled installations.

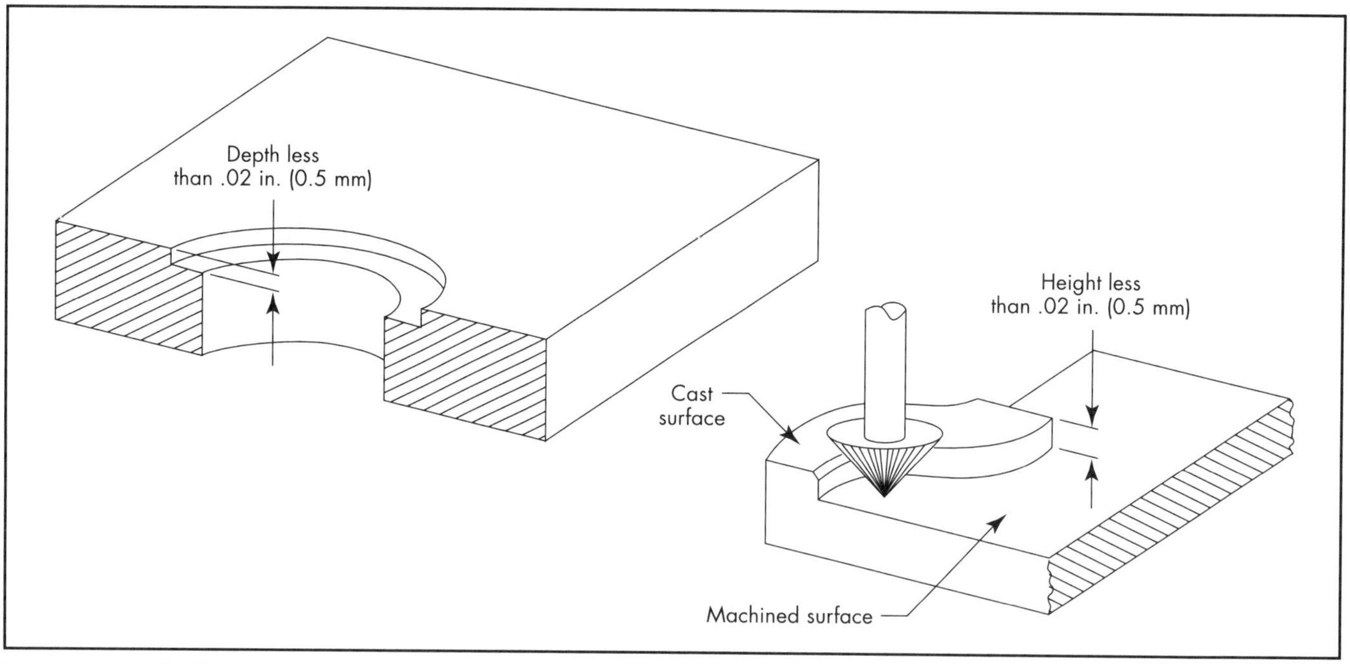

Figure 16-27. Shallow counterbores and shallow steps are poor choices for mechanical deburring by cutting processes.

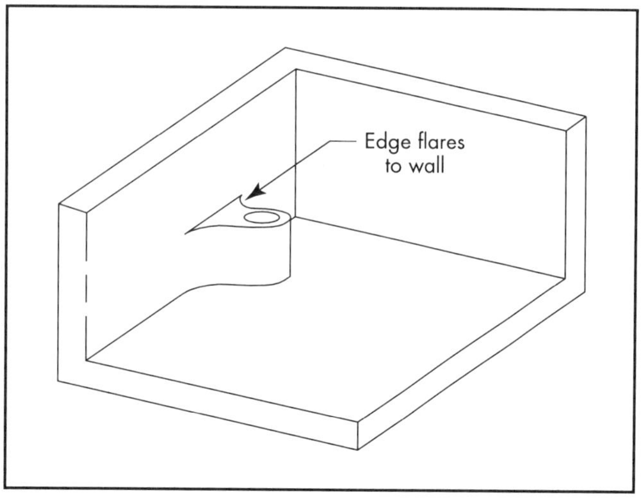

Figure 16-28. Edges flaring into walls will not be totally deburred with cutters unless the wall is also eroded.

If grinding wheels are pinched or twisted as they bind, the wheels will fly apart. A walled area is desirable in all installations involving fettling to prevent injury from flying debris.

Off-line programming provides some safety advantages in that the programmer is not near the robot during initial programming. If the programming is not accurate, however, the robot can be damaged in the first test of the program.

Debris from finishing operations can cause breathing hazards for personnel in the area. Debris can buildup on fixtures and contaminate electrical components. Grinding can generate hot spots and chips; and if swarf is allowed to stick to workpieces or stay in the area, it can melt rubber boots on machines or cause local fires. Hydraulic hoses often break, and nearby staff may be injured by fluids that are released.

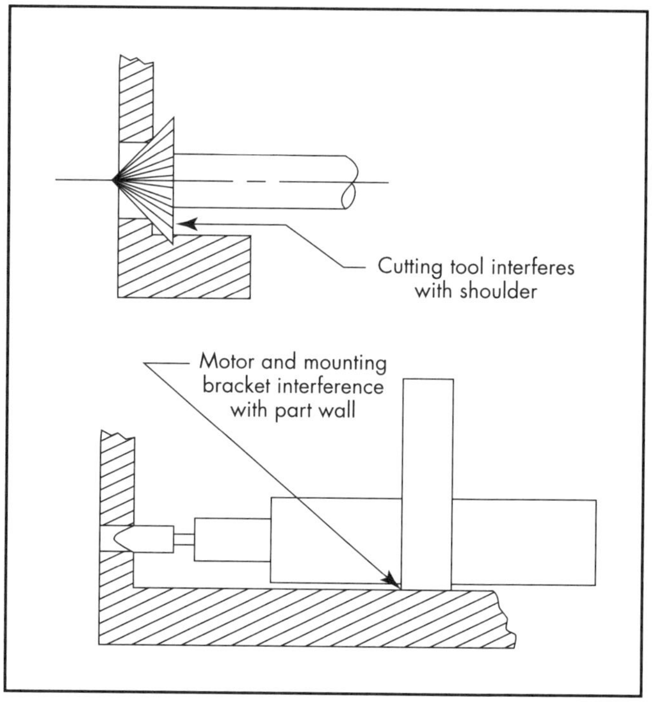

Figure 16-29. Walls adjacent to deburred edges may interfere with cutters or cutter spindles.

Unattended buffing wheels can potentially catch fire. Sanding-type processes can generate enough dust, glue, paper, cloth, or metal particles to cause fires. When fine

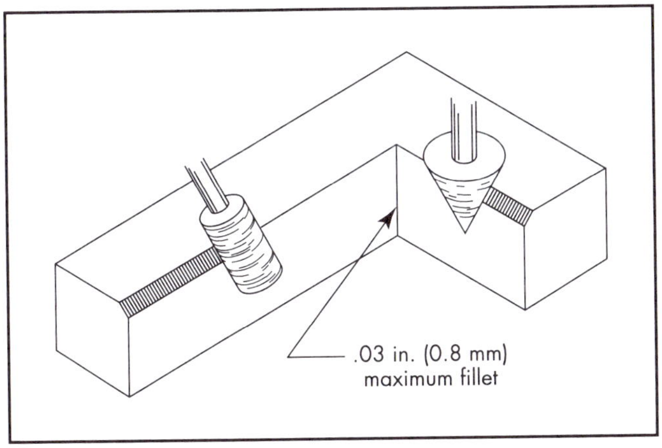

Figure 16-30. Producing small chamfers (for example, .003 in. [76.2 μm]) at internal intersections is difficult.

aluminum or molybdenum particles are present, the potential for an explosion exists. Sanding titanium can generate sparks, and fine chips will burn.

The nickel in stainless steel is a suspected carcinogen. A large volume of dust in the air from stainless steel sanding or grinding may cause long-term health problems. Sampling of air-borne particles may be necessary.

In short, although robots will generally perform safely and consistently, periodic safety checks are advisable to assure that the other elements of the process are also safe.

References

DeMotte, Donald D. 1990. "Simplified Programming for Robotic Deburring." Technical paper MS90-277. Dearborn, MI: Society of Manufacturing Engineers (SME).

Dornfeld, D. A., and Erickson, E. 1989. "Robotic Deburring with Real-time Acoustic Emission Feedback Control." *Mechanics of Deburring and Surface Finishing Processes*. Stango, R.J. and Fitzpatrick, P.R., eds. PED Vol. 38. New York: ASME: 13–26.

Gillespie, LaRoux K. 1987. *Robotic Deburring*. Dearborn, MI: Society of Manufacturing Engineers (SME).

Grace, William H. 1984. "Robotic Precision End Finishing." Technical paper 84-132. Dearborn, MI: Society of Manufacturing Engineers (SME).

Graf, Tim. 1990. "Robotic Deburring and Finishing, Methods and Applications." Technical paper MS90-679. Dearborn, MI: Society of Manufacturing Engineers (SME).

Indge, John H. 1991. "Edge Round with Flexible Bonded Abrasives." Technical paper MR91-128. Dearborn, MI: Society of Manufacturing Engineers (SME).

Koelsch, James R. 1991. "Taking on Tough Burrs." *Manufacturing Engineering,* Vol. 107, (October):35–38.

Loucks, Clifford S., Selleck, Colin B., and Thunborg, Siegfried. 1989. "Cad-directed Robotic Edge Finishing." Technical paper MR89-114. Dearborn, MI: Society of Manufacturing Engineers (SME).

Matsunaga, Masahisa, Osada, Tomoaki, Ioi, Toshihiro, and Endo, Makoto. 1985. "High-speed Adaptive Control and High-performance Grinding Conditions for Deburring Robots." Technical paper MR85-829. Dearborn, MI: Society of Manufacturing Engineers (SME).

McCoy, Danny C., and Hsie, Atlas. 1984. "Robot Applications in Buffing, Polishing, Grinding, and Deburring." Technical paper MS84-360. Dearborn, MI: Society of Manufacturing Engineers (SME).

Nicoletti, Guy M. 1988. "Optoelectronic Proximity Sensors for Automated Deburring." Technical paper AD88-239. Dearborn, MI: Society of Manufacturing Engineers (SME).

Schraft, R.D., Schweizer, M., Abele, E., and Sturz, W. 1983. "Application of Sensor-controlled Robots for Fettling of Castings." Technical paper MR83-379. Dearborn, MI: Society of Manufacturing Engineers (SME).

Bibliography

Abele, E., Sturz, W., and Boley, D. 1985. "Sensor and Tool Development for Robotic Deburring." Technical paper MR85-840. Dearborn, MI: Society of Manufacturing Engineers (SME).

Arnold, Kerry. 1987. "Implementing Robotic Deburring." *Manufacturing Engineering,* Vol. 99, (July):35–37.

Bausch, John J., III, and Gold, Frederick M. 1995. "Design and Application of a Brush-based Active Robotic Deburring System." Technical paper MR95-224. Dearborn, MI: Society of Manufacturing Engineers (SME).

Elbestawl, M. A., Yuen, K. M., Srivastava, A. K., and Dai, H. 1991. "Adaptive Force Control for Robotic Disk Grinding." *CIRP Annals,* 40 (1):391–94.

FitzPatrick, Peter R., and Paul, Frank W. 1987. "Robotic Finishing Using Brushes—Material Removal Mechanics." Technical paper MR87-156. Dearborn, MI: Society of Manufacturing Engineers (SME).

Flores, Gerhard. 1992. "Mechanical Deburring: Process, Tools, Machines, and Applications." Technical paper MR95-272. Dearborn, MI: Society of Manufacturing Engineers (SME).

Gayman, David. 1985. "Defining 'Special' at General Machine." *Manufacturing Engineering,* Vol. 95, (December): 39–40.

Gillespie, LaRoux K. 1996. *Guide to Deburring, Deflashing, and Trimming Equipment, Supplies, and Services.* Kansas City, MO: Deburring Technology International, Inc.

Hickman, Peter K., and Judd, Robert P. 1987. "Robotic Deburring Using an Actively Controlled End Effector." Technical paper MR87-155. Dearborn, MI: Society of Manufacturing Engineers (SME).

Kanda, Yuichi, Fukuda, Yoshiro, and Ioi, Toshihiro. 1985. "Aspects of Robotic Deburring and Finishing in Japan." Technical paper MR85-831. Dearborn, MI: Society of Manufacturing Engineers (SME).

Kanda,Yuichi, Ioi, Toshihiro, Fukuda, Yoshiro, Nishina, Kenji, Takazawa, Koya, and Kinoshita, Naoharu. 1989. "An Expert System for Process Planning in Deburring." Technical paper MR89-239. Dearborn, MI: Society of Manufacturing Engineers (SME).

Koelsch, James R. 1990. "Banish Manual Deburring." *Manufacturing Engineering,* Vol. 105, (July):71–75.

Kramer, B., Gott, R., and Bausch, J. 1985. "Deburring Strategies Using Industrial Robots." Technical paper MR85-843. Dearborn, MI: Society of Manufacturing Engineers (SME).

Lambrechts, R.J., Clark, M.D., Kanies, T.A., Mathews, S.R., and McMillan, D.E. 1986. "A Robotic System for Deburring Machined Castings." Kansas City, MO: Allied (formerly Bendix) Corporation, Division report BDX-613-3522; available from NTIS.

McCain, Harry G., Kilmer, Roger, and Murphy, Karl N. 1984. "Development of a Cleaning and Deburring Workstation for the AMRF." Technical paper MR85-824. Dearborn, MI: Society of Manufacturing Engineers (SME).

Nanlawala, Michael, and Engel, Thomas. 1997. "Automated Deburring of Aerospace Gears." *Proceedings of Gear Manufacturing in a Global Economy.* INFASC Industry Briefing. Chicago, IL: IITRI.

Proctor, Frederick M., Murphy, Karl N., and Norcross, Richard J. 1989. "Automating Robot Programming in Cleaning and Deburring." Technical paper MR89-138. Dearborn, MI: Society of Manufacturing Engineers (SME).

Przyklenk, Klaus. 1985. "Deburring in Europe—Present State and Future Prospects." Technical paper MR85-839. Dearborn, MI: Society of Manufacturing Engineers (SME).

Scoville, John H. 1984. "Robotic Abrasive Process System." Technical paper MS84-361. Dearborn, MI: Society of Manufacturing Engineers (SME).

Spinelli, Richard. 1984. "Aircraft Tube-end Finishing." Technical paper MF84-136. Dearborn, MI: Society of Manufacturing Engineers (SME).

Takeuchi, Yoshimi, Sakamoto, Masafumi, and Kouda, Katsuya. 1987. "Automation of Deburring Operations by Means of Color Information." *CIRP Annals,* 36 (1):293–96.

Van Sickle, Charles, and Flores, Gerhard. 1977. "How to Pick the Right Deburring Process." *Manufacturing Engineering,* Vol. 78, (February):56–62.

Wan, C. Andrew. 1982. "A Rational Approach to Deburring for Flexible Manufacturing Systems." Technical paper MS82-424. Dearborn, MI: Society of Manufacturing Engineers (SME).

17 Manual Deburring

Hand deburring, also called handwork, benchwork, benching, or manual deburring, is any operation in which hand-held deburring tools are used or hand-held workpieces are held against a fixtured tool. The process employs various cutters and motorized tools. Typical tools used for hand deburring are listed in Table 17-1. An estimated 10,000 hand-deburring tools are commercially available, and one tool manufacturing company lists over 1,000 rotary burs in its catalog. Because of the many types of tools used in hand deburring, it overlaps other processes such as brush deburring and coated-abrasive finishing. Hand deburring is the most widely used method for removing burrs, and it is performed not only by those who have the primary responsibility for burr removal, but also by machine operators, inspectors, maintenance workers, and assemblers.

Table 17-1. Typical tools used for hand deburring

Abrasive-filled products
Cork products (bullets, ball-nose cylinders)
Cotton products (cylinders, balls, bars)
Nylon synthetic products
Rubber products (bullets, cylinders, flat bars, disks, cups, dental bullets)
Abrasive paper products (disks, rolls, sheets, cord)
Brushes (wheel, end and cup, tube, cross hole, side action)
Burs, bur balls, rotary files
Countersinks
Drills and reamers
Felt bobs (disks, bullets, cylinders)
Files (large, miniature, round, half-round, triangular, curved, bent)
Hand stones (bars, triangles, cones, points)
Knives (triangular, oval, special shapes, scalpel blades)
Lapping compounds
Mandrels for tools
Peening tools (ball peening, blade peening)
Motorized tools (bench motors, air motors, dental tools, belt sanders, reciprocating files, jitterbug sanders)
Mounted points (balls, disks, cylinders, cones, special shapes)
Picks
Pin vises (dog nose, collet)
Scrapers
Miscellaneous tools (swivel blades, back-side cutters, special designs, vacuum probes)

17.1 WHY HAND DEBURRING IS USED

Despite being labor intensive and often providing less consistent results than desired, hand deburring continues to be widely used because it is versatile and requires minimal floor space and capital investment. Depending on the application, hand deburring may be the least expensive and most efficient approach to deburring. Some of the factors involved in the decision to use hand deburring are shown in Table 17-2.

Hard-to-reach burrs are not in themselves a valid reason for using hand deburring, since some of the processes discussed earlier in this book are capable of accessing remote areas. The quantity of workpieces needing to be processed may not always justify a machine setup for deburring. Usually, a combination of burr accessibility and size, edge and surface requirements, available equipment, and cost dictates the selection of a specific process. Often, mechanized deburring combined with some hand deburring will be the most economical choice.

The term *hand deburring* frequently does not adequately describe the work being performed by the deburring staff. For example, operators may be expected to smooth surfaces, remove broken tools, sand to a specific thickness, and in-

Table 17-2. Reasons for employing hand deburring

Flexible process
Burrs are hard to reach by any other process
Only a small number of workpieces are being processed or a short lead time is required, making hand deburring the least expensive method
Workpieces are delicate or require a precision finish
Workpieces require close tolerances
Required edge radii are large, .060 in. (1.52 mm) or more; or allowable edge radii are small, .002 in. (0.05 mm) or less
Burr size is highly variable, prohibiting use of other processes
Deburring or finishing machine will not accept workpiece size required, or machine is out of commission
Hand deburring can be done while other machining operations are being performed
To remove or reduce the size of burrs or extruded metal left by mechanized finishing processes
To prevent media impregnation, oxide formation, or other undesirable side effects from other finishing processes
Inability to change the process because it has been accepted in conformance with a military specification

spect for nicks, dings, scratches, pits, voids, and stains. Consequently, when efforts are made to reduce hand-deburring costs, the costs of these other tasks may be overlooked. When the full extent of the work involved in hand deburring is taken into consideration, hand deburring is often the most cost-effective process as well as the most timely. There are many industrial instances, however, where hand deburring has been ignored. To make it a viable alternative, significant improvements can be made by a few simple hand tool improvements.

17.2 LIMITATIONS

Consistently maintaining precise edge breaks is difficult with hand deburring, especially when workpieces require many edge breaks. It is almost impossible to consistently produce edge breaks of .002–.003 in. (0.05–0.08 mm) on all portions of every edge of intricate workpieces. However, one company has been able to achieve exacting tolerances by carefully training workers and using microscopes, though the process is time consuming (Gillespie 1980a). Larger breaks, such as .005–.010 in. (0.13–0.25 mm), generally must be specified to ensure that most workpieces will be acceptable.

In the aerospace industry, edge radii to .062 in. (1.57 mm) are sometimes specified. However, hand deburring is very time consuming when edge breaks or radii must be greater than .015 in. (0.38 mm).

17.3 COSTS

An analysis of hand deburring costs is relatively straightforward (a typical form used for analyzing costs is presented in Table 17-3). Cost, however, should not be the only criterion for determining whether special modifications are required to improve manual efforts.

17.3.1 Judging Abnormal Costs

One indicator that overall hand deburring costs may be excessive is when the labor-hours or costs of hand deburring are more than 5% of the total machining time or costs. Another indicator is when most of the production workpieces are being delayed at the deburring benches. A third indicator is when a manufacturing firm is spending considerably less of their capital equipment dollars on finishing equipment than the industry average, which is currently 5–7%.

17.3.2 Independent Cost Assessments

An inexpensive and effective approach to determining if hand deburring costs are excessive is to have a deburring job shop finish a representative sample of workpieces. If the quality of the finished workpieces is acceptable and the cost is less than the in-house cost, in-house hand-deburring methods may require careful scrutiny.

Table 17-3. Typical form used for analyzing deburring costs

Item	Estimated number of h/yr for deburring	Direct salary ($)	Overhead ($/h)	Cost/yr ($)
Hand deburring personnel				
Deburring machine personnel				
Deburring supervisor				
Engineering or management personnel				
Inspection personnel				
Deburring equipment depreciation	—	—		
Equipment maintenance				
Sharpening of deburring tools				
Deburring supplies		—		
Energy costs		—		
Deburring on machine cycle				
Scrap due to inadequate deburring	—			
Warranty work caused by burrs				
Floor space costs	—	—		
Total				

Note: h × direct salary = yearly salary; overhead × salary = yearly overhead; yearly costs = yearly salary and yearly overhead; depreciation only goes in yearly costs; energy costs = h × overhead cost = yearly costs; deburring supplies = h × overhead cost = yearly costs; floor space only goes in yearly costs

17.3.3 Rejected Workpieces

Although costs are a major concern for all manufacturing firms, the rate at which workpieces are being rejected for burr-related reasons is also important. Continual workpiece rejection at later stages indicates an out-of-control situation and implies that deburring operations are not being satisfactorily performed or that workpieces still containing burrs are passing inspection.

17.3.4 Effect of Burr Size

Both burr size and workpiece size have a considerable effect on deburring time and costs. Figure 17-1 shows the effect of burr thickness on hand-deburring time. Washer-like workpieces with burrs as thick as .005 in. (0.13 mm) require four times longer to hand deburr than similar work-pieces with burr thicknesses of .0015 in. (0.038 mm). Figure 17–2 shows typical edge breaks produced by a variety of deburring tools that are discussed in this chapter. Removing heavy burrs by machining frequently costs less than removing them by hand. Since mechanized deburring is generally 80–90% efficient, it should be used whenever possible.

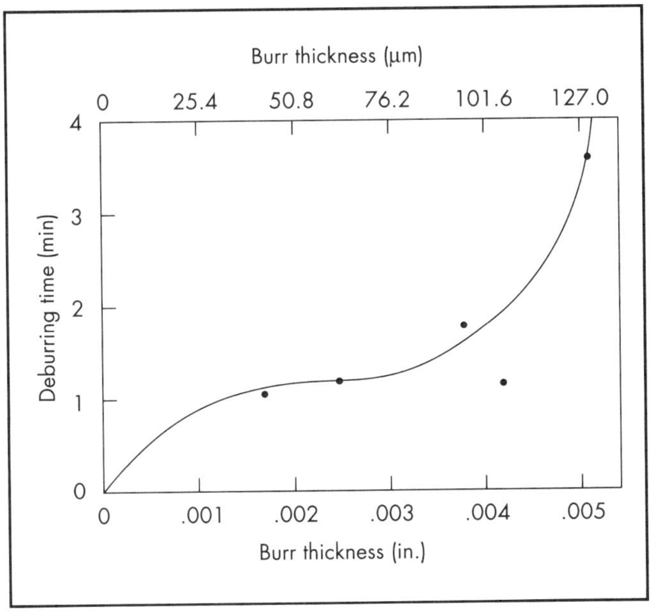

Figure 17-1. Effect of burr thickness on hand deburring time.

17.4 IMPROVING HAND DEBURRING

The capabilities and limitations of most mechanized deburring processes have been well researched and defined, but this is not case for hand deburring methods, which is a major problem in attempts to improve the process. For example, data needs to be made available on coated-abrasive products used in hand deburring to determine which grain size, structure, and bond can be used without damaging a specific surface finish or reducing a diameter more than an allowable amount in a certain time.

17.4.1 Quality Requirements

Clearly defining the quality requirements for various edges and surfaces is an important part of improving the process of hand deburring and reducing its costs. Specific instructions and codes need to be included on workpiece drawings (see Chapter 4) so that every worker knows the exact requirements, the tools employed, and the sequence in which the tools should be used. One method for ensuring quality involves painting a sample workpiece white and then color coding the edges to indicate which tools should be used to finish it (Figure 17-3); when more than one tool is required to finish an edge, the color closest to the edge indicates the first tool to be used, the next closest color, the second tool, and so on.

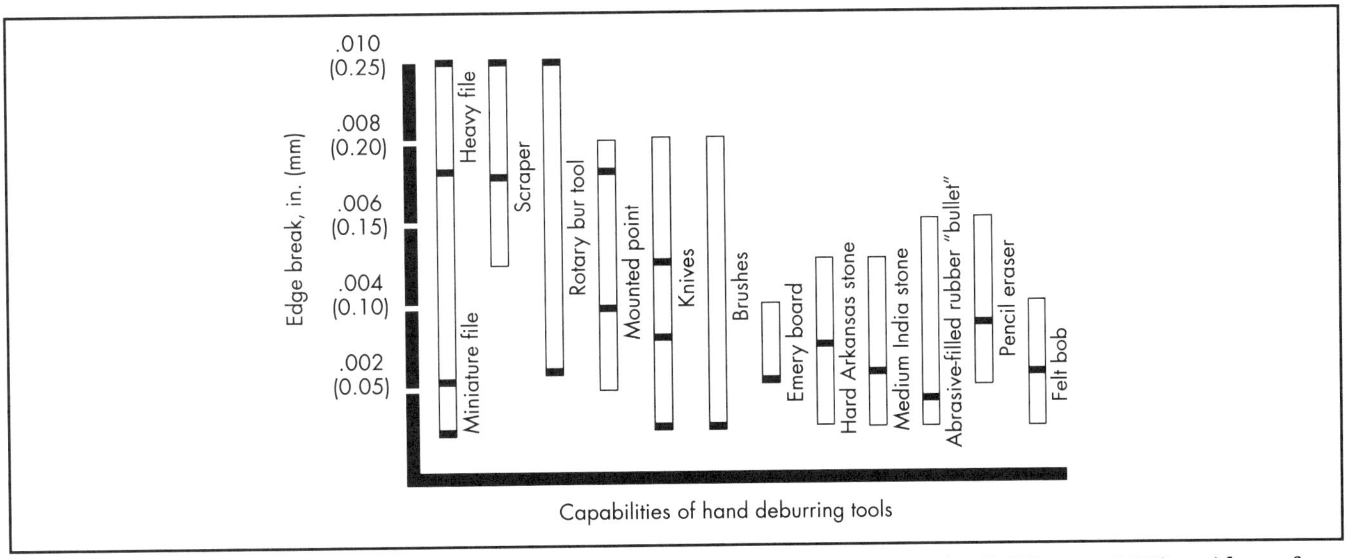

Figure 17-2. Basic edge-break capabilities using specific tools to remove .003 in. × .003 in. (0.076 mm × 0.076 mm) burrs from 303 Se stainless steel (horizontal bars indicate measured results by one operator).

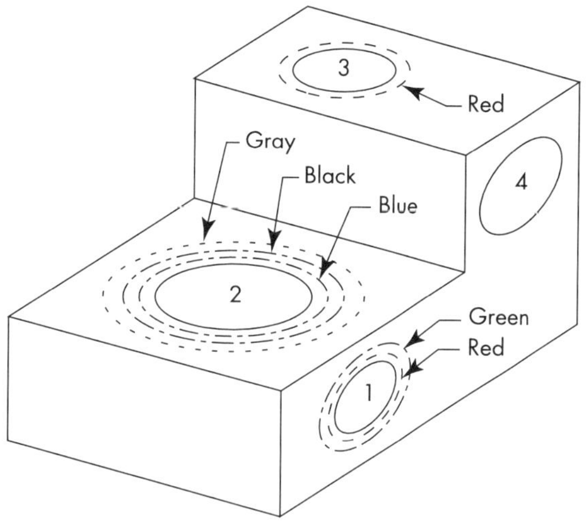

Hole number	Color band	Hand deburring approach
1	Red	Flat sand
1	Green	MX wheel in air motor
2	Blue	Swivel blade knife
2	Black	Triangular knife
2	Gray	Abrasive-filled rubber in motor
3	Red	Flat sand
4	None	None

Figure 17-3. Color-coded sample workpiece indicating finishing requirements for various edges.

17.4.2 Training Programs

Formal training programs are an effective method for providing operators with the skills and expertise necessary for precision hand deburring. Hand deburring by an unskilled operator often results in the original burr being replaced by a chamfer and two new burrs. Training manuals are available for hand deburring (Gillespie 1979; 1980a, b; 1981).

17.5 LOCATING BURRS

Many burrs, especially minute ones, are difficult to see. Cleanliness of the workpieces is essential in locating burrs; on small features, for example, it is practically impossible to tell if a particle at an edge is dirt or a burr unless the workpiece is clean. The use of special lighting as well as microscopes and borescopes may be necessary.

17.5.1 Looking in the Right Direction

Many burrs are almost impossible to see unless they are viewed from a specific direction. For example, burrs on the top edges of holes may not be visible if viewed only from the bottoms of the holes; many of these burrs, however, can be located by looking through the holes from the top to the bottom surfaces of the workpieces. To verify that all burrs have been removed, first look at the top surface and then through the hole; next, turn the workpiece over and look at that surface and then through the hole. To see burrs correctly on most workpieces, tip the workpiece at an angle to the line of eyesight. Observing edges at an angle of 30–60° is generally preferable. A flexible or miniature-lens borescope connected to a video system is helpful. It is also important to have proper lighting at the correct angle. Figure 17-4 illustrates effective lighting angles.

17.5.2 Optimal Lighting

Various lighting techniques are used for microscopic deburring and inspecting. The most common techniques are lights or illuminators placed on or near the microscope, ring lights (fluorescent tubes around the lens), fiber-optic ring lights, coaxial illuminators (through the microscope), back lighting, polarizing accessories, color filters, and miniature illumination systems with flexible probe cords that enter openings as small as .11 in. (2.8 mm) in diameter.

17.5.3 Probing for Burrs

Fingernails and toothpicks are two of a wide variety of items used for determining tactilely if burrs exist. Dental picks are excellent tools for this purpose because of their small diameter and bent ends; electronic probes are also used. Miniature borescopes with video hookups are useful for searching for hidden burrs. Size (gage) wires are being used as probes, particularly for probing through the teeth of small gears. For some applications, pipe cleaners are dragged over edges. Loose pieces of cotton can also indicate rough edges or burrs.

Tactile probing to detect burrs has three major problems:

1. The probe may scratch precision workpieces.
2. The process is time consuming when the entire workpiece must be probed.
3. Tactile probing is not foolproof; for example, it is not satisfactory for probing in a direction parallel to the burr.

Gages are sometimes used for determining whether burrs exist or are of excessive size on workpieces, particularly on workpieces with holes. If a go-type plug gage fails to enter a hole, this indicates that the hole is undersized or that a burr has been thrown over into the hole. Similarly, if the gage enters the hole but does not leave the opposite end, the hole is tapered or undersized, or a burr has been thrown into the exit end of the hole. However, plug gages do not always indicate that a burr is present, and burrs can wear or scratch the gages, rendering them unstable. Furthermore, burrs may lock the gage in the workpiece, and forcing the gage through the hole or pulling it back out can scratch both the gage and the workpiece.

17.6 HAND DEBURRING TOOLS

There are basically 19 types of hand-deburring tools, with 60 subcategories of these tools in everyday use for hand deburring (many are listed in Table 17-1). Tools that are commonly used include knives, files, rotary burs, brushes, and abrasive products.

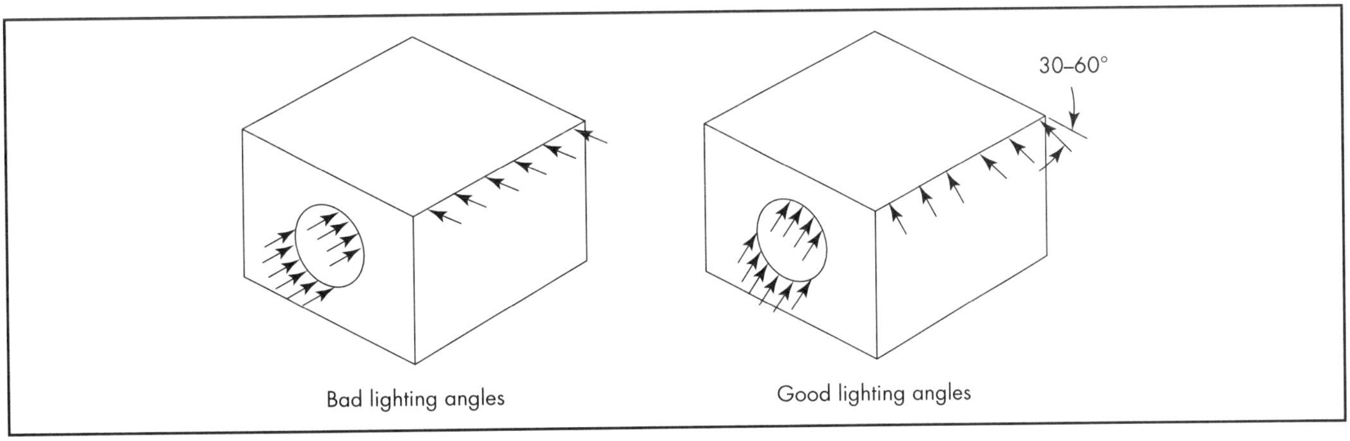

Figure 17-4. Effect of lighting angles on burr detection.

17.6.1 Knives

A wide variety of knives are used for hand deburring. Although knives are inherently hazardous, they do not generally present a safety problem with careful use. Most knives are specially designed for a specific application. They are typically made from American Iron and Steel Institute (AISI) Type M2 high-speed steel and generally have a hardness of R_c6I.

Triangular knives are typically used for cutting and scraping burrs from straight edges. Oval knives are typically used for hole edges or hard-to-reach areas. Specially shaped knives are designed to remove burrs from specific workpieces but are often useful for other applications as well. Scalpel blades are used extensively to trim flash from workpieces made of plastics, but are rarely successful for removing metal burrs.

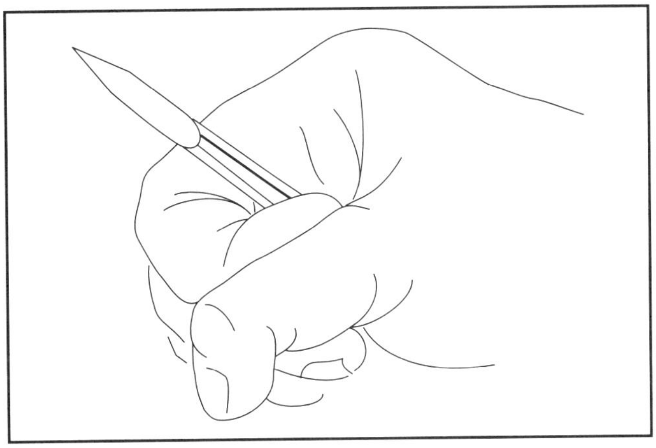

Figure 17-5. Fist position for knife deburring.

Knife-holding Methods

Knives can be used as fist tools or as finger tools. Fist tools, scrapers, large knives, and many other large tools are used with power supplied by the operator's hand and arm (Figure 17-5). The fist-holding position is used extensively for deburring large workpieces, but it is not a successful approach for deburring small or precision workpieces because it does not permit a sense of feel for the burrs. With a knife held in the fist, it is difficult to stop the momentum or control the amount of edge break produced on a workpiece. This method can result in the operator being cut and the workpiece being gouged. Consequently, knives should not generally be used in the fist position.

For small and precision workpieces, the finger is used as a tactile indicator of the actual conditions of the workpiece. There are two approaches to the finger position, and both approaches are widely used for hand deburring because they are safer than the fist position. In the overhand or pencil position, the knife is held with two fingers and the thumb, much like holding a pen or pencil (Figure 17-6). In this position, the knife can be rotated to accommodate changes in intersecting planes on the workpiece. The middle finger can be used as a surface feeling probe. By using this method under a microscope, small burrs can be removed easily.

Figure 17-6. Overhand or pencil position for deburring with knives.

In the underhand position (Figure 17-7), the small and ring fingers are used for surface support and the index

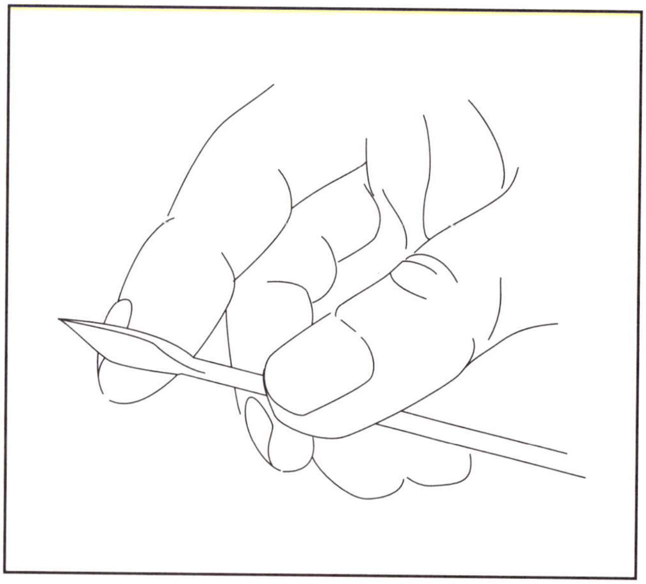

Figure 17-7. Underhand finger position for deburring with knives.

finger as a surface-feeling probe. The knife can be rotated for easy burr removal. This method is particularly useful for deburring holes. It permits rotating the knife around contours and provides uniform force against edges. Interchangeable handles and blade holders, including adjusting and telescoping types for reaching into deep recesses and holes, are available for different knife blades.

Triangular Knives

As implied by its name, the triangular knife has three cutting edges. These knives can be ground either concave or convex (Figure 17-8). Concave or hollow-ground knives have very sharp edges, but they dull quickly and are more likely to chip. Convex-ground tools have a more rigid body behind each edge and, as a result, generally last longer. In addition, they can be reground much faster with conventional equipment than concave tools.

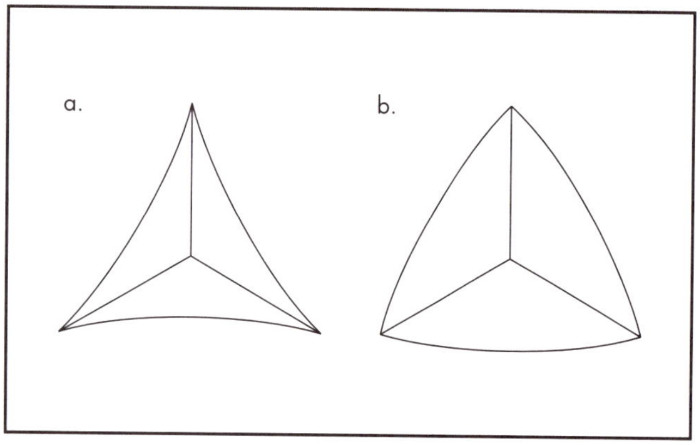

Figure 17-8. Triangular knife, (a) concave grind; and (b) convex grind (concavity and convexity exaggerated for clarity).

Oval Knives

Oval knives (Figure 17-9) have a cylindrical body and a flat surface ground at an angle, which provides a working edge with an oval shape. These knives are designed primarily for removing burrs from holes, but are often used for chiseling out burrs and accessing difficult-to-reach areas.

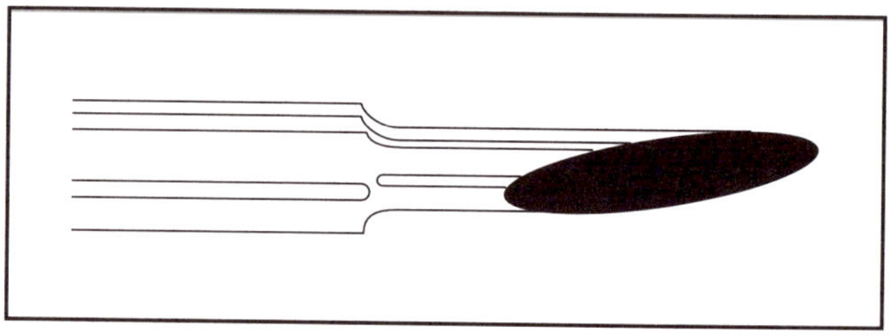

Figure 17-9. Oval knife.

Angle-hooked Knives

Angle-hooked knives (Figure 17-10) have two cutting edges at 90° to the shank. This design allows the user to remove the edge burrs along a slotted groove, sometimes removing burrs from both edges of the slot at the same time. Special angle-hook knives are designed to clean out and deburr threads or to scrape along a single edge.

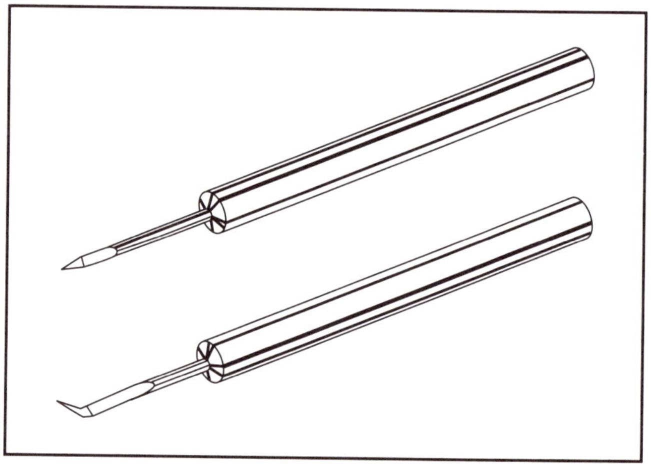

Figure 17-10. Top and side views of angle-hooked knife often used for deburring slots.

Flat-pointed Knives

Flat-pointed knives, also called chisels, are generally cylindrically shaped with a flat ground on each side, resulting in a flat, sharp edge similar to a screwdriver. The tool is used with a gentle pushing motion to scrape burrs. A typical application is cleaning chips from the bottoms of blind-tapped holes.

Cone-end Tools

Cone-end tools have a double cone (back-to-back) with a flat surface ground to the centerline (Figure 17-11). This design enables the tool to remove burrs from hard-to-reach

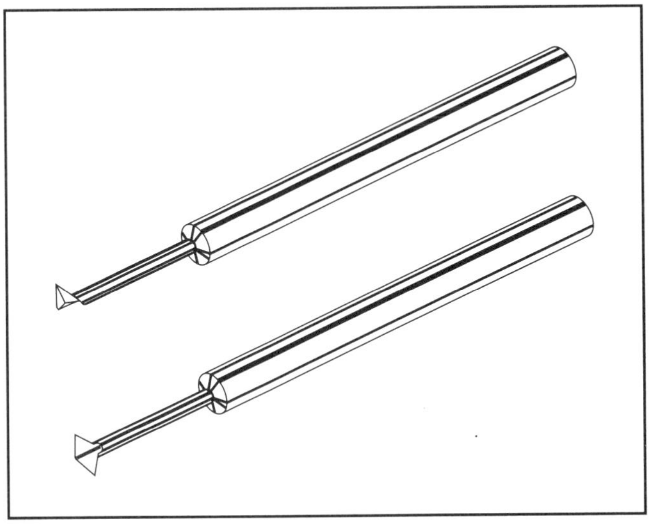

Figure 17-11. Top and side views of cone-end knife/scraper.

areas such as blind holes. The tools are often used to clean and deburr threads and to deburr slots in the walls of tubes. Some users prefer this type of tool to an angle-hook tool for deburring large-sized threads or tapped holes.

Knife/Workpiece Relationship

Normally, workers should not hold workpieces down on a table or microscope base. It is important for both personal safety and workpiece quality to hold workpieces above a solid surface. If necessary, wrists or arms may rest on a tabletop. When using a table to support workpieces, any sudden lunge with the hand holding the knife may gouge the workpiece or cut the other hand. Abrupt changes in burr sizes and locations can also cause lunges. By holding the workpiece in the air, the workholding hand can quickly and easily respond to sudden motions of the knife-holding hand.

17.6.2 Files

More than 2,700 different files in a variety of sizes are commercially available. Common types include Swiss escapement, parallel machine, needle, and riffler. Files, available in a variety of miniature sizes, are rectangular, triangular, round, curved, or bent. Some are made from high-speed steels and others have carbide or diamond particles bonded onto steel shanks. Most files can be used on any workpiece, though it is easy to exceed .003 in. (0.08 mm) edge breaks with many of these tools. Needle files have almost-sharp points, generally with a radius of about .030 in. (0.76 mm). The narrowest rectangular file is about .015 in. (0.38 mm) thick. Some triangular files have knife-like edges thinner than .015 in. (0.38 mm). All files can be further thinned if required.

17.6.3 Rotary Burs

Rotary burs, bur balls, and related tools, sometimes called rotary files, are available in a variety of diameters, shapes, tooth coarseness, and materials (Figure 17-12). Shapes include round, oval, cone, inverted cone, cup, barrel, cylinder, wheel, tree, flame, knife edge, and round edge. Although these tools are sometimes called *burrs,* it is preferable to refer to them as *burs* or *rotary burs* to eliminate confusion between the tools and the material being removed. For hand deburring of most miniature workpieces, a bur tool is placed in pin vises and rotated by hand. For faster action, tools are inserted in bench-mounted electric motors, air motors, and dental motors. They are also used on various machine tools.

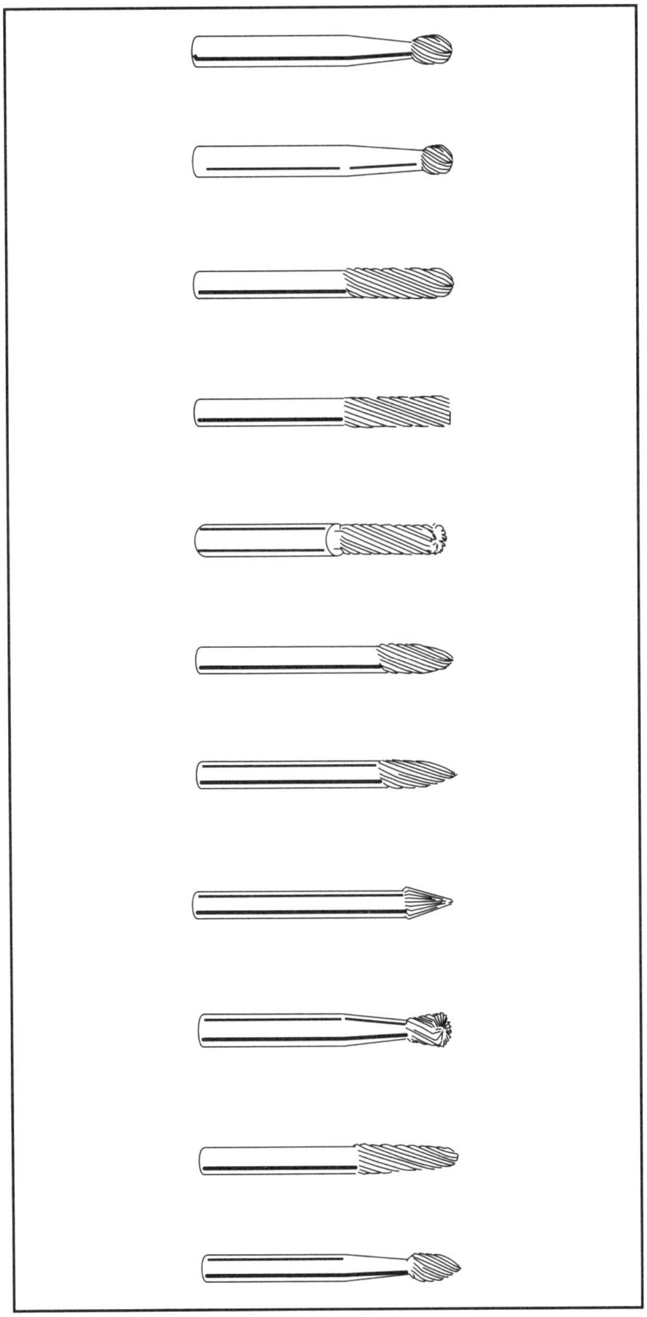

Figure 17-12. Rotary files (burs) used for hand deburring.

The smallest commercially available rotary bur has a .004-in. (0.102-mm) diameter ball, but the shanks are normally much larger, which may present a problem for internal deburring. Rotary burs tend to chatter or cut too deeply unless very fine teeth are used. Diamond-plated burs are being used with good results. Several shapes of these tools are used for deburring the back sides of holes that may be inaccessible by other means.

17.6.4 Brushes

The seven basic types of brushes used for hand deburring are radial, cup, end, tube, cross hole, side action, and toothbrush. The sizes of commercially available brushes used for hand deburring are shown in Table 17-4.

Radial Brushes

Radial brushes, which resemble wheels with spokes extending from them, are the most commonly used brushes. They are generally used in a bench motor but may be used by hand. They should never be used with high-speed air motors because fibers can be thrown from the brushes, and small-diameter shanks may bend.

Table 17-4. Size ranges for commercially available standard brushes used in hand deburring

Brush type	Filler material	Size range, in. (mm)
Radial	Stainless steel	.75–15 (19–381)
	Brass	.75–6 (19–152)
	Nylon	.75–2 (19–51)
	Abrasive-filled nylon	1.38–14 (35–356)
	Nonwoven abrasive-filled nylon	2–6 (51–152)
	Tampico	2–16 (51–406)
	Cord	1–16 (25–406)
	Abrasive tipped	.50–2 (13–51)
	Elastomer	1.25–12 (32–305)
	Flap wheel	1–16 (25–406)
	Felt bob	.25–3 (6–76)
	Abrasive-filled muslin	.19–1 (5–25)
Cup	Stainless steel	.75–6.25 (19–159)
	Brass	.75–3.25 (19–83)
	Elastomer	4–6 (102–152)
End	Stainless steel	.25–1 (6–25)
	Elastomer	.50–1 (13–25)
Cross hole or tube	Stainless steel	.03–3 (0.8–76)
	Brass	.25–2 (6–51)
	Nylon	.25–2 (6–51)
	Elastomer	.50–1 (13–25)
Side-action or sibot	Stainless steel	.25–1.25 (6–32)
	Brass	.25–.88 (6–22)
	Nylon	.25–.88 (6–22)
	Abrasive-filled nonwoven nylon	.50–3 (13–76)
Flare	Steel	.50–4 (13–102)

Cup Brushes

Inverted cup brushes are useful for deburring the edges of relatively large gears, workpieces from which short shafts extend, and difficult-to-reach areas where a soft action is required. They are also useful for deburring workpieces that can be held in the hand and larger workpieces or features with diameters of .5 in. (13 mm) or more. Figure 17-13 illustrates several applications for cup brushes.

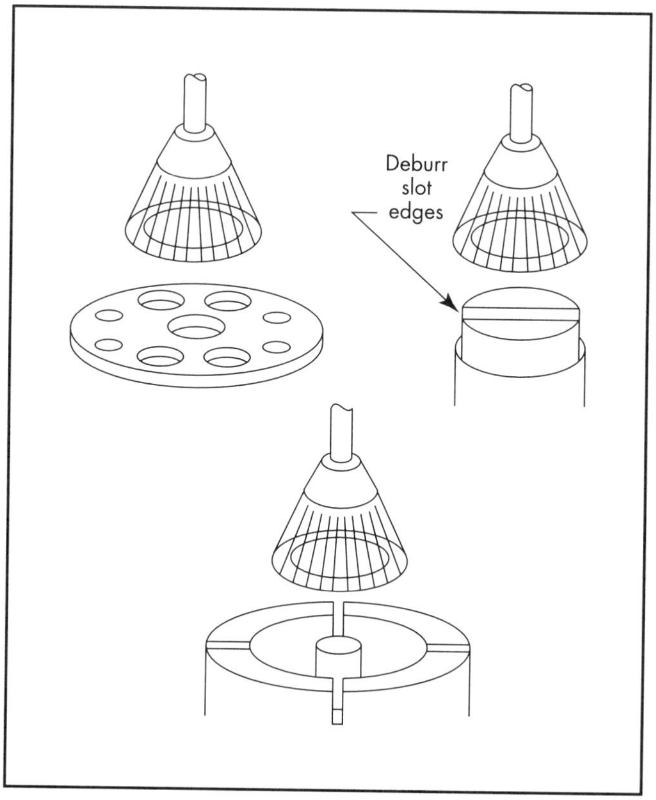

Figure 17-13. Typical applications for cup brushes.

End brushes

On end brushes, the fibers all extend axially outward from the ends of the brushes. End brushes are relatively stiff and are often used in hard-to-reach places. A typical application is shown in Figure 17-14. End brushes are also used for removing recast metal from EDM operations, as well as for cleaning the workpieces.

Tube Brushes

Tube brushes are often used for deburring the insides of tubes, through holes, or relatively deep blind holes. They are available in a variety of stiffnesses, materials, and diameters. In general, they are designed to be softer and more flexible than side-action brushes.

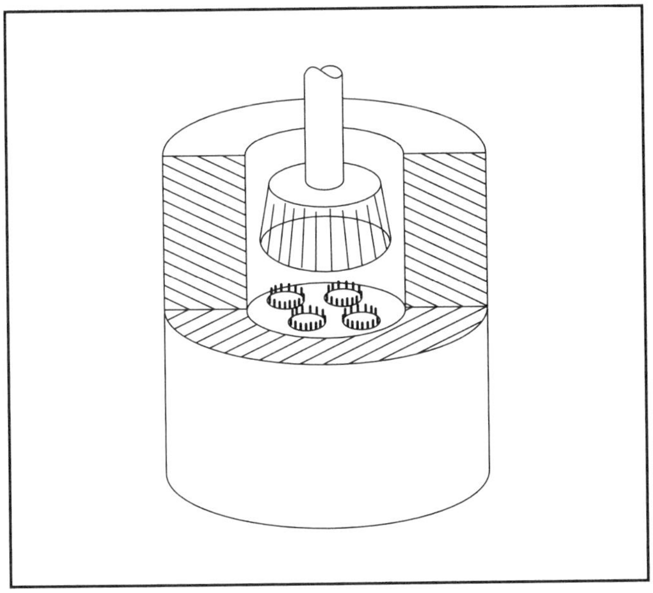

Figure 17-14. An application for end brushing.

Cross hole Brushes

Cross hole brushes have their stainless steel or nylon fibers extending outward in a helical path around the shanks of the brushes. They are available in sizes as small as .024 in. (0.61 mm) in diameter and can be made smaller. Although cross hole brushes look the same as tube brushes, they are considerably smaller and are therefore classified separately. Because of their small sizes, they are fairly soft acting and therefore are not capable of removing large burrs. They are most commonly used for cleaning and removing fine burrs from small-diameter plain and threaded holes.

If the wrong size of brush is used in a threaded hole, the brush fibers can break loose and become lodged in the hole. In many cases, removing these small fibers is a difficult task. Table 17-5 presents recommended brush sizes for various hole and thread sizes.

Side-action and Related Brushes

Side-action (butterfly) brushes are similar to sibot (square trim) brushes (Figure 17-15). An advantage of side-action brushes is that they are available in sizes as small as .25 in. (6.4 mm) in diameter. They are very aggressive and are typically used for deburring plain or threaded holes. When precision dimensions are involved, the brushes work satisfactorily on stainless steels but not on aluminum alloys. Although these brushes can be used by hand, they are normally employed with some type of motorized or powered tool.

Brush Filaments

A variety of bristle materials are used in the manufacture of brushes. However, only the following eight are extensively used for deburring: carbon steel, stainless steel, brass, tampico fibers, nylon, abrasive-filled nylon, nonwoven synthetic material, and elastomer-filled wire bristles.

Carbon steel brushes are still used because of their low cost and long life; however, they are not acceptable for use on many workpieces because the steel rusts easily, and steel brush particles deposited onto workpieces will cause rust damage. Stainless steel brushes require careful use because they can form large burrs on precision parts and may weld to stainless steel workpieces, resulting in rough surfaces.

Brushes containing miniature brass bristles are excellent tools for producing small edge breaks, yet they are soft enough to prevent new burrs from forming. However, brass easily deposits on workpieces and can cause discoloration. Discoloration is not allowable on most finished workpieces. It can be removed by employing proper cleaning procedures.

Brushes made with tampico fibers are too soft to produce much radiusing action, but they do provide good finishes. If tampico brushes are used with a deburring compound containing abrasive particles, they can produce fine edge finishes or aggressive deburring if required. Such compounds, however, are hard to remove from the workpieces.

Nylon bristle brushes are too soft for removing large burrs, but they are useful for cleaning operations. However, nylon brushes with either aluminum oxide or silicon carbide abrasive embedded in the fibers are among the most widely used brushes for deburring and edge finishing precision miniature workpieces. As the nylon wears, new cutting edges are continually exposed. With gentle use, these brushes will provide no more than a .005 in. (0.13 mm) edge break.

Nonwoven synthetic brushes filled with abrasive are used extensively in the aircraft industry. They can provide either gentle or aggressive action and leave an excellent surface finish. For more aggressive action without undue scratching of the surfaces, some firms use plastics-filled

Table 17-5. Cross hole brush sizes used for deburring various-sized holes and threads

Hole diameter, in. (mm)	Brush size to use, in. (mm)
.0197 (0.5)	.024 (0.61)
.0315 (0.8)	.032 (0.81)
.0394 (1)	.032 (0.81)
.0472 (1.2)	.047 (1.19)
Nominal thread size	**Threads per in. (mm)**
0–80	.047 (1.19)
2–56	.079 (2.01)
4–40	.097 (2.46)
5–40	.109 (2.77)
6–32	.125 (3.18)
6–40	.125 (3.18)
8–32	.142 (3.61)
10–24	.156 (3.96)
10–32	.156 (3.96)

Figure 17-15. Side-action and square-trim brushes used for hand deburring.

metal brushes. Plastics keep the metal fibers from spreading apart and reduce wear, thus making the brush a stiffer tool. These brushes are not acceptable for precision or miniature workpieces.

Additional information about brushes and their operation, factors affecting aggressiveness, side effects of brushing, and safety considerations is presented in Chapter 11.

17.6.5 Abrasive Tools

A variety of abrasive tools are used for deburring. The most common include abrasive-filled rubber, cotton, and cork products; abrasive paper; and hand stones.

Abrasive-filled Rubber Products

For deburring small holes, bullet-shaped rubber tools are used extensively. They are available in four levels of aggressiveness, the degree of which is determined by the size and number of silicon carbide particles molded into the rubber. These products are color coded to indicate their degree of aggressiveness, though color codes vary depending on the manufacturer. If they are used on machine tools, the particles that are worn off may score or scratch the machine ways. If the particles are blown into chuck jaws or collets, they can indent workpieces that are subsequently processed.

Short-shank dental tools are less aggressive than industrial tools of this type, and they can be used on smaller features of workpieces. The small size of dental tools allows them to be used crosswise, with dental motors, in holes as small as .750 in. (19 mm) in diameter.

Abrasive-filled Cotton Products

When more aggressive cutting is required than provided by abrasive-filled rubber, abrasive-filled cotton or muslin tools are often used. By varying the size of the abrasive and the type of bonding agent, it is possible to produce smooth radii and quick deburring action with these tools. Although these tools are relatively unknown in the general metalworking industry, they are used extensively in the aircraft industry. Figure 17-16 shows a variety of abrasive-filled cotton and rubber products.

Abrasive-filled Cork Products

Cork and abrasive particles bonded together with special glues or resins are used for fast cutting with smooth radiusing.

Abrasive Paper Products

A variety of sandpaper-like products are used for deburring and finishing. Miniature disks, with adhesive backs for holding them to mandrels, are used for small workpieces. The smallest commercially available disk has a diameter of .5 in. (12.7 mm), but smaller sizes can be fabricated. Abrasive belts for deburring and finishing are used by almost all manufacturing plants.

Hand Stones

A wide variety of hand-held abrasive stones are used for removing burrs and improving finishes. They are available in sizes as small as .010 in. (0.254 mm) thick or less. Stones are typically used only on small burrs and precision surfaces.

Figure 17-16. Abrasive-filled rubber and cotton products.

Mounted Points

Miniature grinding wheels, known as mounted points, are used for edge finishing by many firms. They are available in a wide variety of sizes, shapes, and materials. Mounted points, however, are too aggressive for many applications and often tend to chatter in operation.

The mounted points used in the dental industry, including diamond-coated types, quickly break edges in hard-to-reach areas, without chatter, and are employed extensively for finishing miniature workpieces. Mounted points made of cotton or muslin fibers, rather than bonded abrasives or ceramics, are softer-acting tools and are used primarily in the aircraft industry.

17.6.6 Other Tools

There are many other tools used for deburring and finishing. These include countersink tools, drills and reamers, felt bobs and lapping compounds, pin vises, motorized tools, swivel-blade tools, peening tools, and specially designed tools.

Countersink Tools

Countersinks are used in almost every deburring facility. One source for these tools lists over 70 standard commercial designs of the basic countersink shape. When the various cutting angles and dimensions are considered, at least 1,500 different combinations of countersink tools are available. Some of these come with 6-in. (152-mm) long shanks to allow access to deep holes and hard-to-reach areas.

Drills and Reamers

These tools are frequently used for deburring small intersecting holes. Reamers are generally used first to remove the burrs and, if possible, to leave a sharp edge. When burrs are too thick to be removed by hand using reamers, drills are used.

Felt Bobs and Lapping Compounds

Occasionally, specially formed felt shapes, called *bobs*, are used with a diamond lapping compound to remove burrs. Usually, however, abrasive-filled rubber products will cross hole brush faster without damaging workpieces. Although lapping compounds are not typically used for removing burrs, they can remove small burrs and produce smooth, lustrous surfaces. They are most easily used on flat surfaces and in holes.

Pin Vises

A variety of pin vises are used for holding workpieces and tools for deburring and finishing. Some pin vises have brass collets or jaws to prevent the workpieces from being indented while they are being grasped. Some vises are located on the outer diameter of workpieces, whereas others grasp the inner diameter. When necessary, special vises can be machined to hold small-diameter or specially shaped workpieces.

Motorized Tools

More than 500 motorized tools are available for deburring and finishing. One supplier offers 15 different dental-tool motors. All can be obtained with belt drive, integral electric, or air-powered motors, as well as flexible-shaft drives, and operate at speeds of 2,000–400,000 r/min. For most applications, motors that operate at 20,000 r/min or less are designed to remove burrs by applying a measurable force to the workpieces; faster motors require only a light touch. Dental motors offer more precision, but generally cannot take the extensive use required in industry.

Swivel-blade Tools

Several vendors offer swivel-blade tools that provide fast deburring of holes and edges. A typical set is packaged with 4–15 different tools, most with replaceable blades for general use (Figure 17-17). Swivel blades are the most widely used tools for hand deburring and are found worldwide in similar sets.

Peening Tools

Special tools have been commercially produced to dull the edges of sheet metal. One of these is a V-shaped clothespin-like tool that is vibrated against sheet metal edges. Another device fits into a drill press and straddles the sheet

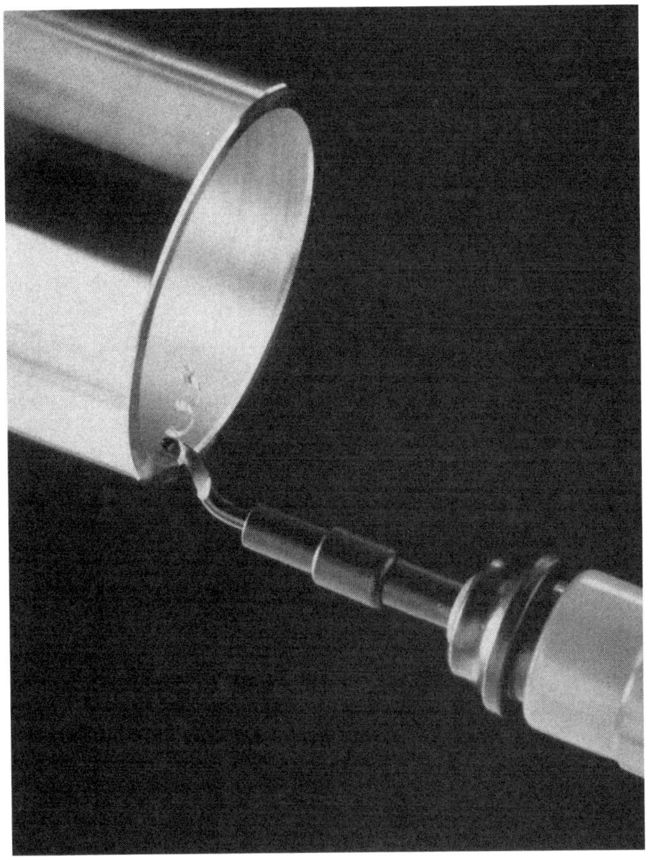

Figure 17-17. Swivel-blade tool. ***(Courtesy Royal Products)***

metal edges. As the tool rotates a cam, small ball-bearing pieces are forced against workpiece edges. The scrapers offered in swivel-blade tool kits provide a more useful edge, but the peening action provided by the drill press tool may be sufficient for some products.

Specially Designed Tools

Whenever conventional tools are not satisfactory for removing burrs, specially designed tools should be considered. One company has designed and used over 60 small, hand-held cutting tools for deburring.

17.7 ENVIRONMENTAL, HEALTH, AND SAFETY CONCERNS

There is little data on hand deburring safety, but knife cuts are one of the most common hazards of hand deburring. When plastics are being deflashed, arms, hands, and fingers can be burned. Workers may need to wear special gloves when handling hot workpieces and when using surgical blades to deflash. Foundry deflashing, cleaning, and snagging, as well as hand sanding and brushing often create very dusty environments. These processes require that appropriate exhaust systems be provided, particularly when stainless steels, nickel, or other suspected or known carcinogens are involved. Some plastics may also produce toxic or unhealthy fumes or dust.

Brushes can throw hand-held workpieces, so normal shielding and related preventive measures may be needed. Rotating motors used in the deburring process pose a potential hazard to operators' eyes, so safety glasses are generally required. Microscope use can result in eye fatigue for some operators; however, most full-time operators who use microscopes do not object to their use because they get used to them. Microscopes require cleaning and adjustments every 3–6 months.

Ergonometric issues are a concern in hand deburring. Using the same tool in the same way for hours, days, and weeks can cause repetitive-motion injuries, which can be very serious and debilitating to operators and can result in massive amounts of lost labor hours. Lifting or moving very heavy or bulky workpieces may lead to back and back-related injuries.

References

Gillespie, LaRoux K. 1979. "Training Manual for Hand Deburring, Part 1." Division report BDX-613-2245. Kansas City, MO: Bendix Corporation, available from NTIS.

———. 1980a. "Training Manual for Hand Deburring, Part 2." Division report BDX-613-2534. Kansas City, MO: Bendix Corporation, available from NTIS.

———. 1980b. "Training Manual for Hand Deburring, Part 3." Division report BDX-613-2572. Kansas City, MO: Bendix Corporation, available from NTIS.

———. 1981. "Training Manual for Hand Deburring, Part 4." Division report BDX-613-2582. Kansas City, MO: Bendix Corporation, available from NTIS.

Bibliography

Gillespie, LaRoux K. 1978. "Hand Deburring: A Plague or the Most Economical Way?" *Machine and Tool Blue Book*. Wheaton, IL: Hitchcock Publishing Co.: 102–13.

———. 1976. "Deburring Capabilities for Miniature Precision Parts." Division report BDX-613-1604. Kansas City, MO: Bendix Corporation, available from NTIS.

———. 1977. "Control of Plant-wide Deburring Costs." Technical paper MRR77-05. Dearborn, MI: Society of Manufacturing Engineers (SME).

———. 1982. "The Management of Deburring." Technical paper MRR82-04. Dearborn, MI: Society of Manufacturing Engineers (SME).

———. 1996. *Guide to Deburring, Deflashing, and Trimming Equipment, Supplies and Services*. Kansas City, MO: Deburring Technology International.

———. 1998. *25 Special Tools for Hand Deburring*. Kansas City, MO: Deburring Technology International.

———. 1999. *Inspecting for Burrs*. Kansas City, MO: Deburring Technology International.

Gillespie, LaRoux K., and McMillen, Jack. 1978. "Hand Deburring, a Necessity that can be Improved." Technical paper MRR78-03. Dearborn, MI: Society of Manufacturing Engineers (SME).

Ostwald, Phillip. 1985/1986. "Deburring." *American Machinist Cost Estimator.* New York: American Machinist.

18 Sheet Metal Deburring

Some deburring processes are particularly suited to sheet metal or pressworked parts. Fully automated, semi-automatic, and manual equipment are available, and several different mechanisms are used. Millions of workpieces are deburred annually using sheet metal deburring techniques, particularly those made from harder materials. Both strip and punched workpiece configurations can be deburred.

Sheet metal deburring processes include press trimming, burr roll back, skiving, edge grinding, pinch rolling, swaging, coining, burnishing, broaching, edge rolling, flanging, beading, peening, flat sanding, brushing, and many of the processes described in other chapters.

As noted in Chapter 3, there are pressworking processes that do not generate burrs, such as opposed punch push-back and blanking opposed die. Although these processes are not generally available, the burr-free edges they produce make them potentially desirable and may warrant the additional effort of cheaper machine design to make them commercially viable.

18.1 STANDARD EDGE CONFIGURATIONS PRODUCED IN STRIP ROLLING

When a roll of metal sheet is pulled under a rolling cutter, strips are produced. In this strip rolling process, burrs traditionally form on both sides of the strip. The contour of the edge of a strip depends on the process used for producing the strip. Five contours are recognized:

1. Sheared strips (strips produced in a shear) have burnished bands along the edges on *opposite* sides of the strip (Figure 18-1a). If burrs are produced because of dull cutting edges, they will also occur on opposite sides of the strip. In addition, sheared strips often become spiraled or curved because the upper blade of the shear is at an angle to the lower blade. This makes the strips difficult to feed through the die unless they are first straightened.
2. Strips produced in the slitter have the burnished bands on the same side of the strip (Figure 18-1b). Blanks produced from these strips in cut-off dies have a better appearance and are fed more easily because they are straighter. Sheared or slit strips may be either produced in the shear department of the plant or ordered directly from the mill.
3. Mill-edge strips have a radius at each corner (Figure 18-1c). They are produced by rolling sheared or slit strips at the mill. Mill-edge strips are used for long stampings, such as handles, shelf brackets, and other parts where sharp edges would be objectionable.
4. Rolled-edge strips have a full radius at each side and are rolled at the mill (Figure 18-1d). They are used for parts where appearance is a factor, such as ornamental grills and gratings.
5. Square-edge strips are ordered from the mill when the sides of the strips must be square and smooth (Figure 18-1e). The widths of these strips are held very accurately. Square-edge strips are specified when blanks are to be bent or formed edgewise. The square edges prevent cracking or splitting in bending and forming operations.

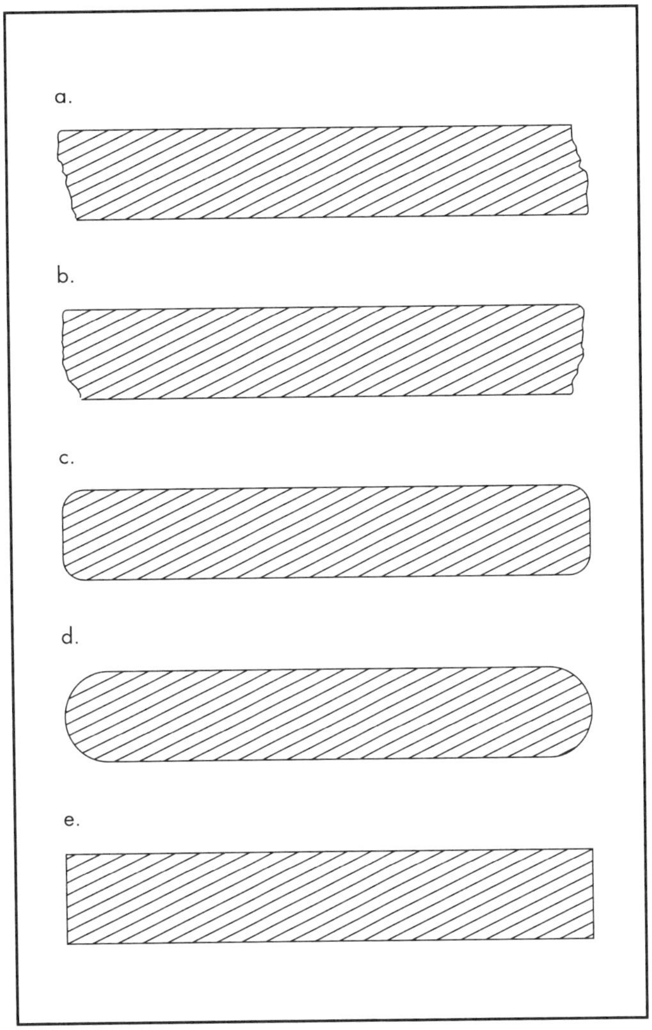

Figure 18-1. Standard edge configurations for strip metals (Paquin and Crowley 1987).

Cold-rolled strip is commonly produced with an as-slit edge (number 3 edge). The definition of the common edges are presented in the following paragraphs (Colombier 1990).

ASTM Standard A 109-93 lists six standards for the edges of carbon cold-rolled steel strip:

1. Number 1 edge (corresponds to items 4 and 5 in the preceding list) is either round or square as specified and is produced by rolling or filing. This edge is often specified when electroplating or when very accurate widths are required.
2. Number 2 edge has no edge processing and is a natural mill edge.
3. Number 3 edge (corresponds to item 2 in preceding list) is an approximately square edge produced by slitting, but the burr is not removed. Because of coiling and stacking of sheets, the burr may be horizontal or vertical.
4. Number 4 edge (corresponds to item 3 in preceding list) is rounded by edge rolling. It is produced from hot-rolled strip or slit-edge strip. This edge has more width tolerance than the number 2 edge.
5. Number 5 edge (corresponds to item 5 in preceding list) is an approximately square edge produced by rolling or filing after slitting. The burr is removed.
6. Number 6 edge (corresponds to item 5 in preceding list) is a square edge from hot-rolled strip or slit-edge strip. This edge has more width tolerance than the number 1 edge.

The round edge is not merely a safety and aesthetic feature. Some high-performance applications require that the brittle material at an edge be removed before the product is sold to prevent the possibility of strip breakage emanating from edge cracks (Bowman 1987).

In many instances, sheet metal burrs are allowable if they do not exceed a certain percentage of the metal thickness; Figure 18-2 shows how the calculation is made. Burr height is determined by subtracting metal thickness from the apparent thickness, including the burr. For example, a burr of .0015 in. (0.038 mm) on a .020 in. (0.51 mm) thick strip represents a burr of 7.5%. Although no industry standard exists for "acceptable" burrs, even in the sheet metal industry, common practice is to accept burrs as long as they are no taller than the following values (Strasser 1983):

- 10% of material thickness for light-gage stock.
- 5% of material thickness for heavy-gage stock.
- No taller than .006 in. (0.15 mm).
- Burrs that are .001 in. (0.03 mm) tall are always acceptable, even if they constitute more than 10% of stock thickness.

It is fairly common with sheet metal workpieces to leave the burr and simply reorient it. Figure 18-3 shows examples of sheet metal edge conditions. In one approach to reorienting a burr (Figure 18-3b), a series of rolls flattens the burr by bending it 90° from a vertical to a horizontal plane.

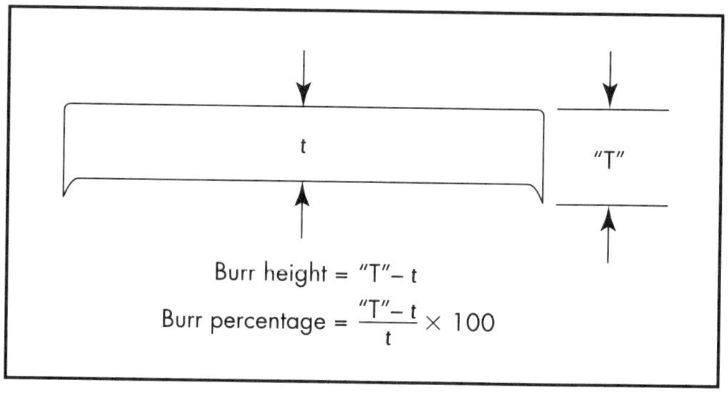

Figure 18-2. Calculating burr size as a percentage of metal thickness (Bowman 1987).

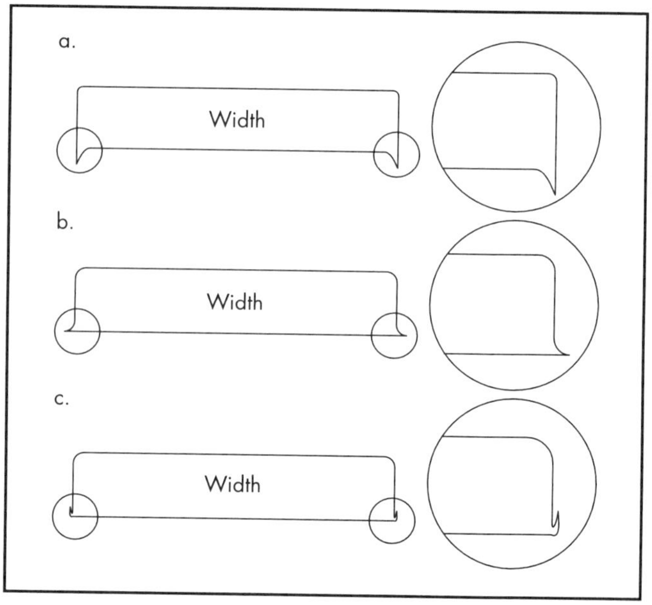

Figure 18-3. Sheet-metal edge conditions: (a) unconditioned edge, (b) conditioned edge where burr has been rolled 90°, and (c) conditioned edge where burr has been rolled and mashed (Bowman 1987).

A truly deburred edge would have a small chamfer on the formerly burr-laden edge.

18.2 SHEET-METAL DEBURRING PROCESSES

18.2.1 Trimming on Presses

Mechanical, hydraulic, or pneumatic presses are used extensively for removing excess material from stampings and drawn workpieces, as well as for deflashing plastic workpieces. Whereas the stock trimmed from sheet metal workpieces are not burrs, the flash from plastic workpieces has many of the same characteristics as burrs.

The sheet metal industry has led the way for innovative die designs to help trimming. Trimming requires the use of cutting dies, such as pinch-trim, Brehm® (shimmy), and notching dies (Figure 18-4). Several variations are available

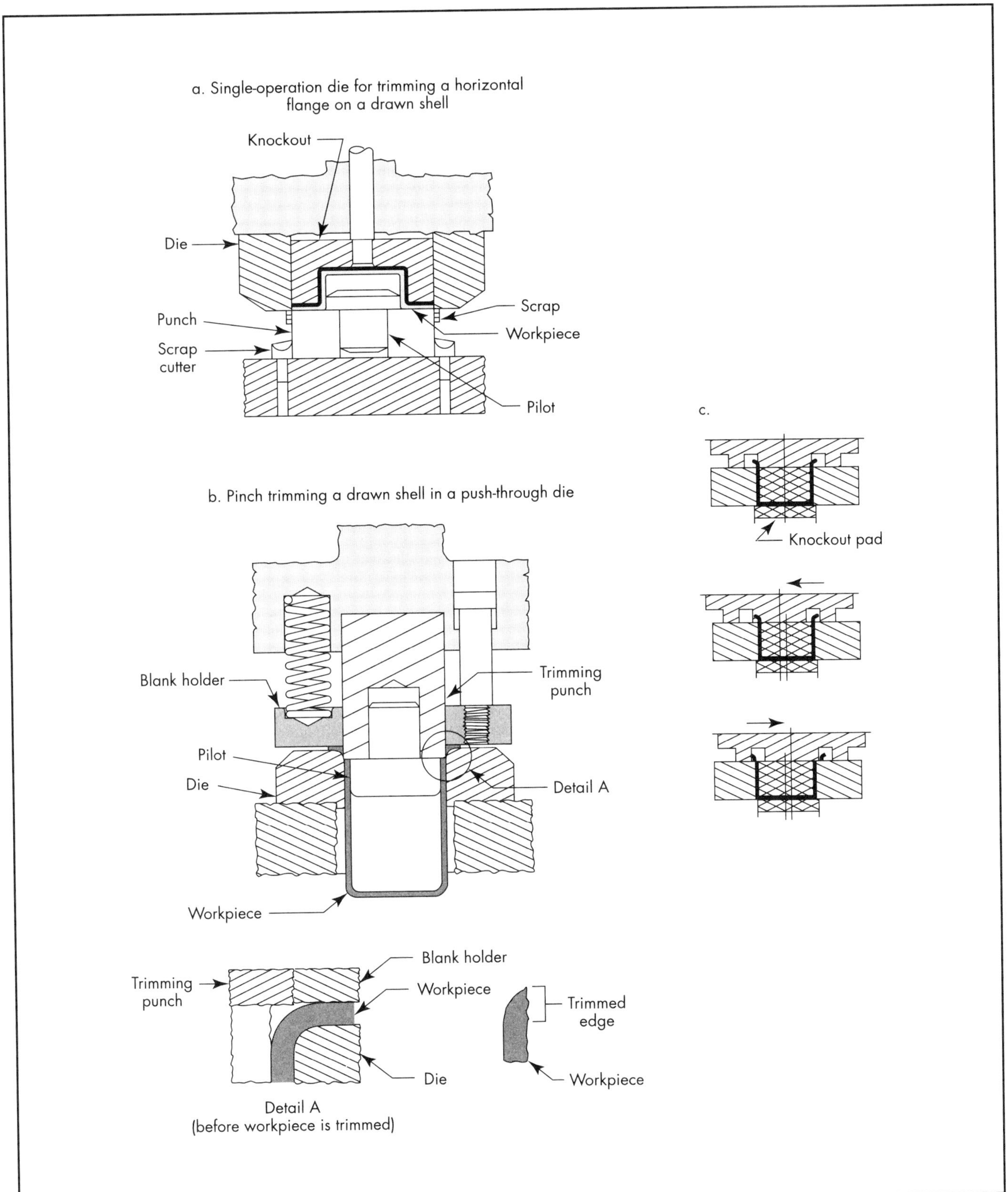

Figure 18-4. Trimming dies: (a) standard approach, (b) pinch trimming, and (c) shimmy trimming (Wick and Veilleux 1985; Lyman 1969).

(Smith and Bakerjian 1990). Trimming can involve standalone manual machines or integrated lines, or it can be part of transfer die lines. Trimming is also often combined with flanging or curling edges to produce a truly finished edge. Shimmy dies move horizontally in four directions to trim. Shaving dies and broach tools are employed for more precise requirements.

Mechanical presses provide 90–300 strokes per min, and workpiece output is typically 600 workpieces per h. Trim accuracies of .005 in. (0.13 mm) are possible. Automobile tire rims are end trimmed or nipped to tolerances of .010 in. (0.25 mm) maximum protrusion left (Kucklick 1996). Brushes and planishing rolls may be required to further finish the edges. Several presses can be equipped with rotary load-unload tables for rapid handling of small workpieces. Small benchtop presses are also available for trimming workpieces that require 4 tons (35.6 kN) of force or less. Broaches are used when conventional trim dies are not adequate.

18.2.2 Presswork Countersinking

Countersinking with punches and dies, also known as *dimpling*, is often used when the sheet metal is too thin for true countersinking with a cutter. To remove burrs from holes, holes can be countersunk as shown in Figure 18-5. Even if the burr is not completely removed, countersinking at least places it in an unobtrusive location.

18.2.3 Edge Curling

Curling dies roll the edges of workpieces (Figure 5-1). For many workpieces, this process saves a step because it simultaneously hides the burr and provides an aesthetic edge (Wilson 1965). This process is called *wiring* when the sheet is curled over a length of wire.

18.2.4 Edge Folding

Edge folding, also known as *hemming* and *folding*, involves mashing and folding over edges containing burrs, as shown in Figure 18-3. The burr still exists, but it is folded out of the way. A variety of edge-folding machines are available. For straight sides, simple folding presses are used, and in Europe some rotary machines are being used for edge folding (for example, burring or flanging machines and Jennies) (Langton 1963).

18.2.5 Edge Machining (Edging)

For some applications, trimming is performed by roller cutters (edging) rather than a press-mounted die. This approach generally requires less-expensive tooling than when trim dies are used. Roller dies can accommodate stock thicknesses of up to .5 in. (12.7 mm) and production rates of up to 500 workpieces per hour. These tools utilize rollers that have a cutter at their center (Figure 18-6) (Wilson 1959).

Another sheet metal edger tool uses two adjustable rotary burs set face to face to chamfer both the top and bottom of sheet metal (Figures 16-23b and 18-7). By loosening a set screw in one cutter, the cutter can be set for different sheet widths. These tools can be used in a drill press or similar hand motor, as well as in more mechanized equipment.

Some sheet metal edgers use small grinding wheels to remove burrs. Edgers employing grinding wheels can generally be adjusted to vary the chamfer produced, and some will deburr both the top and bottom edges on one side of the sheet (Figure 18-8). Such machines are available for sheet stock with thicknesses of .023–.250 in. (0.58–6.35 mm) and widths of .5 in. (12.7 mm) or more. Linear feed rates of 60 ft/min (18.3 m/min) are possible. For heavier grade materials, the machine and cutters shown in Figures 16-1 and 16-2 are used.

18.2.6 Edge Rolling

Edging machines using pinch rolls to remove burrs or reroll the material into the edges can deburr both sides, as well as the top and bottom, of sheet stock in a single pass (Figure 18-9). These machines can process sheets ranging in size from .023 to .120 in. (0.58–3.05 mm) thick and 48 in. (1,219 mm) wide. Feed rates of up to 100 ft/min (30.5 m/min) are possible (Barnes 1996).

On some machines, the stock passes through a set of wide pinch rolls that feed the stock; the next set of rolls, mounted on a vertical axis, sets up the edges. Subsequent sets of power-driven rolls perform the edging and finishing operations. Either chamfered or radiused edges can be produced. The rolls, similar to small V-belt pulleys, are hardened and ground. Pressure exerted by the rolls progressively forces the burrs and sharp edges over, under, or into the stock. Although this method works well with steel, stainless steel, and most aluminum sheets, it will not deburr dead-soft copper or aluminum, which tends to buckle under roll pressure (Langton 1963).

Edge rolling reportedly work-hardens the edge, which can cause forming problems (Kelley 1983). Edge rolling is typically used to produce a fully round edge, commonly called a number 4 edge (Figure 18-1d).

18.2.7 Edge Skiving

In skiving, a small scraper-like tool is used to scrape burrs off strip stock. This tool provides good edge quality at constant feed rates of 400 ft/min (122 m/min); erratic feed will result in chatter or other edge imperfections. Since it generates swirls of fine burrs or wire, the swarf must be planned for removal. The edge produced in skiving is basically square and is commonly referred to as a number 5 edge (Figure 18-1e) (Barnes 1996).

18.2.8 Edge Shaving

Radii can be produced by shaving tools (Kelley 1983). Shaved edges will form better than rolled edges because they are not work-hardened.

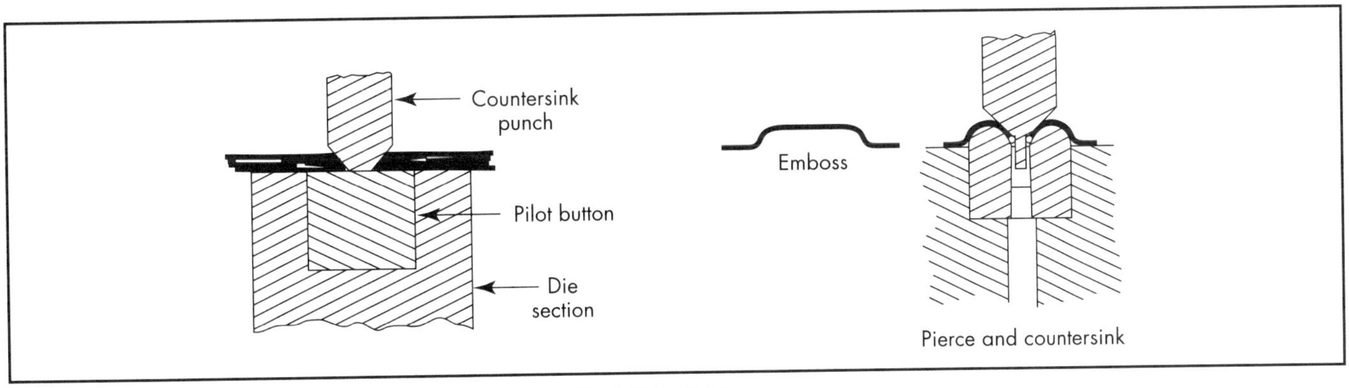

Figure 18-5. Countersinking with sheet metal dies (Kucklick 1996).

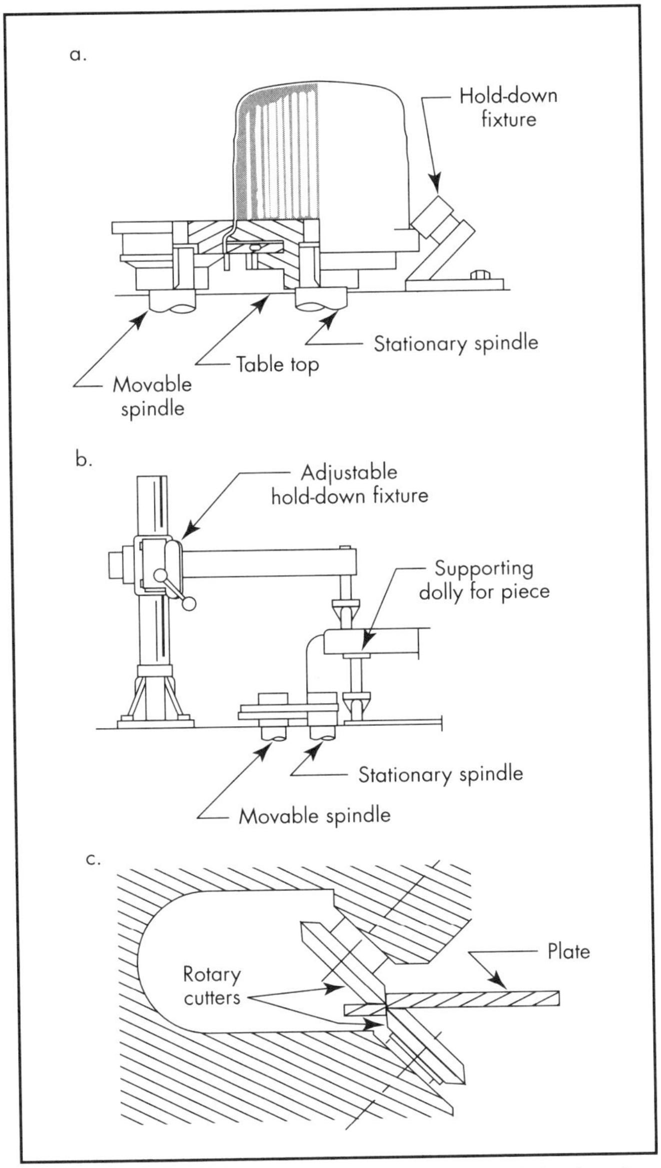

Figure 18-6. Roller-based sheet metal edger: (a) trimming deep drawn housing, (b) trimming steel cabinet top, and (c) close-up view (Wilson 1959).

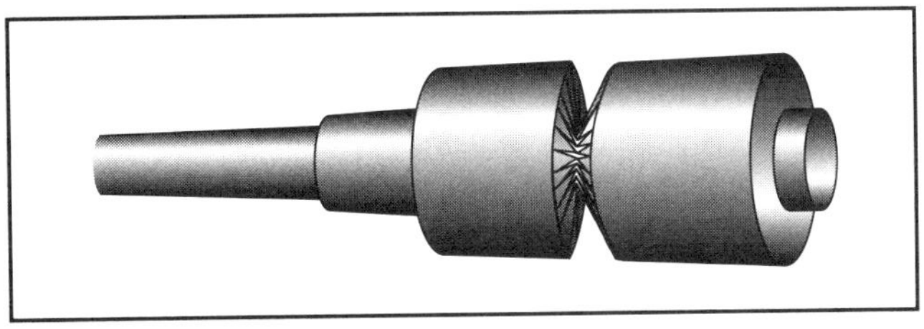

Figure 18-7. Rotary bur-based sheet metal edger. ***(Courtesy Severance Tool Industries)***

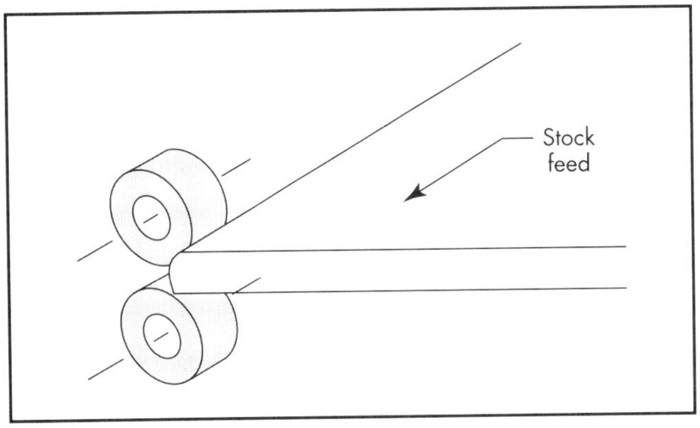

Figure 18-8. Small grinding wheels for deburring and chamfering top and bottom edges of sheet stock at feed rates of up to 60 ft/min (18.3 m/min).

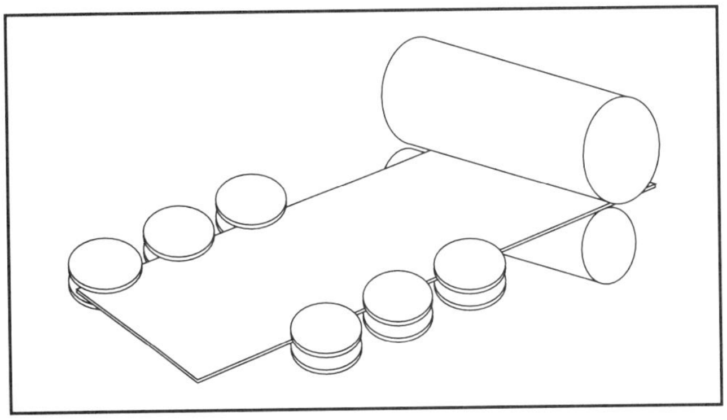

Figure 18-9. Pinch rolls for removing burrs from sheet stock with thicknesses of .023–.120 in. (0.58–3.05 mm).

18.2.9 Edge Swaging

Although deburring sheet metal on the same machine that produced the burr may not seem possible, press equipment is being used in this way by employing swaging dies (Figure 18-10) (Kucklick 1996). Much more can be done with coordinated efforts in this arena. Eliminating separate deburring operations will result in fewer errors, less scrap, and cost and space savings.

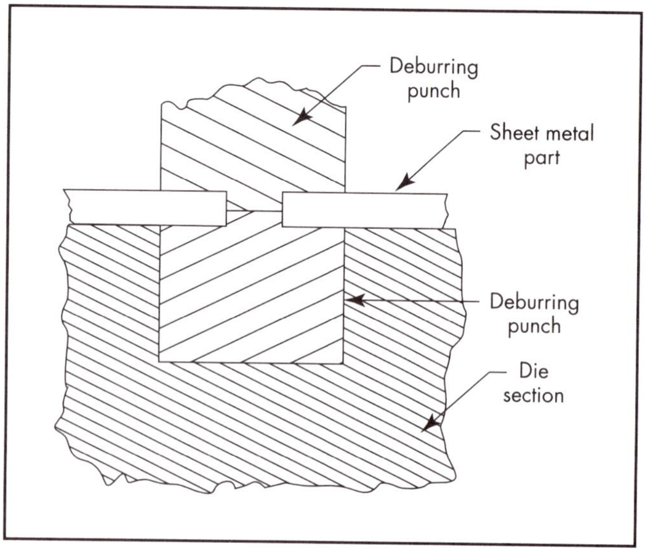

Figure 18-10. Deburring sheet metal in swaging dies (Dallas 1962).

18.2.10 Edge Coining

Any edge configuration desired can be coined on many workpieces. For example, if a large chamfer is desired, it can be coined at high production rates. A radius or combinations of flats and radii can be coined. This technique is not widely used, but it is a viable method for high-speed production and requires a specific coining die.

18.2.11 Edge Burnishing

When hard rolls are rubbed against an edge containing fine burrs, the burrs are rubbed off and the surface is *burnished*. The process can be used for aluminum and copper, as well as for harder materials. The process requires only the use of a wheel, making it very flexible.

18.2.12 Wheel Blending

Wheel blending is similar to edge burnishing, but wheel blending employs a hard, abrasive-filled rubber or cork wheel to smooth edges. Wheel blending is generally used for glass edges but also can be used on metal edges.

18.2.13 Mechanical Peening (Deburring)

Two different tools have been used for peening edges in manual deburring of sheet metal. The first tool employs a bullet-nosed device with a slot cut in its end that is placed in a drill press or drill. The rotation of the drill produces an axial motion that causes the tool to pound against the sheet metal edge, thus dulling it. The second tool employs a rotary disk about 2 in. (51 mm) in diameter that contains a set of small balls sandwiched in the middle, like an Oreo® cookie. The sheet metal workpiece is pushed against the ball-filled center of the disk, and when the drill motor is turned on, a cam in the middle rotates, pushing the small balls against the edges of the workpiece. The continual pounding action of the balls against the sheet metal edges dulls the edges and beats over any burrs (McMaster-Carr 1977). Three sizes are available to accommodate sheet metal that is .010–.125 in. (0.25–3.18 mm) thick.

18.2.14 Bonded Abrasive and Brush Deburring

Figure 18-11 shows an automated machine for deburring flat sheet-metal workpieces using a bonded abrasive. The machine consists of two modules, each containing two media wheels, which are webs of abrasive-impregnated fibers, and a hot air dryer. The four counter-rotating wheels rotate at a speed of 4,000–5,000 surface ft/min (1,219–1,524 m/min) and oscillate along their axis of rotation. Water is applied for cooling and rinsing and is then filtered before being recirculated. Workpieces are fed through the machine by a chain-driven conveyor with rubber-covered rolls. For a one-pass operation, conveyor speed is normally 20–30 ft/min (6.1–9.1 m/min).

Flat-belt sanding machines and brushing machines for general shop use, and wide-belt machines handle piece parts and strips. The soft elastomer backing of the rolls used in sanding allows the belts to reach down below workpiece surfaces and generate a small radius or break. These machines are widely used for metal strip and printed circuit board deburring.

As discussed in Chapter 11, some machines utilize brushes in place of bonded-abrasive products, and some machines combine both types of tools for more effective deburring. Brushes are low cost and are available in hundreds of combinations of aggressiveness, brush material, configuration, and size, and they can be built into an in-line system to quickly remove burrs before workpieces are individually handled.

Figures 18-12 through 18-14 are photomicrographs showing the results of deburring clad sheet aluminum in a machine for deburring printed circuit boards using three-dimensional, nonwoven abrasive-flap wheels. This aluminum is very soft, and maintaining the .0016-in. (0.041-mm) clad thickness is critical for both .032-in. (0.813-mm) and .064-in. (1.63-mm) thick sheets. The photomicrographs confirm that no thickness was removed by any of the abrasive brushes used for deburring. The flap brush (Figure 18-15a) produced smoothly rounded edges, but it did not last long enough for extensive use. The disk brush (Figure 18-15b) lasted for approximately 10,000 workpieces,

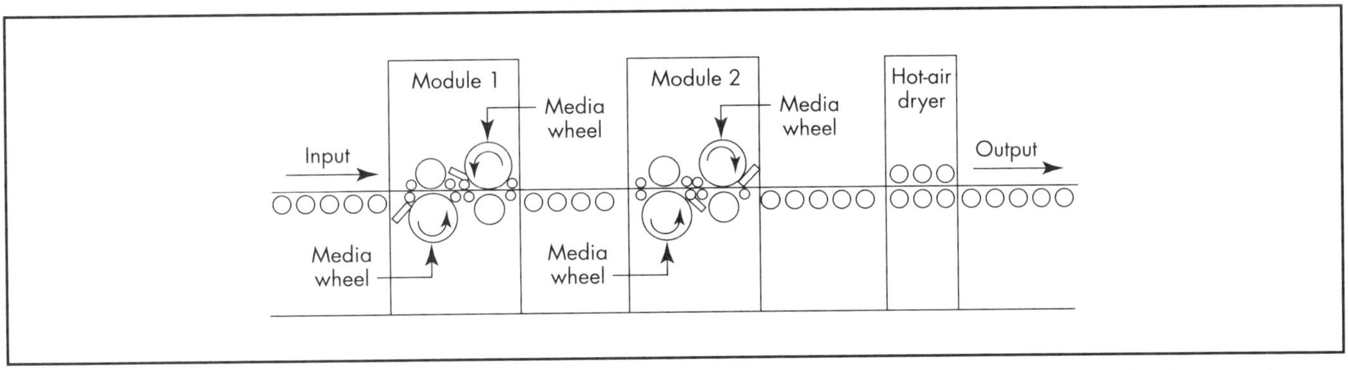

Figure 18-11. Automated machine for deburring flat sheet-metal workpieces. ***(Courtesy Lockheed-Georgia Company)***

rounded the edges appropriately, and did not remove cladding except at the edges of the workpiece. The photomicrographs in Figure 18-14 show the results of deburring with a conformal brush (convolutely wound wheel); the photomicrographs are cross sections of number 40 holes (.098 in. [2.49 mm] in diameter). The ability of this brush to penetrate into these small holes is particularly noteworthy (Behringer 1981). Figure 18-15 shows the various abrasive brushes made from three-dimensional abrasive sheets used in this study.

When a wide brush is used (such as 36 in. [0.9 m]), it is possible to deburr up to 18 pieces per pass and achieve production rates of up to 2,400 workpieces per hour for workpieces 1 in. (25.4 mm) in diameter or length with a feed rate of 10 ft/min (3.1 m/min).

A variety of other mechanical brushing applications using abrasive-filled nylon brushes are possible (for examples, see Flores 1992).

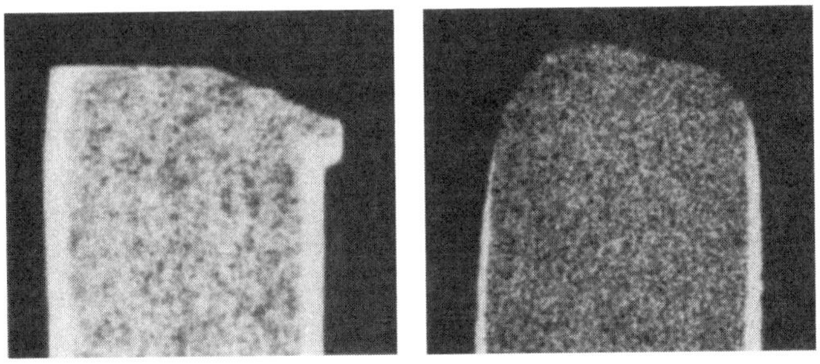

Figure 18-12. Photomicrograph (85×) of 2024-0 clad sheet aluminum before and after deburring by three-dimensional abrasive sheets made into a flap brush.

18.2.15 Other Applicable Deburring Processes

A wide variety of common deburring processes are applicable to sheet metal workpieces (Table 18-1). Electrochemical deburring (ECD), for example, is ideal for finishing multiple hole patterns, small burrs, and precision workpieces. The burrs dissolve in 5–10 s with little effect on the rest of the workpiece. Although stampings are not normally processed with ECD, most metals other than titanium can be successfully deburred using this process. The scrap can be used as the electrode, and there is no potential for the burr to short out the tool as can happen with long machining burrs.

Abrasive flow processes allow inexpensive tooling to be used and produce smooth, burr-free edges. The processes work best when edges are open to allow media to flow over them—stamped workpieces meet this criterion.

Abrasive blasting cuts burrs or beats them over, depending on workpiece material, burr size, media used, and process parameters. Abrasive blasting is fast, easy-to-use, and effective for workpieces as small as semiconductors or as large as ship side plates.

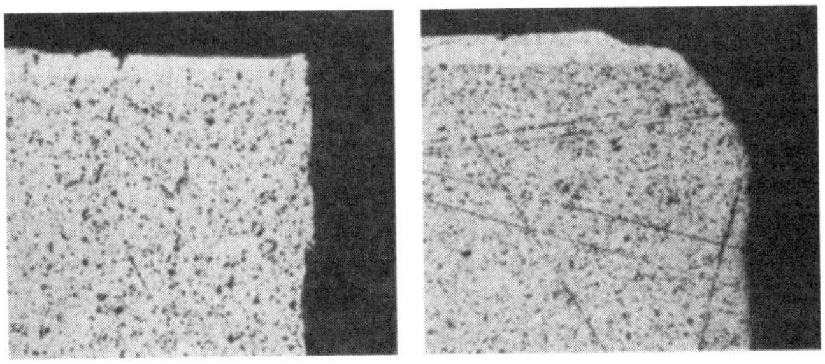

Figure 18-13. Photomicrograph (85×) of 2024-0 clad sheet aluminum before and after deburring by three-dimensional abrasive sheets made into a disk brush.

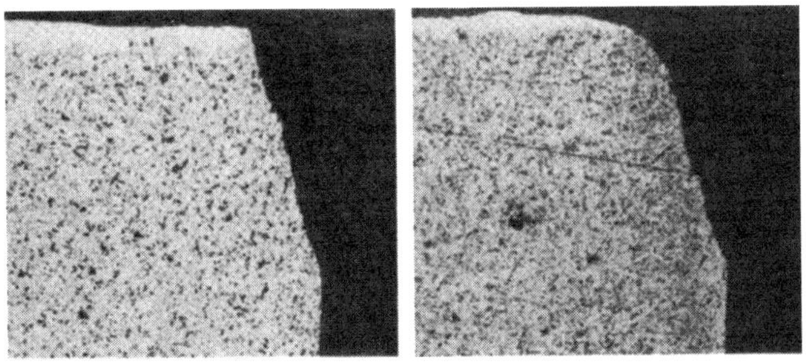

Figure 18-14. Photomicrograph (85×) of 2024-0 clad sheet aluminum before and after deburring by three-dimensional abrasive sheets made into a convolutely wound wheel.

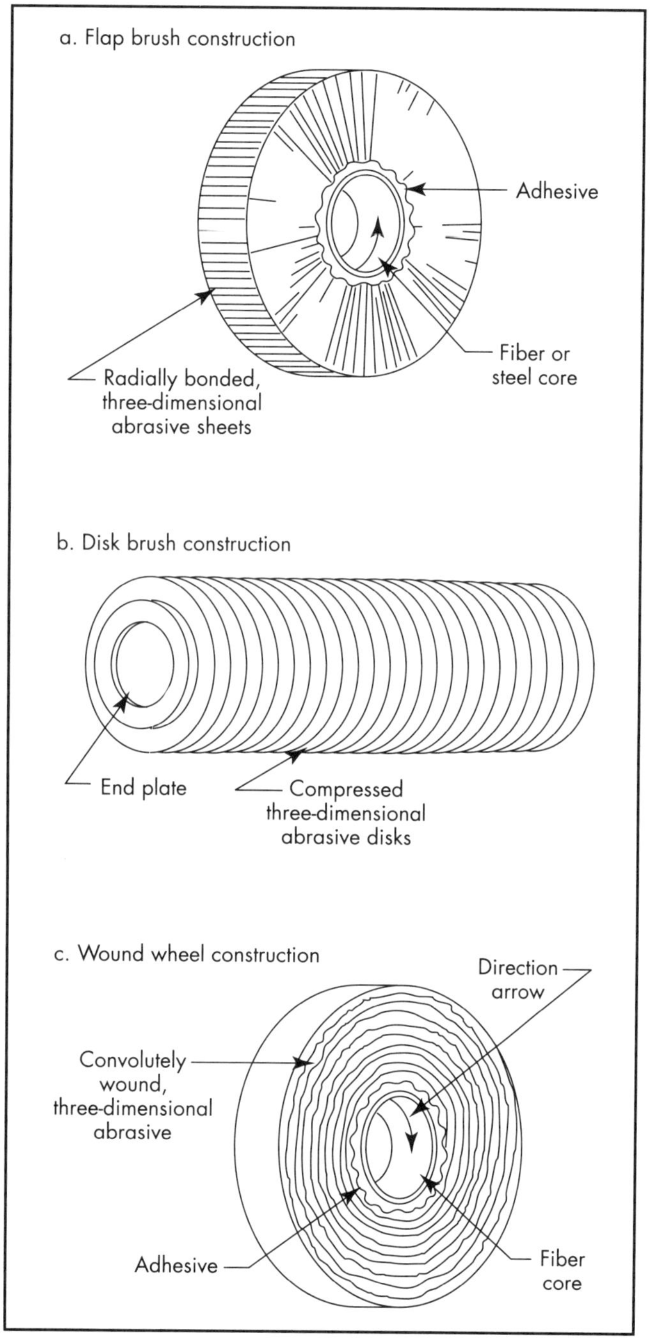

Figure 18-15. Three-dimensional, nonwoven abrasive sheets made into various abrasive brushes.

Centrifugal barrel finishing is the fastest mass-finishing process, but it is a batch process, whereas processes, such as vibratory deburring, can be continuous-flow processes. Adding acidic chemicals to the mass finishing processes will increase burr removal rates. Millions of sheet metal workpieces are deburred annually with loose-abrasive deburring processes.

As noted in Chapter 15, the thermal energy method (TEM) is not generally a good application for sheet metal workpieces because the ratio of burr size to metal thickness and the short blocky nature of sheet metal burrs do not allow the appropriate heat transfer. TEM will remove burrs and is used for this purpose, but it is not ideal for thin sheet metal workpieces.

A new thermal process has been developed for rounding the edges of sheet stock of gages 22–40 (Sonego 1983). This process employs a torch-like heating element to round edges, much like drinking glass edges are rounded.

Electropolish deburring and chemical deburring are generally not used on sheet metal. For these processes to remove burrs, they must also remove surface material, which would not be acceptable with sheet metal. With a normal blanking burr of .002 in. (0.05 mm), .001–.002 in. (0.03–0.05 mm) of material will be removed from the workpiece thickness in the process of removing the burrs; sheet metal is just too thin for this process. However, these processes are often used for removing very thin burrs on highly polished surfaces and are particularly good for stainless steels.

18.3 GENERAL CONSIDERATIONS FOR SHEET METAL

18.3.1 Deburring After Coil Slitting

Both the shearing of narrow strips and edge trimming with a slitting machine are important processes because they are economical and precise. With conventional slitting, however, burr formation at the slit edge is inevitable. To eliminate this problem on conventionally slit metal strips, relatively expensive secondary treatments, such as finish milling, mash rolling, and grinding, must be used.

Mash rolling is often used for most kinds of slit products, but recoiling capacity may be reduced as a result, and edges may be more likely to fracture in subsequent cold-rolling operations. Finish milling may be needed when slit strips are subjected to further welding processes. Deburring by grinding is necessary for insulated strips (particularly oriented silicon steel strips) used in the electrical industry.

18.4 ENVIRONMENTAL, HEALTH, AND SAFETY CONCERNS

The presses used in sheet metal deburring processes require special safety features to ensure that body parts are not caught under the press during operation. Manually operated sheet-metal edgers pose a hazard to operators' fingers and hands; guides or other safety devices will minimize this problem. Grinders that generate dust may cause breathing problems for operators. Venting or exhaust systems are needed. Nickel dust is a suspected carcinogen, and stainless steel containing considerable nickel must be treated as a carcinogen as well; exhaust systems or wet

Table 18-1. Applicability of deburring processes to sheet metal

Process and code	Applicability
Abrasive finishing (A)	
Barrel tumbling (Al)	Yes
Vibratory finishing (A2)	Yes
Vibratory shaker mixer finishing (A2s)	Yes
Roll-flow (centrifugal disk) finishing (A3)	Yes
Centrifugal barrel finishing (A4)	Yes
Spindle finishing (A5)	No
Fluidized bed spindle finishing (A5a)	No
Recipro finishing (A6)	No
Orboresonant finishing (A7)	No
Flow finishing (A8)	No
Cascading media (A9)	No
Chemical loose-abrasive finishing (AC)	
Chemical barrel tumbling (AC I)	Yes
Chemical vibratory finishing (AC2)	Yes
Chemical roll-flow (centrifugal disk) finishing (AC3)	Yes
Chemical centrifugal barrel finishing (AC4)	Yes
Chemical spindle finishing (AC5)	No
Chemical fluidized bed spindle finishing (AC5a)	No
Chemical Recipro finishing (AC6)	Yes
Chemical orboresonant finishing (AC7)	Some
Chemical-flow finishing (AC8)	Some
Cryogenic loose-abrasive finishing (ACRY)	
Cryogenic barrel tumbling (ACRYL)	Yes
Cryogenic vibratory finishing (ACRY2)	Yes
Cryogenic vibratory shaker-mixer finishing (ACRY2s)	Yes
Cryogenic roll-flow (centrifugal disk) finishing (ACRY3)	Yes
Cryogenic centrifugal barrel finishing (ACRY4)	Yes
Cryogenic spindle finishing (ACRY5)	No
Cryogenic fluidized bed spindle finishing (ACRY5a)	No
Cryogenic recipro finishing (ACRY6)	Some
Cryogenic orboresonant finishing (ACRY7)	Some
Cryogenic flow finishing (ACRY8)	Some
Magnetic loose-abrasive finishing (AM)	
Magnetic abrasive barrel finishing (Aml)	Yes
Magnetic abrasive vibratory finishing (AM2)	Yes
Magnetic abrasive spindle finishing (AM5)	No
Magnetic abrasive cylindrical finishing (AM5a)	No
Magnetic abrasive tube-ID finishing (AM5b)	No
Magnetic abrasive ball finishing (AM5c)	No
Magnetic abrasive special-shape finishing (AM5d)	No
Magnetic abrasive prismatic finishing (AM7)	No
Mixed metal fiber magnetic finishing (AM8)	No
Chemical magnetic loose-abrasive finishing (AMC)	
Chemical magnetic abrasive barrel finishing (AMCL)	Yes
Chemical magnetic abrasive vibratory finishing (AMC2)	Yes
Chemical magnetic abrasive spindle finishing (AMC5)	No
Chemical magnetic abrasive cylindrical finishing (AMC5a)	No
Chemical magnetic abrasive tube-ID finishing (AMC5b)	No
Chemical magnetic abrasive ball finishing (AMC5c)	No
Chemical magnetic abrasive special-shape finishing (AMC5d)	No
Chemical magnetic abrasive prismatic finishing (AMC7)	No
Electrochemical loose-abrasive finishing (EAC)	
Electrochemical barrel tumbling (EACL)	Yes
Electrochemical vibratory finishing (EAC2)	Yes
Electrochemical roll-flow (centrifugal disk) finishing (EAC7)	Yes
Electrochemical centrifugal barrel finishing (EAC3)	Yes
Electrochemical spindle finishing (EAC4)	No
Electrochemical fluidized-bed spindle finishing (EAC4a)	No
Electrochemical recipro finishing (EAC6)	Yes
Electrochemical orboresonant finishing (EAC7)	Yes
Electrochemical flow finishing (EAC8)	Yes
Lapping (A9)	Yes
Abrasive jet (AB1)	Yes
Cryogenic abrasive jet (ABCl)	Yes
Liquid hone abrasive-flow deburring (AB2)	Yes
Abrasive-flow finishing (AB3)	Yes
Abrasive-flow orbital finishing (AB3O)	Yes
Abrasive-flow stream (AB3s)	No
Ultrasonic liquid deburring (AUl)	Yes
Ultrasonic abrasive-flow deburring (AB3U)	Yes
Ultrasonic slurry finishing (AU2)	Yes
Chemical deburring (Cl)	Yes
EDM deburring (El)	Yes
Electrochemical deburring processes (EC)	
Electrochemical deburring (salt) (ECl)	Yes
Electrochemical rotary electrode (EC1R)	Some
Electrochemical deburring (glycol) (EC2)	Yes
Electrochemical mesh deburring (EC5)	Yes
Electrochemical moving electrode deburring (EC4)	Yes
Electrochemical brush deburring (EC3)	Yes
Electrochemical nonwoven abrasive magnetic finishing (ECM6)	No
Electropolish (EL3)	Yes
Manual deburring (Ml)	Yes
Mechanized cutting (M2)	Yes
Skiving (M2s)	Yes
Robotic deburring (M2R)	Yes
Brushing (M3)	Yes
Buffing (M3b)	Yes
Wheel blending (M4)	Yes
Sanding (M5)	Yes
Vibratory conveyor deburring (M5a)	Yes
Edge rolling (M7)	Yes
Edge burnishing (M8)	Yes
Edge peening (M13)	Yes
Trimming (presswork) (M9)	Yes
Edge coining (M10)	Yes
Water jet deburring (M6)	Yes
Ballizing (M11)	Some
Tearing (M12)	No
Abrasive chemical finishing (MC1)	Yes
Torch cut deburring (T1)	Some
Plasma deburring (T3)	Some
Plasma glow deburring (T5)	No
Thermal energy method (T2)	Some
Hot wire deburring (T4)	No
Hot blade deburring (T4b)	No

Table 18-1. Applicability of deburring processes to sheet metal *(continued)*

Process and code	Applicability
Hot tool deburring (T4c)	No
Resistance heat deburring (T6)	Some
Laser deburring (T7)	Yes
Chlorine deburring (TC1)	Yes

grinding will eliminate or minimize this hazard. Sanding equipment must be regularly maintained to remove dust particles that may be hazardous to operators' health as well as a source of fires.

References

Barnes, Austin B. ca. 1996. "The State of the Art of Edge Finishing and Decambering—A Review." Buffalo, NY: Barnes Advanced Technology.

Behringer, Brian J. 1981. "Automated Deburring of Flat Sheet Metal." Technical paper MR81-387. Dearborn, MI: Society of Manufacturing Engineers.

Bowman, Hamilton B. 1987. *Handbook of Precision Sheet, Strip, and Foil*. Materials Park, OH: American Society of Metals: 104–11.

Colombier, Louis. 1990. "Stainless and Heat Resisting Steels." *Steel Products Manual*. London, England: Iron and Steel Society: 221.

Dallas, Daniel. 1962. *Progressive Dies: Design and Manufacture*. New York: McGraw-Hill: 19–20.

Flores, Gerhard. 1992. "Mechanical Deburring: Process, Tools, Machines, and Applications." Technical paper MR95-272. Dearborn, MI: Society of Manufacturing Engineers.

Kelley, Harold. 1983. "Special Processing—What Should You Expect from Your Service Center." Technical paper MF83-554. Dearborn, MI: Society of Manufacturing Engineers.

Kucklick, Frederick C. 1996. "Flashless Welding Process Sparks Automotive Production." *American Machinist,* Vol. 140, (October):80–81.

Langton, John W. 1963. *Sheet Metal Working Machinery*. London, England: Sir Isaac Pitman & Sons.

Lyman, Taylor. 1969. *Metals Handbook,* 8th ed., Vol. 4. *Forming*. Materials Park, OH: American Society for Metals: 127.

McMaster-Carr. 1977. *McMaster-Carr Catalog*. Chicago, IL: McMaster-Carr: 83:1,219.

Paquin, J.R., and Crowley, R.E. 1987. *Die Design Fundamentals,* 2nd ed. New York: Industrial Press.

Smith, David A., and Bakerjian, Ramon. 1990. *Die Design Handbook*, 3rd ed. Dearborn, MI: Society of Manufacturing Engineers (SME).

Sonego, Ronald. 1983. "Deburring with High Technology." Technical paper MR83-918. Dearborn, MI: Society of Manufacturing Engineers (SME).

Strasser, Frederico. 1983. *Metal Stamping Plant Productivity Handbook*. New York: Industrial Press: 238–40.

Wick, Charles, and Veilleux, Raymond F. 1985. *Tool and Manufacturing Engineer's Handbook*, 4th ed., Vol. 2. *Forming*. Dearborn, MI: Society of Manufacturing Engineers (SME): 4-5, 4-9, 6-27–28.

Wilson, Frank W., ed. 1959. *Tool Engineer's Handbook*, 2nd ed. New York: McGraw-Hill: 63-7–12.

——. 1965. *Die Design Handbook*, 2nd edition. New York: McGraw-Hill: 9–21.

Bibliography

Carlson, Richard F. 1961. *Metal Stamping Design*. Englewood Cliffs, NJ: Prentice-Hall: 53, 58–74, 119–122.

Gillespie, LaRoux K. 1976. *Guide to Deburring, Deflashing, and Trimming Equipment, Supplies, and Services*. Kansas City, MO: Deburring Technology International, Inc.

Reid, David T., ed. 1991. *Fundamentals of Tool Design*, 3rd ed. Dearborn, MI: Society of Manufacturing Engineers (SME).

Strasser, Frederico. 1959. *Practical Design of Sheet Metal Stampings*. Philadelphia, PA: Chilton Books.

——. 1971. *Functional Design of Metal Stampings*. Dearborn, MI: Society of Manufacturing Engineers (SME).

19 Lesser Known Deburring Processes

Many of the processes developed for deburring do not have widespread commercial application; they may have been developed for special needs, are the results of ongoing research, or are not economically feasible for industrial use. Some are in their infancy and may become more widely used in time. Processes like buffing are generally considered polishing processes, but they remove burrs as readily and as consistently as some of the more well-known processes. The processes described in this chapter will provide the reader with a more complete understanding of the current research, capabilities, and limitations of deburring technology. In today's world, a single innovation can save millions of dollars, and one of the processes described in this chapter may well be the key to that innovation.

The deburring processes discussed in this chapter include abrasive chemical deburring, ballizing, buffing, chemical deburring, chemical magnetic loose-abrasive finishing, chlorine gas deburring, electrical discharge machining, electrochemical nonwoven-abrasive magnetic finishing, hot knife deflashing, hot tool deflashing, hot wire deflashing, lapping, laser deburring, mixed metal-fiber magnetic finishing, plasma arc deburring, plasma-glow deburring, resistance heating, tearing, torch deburring, ultrasonic liquid deburring, ultrasonic slurry deburring, vibratory conveyor deburring, and water-jet deburring.

19.1 ABRASIVE CHEMICAL DEBURRING

In abrasive chemical deburring, a chemical accelerator is introduced in film lapping to improving cutting and polishing action. The process removes microscopic burrs while polishing workpieces. Cylinders are produced with a roughness of about 40 nanometers using fine diamond grain on ceramic parts. The chemical reaction involved in the process is defined as a solid-phase reaction, which prevents grinding scratches (Suzuki and Uematsu 1994).

19.2 BALLIZING (DEBURRING)

Ballizing relies on forcing a ball through a hole in a workpiece to improve its surface finish or to change the size of the hole. It has been used for removing large rollover burrs caused by drilling (Gazan 1977). Figure 19-1 illustrates its use for removing burrs in mismatched holes.

19.3 BUFFING (DEBURRING)

Buffing wheels containing abrasive compounds have been used for many years to polish silverware, pots and pans, and many everyday items. Although the process is not normally used for removing large burrs, it can be that aggressive. It is typically used for removing flash and fine and relatively small burrs, such as those found in grinding. It will remove or smooth larger burrs if they are accessible and the process has been specifically designed for this purpose. The process is used, for example, for deburring and radiusing fir tree slots for jet aircraft turbine rotors, shuttles for high-speed weaving machines, and large aluminum aircraft wing skins. Heat-exchanger tube support plates containing 10,000 holes per side also have been finished with this process.

Production rates vary from 150–2,000 workpieces per h, with costs as low as $.05 per workpiece. The abrasive compound applied to the cloth or felt wheels performs the stock removal action. The compound is tenacious and special cleaning may be required to remove it from workpieces. The process is dirty and dusty, making automation attractive. Both manual and fully automated machines are produced by several companies. The main advantage of this process is its ability to produce high surface luster while

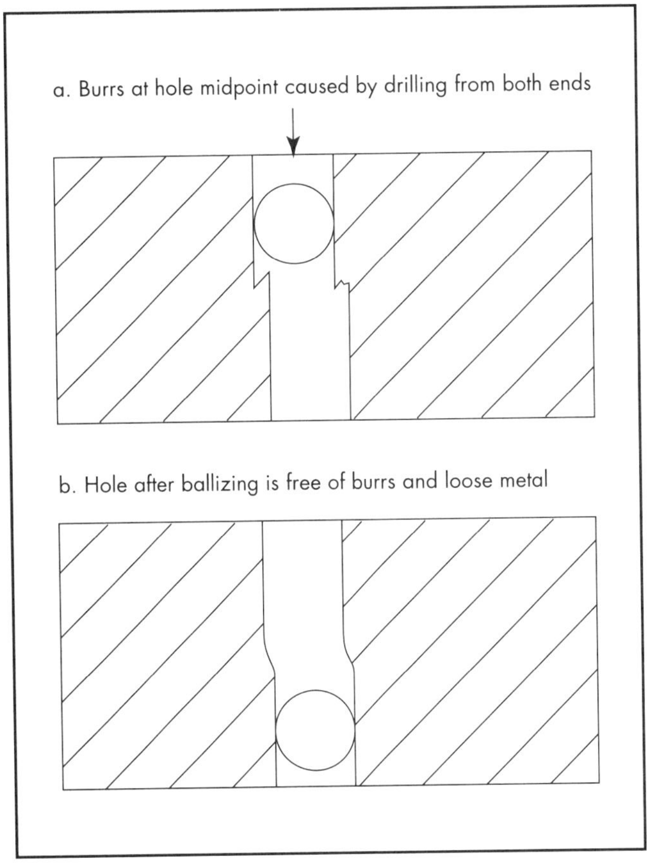

Figure 19-1. Ballizing to remove burrs at mismatched hole intersections.

removing burrs. Many companies employ multiple processes for finishing and deburring in a single dedicated line and use buffing for final steps.

Buffing equipment is configured for manual stands, semiautomatic machines, dedicated lines, CNC machines, and machines in which a robot holds the workpieces against the various buffing heads (Tulinski 1985; Andrew 1981). Much variation exists in equipment buffing wheel designs and materials, as well as in abrasive greases and compounds (McAleer 1975; Wick and Veilleux 1985a).

Some plastic workpieces are finished with a buffing wheel and pumice ash. "Ashing" with pumice is said to be a faster cutting process than using grease- or wax-based compounds and provides faster cooling for plastic workpieces (Richardson 1989). Actual data on the sizes of burrs removed by buffing is nonexistent in the literature, but relatively large radii can be generated if desired. A typical edge radii after buffing appears to be .005–.020 in. (0.13–0.51 mm).

19.4 CHEMICAL DEBURRING

Chemical deburring, a specific use of chemical polishing, is used for a few industrial applications, such as automotive or high-production workpieces requiring generous tolerances. It can be used on a variety of metals to remove small burrs. It is not applicable to thick burrs, because the overall workpiece dimensions are reduced in thickness by almost the same amount as the burr thickness. The time varies for burr removal, but hundreds or thousands of workpieces can be finished in barrels in a day.

19.4.1 Process Details

The terms *chemical polishing* and *bright dipping* are incorrectly used interchangeably. Bright dipping evolved from pickling in much the same way that chemical polishing followed bright dipping. Although bright dipping is still in wide use, mainly because of its lower cost, it is less sophisticated and generally does not have the surface leveling or microdeburring effect achieved through chemical polishing. Chemical polishing selectively removes stock, somewhat like electropolishing, whereas bright dipping does not.

Chemical polishing will brighten the surface of metals, but it is more selective in nature than bright dipping. Thus, it is able, to some extent, to smooth or level the surface and also cause a deburring effect. Chemical polishing is somewhat electrochemical in nature because of the oxidizing agent in the solution.

In almost all bright-dipping and chemical-polishing solutions, an oxidizing agent must be present, usually furnished by the use of nitric acid, chromic acid, dichromates or peroxides, and persulfates. Unfortunately, regardless of the oxidizing agent used, the process has disadvantages. The use of nitric acid produces fumes that are toxic and must be exhausted. Chromic acid and dichromates are reduced from the hexavalent to the trivalent state and therefore become worn out, and the solution must be dumped. Hexavalent chromium wastes must be disposed of in a controlled manner because they are a carcinogen.

Chemical-polishing solutions fall into two categories, either dilute or concentrated acid systems. The concentrated acid systems can tolerate more metal concentrations and can usually be rejuvenated, thus extending their useful life, but they are initially more expensive . Most dilute bright dips operate at ambient temperature. Concentrated acid systems may operate from ambient to elevated temperatures, typically to about 200° F (93° C), depending on the metal or alloy being processed.

Nearly all bright-dipping and chemical-polishing solutions liberate fumes, and hence some degree of exhaust may be required. Also, since most systems are acidic in nature and contain metal salts, the disposal of solution and rinses must comply with local, state, and federal regulations.

Of the various metals and alloys that are bright dipped or chemically polished, aluminum is by far the most popular, followed by copper and its alloys. The brightening of zinc and cadmium on plated workpieces accounts for the greatest tonnage of bright-dip chemical consumption. Nickel and its alloys, carbon steels, and stainless steels are also applicable, but to a much lesser extent.

Grease, oil, soil, or combinations of these must be removed before processing either by solvent cleaning, degreasing, or aqueous alkaline cleaning (soak cleaning).

19.4.2 Equipment

Because most bright-dipping and chemical-polishing solutions are acidic, they require corrosion-resistant equipment. Solution and rinse tanks are usually constructed of austenitic stainless steel, fiberglass, rubber, and plastics, such as polyethylene, polypropylene, and polyvinyl chloride. Because of the variety of solution compositions and temperatures used, the various materials of construction must be evaluated to determine which are best suited for the particular application.

Workpieces may be processed in several ways, depending on their size and shape. Those that are large or have flat surfaces are usually racked individually. Small workpieces that do not tangle, nest, or are not fragile can be processed in bulk in either a dipping basket or a tumbling barrel. When workpieces cannot be bulk processed or when individual hand racking is expensive, compartmentalized or partitioned baskets or containers can be used.

Racks and fixtures are usually made of metal, such as austenitic stainless steel, aluminum, or titanium. Dipping baskets, compartmentalized containers, tumbling barrels, and exhaust equipment are made either of metal or one of the plastics used for tanks.

When either bright dipping or chemical polishing workpieces in production, a means of cooling or heating may be required, depending on the particular solution involved.

19.4.3 Operating Considerations

The degree of surface brightness and, in some instances, surface smoothing depend on the processing time and temperature of the solution. The rate of stock removal increases as processing time and temperature increase up to the point at which increases in either do not improve brightness or surface smoothness. Excessive immersion time can cause a deterioration of brightness or surface smoothing and contribute to shortened solution life because of excessive metal concentration.

Agitation of the solution is desirable, regardless of the type of processing, and mechanical agitation is preferred when possible; methods for agitation include mechanical mixers, pumps, and air. Agitation provides fresh solution at the metal surface for improved chemical polishing; it can also eliminate gas streaks on the processed workpieces and prevent temperature stratification.

Suitable ventilation should be provided to remove mist and/or acid fumes. Exhausting should be sufficient to draw off all mist and fumes during transfer to the first water rinse. Stainless steel, fiberglass, rigid polyvinyl chloride, or other plastic corrosion-resistant coatings are recommended for ducts and blowers, depending on the nature of the solution and the operating temperature.

Several rinses are usually required; if the workpiece is too large to be rinsed by immersion, it must be hosed off. The number and type of rinses are dictated by the metal or alloy to be rinsed and the intended end use of the workpiece. Acid or alkaline neutralizing rinses may be necessary between water rinses. Overflow rinsing tanks are being replaced by more efficient counterflow rinsing tanks, and spray rinsing has gained wide use. In the case of static or overflow rinsing, agitation is beneficial, either by air or mechanical agitation of the solution or the workpieces (Wick and Veilleux 1985b).

Data on burr sizes removed and edge radii produced are nonexistent in the literature. Theoretically, it could be expected that the process could be used for removing burrs on the order of .001 in. (0.03 mm) thick and producing edge radii of .001–.003 in. (0.03–0.08 mm); however, large radii would not be producible.

19.5 CHEMICAL MAGNETIC LOOSE-ABRASIVE FINISHING

Chemical, magnetic, loose-abrasive finishing processes combine the properties of magnetic loose-abrasive finishing with those of chemically accelerated, loose-abrasive finishing. Magnetic loose-abrasive finishing is described in Chapter 8 and in many of the publications cited there. The use of chemical accelerators for this process is not discussed in the literature but should increase deburring rates, although the effect on edges would be relatively small. Expected edge breaks are in the range of .0005–.005 in. (0.013–0.13 mm). The advantage of the process is faster finishing than conventional loose-abrasive finishing.

19.6 CHLORINE GAS DEBURRING

In *chlorine gas deburring*, still only being used on an experimental basis, burrs are removed from heated steel workpieces by exposing them to a chlorine atmosphere in a closed chamber. The workpieces are heated to 600° F (316° C) or more before being exposed to the atmosphere. The chlorine combines with iron in the steel to form iron chloride, which vaporizes below 600° F (316° C). Because of the thin cross section and high surface-area-to-mass ratio of the burrs, the burrs become hotter than the parent metal and are therefore removed faster than material from the workpiece surfaces. The burr-removal cycle time ranges from 20–60 s (Anonymous 1963; Anonymous 1964).

Chlorine gas deburring has a number of limitations. First, it is effective only on iron-based alloys. Second, it changes the dimensions of the workpiece; for example, a 1 min cycle removes .004–.010 in. (0.10–0.25 mm) of stock from the workpiece surface, depending on the temperature of the workpiece. Third, special safety precautions are required because of the toxicity of chlorine; no chlorine gas deburring units are currently commercially available, perhaps because of safety considerations. Fourth, the edge radii produced by the process would be small because it attacks the parent metal at about the same rate as the edge. Edges would be a little hotter and therefore etch at a somewhat higher rate. Expected maximum edge radii would be on the order of .005–.010 in. (0.13–0.25 mm).

19.7 ELECTRICAL DISCHARGE MACHINING

Burrs can be removed using *electrical discharge machining* (EDM). An electrode and the workpiece are submerged in an electrolyte, and the electrode is advanced to within .001–.003 in. (0.03–0.08 mm) of the burr-laden edge. With power on, sparks melt the burrs, and the electrode continues to advance until it is below the burred edge. For circular workpieces with burrs projecting from their diameters, rod or tube electrodes can be used.

EDM removes heavy burrs fairly quickly, but it typically leaves a rough, recast surface at the edges about 30–100 μin. (0.76–2.54 μm) thick. The recast surfaces must be removed if the workpieces are to be used for highly stressed, fatigue-prone applications. Most edges treated by this process will have a small radius, about .003 in. (0.08 mm), but larger radii can be produced with special tooling (News Briefs 1976).

19.8 ELECTROCHEMICAL NONWOVEN-ABRASIVE MAGNETIC FINISHING

Electrochemical, nonwoven-abrasive magnetic finishing is a variation of magnetic loose-abrasive finishing, described in Chapter 8. The loose abrasive is replaced by a nonwoven abrasive-laden pad. Electrolyte is introduced during the polishing, and a cathode is placed above the rotating cylinder. The combination of electrolytic action, oscillating workpieces, and abrasive-laden pad scrubs and chemically removes material at the surfaces. This process is used for fine finishing and small burrs. In one application, a burr .0039 in. (100 μm) tall was removed in 3 min with average surface finishes reduced from 200 to 4 μin. (5 to 0.1 μm) (Yuquan 1994; Yuquan, Guangbin, and Dongxie 1996).

19.9 HOT KNIFE (BLADE) DEFLASHING

Several plastics companies use a thermally hot or warm knife to cut off flash from plastic workpieces. This would work on thermoplastic workpieces but not on thermosetting workpieces. The literature contains no data on the thickness of flash that can be removed (Stoeckhert 1983).

19.10 HOT TOOL DEFLASHING

Trim dies and other trimming tools work better on some materials when they are hot, and a few manufacturers utilize heat in their tooling approach. Recently introduced, *hot tool deflashing* uses an inductively heated, flat ceramic spatula to melt plastic flash on polycarbonate and polyacetal workpieces. Higher temperatures allow faster removal and provide better finishes. Finishes of 59 μin. (1.5 μm) have been obtained at speeds of 5.9 in./min (150 mm/min). A small secondary burr (molten or softened material) forms as the tool removes the primary flash. With polyacetal, the secondary burr was up to .0118 in. (0.300 mm) thick when the tool was traversed at 11.8 in./min (300 mm/min). Slower speeds resulted in smaller secondary burr thickness (Kanda and Shimada 1995, 1996).

19.11 HOT WIRE DEFLASHING

External edges and the surfaces of large cutouts of thermoplastic workpieces can be deflashed with an electrically heated thin wire that is moved parallel to the edges and surfaces. The heat in the thin wire melts and removes the flash. Cycle time is proportional to the linear distance to be trimmed. With some plastics, it should be possible to deflash at a rate of 24 in./min (610 mm/ min) or more.

In *hot wire deflashing*, the heated wire passes over tensioning and drive pulleys (Figure 19-2). On most workpieces, a radius will not be produced on the edges; typically, a small projection of flash is left. It is possible, however, to produce a chamfer if required. One problem with this process is that feed rates must be controlled so the melted plastic does not adhere to the wire and melted material does not reapply to the workpiece (Kanda and Shimada 1995; 1996). Although this process has been in commercial use since at least 1977, no formal data have been published about its capabilities.

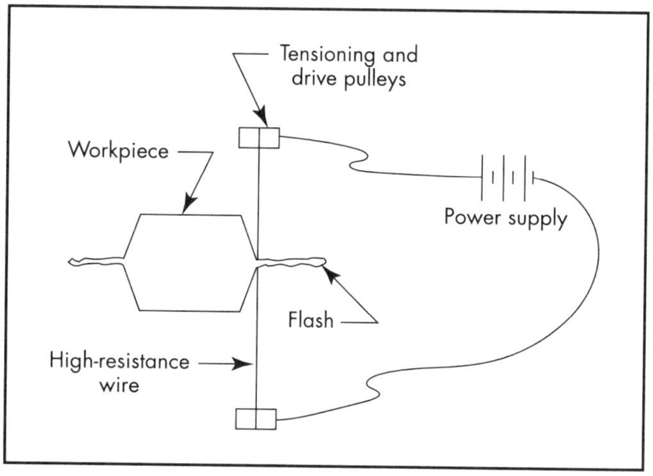

Figure 19-2. Schematic of hot wire process used for deflashing thermoplastic workpieces.

19.12 LAPPING (DEBURRING)

Lapping is not normally considered a deburring process, but some companies use it for that purpose. It removes small burrs left from machining. *Flat lapping* uses a plate with an abrasive-laden lapping compound between the plate and the workpieces to be deburred. With weights on the workpieces, the circular motion of the table provides a gentle material-removal action to take off a thickness of .50 μin. (1.3 μm) or more. Since burrs are at the edges, if they are laying against the table surface they will be removed before the thickness is affected. Burrs must be small for this process to be effective. Hole lapping and honing provide a similar action with the effect concentrated at the hole. Laps are available in sizes smaller than .125 in. (3.18 mm).

An edge rounding occurs when an elastic upper and lower plate are used. Both cylindrical and flat workpieces are edge finished by this process. Carbide cutting-tool inserts are finished in this way, as are some valve spools (Indge 1991).

Orbital vibrational lapping equipment was used in the 1970s (McMaster-Carr 1977). High-speed vibration has taken the place of rotation in this variation. This variation might be more effective than traditional lapping.

19.13 LASER DEBURRING

Laser deburring promises fast, selective edge finishing, but the literature is absent of any reference to actual applications. In fact, some laser deburring patents lie quietly in the patent office collecting dust waiting for potential commercial use. Clearly, the concept has merit for some workpieces, particularly metals and plastics. Loose fiber ends resulting from machining fiberglass-reinforced products are being removed by laser beams. The cost of maintaining laser equipment can be high and requires specialists, but laser machining job shops have found ways to make profits using the basic technology and equipment.

19.14 MIXED METAL-FIBER MAGNETIC FINISHING

For mixed metal-fiber magnetic finishing, magnetic abrasives similar to those used in magnetic loose-abrasive finishing are mixed with short metal fibers. The fibers are produced through chatter machining. In one application, a .8-in. (20.3-mm) diameter tool is rotated with the media and produces surface finishes of 11.8 μin. (0.3 μm) on a die steel. The mixture of media provides more finishing, better media life, and uses much smaller magnetic exciting current (one-tenth) than conventional media. In addition, the allowable gap was about ten times that allowed with conventional media. This mixed-metal approach appears to broaden the potential for this process to many commercial applications for fine burr removal and surface finishing.

19.15 PLASMA-ARC DEBURRING

In *plasma-arc deburring*, burrs are removed by placing the edges of workpieces in a plasma flame (Figure 19-3). Thermal energy from the flame is concentrated at the corners and edges of the workpieces, and the high temperature quickly melts the burrs (Sonego 1983; Rolls Royce Ltd. 1974; Cherepanov, Zinin, and Borovskii 1976). This process is limited, however, because both small and thin precision workpieces may be distorted. It is used for removing fins, sprues, gates, and runners from castings and can remove large burrs from large workpieces. It is not widely used.

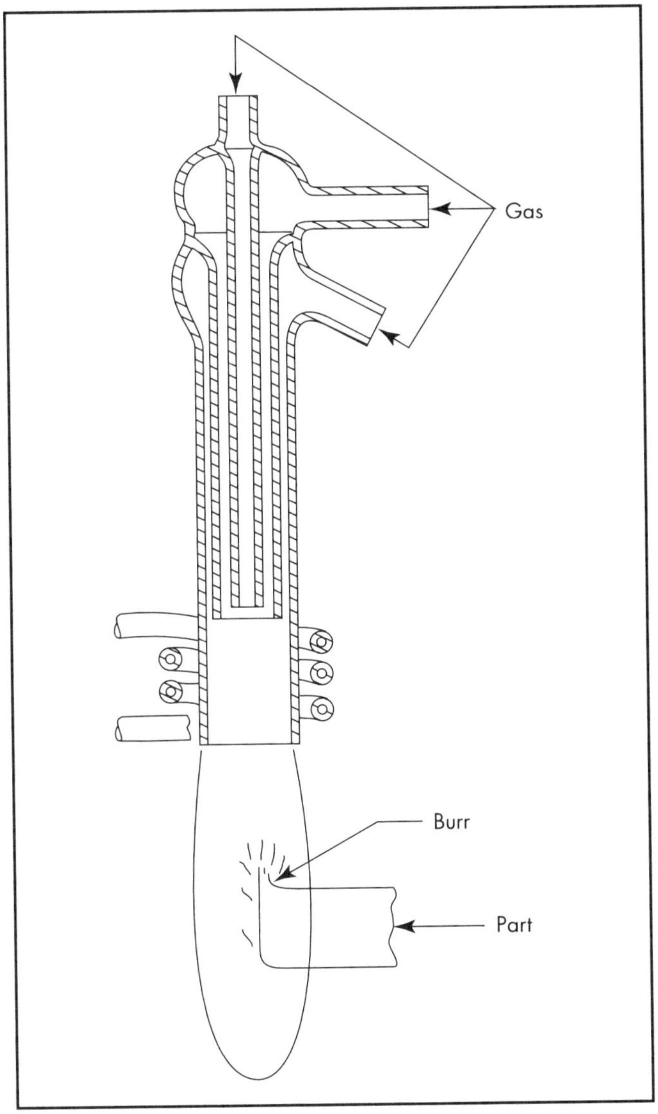

Figure 19-3. Schematic of plasma arc deburring.

19.16 PLASMA-GLOW DEBURRING

For very thin burrs and hairlike projections (such as those 5 μin. [1.27 μm] thick) on plastics, *plasma-glow deburring* (a low-temperature plasma using argon or oxygen gas) can be effective. This process is used primarily for cleaning aerospace and electronic components; deburring is a secondary application. It also has been used for printed circuit board defuzzing. Unlike plasma arc deburring, in this process several workpieces are placed in a chamber and the entire workpiece is subjected to the plasma. Shields can be designed to protect portions of the workpiece if necessary. Plasma cleaning is a high-technology process with many applications.

19.17 RESISTANCE HEATING DEBURRING

In *resistance heating deburring*, heat generated by a current passing through a burr is sufficient to burn away the burr. Workpieces are attached to a power supply that provides low-voltage, high-amperage current (Figure 19-4). The workpiece is attached to the negative side of the electrical source, and a positively charged electrode is placed in contact with the burr. The process can be used for workpieces of any shape, provided that electrical current connections can be made. In one application, burrs are removed from gear teeth by using a meshing gear electrode (Cherepanov, Zinin, and Borovskii 1976).

Possible limitations of the resistance heating process include the need for the workpieces to be electrically conductive and for the electrode to be movable as the burrs shorten. In addition, the larger the burrs, the higher the current required.

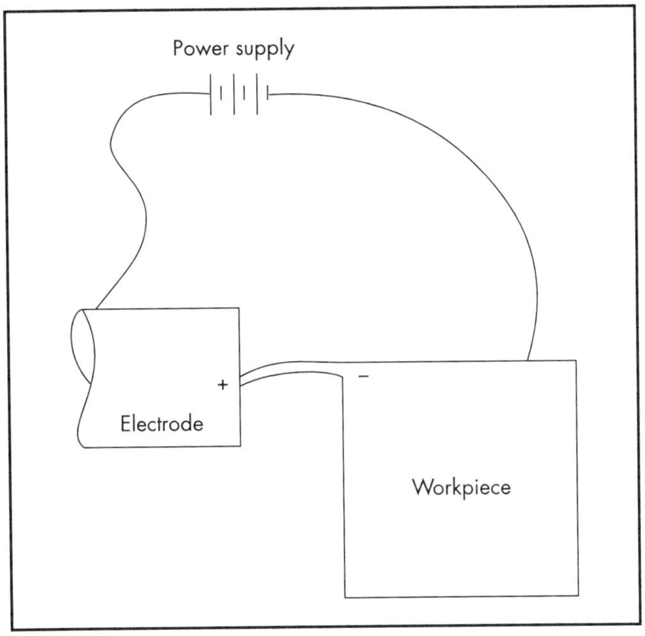

Figure 19-4. Schematic of resistance heating method of deburring.

19.18 TEARING (FLASH)

Some flash is removed from the parent plastic workpiece with mechanical grippers by actually tearing it from the workpiece. Although this method may work adequately for some materials, it is not a normal practice for most materials (Stoeckhert 1983).

19.19 TORCH (FLAME) DEBURRING

The flames from oxyacetylene or similar torches can be used to melt burrs and the edges of workpieces (Sonego 1983). *Torch (flame) deburring* has been used for trimming flash, sprues, and gates from large castings. Torch traverse rates of 20–40 in./min (508–1,016 mm/min) are possible. Most edges exposed to this process will have a radius of .005 in. (0.13 mm) or more, but smaller radii are possible. Possible limitations of torch flame deburring include a heat-affected zone, resolidified droplets around the edges, the need for workholding fixtures, and some means for traversing the torches. In some cases, electric arcs are being used rather than flames. Robots are currently being applied to automate this process in foundries.

19.20 ULTRASONIC LIQUID DEBURRING

Ultrasonic liquid deburring is a little-used process that initially came into being in 1958 to remove fine burrs from microwave and other precision workpieces (Wakefield 1976; Bullen 1976; Wright 1957; Anonymous 1958; Anonymous 1962; Anonymous 1962; Korshunov 1969). Ultrasonic deburring uses ultrasonic cavitation and a specially compounded acid slurry (Figure 19-5). The slurry consists of weak etching solutions and small quantities of abrasive particles. Ultrasound provides high-energy shock waves that cause the abrasive to remain suspended and to constantly bombard the workpiece; the shock waves also increase the etching action. In theory, edges of the workpieces are attacked more than the surfaces because of strain-hardening microstructural changes in the areas of the burrs. Finishes produced are bright and free of oxides. This process is limited to small lots of specialized workpieces containing small burrs, and it is not recommended for high-volume requirements.

Little information is available on the ultrasonic liquid deburring process primarily because the compositions of the slurries are considered trade secrets. Apparently, fine-grained abrasives, such as silicon carbide, .0012–.0079 in. (0.031–0.201 mm) in diameter, are used. Other slurry constituents that have been identified include calcinated soda, stearic acid, sulfuric acid, and glycerin, but the exact quantities are not known. Vibration frequencies to 30 kHz have been employed. Because of the use of acids, tank linings are required.

Ultrasonic liquid deburring is used primarily for thin burrs with thicknesses of about .0005 in. (0.013 mm). It can only generate small edge radii, normally not larger than .005 in. (0.13 mm). The process can be used on a variety of workpiece materials, such as aluminum, beryllium copper, chrome, copper, nickel, mild steels, and stainless steels. It is generally limited to small workpieces because cavitation

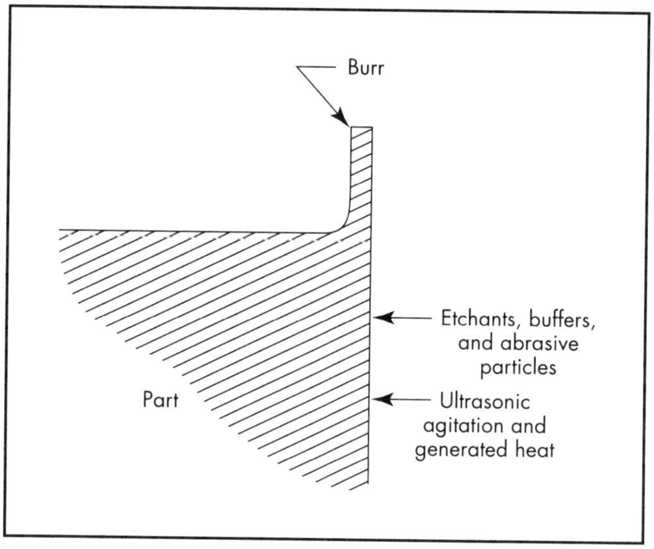

Figure 19-5. Schematic of ultrasonic deburring.

is directly proportional to the size of the transducer, and large workpieces require large tanks and powerful transducers. Cycle times are directly proportional to burr sizes and have been reported to range from 10–25 min. Workpieces should be rinsed with water after processing.

When workpieces are racked, which allows more penetration of the slurry flow, results are often improved. Racking increases production costs, but close-tolerance workpieces requiring this type of finishing generally have to be handled delicately.

The ultrasonic liquid deburring process is especially suitable for reaching inaccessible areas of workpieces, such as removing flash from complexly shaped workpieces containing through or blind holes. Other applications include deburring the teeth of precision gears, the ends of needles, and small precision stampings.

19.21 ULTRASONIC SLURRY DEBURRING

An ultrasonic polisher was introduced in the early 1990s that is used for removing very fine burrs from critical workpieces. The 20,000-cycles/s vibration polishes surfaces and edges. Although only small burrs are removed with this polishing process, unlike ultrasonic machining, it requires no special tools. A blank stick of graphite and a slurry of 200–800 grit abrasive act somewhat like a gentle jitterbug sander (Figure 19-6). The total stock removed by the process is typically .0005 in. (0.013 mm), so the process would be most applicable for removing burrs produced by grinding, honing, lapping, and similar operations.

Workpieces of any size can be deburred by ultrasonic slurry deburring; there are machines with tables measuring 18 × 24 × 12 in. (457 × 610 × 305 mm) to accommodate large parts. Silicone carbide, boron carbide, and diamond are the abrasives used in automated equipment in sizes ranging from 200–800 mesh. The process is adaptable to metals and ceramics. Surface finishes range from

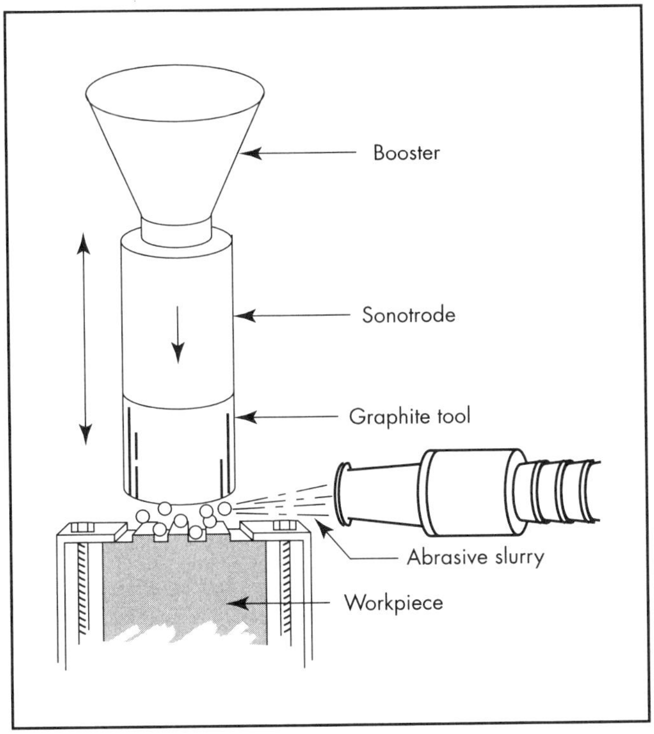

Figure 19-6. Schematic of ultrasonic slurry deburring. ***(Courtesy Extrude Hone)***

8–60 μin. (0.2–1.5 μm); surface thickness may change, but reportedly can be controlled to ±.0002 in. (±5 μm). A typical stock loss is reported as .0005 in. (0.013 mm). The radii produced are very small (Rhoades 1994, 1995).

19.22 VIBRATORY CONVEYOR DEBURRING

In *vibratory conveyor deburring*, workpieces are deburred when they are placed on a conveyor that has adhesive-backed sandpaper attached to the bottom and sides; the conveyor moves the workpieces from presses to other finishing operations, and they are deburred as they move across the sandpaper. No data exist on the size of burrs that can be removed, but the results would be consistent with sanding processes and sheet metal burr sizes (Aoki, Iida, and Jimbo 1982; 1983).

19.23 WATER-JET DEBURRING AND DEFLASHING

In *water-jet deburring and deflashing*, high-pressure fine-diameter streams of water break burrs from the parent material. By oscillating the stream over the burr, the high kinetic energy of the water cuts and breaks the burr. These streams of water cut flash from plastic and metal castings and moldings, and remove thin burrs from machined workpieces. Pressures of 1,000 bar (100,000 kPa) or less are used for soft materials. Since the water is a cleaning agent as well as a deburring tool, completed workpieces are clean.

Water-jet deburring has three variations. In the first variation, a narrow stream is used to concentrate all the energy locally. In the second, a fan of relatively high pressure is used to cover a large area without accurately following each edge. In the third, abrasives are added to the stream to provide faster cutting. Large burrs will not be removed by the process, but all thin and lightly attached burrs will be removed. For many automotive workpieces, this may be adequate, since the primary requirement for some workpieces is that burrs not come off in operation; the assumption is that if they will not come off under high pressure, they will not in normal use. Some burrs in intersecting holes may be removed by the process if the water jet can reach into the feature (Henning 1996; Przyklenk 1994; Labus 1984).

Most deburring applications involve pressures above 3,000 psi (20,684 kPa) but below 20,000 psi (137,891 kPa). One user suggests that for water-jet deburring (as opposed to trimming flash and other fine cutting), the size of the burr that can be removed is determined by extending a .02-in. (0.5-mm) pencil lead 2 in. (51 mm) from the end of the pencil and then pushing the burr. If the burr breaks off, water-jet deburring will remove it; if the pencil lead breaks, the burr will not be removed by the process (Cook and Weightman 1988).

Effective deburring requires the correct nozzle design at the best pressure and at the correct attitude and distance from the burrs. Systems are designed as either fan-shaped patterns to spray for contacting exteriors of workpieces in a general area, lance nozzles that are positioned accurately into bores or specific areas, or multihead nozzles that focus action in a zone but in more limited areas than fan systems.

High-pressure fluid jets (70,000 psi [483 MPa] for example) will sever fingers, hands, or even arms if they are placed under the thin stream of water. One researcher notes that at 45,000 psi (310 MPa), water coming out of a .006 in. (0.15 mm) nozzle is traveling at 1,860 mi/h (2,993 km/h) and the resulting force is similar to a 30-30 rifle bullet (Cook and Weightman 1988). When hoses or pipes develop a leak, they, too, may have enough pressure to injure operators or damage nearby equipment because of the high pressures. Noise levels in the nozzle area may approach 110 dB (Yunker 1991). If robots are used to move the nozzle, standard robot safety issues must be addressed.

19.24 MISCELLANEOUS DEBURRING CONCEPTS

The breadth of deburring techniques, their variations, and supporting equipment is vast, and research and innovation is ongoing; needless to say, space does not allow all potentially useful processes currently under investigation to be covered in this handbook. Examples of processes that are currently under investigation are variations of torch or thermal deburring, magnetic-assisted deburring, and even an induction deburring method (Beier 1990). These and other innovative processes are not in commercial use and are not generally known beyond the research community.

In some instances, newly developed technologies are simply tools for improving existing processes rather than new processes in themselves. Devices have been developed, for example, to improve deburring capability inside hard-to-reach areas, provide portability not found in commercial deburring equipment, and improve sanding processes by increasing the life of abrasives and by increasing the adherence of abrasives to the backing. An estimated five to ten patents are issued each month related to deburring and edge finishing, and deburring staffs rarely know of many of these. Some of these innovations are intriguing and show promise, and the following two examples provide a taste of what has been patented but is not yet commercially available.

U.S. Patent 4,581,853 (Marcus 1982) is a vibratory finishing device; it employs a horizontal-hinged or flexible arm that holds a vibration-inducing tool to create amplitudes similar to those in conventional vibratory finishing. Attached to the vibrating arm is a vertical member on which workpieces (or fixtures for holding workpieces) are attached. The vibrating workpieces are either lowered into a bed of abrasives or the bed of abrasives is raised to encompass the workpieces. The vibrating frequency is variable, generally 500–3,600 vibrations per min; lower frequencies are used during submerging, emerging, and rinsing operations; higher frequencies are used for deburring while the workpieces are submerged in the abrasive. Typically, small abrasive particles, such as 80-grit aluminum oxide or silicon carbide, are used. The abrasive particles travel through interior sections of the workpieces, removing interior burrs and finishing the interior passages. Typical internal deburring times are 30 s for nonferrous workpieces and 2 min for ferrous workpieces.

U.S. Patent 4,823,513 (Marcus and Mercurio 1987) is also a vibratory finishing device. Workpieces are fixtured onto a housing that contains a vibrating device on the inside. The housing is typically a self-contained vibrating motor with a variable-speed drive, though it can be shaft-driven by an external power source. The vibrating housing and attached workpieces are either lowered into a bed of abrasive or the bed of abrasive is raised to encompass the workpieces. The vibrating frequency is variable, usually 500–3,600 vibrations per min. Lower frequencies are used during submerging, emerging, and rinsing operations; higher frequencies are used for deburring workpieces while they are submerged in the abrasive. Typically, small abrasive particles, such as 80-grit aluminum oxide or silicon carbide, are used. The abrasive particles travel through interior sections of the workpieces, removing interior burrs and finishing the interior passages. Typical internal deburring times are 30 s for nonferrous workpieces and 2 min for ferrous workpieces.

References

Andrew, Thomas C. 1981. "New Processes for Buff, Brush, Polish, and Cleaning of Functional and Decorative Components." Technical paper MR81-397. Dearborn, MI: Society of Manufacturing Engineers (SME).

Anonymous. 1958. "Deburring with Ultrasound." *Steel,* 142 (7 April):102–3.

——. 1962. "Ultrasonic Deburrer Cuts Production Costs." *Engineering,* 193 (29 June):853.

——. 1962. "Ultrasonic Deburring." *Aircraft Production,* 24 (11):376.

——. 1963. "New Field Opened by Chlorine Deburring." *Steel,* (4 November):115–116.

——. 1964. "What's New in Materials?" *Materials in Design Engineering,* 59 (January):111–13.

Aoki, Isamu, Iida, Kisuke, and Jimbo, Yasuo. 1982. "Development of New Deburring Method with Vibration." *Bulletin of the Japan Society of Precision Engineering,* 16 (3):191–92.

——. 1983. "New Deburring Method of Blanked Products." *Bulletin of the Japan Society of Precision Engineering,* 17 (3):179–84.

Beier, Hans-Michael. 1990. *Industrielles Entgraten.* Berlin: Verlag Technik.

Bullen, James. 1976. "Capabilities and Limitations of Ultrasonic Chemically Assisted Deburring." Technical paper MR76-131. Dearborn, MI: Society of Manufacturing Engineers (SME).

Cherepanov, V. A., Zinin, Yu. N., and Borovskii, Yu. F. 1976. "The Economics of Flash Removal with a Manual Plasmatron" (in Russian). *Liteinoe Proizvod,* 1 (January):26–27.

Cook, Robert J., and Weightman, David F. 1988. "Automotive Applications of Water-jet Cutting." Technical paper MR88-140. Dearborn, MI: Society of Manufacturing Engineers (SME).

Gazan, George. 1977. "How to Make a Good Hole Better." Technical paper MR77-461. Dearborn, MI: Society of Manufacturing Engineers (SME).

Henning, A. 1996. "Deburring with High-pressure Water Jet." *Proceedings of the 4th International Conference on Precision Finishing and Burr Technology.* Kansas City, MO: Deburring Technology International: 174–93.

Indge, John H. 1991. "Edge Round with Flexible Bonded Abrasives." Technical paper MR91-128. Dearborn, MI: Society of Manufacturing Engineers (SME).

Kanda, Yuichi, and Shimada, H. 1995. "Study on the Micro Deburring System" (in Japanese). Tokyo, Japan: Technical Research Institute, Tokyo University.

——. 1996. "Micro Deburring System for Engineering Plastics." *Proceedings of the 4th International Conference on Precision Finishing and Burr Technology*. Kansas City, MO: Deburring Technology International: 92–100.

Korshunov, B. S. 1969. "Ultrasonic Flash Removal from Small Components." *Russian Engineering Journal,* 49 (12):45–46.

Labus, Thomas J. 1984. "Plant Design and Operational Considerations for Fluid Jet Systems." Technical paper MS88-134. Dearborn, MI: Society of Manufacturing Engineers (SME).

Marcus, R. Steven, and Mercurio, Richard. 1987. U.S. Patent No. 4,823,513, Oct. 13. "Apparatus and Process for Vibratory Finishing of Parts." Washington, DC: U.S. Patent and Trademark Office.

Marcus, Ralph S. 1982. U.S. Patent No. 4,581,853, Feb. 1. "Apparatus for Internal Finishing of Metal Parts." Washington, DC: U.S. Patent and Trademark Office.

McAleer, Howard J. 1975. "Polish Up Your Buffing Data." Technical paper MR79-755. Dearborn, MI: Society of Manufacturing Engineers (SME).

McMaster-Carr. 1977. *McMaster-Carr Catalog*. Chicago, IL: McMaster-Carr: 83:1,243.

News Briefs. 1976. *Tooling and Production,* 41 (November):117.

Przyklenk, Klaus. 1994. "Some Results with Abrasive-flow Machining and Jet Cutting." *Proceedings of the 4th International Conference on Precision Finishing and Burr Technology.* Kansas City, MO: Deburring Technology International: 172–82.

Rhoades, "Lawrence J. 1994. Automatic Finishing of EDM'd Surfaces." Technical paper MR94-250. Dearborn, MI: Society of Manufacturing Engineers (SME).

——. 1995. "Abrasive Flow and Ultrasonic Machining and Polishing." Technical paper MR95-190. Dearborn, MI: Society of Manufacturing Engineers (SME).

Richardson, Terry L. 1989. *Industrial Plastics: Theory and Application*, 2nd edition. Albany, NY: Delmar.

Rolls Royce Ltd. 1974. West German Patent 2401969. London, England: Rolls Royce Ltd.

Sonego, Ronald A. 1983. "Deburring with High Technology." Technical paper MR83-918. Dearborn, MI: Society of Manufacturing Engineers (SME).

Stoeckhert, Klaus. 1983. *Mold-making Handbook*. New York: Society of Plastics Engineers, Hanser Publishers: 99.

Suzuki, Kiyoshi, and Uematsu, Tetsutaro. 1994. "Development of New Surface Finishing Methods." *Proceedings of the 3rd International Conference on Precision Finishing and Burr Technology.* Kansas City, MO: Deburring Technology International: 116–25.

Tulinski, Edward H. 1985. "Flexible Mechanical Systems." Technical paper MR85-822. Dearborn, MI: Society of Manufacturing Engineers (SME).

Wakefield, Brian D. 1976. "Hybrid Line Cuts Assembly Costs." *Production,* 59 (September):83–85.

Wick, Charles, and Veilleux, Raymond, F. 1985a. *Tool and Manufacturing Engineers Handbook,* 4th edition. Vol. 3, *Materials, Finishing, and Coating*. Dearborn, MI: Society of Manufacturing Engineers (SME): 16-19–30.

——. 1985b. "Chemicals Polishing and Bright Dipping." *Tool and Manufacturing Engineers Handbook,* 4th edition. Vol. 3, *Materials, Finishing, and Coating*. Dearborn, MI: Society of Manufacturing Engineers (SME): 17-7–10.

Wright, John P. 1957. "Ultrasonic Deburring." *American Machinist,* 101 (11 February):129–36.

Yunker, Jeff. 1991. "High-Pressure Water Deburring." Technical paper MR91-140. Dearborn, MI: Society of Manufacturing Engineers (SME).

Yuquan, Chen. 1994. "Study on the Electrolytic-magnetic Nonwoven Finishing." *Proceedings of the 3rd International Conference on Precision Finishing and Burr Technology*. Kansas City, MO: Deburring Technology International: 56–62.

Yuquan, Chen, Guangbin, Peng, and Dongxie, Jin. 1996. "Magnetic Field is Helpful to Electrolysis-abrasive Polishing of Pure Titanium." *Proceedings of the 4th International Conference on Precision Finishing and Burr Technology*. Kansas City, MO: Deburring Technology International: 52–57.

Bibliography

Andrew, Thomas C. 1981. "New Processes for Buff, Brush, Polish, and Cleaning of Functional and Decorative Components." Technical paper MR81-397. Dearborn, MI: Society of Manufacturing Engineers (SME).

Bostock, Ray. 1981. "Deburring Costs—Have You Examined Yours Lately?" Technical paper MR81-385. Dearborn, MI: Society of Manufacturing Engineers (SME).

Carlson, Glen A. 1991. "Flexible Computer-integrated Systems Offer Improved Deburring and Finishing Technology."

Technical paper MS91-242. Dearborn, MI: Society of Manufacturing Engineers (SME).

Dargis, Raymond. 1980. "Chemical Deburring and Finishing of Small Parts." Technical paper MR80-918. Dearborn, MI: Society of Manufacturing Engineers (SME).

——. 1985. "Chemical Deburring of Zamac 3 Castings as a Pretreatment for Chromate Coatings or Electroplating." Technical paper MR85-837. Dearborn, MI: Society of Manufacturing Engineers (SME).

Dickman, Alexander. 1991. "Polishing and Buffing." Technical paper MR91-137. Dearborn, MI: Society of Manufacturing Engineers (SME).

Gillespie, LaRoux K. 1974. "Deburring: An Annotated Bibliography." Vol. 1, SME Technical Report. Dearborn, MI: Society of Manufacturing Engineers (SME).

——. 1975. "Deburring: An Annotated Bibliography." Vol. 2, SME Technical Report MRR 75-04. Dearborn, MI: Society of Manufacturing Engineers (SME).

——. 1976. "Deburring: An Annotated Bibliography." Vol. 3, SME Technical Report MRR 76-07. Dearborn, MI: Society of Manufacturing Engineers (SME).

——. 1977. "Deburring: An Annotated Bibliography." Vol. 4, SME Technical Report MRR 77-04. Dearborn, MI: Society of Manufacturing Engineers (SME).

——. 1977. "Deburring: An Annotated Bibliography." Vol. 5, SME Technical Report MRR 78-07. Dearborn, MI: Society of Manufacturing Engineers (SME).

——. 1979. "Deburring: An Annotated Bibliography." Vol. 6, SME Technical Report MRR 79-05. Dearborn, MI: Society of Manufacturing Engineers (SME).

——. 1981. "Progress in the Battle with the Burr." Technical paper MRR81-07. Dearborn, MI: Society of Manufacturing Engineers (SME).

——. 1984. "Deburring: An Annotated Bibliography." Vol. 7. SME Technical Report MRR 84-06. Dearborn, MI: Society of Manufacturing Engineers (SME).

Meehan, James B. 1983. "A New Mass-finishing Process for Internal Burr Removal." Technical paper MR83-188. Dearborn, MI: Society of Manufacturing Engineers (SME).

Sax, Stanley P. 1997. "Buffing Wheels and Equipment." *Metal Finishing Guidebook and Directory*. Tarrytown, NY: Elsevier Science, Inc.: 48–69.

Index

C

D

E

M

N

R

U

V

W